Birkhäuser

Trends in Mathematics

Trends in Mathematics is a series devoted to the publication of volumes arising from conferences and lecture series focusing on a particular topic from any area of mathematics. Its aim is to make current developments available to the community as rapidly as possible without compromise to quality and to archive these for reference.

Proposals for volumes can be submitted using the Online Book Project Submission Form at our website www.birkhauser-science.com.

Material submitted for publication must be screened and prepared as follows:

All contributions should undergo a reviewing process similar to that carried out by journals and be checked for correct use of language which, as a rule, is English. Articles without proofs, or which do not contain any significantly new results, should be rejected. High quality survey papers, however, are welcome.

We expect the organizers to deliver manuscripts in a form that is essentially ready for direct reproduction. Any version of TeX is acceptable, but the entire collection of files must be in one particular dialect of TeX and unified according to simple instructions available from Birkhäuser.

Furthermore, in order to guarantee the timely appearance of the proceedings it is essential that the final version of the entire material be submitted no later than one year after the conference.

More information about this series at http://www.springer.com/series/4961

Emily Clader • Yongbin Ruan
Editors

B-Model Gromov-Witten Theory

Birkhäuser

Editors
Emily Clader
San Francisco State University
San Francisco, CA, USA

Yongbin Ruan
University of Michigan
Ann Arbor, MI, USA

ISSN 2297-0215 ISSN 2297-024X (electronic)
Trends in Mathematics
ISBN 978-3-319-94219-3 ISBN 978-3-319-94220-9 (eBook)
https://doi.org/10.1007/978-3-319-94220-9

Library of Congress Control Number: 2018960402

Mathematics Subject Classification (2010): 81S02, 14H81, 14D21

This book is published under the imprint Birkhäuser, www.birkhauser-science.com, by the registered company Springer Nature Switzerland AG.
The registered company address is: Gewerbestrasse 11, 6330 Cham, Switzerland

Preface

This book is the product of a special semester on B-model Gromov-Witten theory held at the University of Michigan in winter 2014. The goal of the semester was to bring together experts (including both mathematicians and physicists) on various aspects of mirror symmetry in order to better appreciate how their different perspectives fit together into a coherent—yet still not entirely well-understood—story.

Mirror symmetry, an equivalence between two versions of string theory, emerged in the 1980s as a duality in theoretical physics. From a physical perspective, the duality between the A-model of a manifold X and the B-model of its mirror dual $X^\vee$ was natural to expect because the two theories encode the same physics. Mathematically, though, it has taken years to specify the precise data that the A- and B-models should capture, and the correspondences that have resulted have been striking and unexpected. The A-model associated with X, for example, can be viewed mathematically as encoding the enumerative geometry of curves inside of X, leading to the advent of Gromov-Witten theory. When one restricts to genus-zero curves, the B-model can be understood in terms of period integrals on $X^\vee$, which have been classically studied and often explicitly computed. In this way, genus-zero mirror symmetry has led (as in the celebrated example of the quintic threefold) to beautifully explicit answers to some of enumerative geometry's long-standing questions. These answers were mathematically only conjectural when they were first proposed (by physicists Candelas et al. [2]), but a mathematical proof of genus-zero mirror symmetry followed in the 1990s from the work of Givental [4, 5] and Lian–Liu–Yau [12]. This was a major event, leading to the birth of mirror symmetry as a mathematical subject.

Motivated by this success, one would hope to develop an analogous correspondence in all genera, but the pursuit of higher-genus mirror symmetry has been a very difficult task. On the A-side, while Gromov-Witten theory has a firm mathematical foundation in all genera, its computation in genus beyond zero remains one of the hardest problems in geometry and physics. On the B-side, even the theoretical foundations of the higher-genus theory are mathematically not fully understood. One can attempt to forge ahead, nevertheless, using physical methods; indeed, as early as 1993, Bershadsky et al. [1] formulated a higher-genus B-model theory (now known as BCOV theory) in physical language and proposed many of its key properties, such as the famous "holomorphic anomaly equation." Furthermore, they calculated the B-model generating function explicitly in genera one and two, which implied, by way of mirror symmetry, a conjectural closed formula for the generating functions F_g of genus-g Gromov-Witten invariants of the quintic threefold when $g \in \{1, 2\}$. It took ten years before a mathematical proof of the BCOV formula for F_1 was given by Zinger [15], and another ten years for the analogue in genus two, by Guo–Janda–Ruan [8].

Meanwhile, physicists have continued their B-model calculations with great success. A B-model formula for F_3 was calculated by Katz–Klemm–Vafa in 1999 [10], and a number of structural results on the generating functions F_g were predicted on physical grounds. For example, a fundamental physical result of Yamaguchi–Yau [14] states that F_g is a polynomial in five generators, constructed explicitly from the period integrals of the mirror, of which four are holomorphic limits of certain non-holomorphic objects in the B-model; the holomorphic anomaly equation of BCOV theory can be recast into equations relating these four generators. Using the holomorphic anomaly equation, together with other physical predictions regarding the structure of F_g (the conifold gap condition, for example, and orbifold regularity), Huang et al. [9] pushed the physical calculation of F_g to all $g \leq 51$ for the quintic threefold and to other large bounds for targets such as "one-parameter models" and elliptically fibered Calabi–Yau threefolds.

Compared to this stunning success in physics, mathematical progress has been frustratingly slow. One of the reasons is that mathematical understanding of the BCOV B-model theory in higher genus, including the non-holomorphicity of the B-model generating function, remains limited. We hope, in this book, to help change this state of affairs by collecting some of what is known and providing a reference for future study.

The organization of the book is as follows. Chapter "Mirror Symmetry Constructions" (contributed by Emily Clader and Yongbin Ruan) outlines the various ways in which mirror pairs have been constructed. These include the Batyrev construction, which produces a mirror pair of Calabi–Yau hypersurfaces in toric varieties; the Hori–Vafa construction, in which the mirror to a semi-Fano complete intersection in a toric variety is produced as a "Landau–Ginzburg model" (a variety X together with a polynomial function $X \to \mathbb{C}$ known as the superpotential); and the Berglund–Hübsch–Krawitz construction, which produces a mirror pair of Landau–Ginzburg models. For each of these constructions, mirror symmetry is discussed in its most basic manifestation: the "state space correspondence," an isomorphism between bi-graded vector spaces associated with each theory. For the geometric theory, the state space is simply the (orbifold) cohomology of the hypersurface or complete intersection, with different bi-gradings on the A- and B-sides, while in the Landau–Ginzburg theory, it is a certain orbifold Jacobian ring of the superpotential. Mirror symmetry provides a grading-preserving isomorphism between the A- and B-model state spaces. The upshot, for example, in Batryev mirror symmetry is that the Hodge diamonds of a mirror pair of hypersurfaces are related by rotation, which, in the case of Calabi–Yau threefolds, reduces to the statement that mirror symmetry exchanges the two Hodge numbers $h_{1,1}$ and $h_{2,1}$. These are the dimensions, respectively, of the space of infinitesimal deformations of the Kähler structure and the space of infinitesimal deformations of the complex structure, providing a first hint that the state space correspondence is a shadow of a deeper duality.

The bulk of the book is chapter "The B-model Approach to Topological String Theory on Calabi-Yau n-Folds" (contributed by Albrecht Klemm). We focus on a particular mirror symmetry construction—the Batyrev construction—but upgrade our perspective from the state space correspondence to the much richer data of the topological A- and B-models, which are constructed physically out of two different topological twists of an $N = 2$ two-dimensional supersymmetric quantum field theory. In particular, a major part of this long chapter is devoted to providing a full (physical) account of BCOV B-model theory and its key properties.

According to the physical technique of supersymmetric localization, the A-model path integral of a compact Calabi–Yau threefold X localizes to the space of holomorphic curves of prescribed area, so it depends on the complexified Kähler structure (not on the complex structure) of X. The resulting mathematical theory is Gromov-Witten theory. For the B-model, the path integral localizes to the space of constant maps, so the complication lies in the contribution of the transverse directions; in particular, it depends only on the complex structure (not on the Kähler structure) of the target. Thus, for a threefold X with mirror $X^\vee$, the goal of mirror symmetry may be viewed as the construction, at least locally,[1] of a function

$$t_* : \mathcal{M}_{\mathrm{cs}}(X^\vee) \to \mathcal{M}_{\mathrm{cks}}(X)$$

from the moduli space of complex structures on $X^\vee$ to the moduli space of complexified Kähler structures on X, such that t_* relates the A- and B-model amplitudes. Given such a function, the classical duality interchanging $h_{1,1}$ and $h_{2,1}$ (as discussed in chapter "Mirror Symmetry Constructions") can be recovered by taking tangent spaces on both sides.

More specifically, the generating function of Gromov-Witten invariants can be viewed (modulo certain convergence issues) as a holomorphic function defined within the "large radius" region $V \subseteq \mathcal{M}_{\mathrm{cks}}(X)$, and one goal of BCOV B-model theory is to define a similar function on $\mathcal{M}_{\mathrm{cs}}(X^\vee)$ and to prove that these two functions agree under the mirror map t_*. Locally, the study of the complex structure of $X^\vee$ is equivalent to its variation of Hodge structure. For genus-zero worldsheets, this reduces to studying the dependence of period integrals on complex structure, a classical subject in algebraic geometry that can be handled explicitly via the Picard–Fuchs differential equations and monodromy techniques. In higher genus, the entire

[1] Indeed, this correspondence can only be local, because the moduli space of complexified Kähler structures is a ball centered around a point known as the "large radius limit", whereas the moduli space of complex structures carries nontrivial topology. The effort to extend the moduli space of complexified Kähler structures leads to other subjects, such as the gauged linear sigma model, which are outside the scope of this book.

BCOV theory comes into play, and it is much less well known to the mathematical community. Much of chapter "The B-model Approach to Topological String Theory on Calabi-Yau n-Folds" is devoted to explaining these ideas.

In the construction of the B-model generating function, a number of key differences between the A- and B-model theories become apparent. First, the A-model generating function is always holomorphic, whereas the B-model generating function is not; its non-holomorphic dependence is the subject of the holomorphic anomaly equation. Second, in light of the nontrivial topology on the moduli space of complex structures, the B-model generating function can be viewed as a global object—more precisely, a section of a certain line bundle. This naturally leads to connections between the B-model theory and the theory of modular forms, which are holomorphic sections of the same bundle. These two aspects of the theory are intimately related; in dimension one, for example, the holomorphic anomaly equation implies that the B-model generating function is a quasi-modular form. On Calabi–Yau threefolds, moreover, the holomorphic anomaly equation is a powerful computational tool, leading to such structural features as Yamaguchi–Yau's prediction that F_g is a polynomial in certain canonical generators constructed from period integrals. One of the key issues discussed in chapter "The B-model Approach to Topological String Theory on Calabi-Yau n-Folds" is the comparison between the A- and B-model generating function, which involves the delicate procedure of taking a "holomorphic limit" of the B-model generating function using the geometry of the moduli space of complex structures.

The account of BCOV theory provided in chapter "The B-model Approach to Topological String Theory on Calabi-Yau n-Folds" is physical in nature, whereas a mathematical construction of the theory—though still far from complete—has been initiated by Costello and Li [3, 11], whose work we encourage interested readers to consult. A fundamental feature of both the physical and mathematical constructions is that the higher-genus theory is defined as a quantization from genus zero. In order to explain this in mathematical terms, we digress slightly in chapter "Geometric Quantization with Applications to Gromov-Witten Theory" (contributed by Emily Clader, Nathan Priddis, and Mark Shoemaker) to explain the topic of geometric quantization. This is a procedure for producing a Hilbert space $\mathbb{H}_V$ from a polarized symplectic vector space V (finite- or infinite-dimensional), which is functorial in the sense that a symplectic linear transformation $T : V \to W$ gives rise to an operator $\widehat{T} : \mathbb{H}_V \to \mathbb{H}_W$. The physical meaning of this procedure, and the explanation for its name, comes from the passage from a classical theory, in which V represents the space of states, to the associated quantum theory with state space $\mathbb{H}_V$.

Quantization also has deep mathematical significance, encoding the relationship between the genus-zero Gromov-Witten theory of certain targets X and their higher-genus theory. In this case, one sets $V = V_X = H^*(X; \Lambda)((z^{-1}))$, where Λ is a Novikov ring and z a formal parameter. The genus-zero Gromov-Witten invariants of X can be packaged into a Lagrangian submanifold $\mathcal{L}_X \subseteq V_X$, and the higher-genus invariants into a total descendant potential $\mathcal{D}_X \in \mathbb{H}_{V_X}$. If the targets satisfy the condition of "semisimplicity", then a symplectic transformation $T : V_X \to V_Y$ for which $T(\mathcal{L}_X) = \mathcal{L}_Y$ has $\widehat{T}(\mathcal{D}_X) = \mathcal{D}_Y$. In particular, if one can produce a symplectic transformation taking the Gromov-Witten theory of X to the theory of a finite collection of points, then the all genera Gromov-Witten theory of X can be deduced via quantization from that simplest of targets. These ideas, which were developed in the deep foundational work of Givental [6, 7] and Teleman [13], are discussed in the second half of chapter "Geometric Quantization with Applications to Gromov-Witten Theory".

Equipped with the quantization machinery, in chapter "Some Classical/Quantum Aspects of Calabi-Yau Moduli" (contributed by Si Li) we turn to a mathematical perspective on the higher-genus B-model developed in [3]. The main idea is that one should work on the chain level rather than in cohomology. More precisely, the moduli space $\mathcal{M}_{cs}(X)$ of complex structures can be described mathematically in terms of the chain complex of "polyvector fields" on X. From here, via Kyoji Saito's theory of primitive forms, one obtains a Frobenius manifold structure on $\mathcal{M}_{cs}(X)$ and, in particular, a potential function $\mathcal{F}$. Within the "large complex structure" region $U \subseteq \mathcal{M}_{cs}(X)$, this potential function has been identified in a large class of examples with the generating function of genus-zero Gromov-Witten invariants on the mirror $X^\vee$. To obtain the higher-genus B-model, then, one applies the quantization procedure described in chapter "Geometric Quantization with Applications to Gromov-Witten Theory". This is simply a definition, but at the end of chapter "Some Classical/Quantum Aspects of Calabi-Yau Moduli", we discuss the case where X is an elliptic curve, in which one can prove that the higher-genus B-model correlation functions produced via quantization indeed agree in the large complex structure limit with the higher-genus Gromov-Witten invariants of X.

An alternative mathematical development of the higher-genus B-model is presented in chapter "Eynard-Orantin B-model and Its Application in Mirror Symmetry" (contributed by Bohan Fang), by way of the Eynard–Orantin topological recursion. Specifically, one defines a "spectral curve" as an affine algebraic curve $\Sigma \subseteq (\mathbb{C}^*)^2$ equipped with a certain extra structure, and the Eynard–Orantin formalism recursively defines a collection of symmetric meromorphic differential forms $\omega_{g,n}$ on the n-fold product $\overline{\Sigma}^n$ of a compactification $\overline{\Sigma}$. (In fact, formal spectral curves can be defined in a neighborhood of each semisimple point of a generically semisimple Frobenius manifold, and it has been shown that the Eynard–Orantin recursion is equivalent in this case to Givental's quantization.) For certain target spaces X, an associated spectral curve Σ can be defined, and the Eynard–Orantin invariants of Σ can be viewed as a higher-genus B-model of Σ. In particular, the mirror symmetry prediction that the A-model of X and the B-model of Σ agree has been verified in a number of cases, including all toric Calabi–Yau threefolds. One application of this view on the B-model is that it confirms the modularity properties predicted by physicists, as these properties hold in the Eynard–Orantin theory.

Chapter "The Total Ancestor Potential in Singularity Theory" (contributed by Todor Milanov) addresses the question of when the B-model potential function, defined via quantization as in chapter "Some Classical/Quantum Aspects of Calabi-Yau Moduli", is analytic. This is a particularly desirable property that is relevant, for example, in the Landau–Ginzburg/Calabi–Yau correspondence, which relates the

A-model in two different regions of $\mathcal{M}_{\text{cks}}$ (the large radius region, corresponding to Gromov-Witten theory, and the orbifold region, corresponding to the Fan–Jarvis–Ruan–Witten theory of an associated singularity) by passing through a mirror symmetric comparison of both theories to a global B-model. Specifically, chapter "The Total Ancestor Potential in Singularity Theory" focuses on Frobenius manifolds arising via the universal unfolding of a polynomial function $f : \mathbb{C}^{n+1} \to \mathbb{C}$, in which the Frobenius structure corresponds to the choice of a primitive form; this is the analogue in the orbifold region of the Frobenius manifold of chapter "Some Classical/Quantum Aspects of Calabi-Yau Moduli". In contrast to the Calabi–Yau setting considered in chapter "The B-model Approach to Topological String Theory on Calabi-Yau n-Folds", this Frobenius manifold is generically semisimple, which makes the quantization operation much better behaved. In particular, the B-model generating function is always analytic at semisimple points, so the question is whether it extends analytically across the non-semisimple locus. For the Frobenius manifolds discussed in chapter "The Total Ancestor Potential in Singularity Theory", this is indeed the case. Furthermore, when f is quasi-homogeneous and satisfies a condition known as "invertibility," the ancestor potential of the above Frobenius manifold can be identified with the generating function of all genera Fan–Jarvis–Ruan–Witten invariants of f.

Finally, in chapter "Lecture Notes on Bihamiltonian Structures and Their Central Invariants" (contributed by Si-Qi Liu), we turn to a rather different aspect of the B-model: its connection to integrable systems. In the early days of Gromov-Witten theory, Witten conjectured that the generating function of certain intersection numbers on the moduli space of curves is a tau-function of the KdV hierarchy. This conjecture was soon proven by Kontsevich, work for which he was awarded a Fields Medal and that generated a great deal of interest in the interplay between Gromov-Witten theory and integrable systems. A prominent approach to this subject, due to Dubrovin and Zhang, is to develop an axiomatization of the integrable hierarchies arising in geometry. Through this technique, Dubrovin and Zhang proved that the higher-genus generating function associated which a semi-simple Frobenius manifold is uniquely determined by the condition that it is a deformation of the genus-zero generating function and it satisfies a system of differential equations known as the Virasoro constraints. This observation directly motivated Givental's work on quantization. Dubrovin and Zhang pushed their theory much further, however, classifying all possible integrable systems for the higher-genus theory up to "gauge transformations." Since then, their technique has become a standard method for studying the integrable systems associated with semisimple Frobenius manifolds. Chapter "Lecture Notes on Bihamiltonian Structures and Their Central Invariants" gives an account of this fascinating story.

The material presented in this book is by no means complete. For example, it only scratches the surface of Costello–Li's B-model and does not cover any aspects of homological mirror symmetry or the derived category. The reader should take this book as a starting point rather than a definitive reference. As the editors, we wish to thank all the people who contributed to the book, without whose efforts it would not have been possible.

San Francisco, CA, USA Emily Clader
Ann Arbor, MI, USA Yongbin Ruan

References

1. Bershadsky, M., Cecotti, S., Ooguri, H., Vafa, C.: Kodaira-Spencer theory of gravity and exact results for quantum string amplitudes. Commun. Math. Phys. **165**(2), 311–427 (1994)
2. Candelas, P., de la Ossa, X.C., Green, P.S., Parkes, L.: A pair of Calabi-Yau manifolds as an exactly soluble superconformal theory. In: Essays on Mirror Manifolds, pp. 31–95. International Press, Hong Kong (1992)
3. Costello, K., Li, S.: Quantum BCOV theory on Calabi-Yau manifolds and the higher genus B-model (2012). arXiv:1201.4501
4. Givental, A.: A mirror theorem for toric complete intersections. In: Topological Field Theory, Primitive Forms and Related Topics (Kyoto, 1996). Progress in Mathematics, vol. 160, pp. 141–175. Birkhäuser, Boston (1998)
5. Givental, A.B.: Equivariant Gromov-Witten invariants. Int. Math. Res. Not. **13**, 613–663 (1996)
6. Givental, A.B.: Gromov-Witten invariants and quantization of quadratic Hamiltonians. Mosc. Math. J. **1**(4), 551–568, 645 (2001) Dedicated to the memory of I. G. Petrovskii on the occasion of his 100th anniversary
7. Givental, A.B.: Symplectic geometry of Frobenius structures. In: Frobenius Manifolds. Aspects of Mathematics, vol. E36, pp. 91–112. Friedr. Vieweg, Wiesbaden (2004)
8. Guo, S., Janda, F., Ruan, Y.: A mirror theorem for genus two Gromov-Witten invariants of quintic threefolds (2017). arXiv:1709.07392
9. Huang, M.-x., Klemm, A., Quackenbush, S.: Topological string theory on compact Calabi-Yau: modularity and boundary conditions. In: Homological Mirror Symmetry. Lecture Notes in Physics, vol. 757, pp. 45–102. Springer, Berlin (2009)
10. Katz, S., Klemm, A., Vafa, C.: M-theory, topological strings and spinning black holes. Adv. Theor. Math. Phys. **3**(5), 1445–1537 (1999)
11. Li, S.: BCOV theory on the elliptic curve and higher genus mirror symmetry (2011). arXiv:1112.4063
12. Lian, B.H., Liu, K., Yau, S.-T.: Mirror principle. I [MR1621573 (99e:14062)]. In: Surveys in Differential Geometry: Differential Geometry Inspired by String Theory. Surveys in Differential Geometry, vol. 5, pp. 405–454. International Press, Boston (1999)

13. Teleman, C.: The structure of 2D semi-simple field theories. Invent. Math. **188**(3), 525–588 (2012)
14. Yamaguchi, S., Yau, S.-T.: Topological string partition functions as polynomials. J. High Energy Phys. **7**, 047, 20 (2004)
15. Zinger, A.: The reduced genus 1 Gromov-Witten invariants of Calabi-Yau hypersurfaces. J. Am. Math. Soc. **22**(3), 691–737 (2009)

Contents

Mirror Symmetry Constructions

Emily Clader and Yongbin Ruan

Contents

Abstract Mirror symmetry, in general, is a correspondence between objects of a certain type (manifolds, for example, or polynomial functions) and objects of a possibly different type that exchanges the "A-model" of each object with the "B-model" of its image. This equivalence has many manifestations in both mathematics and physics, but in order to discuss any of them, one must first understand how mirror pairs are constructed. We review three such constructions—the Batyrev construction, the Hori–Vafa construction, and the Berglund–Hubsch–Krawitz constructions—and, in each case, describe the A-model and B-model state spaces that mirror symmetry interchanges.

The following notes are based on lectures by Yongbin Ruan during a special semester on the B-model at the University of Michigan in Winter 2014. No claim to originality is made for anything contained within.

E. Clader (✉)
Department of Mathematics, San Francisco State University, San Francisco, CA, USA
e-mail: eclader@sfsu.edu

Y. Ruan
Department of Mathematics, University of Michigan, Ann Arbor, MI, USA
e-mail: ruan@umich.edu

E. Clader, Y. Ruan (eds.), *B-Model Gromov-Witten Theory*, Trends in Mathematics,
https://doi.org/10.1007/978-3-319-94220-9_1

1 Preface: The Idea of Mirror Symmetry

The term "mirror symmetry" is used to refer to a wide array of phenomena in mathematics and physics, and there is no consensus as to its precise definition. In general, it refers to a correspondence that maps objects of a certain type—manifolds, for example, or polynomials—to objects of a possibly different type in such a way that the "A-model" of the first object is exchanged with the "B-model" of its image. The phrases "A-model" and "B-model" originate in physics, and the various definitions of mirror symmetry arise from different ideas about the mathematical data that these physical notions are supposed to capture.

The Calabi-Yau A-model, for example, encodes deformations of the Kähler structure of a Calabi-Yau manifold, while the Calabi-Yau B-model encodes deformations of its complex structure. There is also a Landau-Ginzburg A-model and B-model, which are associated to a polynomial rather than a manifold, and which are somewhat less geometric in nature. The versions of mirror symmetry considered in this course are:

- The Batyrev construction [4, 5], which interchanges the Calabi-Yau A-model of a manifold and the Calabi-Yau B-model of its mirror manifold;
- The Hori-Vafa construction [17, 18, 20], which interchanges the Calabi-Yau (or, more generally, semi-Fano) A-model of a manifold and the Landau-Ginzburg B-model of its mirror polynomial;
- The Berglund-Hübsch-Krawitz construction [6, 23], which interchanges the Landau-Ginzburg A-model of a polynomial and the Landau-Ginzburg B-model of its mirror polynomial.

In each case, mirror symmetry is a conjectural equivalence between the sets of data encoded by the two models. In full generality it remains a conjecture, but many cases are known to hold. The Calabi-Yau/Calabi-Yau mirror symmetry, for example, has been proven whenever the Calabi-Yau manifold X is a complete intersection in a toric variety, and in some cases when X is a complete intersection in a more general GIT quotient.

We should note that, in these notes, mirror symmetry is only discussed as an interchange of cohomology groups (or "state spaces") on the A-side and B-side. At least in the Calabi-Yau case, however, both the A-model and the B-model are understood to capture much more data than these vector spaces alone. The Calabi-Yau A-model, for example, can be encoded in terms of Gromov-Witten theory.

The structure of the notes is as follows. In Sect. 2, we review the fundamentals of toric geometry, which are necessary to explain the Batyrev construction. Sections 3, 4, and 5 develop the three forms of mirror symmetry outlined above. The Appendix reviews the basics of Chen-Ruan cohomology, a cohomology theory for orbifolds that is needed in order to define the state spaces of the Calabi-Yau A- and B-model, and that also provides a useful parallel to the definition of the states spaces in Landau-Ginzburg theory.

2 Toric Geometry

Toric geometry is the study of a class of algebraic varieties whose structure is entirely encoded by combinatorial data. Due to their simplicity, toric varieties provide a natural testing ground for many algebro-geometric ideas, and furthermore, they allow the statement of mirror symmetry to be expressed combinatorially.

The contents of this section are based heavily on Chapter 7 of [21]. Other good references for the basics of toric geometry include [14] and [16].

2.1 *Toric Varieties and Fans*

Definition 1 A **toric variety** is a complex variety X containing an algebraic torus $T := (\mathbb{C}^*)^r$ as an open dense subset, for which the action of T on itself by multiplication extends to an action of T on all of X.

For example, complex projective space $\mathbb{P}^r$ is a toric variety. The open dense torus is

$$\{[x_0 : \cdots : x_r] \mid x_i \neq 0 \;\forall\, i\} \subset \{[x_0 : \cdots : x_r]\} = \mathbb{P}^r.$$

When the toric variety is normal (which is always the case in what follows), it can be constructed from a combinatorial object known as a fan.

Let N be a lattice,[1] a discrete subgroup of $\mathbb{R}^r$ for some r. It follows that $N \cong \mathbb{Z}^r$, but by referring to N abstractly as a lattice, we are *not* fixing an isomorphism. Denote

$$N_{\mathbb{R}} := N \otimes_{\mathbb{Z}} \mathbb{R} \cong \mathbb{R}^r.$$

Definition 2 A **convex rational polyhedral cone** $\sigma \subset N_{\mathbb{R}}$ is a set of the form

$$\sigma = \{a_1 v_1 + \cdots + a_k v_k \mid a_i \geq 0\} \tag{1}$$

for $v_1, \ldots, v_k \in N$. (We sometimes say σ is **generated** by the vectors $\{v_1, \ldots, v_k\}$.) A convex rational polyhedral cone is called **strongly convex** if, furthermore, $\sigma \cap (-\sigma) = \{0\}$; that is, σ does not contain any hyperplanes.

Since we deal exclusively with strongly convex rational polyhedral cones in this course, they are referred to in what follows simply as "cones".

[1] In the general theory of toric varieties, N is allowed to be any abelian group of finite rank. However, in these notes, we assume that N has no torsion, and hence is a lattice.

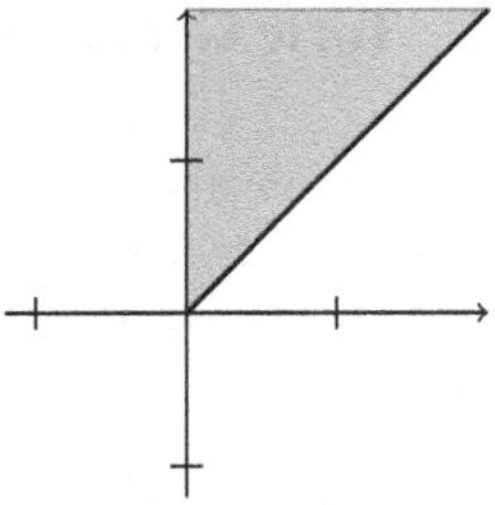

Fig. 1 The cone generated by $(1, 1)$ and $(0, 1)$ in $N \cong \mathbb{Z}^2$

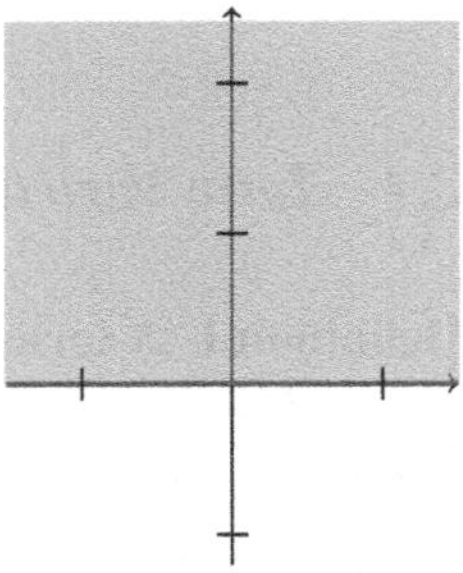

Fig. 2 The cone generated by $(1, 0)$ and $(-1, 0)$ is not strongly convex

See Figs. 1 and 2 for examples. From these illustrations, the notion of a "face" of a cone should be intuitively clear. To put it precisely, a **face** of a cone σ defined as in (1) is a subset given by setting some collection of the a_i's to zero.

Definition 3 A **fan** is a collection Σ of strongly convex rational polyhedral cones satisfying:

(1) Each face of a cone in Σ is also a cone in Σ.
(2) The intersection of any two cones in Σ is a face of each of them.

Example 4 Let $N \cong \mathbb{Z}^2$, and define the following vectors in $N_{\mathbb{R}} \cong \mathbb{R}^2$:

$$v_1 := (-1, -1)$$
$$v_2 := (1, 0)$$
$$v_3 := (0, 1).$$

Then there is a fan Σ whose cones are generated by every proper subset of $\{v_1, v_2, v_3\}$, where by convention, the empty set of vectors generates the 0-dimensional cone $\{0\}$. This fan is illustrated in Fig. 3.

Example 5 Again, let $N \cong \mathbb{Z}^2$. Then there is a fan Σ_n whose cones are generated by proper subsets of $\{(1, 0), (-1, n), (0, 1), (0, -1)\}$. This fan is pictured in Fig. 4.

There are several constructions that yield a toric variety from the data of a fan. Perhaps the most standard (see [16]) involves defining an affine variety for each cone in the fan and using the intersections of the cones to describe how to glue these affine varieties together. This procedure is analogous to the way that one obtains

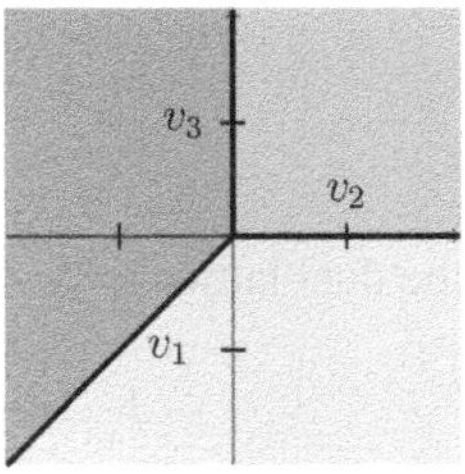

Fig. 3 The fan Σ from Example 4

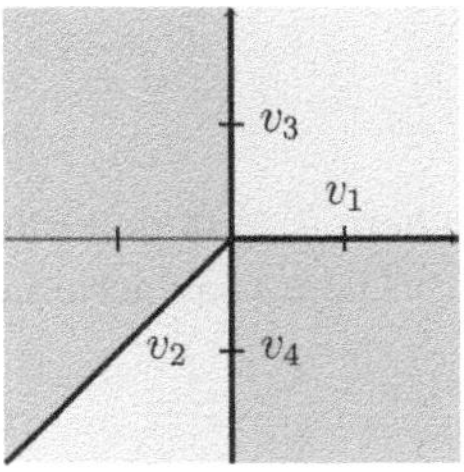

Fig. 4 The fan Σ_n from Example 5

projective space by gluing together the affine subsets on which a given coordinate is nonzero. We return to this perspective in Sect. 2.4.

For the present, we take a different approach to defining toric varieties from fans, which, in the case of projective space, yields the quotient presentation

$$\mathbb{P}^r = (\mathbb{C}^{r+1} \setminus \{0\})/\mathbb{C}^*$$

as opposed to the decomposition into affines.

Fix a fan Σ, and let $\Sigma(1)$ denote the set of 1-dimensional cones. These are sometimes called "edges" or "rays"; explicitly, they are simply the cones generated by a single nonzero vector in N.

For each cone $\rho \in \Sigma(1)$, there is a primitive generator $v_\rho \in N$. That is, v_ρ generates ρ in the sense of Definition 2, and for all integers $k > 1$ we have $\frac{1}{k}v_\rho \notin N$. For convenience, choose an ordering $v_1, \ldots, v_n$ of these vectors, where $n = |\Sigma(1)|$. By abuse of notation, we often write $\Sigma(1) = \{v_1, \ldots, v_n\}$, identifying 1-dimensional cones with their primitive generators.

Consider an affine space $\mathbb{C}^n$ with a coordinate x_ρ for each $\rho \in \Sigma(1)$. In accordance with the above ordering, we sometimes write these coordinates as $x_1, \ldots, x_n$. Inside this affine space, the **discriminant locus** is defined as

$$Z(\Sigma) = \bigcup_{\substack{S \subset \Sigma(1) \\ S \text{ does not span a cone in } \Sigma}} V(I_S),$$

where I_S is the ideal

$$I_S = (\{x_\rho \mid \rho \in S\}) \subset \mathbb{C}[x_1, \ldots, x_n].$$

Let

$$M = \mathrm{Hom}(N, \mathbb{Z}),$$

the dual lattice to N. Then there is a homomorphism

$$\phi : \mathrm{Hom}_{\mathbb{Z}}(\mathbb{Z}^{\Sigma(1)}, \mathbb{C}^*) \to \mathrm{Hom}_{\mathbb{Z}}(M, \mathbb{C}^*) = N \otimes_{\mathbb{Z}} \mathbb{C}^*$$

$$f \mapsto \left(m \mapsto \prod_{v \in \Sigma(1)} f(v)^{\langle m, v \rangle} \right),$$

where $\langle \cdot, \cdot \rangle$ denotes the natural pairing $M \otimes N \to \mathbb{Z}$. Let $G = \ker(\phi)$.

There is an action of G on $\mathbb{C}^n$ defined by

$$g(x_1, \ldots, x_n) = (g(v_1)x_1, \ldots, g(v_n)x_n)$$

for $g \in G$, where we identify $v_\rho \in \Sigma(1)$ with the corresponding standard basis vector for $\mathbb{Z}^{\Sigma(1)}$. It is straightforward to check that this action preserves $Z(\Sigma)$. Thus, one can define

$$X_\Sigma := (\mathbb{C}^n \setminus Z(\Sigma))/G.$$

This is the definition of the toric variety associated to a fan Σ.

Note that X_Σ is indeed toric; the torus that acts is

$$T := (\mathbb{C}^*)^n/G \cong \mathrm{Hom}_{\mathbb{Z}}(M, \mathbb{C}^*) = N \otimes_{\mathbb{Z}} \mathbb{C}^* \cong (\mathbb{C}^*)^r,$$

acting by the quotient of the usual diagonal action of $(\mathbb{C}^*)^n$ on $\mathbb{C}^n$. The first isomorphism in this chain is given by ϕ.

For computations, it is helpful to describe ϕ concretely in coordinates. Under the ordering of $\Sigma(1)$ as $\{v_1, \ldots, v_n\}$, one obtains coordinates $(t_1, \ldots, t_n)$ for $\mathbb{Z}^{\Sigma(1)} \cong \mathbb{Z}^n$. Furthermore, choosing a basis $\{e^1, \ldots, e^r\}$ for N with dual basis $\{e_1, \ldots, e_r\}$ gives an identification $\mathrm{Hom}(M, \mathbb{C}^*) \cong (\mathbb{C}^*)^r$. In these coordinates,

$$\phi(t_1, \ldots, t_n) = \left(\prod_{i=1}^n t_i^{v_{i1}}, \ldots, \prod_{i=1}^n t_i^{v_{ir}} \right),$$

where

$$v_i = \sum_{i=1}^r v_{ij} e^j.$$

One consequence of this description is that G can easily be computed by determining the linear relations satisfied by the vectors $v_1, \ldots, v_n$. We carry this out explicitly in Sect. 2.2 of this chapter.

Let us compute the toric varieties associated to the two fans described above.

Example 6 If Σ is the fan from Example 4, then $r = 2$ and $n = 3$. The discriminant locus is

$$Z(\Sigma) = V(x_1, x_2, x_3) = \{0\},$$

since the only subset of the vectors v_1, v_2, v_3 that does not span a cone is the entire set. In the standard basis $\{(1, 0), (0, 1)\}$ for $N \cong \mathbb{Z}^2$, the homomorphism ϕ is

$$\phi : (\mathbb{C}^*)^3 \to (\mathbb{C}^*)^2$$

$$(t_1, t_2, t_3) \mapsto (t_1^{-1}t_2, t_1^{-1}t_3).$$

Thus,

$$G = \{(t, t, t) \mid t \in \mathbb{C}^*\} \cong \mathbb{C}^*.$$

We obtain

$$X_\Sigma = (\mathbb{C}^3 \setminus \{0\})/\mathbb{C}^* = \mathbb{P}^2.$$

To see how the torus $T = (\mathbb{C}^*)^2$ sits inside $\mathbb{P}^2$, reverse the homomorphism ϕ to write

$$(\mathbb{C}^*)^2 \cong (\mathbb{C}^*)^3/G$$

$$(\lambda_1, \lambda_2) \mapsto [1, \lambda_1, \lambda_2].$$

Thus, $T \subset \mathbb{P}^2$ as $\{[1 : y : z] \mid y, z \neq 0\}$.

Example 7 Let Σ_n be the fan from Example 5, in the special case where $r = 2$ and $n = 4$. Then the discriminant locus is

$$Z(\Sigma_n) = V(x_1, x_2) \cup V(x_3, x_4),$$

and ϕ is given in the standard basis for $N = \mathbb{Z}^2$ by

$$(t_1, t_2, t_3, t_4) \mapsto (t_1 t_2^{-1}, t_2^{-n} t_3 t_4^{-1}).$$

The kernel is

$$G = \{(\lambda_1, \lambda_1, \lambda_1^n \lambda_2, \lambda_2)\} \cong (\mathbb{C}^*)^2.$$

To understand the variety

$$X_{\Sigma_n} = \Big(\mathbb{C}^4 \setminus (\{x_1 = x_2 = 0\} \cup \{x_3 = x_4 = 0\})\Big)/(\mathbb{C}^*)^2,$$

first take the quotient by the λ_1 factor inside $(\mathbb{C}^*)^2$. This yields the quotient of $\mathbb{C}^4 \setminus \{0\}$ by $\mathbb{C}^*$ as

$$\frac{\mathbb{C}^4 \setminus \{0\}}{\lambda(x_1, x_2, x_3, x_4) \sim (\lambda x_1, \lambda x_2, \lambda^n x_3, x_4)},$$

which is the complement of the zero section in the total space of $\mathcal{O}_{\mathbb{P}^1}(n) \oplus \mathcal{O}_{\mathbb{P}^1}$. Now, taking the quotient by the λ_2 factor in $(\mathbb{C}^*)^2$ projectivizes this bundle. Thus,

$$X_{\Sigma_n} = \mathbb{P}(\mathcal{O}_{\mathbb{P}^1}(n) \oplus \mathcal{O}_{\mathbb{P}^1}),$$

which is the Hirzebruch surface F_n. Via the isomorphism

$$T = (\mathbb{C}^*)^2 \cong (\mathbb{C}^*)^4/G$$

$$(\lambda_1, \lambda_2) \mapsto (\lambda_1, 1, \lambda_2, 1)$$

induced by ϕ, one sees that T sits inside F_n as

$$\{(x, 1, z, 1) \mid x, z \neq 0\},$$

where the first two coordinates are the base coordinates and the second two are the fiber coordinates.

Example 8 Consider a fan Σ in $\mathbb{Z}^3$ with

$$\Sigma(1) = \{(1, 0, 1), (0, 1, 1), (-1, -1, 1), (0, 0, 1)\}.$$

Specifically, each subset of $\Sigma(1)$ of size 1 or 2 generates a cone, and the 3-dimensional cones are generated by all subsets of size 3 except for $\{(-1, -1, 1), (1, 0, 1), (0, 1, 1)\}$.

The homomorphism ϕ is easily computed in the standard basis:

$$\phi(t_1, t_2, t_3, t_4) = (t_1 t_3^{-1}, t_2 t_3^{-1}, t_1 t_2 t_3 t_4).$$

Thus,

$$G = \{(t, t, t, t^{-3})\} \cong \mathbb{C}^*,$$

and one obtains

$$X_\Sigma = \mathcal{O}_{\mathbb{P}^2}(-3).$$

Note that this is a noncompact toric variety. In combinatorial terms, the noncompactness of X_Σ is reflected by the fact that the fan does not fill out the entire ambient vector space $\mathbb{R}^3$, leaving some directions "free" from the constraints imposed by the G-action.

In general, given a fan Σ, the resulting toric variety X_Σ is compact if and only if Σ spans $\mathbb{R}^n$. Such fans are referred to as **complete**.

2.2 *The Charge Matrix*

As mentioned previously, there is a straightforward way to read off the group G, and in particular the weights of the G-action on $\mathbb{C}^n$, from the equations satisfied by vectors in $\Sigma(1)$.

To explain this, we need to assume that G does not contain any finite groups as summands; thus, $G \cong (\mathbb{C}^*)^s$ for some s. Then the embedding

$$(\mathbb{C}^*)^s \subset (\mathbb{C}^*)^n$$

induced by viewing $(\mathbb{C}^*)^s$ as the kernel of ϕ can be written in coordinates as

$$(t_1, \ldots, t_s) \mapsto \left(\prod_{a=1}^{s} t_a^{Q_{1a}}, \ldots, \prod_{a=1}^{s} t_a^{Q_{na}} \right) \tag{2}$$

for a matrix Q.

Definition 9 The matrix Q_{ij} in Eq. (2) is called the **charge matrix** for X_Σ.

The terminology relates to a physical connection with the gauged linear sigma model, which is discussed in Sect. 4. It should be noted that the representation of Q as a matrix depends on an identification of G with $(\mathbb{C}^*)^s$ and hence is not canonical.

By the definition of G as the kernel of ϕ, we have

$$\sum_{i=1}^{n} Q_{ia} v_{ik} = 0$$

for all $a = 1, \ldots, s$ and all $k = 1, \ldots, r$. In other words, the charge matrix gives s linear relations

$$\sum_{i=1}^{n} Q_{ia} v_i = 0$$

satisfied by the vectors $v_1, \ldots, v_n$.

It follows that if Λ is the lattice of linear relations on $\Sigma(1)$, then a representation of G as $(\mathbb{C}^*)^s$ is equivalent to a choice of basis for Λ. Having made such a choice, one has

$$X_\Sigma = (\mathbb{C}^n \setminus Z(\Sigma))/(\mathbb{C}^*)^s,$$

in which $(\mathbb{C}^*)^s$ acts by

$$(\lambda_1, \ldots, \lambda_s) \cdot (x_1, \ldots, x_n) = \left(\prod_{a=1}^{s} \lambda_a^{Q_{1a}} \cdot x_1, \ldots, \prod_{a=1}^{s} \lambda_a^{Q_{na}} \cdot x_n \right).$$

Thus, searching for a basis of linear relations among the 1-dimensional cones gives a quick way to read off the toric variety from its fan.

Example 10 Consider the fan Σ from Example 4, for which $X_\Sigma = \mathbb{P}^2$. Then the linear relations among the vectors $\{v_1, v_2, v_3\} = \{(-1,-1), (1,0), (0,1)\}$ are generated by the single relation

$$v_1 + v_2 + v_3 = 0.$$

Under the corresponding identification of G with $\mathbb{C}^*$, then, we have

$$Q = \begin{pmatrix} 1 \\ 1 \\ 1 \end{pmatrix},$$

and indeed,

$$\mathbb{P}^2 = (\mathbb{C}^3 \setminus \{0\})/\mathbb{C}^*,$$

in which $\mathbb{C}^*$ acts with weight 1 in each factor.

Example 11 For the Hirzebruch surface F_n described by the fan in Example 5, the linear relations among $v_1, \ldots, v_4$ are generated by

$$v_3 + v_4 = 0$$

and

$$v_1 + v_2 - n \cdot v_4 = 0.$$

This choice determines an identification of G with $(\mathbb{C}^*)^2$ in which

$$Q = \begin{pmatrix} 0 & 1 \\ 0 & 1 \\ 1 & 0 \\ 1 & -n \end{pmatrix}.$$

Note that, again, the columns give the weights of the two $\mathbb{C}^*$ actions in the quotient

$$X_{\Sigma_n} = \frac{\mathbb{C}^4 \setminus Z(\Sigma_n)}{(\mathbb{C}^*)^2}.$$

2.3 Divisors on X_Σ

Associated to each cone $\sigma \in \Sigma$, there is a subvariety

$$Z_\sigma = \{x_{\rho_1} = \cdots = x_{\rho_k} = 0\} \subset X_\Sigma,$$

where $\rho_1, \ldots, \rho_k$ are the generators of σ. It is easy to check that Z_σ is T-invariant. In fact, all of the T-invariant subvarieties of X_Σ are of the above form.

The codimension of Z_σ is equal to the dimension of σ, so elements $\sigma \in \Sigma(1)$ yield T-invariant *divisors*. If σ has primitive generator $\rho \in N$, we write D_ρ for the T-invariant divisor associated to σ.

The divisors D_ρ in fact generate the group $A_{r-1}(X_\Sigma)$ of Weil divisors modulo linear equivalence; in other words, any divisor on X_Σ is linearly equivalent to a linear combination of T-invariant divisors. Indeed, one can say even more: it is possible to use the combinatorics of the fan to determine when two linear combinations of the divisors D_ρ are linearly equivalent. The criterion involves the notion of the principal divisor associated to a character $m \in M$, which we explain below.

Recall that the torus T is defined as $\mathrm{Hom}(M, \mathbb{C}^*)$. Thus, an element $m \in M$ defines a holomorphic function on T by evaluation. Since X_Σ is a compactification of T, this holomorphic function extends to a meromorphic function f_m on all of X_Σ. Let (m) be the divisor of zeroes and poles of f_m. One can check that (m) is also given by the explicit combinatorial formula

$$(m) = \sum_{\rho \in \Sigma(1)} \langle m, v_\rho \rangle D_\rho \in \mathbb{Z}^{\Sigma(1)} \cong A_{r-1}(X_\Sigma).$$

We will not prove this formula, but let us check it in an example.

Example 12 Let Σ again be the fan from Example 4, whose associated toric variety is $\mathbb{P}^2$. Choose

$$m = (a, b) \in M \cong \mathbb{Z}^2.$$

A point $(\lambda_1, \lambda_2) \in T = (\mathbb{C}^*)^2$ can be viewed as the homomorphism

$$M \to \mathbb{C}^*$$

$$(p, q) \mapsto \lambda_1^p \lambda_2^q.$$

Thus, the function on T associated to the point (a, b) is

$$(\lambda_1, \lambda_2) \mapsto \lambda_1^a \lambda_2^b.$$

Recalling from Example 6 that T sits inside $\mathbb{P}^2$ as $\{[1 : y : z]\}$, it is clear that the unique meromorphic extension f_m of this function to all of $\mathbb{P}^2$ is

$$(x, y, z) \mapsto \left(\frac{y}{x}\right)^a \left(\frac{z}{x}\right)^b.$$

In particular, this confirms that f_m has a zero of order $\langle m, v_2 \rangle = a$ at $D_2 = \{y = 0\}$ (or a pole of order $-a$, if a is negative), and similarly for the other vectors in $\Sigma(1)$.

Using these principal divisors, we obtain an explicit description of $A_{r-1}(X_\Sigma)$.

Theorem 13 *Given Weil divisors* $D = \sum a_\rho D_\rho$ *and* $D' = \sum a'_\rho D_\rho$ *on* X_Σ*, the following are equivalent:*

(1) *D and D' are linearly equivalent;*
(2) *D and D' are homologically equivalent (that is, they define the same element of* $H_{2(r-1)}(X_\Sigma; \mathbb{Z})$*);*
(3) *D and D' have the same associated line bundle;*
(4) *D and D' differ by (m) for some $m \in M$.*

The upshot of everything we have said in this section, then, is that we have a short exact sequence

$$0 \to M \to \mathbb{Z}^{\Sigma(1)} \to A_{r-1}(X_\Sigma) \to 0, \tag{3}$$

where $A_{r-1}(X_\Sigma) \cong H_{2(r-1)}(X_\Sigma; \mathbb{Z})$ is the Chow group of divisors modulo linear equivalence. The first map in the sequence is

$$m \mapsto \Big(\langle m, v_\rho \rangle\Big)_{\rho \in \Sigma(1)},$$

while the second sends $(a_\rho)_{\rho \in \Sigma(1)}$ to $\sum a_\rho D_\rho$. This induces

$$0 \to \mathrm{Hom}(A_{r-1}(X_\Sigma), \mathbb{C}^*) \to \mathrm{Hom}(\mathbb{Z}^{\Sigma(1)}, \mathbb{C}^*) \to \mathrm{Hom}(M, \mathbb{C}^*) \to 0, \tag{4}$$

and the rightmost arrow in this sequence is precisely ϕ.

There are a number of additional consequences of the exact sequences (3) and (4) that we should mention. First, there is now a canonical, basis-independent description of G as

$$G \cong \mathrm{Hom}(A_{r-1}(X_\Sigma), \mathbb{C}^*).$$

In case X_Σ is compact, this is equivalent via Poincaré duality to $G = H_2(X; \mathbb{Z}) \otimes_{\mathbb{Z}} \mathbb{C}^*$.

Second, the entries of the charge matrix can now be described in a more geometrically-motivated way. Rather than expressing it in terms of linear relations on the fan Σ, the sequence (3) shows that such relations can be viewed as elements of $A_{n-1}(X_\Sigma)$. Thus, a basis $C_1, \ldots, C_n$ for the Mori cone of effective curve classes yields a basis for the lattice Λ of relations. Associating the divisors $D_1, \ldots, D_n$ to the generators $v_1, \ldots, v_n$ of $\Sigma(1)$, one can show that

$$Q_{ia} = D_i \cdot C_a,$$

in which $\cdot$ denotes the intersection product.

2.4 *Characterizing Toric Orbifolds*

An orbifold is a variety that is locally given as the quotient of an affine space by a finite group action. In keeping with the idea that all of the structure of toric varieties is represented combinatorially, one can read off from the fan whether a particular toric variety is an orbifold. To do so, we need to briefly describe the construction of X_Σ in terms of charts rather than quotients.

For each top-dimensional cone σ in Σ, define

$$X_\sigma = \{x \in \mathbb{C}^n \setminus Z(\Sigma) \mid x_\rho \neq 0 \text{ for } \rho \notin \sigma\}/G \subset X_\Sigma.$$

In other words, $X_\sigma = X_{\Sigma_\sigma}$, where Σ_σ is the fan consisting of σ and its faces. We claim that

$$X_\Sigma = \bigcup_{\sigma \text{ top-dimensional}} X_\sigma. \tag{5}$$

To see this, choose a point $x \in X_\Sigma$, and let

$$S_x = \{\rho \mid x_\rho = 0\}.$$

Then S_x must span a cone, for otherwise x would lie in $Z(X_\Sigma)$. If σ is the maximal cone containing the cone spanned by S_x, then $x \in X_\sigma$, which proves the claim.

Since G is infinite, the decomposition (5) does not immediately present X_Σ as an orbifold, despite the fact that each of the local patches X_σ is the quotient of an affine variety by a subgroup of G. We require a criterion to ensure that the stabilizer of the G-action on each X_σ is finite.

Definition 14 A fan is **simplicial** if each of its top-dimensional cones gives a $\mathbb{Q}$-basis for N.

Fig. 5 A non-simplicial cone

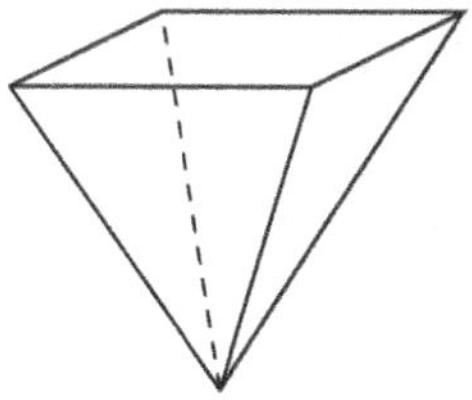

Example 15 The cone in $\mathbb{R}^3$ depicted in Fig. 5 is not simplicial, since it has too many generators. In fact, the toric variety corresponding to this fan is the conifold singularity $xy = uw$.

Theorem 16 *A toric variety is an orbifold if and only if its fan is simplicial. It is a (smooth) manifold if and only if each top-dimensional cone gives a* $\mathbb{Z}$*-basis for* N.

Sketch of proof We discuss only one direction of the proof, since we have explained how to construct a toric orbifold from a fan but not the reverse. See Section 7.5 of [21] for a discussion of the opposite implication.

Let Σ be a simplicial fan, and let σ be a top-dimensional cone. Without loss of generality, we write

$$\Sigma(1) = \{v_1, \ldots, v_n\}$$

and assume that $v_1, \ldots, v_r$ generate the 1-dimensional faces of σ.

We claim that each $x \in X_\sigma$ is equivalent modulo the action of G to a point with $x_\rho = 1$ for all $\rho \notin \sigma$. Indeed, if $\rho \notin \sigma$, then $\rho \cup \sigma$ is linearly dependent, since σ already has maximum dimension. Thus, we have an equation

$$\sum_{i=1}^{r} a_i v_i + a_\rho v_\rho = 0.$$

This implies that

$$(t^{a_1}, \ldots, t^{a_r}, 1, \ldots, t^{a_\rho}, \ldots, 1) \in G$$

for any $t \in \mathbb{C}^*$, where the element t^{a_ρ} is in the spot indexed by v_ρ. Acting by this element, one can rescale x_ρ to 1 without changing any $x_{\rho'}$ for $\rho' \notin \sigma$ other than ρ. Repeating this procedure for each $\rho \notin \sigma$ yields the claim.

It follows that

$$X_\sigma = \mathbb{C}^r/\text{stabilizer}, \tag{6}$$

where the stabilizer is the kernel of the restriction of ϕ to

$$\{x_\rho = 1 \text{ for all } \rho \notin \sigma\} \subset (\mathbb{C}^*)^n.$$

Any element of N yields an element of this stabilizer, since by assumption such an element may be written as

$$\ell_1 v_1 + \cdots + \ell_r v_r$$

with $\ell_i \in \mathbb{Q}$, and it is easily checked that

$$(e^{2\pi i \ell_1}, \ldots, e^{2\pi i \ell_r}, 1, \ldots, 1) \in \ker(\phi),$$

given that $\ell_1 v_1 + \cdots + \ell_r v_r$ is integral. This procedure produces the trivial element of $\ker(\phi)$ exactly when $\ell_i \in \mathbb{Z}$ for each i, so we find that the stabilizer in (6) is isomorphic to $N/\mathbb{Z}\{v_1, \ldots, v_r\}$. This is a finite group, and it is trivial exactly when $\{v_1, \ldots, v_r\}$ forms a $\mathbb{Z}$-basis for N. □

A particularly simple kind of orbifold is a **global quotient**, which is the quotient of a smooth variety by a global finite group action. In terms of fans, toric global quotients are described by passing to a finite-index sublattice.

Fix a fan Σ. Suppose $N \subset N'$ is a sublattice of finite index such that the primitive[2] generators of all of the top-dimensional cones of Σ form integral bases for N.

Since $N_{\mathbb{R}} = N'_{\mathbb{R}}$, either of these lattices can be used with the same fan Σ to define a toric variety. The results, however, are different. Indeed, if

$$T = N \otimes_{\mathbb{Z}} \mathbb{C}^* \text{ and } T' = N' \otimes_{\mathbb{Z}} \mathbb{C}^*,$$

then

$$X_{\Sigma,N} = \frac{X_{\Sigma,N}}{T/T} = \frac{X_{\Sigma,N}}{N/N}.$$

Though we will not prove this fact, let us see how it manifests in an example.

Example 17 Let $X_\Sigma = \mathbb{P}^2$, where the lattice is $N = \mathbb{Z}^2$. Now, suppose we change the lattice to $N' = N + \mathbb{Z}\{(\frac{1}{3}, \frac{2}{3})\}$, which has the effect of adding two additional lattice points to each 1×1 square. For example, the new lattice points in the square whose lower-left vertex is the origin are $(\frac{1}{3}, \frac{2}{3})$ and $(\frac{2}{3}, \frac{1}{3})$.

In the basis $\phi_1 = (\frac{2}{3}, \frac{1}{3})$, $\phi_2 = (\frac{1}{3}, \frac{2}{3})$ for N', the generators of Σ become

$$\{v_1, v_2, v_3\} = \{(2, -1), (-1, 2), (-1, -1)\}.$$

[2] It should be noted that the definition of "primitive" depends on the lattice. In the lattice $\mathbb{Z}^2 \subset \mathbb{R}^2$, for example, the vector $(1, 0)$ is a primitive generator for its ray. The same ray, however, has primitive generator $(\frac{1}{2}, 0)$ if we instead consider the lattice $\frac{1}{2}\mathbb{Z}^2 \subset \mathbb{R}^2$.

As the generators of the one-dimensional cones no longer give integral bases for N', it follows that $X_{\Sigma,N'}$ is an orbifold. Indeed, one can check that $X_{\Sigma,N'} = [\mathbb{P}^2/\mathbb{Z}_3]$.

Conversely, given a global quotient $[X_\Sigma/H]$ of a toric variety, one can reconstruct the inclusion of lattices $N \subset N$ that defines it. To do so, notice that N can be read off from the torus $T = N \otimes_{\mathbb{Z}} \mathbb{C}^*$ as the lattice of 1-parameter subgroups. Although the torus $T \subset X_\Sigma$ descends to a torus $\overline{T} \subset [X_\Sigma/H]$, 1-parameter subgroups of the latter need not lift to 1-parameter subgroups of the former. The new elements of N' that were not present in N are precisely the 1-parameter subgroups of $\overline{T}$ that do not lift. For instance, in Example 17, the subgroups

$$t \mapsto (1, t^{1/3}, t^{2/3})$$

and

$$t \mapsto (1, t^{2/3}, t^{4/3})$$

are well-defined in the quotient but *not* in the original variety $\mathbb{P}^2$, so these give the extra lattice points in N'.

2.5 Toric Resolutions

Now that we have described toric orbifolds in some detail, let us discuss how their singularities can be resolved without leaving the toric setting.

Definition 18 A fan Σ' is said to **subdivide** Σ if

(1) $\Sigma(1) \subset \Sigma'(1)$;
(2) Each cone of Σ' is contained in some cone of Σ.

Suppose Σ' subdivides Σ, and write

$$\Sigma(1) = \{\rho_1, \ldots, \rho_n\},$$

$$\Sigma'(1) = \{\rho_1, \ldots, \rho_n, \rho_{n+1}, \ldots, \rho_m\}.$$

Then the projection $\mathbb{C}^m \to \mathbb{C}^n$ determines a map $X_{\Sigma'} \to X_\Sigma$. One can check that this map restricts to an isomorphism on the tori $T = N \otimes \mathbb{C}^*$ for X_Σ and $X_{\Sigma'}$, so it is birational. We refer to such a map as a **toric resolution**.

Example 19 Consider the toric orbifold $[\mathbb{C}^2/\mathbb{Z}_2]$. By the procedure described in the previous section, this orbifold arises from the fan whose 1-dimensional cones are

$$v_1 = (1, 0) \in \mathbb{Z}^2,$$

$$v_2 = (0, 1) \in \mathbb{Z}^2,$$

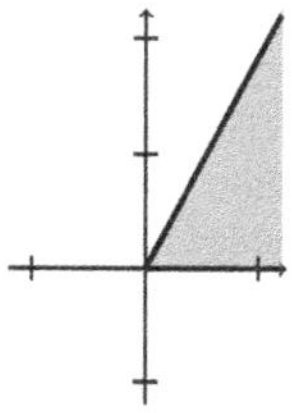

Fig. 6 The fan for $[\mathbb{C}^2/\mathbb{Z}_2]$

where the larger lattice is generated by $(1, 0)$ and $(-\frac{1}{2}, \frac{1}{2})$.

In the basis $\{(1, 0), (-\frac{1}{2}, \frac{1}{2})\}$, the coordinates of the vectors v_i are $(1, 0)$ and $(1, 2)$, respectively. Thus, the fan can equivalently be presented as in Fig. 6. If we add a one-dimensional cone generated by $v_3 = (1, 1)$, subdividing the fan into two 2-dimensional cones, then each of these cones has generators that give an integral basis for the lattice. In other words, we have produced a smooth toric variety birational to $[\mathbb{C}^2/\mathbb{Z}_2]$. In fact, the resulting variety is $\mathcal{O}_{\mathbb{P}^1}(-2)$, which is the blowup of $[\mathbb{C}^2/\mathbb{Z}_2]$ at the origin.

Example 20 Generalizing the above example, consider the global quotient $[\mathbb{C}^2/\mathbb{Z}_{n+1}]$, where the action is by

$$\omega \cdot (z_1, z_2) = (\omega z_1, \omega^n z_2)$$

for $\omega = e^{2\pi i/(n+1)}$.

Starting from the lattice $\mathbb{Z}^2$ that gives the toric variety $\mathbb{C}^2$, the extra lattice points that must be added to obtain $[\mathbb{C}^2/\mathbb{Z}_2]$ are generated by

$$\left(\frac{n}{n+1}, \frac{1}{n+1}\right), \left(\frac{n-1}{n+1}, \frac{2}{n+1}\right), \ldots, \left(\frac{1}{n+1}, \frac{n}{n+1}\right).$$

Denote the corresponding vectors in $\mathbb{R}^2$ by $v_1, \ldots, v_n$; these generate the 1-dimensional cones one must add to the fan for $[\mathbb{C}^2/\mathbb{Z}_2]$ in order to resolve its singularity.

In the refined lattice,

$$\left\{(1, 0), \left(-\frac{1}{n+1}, \frac{1}{n+1}\right)\right\}$$

is an integral basis for $\mathbb{R}^2$, and in this basis, v_i has coordinates $(1, i)$. Therefore, the fan for $[\mathbb{C}^2/\mathbb{Z}_{n+1}]$ and its blowup at the origin can be depicted as in Fig. 7.

Example 21 Consider the global quotient $[\mathbb{C}^3/\mathbb{Z}_3]$, where $\mathbb{Z}_3$ acts with weight 1 in each factor. The fan for $\mathbb{C}^3$ is generated by the 1-dimensional cones $(1, 0, 0), (0, 1, 0), (0, 0, 1)$ in the lattice $\mathbb{Z}^3$. To obtain the quotient, one must add the lattice point $(\frac{1}{3}, \frac{1}{3}, \frac{1}{3})$ and its multiples.

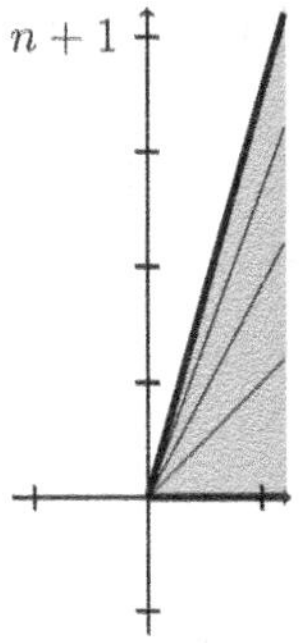

Fig. 7 The fan for $[\mathbb{C}^2/\mathbb{Z}_{n+1}]$ is depicted with bold lines. The lighter lines show the rays that must be added to obtain the toric resolution

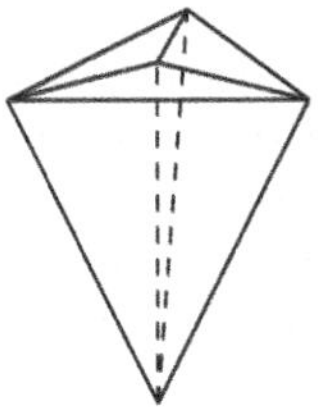

Fig. 8 The fan for $\mathcal{O}_{\mathbb{P}^2}(-3)$, which is a toric resolution of $[\mathbb{C}^3/\mathbb{Z}_3]$

A toric resolution of $[\mathbb{C}^3/\mathbb{Z}_3]$, then, is given by adding a 1-dimensional cone generated by this new lattice point. In the basis

$$\left\{\left(-\frac{1}{3}, \frac{2}{3}, -\frac{1}{3}\right), \left(-\frac{1}{3}, -\frac{1}{3}, \frac{2}{3}\right), \left(\frac{1}{3}, \frac{1}{3}, \frac{1}{3}\right)\right\}$$

for $\mathbb{R}^3$, which is integral under the refined lattice, the coordinates of the 1-dimensional cones in the fan for $[\mathbb{C}^3/\mathbb{Z}_3]$ become

$$(-1, -1, 1), (1, 0, 1), (0, 1, 1),$$

while the coordinates of the lattice point $(\frac{1}{3}, \frac{1}{3}, \frac{1}{3})$ become $(0, 0, 1)$. Thus, the toric resolution can be depicted by the refinement of fans shown in Fig. 8.

Example 22 Toric resolutions need not be unique. For example, consider the (non-orbifold) fan consisting of the single cone shown in Fig. 5, as well as its faces. Specifically, the generators of the 1-dimensional cones are $(1, 0, 0)$, $(0, 1, 0)$, $(0, 0, 1)$, and $(1, 1, -1)$. The resulting toric variety can be embedded as a hypersurface in $\mathbb{C}^4$ by the equation $\{xy = uv\}$, which is called the **conifold singularity**.

There are two toric resolutions, given by adding the 2-dimensional cones shown in Fig. 9. The relationship between the two resulting toric varieties is known as a **flop**. It should be noted that neither is a blowup of the original conifold singularity, since they are given not by introducing new 1-dimensional cones (which would correspond to adding divisors) but rather by adding a 2-dimensional cone.

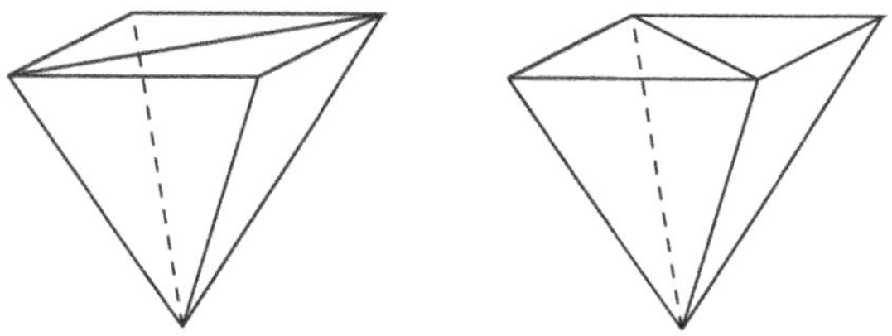

Fig. 9 Two different toric resolutions of the conifold singularity

2.6 Toric Varieties from Polytopes

For projective toric varieties, an alternative combinatorial description can be given; instead of constructing the variety from the data of a fan, it is described via a polytope.

Definition 23 An **integral polytope** in $M_{\mathbb{R}}$ is the convex hull of finitely many points in M.

Given an integral polytope $\Delta \subset M_{\mathbb{R}}$ with integral points

$$\{m_0, \ldots, m_k\} = \Delta \cap M,$$

each m_i can be viewed as a complex-valued function on the torus T, as explained in Sect. 2.3. Let

$$f : T \to (\mathbb{C}^*)^{k+1} \subset \mathbb{P}^k$$

be the map defined by these m_i:

$$f(t) = [m_0(t) : \cdots : m_k(t)].$$

It is easy to check that f is an embedding, assuming that Δ is full-dimensional.

The toric variety associated to Δ is defined as

$$\mathbb{P}_\Delta := \overline{\mathrm{Im}(f)}.$$

Since f is injective, this contains $\mathrm{Im}(f) \cong T$ as a dense open subset, and the fact that the m_i are characters implies that the action of T on itself by multiplication extends to all of $\mathbb{P}_\Delta$. Hence, $\mathbb{P}_\Delta$ is indeed a toric variety.

Two observations can be made right away:

1. Shifting Δ by an integral point $m \in M$ has the effect of multiplying each coordinate of f by the same complex number, which does not change the image in $\mathbb{P}^k$. Thus, $\mathbb{P}_\Delta$ is independent of integral shifts, and in particular, we may assume without loss of generality that $0 \in M \cap \Delta$.
2. If the integral points of Δ satisfy an equation

$$\sum_{i=0}^{k} a_i m_i = 0$$

for integers a_i with $\sum_{i=0}^{k} a_i = 0$, then the homogeneous coordinates y_i of $\mathbb{P}_\Delta$ satisfy the equation

$$\sum_{a_i>0} y_i^{a_i} = \sum_{a_i<0} y_i^{-a_i}.$$

We now have two procedures for obtaining normal toric varieties; the relationship between them is fairly straightforward.

Definition 24 Given an integral polytope Δ, the **normal fan** Σ_Δ has a cone σ_F for each face F of Δ, defined by

$$\sigma_F = \{v \in N_\mathbb{R} \mid \langle m, v\rangle \leq \langle m', v\rangle \text{ for all } m \in F, m' \in \Delta\}.$$

In particular, the 1-dimensional cones of the normal fan are generated by the integral normal vectors v_F to the codimension-1 faces F of Δ, which are determined by the equation

$$F = \{m \mid \langle v_F, m\rangle = 0\}.$$

That is, if N is identified with M via a choice of basis, then v_F is the inward-pointing integral normal vector to F—see Fig. 10.

More generally, if F is a codimension-i face of Δ, then σ_F is an i-dimensional cone. To determine σ_F, one writes F as the intersection of a collection of codimension-1 faces; then σ_F is generated by the integral normal vectors to the faces in this collection.

When $\mathbb{P}_\Delta$ is normal, there is an isomorphism

$$X_{\Sigma_\Delta} \cong \mathbb{P}_\Delta$$

given in homogeneous coordinates by

$$(x_1, \ldots, x_n) \mapsto \left(\prod_{i=1}^{n} x_i^{\langle m_0, v_i\rangle}, \ldots, \prod_{i=1}^{n} x_i^{\langle m_k, v_i\rangle}\right). \tag{7}$$

On the other hand, non-normal toric varieties can arise via polytopes, whereas the normal fan of a polytope is always defined. (In combinatorial terms, this occurs

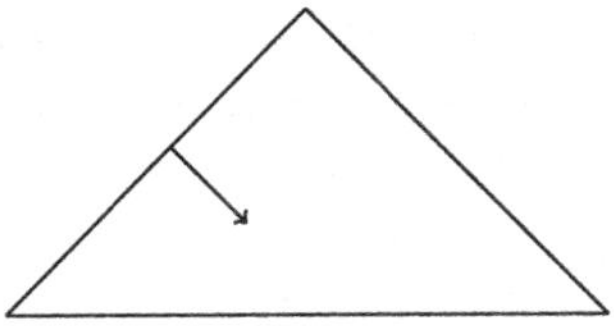

Fig. 10 The inward-pointing normal vector to a face

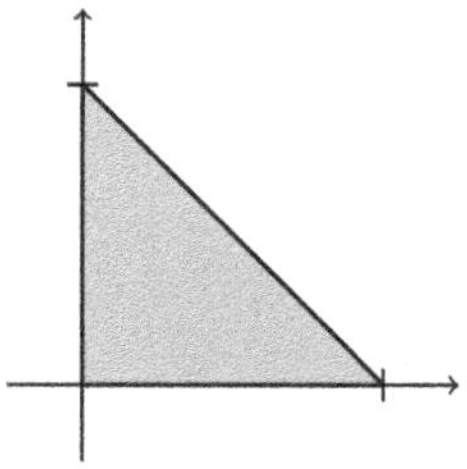

Fig. 11 The toric variety associated to this polytope is $\mathbb{P}^2$

when the lattice points $m_0, \ldots, m_k$ are a proper subset of $\Delta \cap M$ whose convex hull is nevertheless still Δ.) In this situation, the map (7) is still an isomorphism on the torus, and hence is still birational, but it is not an isomorphism on the entire toric varieties; rather, X_{Σ_Δ} is a resolution of $\mathbb{P}_\Delta$.

Example 25 Let $\Delta \subset \mathbb{R}^2$ be the polytope with vertices $(0, 0)$, $(1, 0)$, and $(0, 1)$, as shown in Fig. 11. The resulting map f is

$$f : (\mathbb{C}^*)^2 \to \mathbb{P}^2$$

$$f(t_1, t_2) = [1 : t_1 : t_2],$$

whose image is the standard embedding of the torus in $\mathbb{P}^2$. Thus, $\mathbb{P}_\Delta = \mathbb{P}^2$, and after a shift of Δ, the normal fan is the fan associated to $\mathbb{P}^2$.

Example 26 Scaling a polytope does not change the resulting toric variety (in particular, it does not affect the normal fan of the polytope), but it changes the embedding into projective space. For example, if the polytope in the above example is scaled by a factor of two, then the map f becomes

$$f : (\mathbb{C}^*)^2 \to \mathbb{P}^5$$

$$f(t_1, t_2) = [1 : t_1 : t_1^2 : t_2 : t_2^2 : t_1 t_2].$$

The closure of the image is a copy of $\mathbb{P}^2$ inside $\mathbb{P}^5$.

Example 27 "Cutting off a corner" of a polytope adds a 1-dimensional cone to the normal fan, so it yields a toric resolution. For example, if the top corner of the polytope from the previous example is cut off, the resulting Δ is given in Fig. 12. The associated map is

$$f : (\mathbb{C}^*)^2 \to \mathbb{P}^4$$

$$f(t_1, t_2) = [1 : t_2 : t_1^2 : t_2 : t_2^2].$$

Either via the normal fan or via the polytope construction, it can be verified that the resulting toric variety is F_2.

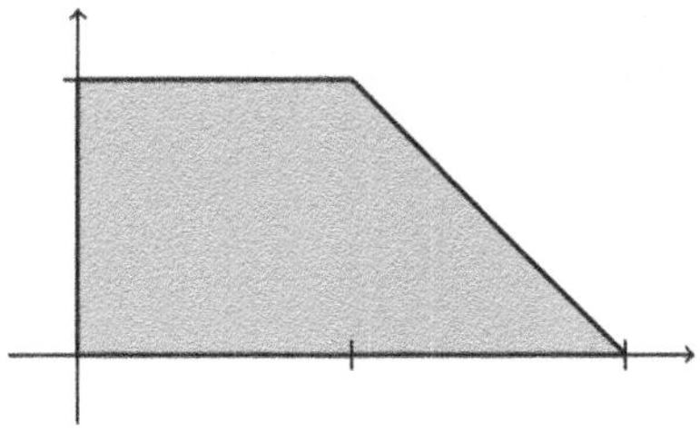

Fig. 12 The toric variety associated to this polytope is F_2

When a toric variety $\mathbb{P}_\Delta$ is constructed out of a polytope, it is automatically equipped with an ample toric divisor D defined as the pullback of the hyperplane class on projective space. Thus, the polytope construction yields strictly more data than the fan method. However, given a toric variety X_Σ *together* with an ample toric divisor, it is possible to reconstruct the polytope Δ that yields this pair.[3]

Let X_Σ be a toric variety equipped with an ample toric divisor D. (The choice only matters up to linear equivalence, since a linearly equivalent divisor will yield a shift of the resulting polytope.) Choosing D amounts to specifying a morphism $X_\Sigma \to \mathbb{P}^k$ for which $\mathcal{O}(D)$ is the pullback of $\mathcal{O}(1)$, and hence the coordinate functions $x_0, \ldots, x_k$ on $\mathbb{P}^k$ yield sections of $\mathcal{O}(D)$. Recall that there is an isomorphism

$$\left\{\text{meromorphic functions } f \text{ on } X_\Sigma \;\middle|\; D + (f) \geq 0\right\} \cong \Gamma(X_\Sigma, \mathcal{O}(D))$$

given by

$$f \mapsto f \cdot s_0,$$

in which s_0 is a global meromorphic function for which $(s_0) = D$. Using this correspondence, the coordinate function x_i yield meromorphic functions f_i on X_Σ. The restriction of each of these to the torus is a character of T, so it can be viewed as an element $m_i \in M$. The polytope Δ is the convex hull of $m_0, \ldots, m_k$.

Example 28 Suppose we begin with the toric variety $\mathbb{P}^2$ and the toric divisor $D_0 = \{x_0 = 0\}$. Then $s_0 = x_0$ is a global meromorphic section whose divisor is D_0, so the functions f_0, f_1, f_2 corresponding to the coordinate sections of D_0 have

$$f_i \cdot x_0 = x_i,$$

and hence $f_i = x_i/x_0$. In terms of the inhomogeneous coordinates $t_1 = \frac{x_1}{x_0}$ and $t_2 = \frac{x_2}{x_0}$ on the torus, these are precisely

$$t_1^0 t_2^0, \; t_1^1 t_2^0, \; t_1^0 t_2^1.$$

[3]The case in which the divisor is ample but not very ample, and hence its associated morphism to projective space is not an embedding, is when the morphism $X_\Sigma \to \mathbb{P}_\Delta$ described in (7) fails to be an isomorphism.

Thus, the polytope Δ is the convex hull of $(0, 0)$, $(1, 0)$, and $(0, 1)$.

Example 29 Repeating the above example with the toric divisor D_3 instead of D_1 yields the convex hull of $(0, 0)$, $(0, -1)$, and $(1, -1)$ as the polytope, which is indeed a shift of Δ.

Example 30 If, instead, the toric divisor is taken as $D_1 + D_2 + D_3$, then the sections of $\mathcal{O}(3)$ are generated by homogeneous degree-three polynomials. These can be dehomogenized by dividing each by $x_1x_2x_3$, at which point they can be expressed in terms of the torus coordinates $t_1 = \frac{x_2}{x_1}$ and $t_2 = \frac{x_3}{x_1}$ to read off the lattice points of the associated polytope. For example, x_1^3 transforms to

$$\frac{x_1^3}{x_1x_2x_3} = \left(\frac{x_2}{x_1}\right)^{-1}\left(\frac{x_3}{x_2}\right)^{-1} = t_1^{-1}t_2^{-1},$$

so the point $(-1, -1)$ lies in the polytope. Repeating this for each of the sections of $\mathcal{O}(3)$ reveals the polytope to be a threefold dilation of the polytope obtained in Example 28.

The above expression for the polytope is expressed more explicitly as follows:

Proposition 31 *The integral points of the polytope associated to a toric variety X_Σ with toric line bundle $\mathcal{O}\left(\sum_\rho a_\rho D_\rho\right)$ (in which $a_\rho \geq 0$) are*

$$\{m \in M \mid \langle m, v_\rho \rangle \geq -a_\rho\}. \tag{8}$$

Proof The previous definition shows that the integer points of the polytope consist of $m \in M$ for which $(m) + D \geq 0$. This means that

$$\sum_\rho \langle m, v_\rho \rangle D_\rho + \sum_\rho a_\rho D_\rho \geq 0,$$

which is clearly equivalent to the description given in the statement of the proposition. □

In other words, the polytope Δ associated to $\left(X_\Sigma, \mathcal{O}(\sum_\rho a_\rho D_\rho)\right)$ is bounded by the affine hyperplanes

$$F_\rho := \{m \in M_\mathbb{R} \mid \langle m, v_\rho \rangle = -a_\rho\}.$$

This fills out the correspondence between the fan and polytope perspectives on toric varieties.

3 Batyrev Mirror Symmetry

The Batyrev mirror symmetry construction applies to toric varieties constructed out of a particular type of polytope, which we will discuss in the first subsection below. Our presentation is based on Section 7.10 of [21]; other key references for this material include [4] and [5].

3.1 Reflexive Polytopes

Definition 32 A full-dimensional integral polytope Δ is **reflexive** if there exist vectors $v_F \in N$ associated to each codimension-1 face F of Δ such that

$$\Delta = \{m \in M_{\mathbb{R}} \mid \langle m, v_F \rangle \geq -1 \text{ for all } F\},$$

and if, furthermore, $0 \in \text{Int}(\Delta)$.

A consequence of reflexivity is that 0 is the only interior integral point of Δ.

Example 33 The polytope Δ appearing in Example 30 of the previous section, depicted in Fig. 13 below, is reflexive. The vectors v_F associated to the edges are shown.

Proposition 31 above shows that in the reflexive case, the divisor in X_{Σ_Δ} determined by the polytope is

$$D = \sum_{\rho \in \Sigma_\Delta(1)} D_\rho.$$

More explicitly, one can check (Theorem 8.2.3 of [14]) that D is the anticanonical divisor of $\mathcal{X}_{\Sigma_\Delta}$. By exploiting this connection to the canonical divisor, one can prove the geometric meaning of reflexivity.

Theorem 34 *A full-dimensional lattice polytope Δ is reflexive if and only if $\mathbb{P}_\Delta$ is Gorenstein and Fano.*

Fig. 13 A reflexive polytope

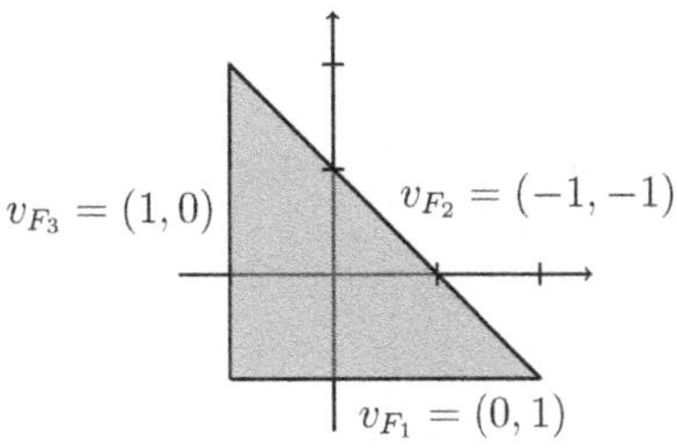

Recall that the Gorenstein condition implies that the canonical bundle extends across the singularities of $\mathbb{P}_\Delta$, and hence forms an honest line bundle on the entire variety. Given this, the Fano condition says that the resulting bundle is ample.

Rather than prove the theorem, we trace its manifestation in the particular case of weighted projective space.

Example 35 By definition, weighted projective space is the quotient

$$\mathbb{P}(c_0, \dots, c_n) = \frac{(\mathbb{C}^{n+1} \setminus \{0\})}{\mathbb{C}^*},$$

where $\mathbb{C}^*$ acts by

$$\lambda(z_0, \dots, z_n) := (\lambda^{c_0} z_0, \dots, \lambda^{c_n} z_n).$$

Weighted projective space is easily seen to be a toric variety. In order to construct its fan, one must find $v_0, \dots, v_n$ satisfying a relation

$$c_0 v_0 + \dots + c_n v_n = 0.$$

Assume, for simplicity, that $c_0 = 1$. Then the vectors v_i can be taken to be the following:

$$\begin{pmatrix} v_1 \\ \vdots \\ v_n \\ v_0 \end{pmatrix} = \begin{pmatrix} 1 & 0 & \cdots & 0 \\ 0 & 1 & \cdots & 0 \\ \vdots & \vdots & & \vdots \\ 0 & 0 & \cdots & 1 \\ -c_1 & -c_2 & \cdots & -c_n \end{pmatrix}.$$

For example, the fan for $\mathbb{P}(1, c_1, c_2)$ is shown in Fig. 14.

Let Δ be the polytope Δ associated to $\mathbb{P}(1, c_1, \dots, c_n)$ and its anticanonical divisor. If Δ is reflexive, then it is defined by the inequalities $\langle m, v_i \rangle \geq -1$ for $0 \leq i \leq n$. In particular, vertices of Δ would correspond to points where all but one of these is an equality:

$$\langle m, v_j \rangle = -1 \text{ for } j \neq i.$$

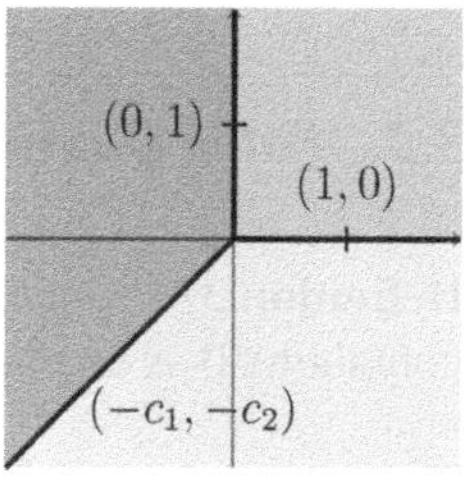

Fig. 14 The fan for $\mathbb{P}(1, c_1, c_2)$

These equations for $i \neq 0$ imply that any point $m = (m_0, \ldots, m_n) \in \Delta$ satisfies

$$m_j = -1 \text{ for } j \neq i,$$

$$-c_1 m_1 - \cdots - c_n m_n = -1,$$

which implies that

$$c_i x_i = 1 + \sum_{j \neq 0, i} c_j = \sum_{j \neq i} c_j.$$

This equation must have an integral solution in order for such a polytope Δ to exist; thus, one must have

$$c_i \,\Big|\, \sum_{j=0}^{n} c_j$$

We have only proved one implication of Theorem 34, and we have made the simplifying assumption that $c_0 = 1$, but in fact, the analysis holds more generally:

Proposition 36 *The following are equivalent:*

(1) $\mathbb{P}(c_0, \ldots, c_n)$ *is Gorenstein.*
(2) The polytope Δ *associated to* $\mathbb{P}(c_0, \ldots, c_n)$ *with its anticanonical bundle is reflexive.*
(3) $c_i \,\big|\, \sum_{j=0}^{n} c_j$ *for each* i.

Moreover, when these are satisfied, the polynomial

$$x_0^{d/c_0} + \cdots + x_n^{d/c_n}$$

with $d := \sum_{j=0}^{n} c_j$ *defines a Calabi-Yau hypersurface of* $\mathbb{P}(c_0, \ldots, c_n)$.

The last statement follows from the adjunction formula. Polynomials of the form $x_0^{a_0} + \cdots + x_n^{a_n}$ are referred to as **Fermat**. Thus, the result implies that a weighted projective space is Gorenstein if and only if it contains a Calabi-Yau hypersurface defined by a Fermat polynomial.

3.2 Polar Duality

Definition 37 Suppose that $\Delta \subset M_{\mathbb{R}}$ is a full-dimensional lattice polytope containing 0 as an interior point. Then the **polar dual** of Δ is

$$\Delta^\circ := \{n \in N_{\mathbb{R}} \mid \langle m, n \rangle \geq -1 \text{ for all } m \in \Delta\}.$$

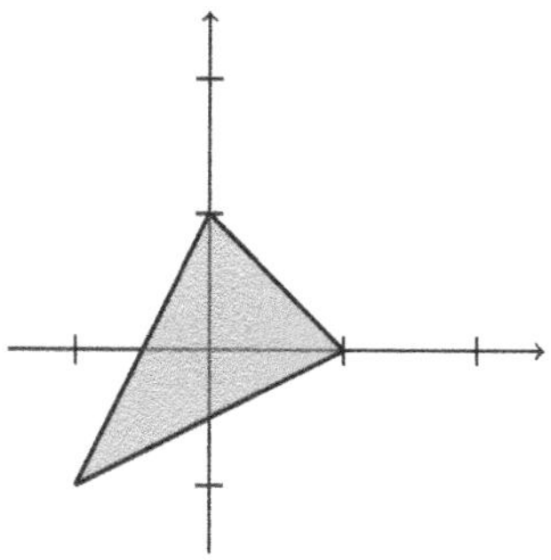

Fig. 15 The polar dual of the polytope in Fig. 13

It is straightforward to check that $(\Delta^\circ)^\circ = \Delta$, justifying the name "duality", and furthermore, that Δ is reflexive if and only if Δ° is reflexive.

Example 38 The polar dual of the polytope Δ in Fig. 13 is shown in Fig. 15. Notice that the vectors v_F for Δ become the vertices of Δ°.

One can see from this example that the normal fan Σ_Δ of a polytope Δ can also be obtained by taking cones over faces of Δ°. This is indeed a general feature of polar duality, and will be important when understanding the relationship between $\mathbb{P}_\Delta$ and $\mathbb{P}_{\Delta^\circ}$ encoded by Batyrev mirror symmetry.

3.3 Batyrev Mirror Symmetry

If $\mathbb{P}_\Delta$ is the toric variety associated to a reflexive polytope Δ, then any hypersurface defined by the vanishing of a generic section of the anti-canonical bundle $\mathcal{O}\left(\sum_\rho \mathcal{D}_\rho\right)$ will automatically be Calabi-Yau by the adjunction formula. Varying the section yields a family of Calabi-Yau hypersurfaces, denoted by $X \subset \mathbb{P}_{\Delta^\circ}$.

Definition 39 Let Δ be a reflexive polytope The **Batyrev mirror** of the family of Calabi-Yau hypersurfaces $X \subset \mathbb{P}_\Delta$ as above is the family $X^\circ \subset \mathbb{P}_{\Delta^\circ}$ of hypersurfaces defined by the vanishing of a generic section of the anticanonical bundle of $\mathbb{P}_{\Delta^\circ}$.

In what sense are these two families of Calabi-Yau hypersurfaces mirror to one another? One fairly straightforward yet important fact is the **monomial-divisor correspondence**, which is a bijection

$$\left\{\begin{array}{c}\text{monomials in}\\ \text{coordinates of } \mathbb{P}_\Delta\end{array}\right\} \leftrightarrow \left\{\begin{array}{c}\text{toric orbifold}\\ \text{divisors in } \Sigma_{\Delta^\circ}\end{array}\right\}.$$

To see why such a correspondence should hold, note that a nonzero element $m \in M \cap \Delta$ can yield two different objects. On the one hand, since $\mathbb{P}_\Delta$ is defined as the closure of the image of the map

$$f : T \to \mathbb{P}^k$$

$$f(t) = [m_0(t) : \cdots : m_k(t)]$$

in which $\{m_0, \ldots, m_k\} = M \cap \Delta \setminus \{0\}$, the element m gives a coordinate function of $\mathbb{P}_\Delta$. On the other hand, each such m generates a ray in a subdivision of the fan Σ_{Δ° given by cones on faces of Δ. Thus, m yields a toric divisor in a resolution of $X_{\Sigma_{\Delta^\circ}}$, which can be viewed as an orbifold divisor in $X_{\Sigma_{\Delta^\circ}}$. The full monomial-divisor correspondence comes from taking linear combinations of the elements $m \in M \cap \Delta$.

More generally, when X and X° are families of Calabi-Yau threefolds, Batyrev's theorem can be phrased in modern terminology as follows:

Theorem 40 (Batyrev) *There are isomorphisms*

$$H^{1,1}_{CR}(X) \cong H^{2,1}_{CR}(X^\circ)$$

and

$$H^{2,1}_{CR}(X) \cong H^{1,1}_{CR}(X^\circ).$$

Note that one must use Chen-Ruan cohomology [8] to account for the presence of orbifold divisors; for a review of Chen-Ruan cohomology, see the Appendix.

To put it another way, one can define a state space for the Calabi-Yau A-model as $H^*_{CR}(X)$, and define a state space for the Calabi-Yau B-model as the same vector space but with a different bi-grading. Theorem 40 is then a bi-degree preserving isomorphism between the A-model of X and the B-model of X°.

For smooth threefolds, toric divisors give elements of $H^{1,1}_{CR}(X^\circ)$, while monomials in the coordinate functions give sections of the anticanonical bundle and these generate $H^{2,1}_{CR}(X)$. Hence, Theorem 40 generalizes the monomial-divisor correspondence.

Example 41 Let us explore the Batyrev mirror symmetry construction for the quintic threefold $X \subset \mathbb{P}^4$, which is defined by the vanishing of a section of the anticanonical bundle $\mathcal{O}_{\mathbb{P}^4}(5)$.

Generalizing Example 6, the fan for $\mathbb{P}^4$ is generated by the rows of the matrix

$$A := \begin{pmatrix} 1 & 0 & 0 & 0 \\ 0 & 1 & 0 & 0 \\ 0 & 0 & 1 & 0 \\ 0 & 0 & 0 & 1 \\ -1 & -1 & -1 & -1 \end{pmatrix}.$$

If Δ is the polytope associated to $\mathbb{P}^4$ and its anticanonical divisor, then taking cones on the faces of Δ° should yield the above fan. From here, it is easy to see that Δ° must be the convex hull of the rows of A, and hence the integral points of Δ° are

$$(0,0,0,0), (-1,-1,-1,-1), (1,0,0,0), (0,1,0,0), (0,0,1,0), (0,0,0,1).$$

Thus, $\mathbb{P}_{\Delta^\circ}$ is the closure of the image of the morphism

$$f^\circ : (\mathbb{C}^*)^4 \to \mathbb{P}^5$$

$$[z_1 : z_2 : z_3 : z_4] \mapsto [1 : z_1^{-1}z_2^{-1}z_3^{-1}z_4^{-1} : z_1 : z_2 : z_3 : z_4].$$

More explicitly, this shows that $\mathbb{P}_{\Delta^\circ} \subset \mathbb{P}^5$ is defined by the equation

$$y_0^5 = y_1 y_2 y_3 y_4 y_5.$$

Let $(\mathbb{Z}_5)^3$ act on $\mathbb{P}^4$ diagonally, where we view

$$(\mathbb{Z}_5)^3 = \left\{ (\omega_0, \omega_1, \omega_2, \omega_3, \omega_4) \in (\mathbb{C}^*)^5 \mid \omega_i^5 = 1 \text{ for all } i, \prod_{i=0}^{4} \omega_i = 1 \right\} / G$$

and $G = \{(\omega, \omega, \omega, \omega, \omega)\}$ is the subgroup of elements that act trivially on $\mathbb{P}^4$. Then there is a map

$$[\mathbb{P}^4/(\mathbb{Z}_5)^3] \to \mathbb{P}^5$$

$$[\hat{x}_1 : \cdots : \hat{x}_5] \mapsto [\hat{x}_1\hat{x}_2\hat{x}_3\hat{x}_4\hat{x}_5 : \hat{x}_1^5 : \cdots : \hat{x}_5^5].$$

This map is an isomorphism onto $\mathbb{P}_{\Delta^\circ} \subset \mathbb{P}^4$.

The family of anti-canonical hypersurfaces in $\mathbb{P}_{\Delta^\circ}$ are given by the restriction of linear functions in the coordinates on $\mathbb{P}^5$. From the perspective of $[\mathbb{P}^4/(\mathbb{Z}_5)^3]$, these correspond to the $(\mathbb{Z}_5)^3$-invariant quintics

$$a_1\hat{x}_1^5 + \cdots + a_5\hat{x}_5^5 + a_0\hat{x}_1 \cdots \hat{x}_5 = 0.$$

Via a change of variables that rescales the coordinates, any such quintic can be expressed as

$$\{x_1^5 + \cdots + x_5^5 + \psi x_1x_2x_3x_4x_5 = 0\} \subset [\mathbb{P}^4/(\mathbb{Z}_5)^3]$$

for a constant ψ. The family of these hypersurfaces (as ψ varies) is the most common expression of the mirror family to the quintic threefold.

4 Hori-Vafa Mirror Symmetry

A few definitions are required in order to set the stage for the Hori-Vafa construction.

4.1 Basics of Symplectic Geometry

References for the material of this section include [7] and [25].

Definition 42 A **symplectic manifold** is a smooth manifold with a closed, nondegenerate 2-form, referred to as a **symplectic form**.

Let M be a symplectic manifold, and let G be a Lie group acting on M that preserves the symplectic form ω. Any element v in the Lie algebra $\mathfrak{g}$ defines a vector field X_v on M giving the infinitesimal action of v—that is,

$$X_v\Big|_x = \frac{d}{dt}\Big|_{t=0} \exp(tv) \cdot x$$

for any point $x \in M$, where $\exp : \mathfrak{g} \to G$ is the exponential map.

Definition 43 The action of G on M is **Hamiltonian** if

(1) there exists a **moment map**

$$\mu : M \to \mathfrak{g}^*,$$

defined by the property that

$$\omega(X_v, \cdot) = d\Big(\mu(\cdot)(v)\Big)$$

for all $v \in \mathfrak{g}$ (here, $\mu(\cdot)(v)$ is a smooth function $M \to \mathbb{R}$, so its differential is a 1-form on M);

(2) the map

$$\mathfrak{g} \to C^\infty(M)$$

$$v \mapsto \mu(\cdot)(v)$$

is a Lie algebra homomorphism, where the Lie bracket on $C^\infty(M)$ is defined to be the Poisson bracket.

Suppose that (M, ω) is a symplectic manifold equipped with a Hamiltonian action by a group G. Let $s \in \mathfrak{g}^*$ be a regular value of μ. Then $\mu^{-1}(s)/G$ has the structure of a symplectic orbifold. Toric varieties can be constructed as symplectic orbifolds in this way via a method known as **symplectic reduction**, which we describe below.

Start with the symplectic manifold $M = \mathbb{C}^n$, with the standard symplectic structure:

$$\omega = \frac{1}{2}\sum_{i=1}^{n} dx_i \wedge dy_i = -\frac{1}{2}\mathrm{Im}\left(\sum_{i=1}^{n} dz_i \wedge d\overline{z_i}\right).$$

Let G be the torus $(S^1)^r$. An action of G on M is specified by a charge matrix $Q = (Q_{ij})$, where

$$g \cdot x := gQx$$

for a row vector $g \in G$ and a column vector $x \in M$. The moment map for this action is

$$\mu(z_1, \ldots, z_n) = \left(\frac{1}{2}\sum_{i=1}^{n} Q_{1i}|z_i|^2, \ldots, \frac{1}{2}\sum_{i=1}^{n} Q_{ri}|z_i|^2\right).$$

Any of the symplectic orbifolds obtained as $\mu^{-1}(s)/G$ for a regular value s of this moment map will be toric; namely, they will be of the form

$$X_\Sigma = \frac{\mathbb{C}^n \setminus Z(\Sigma)}{(\mathbb{C}^*)^r}$$

for some discriminant locus $Z(\Sigma)$. The discriminant locus depends on s—it can be thought of as the complement of the image of $\mu^{-1}(s)$ under the action of $(\mathbb{C}^*)^r$—but only in a rather coarse way. In particular, X_Σ is independent of s within "chambers", connected regions of regular values; only when s crosses a "wall" at a critical value of μ will the toric variety change.

Example 44 Consider a toric variety of the form

$$X_\Sigma = \frac{\mathbb{C}^{n+1} \setminus Z(\Sigma)}{\mathbb{C}^*},$$

in which the action of $\mathbb{C}^*$ on $\mathbb{C}^{n+1}$ has charge matrix $(1, 1, \ldots, 1, -d)$. Such a variety arises via symplectic reduction on the symplectic manifold $M = \mathbb{C}^{n+1}$, where $G = S^1$ and the moment map is given in coordinates $z_1, \ldots, z_n, p$ on M by

$$\mu = \frac{1}{2}\left(\sum_{i=1}^{n} |z_i|^2 - d|p|^2\right).$$

The only place where all of the partial derivatives of μ vanish is $z_1 = \cdots = z_5 = p = 0$, so the only critical value is $s = 0$. Thus, there are two chambers:

- If $s > 0$, then the equation $\mu(z_1, \ldots, z_n, p) = s$ implies

$$\sum_{i=1}^{n} |z_i|^2 = d|p|^2 + \frac{1}{2}s,$$

which can only occur if $\sum_{i=1}^{n} |z_i|^2 \neq 0$. The discriminant locus, then, is

$$Z(\Sigma) = \{z_1 = \cdots = z_n = 0\},$$

and we obtain

$$\mu^{-1}(s)/S^1 = \frac{\mathbb{C}^{n+1} \setminus \{z = 0\}}{\mathbb{C}^*} = \mathcal{O}_{\mathbb{P}^{n-1}}(-d).$$

- If $s < 0$, then the equation $\mu(z_1, \ldots, z_n, p) = s$ forces that $p \neq 0$, so

$$Z(\Sigma) = \{p = 0\},$$

and

$$\mu^{-1}(s)/S^1 = \frac{\mathbb{C}^{n+1} \setminus \{p = 0\}}{\mathbb{C}^*} = [\mathbb{C}^n/\mathbb{Z}_d].$$

4.2 *Gauged Linear Sigma Models*

Using the notion of symplectic reduction, one can define gauged linear sigma models, which are the objects that will be the Hori-Vafa mirrors of varieties.

Definition 45 A **gauged linear sigma model**, or **GLSM**, consists of the following data:

(1) an $r \times n$ charge matrix, which can be viewed as the definition of an action of $(\mathbb{C}^*)^r$ on $\mathbb{C}^n$;
(2) a moment map $\mu : \mathbb{C}^n \to \mathbb{C}^r$;
(3) an R-charge, which is an action of $\mathbb{C}^*$ on $\mathbb{C}^n$ (denoted $\mathbb{C}^*_R$ to distinguish it from the action in (1));
(4) a superpotential, which is a map $W : \mathbb{C}^n \to \mathbb{C}$ satisfying:

   (a) W is invariant under the action of $(\mathbb{C}^*)^r$;
   (b) W is homogeneous of degree 2 under the action of $\mathbb{C}^*_R$;
   (c) the critical locus of W is compact inside $\mu^{-1}(s)/(\mathbb{C}^*)^r$ for any regular value s of μ.

The first two parts of the definition amount to the definition of a toric variety by the technique of symplectic reduction. The superpotential is constructed so it gives a well-defined map with compact critical locus out of this toric variety.

Example 46 The two toric varieties considered in Example 44 are part of the same GLSM, in which the charge matrix and moment map are as specified above. The superpotential is

$$W = p(z_1^d + \cdots + z_n^d),$$

which is invariant under the action of $\mathbb{C}^*$ on $\mathbb{C}^{n+1}$ and hence defines a map out of either of the toric varieties constructed above. Its critical locus in $\mu^{-1}(s)/\mathbb{C}^*$ when $s > 0$ is the hypersurface $\{z_1^d + \cdots + z_n^d = 0\}$ inside the zero-section of $\mathcal{O}_{\mathbb{P}^{n-1}}(-d)$, while its critical locus when $s < 0$ is $\{0\} \subset [\mathbb{C}^n/\mathbb{Z}_d]$. In either case, this locus is indeed compact.

The R-charge of this GLSM can have weights $(0, \ldots, 0, 2)$, but this choice is not unique. For example, another possibility is that the R-charge could be $(1, \ldots, 1, -3)$, and these different choices yield possibly different physical theories.

Definition 47 A **Landau-Ginzburg model** is a GLSM with a choice of chamber but *without* a choice of R-charge. In other words, it consists of a toric variety X_Σ and a map

$$W : X_\Sigma \to \mathbb{C}$$

whose critical locus is compact.

Example 48 Associated to the GLSM considered in Example 46, there are two Landau-Ginzburg models, one for each of the chambers. When $s > 0$, one has the Landau-Ginzburg model

$$W : \mathcal{O}_{\mathbb{P}^{n-1}}(-d) \to \mathbb{C}$$

$$W = p \sum_{i=1}^{5} z_i^d,$$

while for $s < 0$, the Landau-Ginzburg model is

$$W : [\mathbb{C}^n/\mathbb{Z}_d] \to \mathbb{C}$$

$$W = \sum_{i=1}^{n} z_i^d,$$

since the p-coordinate is rescaled to 1 in this presentation.

The various Landau-Ginzburg models associated to a single GLSM are sometimes called **phases**.

Although a Landau-Ginzburg model need not have an R-charge, one can often be chosen, as the above examples illustrate. A general principle of mirror symmetry is that Calabi-Yau models should have quasihomogeneous mirrors; in other words, the Landau-Ginzburg mirror of a Calabi-Yau hypersurface or complete intersection should have a chosen R-charge.

4.3 The Jacobian Ring

Hori-Vafa mirror symmetry is an isomorphism of the Calabi-Yau A-model state space associated to a variety (which, recall, is simply its cohomology) with the Landau-Ginzburg B-model state space of a mirror Landau-Ginzburg model. Specifically, the B-model state space associated to a Landau-Ginzburg model (Y, W), in which Y is a toric variety of dimension n, is the **Jacobian ring** (sometimes called the **Milnor ring**):

$$\mathrm{Jac}(W) := \frac{\mathbb{C}[x_1, \ldots, x_n]}{(\partial_{x_1} W, \ldots, \partial_{x_n} W)}.$$

The situation is especially simple when W is a Morse function. Indeed, the Jacobian ring can be "localized" in the sense that

$$\mathrm{Jac}(W) \cong \prod_{x_0} \mathrm{Jac}(W)|_{\text{neighborhood of } x_0},$$

where x_0 ranges over critical points of W. If W is Morse, then it takes the form $W = \sum_i y_i^2$ in appropriate coordinates in a neighborhood of each critical point. It follows that, in the Morse case, the restriction of $\mathrm{Jac}(W)$ to a neighborhood of each critical point is one-dimensional, so the dimension of $\mathrm{Jac}(W)$ equals the number of critical points. Hori-Vafa mirror symmetry, in this situation, is the claim that if X is a variety with Hori-Vafa mirror (Y, W), then the number of critical points of W is equal to the dimension of $H^*_{CR}(X)$.

Example 49 The A_n-singularity is defined as the polynomial $W = x^{n+1}$, viewed as a function $\mathbb{C} \to \mathbb{C}$. This has a single critical point at $x = 0$, and its Jacobian ring is

$$\mathrm{Jac}(W) = \mathrm{Span}_{\mathbb{C}}\{1, x, x^2, \ldots, x^{n-1}\}.$$

Thus, the dimension of the Jacobian ring equals the number of critical points if and only if $n = 1$, which is indeed the only case in which the A_n-singularity is Morse.

Example 50 Let $W : (\mathbb{C}^*)^n \to \mathbb{C}$ be defined by

$$W = x_1 + \cdots x_n + \frac{e^t}{x_1 \cdots x_n},$$

where t is an unspecified parameter. Then

$$\partial_{x_i} W = 1 - e^t x_1^{-1} x_2^{-1} \cdots x_n^{-1} x_i^{-1},$$

which vanishes only when

$$x_i = e^t x_1^{-1} \cdots x_n^{-1}. \tag{9}$$

Since the right-hand side is independent of i, there can only be a critical point when $x_1 = \cdots = x_n = \lambda$ for some $\lambda \in \mathbb{C}$. Equation (9) implies that $\lambda^{n+1} = e^t$, so the $n+1$ solutions to this equation will yield $n+1$ critical points. It is straightforward to check that all of these critical points are nondegenerate, so W is Morse and the dimension of $\mathrm{Jac}(W)$ equals $n+1$.

4.4 Hori-Vafa Mirrors of Compact Toric Varieties

We are finally equipped to describe the cohomological statement of Hori-Vafa mirror symmetry. The material of this section, and of the remainder of the section, is based on [20].

Definition 51 Let

$$X_\Sigma = \frac{\mathbb{C}^n \setminus Z(\Sigma)}{(\mathbb{C}^*)^r}$$

be a compact toric variety with charge matrix $Q = (Q_{ij})$. Then the **Hori-Vafa mirror** is the Landau-Ginzburg model on the toric variety

$$\{x \in \mathbb{C}^n \mid x_1^{Q_{1j}} \cdots x_n^{Q_{nj}} = e^{t_j} \text{ for all } 1 \le j \le r\}$$

with superpotential given by the restriction of

$$W = x_1 + \cdots + x_n.$$

One should check that the above subset of $\mathbb{C}^n$ is indeed toric, and that the critical locus of W is compact. Having done this, the above constitutes a Landau-Ginzburg model, which should be mirror to X_Σ in the sense that

$$H^*_{CR}(X_\Sigma) \cong \mathrm{Jac}(W)$$

whenever X_Σ is semi-Fano. Let us verify this in some easy examples.

Example 52 The charge matrix of $\mathbb{P}^n$ is $(1, 1, \ldots, 1)$, so the underlying toric variety of the Hori-Vafa mirror is the subset of $\mathbb{C}^{n+1}$ defined by the equation

$$x_0 = \frac{e^t}{x_1 \cdots x_n},$$

which is isomorphic to $(\mathbb{C}^*)^n$. The superpotential is the polynomial considered in Example 50. We saw in that example that $\mathrm{Jac}(W) \cong \mathbb{C}^{n+1}$, so it does match $H^*(\mathbb{P}^n)$ as a vector space.

Example 53 A similar computation yields the Hori-Vafa mirror of weighted projective space. Assume for simplicity that $c_0 = 1$, so $\mathbb{P}(c_0, \ldots, c_n)$ has charge matrix $(1, c_1, \ldots, c_n)$. Then the superpotential of the Hori-Vafa mirror is

$$W = x_1 + \cdots + x_n + \frac{e^t}{x_1^{c_1} \cdots x_n^{c_n}}.$$

To confirm the statement of mirror symmetry, we compute

$$\partial_{x_i} W = 1 - c_i e^t x_1^{-c_1} \cdots x_n^{-c_n} x_i^{-1},$$

so the critical points occur when

$$\frac{x_1}{c_1} = \cdots = \frac{x_n}{c_n} = e^t x_1^{-c_1} \cdots x_n^{-c_n}.$$

If this common value is denoted λ, then we have

$$\lambda \prod_{i=1}^{n} (c_i \lambda)^{c_i} = e^t,$$

yielding an equation $\lambda^{1+c_1+\cdots+c_n} = Ke^t$ for a constant K. It follows that there are $1 + c_1 + \cdots + c_n$ critical points. All of these are Morse, so

$$\mathrm{Jac}(W) \cong \mathbb{C}^{1+c_1+\cdots+c_n},$$

which is indeed isomorphic to $H^*_{CR}(\mathbb{P}(c_0, c_1, \ldots, c_n))$.

The failure of Hori-Vafa mirror symmetry in the non-semi-Fano case can be observed explicitly for Hirzebruch surfaces F_n, which are semi-Fano if and only if $n = 1$.

Example 54 Let F_n be the Hirzebruch surface, given by

$$F_n = \frac{\mathbb{C}^4 \setminus Z(\Sigma)}{(\mathbb{C}^*)^2}$$

with charge matrix

$$Q = \begin{pmatrix} 1 & 1 & n & 0 \\ 0 & 0 & 1 & 1 \end{pmatrix}.$$

The constraints defining the toric variety of the mirror Landau-Ginzburg model, then, are

$$x_1 x_2 x_3^n = e^{t_1},$$

$$x_3 x_4 = e^{t_2}.$$

Denoting $q_i = e^{t_i}$ for $i = 1, 2$, the superpotential is

$$W = x_2 + x_3 + \frac{q_1}{x_2 x_3^n} + \frac{q_2}{x_3}.$$

An elementary computation shows that critical points occur only when

$$n^2 q_1 x_3^n = (x_3^{n+1} - x_3^{n-1} q_2)^2.$$

This is a polynomial of degree $2n + 2$, so for generic values of the parameters q_1 and q_2, it has $2n + 2$ solutions.

On the other hand, $H^*(F_n)$ is 4-dimensional. Thus, the number of critical points matches the rank of the cohomology only when $n = 1$, which is the semi-Fano case.

Remark 55 In recent years, some attempts have been made to adapt Hori-Vafa mirror symmetry to the non-semi-Fano case. This involves adding higher-order terms to W to ensure that the number of its critical points coincides with the rank of the cohomology of the non-semi-Fano variety. These correction terms have an interpretation in terms of counting holomorphic discs in X_Σ.

Beyond the state space level, the statements of mirror symmetry at the level of rings and quantum D-modules have also been extended to non-semi-Fano toric varieties, by passing to certain q-adic completions. See the work of Iritani [22] and Gonzalez-Woodward [19] for details.

4.5 Hori-Vafa Mirrors of Noncompact Toric Varieties

A simple example suffices to show that Hori-Vafa mirror symmetry, in the form stated above, fails when X_Σ is noncompact:

Example 56 Let $X_\Sigma = \mathbb{C}$. Then the Landau-Ginzburg mirror is $W = x : \mathbb{C} \to \mathbb{C}$. This has no critical points, so the duality between $H^*(X_\Sigma)$ and the Jacobian ring fails.

More heuristically, the reason why the above procedure requires compactness involves the aspects of mirror symmetry beyond cohomology. Roughly speaking, genus-zero mirror symmetry is a correspondence between the quantum cohomology of X_Σ, encapsulated by the J-function, and the oscillatory integrals

$$\int_\Delta e^{-W/z} d\log(x_1)\cdots d\log(x_n)$$

against cycles Δ. When W satisfies certain growth properties, the cycles Δ are in bijection with critical points of W, which, in turn, are supposed to correspond via mirror symmetry to the cohomology of X_Σ. In particular, the J-function and the oscillatory integrals depend on the same number of parameters. If, however, X_Σ is noncompact, then the J-function is not well-defined, and the requisite growth properties of W fail to hold. Thus, the statement of mirror symmetry breaks down on both sides.

On the other hand, noncompact toric varieties *do* have equivariant J-functions, so one might expect that there is an equivariant version of the Hori-Vafa mirror for which the mirror symmetry statement is still valid. This is indeed the case.

Definition 57 Let $X_\Sigma = (\mathbb{C}^n \setminus Z(\Sigma))/(\mathbb{C}^*)^r$ be a toric variety with charge matrix Q. Then its **equivariant Hori-Vafa mirror** is the Landau-Ginzburg model with superpotential

$$W = x_1 + \cdots + x_n - \sum_{i=1}^{n} \lambda_i \log(x_i)$$

on the subset of $\mathbb{C}^{n+1}$ defined by the constraints

$$\prod_{i=1}^{n} x_i^{Q_{ib}} = q_b$$

for nonzero parameters q_b. Here, λ_i is a constant, viewed as the equivariant parameter for the ith $\mathbb{C}^*$ action.

Example 58 Let $X_\Sigma = \mathbb{C}$. Then the superpotential of the equivariant Hori-Vafa mirror is

$$W = x - \lambda \log(x),$$

so

$$\partial_x W = 1 - \frac{\lambda}{x}.$$

Unlike the nonequivariant case, this now has a single critical point, so we recover the correspondence between the dimension of $H^*(X_\Sigma)$ and the number of critical points.

Example 59 Let X_Σ be the total space of the bundle $\mathcal{O}_{\mathbb{P}^n}(-d)$, which has charge matrix $(1, 1, \ldots, 1, -d)$. Then

$$W = x_0 + x_1 + \cdots + x_{n+1} - \sum_{i=0}^{n+1} \lambda_i \log(x_i)$$

and the constraint defining the toric variety is

$$x_0 \cdots x_n x_{n+1}^d = q.$$

Thus, we obtain

$$W = x_1 + \cdots + x_{n+1} + \frac{x_{n+1}^d}{x_1 \cdots x_n} - \sum_{i=0}^{n+1} \lambda_i \log(x_i).$$

It is straightforward to check that the number of critical points of this superpotential indeed matches the dimension of $H^*(X_\Sigma)$.

Example 60 It is not always necessary to modify W by all of the terms $\lambda_i \log(x_i)$ in order to achieve mirror symmetry. For example, consider the toric variety $\mathcal{O}_{\mathbb{P}^1}(-1) \oplus \mathcal{O}_{\mathbb{P}^1}(-1)$, which has charge matrix $Q = (1, 1, -1, -1)$. Its nonequivariant Hori-Vafa mirror is

$$W = x_2 + x_3 + x_4 + q\frac{x_3 x_4}{x_2}$$

defined over $(\mathbb{C}^*)^3$, while the equivariant mirror would subtract the terms $\sum_{i=1}^4 \lambda_i \log(x_i)$ from the above.

Even the partial modification

$$\widetilde{W} = x_2 + x_3 + x_4 + q\frac{x_3 x_4}{x_2} - \lambda_3 \log(x_3) - \lambda_4 \log(x_4)$$

of W upholds mirror symmetry, though. This makes sense from the perspective of the J-function; only the two $\mathbb{C}^*$ actions in the noncompact fiber directions of $\mathcal{O}_{\mathbb{P}^1}(-1) \oplus \mathcal{O}_{\mathbb{P}^1}(-1)$ are necessary in order to make the equivariant Gromov-Witten theory well-defined.

Despite the failure of nonequivariant Hori-Vafa mirror symmetry for noncompact toric varieties, the noncompact mirrors are still worth remembering, as they play a role in the hypersurface version of mirror symmetry considered below.

4.6 The Orbifold Jacobian Ring

The Hori-Vafa construction can be adapted to give the mirror of a smooth semi-Fano hypersurface in a toric variety. As we will see, however, the resulting Landau-Ginzburg model will have a nontrivial symmetry group, and the definition of its B-model state space must be modified accordingly.

Consider a Landau-Ginzburg model of the form $(\mathbb{C}^N, W)$, where $W : \mathbb{C}^N \to \mathbb{C}$ is a superpotential. Let $G \subset (\mathbb{C}^*)^N$ be a (finite) group of diagonal matrices preserving W; this is referred to as a symmetry group of the Landau-Ginzburg model.

We will define an "orbifolded" version of the Jacobian ring $\text{Jac}(W)$ that takes the data of G into account. This is modelled on the definition of the Chen-Ruan cohomology of a global quotient, which is described in the Appendix. As in the case of Chen-Ruan cohomology, a certain cohomology group is attached to each fixed point of the G-action, and the state space is formed by taking the G-invariant part of the direct sum of the contributions from all of the fixed points.

Definition 61 Given a polynomial $W : \mathbb{C}^N \to \mathbb{C}$ and a group of symmetries G of the associated Landau-Ginzburg model, the **B-model Landau-Ginzburg cohomology** of $(\mathbb{C}^N, W, G)$ is the vector space

$$\text{Jac}(W, G) := \left(\bigoplus_{g \in G} \text{Jac}(W_g) \right)^G,$$

where

$$W_g = W|_{\text{Fix}(g)}$$

and the G-invariant part is taken with respect to the action of G on $\bigoplus_{g \in G} \text{Jac}(W_g)$ that sends

$$\text{Jac}(W_g) \to \text{Jac}(W_{h^{-1}gh})$$

$$\phi \mapsto \det(h) \cdot h^* \phi.$$

for each $h \in G$.

This state space will be explained further in Sect. 5.5 of the next section.

4.7 Hori-Vafa Mirrors of Hypersurfaces in Toric Varieties

Let us begin by studying the case of Fermat hypersurfaces in projective space.

Example 62 Let $X_d \subset \mathbb{P}^{N-1}$ be a smooth degree-d hypersurface defined by the vanishing of the polynomial $A = x_1^d + \cdots + x_N^d$. Explicitly, the semi-Fano condition corresponds to the requirement that $d \leq N$.

To form the Hori-Vafa mirror, one first constructs a GLSM on a noncompact toric variety for which X_d is the critical locus of the superpotential in a particular phase. Namely, let $\mathbb{C}^*$ act on $\mathbb{C}^{N+1}$ with charge matrix

$$(1, \ldots, 1, -d),$$

so that the first N factors give precisely the charge matrix for $\mathbb{P}^{N-1}$. Let the superpotential be

$$W = p \cdot A(x_1, \ldots, x_N) : \mathbb{C}^{N+1} \to \mathbb{C}$$

in coordinates $(x_1, \ldots, x_N, p)$ on $\mathbb{C}^{N+1}$, and let the moment map be

$$\mu = \frac{1}{2}\left(\sum_{i=1}^{N} |z_i|^2 - d|p|^2\right).$$

(This generalizes the GLSM appearing in Examples 44 and 46.) As we have seen previously, in the $s > 0$ phase of this GLSM, the critical locus of W is precisely the hypersurface X_d.

Next, construct the (non-equivariant) Hori-Vafa mirror of this noncompact toric variety containing X_d. This is sometimes called the **pre-Hori-Vafa mirror** of X_d. In this case, it is the Landau-Ginzburg model with superpotential

$$\widetilde{W} = x_1 + \cdots + x_N + x_{N+1}$$

on the subset of $\mathbb{C}^{N+1}$ satisfying the constraint

$$x_1 \cdots x_N x_{N+1}^{-d} = e^t.$$

In other words, setting

$$x_i = u_i^d \text{ for } 1 \leq i \leq N$$

and

$$x_{N+1} = u_{N+1},$$

the pre-Hori-Vafa mirror becomes

$$\widetilde{W} = u_1^d + \cdots + u_N^d + e^{-t/d} u_1 \cdots u_N$$

on the toric variety $(\mathbb{C}^*)^N$.

This Landau-Ginzburg model has a nontrivial symmetry group.[4] Namely, an automorphism of $(\mathbb{C}^*)^N$ of the form

$$u_i \mapsto \omega_d^{p_i} u_i,$$

for which ω_d is a dth root of unity and

$$\omega_d^{p_1+\cdots+p_N} = 1,$$

will preserve the superpotential $\widetilde{W}$. The group of such symmetries is denoted $\mathrm{SL}(W_0)$, since if $W_0 = u_1^d + \cdots + u_N^d$, the automorphisms in question are precisely

$$\left\{ \begin{pmatrix} \lambda_1 & & \\ & \ddots & \\ & & \lambda_N \end{pmatrix} \;\middle|\; W_0(\lambda_1 x_1, \ldots, \lambda_N x_N) = W_0(x_1, \ldots, x_N) \right\} \cap \mathrm{SL}_N(\mathbb{C}).$$

Finally, to form the Hori-Vafa mirror of X_d, one "compactifies" the pre-Hori-Vafa mirror by adding the point $u_1 = \cdots = u_N = 0$ to the domain $\mathbb{C}^N$ of the Landau-Ginzburg model, and then takes the quotient by the above symmetry group. This yields

$$\widetilde{W} : [\mathbb{C}^N/\mathrm{SL}(W_0)] \to \mathbb{C}$$

$$\widetilde{W} = u_1^d + \cdots + u_N^d + e^{-t/d} u_1 \cdots u_N$$

as the mirror. The claim, then, is:

$$H^*_{CR}(X_d) \cong \mathrm{Jac}(\widetilde{W}, \mathrm{SL}(W_0)).$$

That is, the Calabi-Yau A-model cohomology of X_d is isomorphic to the Landau-Ginzburg B-model cohomology of $(\mathbb{C}^N, \widetilde{W}, \mathrm{Aut}(\widetilde{W}))$.

[4] An alternate explanation for the appearance of this automorphism group can be given in terms of the data of the Landau-Ginzburg B-model beyond cohomology. Specifically, as we have mentioned previously, the full Landau-Ginzburg B-model in genus zero can be viewed as encoding certain oscillatory integrals

$$\int_\Delta e^{-W/z} \omega,$$

where ω is a "primitive form". In the toric case, we have mentioned that $\omega = d\log(x_1) \wedge \ldots \wedge d\log(x_n)$. On the other hand, the primitive form for a hypersurface is $\omega = dx_1 \wedge \ldots \wedge dx_n$. From this perspective, the appearance of the automorphism group $\mathrm{SL}(W_0)$ is explained by the fact that only automorphisms with determinant 1 will preserve this form.

The same basic procedure computes the Hori-Vafa mirror of a hypersurface in a more general weighted projective space. It is necessary, however, to restrict to a certain class of polynomials.

Definition 63 A quasihomogeneous polynomial is **invertible** if the number of monomials equals the number of variables. That is, after rescaling the variables to absorb any coefficients, the polynomial can be written in the form

$$A = \sum_{i=1}^{N} \prod_{j=1}^{N} x_j^{m_{ij}}.$$

This condition implies that the exponent matrix $E_A = (m_{ij})$ of A is square. Assuming that $\{A = 0\}$ also defines a smooth orbifold in weighted projective space, it follows moreover that E_A is invertible, which explains the terminology.

Example 64 Consider a smooth degree-d hypersurface X in weighted projective space $\mathbb{P}(c_1, \ldots, c_N)$ defined by the vanishing of an invertible polynomial A with exponent matrix (m_{ij}). In this case, the semi-Fano condition is $d \leq \sum_{i=1}^{N} c_i$.

Again, one begins by constructing a GLSM in which X is the critical locus of the superpotential in some phase. Namely, the GLSM will have charge matrix $(c_1, \ldots, c_N, -d)$ and the superpotential will be $W = p \cdot A(x_1, \ldots, x_N)$. It follows that the pre-Hori-Vafa mirror of X is the nonequivariant Hori-Vafa mirror of the toric variety $\mathcal{O}_{\mathbb{P}(c_1,\ldots,c_N)}(-d)$ of this GLSM, which is

$$\widetilde{W} = x_1 + \cdots x_N + x_{N+1}$$

defined over the subset of $\mathbb{C}^{N+1}$ with constraint

$$x_1^{c_1} \cdots x_N^{c_N} x_{N+1}^{-d} = e^t. \tag{10}$$

The constraint can be "solved"—that is, expressed in terms of only N variables—by a change of coordinates:

$$x_i = \prod_{j=1}^{N} u_j^{m_{ji}}, \quad i = 1, \ldots, N, \tag{11}$$

$$x_{N+1} = u_{N+1}.$$

Then (10) becomes

$$\prod_{j=1}^{N} u_j^{\sum_i m_{ji} c_i} u_{N+1}^{-d} = e^t.$$

The fact that A is quasihomogeneous means that

$$\sum_{j=1}^{N} m_{ij} c_j = d$$

for each i, so the constraint is in fact

$$u_{N+1} = e^{-t/d} u_1 \cdots u_N.$$

It follows that the pre-Hori-Vafa mirror has superpotential

$$\widetilde{W} = \sum_{i=1}^{N} \prod_{j=1}^{N} u_j^{m_{ji}} + e^{-t/d} u_1 \cdots u_N : (\mathbb{C}^*)^N \to \mathbb{C}.$$

To form the mirror itself, we add $u_1 = \cdots = u_N = 0$ and take the quotient by $\mathrm{SL}(W_0)$, where $W_0 = \sum_{i=1}^{N} \prod_{j=1}^{N} u_j^{m_{ji}}$. Thus, the mirror is

$$\widetilde{W} : [\mathbb{C}^N / \mathrm{SL}(W_0)] \to \mathbb{C}$$

with $\widetilde{W}$ as above.

It is interesting to note that the term W_0 in the superpotential of the mirror is the transpose of the defining polynomial A of the hypersurface—that is, the exponent matrix of W_0 is the transpose of the exponent matrix for G. The idea that the transpose polynomial should appear in mirror symmetry was actually suggested before Hori and Vafa's work, by physicists Berglund and Hübsch, as we will discuss in the next section. At the time when Hori and Vafa proposed their mirror symmetry construction, however, this connection to previous work was not realized.

More generally, the hypersurfaces in toric varieties for which the Hori-Vafa mirror is defined are as follows:

Definition 65 Let $X \subset X_\Sigma := (\mathbb{C}^N \setminus Z(\Sigma))/(\mathbb{C}^*)^r$ be a hypersurface defined by the vanishing of a polynomial $A(x_1, \ldots, x_M)$, where $M \leq N$ and A is homogeneous of degrees $d_1, \ldots, d_r$ with respect to the r actions of $\mathbb{C}^*$. Let Q be the charge matrix of X_Σ. We say that X is **invertible** if

(1) $N - M + 1 = r$;
(2) the $r \times r$ matrix given by taking the last $N - M$ rows of Q and appending the row $(-d_1, \ldots, -d_r)$ is invertible;
(3) the exponent matrix of A is square.

When $X \subset X_\Sigma$ is a compact invertible semi-Fano hypersurface, its Hori-Vafa mirror can be constructed by exactly the same procedure as in Example 64. To summarize, one first forms the Hori-Vafa mirror of the toric variety

$$\frac{(\mathbb{C}^N \setminus Z(\Sigma)) \times \mathbb{C}}{(\mathbb{C}^*)^r},$$

in which the charge matrix is given by appending the row $(-d_1, \ldots, -d_r)$ to the charge matrix Q of X_Σ. This has superpotential

$$\widetilde{W} = x_1 + \cdots + x_{N+1}$$

and is defined on the subset of $\mathbb{C}^{N+1}$ satisfying the constraints

$$\prod_{i=1}^{N} x_i^{Q_{ij}} x_{N+1}^{-d_j} = e^{t_j}$$

for $j = 1, \ldots, r$. Then, one uses invertibility to express this subset of $\mathbb{C}^{N+1}$ as $(\mathbb{C}^*)^M$ in variables $u_1, \ldots, u_M$. Namely, if $(m_{ij})_{i,j=1,\ldots,M}$ is the exponent matrix of A and $(D_{ij})_{i,j=1,\ldots,r}$ is the invertible matrix described by condition (2) of Definition 65, then

$$x_i = \prod_{j=1}^{M} u_j^{m_{ji}}$$

for $i = 1, \ldots, M$ and

$$x_{M+b} = e^{\sum_{j=1}^{r} t_j D^{jb}} \prod_{i,j,k} u_k^{-m_{ki} Q_{ij} D^{jb}}$$

for $b = 1, \ldots, r$, where (D^{ij}) is the inverse of (D_{ij}).

The Hori-Vafa mirror is then

$$\widetilde{W} : [\mathbb{C}^M / \mathrm{Aut}(\widetilde{W})] \to \mathbb{C}.$$

Note that the symmetry group $\mathrm{Aut}(\widetilde{W})$ is in general smaller when $N - M$ is larger, since there are more constraints on the variables $u_1, \ldots, u_M$. This situation occurs when the original polynomial $W = p \cdot A$ has a larger symmetry group. We will explore the duality between the symmetries of W and those of $\widetilde{W}$ much further in the next section. For now, let us simply look at an example.[5]

Example 66 Consider the ambient toric variety

$$X_\Sigma = \frac{(\mathbb{C}^3 \setminus \{0\}) \times (\mathbb{C} \setminus \{0\})}{(\mathbb{C}^*)^2}$$

with charge matrix

$$Q = \begin{pmatrix} 1 & -2 \\ 1 & -1 \\ 1 & 0 \\ 0 & 3 \end{pmatrix}.$$

It is straightforward to check that $X_\Sigma = [\mathbb{P}^2/\mathbb{Z}_3]$. Let X be the hypersurface defined by the polynomial

$$A = x_1^3 + x_2^3 + x_3^3.$$

To form the Hori-Vafa mirror, first consider the toric variety

$$\frac{(\mathbb{C}^3 \setminus \{0\}) \times (\mathbb{C} \setminus \{0\}) \times \mathbb{C}}{(\mathbb{C}^*)^2}$$

with charge matrix

$$\begin{pmatrix} 1 & -2 \\ 1 & -1 \\ 1 & 0 \\ 0 & 3 \\ -3 & 0 \end{pmatrix}.$$

Its Hori-Vafa mirror (the pre-Hori-Vafa mirror of X) is

$$\widetilde{W} = x_1 + x_2 + x_3 + s + p$$

on the subset of $\mathbb{C}^5$ defined by the constraints

$$x_1 x_2 x_3 p^{-3} = e^{t_1}, \quad x_1^{-2} x_2^{-1} s^3 = e^{t_2}.$$

[5] We are grateful to Mark Shoemaker for pointing out this example, and for correcting our definition of invertibility in light of it.

Setting $x_i = u_i^3$ for $i = 1, \ldots, 3$, we find that

$$\widetilde{W} = u_1^3 + u_2^3 + u_3^3 + e^{-t_1/3} u_1 u_2 u_3 + e^{t_2/3} u_1^2 u_2.$$

The symmetry group of $\widetilde{W}$ is not all of $\mathrm{SL}(A^T)$, but instead is isomorphic to $\mathbb{Z}_3$. Thus, the Hori-Vafa mirror of X is

$$\widetilde{W} : [\mathbb{C}^3/\mathbb{Z}_3] \to \mathbb{C}$$

for $\widetilde{W}$ as above.

4.8 *Hori-Vafa Mirrors of Complete Intersections in Toric Varieties*

A similar procedure yields the Hori-Vafa mirror of a complete intersection in a toric variety, but the notion of invertibility must once again be adapted.

Definition 67 Let $X \subset X_\Sigma := (\mathbb{C}^N \setminus Z(\Sigma))/(\mathbb{C}^*)^r$ be a complete intersection defined by the vanishing of polynomials

$$A_1(x_1, \ldots, x_M), \ldots, A_k(x_1, \ldots, x_M),$$

where $M \le N$ and A_b is homogeneous of degree d_{ba} with respect to the ath $\mathbb{C}^*$ action. Let Q be the charge matrix of X_Σ. We say that X is **invertible** if

(1) $N - M + k = r$;
(2) the $r \times r$ matrix given by taking the last $N - M$ rows of Q and appending the matrix $(-d_{ba})$ is invertible;
(3) there exists a collection of N monomials

$$\prod_{j=1}^{N} x_j^{m_{ij}}, \quad i = 1, \ldots, N$$

such that

$$A_b = \sum_{i=1}^{N} n_{ib} \prod_{j=1}^{N} x_j^{m_{ij}}$$

with each $n_{ib} \in \{0, 1\}$ and $\sum_{b=1}^{k} n_{ib} = 1$. (In other words, there are no repeated monomials among the A_b, and the total number of monomials appearing is equal to the number of variables.)

Example 68 A complete intersection in $\mathbb{P}^{N_1-1} \times \cdots \times \mathbb{P}^{N_k-1}$ defined by invertible polynomials $\{A_i = 0\} \subset \mathbb{P}^{N_i-1}$ is invertible.

Example 69 A complete intersection in

$$\mathbb{P}^{N-1} \times \mathbb{P}^{M-1}$$

defined by polynomials

$$A_1 = \sum_{i=1}^{N} s_i^{d_i}$$

and

$$A_2 = \sum_{j=1}^{M} s_j t_j^{d_2}$$

with $N \geq M$ is invertible.

As in the case of hypersurfaces, we will proceed by first associating to a compact invertible semi-Fano complete intersection $X \subset X_\Sigma$ a GLSM for which X is the critical locus of the superpotential in a particular phase. From here, a similar set of constraints determined by the defining polynomials of X will yield the pre-Hori-Vafa mirror as a Landau-Ginzburg model. A partial compactification and quotient by symmetries will convert this into the final Hori-Vafa mirror of X.

For simplicity, let us describe the procedure only in the case where $N = M$, so that $k = r$. Then X is given by

$$\{A_1 = \cdots = A_k = 0\} \subset \frac{\mathbb{C}^N \setminus Z(\Sigma)}{(\mathbb{C}^*)^k}$$

in which the ambient toric variety X_Σ has charge matrix $Q = (Q_{ia})$. Suppose that A_b is homogeneous of degree d_{ba} with respect to the ath $\mathbb{C}^*$ action.

The associated GLSM is

$$W : \frac{(\mathbb{C}^N \setminus \{0\}) \times \mathbb{C}^k}{(\mathbb{C}^*)^k} \to \mathbb{C},$$

where $(\mathbb{C}^*)^k$ acts with charge matrix

$$\begin{pmatrix} Q_{11} & \cdots & Q_{1N} \\ \vdots & & \vdots \\ Q_{k1} & \cdots & Q_{kN} \\ -d_{11} & \cdots & -d_{1k} \\ \vdots & & \vdots \\ -d_{k1} & \cdots & -d_{kk} \end{pmatrix}$$

and

$$W = p_1 A_1(x_1, \ldots, x_N) + \cdots + p_k A_k(x_1, \ldots, x_N)$$

in coordinates $x_1, \ldots, x_N, p_1, \ldots, p_k$ on $\mathbb{C}^N \times \mathbb{C}^k$.

The pre-Hori-Vafa mirror is the Landau-Ginzburg model whose domain is the subset of $\mathbb{C}^{N+1}$ defined by the constraints

$$\prod_{i=1}^{N} x_i^{Q_{ia}} \prod_{b=1}^{k} x_N^{-d_{ba}} = e^{t_a}$$

for $a = 1, \ldots, a_k$ and whose superpotential is the restriction of

$$\widetilde{W} = x_1 + \cdots + x_N + x_{N+1} + \cdots + x_{N+k}$$

to this subset. If m_{ij} and n_{ib} are defined as in Definition 67, then one can check that the change of variables

$$x_i = \prod_{j=1}^{N} u_j^{m_{ji}}, \quad i = 1, \ldots, N,$$

$$x_{N+b} = e^{-\sum_{i=1}^{k} d^{bc} t_c} \prod_{j=1}^{N} u_j^{n_{jb}}, \quad b = 1, \ldots, k$$

solves the constraint. Here, d^{bc} is the inverse of the matrix d_{bc} and the expression $-d^{bc} t_c$ uses the Einstein summation convention, and hence should be understood as a sum over c. After this coordinate change, the superpotential of the pre-Hori-Vafa mirror is

$$\widetilde{W} = \sum_{i=1}^{N} \prod_{j=1}^{N} u_j^{m_{ji}} + \sum_{b=1}^{k} e^{-d^{bc} t_c} \prod_{j=1}^{N} u_j^{n_{jb}}.$$

After adding $u_1 = \cdots = u_N = 0$ and taking the quotient of the domain by symmetries preserving $\widetilde{W}$, one obtains the Hori-Vafa mirror.

4.9 An Alternative Description

For Calabi-Yau hypersurfaces in toric varieties, we now have two notions of the mirror: the Batyrev mirror and the Hori-Vafa mirror. Although the two constructions appear unrelated, a different description of the Hori-Vafa mirror can be used to show that it coincides in the Calabi-Yau case with Batyrev's definition.

Recall that for a degree-d semi-Fano hypersurface $X_d = \{A_d = 0\} \subset \mathbb{P}^{N-1}$, the first step in defining the Hori-Vafa mirror is to associate to X_d a GLSM with charge matrix $(1, \ldots, 1, -d)$ and superpotential $W = p \cdot A_d$. From here, the pre-Hori-Vafa mirror is defined as the Landau-Ginzburg model on the subset of $\mathbb{C}^{N+1}$ satisfying the constraint

$$x_1 \cdots x_N p^{-d} = e^t$$

with superpotential $\widetilde{W} = x_1 + \cdots + x_N + p$.

Previously, we expressed this Landau-Ginzburg model in terms of coordinates $u_1, \ldots, u_N$. Suppose, however, that we instead used the change of variables

$$p = \tilde{p}$$

$$x_i = \tilde{u_i}\tilde{p} \text{ for } 1 \leq i \leq d,$$

$$x_i = \tilde{u_i} \text{ for } d+1 \leq i \leq N.$$

Note that the semi-Fano condition $d \leq N$ is necessary for this to be well-defined.

In these coordinates, the pre-Hori-Vafa mirror is

$$\widetilde{W} = \tilde{p}(\tilde{u_1} + \cdots + \tilde{u}_{d+1}) + \sum_{i=d+1}^{N} \tilde{u_i}$$

on the subset of $\mathbb{C}^{N+1}$ satisfying the constraint

$$\prod_{i=1}^{N} \tilde{u_i} = e^t.$$

This GLSM, however, is "equivalent" to the GLSM with superpotential

$$\widetilde{W} = -\sum_{i=d+1}^{N} \tilde{u_i}$$

subject to the constraints

$$\sum_{i=1}^{d} u_i = -1, \quad \prod_{i=1}^{N} u_i = e^t. \tag{12}$$

Intuitively, it makes sense that these two theories would be equivalent, since the first of these new constraints kills the first term of the old superpotential. More explicitly, "equivalent" means that the two theories yield the same oscillatory integrals. In particular, since mirror symmetry is defined as a correspondence between the J-function and the generating function of such integrals, a theory equivalent to the mirror of X_d will still be mirror.

In the Calabi-Yau case, this new presentation of the mirror agrees with the Batyrev construction.

Example 70 The Calabi-Yau condition on $X_d \subset \mathbb{P}^{N-1}$ is $d = N$. In this case, the new version of the pre-Hori-Vafa mirror described above has no superpotential; it is simply the open manifold defined by (12). The second equation of (12) implies that the coordinates $\tilde{u_i}$ can be expressed in terms of new variables z_i as

$$\tilde{u_i} = e^{t/N} \frac{z_i^N}{z_1 \cdots z_N},$$

after which the second equation of (12) becomes

$$z_1^N + \cdots + z_N^N + e^{t/N} z_1 \cdots z_N = 0.$$

After the compactification adding $z_1 = \cdots = z_N = 0$ and the quotient by symmetries of $\widetilde{W}$, this is precisely the Batyrev mirror, as we computed in the case $d = N = 5$ in Example 41.

5 Berglund-Hübsch-Krawitz Mirror Symmetry

As we have seen, Hori-Vafa mirror symmetry for hypersurfaces reveals an interesting duality between certain polynomials and their transposes: a semi-Fano hypersurface $X = \{W = 0\}$ in weighted projective space is Hori-Vafa mirror to the Landau-Ginzburg orbifold

$$\widetilde{W} = W^T + e^t x_1 \cdots x_N \tag{13}$$

modulo the group $\mathrm{SL}(W^T)$.[6] Given that both W and W^T give rise to Landau-Ginzburg models, this suggests an LG-to-LG version of mirror symmetry. Our discussion of the resulting statement, known as Berglund-Hübsch-Krawitz Mirror Symmetry, is based on [6] and [23].

[6] We have changed notation from the previous section, denoting the defining polynomial of the hypersurface by W rather than A, to be consistent with the literature on LG-to-LG mirror symmetry.

5.1 Phases of the GLSM

In order to cast Hori-Vafa's Fano-to-LG mirror symmetry in this new framework, we will need to replace the geometric Fano model of the hypersurface X by a Landau-Ginzburg model. We have already seen how to do this; the trick involves passing from X to an associated gauged linear sigma model.

Suppose $X \subset \mathbb{P}(c_1, \ldots, c_N)$ is a hypersurface in weighted projective space defined by the vanishing of an invertible quasihomogeneous polynomial W of degree d. Then, as observed in the construction of the Hori-Vafa mirror, there is a GLSM for which X is the critical locus in one phase. Namely, let

$$\overline{W} := p \cdot W : \frac{\mathbb{C}^N \times \mathbb{C}}{\mathbb{C}^*} \to \mathbb{C},$$

where the charge matrix for the action of $\mathbb{C}^*$ is $(c_1, \ldots, c_N, -d)$. The moment map for this GLSM is

$$\mu = \frac{1}{2}\left(\sum_{i=1}^{N} c_i |x_i|^2 - d|p|^2\right).$$

The only critical value is $s = 0$, so the GLSM has two phases. The important fact for Hori-Vafa mirror symmetry is that X is the critical locus of $\overline{W}$ in the phase $s > 0$, which is the Landau-Ginzburg model

$$\overline{W} : \mathscr{O}_{\mathbb{P}(c_1,\ldots,c_N)}(-d) \to \mathbb{C},$$

as computed in Example 48 of the previous section. This is sometimes referred to as the "geometric phase", since it arises out of a hypersurface. The $s < 0$ phase, sometimes called the "Landau-Ginzburg phase", is

$$W : [\mathbb{C}^N/\mathbb{Z}_d] \to \mathbb{C},$$

where the generator of $\mathbb{Z}_d$ acts on $\mathbb{C}^N$ via the matrix

$$J := \begin{pmatrix} e^{2\pi i \frac{c_1}{d}} & & \\ & \ddots & \\ & & e^{2\pi i \frac{c_N}{d}} \end{pmatrix} \in U(1). \tag{14}$$

A Landau-Ginzburg model admits two types of cohomology. The one that we have seen thus far, the orbifold Jacobian ring of the superpotential, is the **Landau-Ginzburg B-model cohomology**. On the other hand, there is a **Landau-Ginzburg A-model cohomology** [15], which we will define later in this section.

When applied to the geometric phase of the above GLSM, the Landau-Ginzburg A-model cohomology is simply the cohomology of the hypersurface X. Thus, the Hori-Vafa mirror symmetry statement

$$H^*_{CR}(X) \cong \mathrm{Jac}(\widetilde{W}, \mathrm{SL}(W^T))$$

can be phrased as an exchange of the Landau-Ginzburg A-model cohomology of the $s > 0$ phase of the GLSM with the Landau-Ginzburg B-model cohomology of the model $(\mathbb{C}^N, \widetilde{W}, \mathrm{SL}(W^T))$ defined in the previous section.

To be more precise, the Hori-Vafa mirror $\widetilde{W}$ of X depended on a parameter t, and in fact, X is only mirror to this Landau-Ginzburg model when t is near ∞—this is the **large complex structure limit** of the family of Landau-Ginzburg models. On the other hand, if t is near $-\infty$, then the B-model Landau-Ginzburg cohomology of $(\mathbb{C}^N, \widetilde{W}, \mathrm{SL}(W^T))$ will instead match the A-model of the $s < 0$ phase of the GLSM.[7]

In the special case where X is Calabi-Yau, the Landau-Ginzburg B-models of $\widetilde{W}$ for different values of the parameter t can be related to one another. We thus obtain a diagram:

$$\begin{array}{ccc}
\text{LG A-model of } \overline{W}|_{s>0} & & \text{LG A-model of } \overline{W}|_{s<0} \\
\text{Hori-Vafa} \updownarrow & & \updownarrow \text{Hori-Vafa} \\
\text{LG B-model of } \widetilde{W}|_{t\to\infty} & \leftarrow - \rightarrow & \text{LG B-model of } \widetilde{W}|_{t\to-\infty},
\end{array} \quad (15)$$

where each Landau-Ginzburg model should be understood as orbifolded with respect to its symmetry group. The dotted horizontal arrow is a special case of a conjecture known as the Landau-Ginzburg/Calabi-Yau (LG/CY) correspondence, which relates the various phases of a GLSM to one another [26]. The conjecture has been proven in many instances; see [1, 9–13], among many other references, for more information.

Given the discussion of the previous paragraphs and the fact that $\widetilde{W} \to W^T$ as $t \to \infty$, the diagram (15) can be re-expressed as:

$$\begin{array}{ccc}
\text{CY A-model of } X & & \text{LG A-model of } (W, \mathbb{Z}_d) \\
\updownarrow & \swarrow\!\!\nearrow \text{(dashed)} & \updownarrow \\
\text{LG B-model of } (W^T, \mathrm{SL}(W^T)) & \leftarrow - \rightarrow & \text{LG B-model of } (\widetilde{W}|_{t\to-\infty}, \mathrm{Aut}(\widetilde{W})).
\end{array} \quad (16)$$

[7] Roughly, t is the real part of the parameter s in the GLSM.

The arrow from the bottom-left to the top-right, which is implied by the LG/CY correspondence, is Berglund-Hübsch-Krawitz mirror symmetry.

More generally, Berglund-Hübsch-Krawitz mirror symmetry can be extended by replacing the group $\mathbb{Z}_d$ with any group G of diagonal symmetries of $\mathbb{C}^N$ that preserves W and contains the element J defined in (14). The construction yields a "mirror group" G^T associated to any such G. The generalization of the diagonal arrow in (16), then, is

$$\text{LG A-model of } (W, G) \cong \text{LG B-model of } (W^T, G^T),$$

the latter of which is Hori-Vafa mirror to the orbifold $[X/(G/\langle J \rangle)]$.

All three forms of mirror symmetry discussed in these notes can now be schematically related. Let $X_W := \{W = 0\}$ be a Calabi-Yau hypersurface. For a group G of symmetries as above, let $\widetilde{G}$ denote $G/\langle J \rangle$. Then we have a diagram:

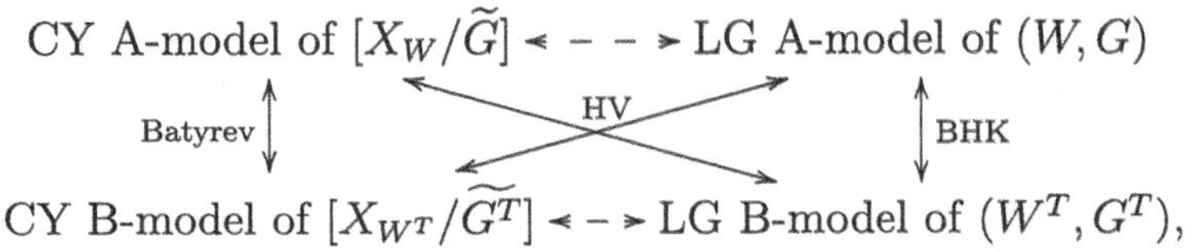

in which the horizontal arrows are again the LG/CY correspondence.

It should be noted that, although we have used the Calabi-Yau assumption to motivate the appearance of Berglund-Hübsch-Krawitz mirror symmetry, the state space isomorphism between (W, G) and (W^T, G^T) holds even when the Calabi-Yau assumption fails. The remainder of this section will be devoted to making the specific assumptions and results of Berglund-Hübsch-Krawitz mirror symmetry precise.

5.2 *Classification of Nondegenerate Singularities*

The polynomial W will be required to be invertible, quasihomogeneous, and to satisfy a certain nondegeneracy condition.

Definition 71 A quasihomogeneous polynomial W is **nondegenerate** if

(1) zero is the only critical point;
(2) there are no monomials of the form $x_i x_j$ with $i \leq j$.

A few remarks about this definition are in order. First, the condition that zero is the only critical point is equivalent to the existence of only isolated critical points, since quasihomogeneity implies that W and its derivatives vanish at $\mathbf{x}$ if and only if the same is true for any scalar multiple $\lambda\mathbf{x}$.

Second, suppose that we define:

Definition 72 The **exponent matrix** of an invertible quasihomogeneous polynomial

$$W = \sum_{i=1}^{N} a_i \prod_{j=1}^{N} x_j^{m_{ij}}$$

is the matrix $E_W := (m_{ij})$.

Definition 73 The **charges** of a quasihomogeneous polynomial with weights $c_1, \ldots, c_N$ and degree d are the rational numbers

$$q_j = \frac{c_j}{d}.$$

Then the definition of nondegeneracy has the following elementary consequences:

1. The charges q_j are unique.
2. $q_j \leq 1/2$ for all j.
3. The exponent matrix of W is nonsingular.

Polynomials satisfying the above conditions admit a very nice classification.

Theorem 74 (Kreuzer-Skarke [24]) *Suppose that W is a nondegenerate quasihomogeneous invertible polynomial. Then W can be written as a sum of polynomials*

$$W = \sum_{s=1}^{k} W_s$$

in disjoint sets of variables, where each W_s is of one of the following types:

(1) Fermat: x^a for some $a \geq 2$;
(2) Loop: $x_1^{a_1}x_2 + x_2^{a_2}x_3 + \cdots + x_N^{a_N}x_1$;
(3) Chain: $x_1^{a_1}x_2 + x_2^{a_2}x_3 + \cdots x_N^{a_N}$.

This is extremely useful for computations in the Landau-Ginzburg model, because the model decomposes as a "product", in a precise sense, whenever the superpotential breaks up as a disjoint sum. As a result, the study of the LG model reduces to the study of the model associated to Fermat, loop, and chain polynomials separately. Moreover, the operation of transpose preserves the types, so mirror symmetry can also be studied in the three cases individually.

Let us mention a few specific examples of nondegenerate invertible quasihomogeneous polynomials, following [3]. An important invariant of such a polynomial is its **central charge**

$$c_W := \sum_{i=1}^{N}(1 - 2q_i). \tag{17}$$

When $\sum q_i = 1$—or, equivalently, $c_W = N - 2$—the polynomial is said to be **Calabi-Yau**, since it defines a Calabi-Yau hypersurface in weighted projective space.

The only polynomials with $c_W < 1$ are the **ADE**-singularities

- $A_n = \frac{1}{n+1}x^{n+1}$;
- $D_n = x^{n-1} + xy^2, n \geq 4$;
- $E_6 = x^3 + y^3$;
- $E_7 = x^3 + xy^3$;
- $E_8 = x^3 + y^5$.

These all have $\sum q_i < 1$, so none of them is Calabi-Yau. Note, furthermore, that A-type and E-type polynomials are self-mirror under the operation of transpose, whereas D-type polynomials are not self-mirror.

Among the Calabi-Yau examples, those with $c_W = 1$ are the **elliptic singularities**

- $P_8 = x^3 + y^3 + z^3$,
- $X_9 = x^2 + y^4 + z^4$,
- $J_{10} = x^2 + y^3 + z^6$.

The Calabi-Yau examples with $c_W = 2$ have four variables. These are the **K3 singularities**, of which there are 95 types. When $c_W = 3$, we have the Calabi-Yau threefolds, for which there are thousands of examples. As one can see, classification becomes unwieldy beyond this point.

5.3 The Maximal Diagonal Symmetry Group

Let W be a nondegenerate quasihomogeneous invertible polynomial. The **maximal diagonal symmetry group** of W is the group of diagonal matrices preserving W; that is,

$$G_{max} = \left\{ g = \begin{pmatrix} e^{2\pi i g_1} & & \\ & \ddots & \\ & & e^{2\pi i g_N} \end{pmatrix} \;\middle|\; W(g \cdot \mathbf{x}) = W(\mathbf{x}) \right\}.$$

The groups G for which the Landau-Ginzburg A-model and B-model are defined will always be subgroups of G_{max}.

There is a convenient description of this group in terms of the exponent matrix of W. By the definition of invertibility, W is of the form

$$W = \sum_{i=1}^{N} \prod_{j=1}^{N} x_j^{m_{ij}},$$

for a matrix $E_W = (m_{ij})$. Let

$$\rho_k = \begin{pmatrix} \rho_1^{(k)} \\ \vdots \\ \rho_N^{(k)} \end{pmatrix}, \quad k = 1, \ldots, N$$

be the columns of the inverse matrix E_W^{-1}. Convert each into a diagonal matrix

$$\overline{\rho_k} := \begin{pmatrix} e^{2\pi i \rho_1^{(k)}} & & \\ & \ddots & \\ & & e^{2\pi i \rho_N^{(k)}} \end{pmatrix}.$$

Proposition 75 *The maximal diagonal symmetry group G_{max} is generated by the matrices $\overline{\rho_1}, \ldots, \overline{\rho_N}$.*

Proof The first thing to check is that each $\overline{\rho_k}$ lies in G_{max}. Indeed,

$$\begin{aligned} W(\overline{\rho_k}\mathbf{x}) &= W(e^{2\pi i \rho_1^{(k)}} x_1, \ldots, e^{2\pi i \rho_N^{(k)}} x_N) \\ &= \sum_{i=1}^{N} \prod_{j=1}^{N} e^{2\pi i \rho_j^{(k)} m_{ij}} x_j^{m_{ij}} \\ &= \sum_{i=1}^{N} \left(e^{2\pi i \sum_j \rho_j^{(k)} m_{ij}} \right) \prod_{j=1}^{N} x_j^{m_{ij}}. \end{aligned}$$

The exponent $\sum_j \rho_j^{(k)} m_{ij}$ is precisely the ith entry in the vector $E_W \cdot \rho_k$, which is the (i, j)th entry in the matrix $E_W \cdot E_W^{-1}$. Hence, it equals δ_{ij}, so we obtain

$$W(\overline{\rho_k}\mathbf{x}) = \sum_{i=1}^{N} \prod_{j=1}^{N} x_j^{m_{ij}} = W(\mathbf{x}),$$

as required.

Now, to see that $\overline{\rho_1}, \ldots, \overline{\rho_N}$ generate G_{max}, notice that for a diagonal matrix

$$g = \begin{pmatrix} e^{2\pi i g_1} & & \\ & \ddots & \\ & & e^{2\pi i g_N} \end{pmatrix} \in G_{max}, \tag{18}$$

we have

$$E_W \begin{pmatrix} g_1 \\ \vdots \\ g_N \end{pmatrix} = g_1 E_W \rho_1 + \cdots + g_N E_W \rho_N = E_W(g_1\rho_1 + \cdots + g_N\rho_N),$$

since the column vectors ρ_k are defined so that $E_W\rho_k$ is a vector with a 1 in the kth entry and zeroes elsewhere. Cancelling E_W from both sides of (18) shows that

$$g = g_1\overline{\rho_1} + \cdots + g_N\overline{\rho_N},$$

so g can be expressed in terms of $\overline{\rho_1}, \ldots, \overline{\rho_N}$. □

Note, in particular, that the distinguished element $J \in G_{max}$ defined in (14) is presented in these generators as

$$J = \overline{\rho_1} \cdots \overline{\rho_k},$$

since it is the diagonal matrix corresponding to the column vector $(q_1, \ldots, q_N)^T$ and

$$E_W \cdot \begin{pmatrix} q_1 \\ \vdots \\ q_N \end{pmatrix} = \begin{pmatrix} 1 \\ \vdots \\ 1 \end{pmatrix}.$$

5.4 The Mirror Group

Given a quasihomogeneous invertible polynomial W, denote its maximal diagonal symmetry group by $G_{W,max}$, and denote the maximal diagonal symmetry group of the transpose polynomial by $G_{W^T,max}$; recall, W^T is defined by the relationship

$$E_{W^T} = (E_W)^T.$$

For any subgroup $G \subset G_{W,max}$ containing J, we will construct a "mirror" subgroup $G^T \subset G_{W^T,max}$. Before doing so, two simple observations are useful.

First, whereas $G_{max,W}$ consists of matrices $\rho = \rho_1^{k_1} \cdots \rho_N^{k_N}$, where ρ_i are the diagonal matrices associated to columns of W, the group G_{max,W^T}, consists of matrices $h = h_1^{r_1} \cdots h_N^{r_N}$, in which h_i are associated to the *rows* of W.

Second, the same proof that showed that $\rho_k \in G_{max}$ implies that a matrix

$$g = \begin{pmatrix} e^{2\pi i g_1} & & \\ & \ddots & \\ & & e^{2\pi i g_N} \end{pmatrix}$$

lies in the maximal diagonal symmetry group G_{max} of W if and only if

$$E_W^{-1}\begin{pmatrix} g_1 \\ \vdots \\ g_N \end{pmatrix} \in \mathbb{Z}^N.$$

A similar observation holds for the maximal diagonal symmetry group of W^T, but replacing right multiplication by a column with left multiplication by a row.

With these facts in mind, we define G^T as follows.

Definition 76 Given a subgroup $G \subset G_{W,max}$ containing J, the **mirror (or transpose) group** is

$$\left\{ h_1^{r_1} \cdots h_N^{r_N} \;\middle|\; (r_1 \cdots r_N)\, E_W^{-1}\begin{pmatrix} k_1 \\ \vdots \\ k_N \end{pmatrix} \in \mathbb{Z}^N \text{ for any } \rho_1^{k_1} \cdots \rho_N^{k_N} \in G \right\}.$$

To put it another way, one can define a pairing between $G_{max,W}$ and G_{max,W^T} by

$$\langle \rho, h \rangle = (r_1 \cdots r_N)\, E_W^{-1}\begin{pmatrix} k_1 \\ \vdots \\ k_N \end{pmatrix} \mod \mathbb{Z},$$

and in terms of this pairing, G^T is the orthogonal complement of G.

The following are some useful observations regarding the operation of transpose on both polynomials and groups:

1. If $G_1 \subset G_2$, then $G_2^T \subset G_1^T$.
2. $(G^T)^T = G$.
3. $\{1\}^T = G_{max}$.
4. $\langle J \rangle^T = \mathrm{SL}(W^T)$.

In particular, these facts imply that the transpose of groups preserves the **Calabi-Yau condition**

$$\langle J \rangle \subset G \subset \mathrm{SL}(W)$$

for subgroups G. This is important for aspects of LG-to-LG mirror symmetry that will not be discussed in these notes; namely, there are ring structures on the A-model and B-model LG state spaces, and mirror symmetry gives a ring isomorphism only in the Calabi-Yau case.

5.5 B-Model LG Cohomology

We have now defined the mirror (W^T, G^T) of a Landau-Ginzburg model (W, G). In order to make the statement of Berglund-Hübsch-Krawtiz mirror symmetry precise, we must carefully define the cohomology groups on the A-side and the B-side that will be exchanged.

The Landau-Ginzburg B-model cohomology has already appeared in the context of Hori-Vafa mirror symmetry, where we viewed it as an orbifold Jacobian ring. Let us recall and expand upon this definition.

It is convenient to give a somewhat different presentation of the Jacobian ring. Let

$$\Omega_W = \Omega^N(\mathbb{C}^N)/\big(dW \wedge \Omega^{N-1}(\mathbb{C}^N)\big).$$

Then the map

$$\mathrm{Jac}(W) \to \Omega_W$$

$$\phi \mapsto \phi \cdot dx_1 \cdots dx_n =: \phi d\underline{x}$$

is an isomorphism of vector spaces.[8]

There is a pairing on Ω_W defined by

$$\begin{aligned}\langle \phi d\underline{x}, \phi' d\underline{x}\rangle &= \mathrm{Res}\left(\frac{\phi\phi' d\underline{x}}{\partial_1 W \cdots \partial_N W}\right)\\ &= \frac{1}{2\pi i}\int_{|\partial_i W|=\epsilon} \frac{\phi\phi' d\underline{x}}{\partial_1 W \cdots \partial_N W}.\end{aligned}$$

Furthermore, there is a grading

$$\deg(\phi d\underline{x}) = \deg(\phi) + \sum_{i=1}^{N} q_i$$

on W, in which the **charges** q_i are defined by

$$q_i = \frac{c_i}{d}.$$

[8] Some care should be taken here, because both sides admit both a grading and a G-action, and the isomorphism does not preserve these aspects. We will be careful to specify these in what follows, but the basic principle is that they are inherited from Ω_W, not from $\mathrm{Jac}(W)$.

Under this grading, the pairing can equivalently be computed by using the fact that, like the cohomology of a manifold, Ω_W has a unique generator in the top degree. This top degree is the central charge c_W defined in (17), and the generator is the Hessian Hess(W). Just as one computes the pairing on the cohomology of a manifold via the volume form, then, one can compute

$$\langle \phi, \phi' \rangle = \lambda,$$

where

$$\phi \cdot \phi' = \frac{\lambda}{\mu}\mathrm{Hess}(W) + \text{ lower-order terms.}$$

Here, the normalizing constant is the Milnor number μ, defined by

$$\mu = \dim(\mathrm{Jac}(W)) = \prod_{i=1}^{N} \left(\frac{1}{q_i} - 1 \right).$$

The B-model Landau-Ginzburg cohomology can now be built by orbifolding, as was previously described.

Definition 77 Given a nondegenerate quasihomogeneous invertible polynomial W and a subgroup $G \subset G_{W,max}$ containing J, the **B-model Landau-Ginzburg cohomology** of (W, G) is the vector space

$$\Omega_{W,G} = \left(\bigoplus_{g \in G} \Omega_{W_g} \right)^G, \tag{19}$$

where

$$W_g = W|_{\mathrm{Fix}(g)}.$$

Here, the G-invariant part is taken with respect to the action of G on $\bigoplus_{g \in G} \Omega_{W_g}$ that sends

$$\Omega_{W_g} \to \Omega_{W_{h^{-1}gh}}$$

via pullback under multiplication by h.

We should note that the restriction of W to Fix(g) is still nondegenerate, so each Ω_{W_g} is of the form described previously.

The pairings on the vector spaces Ω_{W_g} defined above can be combined to give a pairing on $\Omega_{W,G}$, again in much the same way that the Poincaré pairing is defined in Chen-Ruan cohomology. Specifically, one pairs

$$\Omega_{W_g}^G \otimes \Omega_{W_{g^{-1}}}^G \to \mathbb{C}$$

with the previously-defined residue pairing, using the fact that $\mathrm{Fix}(g) = \mathrm{Fix}(g^{-1})$.

As in Chen-Ruan cohomology, a shift in the grading on $\Omega_{W,G}$ is necessary in order to make the pairing behave like a Poincaré pairing. First, define an unshifted (or "internal") bigrading on Ω_W by doubling the single grading:

$$\Omega_{W_g}^{p,q} = \begin{cases} (\Omega_{W_g})^p & \text{if } p = q \\ 0 & \text{if } p \neq q. \end{cases}$$

Then, define a **degree shift**:

$$(Q_-^B, Q_+^B) = (\iota_{(g^{-1})}, \iota_{(g)}) - \left(\sum_{i=1}^N q_i, \sum_{i=1}^N q_i\right).$$

Here, ι_g is the age shift from Chen-Ruan cohomology, defined by

$$\iota_{(g)} = \sum_{i=1}^N \frac{m_i}{m},$$

where $m = \mathrm{ord}(g)$ and g acts on the tangent space T_xX to a point $x \in \mathrm{Fix}(g)$ by

$$g = \begin{pmatrix} e^{2\pi i \frac{m_1}{m}} & & \\ & \ddots & \\ & & e^{2\pi i \frac{m_n}{m}} \end{pmatrix}.$$

The grading on $\Omega_{W,G}$ is then defined as

$$\Omega_{W,G}^{p,q} = \left(\bigoplus_{g \in G} \Omega_{W_g}^{p-Q_-^B, q-Q_+^B}\right)^G.$$

The purpose of this shift is to ensure that the following result holds:

Lemma 78 *If*

$$\langle \phi_1 d\underline{x}_g, \phi_2 d\underline{x}_{g^{-1}} \rangle \neq 0,$$

then

$$\deg^{\pm}_W(\phi_1 d\underline{x}_g) + \deg^{\pm}_W(\phi_2 d\underline{x}_{g^{-1}}) = c_W.$$

Proof We will require the fact that, if

$$g = \begin{pmatrix} e^{2\pi i \Theta_g^1} & & \\ & \ddots & \\ & & e^{2\pi i \Theta_g^N} \end{pmatrix},$$

then

$$\iota_{(g)} = \sum_{\Theta_g^i \neq 0} \Theta_g^i,$$

whereas

$$\iota_{(g^{-1})} = \sum_{\Theta_g^i \neq 0} (1 - \Theta_g^i).$$

Using this, we have

$$\begin{aligned}
&\deg^{\pm}_W(\phi_1 d\underline{x}_g) + \deg^{\pm}_W(\phi_2 d\underline{x}_{g^{-1}}) \\
&= \deg(\phi_1 d\underline{x}_g) + Q^B_{\pm}(g) + \deg(\phi_2 d\underline{x}_{g^{-1}}) + Q^B_{\pm}(g^{-1}) \\
&= \deg(\phi_1) + \sum_{\Theta_g^i = 0} q_i + Q^B_{\pm}(g) + \deg(\phi_2) + \sum_{\Theta_g^i = 0} q_i + Q^B_{\pm}(g^{-1}) \\
&= \deg(\phi_1) + \deg(\phi_2) + \sum_{\Theta_g^i = 0} (1 - 2q_i) + 2 \sum_{\Theta_g^i = 0} q_i + \sum_{\Theta_g^i \neq 0} 1 - 2\sum_{i=1}^{N} q_i \\
&= c_{W_g} + \sum_{\Theta_g^i \neq 0} (1 - 2q_i) \\
&= \sum_{i=1}^{N} (1 - 2q_i) \\
&= c_W.
\end{aligned}$$

□

The upshot of Lemma 78 is that $\Omega_{W,G}$ behaves like the cohomology of a complex manifold of dimension c_W—despite the fact that c_W may be fractional. This is referred to as a “manifold of dimension c_W” in the physics literature to make sense of the notion of fractional dimension.

The components of the decomposition (19) are referred to as **sectors**. There are a number of special cases that yield particularly important sectors.

First, when $g = 1$, the component of $\Omega_{W,G}$ is called the **nontwisted sector**. It is isomorphic to Ω_W but is graded via the shift

$$(Q_-^B, Q_+^B) = -\left(\sum_{i=1}^N q_i, \sum_{i=1}^N q_i\right).$$

Thus,

$$\deg_W^{\pm}(\phi d\underline{x}) = \deg(\phi).$$

The element $1d\underline{x}$ of bidegree $(0,0)$ and the element $\mathrm{Hess}(W)d\underline{x}$ of bidegree (c_W, c_W) both lie in the nontwisted sector.

The J-sector, for which $g = J$, is 1-dimensional:

$$\Omega_{W_J}^G \cong \mathbb{C},$$

since $\mathrm{Fix}(J) = \{0\}$. It degree shift is

$$(Q_-^B, Q_+^B) = \left(\sum_{i=1}^N (1-q_i), \sum_{i=1}^N q_i\right) - \left(\sum_{i=1}^N q_i, \sum_{i=1}^N q_i\right) = (c_W, 0).$$

The J^{-1}-sector is also one-dimensional, but its degree shift is

$$(Q_-^B, Q_+^B) = (0, c_W),$$

by an analogous computation.

Thus, the components of various bidegrees in the Landau-Ginzburg B-model cohomology can be compiled into the following rough outline of a Hodge diamond:

$$\begin{array}{ccc} & H^{c_W,c_W} & \\ \mathbb{C} \cong \Omega_{W_J} \cong H^{c_W,0} & & H^{0,c_W} \cong \Omega_{W_{J^{-1}}} \cong \mathbb{C} \\ & H^{0,0}, & \end{array}$$

where the middle column is the nontwisted sector.

5.6 A-Model LG Cohomology

We have actually mentioned the A-model Landau-Ginzburg cohomology before briefly, as well. Analogously to the way in which the B-model cohomology is built out of Jacobian rings, the building blocks of the A-model cohomology are the vector spaces

$$H^N(\mathbb{C}^N, W^{+\infty}; \mathbb{C}), \tag{20}$$

where

$$W^{+\infty} = (\mathrm{Re}W)^{-1}(\rho, \infty)$$

for $\rho \gg 0$.[9]

Just as in the B-model case, we will define the A-model by taking a direct sum of middle cohomology groups as in (20) for restrictions of W to the fixed-point sets of $g \in G$. There will be a degree shift, defined so that a certain pairing behaves like a Poincaré pairing. Towards this end, we will need a pairing on (20).

The first step in that direction is to define a pairing

$$H^N(\mathbb{C}^N, W^{-\infty}; \mathbb{Q}) \otimes H^N(\mathbb{C}^N, W^{+\infty}; \mathbb{Q}) \to \mathbb{Q}, \tag{21}$$

where

$$W^{-\infty} = (\mathrm{Re}W)^{-1}(-\infty, -\rho)$$

for $\rho \gg 0$. To do so, we use the fact that $H^N(\mathbb{C}^N, W^{+\infty}, \mathbb{Q})$ is dual to the homology group $H_N(\mathbb{C}^N, W^{+\infty}; \mathbb{Q})$. The latter has a basis consisting of the preimages in $\mathbb{C}^N$ of a collection of nonintersecting paths in $\mathbb{C}$ that begin at the critical values of W and move in the direction of $\mathrm{Re}(z) = \infty$, eventually becoming horizontal. These subsets of $\mathbb{C}^N$ are called **Lefschetz thimbles**; see Fig. 16. There is an analogous basis for $H_N(\mathbb{C}^N, W^{-\infty}; \mathbb{C})$, consisting of "opposite" thimbles.

The two types of thimbles can intersect one another, so there is an intersection pairing

$$H_N(\mathbb{C}^N, W^{-\infty}; \mathbb{Q}) \otimes H_N(\mathbb{C}^N, W^{+\infty}; \mathbb{Q}) \to \mathbb{Q}.$$

[9] Sometimes $W^{+\infty}$ is replaced with a "Milnor fiber" $W^{-1}(z_0)$ of W, where z_0 is a sufficiently large real number. This does not affect the vector space; however, in other parts of the theory it is necessary to use the Hodge structure on cohomology, and in order for a natural Hodge structure to be defined, it is necessary to use $W^{+\infty}$ and not a single fiber.

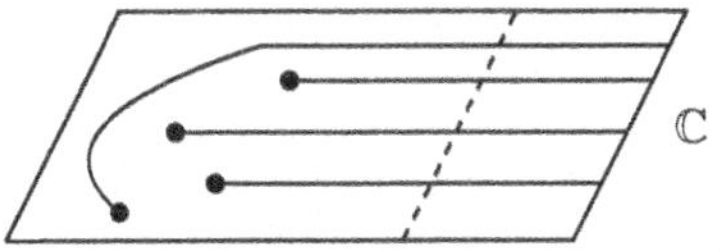

Fig. 16 The preimages in $\mathbb{C}^N$ of these paths are closed at one end and open at the other, giving them the appearance of infinite "thimbles"

In fact, it is a perfect pairing, and (21) is obtained by dualization. To be more explicit, if δ_i is a basis for $H_N(\mathbb{C}^N, W^{-\infty}; \mathbb{Q})$ and $\delta_i^\vee$ is a basis for $H_N(\mathbb{C}^N, W^{+\infty}; \mathbb{Q})$, then

$$\langle \alpha, \beta \rangle = \sum_{i,j} \langle \alpha, \delta_i \rangle \langle \delta_i, \delta_j^\vee \rangle \langle \beta, \delta_j^\vee \rangle,$$

where the three pairings on the right-hand side are given by intersection in $H_N(\mathbb{C}^N, W^{+\infty}; \mathbb{Q})$, $H_N(\mathbb{C}^N; \mathbb{Q})$, and $H_N(\mathbb{C}^N, W^{-\infty}; \mathbb{Q})$, respectively.

From here, (21) can be converted into a pairing on (20) by using the morphism

$$I : \mathbb{C}^N \to \mathbb{C}^N$$

$$(x_1, \ldots, x_N) \mapsto (\xi^{c_1} x_1, \ldots, \xi^{c_N} x_N)$$

for a chosen ξ satisfying $\xi^d = -1$. This morphism has

$$W(I(x_1, \ldots, x_N)) = -W(x_1, \ldots, x_N),$$

and hence it interchanges $W^{+\infty}$ with $W^{-\infty}$. The pairing on (20) is defined as

$$\langle \alpha, \beta \rangle := \langle \alpha, I^* \beta \rangle.$$

Although the resulting pairing appears to depend on a choice of ξ, we will always look at its restriction to

$$H^N(\mathbb{C}^N, W^{+\infty}; Q)^G \tag{22}$$

for G a subgroup of G_{max} containing J. One can check that the pairing on $H^N(\mathbb{C}^N, W^{+\infty}; \mathbb{Q})^{\langle J \rangle}$ is well-defined and independent of ξ, so the same is true for the restriction to the smaller spaces (22).

The orbifolding construction is very similar to what we have seen previously.

Definition 79 If W is a nondegenerate quasihomogeneous invertible polynomial and $G \subset G_{W,max}$ is a subgroup containing J, then the **A-model Landau-Ginzburg cohomology** of (W, G) is the vector space

$$\mathcal{H}_{W,G} := \left(\bigoplus_{g \in G} H^{N_g}(\mathbb{C}_g^N, W_g^{+\infty}; \mathbb{Q}) \right)^G, \tag{23}$$

where $\mathbb{C}_g^N$ is the fixed locus of g, $W_g^{+\infty} = \mathrm{Re}(W_g^{-1}(\rho, \infty))$ for $\rho \gg 0$, $W_g = W|_{\mathbb{C}_g^N}$, and N_g is the complex dimension of $\mathbb{C}_g^N$. The action of G takes the sector indexed by g to the sector indexed by $h^{-1}gh$ via pullback under multiplication by h.

This has a pairing, given by mapping

$$H^{N_g}(\mathbb{C}_g^N, W_g^{+\infty}; \mathbb{Q})^G \otimes H^{N_{g^{-1}}}(\mathbb{C}_{g^{-1}}^N, W_{g^{-1}}^{+\infty}; \mathbb{Q})^G \to \mathbb{Q}$$

via the pairing defined previously; note that $\mathbb{C}_g^N = \mathbb{C}_{g^{-1}}^N$. There is also a degree shift, defined so that this pairing behaves like a Poincaré pairing. It is:

$$\mathcal{H}_{W,G}^{p,q} = \left(\bigoplus_{g \in G} \Omega_{W_g}^{p-Q_-^A, q-Q_+^A} \right)^G,$$

where the shift is

$$(Q_-^A, Q_+^A) := (\iota_g, \iota_g) - \left(\sum_{i=1}^N q_i, \sum_{i=1}^N q_i \right)$$

and the internal bi-grading is given by the Hodge structure on the vector space $H^{N_g}(\mathbb{C}_g^N, W_g^{+\infty}; \mathbb{Q})$.

In fact, the A-model and B-model Landau-Ginzburg cohomology of (W, G) are isomorphic to one another, but with different gradings. To prove this, one shows that the map

$$\Omega_W \to H^N(\mathbb{C}^N, W^{+\infty}; \mathbb{C}) = \mathrm{Hom}(H_N(\mathbb{C}^N, W^{+\infty}; \mathbb{Z}), \mathbb{C})$$

$$\phi d\underline{x} \mapsto \left(\Delta \mapsto \int_\Delta e^{-W} \phi d\underline{x} \right)$$

is an isomorphism; it also respects the pairing on either side, up to a constant.[10] It does not respect the grading, though; instead, Ω_W^p maps to the bidegree $(N-p, p)$ part of $H^N(\mathbb{C}^N, W^{+\infty}; \mathbb{C})$.

[10]This is a more computationally practical way to compute the A-model pairing than via Lefschetz thimbles.

Even after the degree shift, the two sides are bigraded differently. An element of $(\bigoplus_g \Omega_{W_g})^G$ with internal bidegree (p, p) and degree shift

$$(Q_B^-, Q_B^+) = (\iota_{g^{-1}}, \iota_g) - \left(\sum q_i, \sum q_i\right)$$

corresponds to an element in $\left(\bigoplus_g H^{N_g}(\mathbb{C}_g^N, W_g^{+\infty}; \mathbb{Q})\right)^G$ with internal degree $(N_g - p, p)$ and degree shift

$$(Q_A^-, Q_A^+) = (\iota_g, \iota_g) - \left(\sum q_i, \sum q_i\right).$$

One can compute, then, that

$$\deg_A^+ = \deg_B^+$$

but

$$\deg^- A = c_W - \deg_B^- .$$

In particular, this says that the A-model Landau-Ginzburg cohomology of (W, G) is obtained from the B-model cohomology of (W, G) by flipping the Hodge-diamond.

Let us use this relationship to the B-model cohomology to study a few of the special sectors in the decomposition (23). The nontwisted sector, indexed by the element $g = 1$, has degree shift

$$(Q_-^A, Q_+^A) = \left(-\sum_{i=1}^N q_i, -\sum_{i=1}^N q_i\right)$$

and internal grading

$$\deg(\phi d\underline{x}) = \deg(\phi) + \sum_{i=1}^N q_i.$$

In particular, the A-model image of the element $1 d\underline{x}$ has bidegree

$$\begin{aligned}
&\left(\deg_A^-(d\underline{x}), \deg^- A(d\underline{x})\right)\\
=&(N - \sum q_i, \sum q_i) + (-\sum q_i, -\sum q_i)\\
=&(c_W, 0),
\end{aligned}$$

and the A-model image of the element $\mathrm{Hess}(W)d\underline{x}$ has bidegree

$$\begin{aligned}&\left(\deg_A^-(\mathrm{Hess}(W)d\underline{x}), \deg_A^-(\mathrm{Hess}(W)d\underline{x})\right)\\ =&(N - c_W - \sum q_i, c_W + \sum q_i) + (-\sum q_i, -\sum q_i)\\ =&(0, c_W).\end{aligned}$$

The sector $g = J$ has $\mathbb{C}_J^N = \{0\}$, so the middle-dimensional relative cohomology is generated by the constant function 1, which we denote e_J. This has

$$\left(\deg_A^-(e_J), \deg_A^+(e_J)\right) = \left(\sum q_i, \sum q_i\right) + \left(-\sum q_i, -\sum q_i\right) = (0, 0).$$

The sector $g = J^{-1}$ also has a single generator $e_{J^{-1}}$, with bidegree (c_W, c_W). Comparing with the sketch of the B-model Hodge diamond computed in the previous section, we again see that passing between the A-grading and the B-grading corresponds to a rotation.

5.7 *Berglund-Hübsch-Krawitz Mirror Symmetry*

The result that makes the association $(W, G) \mapsto (W^T, G^T)$ qualify as mirror symmetry is the following:

Theorem 80 (Krawitz [23]) *There is a bidegree-preserving isomorphism*

$$\Omega_{W,G} \cong \mathcal{H}_{W^T,G^T}.$$

Alternatively, the results of the previous section show that this can be expressed as an isomorphism between the B-model of (W, G) and the B-model of (W^T, G^T) that rotates the bigrading.

For $g \in G$, let $F_g \subset \{1, \ldots, N\}$ be the indices of the coordinate directions fixed by g, so that

$$\mathrm{Fix}(g) = \mathrm{Spec}\left(\mathbb{C}\left[\{x_i \mid i \in F_g\}\right]\right).$$

An element of $\Omega_{W,G}$ will be written as

$$\sum_{i \in F_g} x_i^{r_i} dx_i \left| \prod_{j=1}^{N} \rho_j^{s_j+1} \right\rangle$$

if it is drawn from the sector indexed by

$$g = \prod_{j=1}^{N} \rho_j^{s_j+1} \in G.$$

Under this notation, the mirror map in Theorem 80 is given in terms of the B-model Landau-Ginzburg cohomology by

$$\Omega_{W,G} \to \Omega_{W^T,G^T}$$

$$\prod_{i \in F_g} x_i^{r_i} dx_i \left| \prod_{j=1}^{N} \rho_j^{s_j+1} \right\rangle \mapsto \prod_{j=1}^{N} y_j^{s_j} dy_j \left| \prod_{i \in F_g} \rho_i^{r_i+1} \right\rangle.$$

That is, it exchanges group elements with monomials in the coordinates.

Example 81 Let

$$W = x^3 y + x y^5,$$

for which $W^T = W$. This has weights $c_x = 2$ and $c_y = 1$ and degree 7, so

$$J = \begin{pmatrix} e^{2\pi i \frac{2}{7}} & 0 \\ 0 & e^{2\pi i \frac{1}{7}} \end{pmatrix}.$$

Let

$$G = \langle J \rangle,$$

for which

$$G^T = \mathrm{SL}(W^T) = \left\langle \begin{pmatrix} e^{2\pi i \frac{1}{2}} & 0 \\ 0 & e^{2\pi i \frac{1}{2}} \end{pmatrix} \right\rangle.$$

By computing the inverse of the exponent matrix, it is easy to verify that the generators of $G_{\max,W}$ are

$$\rho_x = \begin{pmatrix} e^{2\pi i \frac{5}{14}} & 0 \\ 0 & e^{2\pi i \frac{-1}{14}} \end{pmatrix}$$

and

$$\rho_y = \begin{pmatrix} e^{2\pi i \frac{-1}{14}} & 0 \\ 0 & e^{2\pi i \frac{3}{14}} \end{pmatrix}.$$

There are seven elements in the subgroup $\langle J\rangle$. Using the fact that $J = \rho_x\rho_y$ and the identity $\rho_x^3\rho_y = 1$, these can be written as

$$\langle J\rangle = \{\rho_x^0\rho_y^0, \rho_x^1\rho_y^1, \rho_x^2\rho_y^2, \rho_x^3\rho_y^3, \rho_x^1\rho_y^3, \rho_x^2\rho_y^4, \rho_x^3\rho_y^5\}.$$

All but the first of these is narrow, so the corresponding sector of $\Omega_{W,\langle J\rangle}$ will just be

$$\mathbb{Q}\cdot e_{\rho_x^a\rho_y^b},$$

in which $e_{\rho_x^a\rho_y^b}$ denotes the volume form on the sector indexed by $\rho_x^a\rho_y^b$. There is no need to restrict to the $\langle J\rangle$-fixed part on these sectors, since the action is trivial. For the remaining sector, indexed by $\rho_x^0\rho_y^0$, the action of J sends

$$x^iy^jdxdy \mapsto \det(J)\cdot(e^{2\pi i\frac{2}{7}}x)^i(e^{2\pi i\frac{1}{7}}y)^jdxdy = e^{2\pi i\frac{3+2i+j}{7}}x^ay^bdxdy.$$

As a result, the $\langle J\rangle$-invariant part is spanned by

$$\{x^2dxdy, xy^2dxdy, y^4dxdy\}.$$

(Since we are working in the Jacobian ring

$$\mathrm{Jac}(W) = \frac{\mathbb{C}[x,y]}{(3x^2y+y^5, x^3+5xy^4)},$$

powers of y greater than 4 can always be expressed in terms of smaller exponents.)

As for the mirror side, there are two elements in $\langle J\rangle^T$, which can be written as

$$\langle J\rangle^T = \{\rho_x^0\rho_y^0, \rho_x^2\rho_y^3\}.$$

The second of these gives the sector

$$\mathbb{Q}\cdot e_{\rho_x^2\rho_y^3}$$

of $\Omega_{W^T,\langle J\rangle^T}$. The first gives the nontwisted sector. Since the nontrivial element of $\langle J\rangle^T$ acts by

$$x^iy^jdxdy \mapsto \det(\rho_x^2\rho_y^3)\cdot(e^{2\pi i\frac{1}{2}}x)^i(e^{2\pi i\frac{1}{2}}y)^jdxdy = e^{2\pi i\frac{i+j}{2}}x^ay^bdxdy,$$

the $\langle J\rangle^T$-invariant part of the nontwisted sector is spanned by

$$\{1, xy, x^2y^2, y^2, xy^3, x^2y^4, x^2, y^4\},$$

again keeping the powers of y below 5.

These generators are matched up by the mirror map in the following way:

(W, J)	$e_{\rho_x^1\rho_y^1}$	$e_{\rho_x^2\rho_y^2}$	$e_{\rho_x^3\rho_y^3}$	$e_{\rho_x^1\rho_y^3}$
$(W^T, \langle J\rangle^T)$	$1e_{\rho_x^0\rho_y^0}$	$xye_{\rho_x^0\rho_y^0}$	$x^2y^2e_{\rho_x^0\rho_y^0}$	$y^2e_{\rho_x^0\rho_y^0}$

(W, J)	$e_{\rho_x^2\rho_y^4}$	$e_{\rho_x^3\rho_y^5}$	$x^2e_{\rho_x^3\rho_y^1}$	$xy^2e_{\rho_x^0}$	$y^4e_{\rho_x^1\rho_y^5}$
$(W^T, \langle J\rangle^T)$	$xy^3e_{\rho_x^0\rho_y^0}$	$x^2y^4e_{\rho_x^0\rho_y^0}$	$x^2e_{\rho_x^3\rho_y^1}$	$e_{\rho_x^2\rho_y^3}$	$y^4e_{\rho_x^1\rho_y^5}$

Notice that the mirror map sends "narrow sectors"—those indexed by group elements g for which $\mathrm{Fix}(g) = \{0\}$—to elements of the nontwisted sector. In the Calabi-Yau case, there are exactly as many narrow sectors for (W, G) as elements of the nontwisted sector for (W^T, G^T), and the matching is perfect.

Via the Landau-Ginzburg/Calabi-Yau correspondence, this sets up a strong parallel between the Berglund-Hübsch-Krawitz and Batyrev-Borisov mirror symmetry constructions. Indeed, in all known cases of the LG/CY correspondence, the narrow sectors of the Landau-Ginzburg cohomology correspond to the ambient part of the cohomology of X_W, consisting of classes pulled back from the weighted projective space. Such ambient cases are generated by toric divisors. Thus, the exchange between narrow group elements and nontwisted monomials in Berglund-Hübsch-Krawitz mirror symmetry is matched, via the correspondence, with the exchange between toric divisors and monomials in Batyrev mirror symmetry.

Acknowledgements The authors would like to thank Mark Shoemaker for his detailed and valuable comments. Thanks are also due to Kentaro Hori for answering many of the authors' questions, and to all of the students in the course on which these notes are based, without whom many corrections and clarifications would not have been made. The authors were partially supported by NSF RTG grant 1045119.

Appendix: Chen-Ruan Cohomology

Many more details on this topic can be found in [2].

An orbifold, speaking geometrically, is an object that is locally the quotient of a manifold by the action of a finite group. One can make this definition precise in the category-theoretic language of groupoids. From this perspective, an orbifold is a groupoid for which the objects and arrows form manifolds and the structure morphisms (source, target, composition, identity, and inversion) are all smooth.

The case on which we will focus is when X is a complex manifold and G is a finite group acting on X. Then there is an orbifold

$$\mathcal{X} = [X/G].$$

This should be thought of as a version of the quotient that records any isotropy of the original action. The fact that the orbifold "remembers" the data of the action is clear from the groupoid point of view: the orbifold groupoid $[X/G]$ has objects X and arrows $x \to g \cdot x$ for each $x \in X$ and $g \in G$.

The de Rham or singular cohomology of an orbifold can be defined, but it is insufficient for capturing all of the data of $\mathcal{X}$. One way to understand the problem is through Gromov-Witten theory. Ordinarily, in Gromov-Witten theory, there would be evaluation morphisms

$$\mathrm{ev}_i : \overline{\mathcal{M}}_{g,n}(Y, \beta) \to Y$$

to record the images of the various marked points. In the orbifold setting, however, a morphism $f : \mathcal{C} \to \mathcal{X}$ from an *orbifold* curve $\mathcal{C}$ to an *orbifold* $\mathcal{X}$ has more local data around a marked point x_i than simply its image. Namely, $\mathcal{C}$ is of the form $[\mathbb{C}/G_i]$ near x_i for a finite group G_i, and part of the data of f is a homomorphism $G_i \to G$. Thus, the evaluation maps should keep track not only of the images of the marked points but of the homomorphisms on isotropy, and for this reason, they should land not in $\mathcal{X}$ but in a more complicated object.

Definition 82 The **inertia orbifold** of $\mathcal{X} = [X/G]$ is

$$\Lambda[X/G] = \left[\bigsqcup_{g \in G} (X^g \times \{g\})/G \right],$$

where

$$X^g = \{x \in X \mid gx = x\},$$

and G acts on this disjoint union by

$$h \cdot (x, g) = (hx, hgh^{-1}).$$

Equivalently, one can write

$$\Lambda[X/G] = \left[\bigsqcup_{(g) \in G^*} (X^g \times \{g\})/C(g) \right],$$

in which G^* denotes the set of conjugacy classes of G. The advantage of this presentation is that it breaks $\Lambda[X/G]$ into its connected components. In particular, there is a component

$$[(X \times \{1\})/G] \subset \Lambda[X/G]$$

that is isomorphic to $\mathcal{X}$ itself, called the **nontwisted sector**. The other connected components are referred to as **twisted sectors**.

The inertia orbifold is the image of the evaluation maps in Gromov-Witten theory, so, at least from that perspective, its cohomology is the natural object to study.

Definition 83 As a vector space, the **Chen-Ruan cohomology** of $\mathcal{X}$ is defined as the cohomology of $\Lambda[X/G]$.

However, the grading on Chen-Ruan cohomology differs from that of the inertia stack.

Definition 84 The **degree (or age) shift** of an element $g \in G$ is defined as

$$\iota(g) := \sum_{i=1}^{n} \frac{m_{i,g}}{m_g},$$

where $m_g = \mathrm{ord}(g)$ and g acts on the tangent space $T_x X$ to a point $x \in X^g$ by

$$\rho_x(g) := \begin{pmatrix} e^{2\pi i \frac{m_{1,g}}{m_g}} & & \\ & \ddots & \\ & & e^{2\pi i \frac{m_{n,g}}{m_g}} \end{pmatrix}. \tag{24}$$

One can check that ι gives a locally constant function on $\Lambda[X/G]$. The grading on Chen-Ruan cohomology is defined as follows:

$$H^{p,q}_{CR}([X/G]; \mathbb{Q}) := \prod_{(g)\in G^*} H^{p-\iota(g),q-\iota(g)}(X^g/C(g); \mathbb{Q}).$$

There is an involution $I : X^g \times \{g\} \to X^{g^{-1}} \times \{g^{-1}\}$ for any $g \in G$, which is simply the identity on the first component. Using this, one can define a pairing on Chen-Ruan cohomology by

$$\langle\, ,\, \rangle : H^{p,q}(X^g \times \{g\}/C(g); \mathbb{Q}) \otimes H^{p',q'}(X^{g^{-1}} \times \{g^{-1}\}/C(g^{-1}); \mathbb{Q}) \to \mathbb{Q}.$$

Proposition 85 *Let $n = \dim_{\mathbb{C}}(X)$, and choose $\alpha \in H^{p,q}(X^g \times \{g\}/C(g))$ and $\beta \in H^{p',q'}(X^{g^{-1}} \times \{g^{-1}\}/C(g^{-1}))$. If $\langle \alpha, \beta \rangle \neq 0$, then*

$$p + \iota(g) + p' + \iota(g^{-1}) = n,$$

$$q + \iota(g) + q' + \iota(g^{-1}) = n.$$

That is, the degrees after shifting must sum to n.

Proof We claim that

$$\iota(g) + \iota(g^{-1}) = \text{rank}(\rho_x(g) - I),$$

with $\rho_x(g)$ as in (24). Indeed, $m_{i,g^{-1}} \equiv -m_{i,g} \mod m_g$, so if $m_{i,g} \neq 0$, then $m_{i,g} + m_{i,g^{-1}} = m_{i,g}$ and hence

$$\frac{m_{i,g}}{m_g} + \frac{m_{i,g^{-1}}}{m_g} = 1.$$

If $m_{i,g} = 0$, however, then $m_{i,g^{-1}} = 0$, and so

$$\frac{m_{i,g}}{m_g} + \frac{m_{i,g^{-1}}}{m_g} = 0.$$

It follows that $\iota(g) + \iota(g^{-1})$ is the number of entries not equal to 1 in $\rho_x(g)$, which is precisely the rank of $\rho_x(g) - I$.

The dimension of the fixed locus in T_xX under the action of g is the same as the dimension of X^g, so the above implies that

$$\iota(g) + \iota(g^{-1}) = n - \dim_{\mathbb{C}}(X^g).$$

Thus, the Proposition follows from the fact that two elements of ordinary cohomology pair nontrivially only if

$$p + p' = q + q' = \dim_{\mathbb{C}}(X^g).$$

□

The role of the degree shift is now clear: it makes the pairing on Chen-Ruan cohomology behave with respect to degrees like the Poincaré pairing on the ordinary cohomology of a manifold. There are further parallels between Chen-Ruan and ordinary cohomology. For example, there is a product structure, with a unit lying in the nontwisted sector whose Poincaré dual is the volume form, also drawn from the nontwisted sector. However, we only need to study the Chen-Ruan cohomology as a graded vector space in these notes.

Example 86 Let $\mathbb{Z}_r$ act on $\mathbb{C}$ via multiplication by rth roots of unity. Then

$$\Lambda[\mathbb{C}/\mathbb{Z}_r] = [\mathbb{C}/\mathbb{Z}_r] \sqcup [\bullet/\mathbb{Z}_r] \sqcup \cdots \sqcup [\bullet/\mathbb{Z}_r],$$

in which there are $r - 1$ twisted sectors in addition to the nontwisted sector. Thus, as a vector space, the Chen-Ruan cohomology of $[\mathbb{C}/\mathbb{Z}_r]$ is $\mathbb{Q}^r$, with generators $e_0, e_1, \ldots, e_{r-1}$ given by the constant function 1 on each of the above components.

The degree-shift on the component generated by e_i is $\frac{i}{r}$, so the r copies of $\mathbb{Q}$ occur in bidegrees $(0, 0)$, $(\frac{1}{r}, \frac{1}{r}), \ldots, (\frac{r-1}{r}, \frac{r-1}{r})$.

References

1. Acosta, P.: Asymptotic expansion and the LG/(Fano, General Type) correspondence (2014). ArXiv preprint, arXiv:1411.4162
2. Adem, A., Leida, J., Ruan, Y.: Orbifolds and Stringy Topology, vol. 171. Cambridge University Press, Cambridge (2007)
3. Arnol'd, V.I.: Singularity Theory, vol. 53. Cambridge University Press, Cambridge (1981)
4. Batyrev, V.: Dual polyhedra and mirror symmetry for Calabi-Yau hypersurfaces. J. Algebraic Geom. **3**, 493–535 (1994)
5. Batyrev, V., Borisov, L.: On Calabi-Yau complete intersections in toric varieties. In: Andreatta, M., Peternell, T. (eds.) Higher-Dimensional Complex Varieties (Trento, 1994). Walter de Gruyter, Berlin (1996)
6. Berglund, P., Hübsch, T.: A generalized construction of mirror manifolds. Nucl. Phys. B **393**, 377–391 (1994)
7. Cannas da Silva, A.: Lectures on Symplectic Geometry. Lecture Notes in Mathematics, vol. 1764. Springer, Berlin (2008)
8. Chen, W., Ruan, Y.: A new cohomology theory of orbifold. Commun. Math. Phys. **248**(1), 1–31 (2004)
9. Chiodo, A., Ruan, Y.: Landau–Ginzburg/Calabi–Yau correspondence for quintic three-folds via symplectic transformations. Invent. Math. **182**(1), 117–165 (2010)
10. Chiodo, A., Ruan, Y.: LG/CY correspondence: the state space isomorphism. Adv. Math. **227**(6), 2157–2188 (2011)
11. Chiodo, A., Iritani, H., Ruan, Y.: Landau-Ginzburg/Calabi-Yau correspondence, global mirror symmetry and Orlov equivalence. Publ. Math. IHÉS, **119**, 1–90 (2012)
12. Clader, E.: Landau-Ginzburg/Calabi-Yau correspondence for the complete intersections $X_{3,3}$ and $X_{2,2,2,2}$ (2013). ArXiv preprint, arXiv:1301.5530
13. Clader, E., Ross, D.: Sigma models and phase transitions for complete intersections. Int. Math. Res. Not. (2015, in press). arXiv:1511.02027
14. Cox, D.A., Little, J.B., Schenck, H.K.: Toric Varieties. American Mathematical Society, Providence (2011)
15. Fan, H., Jarvis, T., Ruan, Y.: The Witten equation, mirror symmetry and quantum singularity theory. Ann. Math. **178**, 1–106 (2013)
16. Fulton, W.: Introduction to Toric Varieties. Annals of Mathematics Studies, vol. 131. Princeton University Press, Princeton (1993)
17. Givental, A.: Equivariant Gromov-Witten invariants. Int. Math. Res. Not. **13**, 613–663 (1996)
18. Givental, A.: A mirror theorem for toric complete intersections. In: Topological Field Theory, Primitive Forms, and Related Topics. Progress in Mathematics, vol. 160, pp. 141–175. Birkhäuser, Basel (1998)
19. Gonzalez, E., Woodward, C.: Quantum cohomology and toric minimal model programs (2012). ArXiv preprint, arXiv:1207.3253
20. Hori, K., Vafa, C.: Mirror symmetry (2000). arXiv preprint hep-th/0002222
21. Hori, K., Katz, S., Pandharipande, R., Thomas, R., Vafa, C., Vakil, R., Zaslow, E.: Mirror Symmetry. American Mathematical Society/Clay Mathematics Institute, Peterborough (2003)
22. Iritani, H.: Convergence of quantum cohomology by quantum Lefschetz. J. Reine Angew. Math. **610**, 29–69 (2007)
23. Krawitz, M.: FJRW rings and Landau-Ginzburg mirror symmetry (2009). ArXiv preprint, arXiv:0906.0796

24. Kreuzer, M., Skarke, H.: On the classification of quasihomogeneous functions. Commun. Math. Phys. **150**, 137–147 (1992)
25. McDuff, D., Salamon, D.: Introduction to Symplectic Topology. Oxford Mathematical Monographs. Oxford University Press, Oxford (1999)
26. Witten, E.: Phases of N = 2 theories in two dimensions. Nucl. Phys. B **403**(1), 159–222 (1993)

The B-Model Approach to Topological String Theory on Calabi-Yau n-Folds

Albrecht Klemm

Contents

Abstract The main goal of these lectures is to explain, to both mathematicians and physicists, some of the new ideas and techniques that have been developed in the course of considering compactifications of string theories from their critical dimension on special holonomy manifolds M (such as Calabi–Yau manifolds) to four dimensions. The physical motivation in this subject is simply to describe the four-dimensional physical theories which arise in this construction, while for mathematicians, the interest is that the physical perspective was effective in discovering unexpected mathematical structures of these manifolds. Physicists and mathematicians alike find interesting structures and simplifications in supersymmetric theories, in particular exactly solvable subsectors, which might shed light on more general mathematical properties of quantum theories of fields and strings. We give an in-depth overview from a physical perspective, covering such topics as the monodromy group and Picard-Fuchs equations, tt* structures, the holomorphic anomaly equation, and modularity properties.

A. Klemm (✉)
Hausdorff Center for Mathematics and Bethe Center for Theoretical Physics (BCTP), Physikalisches Institut, Universität Bonn, Bonn, Germany
e-mail: aklemm@th.physik.uni-bonn.de

E. Clader, Y. Ruan (eds.), *B-Model Gromov-Witten Theory*, Trends in Mathematics,
https://doi.org/10.1007/978-3-319-94220-9_2

1 Physical Motivation and Physical Applications

The main goal of the present lecture is to explain to mathematicians and physicists some of the new ideas and techniques that have been developed in the course of considering compactification of string theories from their critical dimension on special holonomy manifolds M such as Calabi-Yau manifolds to four dimensions. The physical motivation in this subject is simply to describe the four dimensional physical theories which arise in this construction, because starting with superstring theory they can describe all observed interactions and consistently combine them with quantum gravity and are therefore candidates to describe the physical reality at a deeper level and with a higher level of consistency then previously possible. The theories typically carry some amount of supersymmetry and to find the mechanism to break this symmetry at a suitable scale is still one of the biggest open challenges. The interest for mathematicians is that the physical perspective was effective in discovering unexpected mathematical structures of these manifolds and helped to develop powerful new ideas to calculate for example topological information for these manifolds explicitly. Physicists and mathematicians alike find interesting structures and simplifications in supersymmetric theories, in particular exactly solvable subsectors, which might shed light on more general properties of quantum theories of fields and strings.

The purpose of the introduction section is mainly to familiarize mathematicians with the key physical ideas of this setting. String physicist will know this part from books and review, but hopefully both communities can profit from the outline at the end of this section.

1.1 Kaluza-Klein Reduction and Supersymmetry

The original idea of Kaluza and Klein to *geometrize* physical properties by adding compact extra dimensions was developed in the twenties of the last century. It was motivated by the success of *geometrizing* gravity by Einstein's theory of general relativity and the *geometrization* of electromagnetism as a gauge theory. It attempted to unify these two interactions in a five dimensional space time that had to be reduced on a circle to four dimensions. While physically unsuccessful the idea contributed to the development of non abelian gauge theory as explained in the short historical account [272], where also the original references can be found.

Beside the spectacular success of the gauge principle as description of the *electro-weak interactions* by Glashow, Weinberg and Salam in 1967 and the *strong interactions* by the $SU(3)$ gauge theory called quantum chromodynamics in 1973 that culminated in the standard model, a particularly broken non-abelian gauge theory with gauge group $SU(3) \times SU(2) \times U(1)$, the advent of supersymmetry and superstring theory added substantially to the motivation to study Kaluza-Klein reductions on compact manifolds M.

Supersymmetric theories, characterized by an integer $\mathcal{N}$, have a symmetry that transforms bosonic degrees to fermionic degrees of freedom and vice versa. The corresponding supersymmetry generators Q^i_α $i = 1, \ldots, \mathcal{N}$ transform in spinor representations w.r.t. the D dimensional Poincaré group and fulfill the algebra

$$\{Q^i_\alpha, {Q^j}^\dagger_\beta\} = 2\delta^{ij}\sigma^\mu_{\alpha\beta}P_\mu \,. \tag{1.1.1}$$

According to the Coleman-Mandula theorem supersymmetry is the largest symmetry that extends the Poincaré symmetry, introduces only a finite number of spinor fields and allows still for non-trivial S-matrix elements, i.e. for interesting physical processes. It fits the experience in particle physics that at higher energy scales the physics should be more symmetric, even though at the energy scale of present experiments supersymmetry is not yet found and must be broken. The bigger $\mathcal{N}$ the more symmetric, tamer and unrealistic from the low energy point of view is the theory. However solving an interacting field in three and more dimensions is so difficult that solving the simpler supersymmetric version would be a big breakthrough. A strong motivation to consider supersymmetric theories are the cancellations of short distance divergencies that are typical in quantum fields theories and render for example a perturbative point particle description of quantum gravity in four dimensions inconsistent. These cancellations make supersymmetric field theories better behaved at the quantum level than their non-supersymmetric cousins. For example in the standard model of particle physics the leading order correction to the mass of the Higgs at one loop level is quadratically divergent $\delta m = (\Lambda/m)^2$ with the cut-off parameter Λ, while in the supersymmetric version of the standard model it is only logarithmically divergent $\delta m = \log(\Lambda/m)$. This cut-off parameter, here chosen to be an energy scale, is unphysical, but necessary to render the amplitudes finite at small length or high energy scales. In renormalizable quantum field theories it must be possible to absorb the cut-off dependence uniquely using a finite set of counterterms so that the amplitudes calculated including these counterterms depend, after enforcing a finite set of physical renormalization conditions, only on finitely many measurable physical quantities, such as masses and couplings, and are in particular independent of the cut-off parameter. Supergravity theories, the supersymmetric cousins of gravity, are yet not renormalizable in this sense. One possible way out is super string theory in which a perturbative description of super gravity makes sense due to a physical cut-off, which in energy units is the inverse of the string length. What singles this cut-off out among other possible choices is the fact that it is compatible with the Poincaré symmetry and internal symmetries such as gauge symmetries. If there is now a physical cut-off, the milder scale dependence of quantum corrections in supersymmetric theories are often invoked to argue that it might be natural to have large hierarchies as for example between the string scale and the standard model scale. On the other hand consistent string theories require supersymmetry. This, together with the facts that the scale dependent couplings—the strong, the weak and electric one—of a supersymmetrized standard model, unify at a scale close to the

Planck scale, which would be a natural estimate for the string scale and that string theories like the heterotic string can give an unified description of these interactions plus gravity has given a perspective on a physical realization of supersymmetry and string theory in nature.

Spinors have in D dimensions $2^{\lfloor \frac{D+1}{2} \rfloor -1}$ components. As discussed in more detail below, depending on dimension D and the space time Lorentz signature $(t, s = D - t)$, these spinor representations can be real $\mathbb{R}$ also called Majorana spinors, complex $\mathbb{C}$, or quaternionic $\mathbb{H}$.

In any case the degrees of freedom in the spinor representations grow approximately exponentially with D, while bosonic degrees of a given spin grow only polynomially with D. For this reason there is a maximal dimension in which one can supersymmetrize pure gravity, if one insists that the graviton sits in a supersymmetry representation with $J \leq 2$. The maximal dimension is $D = 11$ for the signature[1] $(1, D - 1)$, since it is the maximal dimension in which a $J \leq 2$ supersymmetry representation exists [266]. The corresponding supergravity action was constructed in [80]. Similarly gravity coupled to a gauge theory with signature $(1, D-1)$ can be supersymmetrized in maximal $D = 10$ dimensions, if one insist that the gauge boson sits in a representation with $J \leq 1$. In addition the mixed gravitational and gauge anomalies cancel only for the gauge groups $E_8 \times E_8$, $SO(32)$, or $SO(16) \times SO(16)$ see chapt. 13.5 of [142] for a review. Likewise purely from representation theoretic arguments Nahm concluded [266] that six is the maximal dimensions in which super conformal quantum field theories could exist. In this case there is no Lagrangian description, but the advances of geometrical engineering of quantum field theories, described in Section allow now to explore certain properties of these theories.

The existence of these unique supersymmetric theories in the maximal dimensions leads naturally to the program of a geometric classification of supersymmetric theories in lower dimensions $\mathfrak{d} = D - d$ by Kaluza-Klein reduction on real d dimensional manifolds M, see [103] for a review. In particular in $D = 11$ the supersymmetry generator transforms in one $\mathcal{N} = 1$ real $32 = 2^5$ spinor representation with components Q_α, $\alpha = 1, \ldots, 32$. Lower dimensional theories can have supersymmetry generators in different spinor representations but with maximally 32 real components, iff M has the maximal number of covariant constant spinors as it is the case in particular for compactifications on the d dimensional torus T^d.

The question what principal type of lower dimensional supergravity is realized depends on how the 32 supercharges Q_α fill the lower dimensional spinor representations. If M has less covariant constant spinors one can get supersymmetric theories in $\mathfrak{d} = D - d$ with less, e.g. none, supersymmetries. The number of covariant constant spinors depends on the *holonomy group* of the compactification manifold M. As the symmetric spaces turn out to be too rigid to obtain suitable $\mathfrak{d} = D - d$ physics the *special holonomy manifolds* classified by Berger have become of great

[1] For signature $(2, D - 2)$ the maximal dimension is $D = 12$.

interest to physicists. The mathematical theory is reviewed in [42].[2] In our lectures the Calabi-Yau n-folds, which have holonomy group $SU(n)$ will play a prominent role.

Of course the crucial question in the Kaluza-Klein program is how much about the $\mathfrak{d} = D - d$ theory can be learned from the geometry of the compactifying space M and its symmetries in the supersymmetric case. An important statement concerns the Kaluza-Kein spectrum. If the fields in D dimensions are massless and we denote the Laplacian on a form (or spinor) field γ in m dimensions irrespectively of the signature Δ_m then we get for the decomposition γ into the $\mathfrak{d}$ and d dimensional part[3] $\gamma = \gamma' \cdot \gamma''$

$$\Delta_D \gamma = (\gamma'' \Delta_{\mathfrak{d}} \gamma' + \gamma' \Delta_d \gamma'') = \gamma''(\Delta'_{\mathfrak{d}} + m^2_{\gamma''}(g))\gamma' = 0 \,. \tag{1.1.2}$$

The second equation tells us that the masses in $\mathfrak{d}$ dimensions are given by the spectrum of the Laplacian on the internal manifold M, which implies that the massless fields correspond to harmonic forms on M, while the masses of massive fields depend on the metric g and are in particular very massive if the internal manifold is very small. Typically one assumes it to be close to the Planck scale, which is in energy units at 10^{19} GeV which corresponds in length units to $1.6.10^{-35}$ m. The dependence of the spectrum and couplings between the fields in the $\mathfrak{d}$-dimensional effective Langrangian on those deformations of the metric g, which keep the holonomy and therefore the supersymmetry and maybe even more so on the ones which break supersymmetry, is clearly physically of great interest. The first mentioned setting in which one stays within the supersymmetric theory is the main theme of the lecture. As we mentioned before supergravity are theories alone are not a good starting points because they are not renormalisable, so one rather has to start with string theory in D dimensions. However many properties of the low energy spectrum are unchanged by this, because the string excitations are again of very high energy if the string tension is very high. Frequently one assumes it to be also roughly the Planck-scale, while charged string excitations are experimentally excluded at the scale of roughly 10^4 GeV.

To conclude the section we want to mention already that after including string effects in simple situations, like compactification on a circle of radius $\mathfrak{r}$—dimensionless in string units—denoted as $S^1_{\mathfrak{r}}$ and understanding T-duality to a compactification on $S^1_{\frac{1}{\mathfrak{r}}}$ as well as orbifolds compactifications on $S^1_{\mathfrak{r}}/\mathbb{Z}_2$, the program of studying supergravity theories and their string extensions in lower dimensions from dimensional reduction of eleven dimensional supergravity, whose putative microscopic description is called M-theory and is believed to be a non perturbative completion of the Type IIA string, becomes rather complete and successful.

---

[2] See in particular Theorem 10.90 (Berger and Simons) and Corollary 10.92.

[3] Here we assume the d-dimensional internal space and the $\mathfrak{d}$-dimensional space time to be a direct product.

1.1.1 The String Perspective

As we mentioned above, beside the unique supergravity theories in maximal dimensions another strong motivation to reconsider Kaluza-Klein compactifications is superstring theory whose critical dimension is $D = 10$. String theory has a perturbative definition, which expands a *functional integral* in which one varies over the map

$$X : \Sigma \to M_{st}, \tag{1.1.3}$$

from the two dimensional world-sheet Σ to the space time M_{st}. In the spirit of the Kaluza-Klein approach one views the latter as product, or maybe a warped product, of a large four dimensional space U_4, our Universe as it is visible with low energy experiments, and an internal six dimensional space M, in the simplest case $M_{st} = U_4 \times M$. The bosonic part of the weight functional associated to this map, called the action

$$S(X, h, \psi_{ferm}, G, B, \phi) = \int_\Sigma \mathrm{d}^2\sigma \mathcal{L}(X, h, \psi_{ferm}, G, B, \phi) \tag{1.1.4}$$

is the area of the image curve in M_{st}, also known as the Dirichlet energy of the map. The Lagrangian $\mathcal{L}$ contains as bosonic dynamical fields X and the two dimensional metric h as Lagrange multiplier. The action has to be supersymmetrized by fermions as the pure bosonic string has a negative energy state, called tachyon, and is therefore not consistent. We indicate the fermionic super partners only schematically by ψ_{ferm}. The Lagrangian also contains as background fields the *metric* G of M_{st}, the *Neveu-Schwarz background 2-form field* B in M_{st}, to which the string couples, as well as dilaton whose vacuum expectation value determines the string coupling. The dilaton ϕ multiplies the Euler density of Σ and a constant dilaton gives a topological contribution $\phi\chi(\Sigma)$ to the Lagrangian. The usual approach to string theory is a first quantized one, i.e. one does not vary in the path integral the background fields G, B and ϕ. The "path" integral e.g. for the zero point function is hence

$$Z(G, B, \phi) = \int \mathcal{D}X\mathcal{D}h\mathcal{D}\psi_{ferm} e^{\frac{i}{\hbar}S(X,h,\psi_{ferm},G,B,\phi)} \,. \tag{1.1.5}$$

Classically the world-sheet action depends only on the *conformal structure* of Σ, i.e. it is scaling and reparametrisation invariant. In the critical dimension which is $D = 10$ for the superstring this holds also in the quantum theory, as a potential anomaly of the scaling symmetry cancels. As a consequence the variation $\mathcal{D}h$ can be replaced at each genus by a variation over *conformal structure*, which is also the *complex structure* for a two dimensional surface. Schematically one has the substantial simplification

$$\int \mathcal{D}h \to \sum_{g=0}^{\infty} \int_{\mathcal{M}_{\Sigma_g}} \mu_{3g-3}, \tag{1.1.6}$$

from a functional integral to a discrete sum over finite dimensional integrals in the critical dimension. Here μ_{3g-3} is a measure on the moduli space of complex structures of Σ_g. By the Riemann-Roch theorem it has generically, i.e. for $g \geq 2$, the complex dimension $3g-3$. Riemann surfaces with genus zero or one have three or one complex Killing field(s) respectively. This raises the value of the dimension of the moduli space given by the Riemann-Roch theorem to zero and one. The latter is called τ and lives in the upper complex half-plane.

Consistency requires reparametrization invariance to hold also for those reparametrizations, which are not continuously deformable to the identity. They are called *global reparametrizations* and fall in a discrete number of topological types. Not all,[4] but essential consequences of this general consistency requirement can be studied for the string vacuum amplitude at world sheets of genus *one* Σ_1, which can be interpreted as the worldsheet *partition function* $Z_1(\tau)$ of the string theory,[5] which contains the information about the *spectrum* of the theory. In the genus one case the different topological types of maps from Σ_1 to itself are characterized by the Dehn twists which form the group Sl(2, $\mathbb{Z}$), also known as *modular group*, as it acts projectively on the complex modulus τ of Σ_1. The decisive requirement is then that $Z_1(\tau)$ is modular invariant and that there are no negative energy states, called tachyons, in the spectrum.

It gave much credibility to the importance of string theory as a fundamental microscopic description of field theory that the consistent supergravity theories in $D = 10$ can be all realized as the point particle limit of different superstring theories, that are consistent in the sense described above, as was discovered in the 1980s. Again this is reviewed in [142]. These supergravity theories are: The chiral type IIB theory in which the supercharges transform in two ($\mathcal{N} = 2$) 16 dimensional spinors of one chirality; the non-chiral type IIA theory in which the supercharges transform in two ($\mathcal{N} = 2$) 16 dimensional spinors of opposite chirality; the chiral supergravity with the supercharges transforming in one ($\mathcal{N} = 1$) 16 dimensional spinor representation coupled to a gauge theory, which can be for the requirement of one-loop anomaly cancellation only have the gauge group $SO(32)$ or $E_8 \times E_8$.[6] The corresponding D=10 closed string theories go by similar names: Type IIA/B string theory and heterotic $SO(32)/E_8 \times E_8$ string theory. The $SO(32)$, $\mathcal{N} = 1$ supergravity can also be realized as the point particle limit of an unoriented open string called Type I string, where the consistency conditions imposed by global reparametrization invariance are of course different then the ones described at the end of the last paragraph and are known as tadpole cancellations conditions. Note that the fundamental D=11 supergravity with $32 = 16 + 16$ supercharges had found no microscopic description by a perturbative critical string theory. A Kaluza Klein

---

[4] See [263] for the complete conditions in rational conformal two dimensional field theories.

[5] Note that it will depend on the background fields, which are trivial in $D = 10$ flat space. Also unlike as for Z in (1.1.5) the integration over the complex moduli space, parametrized at genus one by the fundamental region $\mathcal{F} = \mathbb{H}_+/\mathrm{PSL}(2, \mathbb{Z})$ is not performed.

[6] The $SO(16) \times SO(16)$ theory has a dilaton tachyon [142].

reduction of it on a circle S^1_τ leads to the non-chiral Type IIA supergravity. It was proposed to provide a non-perturbative description of Type IIA theory, called *M*-theory, in which the type IIA string coupling $g_s^{IIA} = \tau^{-1}$ is identified with the inverse of the dimensionless radius of the circle S^1_τ [325].

If the string length[7] l_s is very small compared to the typical curvature radii r of M, the Kaluza-Klein point particle approach is a good approximation. One can introduce $1/\tau = l_s/r \ll 1$ as dimensionless parameter and provide a perturbative scheme to include string corrections to higher orders in $g_\sigma = 1/\tau$ to the point particle results. The starting point of this perturbative expansion is the action of the *non-linear* σ*-model*. Its action contains the metric and the expansion is around the flat or infinite volume metric.

If on the other hand l_s is comparable to r this approach breaks down and one has to the consider, e.g. in the closed string, the geometry of the loop space of M to get information about the effective theory in $D - d$ dimensions. In particular one has to take into account the string instantons, which are in the limit discussed before exponentially suppressed by their volume. To find just the contributions of these classical solutions of the string action is already a very complicated problem on a general special holonomy manifold M. The main reason that we can address it at all is that the full string amplitudes exhibit so called *spacetime duality symmetries*, precisely because of the extended nature of the string.

The simplest example[8] is the $\Gamma = \mathbb{Z}_2$ *duality group* that states that string amplitudes on a circle S^1_τ of radius $\tau = r/l_s$ are equivalent to those on a circle of radius $\tau' = 1/\tau = l_s/r$. Here we used again the dimensionless radius τ as the real modulus of the compactifications manifold S^1_τ. This allows at least to relate the very stringy regime $l_s \gg r$, where the instantons that correspond to strings winding multiple times around the circle contribute most to the known point particle regime.

For superstring theories there are for certain amplitudes conditions fulfilled that allow to reconstruct even very non-trivial amplitudes:

- The moduli are *complex parameters*, corresponding to the vacuum expectation values of a complex scalar fields in a flat potential, in supersymmetric theories certain amplitude depend *holomorphically* on them or can be related to *holomorphic* expressions of them.
- The duality symmetries form an infinite discrete group Γ, for example finite index subgroups of $\mathrm{Sl}(2, \mathbb{Z})$ occur, that admits only a finitely generated ring of holomorphic automorphic forms with the appropriate invariance properties under the action of Γ on the moduli.
- The physics of the problem determines the automorphic weight and the boundary behaviour that follows typically from a physical description of the theory in a weak coupling regime and fixes the finite ambiguity in the amplitude.

---

[7]The string tension T_s can be introduced as the only dimensionful quantity $[T] = \text{length}^{-2}$. Other common choices are the Regge slope $\alpha' = \frac{1}{2\pi T}$ or the string length $l_s = 2\pi\sqrt{\alpha'}$.

[8]In this simple example the full string reduction can be performed.

The above condition are typically met for protected amplitudes that arise in compactifications of the type IIA/IIB string on Calabi-Yau 3-folds. In this case the $16 + 16$ supercharges in ten dimensions, see Table 1, are reduced by the nontrivial $SU(3)$ holonomy to 8 which lie in two four component Majorana spinors. The four dimensional supersymmetry is hence $\mathcal{N} = 2$ supergravity and the techniques outlined above carry over to the limit of the type IIA/IIB geometry in which gravity decouples [205, 214, 231] and allow to understand the famous $\mathcal{N} = 2$ gauge theory results of [287, 288] from the point of view of string theory.

The proposal for the non-perturbative description of the type IIA theory of Witten by M-theory [325] predicts that M-theory has as effective eleven dimensional low energy action the unique $D = 11$ $\mathcal{N} = 1$ supergravity action that was discovered by [80, 266]. The proposal does not include a microscopic description of M-theory, like the ones given by the $D = 10$ string theories for the $D = 10$ dimensional super gravities. It merely suggest that the M2-brane plays a similar role then the string and that the M5-brane plays an important role as the dual degree of freedom in $D = 11$. The quantisation of these systems is very difficult, but conjectures of special properties for the M2 brane degrees of freedom have been made in large N-limits [8]. For circle compactifications detailed descriptions for the topological string expansion including non-perturbative properties have been made in [139, 140], using the $D0/D2$ brane spectrum, as we review in Sect. 4.3.3.

1.1.2 The B-Model in Mirror Symmetry

The B-model is concerned with the study of the dependence of the amplitudes of superstring compactifications on the complex moduli, which correspond to the *deformation of the complex structure* of the compactification manifold. This complex moduli space can itself be compactified to a complex manifold, which we call $\mathcal{M}_{cs}(M)$.

An important physical application of the B-model is within the framework of *mirror symmetry*. Mirror symmetry exchanges one Calabi-Yau manifold M with another Calabi-Yau manifold W, so that the complex structure moduli space $\mathcal{M}_{cs}(W)$ of W is exchanged with the moduli space $\mathcal{M}_{cks}(M)$ of areas, complexified with the NS B-field, of M and vice versa. The latter is called the complexified Kähler structure moduli space.

Without going into the technical details let us recapitulate the facts, which makes this setting so powerful for calculations: Mirror symmetry exchanges the topological A-model with the topological B-model. The couplings in these topological theories are of the same structure. They encode on the world sheet sphere the 3-point couplings and on higher genus g world sheets, where the theory has to be coupled to world-sheet gravity, the couplings between the selfdual part of the space time curvature R_+ and the selfdual part F_+ of the graviphoton fields strength $R_+^2 F_+^{2g-2}$.

By supersymmetric localisation one can show that in the A-model, these world-sheets are mapped to holomorphic curves of definite area. In general they depend therefore on the complexified size parameters of the Calabi-Yau space. One the other hand one can show that they do not depend on the complex structure of the Calabi-Yau space.

By supersymmetric localisation in the B-model one can establish that the world-sheets are mapped to points. Hence the resulting theory does not depend on the complexified areas. In fact the resulting theory is a field theory on the Calabi-Yau space, known as Kodaira-Spencer gravity. As the name suggests its amplitudes do depend on the complex structure.

Schematically one has the following picture

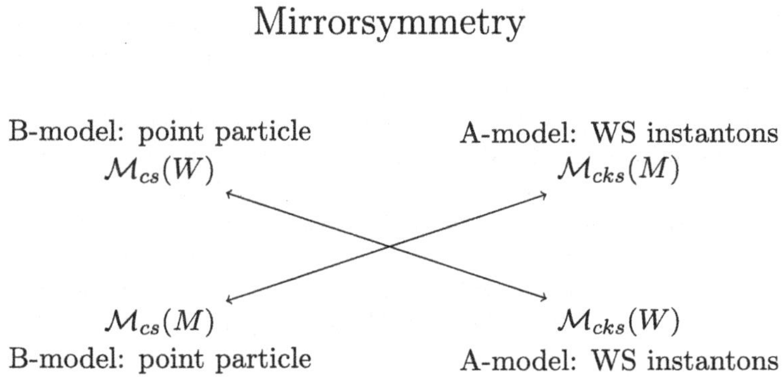

One uses this setting in the following way: One determines the moduli dependent amplitudes on the side where they depend only on the complex structure deformations. Here they can be calculated exactly. In fact they are sections of bundles over the complex moduli space, which can be expanded in convergent power series in various regions of $\mathcal{M}_{cs}(*)$. If the mirror map between the complex moduli and the complexified Kähler structure moduli spaces say

$$t_* : \mathcal{M}_{cs}(W) \to \mathcal{M}_{kcs}(W) \tag{1.1.7}$$

is locally known between the patches $U_* \in \mathcal{M}_{cs}(W)$ and $V_* \in \mathcal{M}_{cks}(M)$, one can relate the individual terms of the expansion of the B-model amplitudes to individual instanton contributions of the A-model amplitudes. This is particular successful in region U_{lr} of $\mathcal{M}_{cs}(W)$, which correspond to a large radius region V_{lr} of $\mathcal{M}_{kcs}(M)$, because here one understands in the A-model the principal form in which individual world sheet instantons of a given genus and degree contribute to the amplitude. Hence one can identify these contributions one by one. In this way the expansion of the B-model amplitudes in the right parametrizations t_{lr}, becomes a generating function of an infinite number of WS-instantons contributions also known as Gromov-Witten invariants, each of which would be very hard to calculate directly. Other regions of $\mathcal{M}_{cks}(M)$ where again the expansion in the A-model is well understood are the orbifold points of M where M looks locally like $\mathbb{C}^3/G$ where G is a discrete subgroup of $SU(3)$. Again if the map $t_{orb} : U_{orb} \to V_{orb}$ is locally known, the B-model amplitudes expanded in t_{orb} become generating functions for "Orbifold Gromov-Witten" invariants. Another interesting aspect of the orbifold points is, that for these values of the moduli one has at least in many cases an exact description of the two dimensional superconformal worldsheet theory. The latter was suggested by Doron Gepner and for this reason these points are also known as Gepner points. There are many other special regions and points corresponding to critical locii in $\mathcal{M}_{cks}(M)$, where M either degenerates of becomes specially symmetric, where the corresponding A-model interpretation of the amplitudes has yet to be found.

The key B-model techniques that we develop in the bulk of the lectures can be summarized as follows. This complex moduli space can be compactified to a complex manifold, which we call $\mathcal{M}_{cs}$. The study of the complex structure of M is at least locally equivalent to the *variation of the Hodge structure* of M. The information of this variation is contained in the dependence of *periods integrals* on the complex structure moduli. The periods integrals have also have a direct interpretation in terms of the physical amplitudes. For example for Calabi-Yau n-folds $\mathcal{M}_{cs}$ is Kähler and the Kähler metric is determined by the periods. The periods are solutions of systems of *differential equations*, often called the Picard-Fuchs equations, which determines up to choice of basis of the solutions the periods integrals and their dependence on the complex moduli. These systems of *differential equations* can geometrically be interpreted as the flatness of a *connection* of a bundle over $\mathcal{M}$ in which the periods are sections. The connection is known as the *Gauss-Manin connection*. The local flatness does not mean that the connection is trivial, rather the periods experience *monodromies*, when they are analytically continued around critical loci which are at complex codimension one in $\mathcal{M}_{cs}$. At these loci in $\mathcal{M}_{cs}$ the manifold M is either singular or exhibits additional discrete symmetries. While the monodromy group $\Gamma \subset \mathrm{Sl}(2, \mathbb{Z})$ for elliptic curves, i.e. Calabi-Yau spaces in complex dimension one, has been intensively studied and found to be related to deep number theoretical questions, not much is known about the monodromy group $\Gamma \subset \mathrm{Sp}(h_3(W), \mathbb{Z})$ of Calabi-Yau threefolds and even less about its automorphic forms. It is this difficult question that can be at least in part addressed by mirror symmetry and the B-model techniques that we discuss in the main part of the lectures.

1.1.3 The Full Physical Mirror Conjecture

One should notice that the mirror conjecture as suggested from the worldsheet point of view of type II string, which can be made rather precise for example at the Gepner point, goes much further then exchanging the complex structure with the complexified Kähler structure with the technical advantages explained above. Generally one has an $(\mathcal{N}, \bar{\mathcal{N}}) = (2, 2)$ superconformal field theory on the worldsheet of the closed string theories propagating on M as well as on W. The surprising prediction is that these two $(\mathcal{N}, \bar{\mathcal{N}}) = (2, 2)$ superconformal field theories are completely identical after some rather trivial field identification, which just flips the relative left and right charge assignment of the composite fields that are constructed from left and the right moving fields.

Already the identification of the *topological sectors* of the $(\mathcal{N}, \bar{\mathcal{N}}) = (2, 2)$ superconformal field theory with the cohomology groups of M and W implies that

$$H^{n-k,k}(M) = H^{k,k}(W)\,, \quad k, n-k \geq 0 \qquad (1.1.8)$$

i.e. for odd dimension n, odd forms are map to even forms and vice versa. The full prediction is that not just the harmonic forms, but for a suitable and possible choice of metrics on M and W, the complete spectrum of the Laplacians can be identified and moreover all non-topological *closed string amplitudes* are also the same.

In odd dimensions mirror symmetry exchanges Type IIA with IIB theories and even branes with odd branes. These branes correspond to possible boundary conditions of the $(\mathcal{N}, \bar{\mathcal{N}}) = (2, 2)$ superconformal field and should be likewise identified together with all possible *open string amplitudes* that are possible with this boundary conditions. Again this should hold if these branes are topological but even if they are not topological. The former statement leads to *homological mirror symmetry conjecture*. Since the even topological branes and their interactions are mathematical described by the *derived category of coherent sheaves* while the odd topological branes and their interactions by the *Fukaya category of special Lagrangians* the two structures are expected to be identified by mirror symmetry.

1.1.4 Comment on the Construction of Mirror Pairs

One obvious question is how to construct the mirror partner W from M. This questions has a very nice conceptual answer outlined by Strominger, Yau and Zaslow, because mirror symmetry can be understood as the T-duality $\tau' = \frac{1}{\tau}$, that we discussed in the last section for one S^1, performed on half of the real dimensions of a Kähler manifold of complex dimension n. However in practice this program is difficult to implement—we explain some difficulties below—, but there are several constructive ways to produce mirror pairs. We list all of them below

- The Batyrev construction [28] defines pairs of reflexive polyhedra $(\Delta, \hat{\Delta})$, which specify pairs of compact toric varieties $(\mathbb{P}_\Delta, \mathbb{P}_{\hat{\Delta}})$, which are almost Fano. One can define therefore pairs of smooth Calabi-Yau manifolds $(M_\Delta, W_{\hat{\Delta}})$ as the canonical bundles in these toric ambient spaces and show that the complex structure and complexified Kähler structure are exchanged in these pairs.
- The Greene-Plesser construction [144] constructs the mirror of a Fermat hypersurfaces in weighted projective spaces by a maximal discrete orbifold [121]. It is inspired by a duality on the worldsheet conformal field theory and yields a subset of the hypersurface pairs within the Batyrev construction.
- The Batyrev-Borisov construction generalizes the Batyrev construction by specifying dual toric ambient spaces together with a dual nef partition that defines

pairs of complete intersections Calabi-Yau manifolds in the toric ambient spaces [29].

- The Berglund-Hübsch construction defines the mirror by transposing the adjacency matrix of the exponents in a hypersurface constraint. It has a considerable but not complete overlap with the Batyrev construction.
- The Strominger-Yau-Zaslow conjecture, which asserts that the D_0 brane or sky-scraper sheaf on M is identified with a special Lagrangian brane L on W. As the moduli space of the D_0 brane is the Calabi-Yau manifold M this particular special Lagrangian brane L must have a moduli space that can also be the Calabi-Yau space M. As the dimension of the geometric deformation space is $b_3(L)$, it can be argued to be a real n-dimensional torus. M is then the total space of the T-dual torus fibred over this deformation space, the real n-dimensional base. This works because the complex deformation space of the brane is the geometrical of L together with a flat gauge connection on L, which is described by the dual torus. This picture not only implies that generically Calabi-Yau spaces have special Lagrangian torus fibrations, it also predicts that the mirror Calabi-Yau W is obtained by T-duality on the corresponding fibre torus. It can be shown deformation space that this construction exchanges the complexified Kähler structure or more general the symplectic structure with the complex structure. It is easy to see that the T-duality on an odd number of circles maps type IIA to type IIB string. The T-duality can be easily defined on the generic fibres. The extension to the degenerate tori, which are necessarily present in actual Calabi-Yau spaces with the full $SU(n)$ holonomy, is harder to describe.

In practice for the examples explained here the Batyrev and Batyrev-Borisov construction are sufficient to construct the mirrors. We should mention further that recent progress in the localisation within the gauge linear σ-model, allows to construct at least at genus topological A- and B- model data without actually constructing the geometrical mirror pairs.

1.1.5 Bergers List

We finish this motivational introduction, which started with basic concepts of Kaluza-Klein reduction, with a section containing a short summary of the spinor representations in various dimensions and the corresponding special holonomy manifolds.

For a vector bundle V with a connection Γ over a path connected manifold M the holonomy group g_Γ is generated by all transformations that a vector $v \in V_p$ at the point p experiences, when it is parallel transported around all possible closed loops in M back to p. In particular for a d dimensional Riemannian manifold M with $\Gamma(g)$ the metric connection and $V = TM$ the tangent bundle one denotes the corresponding holonomy group $\mathrm{Hol}(g)$. Clearly $\mathrm{Hol}(g) \subset O(d)$ in general, $\mathrm{Hol}(g) \subset SO(d)$ if M is oriented, $\mathrm{Hol}(g) = 1$ if g is a flat metric and $\mathrm{Hol}(g) = H$, if M is a symmetric manifold defined by the Lie groups G and H as G/H.

On orbifolds spaces of flat spaces with conical singularities one can have discrete holonomy groups.

Berger classified all possible holonomy groups on simply connected irreducible and non-symmetric manifolds of real dimension d. We reproduce the list with some additional information about the properties of the metric and the number N_+, N_- of complex covariant constant spinors with positive and negative chirality [312] respectively. If d is odd the spinor representation is irreducible and we have just one type of spinor. The last part in the list below comments on the special forms that exist on this manifold. See [155, 203, 312] for more background.

1. $\mathrm{Hol}(g) = SO(d)$, *generic oriented manifold*, not necessarily spin.
2. $d = 2n$ with $n \geq 2$: $\mathrm{Hol}(g) = U(n)$, *Kähler manifold, Kähler*, not nec. spin; ω Kähler form of Hodge type $(1, 1)$.
3. $d = 2n$, $n \geq 2$: $\mathrm{Hol}(g) = SU(n)$, *Calabi-Yau manifold, Ricci-flat, Kähler*, $N_\pm = 1$ for n odd, $N_+ = 2$ for n even; ω Kähler $(1, 1)$-form and Ω holomorphic $(n, 0)$-form.
4. $d = 4n$, $n \geq 2$: $\mathrm{Hol}(g) = Sp(n)$, *Hyperkähler manifold, Ricci-flat, Kähler*, $N_+ = n + 1$; H, I, J $SU(2)$ triplet of $(1, 1)$ forms.
5. $d = 4n, n \geq 2$: $\mathrm{Hol}(g) = Sp(n)\cdot Sp(1)$, *Quaternionic Kähler manifold, Einstein, not Ricci-flat, not Kähler* .
6. $d = 7$: $\mathrm{Hol}(g) = G_2$, *G_2-manifold, Ricci-flat*, $N = 1$; Φ associative 3-form, $*\Phi$ coassociative 4-form.
7. $d = 8$: $\mathrm{Hol}(g) = Spin(7)$, *Spin(7)-manifold, Ricci-flat*, $N_- = 1$; Ψ Cayley 4-form.

1.1.6 Spinors and Supergravity in Various Dimensions

Here we provide a list of spinors in the dimensions which are relevant to study Superstring-, M- and F-theory compactifications. In this sections spinors are irreducible representations V of the $D = t + s$ dimensional Lorentz algebra $\mathfrak{so}(t, s)$. They have generically $2^{\lfloor \frac{D+1}{2} \rfloor - 1}$ components.

However more information is needed, as one has depending on the dimension natural bilinear forms $\langle \cdot, \cdot \rangle : V \times V \to \mathbb{K}$ associated to the representation V, which fulfill $\langle s_1, s_2 \lambda \rangle = \langle s_1, s_2 \rangle \lambda$ and $\langle s_1, s_2 \rangle = \overline{\langle s_2, s_1 \rangle}$. Depending on whether $\mathbb{K}$ is $\mathbb{R}$, $\mathbb{C}$ or $\mathbb{H}$ one calls the spinor *real, complex* or *quaternionic*. In the physics literature real spinors are denoted *Marjorana* spinors and quaternionic spinors make two *symplectic Majorana* spinors. In even dimensions one uses in the physical nomenclature often Weyl spinors.

Of course in physics one wants to have spinors that are irreducible representations under space reflections (parity), time reflections and charge conjugation as for example the Dirac spinor in four dimensions or one wants to put additional constraints to obtain the four dimensional Majorana spinor. These are more detailed or specialized considerations as covered in the present discussions of irreducible representations of $\mathfrak{so}(t, s)$.

Real spinors and complex spinors have the naive number of real components but a nontrivial constraint on the quaternionic spinors eliminates half of their components.

Let us now give a list of the spinor representations. Since it is an useful intermediate information we give also the irreducible representations of the Clifford algebra called pinors below.

Using Bergers list and the list of spinors one can infer how many supersymmetries one gets in the uncompactified dimension $\mathfrak{d} = D - d$ for a compactification on a $d = 2n$- dimensional internal space. The spinor of the uncompactified space decomposes under the representations of the low energy space time and the internal space according to the embedding of the maximal subalgebra

$$\mathfrak{so}(1, D-1) \supset \mathfrak{so}(1, \mathfrak{d}-1) \oplus \mathfrak{so}(d) \tag{1.1.9}$$

The holonomy group acts on the representations of $\mathfrak{so}(d)$ of the internal space and the number of surviving supersymmetries correspond to the number of covariantly constant components of the spinor in the internal space.

For example for the compactification of the maximal supersymmetric type IIA/B theories in $D = 10$ on a Calabi-Yau 3-fold one has according to Bergers list $\mathfrak{Hol}(g) = \mathfrak{su}(3)$. Now (1.1.9) reads

$$\mathfrak{so}(1,9) \supset \mathfrak{so}(1,3) \oplus \mathfrak{so}(6) \cong \mathfrak{so}(1,3) \oplus \mathfrak{su}(4) \tag{1.1.10}$$

where we identified $\mathfrak{so}(6) \sim \mathfrak{su}(4)$. The eight dimensional spinor of $\mathfrak{so}(6)$ decomposes in spinors of positive- and negative chirality, which transform as **4** and $\bar{\mathbf{4}}$ representations of $\mathfrak{su}(4)$. Now we have to study the invariant pieces of these spinors under the holonomy $\mathfrak{su}(3)$. Luckily this is fixed by representation theory. One has

$$\mathbf{4} = \mathbf{3} + \mathbf{1}, \tag{1.1.11}$$

where we have $\mathfrak{su}(3)$ representations on the right hand side. Hence there is one invariant or more precisely covariantly constant component for the positive- and similar one for the negative chiral spinor of $\mathfrak{so}(6)$. That leads to spinors of two chiralities transforming under the Lorentzgroup $\mathfrak{so}(1,3)$ of the uncompactified four dimensional space-time filling the $\mathbb{C}^2$ representation of Table 1. Hence if starts with $\mathcal{N} = 2$, i.e. two 16 component spinors Q^I_{10}, $I = 1,2$ generating the $\mathcal{N} = 2$ supersymmetry algebra in ten dimension, one gets two spinors Q^I_4, $I = 1,2$ generating the $\mathcal{N} = 2$ supersymmetry in four dimensions.

Often in the physics literature one speaks of the components of Q_{dim} as supercharges. Equation (1.1.11) is then interpreted that the $\mathfrak{su}(3)$ holonomy on a Calabi-Yau 3 fold divides the number of supercharges by four.

Similar the compactification of the ten dimensional theory on $K3$ yields the decomposition

$$\mathfrak{so}(1,9) \supset \mathfrak{so}(1,5) \oplus \mathfrak{so}(4) \cong \mathfrak{so}(1,5) \oplus \mathfrak{su}(2) \oplus \mathfrak{su}(2) \tag{1.1.12}$$

Table 1 Pinors, Spinors, Supersymmetry Representations and maximal number of supersymmetries

$D = t + s$	Pinors	Spinors	Q	$\mathcal{N}$
12=2+10	$\mathbb{R}^{64}$	$\mathbb{R}^{32} \oplus \mathbb{R}^{32}$	$\mathbb{R}^{64}$	1
11=1+10	$\mathbb{R}^{32} \oplus \mathbb{R}^{32}$	$\mathbb{R}^{32}$	$\mathbb{R}^{32}$	1
10=1+9	$\mathbb{R}^{32}$	$\mathbb{R}^{16} \oplus \mathbb{R}^{16}$	$\mathbb{R}^{16}$	2
9=1+8	$\mathbb{C}^{16}$	$\mathbb{R}^{16}$	$\mathbb{R}^{16}$	2
8=1+7	$\mathbb{H}^{8}$	$\mathbb{C}^{8}$	$\mathbb{C}^{8}$	2
7=1+6	$\mathbb{H}^{4} \oplus \mathbb{H}^{4}$	$\mathbb{H}^{4}$	$\mathbb{H}^{8}$	2
6=1+5	$\mathbb{H}^{4}$	$\mathbb{H}^{2} \oplus \mathbb{H}^{2}$	$\mathbb{H}^{4}$	4
5=1+4	$\mathbb{C}^{4}$	$\mathbb{C}^{4}$	$\mathbb{C}^{4}$	4
4=1+3	$\mathbb{R}^{4}$	$\mathbb{C}^{2}$	$\mathbb{C}^{2}$	8
3=1+2	$\mathbb{R}^{2} \oplus \mathbb{R}^{2}$	$\mathbb{R}^{2}$	$\mathbb{R}^{2}$	16
2=1+1	$\mathbb{R}^{2}$	$\mathbb{R} \oplus \mathbb{R}$	$\mathbb{R}^{2}$	16
1=1+0	$\mathbb{C} \oplus \mathbb{R}$	$\mathbb{R}$	$\mathbb{R}$	32

and the **16** decomposes as

$$\mathbf{16} = (\mathbf{4}, \mathbf{2}, \mathbf{1}) \oplus (\mathbf{4}, \mathbf{1}, \mathbf{2}) \, . \tag{1.1.13}$$

We can identify the second $\mathfrak{su}(2)$ with the holonomy of the $K3$, which as Calabi-Yau 2-fold has $\mathfrak{Hol}(g) = \mathfrak{su}(2)$. The two invariant combinations in the other $\mathfrak{su}(2)$ give one quaternonic spinor $\mathbb{H}^4$ for each ten dimensional spinor **16**. E.g. $\mathcal{N} = 1$ in ten dimensions leads to $\mathcal{N} = 1$ in six dimensions and the number of supercharges is divided by two by the $\mathfrak{su}(2)$ holonomy.

By the considerations above it is clear that compactifications on flat tori with trivial holonomy leads to no reduction of the number of supercharges. If one starts with the maximal supersymmetry in eleven dimensions one gets upon $T^d \sim (S^1)^d$ compactifications theories with the maximal amount of supersymmetries in all lower dimensions that are indicated in Table 1.

One does not get all possible chiralities though. E.g. by compactifying eleven dimensional supergravity on S^1 one gets the chiral type IIA supergravity in ten dimensions which has one positive- and one negative chiral **16** spinor generating the $\mathcal{N} = 2$ theory. If one wants to express that fact one uses the notation $\mathcal{N} = (1, 1)$ instead of $\mathcal{N} = 2$.

1.1.7 String Dualities and Further Applications of B-Model Techniques

There have been indications that compactifications, which start with different supersymmetry in ten dimensions but end up in the un-compactified $\mathfrak{d}$-dimensional space with the same supergravity theory, are dual to each other, after non-perturbative contributions have been taken into account.

Heterotic/Type II Duality and Geometric Engineering of QFT

The earliest dualities start with the $\mathcal{N} = 1$ heterotic string and the $\mathcal{N} = 2$ type II string in ten dimensions and compactify on the heterotic side on manifolds which smaller holonomy groups, which break less supersymmetry [197]. For example the following dualities between $\frac{1}{2}$—and $\frac{1}{4}$ super symmetric theories[9] are believed to be true

$$E_8 \times E_8 \text{ heterotic string on } T^4 = \text{Type II String on } K3$$

$$E_8 \times E_8 \text{ heterotic string on } K3 \times T^2 = \text{Type II String on } K3 \text{ fibred CY 3-fold} \tag{1.1.14}$$

Convincing evidence for the first case was provided comparing the BPS spectrum on both sides [290]. The most remarkable feature is that in this duality the heterotic string coupling, which is given by a complex combinations of the vacuum expectation value of heterotic dilaton ϕ and axion is identified with a geometric modulus on the type II side and vice versa. For example in the second duality the complexified size of the base $\mathbb{P}^1$ of the K3 fibred Calabi-Yau 3 fold is identified with the heterotic string coupling. Together with mirror symmetry this conjectural setting allows to calculate non-perturbative effects in heterotic string with B-model techniques, e.g. on minons of K3 fibred Calabi-Yau 3-folds [204, 229]. While the heterotic string comes naturally with non-abelian gauge groups, the charged gauge bosons on the Type II side are non-perturbative Ramond-Ramond states that are geometrically realised as solitonic branes configurations near singular limits of the Calabi-Yau manifold. One can focus on the singular geometry and decompactify the rest of the geometry to extract gauge theory quantities. The rigid Kähler potential in this limit gives rise to the exact gauge coupling in suitable singular geometries, while the periods encode the BPS masses. Stability of the BPS branes is related to geodesics on the base [231]. This rich application of quantities calculable in the B-model on specially chosen Calabi-Yau manifolds to $\mathcal{N} = 2$ quantum field theories in $\partial = 4$ dimensions is called geometrical engineering of quantum field theories [214, 221]. It is not confined to theories which have Lagrangian descriptions, but rather has become very useful to find and explore much more general classes of $\mathcal{N} = 2$ theories then previously known.

While the B-model is strong in calculating non-perturbative effects, the perturbative heterotic string makes strong higher genus predictions in the topological sector of the Type II string. The prototypical example is the prediction of Yau and Zaslow [333] relating the count of nodal rational curves on K3 to the one loop partition function of the heterotic string projected on the right moving (super symmetric) ground state, also known as elliptic genus, giving the contributions of

[9]Upon further torus compactifications the first lead to $\mathcal{N} = 4$ while the second lead directly to $\mathcal{N} = 2$ theories in $\partial = 4$.

the 24 transversal bosonic modes hence the inverse of (comp (A4.28))

$$\Delta = q\prod_{n=1}^{\infty}(1-q^n)^{24}\,. \tag{1.1.15}$$

This rational curves can be understood as degenerate higher genus curves, where the points of degeneration are counted by the Euler characteristic of the Hilbert scheme of points on K3. The genus can be made explicit by introducing the topological string coupling $q_\lambda = e^{i\lambda}$ in the formula for the refined Euler characteristic of the Hilbert Scheme, yielding the higher genus counting formula for the topological string partition function [215], see (4.3.22) for a definition

$$Z^{hol}(K3, q_\lambda, q) = \sum_{n\geq 0}\chi_\lambda(\mathrm{Hilb}^n(K3))q^{n-1} = \frac{1}{q\prod_{n=1}^{\infty}(1-q_\lambda q^n)^2(1-q^n)^{20}(1-q_\lambda^{-1}q^n)^2}\,. \tag{1.1.16}$$

This application of heterotic type II duality to curve counting that also yields closed modular formulas for the higher genus B-model amplitudes has many generalisations. For instance the only known formula for all genera (reduced) invariants of a compact threefold is the one for $K3\times T_2$ given in [215], but the idea can be generalised to the $K3$ fibred Calabi-Yau spaces [222, 233, 260], as well as to elliptically fibered Calabi-Yau spaces [195, 235].

M-Theory Unifying String Dualities

Given the uniqueness of the $\mathcal{N}$ supergravity in $D = 11$ dimensions and the potential of M-theory of being, the maybe ultimate, non-perturbative description of string theory it is natural to embed these string theory dualities on various background geometries into M-theory compactifications. A key insight by Horava and Witten that the $D = 10$ heterotic and type I string can be obtained by M-theory compactifications with boundaries [180] in $D = 10$ and $D = 9$ dimensions serves as staring point and a summary how these duality conjectures are related can be found e.g. in [291]. A systematic approach stressing the E_{11} symmetry of $D = 11$ supergravity was developed in [313]. The unique covariant constant spinor in Bergers list for seven manifolds implies that compactifications of $D = 11$ supergravity on G_2 manifolds leads to $\mathcal{N} = 1$ supergravity in $\mathfrak{d} = 4$. Compact G_2 manifolds have been constructed by Joyce [202, 203] and more recently by Kovalev [241]. From the point of view of calculating the $\mathcal{N} = 1$ effective action in $\mathfrak{d} = 4$ this case would be the next level of complexity relative to the Type II string compactifications on Calabi-Yau threefolds discussed mostly in the lecture. Some aspects of the problem have been outlined in [159].

F-Theory and 6D SCFT

The non-perturbative type IIB theory in $\mathfrak{d} \leq 8$ with varying axio-dilaton is easier described from an twelve dimensional starting point, where there is however no supergravity multiplet in the right signature, see Table 1. Nevertheless a very efficient way to describe these general type IIB in $\mathfrak{d} = 2n$ is to consider an $6 - n$ complex dimensional elliptically fibred Calabi-Yau manifold over a $d = 5 - n$ complex dimensional base [307], which needs to have some positivity conditions, rendering the choice finite. In this construction known as F-theory, the complex structure of the elliptic fibre parametrise the axio-dilaton and it has been checked progressively in dimension of the base that the Calabi-Yau condition ensures anomaly free theories in $\mathfrak{d}$ dimensions. This is why elliptically Calabi-Yau fourfolds, which lead to $\mathcal{N} = 1$ theories in $\mathfrak{d} = 4$ are also relevant to physics and indeed mirror symmetry and B-model techniques are of help to evaluate the holomorphic quantities like the superpotential, the gauge kinetic functions as well as some simple non-holomorphic ones as the Kähler potential [232, 262]. F-theory compactifications on elliptic Calabi-Yau threefolds with contractable singularities can be used to conjecturally classify the elusive $\mathfrak{d} = 6$ theories [171, 172], whose existence was conjectured by Nahm [266] as mentioned above in Sect. 1.1. B-model techniques. B-model techniques on the elliptic Calabi-Yau 3-fold are useful to explore these theories [166, 230]. In particular a building block the E-string can be very efficiently solved using modularity and B-model techniques [158].

1.1.8 The Plan of the Lecture

Chapter 2 is devoted to the main object that underlies all studies in this lecture namely deformation families of Calabi-Yau n-folds

$$\pi : M_n \to \mathcal{M}_{cs} \tag{1.1.17}$$

over their complex structure moduli space $\mathcal{M}_{cs}$. This is a very well studied beautiful subject in algebraic geometry.

Physical input from string compactifications on Calabi-Yau manifolds added the perspective of viewing this subject from the $\mathcal{N} = (2, 2)$ superconformal field theory point of view. This theory allow a topological twisting that leads to a topological subsector called the *topological B-model*, whose operators and deformations are directly related to the complex structure deformations.

It describes also in the effective action of *smooth* Calabi-Yau compactifications of Type II string a topological subsector whose metric (2-point amplitudes) and higher point amplitudes depend on the complex the structure moduli and not on the complexified Kähler moduli.

The main new input of physics is of course mirror symmetry which states that the B-model on mirror pairs of Calabi-Yau spaces is strictly equivalent to another topological subsector that depends for smooth Calabi-Yau compactifications on the complexified Kähler moduli, but not on the complex the structure moduli. It is called the *topological A-model*. The names are given in the paper of Witten [319], where the decoupling is explained.

In the next chapters we explore properties of the *topological B-model* amplitudes that on *topological A-model* side correspond to amplitudes that get only corrected by genus zero instantons. Since mirror symmetry is proven at genus zero by Givental [136, 137], Lian-Liu-Yau, [253–255] the statements below that we are deriving for the complex the structure moduli of the manifold M are expected to be strictly correct also for the A-model on the mirror manifold, at least for the intersection Calabi-Yau manifolds in toric ambient spaces.

We start with an exposition about the basic geometric properties of Calabi-Yau manifolds. A possible definition is the following: A Calabi-Yau n-fold M is a complex manifold, which is also a Kähler manifold, and whose holonomy group is the group $SU(n)$.

However this is not the only way to define a Calabi-Yau space. It is very useful to study equivalent properties on complex n dimensional Kähler manifolds M, which can be used to define Calabi-Yau manifolds and highlight different geometric and algebraic aspects of the latter. Before we do this we need to review topological facts about real manifolds in Sect. 2.1. We then turn to introduce first properties and examples of complex manifolds in Sect. 2.2. The notions of Kähler manifolds are introduced in Sect. 2.3. The equivalent definitions of Calabi-Yau manifolds and their physical relevance are discussed in Sect. 2.4.1. Their complex deformation theory is closely related to their middle cohomology whose properties are reviewed in Sect. 2.3.4. The main property of the this deformation theory on Calabi-Yau manifolds is that it is unobstructed as explained in section "First Order Complex Structure Deformations".

Due to the local and global Torelli theorems a key tool to describe these moduli spaces is the variation of Hodge structures that is concretely expressed by the notion of the Picard-Fuchs equation describing the flat Gauss-Manin connection. These concept are first abstractly introduced in Sect. 2.4.3 and exemplified with the Legendre family of genus one curves.

The main focus then is on the stringy moduli space and its properties under duality symmetries including in particular mirror symmetry. This is explained in a short journey from one to higher complex dimensions in Sect. 2.4.4. The main focus of this review are the Calabi-Yau 3 folds. Apart from being Kähler their moduli space enjoys an additional structure, called special Kähler geometry which is central to this review. Its discussion starts in Sect. 2.5.

The monodromy group, which plays an important role, as the string amplitudes of interest are automorphic forms w.r.t. to that monodromy groups is discussed in Sect. 2.6. The behaviour of the periods at the singular loci plays also an important role in fixing the integral symplectic basis; we give a complete characterisations of the latter in section "The $\hat{\Gamma}$ Classes and Homological Mirror Symmetry".

We then turn to a first discussion of explicit examples of local and global Calabi-Yau manifolds and their mirrors in Sect. 2.7. The constructions are based on toric geometry and the work of Batyrev.

Next we discuss explicit representations of the holomorphic (n,0)-forms and their period integrals on Calabi-Yau n-folds in Sect. 2.8. In Sect. 2.9 we describe the differential ideal that annihilates the periods, named the Picard-Fuchs system. From a complete set of generators of the ideal of linear differential one can derive the n-point couplings, as described in Sect. 2.9.2. The rest of the section describes various methods to get these systems of differential operators and properties of their solutions. In the case the operators comes from a Gelfand-Greav-Kapranov-Zelevinsky system their solutions can be given very explicitly in terms of generalised hypergeometric functions.

This is all exemplified in detail for the quintic hypersurface in $\mathbb{P}^4$ and other hypergeometric and more general one parameter families in Sect. 2.10 and a selection of two parameter families exhibiting K3 and elliptic fibrations in Sect. 2.11.

An interesting property of Calabi-Yau Picard Fuchs equations is the integrality of the mirror map for which we give some examples in Sect. 2.11.3. This property gives hints for the role of the automorphic symmetries, which can be most completely used, if the compact part of a non-compact mirror is an elliptic curve. As shown in Sect. 2.12 these local cases which are dual to del Pezzo surfaces can be all treated in an universal way.

Chapter 3 is concerned with the world-sheet aspects of string theory on Calabi-Yau manifolds. It reviews material that were part of earlier lectures of myself and that is essential to get good understanding of Sect. 4 that contains the description of the complete solution of the B-model on local geometries and state of the art of solving it on compact Calabi-Yau manifolds. Firstly we describe the relevant superconformal algebras on the World-Sheet in Sect. 3.1 including the spectral flow Sect. 3.1.2 and the supersymmetric non-linear sigma model that occurs for the string world-sheet in a Calabi-Yau background geometry in Sect. 3.2. This allows to define by the topological twists in Sect. 3.3 the topological A- in Sect. 3.4 and the B-model in Sect. 3.5.

Much of the structure of special geometry can be more generally defined using the tt^* structure. This combination of spacetime and world-sheet techniques is important to understand the derivation of the holomorphic anomaly equation in Sect. 4. We therefore devote two subsections in Sect. 3.6 to develop the tt^* algebra. Section 3.6.2 exploits this tt^* structure to describe the n-point couplings on Calabi-Yau n-folds and the relation to the Picard-Fuchs operators. Section 3.7 falls somewhat out of the context of Sect. 3 as it sketches a space-time approach to the B-model, which however gives results that clarify for example boundary properties of the amplitudes that are important to fix the holomorphic ambiguity.

Chapter 4 contains in Sect. 4.1 the coupling of the B-model to gravity, which leads to the derivation of the holomorphic anomaly equation in Sect. 4.1.1, which have a wave function interpretation on which we comment on in Sect. 4.1.2.

The recursive character of holomorphic anomaly combined with some structures, typical in the rings of almost holomorphic modular forms, allows under some assumptions to directly integrate the holomorphic anomaly so that the resulting amplitudes are polynomials in holomorphic generators the so called *propagators* and meromorphic coefficients as explained in Sect. 4.2.

There are physical and geometrical interpretations of the higher genus amplitudes at various critical points in the moduli space corresponding to singularities in the Calabi-Yau family which require a holomorphic limit and certain gauge choices discussed in Sect. 4.3.1. Most notable is of course the prediction of the Gromov-Witten invariants at the maximal unipotent monodromy or large volume point. More suitable to explain the integrality are the Donaldson-Thomas and Pandharipande-Thomas invariants as the latter can be related to the BPS invariants, whose physical origin comes from the counting of super symmetric ground states of classical D-brane configurations, that make their way into the amplitudes by a Schwinger-Loop calculation. The fact that counting is involved gives the remarkable integrality structures in the expansion of the amplitudes as discussed in Sect. 4.3.

These structures allow to a large extend fix the huge kernel in the holomorphic anomaly equations, called the holomorphic ambiguity. We discuss both the refined integrality which involves the actual counting as well as the unrefined that counts more robust indices that are directly related to the genus expansion of the topological string.

Constrained by modular properties, holomorphicity, or rather restricted meromorphicity, and the physical and geometrical expectations at the boundaries it is possible to fix the refined and unrefined holomorphic anomaly completely for the local models and the latter to some extend for the global models as we explain in Sect. 4.5. We demonstrate this for the local $\mathbb{P}^2$ geometry the $K3 \times T2$ example and the quintic.

Chapter 5 is less self contained and aims at an overview over further and more recent B-model—or closely related modular techniques that allow to obtain impressive higher genus results. They fall into two classes, one in which one evaluated a one-loop amplitude or rather an elliptic genus in a dual theory, like the heterotic theory or a quiver gauge theory, that give all genera results for the certain classes in the Calabi-Yau manifold. This is discussed in Sect. 5.1 for the two cases Sects. 5.1.1 and 5.1.2 in turn.

In section we explain the open string disk calculation in Sect. 5.2.1 with an example for the local $\mathbb{P}^2$ example in section "Disks on Harvey-Lawson Branes in the $\mathcal{O}(-3) \to \mathbf{P}^2$ Geometry". The main objective of this material is to give latter a short account of the remodelling the B-model by matrix model techniques in Sect. 5.2.2, that applies open and closed higher genus amplitudes of local Calabi-Yau spaces and gives impressive analytic results.

2 Geometry of Calabi-Yau Manifolds

2.1 *Some Aspects of (Co)Homology Theory on Real Manifolds*

In this short section we review some basic notions of (Co)Homology theory on real manifolds, the Hodge star operator and harmonic analysis. This is to define notations and stress a few aspects that are used later.

2.1.1 Homology and Cohomology

Before we engage in the discussion of complex manifolds, we review aspects of homology and cohomology on any manifold M. No matter whether we have real or complex manifolds the homology is naturally defined over the ring of integers $\mathbb{Z}$ or eventually over finite fields. Cohomology for complex manifolds is naturally defined over $\mathbb{C}$. On the other hand for a fixed manifold the two concepts are dual, the pairing defined by integration gives rise to the period integrals. One of the key ideas in the study of families of complex manifolds that we discuss latter is to study suitable cohomology groups as they vary with the complex structure using the complex structure dependence of these period integrals.

2.1.2 Homology

Let us review some elementary notations of singular homology theory as can be found in the books [143, 296]. We then study it's relation to de Rham cohomology for smooth manifolds of real dimension d and finally to the cohomology of complex manifolds. Recall that a p-chain C_p is a

$$C_p = \sum_k b_k N_k \tag{2.1.1}$$

sum of p dimensional oriented simplicial subsets N_k of M. Often we can think of them simply as oriented smooth submanifolds of M. However the latter do not in general represent all homology classes to be defined below.

Such linear combinations are particular natural if one thinks of chains as domains over which p-forms γ_p can be integrated over

$$\int_{C_p} \gamma_p = \int_{\sum_k b_k N_k} \gamma_p = \sum_k b_k \int_{N_k} \gamma_p \,. \tag{2.1.2}$$

The coefficients b_k should come from a ring such as $\mathbb{Z}$ or from $\mathbb{R}$, $\mathbb{C}$ or other number fields, such as the p-adic numbers, in which case one speaks of a integer, real, complex, or p-adic chains. The most natural choice for our purposes is $\mathbb{Z}$. Addition of chains is abelian and in the simplest case the C_p are free abelian groups. However torsion, i.e. chains whose multiple is trivial, can occur.

To any oriented simplicial complex or any oriented manifold M of dimension p one can associate its oriented $p-1$ dimensional boundary by the boundary operator ∂. A fundamental fact is that the boundary of a boundary is empty. In terms of the boundary operator ∂ this fact is expressed as

$$\partial\partial M = 0 . \tag{2.1.3}$$

By linearity of the boundary operator it extends immediately to chains and maps p-chains to $(p-1)$-chains

$$\partial C_p = \sum_k b_k \partial N_k \tag{2.1.4}$$

and makes chains into a *chain complex.*

p-*Cycles* Z_p are p-chains which are closed, i.e they have no boundaries or equivalently they are in the *kernel* of ∂

$$\partial Z_p = 0 . \tag{2.1.5}$$

Among them there are the trivial- or exact p-cycles B_p, which in the *image* of ∂, i.e. they are itself boundaries of a $(p+1)$-chains

$$B_p = \{C_p | C_p = \partial N_{p+1}\} . \tag{2.1.6}$$

The so called singular homology groups of the manifold M is then given by the quotients

$$H_p(M) = Z_p / B_p , \tag{2.1.7}$$

describing the corresponding equivalence classes of closed modulo exact p-chains. The dimensions of $H_p(M)$ as vector spaces are called the *Betti numbers*

$$b_p = \dim(H_p(M)) . \tag{2.1.8}$$

As mentioned a natural ring for the homology is $\mathbb{Z}$, because of the possibility to define an integral intersection form between elements of $H_p(M, \mathbb{Z})$ and $H_q(M, \mathbb{Z})$ with $p + q = d$

$$H_p(M, \mathbb{Z}) \times H_{d-p}(M, \mathbb{Z}) \to \mathbb{Z} \tag{2.1.9}$$

By the *theorem of Poincaré duality* for compact, orientable manifolds without boundary[10] this intersection pairing is unimodular, which means that any linear functional $H_{d-p}(M,\mathbb{Z}) \to \mathbb{Z}$ is expressible as intersection with a class $C_p \in H_p(M,\mathbb{Z})$ and any class $C_p \in H_p(M,\mathbb{Z})$ that has intersection 0 with all classes in $H_{d-p}(M,\mathbb{Z})$ is a *torsion class*. So the groups $H_p(M,\mathbb{Z})$ are abelian groups of finite rank (lattices), possibly with *torsion*. Further aspects of (2.1.9) will be discussed in Sect. 2.3.4.

2.1.3 Cohomology

Next we come to cohomology and recall that a p-form $\gamma_p \in \Omega^p(\mathbb{R})$ is given in the local coordinates x^i of a real manifold by

$$\gamma_p = \frac{1}{p!}\gamma(x)_{i_1 i_2 \ldots i_p} dx^{i_1} \wedge dx^{i_2} \wedge \ldots \wedge dx^{i_p} = \gamma(x)_{I_p} dx^{I_p} \tag{2.1.10}$$

with $\gamma(x)_{i_1 i_2 \ldots i_p}$ a skew symmetric tensor that is often enough differentiable for what follows.[11] The dx^i span the cotangent space and transform in the naive way $d\tilde{x}^i(x) = \frac{\partial \tilde{x}^i}{\partial x^k} dx^k$ under coordinate changes.[12] The symbol $dx^{I_p} = dx^{i_1} \wedge dx^{i_2} \wedge \ldots \wedge dx^{i_p}$, with $i_1 < i_2 < \ldots < i_p$ is also skew symmetric, i.e. $dx^i \wedge dx^j = -dx^j \wedge dx^i$, and spans the $\frac{d!}{p!(d-p)!}$ dimensional vector space Λ^p of forms. From the transformation of dx^i follows that $dx^{i_1} \wedge dx^{i_2} \wedge \ldots \wedge dx^{i_p}$ transforms as a skew symmetric covariant tensor. In particular the form of maximal degree $p = d$ transforms with multiplication by the Jacobi-determinant, i.e. like a volume element under coordinate changes. The skew symmetric tensor $\gamma(x)_{i_1 i_2 \ldots i_p}$ transforms contravariant so that γ_p is invariant. Recall that a 0-from is just a differentiable function, the components of a 1-forms are co-vector fields, that Λ^p has the same dimension then Λ^{d-p} and q and p-forms can be multiplied to $p+q$-forms with $\beta_q \wedge \gamma_p = (-1)^{pq} \gamma_p \wedge \beta_q$.

An important operation on forms is the exterior derivative d

$$d : \Omega^p(\mathbb{R}) \to \Omega^{p+1}(\mathbb{R}) \tag{2.1.11}$$

defined on

- $f \in \Omega^0(\mathbb{R})$ as $df = \frac{\partial f}{\partial x^i} dx^i$ with summation over equal indices implied and on general forms
- $\gamma_p = \gamma_{I_p} dx^{I_p} \in \Omega^p(\mathbb{R})$ as $d\gamma_p = d\gamma_{I_p} dx^{I_p}$.

---

[10]Later in the lecture we deal also with non-compact so called local Calabi-Yau manifolds. Often in this situations one can still make partially sense of (2.1.9) by restriction to the compact part.

[11]We will often assume that it is in C^∞, i.e. arbitrarily often differentiable in the coordinates x.

[12]We assume summation over equal indices.

This is a canonical generalisation of the well known three dimensional case where the exterior derivative d on 0-forms is the gradient, on 1-forms is the curl and on 2-forms is the divergence. Note also that by the Leipnitz- and the multiplication rule of forms d is an anti-derivation: $d(\alpha_p \wedge \beta_q) = d\alpha_p \wedge \beta_q + (-1)^p \alpha_p \wedge d\beta_q$.

Since partial derivatives commute the exterior derivative operator is nilpotent $d^2 = 0$ similar as the boundary operator and one can define as before the *closed* p-forms

$$Z^p = \{\gamma_p | d\gamma_p = 0\} \tag{2.1.12}$$

in the kernel of d, the *exact* p-forms

$$B^p = \{\gamma_p | \gamma_p = d\nu_{p-1}\} \tag{2.1.13}$$

in the image of d and the *de Rham cohomology groups*

$$H^p(M) = Z^p / B^p \ . \tag{2.1.14}$$

The operation that makes homology and cohomology roughly dual to each other is integration, which defines the *periods* of M

$$\pi(C_p, \gamma_p) = \int_{C_p} \gamma_p \ . \tag{2.1.15}$$

The fact that makes π a well defined map from $H_p(M) \times H^p(M)$ typically to $\mathbb{C}$ is the *Stokes theorem*

$$\int_{C_p} d\gamma_{p-1} = \int_{\partial C_p} \gamma_{p-1} \ . \tag{2.1.16}$$

The reader should check that this eliminates the dependence on the representative with respect to the equivalence $\gamma_p \sim \gamma_p + d\nu_{p-1}$ defining the classes in H^p as well as w.r.t. $C_p \sim C_p + \partial N_{p+1}$ defining classes in H_p and that the Stokes theorem implies the theorems of three dimensional integral calculus known as Stokes- and Gauss theorem. The latter requires the Hodge star operator $*$ defined below.

Periods play a central rôle in the study of the families of complex manifolds M. In this case they are not numbers, but functions of the complex structure parameters of M. The *singular cohomology groups* are dual to the singular homology groups. For example a natural number field for them will be $\mathbb{R}$ or $\mathbb{C}$ and we have e.g. $H^p(M, \mathbb{C}) = H^p(M, \mathbb{Z}) \otimes_{\mathbb{C}} \mathbb{Z} = Hom(H_p(M; \mathbb{Z}), \mathbb{C})$, because De Rhams theorems

- for a basis $\{Z_i\}$ of H_p and any sets of periods values in the field μ_i with $i = 1, \ldots, b_p$ one can find a closed p-form γ_p such that $\pi(Z_i, \gamma_p) = \mu_i$
- if all the periods of γ_p over the basis $\{Z_i\}$ vanish then γ_p is exact

ensure that the integral (2.1.15) defines such a homomorphism. Explicitly[13] one can hence state $H^*_{sing}(M,\mathbb{C}) = H^*_{\text{de Rahm}}(M,\mathbb{C})$.

2.1.4 The Hodge Star Operator ∗

Obviously the structures apart from the natural ring or number field K, which we take to be real $\mathbb{R}$ in this section, are very similar and one immediately might wonder about the analog of (2.1.9) on the cohomology side. An natural guess is the integral over M, which defines a number in K rather then an integer. Given a metric structure one can define a positive real structure proportional to the volume using the Hodge $*$ operator

$$* : C^\infty(\Lambda^p) \to C^\infty(\Lambda^{d-p}) \,, \tag{2.1.17}$$

whose definition on a basis of Λ^p is

$$*(dx^{m_1} \wedge \ldots \wedge dx^{m_p}) = \frac{1}{(n-p)!}\sqrt{g}g^{m_1 k_1} \cdots g^{m_p k_p} \varepsilon_{k_1 \ldots k_p k_{p+1} \ldots k_d} dx^{k_{p+1}} \wedge \ldots \wedge dx^{k_d} \tag{2.1.18}$$

and that requires as additional structure the metric g_{ij} whose determinant we denote by g. An useful exercise that requires the definition of the determinant using the totally anti-symmetric tensor ε similar as in the transformation of dx^{I_d}, shows that

$$* * \gamma_p = (-1)^{p(d-p)} \gamma_p \,. \tag{2.1.19}$$

The integral allows now to define on oriented manifolds a symmetric, positive definite, bilinear inner product, i.e. a norm on $C^\infty(\Lambda)$

$$(\alpha_p, \beta_p) = \int_M \alpha_p \wedge * \beta_p = \int_M^V \frac{1}{p!}\alpha_{m_1 \ldots m_p} \beta^{m_1 \ldots m_p} \text{vol}_g, \tag{2.1.20}$$

where

$$\text{vol}_g = \sqrt{g} dx^1 \wedge \ldots \wedge dx^d \tag{2.1.21}$$

is the volume form, which in turn allows to define the adjoint operator

$$d^\dagger : C^\infty(\Lambda^p) \to C^\infty(\Lambda^{p-1}) \tag{2.1.22}$$

[13] We dropped the qualifiers "sing" and "de Rham" in the discussion.

to the exterior derivative by

$$(\alpha_p, d\beta_{p-1}) = (d^\dagger \alpha_p, \beta_{p-1}) \, . \tag{2.1.23}$$

Using Stokes theorem on closed manifolds M, that d is an anti-derivation and the definition for the covariant derivative w.r.t to the metric ∇ one can show as exercise that on p-forms

•

$$d^\dagger = \begin{cases} *d* \text{ if } d \text{ odd} \\ (-1)^p * d* \text{ if } d \text{ even} \end{cases} \tag{2.1.24}$$

•

$$d^\dagger \alpha_p = -\frac{1}{(p-1)!} \nabla^k \alpha_{km_2 \ldots m_p} d^{m_2} \wedge \ldots \wedge d^{m_p} \, . \tag{2.1.25}$$

Of course $d^\dagger d^\dagger = 0$. An important feature of d and its adjoint is that one can define a generalisation of the second order Laplace operator that preserves the form degree

$$\Delta_d = dd^\dagger + d^\dagger d \tag{2.1.26}$$

acting in coordinates on p forms as[14]

$$(\Delta \alpha_p)_{m_1 \ldots m_p} = -\nabla^k \nabla_k \alpha_{m_1 \ldots m_p} - p R_{k[m_1} \alpha^k{}_{m_2 \ldots m_p]} - \frac{1}{2} p(p-1) R_{jk[m_1 m_2} \alpha^{jk}{}_{m_3 \ldots m_p]} \, . \tag{2.1.27}$$

A form is called *harmonic* if

$$\Delta \alpha = 0 \tag{2.1.28}$$

which is equivalent to the fact that it is closed $d\alpha = 0$ and *co-closed* $d^\dagger \alpha = 0$. A very important theorem is the *Hodge theorem* that states that on a compact oriented manifold without boundary, any p-form admits an *unique decomposition* into an *harmonic*, exact and co-exact piece as follows

$$\omega_p = \alpha_p + d\beta_{p-1} + d^\dagger \gamma_{p+1} \, . \tag{2.1.29}$$

Comparing with the last section this implies in particular that any element in cohomology $H^p(M)$ can be represented by an harmonic form α_p, because this is closed and if it were exact it would be zero due to the uniqueness of the decomposition.

[14] See [42] for the curvature tensors.

2.2 Complex Manifolds

Consider a real $2n$ dimensional manifold M with a covering by coordinate patches $\mathcal{U}_i$, $i = 1, \ldots, r$, which are homeomorphic to a neighborhood $U_i \in \mathbb{C}^n$. Then we can pick $z_\alpha^{(i)}$, $\alpha = 1, \ldots, n$ complex coordinates on each $\mathcal{U}_i$. M is a complex manifold, if all transition functions

$$f^{(jk)} : z^{(k)} \to z^{(j)} , \tag{2.2.1}$$

defined for all points $p \in \mathcal{U}_j \cap \mathcal{U}_k$, are biholomorphic.

2.2.1 Examples

Obviously $\mathbb{C}^n$ is a non-compact complex manifold with one chart and the identity map as transition function.[15] One may hope to get examples of compact complex manifolds by considering constraints like $f(z_1, \ldots, z_n) = 0$, which are holomorphic in all variables. While this leads indeed to a complex manifold, it fails to define compact ones, because of the maximum modulus theorem, which states that the maximum value of the modulus of a non constant differentiable function on an arbitrary domain D is taken at the boundary of D. If now $f = 0$ is solved for some z_i in a compact domain D of the other variables, z_i takes its maximal modulus on the boundary of D and the construction fails to define a differentiable compact manifold.

A way out is to use identifications on $\mathbb{R}^{2n}$ by discrete shift symmetries, i.e. consider tori $T^{2n} = \mathbb{R}^{2n}/\Gamma_{2n}$, where the lattice $\Gamma_{2n} \cong \mathbb{Z}^{2n}$ is identified with $\mathbb{Z}^{2n}$ as an abelian group. If one chooses a complex structure on $\mathbb{R}^{2n}$ by aligning real and imaginary directions of $T^*\mathbb{R}^{2n} \cong \mathbb{R}^{2n}$ with the basis of Γ_{2n} one gets compact complex tori $T^n_{\mathbb{C}}$. They are flat and have hence trivial holonomy. Dividing by discrete rotations G of the lattice Γ_{2n} leads to orbifold compactifications. For example if G acts as a discrete subgroup of $SU(3)$ in the fundamental representation on the complex coordinates of $T^3_{\mathbb{C}}$ then one gets a complex orbifold with curvature singularities at the fixset of G. The corresponding lattice automorphisms have been classified [108]. Remarkably one can prove that this curvature singularities can be smoothed to get a Kähler manifold with $SU(3)$ holonomy.

An alternative route to construct simple compact complex manifolds is by dividing by $\mathbb{C}^* := \mathbb{C} \setminus \{0\}$ actions. E.g. $\mathbb{P}^n$ is defined as the space of complex lines through the origin in $\mathbb{C}^{n+1}$. This is the space of equivalence classes of $[z_1, \ldots, z_{n+1}]$ in $\mathbb{C}^{n+1} \setminus \{\underline{0}\}$ with the equivalence relation

$$(z_1, \ldots, z_{n+1}) \sim \lambda(z_1, \ldots, z_{n+1}), \tag{2.2.2}$$

[15] It is also Kähler as seen below.

where $\lambda \in \mathbb{C}^*$. For the charts we take

$$\mathcal{U}_i = \{z_i \neq 0 | z_i \in \mathbb{P}^n\}$$

and as their coordinates the non-trivial ratios $z_m^{(i)} = z_m/z_i$. It is convenient to include the trivial ratios and to define on $\mathcal{U}_j \cap \mathcal{U}_k$ with

$$z_m^{(j)} = \frac{z_m}{z_k} / \frac{z_j}{z_k} = \frac{z_m^{(k)}}{z_j^{(k)}}, \qquad (2.2.3)$$

biholomorphic transitions functions between the complex coordinates. $\mathbb{P}^n$ is a obviously compact and as it turns out also a Kähler manifold.

A polynomial hypersurface constraint in $\mathbb{P}^n$ of the type $P(z_1, \ldots, z_{n+1}) = 0$ must be homogeneous of some degree d in the z_i, i.e. $P(\lambda z_1, \ldots, \lambda z_{n+1}) = \lambda^d P(z_1, \ldots, z_{n+1})$, to be well defined on the equivalence classes. It defines a compact complex variety. This is a smooth Kähler manifold if p is transversal, i.e. $\mathrm{d}P \neq 0$ for $P = 0$. We will give an overview about the application of this construction and generalizations to Calabi-Yau manifolds in Appendix 3 and in Sect. 2.7.1 as well as Sect. 2.7.3.

2.2.2 Almost Complex Manifolds

Conceptional it is an important question if and how many complex structures an even dimension real manifold possesses. A necessary prerequisite to have a complex structure is a differentiable endomorphism of the tangent bundle $J : TM \to TM$ with $J^2 = -\mathbf{1}$. J corresponds to multiplication of the tangent bundle by $i = \sqrt{-1}$ and a manifold with this structure is called an *almost complex manifold*.[16] With J we can define projectors

$$P = \frac{1}{2}(\mathbf{1} - iJ)$$

onto the holomorphic sub-bundle—and with

$$\bar{P} = \frac{1}{2}(\mathbf{1} + iJ)$$

[16]A complex manifold is almost complex, because multiplying the basis of TM of a complex manifold with coordinates $z^k = u^k + iv^k$ by $i = \sqrt{-1}$ maps $\begin{pmatrix} \frac{\partial}{\partial u^k} \\ \frac{\partial}{\partial v^k} \end{pmatrix} \mapsto \begin{pmatrix} \frac{\partial}{\partial v^k} \\ -\frac{\partial}{\partial u^k} \end{pmatrix}$, i.e. $J = du^i \otimes \frac{\partial}{\partial v^i} - dv^i \otimes \frac{\partial}{\partial u^i}$. In holomorphic and anti-holomorphic coordinates this means $J^i_j = i\delta^i_j$, $J^{\bar{i}}_{\bar{j}} = -i\delta^{\bar{i}}_{\bar{j}}$ and $J^i_{\bar{j}} = J^{\bar{i}}_j = 0$.

onto the anti-holomorphic sub-bundle of the tangent bundle respectively. The significance of the projectors is that on an almost complex manifold one can already project r-forms Ω with p P's and q $\bar{P}$'s $(r = p+q)$ to (p, q)-forms $\Omega^{p,q}$. However without the notion of an integrable complex structure these projected spaces do not behave naturally under derivations.

2.2.3 Complex Vector Bundles

Let us suppose for the moment complex coordinates $z^k = u^k + i v^k$ with

$$\partial_k := \frac{\partial}{\partial z^k} = \frac{1}{2}\left(\frac{\partial}{\partial u^k} - i\frac{\partial}{\partial v^k}\right), \qquad \partial_{\bar{k}} := \frac{\partial}{\partial \bar{z}^k} = \frac{1}{2}\left(\frac{\partial}{\partial u^k} + i\frac{\partial}{\partial v^k}\right) \tag{2.2.4}$$

can be defined. We then can split $T_{\mathbb{C}}M = T_{\mathbb{R}}M \otimes \mathbb{C}$, which is spanned over $\frac{\partial}{\partial w_k}$, $k = 1, \ldots, 2n$ with complex coefficients V^i as $T_{\mathbb{C}}M = T^{1,0}M \oplus T^{0,1}M$. Here $\{u_k, v_k\} =: \{w_k, w_{k+n}\}$ and each vector V in $T_{\mathbb{C}}M$ decomposes as

$$V = \sum_{k=1}^{2n} V^k \frac{\partial}{\partial w_k} = \sum_{k=1}^{n} \left[(V^k + i V^{n+k})\partial_k + (V^k - i V^{n+k})\partial_{\bar{k}}\right] =: V^{1,0} + V^{0,1}\,. \tag{2.2.5}$$

We call $T^{1,0}M$ $[T^{0,1}]$ spanned by ∂_k, $[\partial_{\bar{k}}]$ the [anti]holomorphic tangent bundle. The corresponding transition functions are [anti]holomorphic and are given by $\partial_k^{(i)} = \frac{\partial z^{(j)l}}{\partial z^{(i)k}}\partial_l^{(j)}$ from (2.2.1) and the complex conjugate. Obviously under complex conjugation $T^{0,1}M = \overline{T^{1,0}M}$. Similarly the cotangent bundle splits $T^*_{\mathbb{C}}M = T^{*1,0}M \oplus T^{*0,1}M$ into a holomorphic and an anti-holomorphic sub-bundle spanned by $\mathrm{d}z^k$ and $\mathrm{d}\bar{z}^k := \mathrm{d}\bar{z^k}$ respectively.[17] Sections of $\wedge^r T^*_{\mathbb{C}}M$ are called r-forms Ω^r and can be decomposed into sections of $\wedge^p T^{*1,0}M \wedge^q T^{*0,1}M$, which are called (p, q)-forms $\Omega^{p,q}$, i.e the space A^r of r forms splits into the space $A^{p,q}$ of (p, q)-forms $A^r = \bigoplus_{r=p+q} A^{p,q}$. We also get natural holomorphic $(1, 0)$ and an-holomorphic $(0, 1)$ derivative operators as

$$\partial f = \sum_{i=1}^{n} \frac{\partial f}{\partial z^i}\mathrm{d}z^i =: \partial_i f \mathrm{d}z^i\,, \qquad \bar{\partial} f = \sum_{\bar{j}=1}^{n} \frac{\partial f}{\partial \bar{z}^{\bar{j}}}\mathrm{d}\bar{z}^{\bar{j}} =: \bar{\partial}_{\bar{j}} f \mathrm{d}\bar{z}^{\bar{j}}\,. \tag{2.2.6}$$

[17] To avoid too complicated notations TM (T^*M) will mean in the following the holomorphic tangent bundle $TM = T^{10}M$ (cotangent bundle $T^*M = T^{*\,1,0}M$).

Integrable Complex Structures

Still we have to establish the existence of complex coordinates. According to a theorem of Niremberg and Newlander a necessary and sufficient[18] condition for the existence of global complex coordinates, i.e. a *complex structure*, is that the Lie bracket (2.4.14) of two holomorphic vector fields X, Y, defined as above by such coordinates, is always a holomorphic vector field again [269] (see [182] and [57] Chap. V. for physicists review). Written with the projectors this condition becomes

$$\bar{P}[PX, PY] = 0. \tag{2.2.7}$$

This integrability condition leads to $[JX, JY] - J[X, JY] - J[JX, Y] - [X, Y] = 0$. In local flat coordinates $J(\partial_b) = J_b^e \partial_e$ and with $J_c^b J_d^c = -\delta_d^b$, i.e. $(\partial_a J_c^b) J_d^c = -J_c^b(\partial_a J_d^c)$ (2.2.7) means that the so called Nijenhuis tensor vanishes identically [269]

$$N_{bd}^c := J_b^a(\partial_a J_d^c - \partial_d J_a^c) - J_d^a(\partial_a J_b^c - \partial_b J_a^c) \equiv 0\,. \tag{2.2.8}$$

2.2.4 Exterior Derivatives on Complex Manifolds

A related perspective on the existence of a complex structure can be given as follows. On an almost complex manifolds due to the dependence of J on the coordinates the exterior derivative has a priori the pieces $\mathrm{d}\Omega^{p,q} = (\mathrm{d}\Omega)^{p-1,q+2} + (\mathrm{d}\Omega)^{p,q+1} + (\mathrm{d}\Omega)^{p+1,q} + (\mathrm{d}\Omega)^{p+2,q-1}$. One may then define $\partial\Omega^{p,q} = (\mathrm{d}\Omega)^{p+1,q}$ and $\bar{\partial}\Omega^{p,q} = (\mathrm{d}\Omega)^{p,q+1}$ as the $(1,0)$ and $(0,1)$ parts of the d operator and search for complex variables such that the latter follow from extending (2.2.6) in the canonical way to forms. One can check that the condition $\bar{\partial}^2 = 0$ is equivalent to $N_{cd}^b \equiv 0$ and implies that the $(\mathrm{d}\Omega)^{p-1,q+2}$ and $(\mathrm{d}\Omega)^{p+2,q-1}$ pieces in the exterior derivative disappear. It follows further by consideration of the (p,q) type that the equation $\mathrm{d}^2 = 0$ on Ω^* implies $\partial^2 = 0$ and $\bar{\partial}\partial + \partial\bar{\partial} = 0$. To summarize if J is integrable one has for

$$\phi = \frac{1}{p!q!}\phi_{i_1\ldots i_p \bar{j}_1\ldots\bar{j}_q}\mathrm{d}z^{i_1}\wedge\ldots\wedge z^{i_p}\wedge \mathrm{d}\bar{z}^{\bar{j}_1}\wedge\ldots\wedge \mathrm{d}z^{\bar{j}_q} \tag{2.2.9}$$

$$\begin{aligned}\partial\phi &= \tfrac{1}{p!q!}\partial\phi_{i_1\ldots i_p \bar{j}_1\ldots\bar{j}_q}\mathrm{d}z^{i_1}\wedge\ldots\wedge z^{i_p}\wedge \mathrm{d}\bar{z}^{\bar{j}_1}\wedge\ldots\wedge \mathrm{d}z^{\bar{j}_q}\\ \bar{\partial}\phi &= \tfrac{1}{p!q!}\bar{\partial}\phi_{i_1\ldots i_p \bar{j}_1\ldots\bar{j}_q}\mathrm{d}z^{i_1}\wedge\ldots\wedge z^{i_p}\wedge \mathrm{d}\bar{z}^{\bar{j}_1}\wedge\ldots\wedge \mathrm{d}z^{\bar{j}_q}\end{aligned} \tag{2.2.10}$$

and the de Rham exterior derivative splits into and $(1,0)$ and an $(0,1)$ piece

$$\mathrm{d} = \partial + \bar{\partial}\,, \tag{2.2.11}$$

[18] This is the nontrivial part.

so that $\mathrm{d}\Omega^{p,q} \in A^{p+1,q} \oplus A^{p,q+1}$. Since $\bar{\partial}$ is nilpotent we can define the cohomology

$$H^*_{\bar{\partial}} = \frac{\mathrm{Kern}\bar{\partial}}{\mathrm{Im}\bar{\partial}} \,. \tag{2.2.12}$$

As in Sect. 2.1.4, better insights into these cohomology groups and its representatives can be drawn, if one has a norm and with it the possibility to define adjoint operators. Therefore we relegate this points after the discussion of the corresponding structure to Sect. 2.3.2.

We discuss further important aspects of different complex structures in section "First Order Complex Structure Deformations". In particular these different complex structures on a real manifold can form *families* parametrized by the complex structure moduli space. This is in particular the case for Calabi-Yau manifolds, which is the reason that we relegate section "First Order Complex Structure Deformations" after the definition of Kähler manifolds in Sect. 2.3 and Calabi-Yau manifolds in Sect. 2.4.1.

2.3 *Kähler Manifolds*

Complex- and Kähler manifolds play a double role in supersymmetric Kaluza-Klein and string compactifications. On the one hand the target space is in many cases—a notable exception are G2 manifolds—a complex- and Kähler manifold. On the other hand their deformations- or moduli spaces, become the physical moduli spaces or vacuum manifolds. For global and local supersymmetric theories these spaces are also complex- and Kähler manifolds and in the case of extended super symmetry or compactifications on special holonomy manifolds they can have additional geometric structures. We therefore recall now some essential aspects of hermitian and Kähler manifolds.

2.3.1 Metric Aspects

A *hermitian metric* is a positive-definite inner product $TM \otimes \bar{T}M \to \mathbb{C}$. Locally it can be given by a covariant tensor[19]

$$\mathrm{d}s^2 = \sum_{i,j}^{n} G_{i\bar{j}}(w)\mathrm{d}z^i \otimes \mathrm{d}z^{\bar{j}} \tag{2.3.1}$$

[19]Note that the first index of $G_{i\bar{j}}$ is only summed over the unbarred $i = 1, \ldots, n$ and the second only over barred $\bar{j} = \bar{1}, \ldots, \bar{n}$ indices respectively.

with the properties that $G_{j\bar{i}} = \overline{G_{i\bar{j}}}$ and $G_{i\bar{j}}$ is positive. Explicitly the positivity means that

$$v^i G_{i\bar{j}} \bar{v}^{\bar{j}} \geq 0, \qquad \forall\, \underline{v} \in \mathbb{C}^n \,. \tag{2.3.2}$$

The equality holds only if $\underline{v} = \underline{0}$, which in particular implies that $\det(G_{i\bar{j}}) > 0$.

To define an hermitian metric, given a real metric, an almost complex structure is sufficient, see Theorem 3.14 of [238]. Hermiticity is the condition $G(X, Y) = G(JX, JY)$ on the real metric, which becomes

$$G_{mn} = J^a_m J^b_n G_{ab} \tag{2.3.3}$$

in coordinates. It does not constraint M further then admitting J and any metric say G', because for any such G' the metric $G_{mn} = \frac{1}{2}(G'_{mn} + J^a_m J^b_n G'_{ab})$ is hermitian. In particular on any complex manifold we can define a *hermitian metric* see [237] Chap 3.5. Multiplying (2.3.3) with J^m_p, defining $J_{nm} = J^a_n G_{am}$ and using $J^m_p J^a_m = -\delta^a_p$ we see that $J_{nm} = -J_{mn}$. Hence we can define a 2-form $\omega = J_{nm} \mathrm{d}w^n \wedge \mathrm{d}w^m$. In complex notation this becomes

$$\omega = i \sum_{i,j=1}^{n} G_{i\bar{j}} \mathrm{d}z^i \wedge \mathrm{d}z^{\bar{j}} \,. \tag{2.3.4}$$

This is a real form $\bar{\omega} = \omega$ of type $(1, 1)$ and is called the *fundamental form* associated to the *hermitian metric*. Because[20] $G := \det(G_{i\bar{j}}) > 0$ one gets by wedging ω n-times

$$\mathrm{vol} = \frac{\omega^n}{n!} = i^n \det(G_{i\bar{j}}) \mathrm{d}z^1 \wedge \mathrm{d}\bar{z}^1 \wedge \ldots \wedge \mathrm{d}z^n \wedge \mathrm{d}\bar{z}^n = 2^n \det(G_{ij})^{\frac{1}{2}} \mathrm{d}w^1 \wedge \ldots \wedge \mathrm{d}w^{2n} \tag{2.3.5}$$

a positive volume form on M, which implies also that M is orientable.

Let us prepare some notions for the harmonic theory on complex manifolds. On (p, q)-forms in $A^{p,q}$

$$\phi = \frac{1}{p!q!} \phi_{i_1,\ldots,i_p,\bar{j}_1\ldots,\bar{j}_q}(z) \mathrm{d}z^{i_1} \wedge \ldots \wedge \mathrm{d}z^{i_p} \wedge \mathrm{d}z^{\bar{j}_1} \wedge \ldots \wedge \mathrm{d}z^{\bar{j}_q} \tag{2.3.6}$$

we have an *local inner product* defined by the hermitian metric

$$(\phi, \psi)(z) = \frac{1}{p!q!} \phi_{i_1\ldots i_p \bar{j}_1\ldots\bar{j}_q} \bar{\psi}^{i_1\ldots i_p \bar{j}_1\ldots\bar{j}_q} \tag{2.3.7}$$

[20] Note in coordinates $x^i, x^{\bar{i}}$ one has the block form $G_{nm} = \begin{pmatrix} 0 & G_{\mu\bar{\nu}} \\ G_{\sigma\bar{\rho}} & 0 \end{pmatrix}$ and e.g. [57] defines $G := \det(G_{nm}) = \det^2 G_{\mu\bar{\nu}}$.

where

$$\bar{\psi}^{i_1\ldots i_p\bar{j}_1\ldots\bar{j}_q} = G^{\bar{k}_1 i_1}\ldots G^{\bar{k}_p i_p}\cdot G^{\bar{j}_1 l_1}\ldots G^{\bar{j}_q l_q}\overline{\psi_{k_1\ldots k_p,\bar{l}_1\ldots\bar{l}_q}}\,. \qquad (2.3.8)$$

With this we can adapt the bilinear (2.1.20) to complex manifolds and define a *global bilinear inner product* $A^{p,q}\times A^{p,q}\to\mathbb{C}$

$$(\phi,\psi) = \int_M (\phi,\psi)(z)\mathrm{vol}\,. \qquad (2.3.9)$$

One a compact manifolds one has the positivity properties

$$(\phi,\psi) = \overline{(\psi,\phi)}, \qquad (\phi,\phi) > 0 \text{ unless } \phi = 0, \qquad (2.3.10)$$

which makes $A^{p,q}$ in a pre-Hilbert space. The Hodge operator $*$ maps now[21] $*: A^{p,q}\to A^{n-q,n-p}$ and is defined as

$$(\phi,\psi)\mathrm{vol} = \phi\wedge *\bar{\psi}\,. \qquad (2.3.11)$$

Here

$$\bar{\psi} = \frac{1}{p!q!}\overline{\psi}_{j_1\ldots j_q,\bar{i}_1\ldots\bar{i}_p}\mathrm{d}z^{j_1}\wedge\ldots\wedge\mathrm{d}^{j_q}\wedge\mathrm{d}z^{\bar{i}_1}\wedge\ldots\wedge\mathrm{d}z^{\bar{i}_p} \qquad (2.3.12)$$

and $\overline{\psi_{i_1\ldots i_p,\bar{j}_1\ldots\bar{j}_q}} = (-1)^{pq}\bar{\psi}_{j_1\ldots j_q,\bar{i}_1\ldots\bar{i}_p}$. Explicitly

$$*\psi = \frac{i^n(-1)^{n(n-1)/2+np}}{p!q!(n-p)!(n-q)!}\det(G)\epsilon^{k_1\ldots k_p}{}_{\bar{j}_1\ldots\bar{j}_{n-p}}\epsilon^{\bar{l}_1\ldots\bar{l}_q}{}_{i_1\ldots i_{n-q}}\psi_{k_1\ldots k_p,\bar{l}_1\ldots\bar{l}_q}$$
$$\mathrm{d}z^{i_1}\wedge\ldots\wedge\mathrm{d}z^{i_{n-q}}\wedge\mathrm{d}z^{\bar{j}_1}\wedge\ldots\wedge\mathrm{d}z^{\bar{j}_{n-p}}\,. \qquad (2.3.13)$$

One checks $*\bar{\psi} = \overline{*\psi}$ and $**\psi = (-1)^{pq}\psi$ for ψ a (p,q)-form. On the middle cohomology $p+q=n$ one has for $A^{p,q}$ in a suitable basis the eigen values[22]

$$*\psi = i^{(p-q)}\psi\,. \qquad (2.3.14)$$

With the norm $(\cdot,\cdot)$ we can define the *adjoint operators* $\partial^*: A^{p,q}\to A^{p-1,q}$ and $\bar{\partial}^*: A^{p,q}\to A^{p,q-1}$ by

$$(\partial^*\psi,\phi) := (\psi,\partial\phi), \qquad \text{and} \qquad (\bar{\partial}^*\psi,\phi) := (\psi,\bar{\partial}\phi) \qquad (2.3.15)$$

[21] Here the conventions are as in [237]. The $*$ operator in [150] maps $*_{gh}: A^{p,q}\to A^{n-p,n-q}$, so it involves an additional complex conjugation $*_{gh}\psi = *_{ko}\bar{\psi}$.

[22] For CY 3-fold formulas are summarized in Chapter 14 of [47] that involve the Kähler class ω.

respectively. On a compact manifold one can calculate from the above that $\bar{\partial}^* = - * \partial *$.

Let us next come to the Kähler condition. If $G = \delta_{i\bar{j}}$ is Euclidean it is an exercise to show that the new Laplacian and the old Laplacian (2.1.26)

$$\Delta_{\bar{\partial}} = \bar{\partial}\bar{\partial}^* + \bar{\partial}^*\bar{\partial}, \qquad \Delta_{\rm d} = d^\dagger d + dd\dagger \tag{2.3.16}$$

are related by

$$\Delta_{\bar{\partial}} = \frac{1}{2}\Delta_{\rm d} , \tag{2.3.17}$$

a fact that is important for the Hodge decomposition described below. On a complex manifold it turns out that for this to be true the metric $G = \delta_{i\bar{j}}$ must approximate the Euclidean metric up to order two in z at each point in M. This is the Kähler condition, which can be given in different equivalent ways, see [149], one of which we state below. Let us denote the $(1, 1)$-form associated to $G_{i\bar{j}}$

$$\omega = i \sum_{i,\bar{j}=1}^{n} G_{i\bar{j}} {\rm d}z^i \wedge {\rm d}\bar{z}^{\bar{j}} . \tag{2.3.18}$$

An hermitian metric $G_{i\bar{j}}$ whose fundamental form is closed ${\rm d}\omega = 0$ is called a Kähler metric and ω is called the *Kähler form*. Since from ${\rm d}\omega = 0$ follows

$$\partial_i G_{j\bar{k}} = \partial_j G_{i\bar{k}}, \qquad \bar{\partial}_{\bar{i}} G_{j\bar{k}} = \bar{\partial}_{\bar{k}} G_{j\bar{i}} , \tag{2.3.19}$$

which are the integration conditions for the local existence of a Kähler potential K with

$$G_{i\bar{j}} = \partial_i \bar{\partial}_{\bar{j}} K = -\frac{1}{2}{\rm d}(\partial - \bar{\partial})K \tag{2.3.20}$$

Note that despite the form above ω cannot be exact. For if $\omega = {\rm d}A$ would have been exact (2.3.5) could not be true, because using Stokes theorem the integral $\int \omega^n$ would be zero. That means that $(\partial - \bar{\partial})K$ is not globally defined. Indeed as far as the definition of ω goes $K(z, \bar{z})$ only needs to be defined up to a Kähler transformation

$$K(z, \bar{z}) \to K(z, \bar{z}) - f(z) - \bar{f}(\bar{z}) , \tag{2.3.21}$$

so e^K will be in general a section of a nontrivial line bundle over M. In general two Kähler forms ω and ω' are in the same class in $H^2(M, \mathbb{R})$, if we can find a smooth global real function ϕ on M and

$$\omega' = \omega + \partial\bar{\partial}\phi(z, \bar{z}) . \tag{2.3.22}$$

In Kähler geometry the non vanishing Christoffel symbols Γ have only pure indices, e.g. they are given by

$$\Gamma^i_{jk} = G^{i\bar{l}}\partial_j G_{k\bar{l}}, \qquad \Gamma^{\bar{i}}_{\bar{j}\bar{k}} = G^{\bar{i}l}\partial_{\bar{j}} G_{\bar{k}l} = \overline{\Gamma^i_{jk}}\,, \tag{2.3.23}$$

which define a torsion free $T^i_{jk} = \Gamma^i_{jk} - \Gamma^i_{kj} = 0$ metric connection $\nabla_i G_{j\bar{k}} = 0$, with $\nabla_i V_k = \partial_j V_k - \Gamma^p_{ik} V_p$ and analogous equations for $\bar{\nabla}_{\bar{i}}$. Further one has the simplified relation for the Riemann tensor[23]

$$R_{i\bar{j}l}{}^k = \bar{\partial}_{\bar{j}}\Gamma^k_{il}, \qquad R_{\bar{i}jl}{}^k = -\bar{\partial}_{\bar{i}}\Gamma^k_{jl}, \qquad R_{i\bar{j}k\bar{l}} = -\partial_i\partial_{\bar{j}}G_{k\bar{l}} + G^{m\bar{n}}(\partial_i G_{k\bar{n}})(\partial_{\bar{j}} G_{m\bar{l}})\,. \tag{2.3.24}$$

This implies an additional symmetry under the exchange of i and l and j and l respectively. The other non-vanishing components are given by complex conjugation

$$R_{\bar{i}j\bar{l}}{}^{\bar{k}} = \overline{R_{i\bar{j}l}{}^k}, \qquad R_{i\bar{j}\bar{l}}{}^{\bar{k}} = \overline{R_{\bar{i}jl}{}^k} \tag{2.3.25}$$

as well as by the usual symmetries of the Riemann tensor [42]. Our sign conventions, which are really conventions about the position of the indices, are so that

$$[\nabla_i, \nabla_{\bar{j}}]V_k = R_{i\bar{j}k}{}^p V_p\,. \tag{2.3.26}$$

The Ricci tensor is defined by contracting

$$R_{i\bar{j}} = R_{i\bar{a}\bar{j}}{}^{\bar{a}} = -\partial_i\bar{\partial}_{\bar{j}} \log\det(G) \tag{2.3.27}$$

is hermitian, i.e. $R_{i\bar{j}} = R_{\bar{j}i}$ and $R_{i\bar{j}} = \overline{R_{j\bar{i}}}$, which follows from the fact that G is hermitian and $\det G$ is real. It can be used to define the *Ricci-form* $\mathcal{R}$ as

$$\mathcal{R} = \frac{i}{2\pi} R_{i\bar{j}} dz^j \wedge d\bar{z}^{\bar{j}} = \frac{i}{2\pi}\partial\bar{\partial}\log\det(G) = \frac{i}{4\pi} \mathrm{d}(\partial - \bar{\partial})\log\det(G)\,. \tag{2.3.28}$$

The latter also represents the first Chern class of TM

$$c_1(TM) = [\mathcal{R}]\,. \tag{2.3.29}$$

It is a fundamental property of the Chern-class of holomorphic vector bundles, that they don't depend on the choice of the connection. So despite the appearance of the metric or its connection in (2.3.28) the first Chern class $c_1(TM)$ is independent of these data.

[23]Here we follow the conventions in which the real Riemann tensor has the index structure $R_{ijk}{}^l = \partial_j\Gamma^l_{ik} - \partial_i\Gamma^l_{jk} + \Gamma^l_{jr}\Gamma^r_{ik} - \Gamma^l_{ir}\Gamma^r_{jk}$.

Example $\mathbb{P}^n$ is a Kähler manifold. This can be established by giving with the *Fubini-Study metric* an explicit form of a Kähler potential for a Kähler metric. In the $\mathcal{U}_i$, $i = 0, \ldots, n$ patches the Kähler potential is given by $K^{(i)}(z^{(i)}, \bar{z}^{(i)}) = \log(1 + |z^{(i)}|^2)$, where $|z^{(i)}|^2 = \sum_{j\neq i} |z_j^{(i)}|^2$. Using (2.2.3) we see that $K^{(i)}(z^{(i)}, \bar{z}^{(i)}) = K^{(j)}(z^{(j)}, \bar{z}^{(j)}) - \log \frac{z_i}{z_j} - \log \frac{\bar{z}_i}{\bar{z}_j}$. The latter two terms are holomorphic and antiholomorphic sections respectively on $\mathcal{U}_i \cap \mathcal{U}_j$. Hence they do not affect the metric $g_{i\bar{j}} = \partial_i \partial_{\bar{j}} K(z, \bar{z})$, which is globally well defined. Dropping the index for the patch we get

$$\omega = i\partial\bar{\partial}\log(1 + |z|^2) = i\left(\frac{\mathrm{d}z^i \wedge \mathrm{d}z^{\bar{i}}}{1 + |z|^2} - \frac{\bar{z}^i \mathrm{d}z^i \wedge z^j \mathrm{d}z^{\bar{j}}}{(1 + |z|^2)^2}\right) . \tag{2.3.30}$$

This defines a positive-definite metric. With $\det(g_{i\bar{j}}) = \frac{1}{(1+|z|^2)^{n+1}}$ one calculates the Ricci tensor $R_{i\bar{j}} = -\partial_i \partial_{\bar{j}} \log \det(g_{i\bar{j}}) = (n + 1)g_{i\bar{j}}$. If the Ricci tensor is proportional to the Kähler metric one calls the metric Kähler-Einstein.

Let us mention briefly further important facts about Kähler manifolds. The property of the Christoffel symbol to have only pure indices leads to the fact that parallel transport of a vector generates only the holonomy group $U(n) \in SO(2n)$ rather then $SO(2n)$, which would be the holonomy of a generic orientable manifold.

2.3.2 Hodge Theorem and Hodge Decompositions

Let us review next simple topological facts. As a consequence of (2.3.17) $\Delta_{\mathrm{d}} = 2\Delta_{\bar{\partial}}$ does not change the (p, q)-type, i.e. $H_d^{p,q}(M) = H_{\bar{\partial}}^{p,q}(M)$ and the same equality is true for the harmonic representatives $\mathcal{H}_d^{p,q}(M) = \mathcal{H}_{\bar{\partial}}^{p,q}(M)$. Taking the harmonic forms as distinguished representatives we get the Hodge decomposition of the de Rham cohomology groups

$$\begin{aligned} H^r(M, \mathbb{C}) &= \bigoplus_{p+q=r} H^{p,q}(M) \\ H^{p,q}(M) &= \overline{H^{q,p}} . \end{aligned} \tag{2.3.31}$$

Forms of Kähler manifolds are related by complex conjugation $\overline{A^{p,q}} = A^{q,p}$, which implies for the cohomology groups $\overline{H^{p,q}(M)} = H^{q,p}(M)$, since complex conjugation commutes with Δ_{d}. The star operator $* : A^{p,q} \to A^{n-q,n-p}$ is another bijection which commutes with Δ_{d} and hence

$$H^{q,p}(M) = H^{p,q}(M) = H^{n-q,n-p}(M) . \tag{2.3.32}$$

The Hodge theorem states that every element $\phi \in A^{p,q}$ has an unique orthogonal decomposition into a harmonic form h, an exact piece $\bar{\partial}\xi$ with $\xi \in A^{p,q-1}$ and a co-exact piece $\bar\partial^*\eta$ with $\eta \in A^{p,q+1}$ i.e.

$$A^{p,q} = \mathcal{H}^{p,q} \oplus \bar\partial A^{p,q-1} \oplus \bar\partial^* A^{p,q+1} \,. \tag{2.3.33}$$

This is in analogy with the de Rham decomposition $A^p = \mathcal{H}^p \oplus \mathrm{d}A^{p-1} \oplus \mathrm{d}^* A^{p+1}$. The usual argument shows that if ϕ is closed, i.e. $\bar\partial\phi = 0$, then the $\bar\partial^*\eta$ piece in the decomposition is zero, because $\bar\partial\phi = \bar\partial\bar\partial^*\eta$ and thus $0 = (\bar\partial\phi, \eta) = (\bar\partial^*\eta, \bar\partial^*\eta)$, which implies $\bar\partial^*\eta = 0$. This in turn means that every $\bar\partial$ closed form can be uniquely decomposed into a harmonic form w.r.t. $\Delta_{\bar\partial}$ and a $\bar\partial$ exact piece, which implies $H^{p,q}_{\bar\partial}(M) \cong \mathcal{H}^{p,q}(M)$.

A central result is the Čech-Dolbeault isomorphism, which follows from the Hodge-de Rham isomorphism see [150] and the $\bar\partial$-Poincaré Lemma. It states for sheaves of forms contracting sheafs of vectors fields F that

$$H^q(M, \Omega^p(F)) \cong H^{p,q}_{\bar\partial}(M, F) \,. \tag{2.3.34}$$

For example $H^q(M, \wedge^p T^* M) \cong H^{p,q}(M, TM) =: H^{p,q}(M)$ and for the holomorphic forms in $H^{p,0}(M) = H^0(M, \Omega^p)$ it follows from the Hodge decomposition that they are harmonic for any Kähler metric on a compact Kähler manifold. Vise versa by consideration of type, we have for every holomorphic $(p, 0)$-form Ω_p that $\bar\partial^*\Omega_p = 0$ as it maps to $A^{p,-1}$ which is trivial. If $\Delta_{\bar\partial}\Omega_p = 0$ then from $\bar\partial^*\bar\partial\Omega_p = 0$ follows $\bar\partial\Omega_p = 0$.

Using $(\bar\partial^*\psi)_{i_1\ldots i_p \bar j_2 \ldots \bar j_p} = (-1)^{p+1}\nabla^{\bar j_1}\psi_{i_1\ldots i_p \bar j_1 \bar j_2 \ldots \bar j_q}$ one can show that the Kähler ω form is harmonic. Hence $h^{1,1}(M) \geq 1$ on a Kähler manifold. Similarly one shows that all ω^m, $m = 1, \ldots, n$ are nontrivial elements in $H^{m,m}(M)$.

2.3.3 Lefshetz Decomposition

On the cohomology of a Kähler manifold with $n = \dim_C(M)$ one can define the exterior product with the standard Kähler form ω as[24] lowering operator S^-, the adjoint operator as raising operator S^+ and the diagonal operator, which associates to each form of degree r the eigenvalue $(n-r)/2$, as H. Then $H, S^\pm$ fulfill the Lie algebra of $sl(2, \mathbb{C})$, $[S^+, S^-] = 2H$, $[H, S^\pm] = \pm S^\pm$ and the cohomology decomposes into its irreducible representations. More precisely the *Hard Lefshetz Theorem* [148] says the following: $(S^-)^k : H^{n-k} \to H^{n+k}$ is an isomorphism and with $P^{n-k} := (\mathrm{Ker}(S^-)^{k+1} : H^{n-k} \to H^{n+k+2}) = (\mathrm{Ker} S^+) \cap H^{n-k}$ the *primitive cohomology* has the *Lefshetz decomposition*

$$H^r(M) = \bigoplus_k (S^-)^k P^{r-2k}(M) \,. \tag{2.3.35}$$

The primitive parts of the cohomology play here the rôle of highest weight vectors.

---

[24] Normalized on a $\mathbb{C}^n$ as $\omega = \frac{i}{2}\sum_i \mathrm{d}z_i \wedge \mathrm{d}\bar z_i$.

Examples The cohomology of $\mathbb{P}^n$ forms a representation $\left(\frac{\mathbf{n}}{\mathbf{2}}\right)$. The cohomology of the two torus ($\dim_R(T^2) = 2$) decomposes as $2(\mathbf{0}) + \left(\frac{\mathbf{1}}{\mathbf{2}}\right)$, where the two $(\mathbf{0})$ representations are dz and $d\bar{z}$ while $[1, dz \wedge d\bar{z}]$ form the $\left(\frac{\mathbf{1}}{\mathbf{2}}\right)$ representation. Check that the cohomology of the T^{2n} torus has the $sl(2,\mathbb{C})$ decomposition

$$\left(2(\mathbf{0}) + \left(\frac{\mathbf{1}}{\mathbf{2}}\right)\right)^{\otimes n} = \bigoplus_{r=1}^{n} \left(\binom{2n}{n-r} - \binom{2n}{n-r-2} \right) \left(\frac{\mathbf{r}}{\mathbf{2}}\right) \bullet \tag{2.3.36}$$

2.3.4 The Middle (Co)homology and the Riemann Bilinear Relations

In the middle homology of an even real dimensional manifold one has an obvious bilinear intersection form (2.1.9). There is a dual intersection form on the cohomology, whose properties are particular interesting on Kähler manifolds and which we discuss below.

The most important invariant of the bilinear form (2.1.9) is its signature, which is given by the Hirzebruchs signature theorem

$$\sigma = h^n_+(M) - h^n_-(M) = \int_{M_n} L_{\frac{n}{2}} \,. \tag{2.3.37}$$

Here L_m is the Hirzebruch L polynomial a multiplicative class defined by

$$Q(x) = \frac{x}{\tanh(x)} = 1 + \sum_{k=0}^{\infty} (-1) \frac{2^{2k}}{(2k)!} B_k x^{2k} \,, \tag{2.3.38}$$

where the Bernoulli numbers are normalized to $B_1 = \frac{1}{6}$, $B_2 = \frac{1}{30}$, $B_3 = \frac{1}{42}$, $B_4 = \frac{1}{30}$, etc. Expanding this in Pontryagin- (p_k) or Chern classes (c_k) yields

$$\begin{aligned} L_{\frac{m}{2}} &= 0, \quad \text{for } m \text{ odd}\,, \\ L_1 &= \frac{1}{3} p_1 = \frac{1}{3}(c_1^2 - 2c_2)|_{CY} = -\frac{2\chi}{3}\,, \\ L_2 &= \frac{1}{45}(7p_2 - p_1^2) = \frac{1}{45}(4c_1^2 c_2 - c_1^4 - 14 c_1 c_3 + 3c_2^2 + 14 c_4)|_{CY} = \frac{\chi}{3} + 32 \\ &\textit{etc.} \end{aligned} \tag{2.3.39}$$

From the first line follows for odd complex dimension, that one can find an symplectic basis for the middle homology with equal positive and negative eigenvalues. For the first even dimensional manifolds, we restricted the result in the last line to Calabi-Yau manifolds. For complex 2-fold Calabi-Yau manifolds, i.e. $K3$, it follows simply from $c_1 = 0$, but for Calabi-Yau 4-folds (and higher ones) one can invoke in addition the Hirzebruch-Rieman-Roch theorem (A2.18), which states CY 4-folds that the first arithmetic genus $\chi_0 = h_{00} + h_{40} = 2 = \frac{1}{720}\int_{M_4} 3(c_2^2 - c_4)$ to simplify the expression. In these even cases classifications of even lattices put further powerful restrictions on the intersection forms. We discuss some of those in section "K3 Surfaces".

On the middle dimensional cohomologies of Kähler manifolds $H^{p,q}$ with $p + q = n$ it is customary to specialize[25] the bilinear form (2.3.9) to the bilinear intersection form $\langle \cdot, \cdot \rangle : H^n \otimes H^n \to \mathbb{C}$, defined by integration

$$\langle \alpha, \beta \rangle = \int_{M_n} \alpha \wedge \beta \,. \tag{2.3.40}$$

This form is obviously odd (even) for n odd (even) and has the property for $\omega^{p,q} \in H^{p,q}$ and $\omega^{r,s} \in H^{r,s}$

$$\langle \omega^{p,q}, \omega^{r,s} \rangle = 0, \qquad \text{unless} \quad p = s \ \text{ and } q = r \,. \tag{2.3.41}$$

Moreover if α is a primitive form in the middle cohomology then one has a positive real structure, i.e.

$$R(\alpha) = i^{p-q} \langle \alpha, \bar{\alpha} \rangle > 0 \,. \tag{2.3.42}$$

Applied to the unique $(n, 0)$ form this real structure determines the Kähler Weil-Peterssen metric on the complex moduli space of Calabi-Yau metrics.

We are interested how the decomposition (2.3.31) of the middle cohomology varies, when move the family of Calabi-Yau n-folds (1.1.17) over $\mathcal{M}$. The theorems of Tian and Todorov states the global unobstructedness of the complex moduli deformations. This insures that $\mathcal{M}$ exists as complex and in fact Kähler manifold. Moreover it states that the complex dimension is given by the Hodgenumber $h = h_{n-1,1}(M_n) = \dim_{\mathbb{C}}(\mathcal{M})$. We parametrize $\mathcal{M}$ by complex coordinates z^i, $i = 1, \ldots, h$.

For n even the question which part of the middle cohomology varies with the complex structure depends on the choice of the polarisation. The part that varies with the complex structure for algebraic embeddings on M_n, which fix the polarisation, is called the (primitive) horizontal cohomology.

For $n = 3$ the whole middle cohomology is horizontal. From now let us denote the decomposition of horizontal cohomology simply by

$$H^n(M) = \bigoplus_{k=0}^{n} H^{n-k,k}(M) \tag{2.3.43}$$

[25] In view of (2.3.14) the relation is $(\alpha, \beta) = i^{p-q} \langle \alpha, \bar{\beta} \rangle$.

even so one strictly would need to call it $H^n_{hor}(M)$ etc. We drop the qualifier horizontal and the n on M_n below.

2.4 Calabi-Yau Manifolds and Their Deformation Spaces

Here we give first equivalent definitions of Calabi-Yau manifolds. We then discuss generic structures of their deformation space. We focus on deformation families of Calabi-Yau manifolds over their complex moduli space $\mathcal{M}_{cs}(W)$, because this is mathematically better explored.

What string theory adds is mirror symmetry, which due to an approximate decoupling between $\mathcal{M}_{cs}(W)$ and $\mathcal{M}_{cks}(M)$ states that this structure will be mirrored into the complexified Kähler deformation space $\mathcal{M}_{cks}(M)$ of the Calabi-Yau manifolds, the Kähler deformation space supplemented by the B-field.

2.4.1 Properties of Calabi-Yau Manifolds

We come now to the description of Calabi-Yau spaces by discussing important, partly equivalent properties, that make a Kähler manifold into a Calabi-Yau manifold M. The precise relation between these properties is discussed below

(a) The canonical class K_M is trivial.
(b) The first Chern class of the tangent bundle vanishes[26] $c_1(TM) = 0$.
(c) In each Kähler class on M there exists an unique Kähler metric[27] g whose Ricci tensor vanishes $R_{i\bar{j}}(g) = 0$.
(d) There exists an—up to a constant—unique nowhere vanishing holomorphic $(n, 0)$ form called Ω.
(e) The holonomy group $\mathrm{Hol}(g)$ of M is a subgroup of $SU(n)$.
(f) M admits a pair of globally defined covariantly constant (parallel) spinors ξ and $\bar{\xi}$ of opposite chirality if n is odd and of the same chirality if n is even.

We see that on a Calabi-Yau manifold one has two important forms: The Kähler $(1, 1)$-form ω and the holomorphic $(n, 0)$ form Ω. The above listed properties are of different relevance in different physical contexts, but only partly equivalent. For these reasons one finds different definitions of Calabi-Yau manifolds in the physics literature. For reference let us give here a definition of a compact Calabi-Yau manifold.

Definition For $n \geq 2$ a Calabi-Yau manifold is the quadruple (M, J, g, Ω), where (M, J) is a compact complex manifold, g a Kähler metric, ω the Kähler form and

[26] We assume that we have a connection without torsion on TM.

[27] In the following g denotes the Kähler metric of the Calabi-Yau manifolds and G the one of its moduli space.

Ω an up to a phase *unique* non-vanishing holomorphic $(n, 0)$ form. The latter forms are linked by the fact that $\Omega \wedge \bar{\Omega}$ is proportional to the volume form as the n-th exterior power of the Kähler form

$$\frac{\omega^n}{n!} = (-1)^{\frac{n(n-1)}{2}} \left(\frac{i}{2}\right)^n \Omega \wedge \bar{\Omega} \bullet \tag{2.4.1}$$

This is a natural normalization which makes $\text{Re}(\Omega)$ a calibration for the *special Lagrangian submanifolds* $L \subset M$ of real dimension n defined by

$$Re(e^{i\theta}\Omega)|_L = \text{vol}_L, \quad \text{and} \quad \omega|_L = 0\,. \tag{2.4.2}$$

Imposing (2.4.1) reduces the freedom in the constant in (e) to the phase θ [203]. Likewise there are the holomorphic submanifolds Z_k of complex dimension k which are calibrated as follows

$$\left.\frac{\omega^k}{k!}\right|_{Z_k} = \text{vol}_{Z_k}, \quad \text{and} \quad \Omega|_{Z_k} = 0\,. \tag{2.4.3}$$

Here vol_Z is the volume form on the corresponding cycle Z and is important to notice that (2.4.2) and (2.4.3) are *minimal volume conditions*. In particular *calibrated cycles* are the cycles of the least volumes in its homology class [203].

In the physics literature calibrated cycles denoted in general by Σ_{p+1} are know as *supersymmetric cycles*, see [129] for a review. It was shown in [33] that the requirement of maximal supersymmetry on the worldvolume of a cycle, defined by the embedding map $\Phi\ :\ \Sigma_{p+1}\ \hookrightarrow\ M$, is ensured if the global ten or eleven dimensional susy transformation on the worldvolume can be undone by a worldvolume κ symmetry. That leads to the conditions

$$P_-\eta = \frac{1}{2}\left(1 - \frac{i}{(p+1)!}h^{-1/2}\varepsilon^{\sigma_1\ldots\sigma_{p+1}}\partial_{\sigma_1}\phi^{m_1}\ldots\partial_{\sigma_{p+1}}\phi^{m_{p+1}}\Gamma_{m_1\ldots m_{p+1}}\right)\eta = 0, \tag{2.4.4}$$

where η is a ten or eleven dimensional constant spinor, $P_\pm$ are projection operators and h is the induced metric on the word-volume of Σ_{p+1}. It was shown in [33] that (2.4.4) specializes to the *calibration conditions* (2.4.2), (2.4.3).

It useful to note that all the forms mentioned in Bergers list in Sect. 1.1.5 define calibrations and equivalently supersymmetric cycles. If the complex dimensions of a Calabi-Yau manifold n is even one can also define mixed calibration conditions, e.g. on a fourfold the calibration with respect to $\frac{\omega^2}{2} + \text{Re}(e^{i\theta}\Omega)$ leads to the so called *Caley cycles* [129].

A general Kähler manifold has holonomy $U(n)$. Under the above condition Yau's theorem states that one can chose a metric G on M_n with $R_{i\bar{j}}(g) = 0$. Since $R_{i\bar{j}}$ generates the $U(1)$ in the holonomy group $U(n)$, a general Calabi-Yau manifold has holonomy $SU(n)$. Unless mentioned otherwise we assume that the holomomy of M_n is the full $SU(n)$. That excludes the complex tori $T^n_{\mathbb{C}}$ which are flat Kähler manifolds and have trivial holonomy or products of lower dimensional Calabi-Yau manifolds, potentially with complex tori, which likewise have reduced holonomy. It also excludes the more exotic cases such as the Enriques Calabi-Yau manifold, which has holonomy $\mathbb{Z}_2 \times SU(2)$.

Let us now discuss the relation between the statements (a)–(f). In order to connect (a)–(d) to (e) and (f) we will assume that M is simply connected and not of product form.

(a) $\leftrightarrow$ (b) follows from (A2.7).

(c) $\rightarrow$ (b) is a simple consequence of the independence of the Chern classes on the choice of the Kähler metric. Once one knows that there exists a Ricci-flat metric clearly $c_1(TM) = 0$ and that holds for all Kähler metrics.

(b) $\rightarrow$ (c) is a corollary to Yau' theorem [331], which proves the conjecture that E. Calabi formulated in (1956). It states that given the data

- (C.a) of a Kähler metric g,a Kähler form ω, a Ricci form $\mathcal{R}$ on M and a real closed $(1, 1)$ form $\mathcal{R}'$, which represents the Chern class $[\mathcal{R}] = [\mathcal{R}'] = 2\pi c_1(TM)$

one can construct

- (C.b) an unique metric g' on M with associated Kähler form ω' such that $[\omega'] = [\omega] \in H^2(M, \mathbb{R})$ and the Ricci form of g' is $\mathcal{R}'$.

In particular $c_1(TM) = 0$ can be represented by $\mathcal{R}' \equiv 0$ and then according to the above there exists a unique metric g' whose Ricci form is $\mathcal{R}'$. Therefore its Ricci tensor vanishes.

One can formulate simpler equivalent versions of (C.a) and (C.b) as requirements on the existence of functions on M as follows. $\mathcal{R} - \mathcal{R}'$ is a $\bar\partial$ exact and d closed real $(1, 1)$ form. By the ∂, $\bar\partial$ Lemma one has a real function f on M so that $\mathcal{R} - \mathcal{R}' = i\partial\bar\partial f$ up to a constant κ. Recalling (2.3.29) how $\mathcal{R}$ is derived from the positive function multiplying $w^1 \wedge \ldots \wedge w^{2n}$ in (2.3.5), which is itself determined by $\frac{\omega^n}{n!}$, we conclude that f must make its appearance also in $e^f \omega^n = (\omega')^n$. In fact the constant κ can be fixed by normalizing the volume $\int_M e^f \omega^n = \int_M \omega^n$. The simplification is that instead of requiring g' to lead to a prescribed $\mathcal{R}'$ one requires that it leads to a prescribed volume form and the statement about $\mathcal{R}$ and $\mathcal{R}'$ can be replaced by a statement about f. Similarly one can formulate the $[\omega'] = [\omega]$ condition in (C.b) as a search for a real function ϕ as in (2.3.22). ϕ can be made unique by requiring $\int_M \phi \mathrm{vol}_g = 0$. So the simplified version of (C.a) and (C.b) is

- (C'.a) that for every given Kähler metric g, Kähler form ω and a real smooth function f on M with $\int_M e^f \omega^n = \int_M \omega^n$

one can construct

- (C'.b) an unique smooth real function ϕ on M such that (i) $\omega + i\partial\bar\partial\phi$ is a positive $(1, 1)$ form ω', (ii) $\int_M \phi \mathrm{vol}_g = 0$ and (iii) $(\omega + i\partial\bar\partial\phi)^n = e^f \omega^n$.

Yau proved that the non-linear p.d.e (iii) on ϕ admits a unique solution which fulfills (i) and (ii). This is an existence proof and up to date no explicit solutions for ϕ and[28] e.g. the Ricci-flat metric on any compact Calabi-Yau manifold has been given.

(c) $\rightarrow$ (e) From the special form of the curvature tensor on Kähler manifolds (2.3.26) one can argue that the holonomy group of a Kähler manifold is generically $U(n)$. Moreover from the definition of the Ricci-tensor (2.3.27) as the trace, one concludes that it is generating the $U(1)$ part of $U(n) \cong SU(n) \times U(1)$ holonomy. On a Ricci-flat manifold this part is not generated and the holonomy is reduced to $SU(n)$.

(e) $\rightarrow$ (d) An $(n, 0)$-form can always locally be written as $\Omega_{i_1,\ldots,i_n} = f(x)\epsilon_{i_1,\ldots i_n}$. It is therefore in the total antisymmetric representation of the holonomy group $SU(n)$, i.e. a singlet invariant under Hol. One can conclude from this that $\nabla\Omega = 0$. Since Γ has no mixed indices $\bar\partial_{\bar i}\Omega = \nabla_{\bar i}\Omega = 0$ and Ω is holomorphic. This implies that $f(x)$ has to be a globally defined holomorphic function over the compact manifold M and hence a constant. Note that ω, locally written as $\omega = \frac{i}{2}\sum_{i=1}^n \mathrm{d}x^i \wedge \mathrm{d}x^{\bar i}$, and g, locally written $g = \sum_{i=1}^n |\mathrm{d}x^i|^2$, are also covariantly constant. The normalization (2.4.1) established at a point requires $|f| = 1$, but since all quantities are covariantly constant (2.4.1) will hold at any point.

Ω is also harmonic $\Delta_{\bar\partial}\Omega = 0$ as beside $\bar\partial\Omega = 0$ also $\bar\partial^*\Omega = -*\partial*\Omega = 0$, because $* : A^{n,0} \rightarrow A^{n,0}$ and $\partial : A^{n,0} \rightarrow A^{n+1,0} = \{0\}$.

(d) $\rightarrow$ (a) We just constructed with Ω a trivial constant section of the canonical bundle $\wedge^n T^{*1,0}M$.

(d) $\rightarrow$ (b) Assume a nowhere vanishing holomorphic $(n, 0)$ exists. We get then a globally well defined scalar function

$$||\Omega||^2 = \frac{1}{n!}\Omega_{i_1\ldots i_n}\bar\Omega^{i_1\ldots i_n}\,, \tag{2.4.5}$$

where the indices are raised by the hermitian metric $g^{i\bar j}$. Locally Ω is given by $\Omega_{i_1,\ldots,i_n} = f(x)\epsilon_{i_1,\ldots i_n}$, where $f(x)$ is a non-vanishing holomorphic function in each patch. We can obtain $\bar\Omega^{i_1,\ldots i_n} = \frac{\bar f}{g}\epsilon^{i_1\ldots i_n}$ and it follows that $g = \det(g_{i\bar j}) = \frac{|f|^2}{||\Omega||^2}$. Inserting in (A2.6) we get $c_1(TM) = -\frac{i}{2\pi}\partial\bar\partial\log|\Omega|^2$ which is exact since $\log||\Omega||^2$ is a scalar, hence $c_1(TM) = 0$ in cohomology.

(f) $\leftrightarrow$ (d) is proven in generality in [312]. This is done using representation theory. Let us just give a simple relevant example namely the threefold case, $n = 3$. We must figure out how many spinors transform as singlets under the holonomy $SU(3)$. Under generic rotations in the internal 6d space, vectors transform by $SO(6)$ and the associated spin group with the same Lie algebra is isomorphic to $SU(4)$. The spinor representation in $6d$ is $2^{\frac{6}{2}} = 8$ dimensional and splits according to the chirality into representations $(\mathbf{4}, \bar{\mathbf{4}})$ of this $SU(4)$. Now the holonomy is reduced

[28] It is not that difficult to find a Kähler metric on a Calabi-Yau manifold, e.g. by constructing the induced metric of the Fubini-Study metric on the quintic in $\mathbb{P}^4$, see [302].

to $SU(3)$ and embedding the $SU(3)$ in $SU(4)$ singles out an $U(1)$, i.e. one has $SU(3)\otimes U(1)\in SU(4)$. The decomposition of the $(\mathbf{4},\bar{\mathbf{4}})$ into the representations of this $U(1)$ and $SU(3)$ is unique $(\mathbf{4},\bar{\mathbf{4}})=(\mathbf{3}^1\otimes\mathbf{1}^{-3},\bar{\mathbf{3}}^{-1}\otimes\mathbf{1}^3)$, where the superscripts are the $U(1)$-charges. Hence we can conclude that there are indeed one invariant and therefore covariantly constant spinor of each helicity. Bilinears of the covariantly constant spinors can be used to build the covariantly constant tensors discussed above. In particular the almost complex structure as $J^a_b=-i\xi^\dagger\Gamma^a_b\xi$, the metric as $g_{i\bar{j}}=ij^\dagger\Gamma_{i\bar{j}}j$ and the (3, 0) form as $\Omega_{ijk}=e^{-i\alpha}\xi^T\Gamma_{ijk}\xi$. In this way one can show (f) → (d) see [57] for details. Furthermore it is easy to see that the eight spinors can be generated from $\xi\in\mathbf{1}^{-3}$ as $\Gamma_i\xi\in\bar{\mathbf{3}}^{-1}$, $\Gamma_{ij}\xi\in\mathbf{3}^1$, $\Gamma_{ijk}\xi\in\mathbf{1}^3$ and decomposed as

$$\eta=\Omega^{0,0}\xi+\Omega^{0,1}_{\bar{i}}\Gamma^{\bar{i}}\xi+\Omega^{0,2}_{\bar{i}\bar{j}}\Gamma^{\bar{i}\bar{j}}\xi+\Omega^{0,3}_{\bar{i}\bar{j}\bar{k}}\Gamma^{\bar{i}\bar{j}\bar{k}}\xi,\quad \text{where } \Omega^{0,n}_{\bar{i}_1\ldots\bar{i}_r}\mathrm{d}z^{\bar{i}_1}\wedge\ldots\wedge\mathrm{d}z^{\bar{i}_r}\in H^{0,r}_{\bar{\partial}}(M)\,. \tag{2.4.6}$$

On $T^3_{\mathbb{C}}$ one has therefore eight covariant constant spinors and on $T^1_{\mathbb{C}}\times K3$ four.

A very general tool in Čech cohomology is Serre duality which states for any sheaf E on M that

$$H^k(E)^*\cong H^{n-k}(E^*\otimes K_M)\,. \tag{2.4.7}$$

Using the Čech-Dolbeault isomorphism $H^k(E)\cong H^k_{\bar{\partial}}(M,E)$, $H^r(M,\wedge^s T^*M)=H^{s,r}(M)$ and $K_M=\mathcal{O}_M$ we relate on a Calabi-Yau manifold the cohomology groups $H^{0,r}(M)\cong H^{0,n-r}(M)$ by taking $E=\mathcal{O}(M)$ or by complex conjugation the cohomology groups $H^{r,0}(M)\cong H^{n-r,0}(M)$. This particular result can be seen also in a more direct way by contracting a $(p,0)$ form $\omega_{i_1\ldots i_p}\mathrm{d}x^{i_1}\wedge\ldots\wedge\mathrm{d}x^{i_p}$ with the unique $(0,n)$ form to define a $(0,n-p)$-form $\hat{\omega}_{\bar{j}_{p+1}\cdots\bar{j}_n}=\frac{1}{p!}\bar{\Omega}_{\bar{j}_1\cdots\bar{j}_n}\omega^{\bar{j}_1\cdots\bar{j}_p}$. One shows easily that this is an invertible map that commutes with Δ, i.e. $H^{p,0}(M)\cong H^{0,n-p}(M)\cong H^{n-p,0}(M)$. As an exercise use the index theorem (A2.11) to argue that $h^{1,0}=h^{2,0}$ on a Calabi-Yau 3-fold.

With $h^{n,0}(M)=h^{0,0}=1$ Eq. (2.4.6) implies that one has at least two covariantly constant spinors on a Ricci-flat manifold. In order to show that one has only this two on a manifold with Hol $=SU(n)$ we shall show that $h^{p,0}=0$ for $0<p<n$. On a compact Kähler manifold harmonicity of $(p,0)$-form implies holomorphicity as argued after (2.3.33) by consideration of type. Specializing (2.1.27) to $R_{ij\bar{k}\bar{l}}=0$ for Kähler- and $R_{i\bar{j}}=0$ for Ricci-flat manifolds harmonicity means $\nabla^\nu\nabla_\nu\omega_{i_1\ldots i_p}=0$. On a compact manifold one can use pairing and partial integration to see that this requires $\nabla_{\bar{j}}\omega_{i_1\ldots i_p}=0$ (and also $\bar{\partial}\omega=0$). From these equations we conclude that all harmonic $(p,0)$ forms are covariantly constant. However that would mean that they are invariant under $SU(n)$, which is impossible for $0<p<n$ as only the trivial and the total antisymmetric representation are invariant.

First Order Complex Structure Deformations

Our aim is to explain in this section the local tangent space of the complex structure moduli space from a point of view and put forward by Kodaira and Spencer [237] and to explain in the next section why the first order deformations on a Calabi-Yau manifold are *unobstructed*.

Consider as in Sect. 2.2 a $2n$ real dimensional manifold M and a covering of it by coordinate patches $\mathcal{U}_i$, $i = 1, \ldots, r$, which are homeomorphic to a neighborhood $U_i \in \mathbb{C}^n$ with coordinates $z_\alpha^{(i)}(p)$, $\alpha = 1, \ldots, n$. As was mentioned M is a complex manifold if the transition functions $f^{(jk)} : z^{(k)}(p) \to z^{(j)}(p)$, defined for $p \in \mathcal{U}_j \cap \mathcal{U}_k$, are biholomorphic. One attempt now to define a family of complex manifolds M_u, by considering a family of transition functions $z_\alpha^{(j)} = f_\alpha^{(jk)}(z^{(k)}, u)$, which depend also holomorphically on the complex parameters u. The difficulty is that some u dependence of $f_\alpha^{(ik)}(z^{(k)}, u)$ corresponds just to different choices of local coordinates systems on the same complex manifold. In order to decide whether the $f^{(jk)}(z^{(k)}, u)$ really induce changes of the complex structure Kodaira [237] considers in every patch U_k an infinitesimal coordinate changes that is characterized by a holomorphic vector field $V^{(k)}(u) = \sum_{\alpha=1}^n \frac{\partial f_\alpha^{(k)}(z^{(k)},u)}{\partial u} \frac{\partial}{\partial z_\alpha^{(k)}}$. Next he forms the composition of transition functions in $\mathcal{U}_i \cap \mathcal{U}_j \cap \mathcal{U}_k$. Per definition the identity

$$f_\alpha^{(ik)}(z^{(k)}, u) = f_\alpha^{(ij)}(f_1^{(jk)}(z^{(k)}, u), \ldots, f_n^{(jk)}(z^{(k)}, u), u) \tag{2.4.8}$$

holds. Differentiation w.r.t. to u gives

$$\frac{\partial f_\alpha^{(ik)}(z^{(k)}, u)}{\partial u} = \frac{\partial f_\alpha^{(ij)}(z^{(j)}, u)}{\partial u} + \sum_{\beta=1}^n \frac{\partial z_\alpha^{(i)}}{\partial z_\beta^{(j)}} \frac{\partial f_\beta^{(jk)}(z^{(k)}, u)}{\partial u} . \tag{2.4.9}$$

Denote general vector fields by

$$A^{(jk)}(u) = \sum_{\alpha=1}^n \frac{\partial f_\alpha^{(jk)}(z^{(k)}, u)}{\partial u} \frac{\partial}{\partial z_\alpha^{(j)}}, \qquad z^{(k)} = f^{(kj)}(z_j, u) . \tag{2.4.10}$$

Note that $A^{(kk)}(u) = 0$ since $f_\alpha^{(kk)} = z^{(k)}$ independently of u. Therefore Eq. (2.4.9) written covariantly in terms of the vector fields (2.4.10) implies $A^{(kj)}(u) = -A^{(jk)}(u)$. For general i, j, k (2.4.9) is a Čech[29] 1-cocycle condition for the $A^{(ij)}$

$$A^{(ij)}(u) + A^{(jk)}(u) + A^{(ki)}(u) = 0 . \tag{2.4.11}$$

[29] Čech cohomology made a prominent physical appearance in topological charge quantization in [15].

The exact 1-cocycles come precisely from the infinitesimal coordinates changes setting $A^{(jk)}(u) = V^{(j)}(u) - V^{(k)}(u)$, while the true changes of complex structure correspond to 1-cocycles, which are not exact, i.e. elements of $H^1(M, A)$, where A are sheaves of vector fields $A = \mathcal{O}(TM)$. The Čech-Dolbeault theorem (2.3.34) with $F = \mathcal{O}(TM)$ implies that complex structure deformations are given by elements in $H^{0,1}(M, TM)$, which we also call A.

2.4.2 Unobstructedness of the Complex Deformation Space

As explained further in [237] the existence of a global complex structure deformation requires the vanishing of higher Čech cohomology groups for vector fields. Tian [301] and Todorov [303] have proven that these higher order conditions are automatically fulfilled on a Calabi-Yau space.

The elements

$$A(u) = A^i_{\bar{j}}(x, u)\mathrm{d}z^{\bar{j}} \frac{\partial}{\partial z^i} \quad \in H^{0,1}(M, TM) \tag{2.4.12}$$

representing first order deformation in the complex moduli space can be used to deform the $\bar{\partial}$ operator to $\bar{\partial}_u = (\bar{\partial} + A(u))$ so that $\bar{\partial}_u f(z) = 0$, defines what a holomorphic function on M is w.r.t. the new complex structure. The requirement that $\bar{\partial}_u^2 = 0$ leads to

$$\bar{\partial} A(u) + \frac{1}{2}[A(u), A(u)] = 0, \tag{2.4.13}$$

where $[.,.]$ is the Lie bracket. For $\phi(z) = \phi^i(z)\partial_{z_i} \in \mathcal{L}^{0,p}(T)$, with $\phi^i = \frac{1}{p!}\phi(z)^i_{\bar{i}_1,\ldots,\bar{i}_p}\mathrm{d}z^{\bar{i}_1} \wedge \ldots \wedge \mathrm{d}z^{\bar{i}_p}$, and $\omega(z) \in \mathcal{L}^{0,q}(T)$ similarly defined one has

$$[\phi, \omega] = (\phi^i \wedge \partial_i \omega^j - (-1)^{pq} \omega^i \wedge \partial_i \phi^j)\partial_j, \tag{2.4.14}$$

giving above a $(0, 2)$ form vector field from two $(0, 1)$-form vector fields. Condition (2.4.13) is equivalent to the vanishing of the Nijenhuis tensor (2.2.8) [237].

The main idea of the proof is that the existence of the holomorphic $(n, 0)$ form induces an isomorphism

$$H^{0,p}(M, TM) \cong H^{n-1,p}(M)\,. \tag{2.4.15}$$

which converts the condition (2.4.13) into a cohomological question, which is solved by the $\partial\bar{\partial}$ lemma. This conversion of the deformation problem to a cohomological question, which is solved by an analog of the $\partial\bar{\partial}$ Lemma extends to deformations of G_2 metrics [177, 202] as well as to the extended moduli space considered in [27].

Contraction with the homolomorphic $(n,0)$ form associates to $A = A^i_{\bar{j}_1,\ldots,\bar{j}_p} dz^{\bar{j}_1} \wedge \ldots \wedge dz^{\bar{j}_p} \frac{\partial}{\partial z^i} \in H^{0,p}(M,TM)$ an $\hat{A} \in H^{n-1,p}(M)$ as

$$\hat{A} = \frac{1}{(n-1)!} A^j_{\bar{j}_1,\ldots,\bar{j}_p} \Omega_{j,i_2,\ldots,i_n} dz^{i_2} \wedge \ldots \wedge dz^{i_n} dz^{\bar{j}_1} \wedge \ldots \wedge dz^{\bar{j}_p} \tag{2.4.16}$$

with the inverse

$$(\hat{A})^\vee = \frac{1}{(n-1)!|\Omega|^2} \bar{\Omega}^{i,i_2,\ldots,i_n} \hat{A}_{i_2,\ldots i_n,\bar{j}_1,\ldots,\bar{j}_p} dz^{\bar{j}_1} \wedge \ldots \wedge dz^{\bar{j}_p} \frac{\partial}{\partial z^i} \tag{2.4.17}$$

where $|\Omega|^2$ is defined in (2.4.5). One checks that A is harmonic iff $\hat{A}$ is harmonic and the operation is invertible i.e. $A = (A^\wedge)^\vee$, which shows (2.4.15).

Since Ω is holomorphic the hat operation (2.4.16) commutes with $\bar{\partial}$ and we get

$$\bar{\partial}\hat{A} = \widehat{\bar{\partial}A} = -\frac{1}{2}\widehat{[A,A]} =: -\frac{1}{2}[\hat{A},\hat{A}], \tag{2.4.18}$$

as equivalent to the condition (2.4.13).

The main technical instrument is the following Lemma (Tian-Todorov)

$$[\hat{A},\hat{B}] := \widehat{[A,B]} = \partial(\widehat{A \wedge B}) - (D\cdot A)\wedge \hat{B} + \hat{A} \wedge (D\cdot B), \tag{2.4.19}$$

where $D \cdot A = (\partial_i A^i_{\bar{j}_1\ldots\bar{j}_p}) z^{\bar{j}_1} \wedge \ldots \wedge dz^{\bar{j}_p}$ is a contraction. The calculation is a straightforward exercise whose solution is made explicit in [301].

Equation (2.4.19) becomes particularly useful, if one can choose "gauge" representatives for A and B so that $(D\cdot A) = (D \cdot B) = 0$. To control this "gauge" condition Tian considers a Taylor expansion $A(u) = A_1 u + A_2 u^2 + \ldots$ with A_i sections of $\Gamma(M, \Omega^{(0,1)}(TM))$ and starting data $\bar{\partial}_0 = \bar{\partial}$, i.e. $A(0) = 0$. To order u (2.4.13) states $\bar{\partial}A_1(z) = 0$ and we already argued that in order to get rid of complex coordinate transformations we should consider $A_1 \in H^{01}_{\bar{\partial}}(M,TM)$ only. One wants now to prove inductively that $\partial A_k + \frac{1}{2}\sum_{i=1}^{k-1}[A_i, A_{k-i}] = 0$ for $k > 1$ which by (2.4.18) is equivalent to

$$\bar{\partial}\hat{A}_k = \frac{1}{2}\sum_{i=1}^{k-1}[\hat{A}_i, \hat{A}_{k-i}], \quad \text{for } k > 1\,. \tag{2.4.20}$$

First step of induction: To first order in u one has simply as above $\hat{A}_1 \in H^{n-1,1}(M)$ and we pick the harmonic representative $\hat{A}_1$. In fact on compact Kähler manifolds it follows from (2.3.10), (2.3.33) that every harmonic representative fulfills $\bar{\partial}A_1 = \bar{\partial}^* A_1 = 0$. Moreover with $\Delta_{\bar{\partial}} = \Delta_\partial$, see Sect. 2.3.1 also $\partial\hat{A}_1 = 0$ holds. This implies $D \cdot A_1 = 0$ and by (2.4.19) $[\hat{A}_1, \hat{A}_1] = \partial(\widehat{A_1 \wedge A_1})$ is ∂-exact. On the other hand for $\hat{A}_1 \in H^{n-1,1}(M)$ hence $\bar{\partial}A_1 = 0$ it is immediate from the definition of the bracket that $\bar\partial[\hat A_1,\hat A_1]=\bar\partial\partial(\widehat{A_1\wedge A_1})=0$. The $\partial\bar\partial$ *Lemma* of Kähler geometry ([150], p. 149) states that if a form $\eta\in\Omega^{p,q}$ is $\bar\partial$ closed and d-, ∂- or $\bar\partial-$ exact, then it can be written as $\eta=\partial\bar\partial\psi$. Applied to the bracket we can write $[\hat A_1,\hat A_1]=\partial\bar\partial\psi_1$ for some $\psi_1\in\Omega^{1,1}$. Identifying $\hat A_2=\frac{1}{2}\partial\psi_1$ we have constructed a solution to $\bar\partial\hat A_2+\frac{1}{2}[\hat A_1,\hat A_1]=0$.

General induction: If for some N one has solved for $\hat A_i$ with $\partial\hat A_i=0$ and $\bar\partial\hat A_i+\frac{1}{2}\sum_{j=1}^{i-1}[\hat A_j,\hat A_{i-j}]=0$, $i=1,\ldots,N$, then

$$\sum_{j=1}^{N}[\hat A_j,\hat A_{N+1-j}]=\partial\sum_{j=1}^{N}(A_j\wedge A_{N+1-j})^{\wedge} \tag{2.4.21}$$

and one also checks that

$$\begin{aligned}\bar\partial\left(\sum_{j=1}^{N}[\hat A_j,\hat A_{N+1-j}]\right)&=\bar\partial\partial\left(\sum_{j=1}^{N}[A_j,A_{N+1-j}]\right)^{\wedge}\\&=\frac{1}{2}\partial\left(\sum_{j=1}^{N}\sum_{k=1}^{j-1}[[A_k,A_{j-k}],A_{N+1-j}]-[A_j,[A_k,A_{N+1-j-k}]]\right)^{\wedge}=0\,.\end{aligned}$$

Here we used first (2.4.19), then the fact that $\bar\partial$ and $\wedge$ commutes, (2.4.21) for A_k with $k\le N$ and the Jacobi identity for (2.4.14). By the ∂, $\bar\partial$ Lemma one can set $\hat A_{N+1}=\frac{1}{2}\partial\psi_N$ and since $\partial\hat A_{N+1}=0$ the induction proceeds. Moreover one has arguments that the series converges in $H^{n-1,1}(M)$ [301].

Hence there exist always a family of Calabi-Yau manifolds with varying complex structure parameters, whose complex dimension is $h^{0,1}(M,TM)=h^{n-1,1}(M)$. Tians and Todorovs result is very important also with respect to the world sheet theory, where is very not-trivial to establish that a deformation of type (4.1.7) is exactly marginal and does lead to family of $N=2$ SCFTs.

2.4.3 The Variation of Hodge Structures

The previous section gave us an understanding of the generic local structure of the complex moduli space, which has the special good property of unobstructedness for Calabi-Yau manifolds. To describe it globally we will discuss the idea of the variation of Hodge structure which leads to the more concrete notion of Picard-Fuchs differential equation and periods, which we discuss as important tools later.

Let us first informally describe the main idea with a short account of its possible pitfalls, which are as we see irrelevant for our most natural applications. An very nice introduction of the concept for Riemann surfaces can be found in [100]. In Sect. 2.9.7 we go more in the technical aspects of the derivations of the Picard-Fuchs equations applicable for general Calabi-Yau hypersurfaces and complete intersections.

Above we introduced the notion of a varying complex structures. Let us introduce now a family $\mathfrak{M}$ of complex manifolds $\{M_{\underline{u}}\}$ parametrized by some coordinates $\underline{u}$ varying smoothly over some base space U, so that there is a projection π : $\mathfrak{M} \to U$ with pre-image $\pi^{-1}(\underline{u}) = M_{\underline{u}}$. As we just saw the complex dimension of U is expected to be $h^{n-1,1}(M)$. It will useful and possible as we will see to consider a compactification of U to the complex moduli space $\mathcal{M}_{cs}$, which in particular contains singular fibres $\pi^{-1}(u_s)$ of $\mathfrak{M}$, which occur at least at complex co-dimension one in $\mathcal{M}_{cs}$. That is generically, away from the singular fibres, the fibres M_u are smooth and diffeomorphic to each other. In particular locally in some neighborhood $V \in \mathcal{M}_{cs}$ away from the singularity the mapping π describes a product and we can pick a k-chain or real k-dimensional sub manifold $C_{\underline{u}} \in M_{\underline{u}}$ that is *locally constant* in the sense that it always represents the same integer homology class $[C_{\underline{u}}] \in H_k(M_{\underline{u}}, \mathbb{Z})$. Because the homology is integer it is clear that class can not jump by smooth variations along $\mathcal{M}_{cs}$ as long as we stay away from the singularities. Elements in the dual cohomology are more naturally defined over $\mathbb{C}$ and can smoothly depend on $\underline{u}$. The latter can be defined as a restriction of a not necessarily closed form ω on $\mathfrak{M}$ to the fibres to yield a k-form $\omega_{\underline{u}} = \omega|_{M_{\underline{u}}}$. If $\omega_{\underline{u}}$ is a closed k-form on each fibre then the definition (2.1.15) extends immediately defining for the family $\{M_{\underline{u}}\}$ a period function by $\pi(\underline{u}) = \int_C \omega_{\underline{u}}$. Here we suppressed the dependence of C on $\underline{u}$, because it is a constant class in $H_k(M_{\underline{u}}, \mathbb{Z})$. Let $\underline{x}$ be coordinates on M, then because of the local product form of $\mathfrak{M}$ derivatives w.r.t ∂_{u^i}, $i = 1, \ldots, h_{n-1,1}$ commute with derivatives ∂_{x_i}, $i = 1, \ldots, \dim_{\mathbb{R}}(M)$. In particular if $\omega_{\underline{u}}$ is closed, so is $\partial^p_{u^i}\omega_{\underline{u}}$, $p = 0, \ldots, b_k(M_{\underline{u}})$. However we can have only $b_k(M_{\underline{u}})$ independent elements in $H^k(M_{\underline{u}}, \mathbb{R})$, hence by the theorem of de Rham we must have a linear combination that is exact:

$$\sum_{i=0}^{b_k} a_i(\underline{u})\partial^i_{u_k}\omega_{\underline{u}} = d\eta_{\underline{u}} \,. \tag{2.4.22}$$

By the Stokes theorem this immediately implies a linear differential equation the so called Picard-Fuchs equation for the period function, which we call often simply the period, which reads

$$\sum_{i=0}^{b_k} a_i(\underline{u})\partial^i_{u_k}\pi(\underline{u}) = 0 \,. \tag{2.4.23}$$

The argument has interesting shortcomings when we really apply it to the context of complex manifolds as intended, see [211] for an account of the problems. Let us pass from the real coordinates x_i, $i = 1, \ldots, 2n$ to complex coordinates $(z_i, z_{\bar{\imath}})$, $i = 1, \ldots, n$. Because of the Hodge theorem (2.3.31) we have the following situation: Let us start with any form in $A^{p,q}$ that is a holomorphic or algebraic (p, q)-form with $p + q = k$. Since we can still locally trivialize π as holomorphic map and since ∂_{u^i} commutes in this trivialization with ∂_{z_i} we would not change the Hodge type (p, q) by derivatives w.r.t. u and do not reach all of $H_k(M_{\underline{u}}, \mathbb{C})$.

A related issue that provides part of the solutions is that $\omega_{\underline{u}}$ vanishes at some point u_0 in $\mathcal{M}_{cs}$. Then taking the derivative w.r.t. u at this point can create poles in $\omega_{\underline{u}}$. The rectification of both problem is wellknown, if M is a elliptic curve. Then we extend the class of holomorphic one forms to include meromorphic one forms which have poles of arbitrary order but no residue at those poles. The latter are known as one forms of the *second kind*. Since they have no residue, their integral is still independent of the class of the cycle and they represent an element in $H_1(\Sigma_u)$, but this class of forms closes under differentiation w.r.t. u.

Example The *Legendre Elliptic Family* is given by

$$y^2 = x(x-1)(x-s)\,. \tag{2.4.24}$$

The (1,0) form differential

$$\omega_s = \frac{dx}{y} \tag{2.4.25}$$

can be differentiated as

$$\mathcal{L}^{(2)}\omega_s = (1+4(2s-1)\partial_s + 4s(s-1)\partial_s^2)\omega_s = -2\,d\left(\frac{\sqrt{x(x-1)}}{(x-s)^{3/2}}\right) \tag{2.4.26}$$

to yield an exact form η as in (2.4.22). Hence

$$\mathcal{L}^{(2)}\pi(s) = 0 \tag{2.4.27}$$

is the Picard Fuchs equation, which is a Hypergeometric system ${}_2F_1(\frac{1}{2},\frac{1}{2},1;s)$. The normalized period over the vanishing cycle at $s=0$ is $\int_A \omega_s = \int_1^\infty \omega_s = 2\pi{}_2F_1(\frac{1}{2},\frac{1}{2},1;s)$.

On higher dimensional manifolds there is an analog of the differentials of the second kind [150], which is however not quite sufficient in general to provide a basis of the cohomology that is suitable to describe the variation of Hodge structure. Algebraic De Rham theory is sufficient to describe the suitable cohomological objects, i.e. the cohomology groups on the left hand side of (2.3.34) with coefficients in the sheaf of p forms, which are algebraically defined, if M is an affine variety. In general one has to pass [156] to Hypercohomology [150].

The primitive horizontal subspace in the cohomology of a Calabi-Yau n-fold M_n comes with a polarized Hodge structure, see [76, 78, 256] for reviews. In order to study its variation we fix an integral structure $H^n(M,\mathbb{Z})$ relative to which we can measure the change of the spaces $H^{p,q}(M_{\underline{z}})$ in $H^n(M,\mathbb{R})$ with the change in complex structure parametrized by $\underline{z}$. The spaces $H^{p,q}(M_{\underline{z}})$ do not fit into holomorphic vector bundles over $\mathcal{M}_{cs}$. The way of capturing the variation of the polarized Hodge of Calabi-Yau n-folds $\pi\ :\ \mathcal{M}_n \rightarrow \mathcal{M}_{cs}$ over their complex moduli space $\mathcal{M}_{cs}$ with fibre $\pi^{-1}(t) = M_t$ is therefore as follows: One captures the Hodge decomposition $H^n(M,\mathbb{C}) = \oplus_{k=0}^{n} H^{n-k,k}$ with $\overline{H^{p,q}(M)} = H^{q,p}(M)$, with respect to the real structure $H^n(M_n,\mathbb{R})$, in terms of Hodge filtrations[30] $F^\bullet = \{F^p\}_{p=0}^{n}$ with

$$F^p = \bigoplus_{l\geq p} H^{l,n-l} \tag{2.4.28}$$

so that

$$H^n = F^0 \supset F^1 \supset \ldots \supset F^n \supset F^{n+1} = 0\,. \tag{2.4.29}$$

One can recover the Hodge decomposition from the $F^\bullet$ in (2.4.28) from the following relations

$$F^p \oplus \overline{F^{n-p+1}} = H^n(M,\mathbb{C}), \qquad H^{p,q}(M) = F^p(M) \cap \overline{F^q(M)} \tag{2.4.30}$$

and

$$H^{p,q}(M) = F^p/F^{p+1}\,. \tag{2.4.31}$$

Together with the lattice $H^n(M_n,\mathbb{Z})$ this defines the *Hodge structure*. Unlike the $H^{p,q}(M_n)$ the $F^p(M_n)$ vary holomorphically with the complex structure and fit into locally free sheaves over $\mathcal{M}_{cs}$ with inclusion $\mathcal{F}^p \subset \mathcal{F}^{p-1}$. This defines a *decreasing varying Hodge filtration*

$$\mathcal{H} = \mathcal{F}^0 \supset \mathcal{F}^1 \supset \ldots \supset \mathcal{F}^n \supset \mathcal{F}^{n+1} = 0\,. \tag{2.4.32}$$

for the locally free sheaves $\mathcal{F}^p$.

In particular $\mathcal{F}^0 = R^n\pi_*\mathbb{C} \otimes \mathcal{O}_{\mathcal{M}_{cs}}$ is the Hodge bundle $\mathcal{H}$, which has a *locally constant subsheaf* $R^n\pi_*\mathbb{C}$ over the moduli space $\mathcal{M}_{cs}$. Taking this as the flat section of $\mathcal{F}^0$ defines a flat connection, called the *Gauss Manin connection* $\nabla^{\mathrm{GM}} : \mathcal{F}^0 \to \mathcal{F}^0 \otimes \Omega^1_{\mathcal{M}_{cs}}$ by

$$\nabla^{\mathrm{GM}}(s \otimes f) = s \otimes \mathrm{d}f\,. \tag{2.4.33}$$

The locally constant subsheaf $\mathcal{H}_{\mathbb{C}} = R^n\pi_*\mathbb{C}$ of flat sections has as subsheaf the sheaf of integer sections $\mathcal{H}_{\mathbb{Z}} = \mathrm{Im}(R^n\pi_*\mathbb{Z} \to R^n\pi_*\mathbb{C})$. The Gauss Manin connection fulfills the Griffiths nilpotency condition w.r.t. to the varying Hodge

[30] Sometimes, especially in the physics literature, $F_p = F^{n-p}$ is used.

filtration

$$\nabla^{GM}\mathcal{F}^p \subset \mathcal{F}^{p-1} \otimes \Omega^1_{\mathcal{M}_{cs}} \,. \tag{2.4.34}$$

In particular if we pick any $\Omega \in \mathcal{H}$ then in view of (2.4.34), (2.4.32) the expression

$$\left(\nabla^{GM}\right)^{n+1} \Omega \tag{2.4.35}$$

for derivatives in any directions in $\mathcal{M}_{cs}$ must be expressible in terms of $\left(\nabla^{GM}\right)^k \Omega$, $k = 0, \ldots, n$ in cohomology. This is equivalent to the Picard-Fuchs differential ideal $\mathcal{I}_{PF}$, which is *the kernel* of the map $\phi(X_1, \ldots, X_r) = \nabla_{X_1} \ldots \nabla_{X_r} \Omega(t)$ from the sheaves of linear differential operators $\mathcal{D}_i$ on M_{cs} to the Hodgebundle $\mathcal{F}^0$, which makes the latter into a *$\mathcal{D}$-module*. Here $X_i = \partial_{z_i}$, $i = 1, \ldots, r = \dim(\mathcal{M}_{cs})$ is a basis of vector fields on $\mathcal{M}_{cs}$. As

$$\partial_{z_i} \int_{\Gamma_t} \Omega(t) = \int_{\Gamma_t} \nabla_{\partial_{z_i}} \Omega(t) \tag{2.4.36}$$

the $\mathcal{D}_i$ annihilate the periods and determine them. The solutions to the Picard-Fuchs equations determine then the above mentioned flat sections. The local flatness implies that the non-trivial essence in this setup is to large extend in the global monodromy group Γ acting on the sections over $\mathcal{M}_{cs}$, which explains the pivotal role of Γ.

In fact one can turn the problem around and given start with the monodromies at critical divisors, which determine the leading behaviour of the sections, which in turn are captured for simple examples by the Riemann-Hilbert symbol, and ask the question whether there is a differential system with regular singularities only, which then fully determines the sections. Problems of these kind are called Riemann-Hilbert problems.

2.4.4 Periods, Torelli Theorems and the Stringy Moduli Space

Based on the signature theorem of Sect. 2.3.4 we review first the possible intersection forms on the middle homology (in Section "The Integral Basis in the Middle (Co)homology"). We assume here based on the general remarks that we can obtain the periods of the complex n-dimensional manifold over an integral basis of real n dimensional cycles in the middle homology. We comment on the general choice of local coordinates on $\mathcal{M}_{cs}$ using the periods, which is possible due to a local Torelli theorem.

Then we explain how the complex structure deformations can be viewed as one part of possible deformations of the Ricci-flat metric. The other part, given geometrically as volume or so called Kähler deformations, complements this to the geometrical moduli space.

In string theory it is very natural to extend the moduli space of Kähler deformations by the Neveu-Schwarz antisymmetric background fields, that are not related to the metric. However as it turns out that the extended moduli spaces have better properties in particular holomorphicity and nicer symmetries. Mirror symmetry is the most famous example for the latter statement.

We discuss in examples in one and two dimensions the important aspect namely that the periods or rather ratios of periods parametrize the moduli space faithfully also globally. Such statements are know as global Torelli theorems in the literature.

We conclude the section with some general remarks on the higher dimensional Calabi-Yau cases and the physical moduli space on Calabi-Yau 3 folds. The detailed study of the moduli spaces of Calabi-Yau 3- and n-folds is discussed in Sect. 2.5.

The Integral Basis in the Middle (Co)homology

For odd dimension n this integral basis can be chosen to be *integral symplectic*, i.e. in homology we pick such a symplectic basis $A^I, B_I \in H_n(M_n, \mathbb{Z})$ and in cohomology we pick a dual basis $\alpha_I, \beta^I \in H^n(M_n, \mathbb{Z})$ with $I = 0, \ldots, \frac{b_n}{2} - 1$ such that the non-zero pairings are as follows

$$A^J \cap B_I = -B_I \cap A^J = \int_{M_n} \alpha_I \wedge \beta^J = -\int_{M_n} \beta^J \wedge \alpha^I = \int_{A^J} \alpha_I = \int_{B_I} \beta^J = \delta_I^J \tag{2.4.37}$$

and we define the periods w.r.t. to this basis as

$$\vec{\Pi} = \begin{pmatrix} F_I(z) \\ X^I(z) \end{pmatrix} = \begin{pmatrix} \int_{B_I} \Omega(z) \\ \int_{A^I} \Omega(z) \end{pmatrix}. \tag{2.4.38}$$

For even dimension n we have an even intersection forms, whose signature can be calculate by the Hirzebruch signature theorem, reviewed in Sect. 2.3.4. It is still of importance for the definition of the inhomogeneous coordinates on the complex moduli space $\mathcal{M}_{cs}$ that there are $h_{n-1,1} = h$ non-intersecting A-cycles A^I, $I = 0, \ldots, h$, such that the inhomogeneous coordinates can be defined as $t_i = X^i/X^0$ see below.

Often as for example for $K3$, discussed below, the intersection form is restricted by the classification of integer lattices. Given an algebraic realization which fixes a polarization, only the horizontal subspace of the middle cohomology can be described by periods integrals fulfilling Picard-Fuchs equations.

Using Griffiths residuum representation of the holomorphic (n,0) the integral over some cycles in the symplectic basis in some regions of the moduli space might be performed explicitly. In the general situation one finds the exact complex moduli dependence on z and an $\mathbb{C}$ basis of the periods using the fact that Π fulfills a systems of linear differential equations in the complex moduli. In order to determine the integral basis one matches the so found solutions to the known leading behaviour of integral cycles at degenerations points. The latter task is very much helped by mirror symmetry. One also might use the fact that the monodromy respects the integer symplectic basis, i.e. it has to be integer symplectic.

Choice of Coordinates

The existence of the holomorphic $(n, 0)$ form on Calabi-Yau spaces relates the infinitesimal complex structure deformations (2.4.12) to the middle cohomology $H^{n-1,1}(M)$ by (2.4.15). One can study the corresponding complex structure deformations at least locally by the variation of the Hodge structures using periods $I = 0, \ldots, h_{n-1,1}(M)$ over suitable cycles in the integral basis. Let us call these periods

$$X^I, \qquad I = 0, \ldots, h_{n-1,1}(M) = h \tag{2.4.39}$$

even so only for three-folds this can coincide with the definition in (2.4.38). These periods constitute *homogeneous coordinates* of the, so called big moduli space, whose dimension is $h_{n-1,1}(M) + 1$. The

$$t^i = \frac{X^i}{X^0}, i = 1, \ldots, h_{n-1,1}(M) . \tag{2.4.40}$$

are at least locally *inhomogeneous coordinates* of the actual complex moduli space $\mathcal{M}_{cs}$. What periods we select to define these coordinates depends on the locus in the moduli space and in particular on the local monodromy as explained in Sect. 2.6.2. For example for the 3-fold case, what we call A-cycles and therefore X^I in (2.4.38) is of course a choice up to integer matrix M transformations, which respects the intersection form Σ defined below in (2.4.37). I.e. for a Calabi-Yau 3-fold or Riemann surfaces of genus g these are $\mathrm{Sp}(h_3(M), \mathbb{Z})$ or $\mathrm{Sp}(2g, \mathbb{Z})$ transformations respectively. As we will see in Sect. 2.6.4, at the points of maximal unipotent monodromies, there is an unique holomorphic period X^0 and exactly $h_{n-1,1}$ single logarithmic periods X^i, $i = 1, \ldots, h_{n-1,1}(M)$. This singles out the ratios (2.4.40) as the mirror map. This and choices at other points will be discussed more later.

The First Order Deformations of the Metric

We known the first order deformations and for the complex structure deformations the fact that they are unobstructed as discussed in (2.4.2). Let us review [58, 60] the linearization approach to the Ricci-flat metric and to get the stringy moduli space to the B–field and further background fields. So we consider deformations $g_{\mu\nu}+\delta g_{\mu\nu}$, which do not change the Calabi-Yau condition[31] $R_{\mu\nu}(g) = 0$, i.e.

$$R_{\mu\nu}(g + \delta g) = 0 \,. \tag{2.4.41}$$

In analyzing this equation we have to eliminate the infinitesimal changes δg, which come from coordinate transformations. Coordinate transformations or equivalently diffeomorphism of M are generated by vectors fields V^μ, compare section "First Order Complex Structure Deformations". An actual change of the metric $\delta g_{\mu\nu}$ is orthogonal to diffeomorphism generated by the vector field in the following sense $\int \sqrt{g}\delta g^{\mu\nu}(\nabla_\mu V_\nu + \nabla_\nu V_\mu)\mathrm{d}^m x = 0$, which is equivalent to the gauge condition $\nabla^\mu \delta g_{\mu\nu} = 0$, compare (3.3.20) and (3.3.21). Expanding with this constraint (2.4.41) to linear order around $R(g) = 0$ one gets

$$\nabla^\rho \nabla_\rho \delta g_{\mu\nu} - 2R_{\mu\ \nu}^{\ \kappa\ \sigma}\delta g_{\kappa\sigma} = 0 \tag{2.4.42}$$

Using the splitting of a Kähler metric in holomorphic and holomorphic indices one can analyze $\delta g_{i\bar{j}}$, and δg_{ij} separately. Note that $\delta g_{i\bar{j}}$ is real, while δg_{ij} with $\overline{\delta g_{ij}} = \delta g_{\bar{i}\bar{j}}$ is complex. From (2.1.27) it follows that $\delta g_{i\bar{j}}$ is Δ_d harmonic and $\delta g^i = \delta g^i_{\bar{j}}\mathrm{d}z^{\bar{j}} = g^{i\bar{k}}\delta g_{\bar{k}\bar{j}}\mathrm{d}z^{\bar{j}}$ is $\Delta_{\bar{\partial}}$ harmonic. In other words the first order deformations factorize and correspond to elements in $H^{1,1}(M)$ and $H^1(M, TM) \sim H^{2,1}(M)$ respectively. As mentioned the last equivalence is due to Serre duality (2.4.7), via the no-where vanishing $(3, 0)$ form. These are also among the deformations of the A- and B-model as mentioned above.

As it is clear from the fact that the deformations $\delta g_{ij}, \delta g_{\bar{i}\bar{j}}$ change the $(i, \bar{\imath})$ type of the metric, the moduli space $H^1(M, TM)$ is associated to complex structure deformations, discussed already in section "First Order Complex Structure Deformations" and Sect. 2.4.2.

Next we discuss the two moduli space associated to $H^{1,1}(M)$. In a basis of $(1, 1)$-forms $\omega^{(k)}$, we expand a Kähler form ω

$$\omega = \sum_{k=1}^{h^{11}} v^k \omega^{(k)} \tag{2.4.43}$$

in terms of the real Kähler parameters $v^k > 0$, which are the *2-volumes of the curves* C_k dual to $\omega^{(k)}$

$$v^k = \int_{C_k} \omega \,. \tag{2.4.44}$$

[31]Strictly speaking one should ask for perturbations, which leave the Ricci-form $\mathcal{R}$ in the $c_1(M) = 0$ cohomology class. Though the representatives of the deformations in the cohomology classes would be different, the counting would be the same, see Sect. 3.3.1.

Therefore the range of v^k is bounded by the inequalities, which ensure positivity of the 2-volumes of curves C and for examples for 3-fold the 4-volumes of divisors D and the 6-volumes of M, i.e.

$$\int_C \omega > 0, \quad \int_D \omega \wedge \omega > 0, \quad \int_M \omega \wedge \omega \wedge \omega > 0 \,. \tag{2.4.45}$$

These conditions describe a real cone in $\mathbb{R}_+^{h^{1,1}}$, which is called the *Kähler cone*. The volume parameters v^k shrink to zero area at the boundaries of the Kähler cone.[32]

The Neveu-Schwarz B-Field and the Complexified Kähler Moduli Space

The string σ-model action (3.4.1) describes the string theory perturbatively in the large volume limit. The relevant bosonic part is

$$\int_\Sigma \mathrm{d}\sigma^2 \left(g_{i\bar{\jmath}}\eta^{ab} + i b_{i\bar{\jmath}}\varepsilon^{ab}\right) \partial_{\sigma^a} x^i \partial_{\sigma^b} x^{\bar{\jmath}} \,. \tag{2.4.46}$$

Here σ^a with $a, b = 0, 1$ are world sheet coordinates, $x^i(\sigma)$, $x^{\bar{\jmath}}(\sigma)$ with $i, \bar{\jmath} = 1, \ldots, n$ are the pullback of the complex coordinates on M to the world-sheet Σ describing the embedding map $X : \Sigma \to M$. η^{ab} and ε^{ab} are the flat worldsheet metric and the worldsheet antisymmetric tensor respectively. Note that the equations of motion for the B-field enforce that $b_{i\bar{\jmath}}$ is a harmonic $(1, 1)$ and therefore a representative of $H^{1,1}(M)$. Combined with the possible Ricci-flat deformations (2.4.42) is is therefore natural to decompose the space time moduli metric element into a complex field for every cohomology element in $H^{2,1}(M)$, while the B field allows to complexify the real Käher modulus corresponding to each element in $H^{1,1}(M)$

$$\mathrm{d}s^2 = \frac{1}{2V}\int_M g^{i\bar{m}} g^{j\bar{n}} \left[\delta g_{ij}\delta g_{\bar{m}\bar{n}} + (\delta g_{i\bar{m}}\delta g_{j\bar{n}} + \delta b_{i\bar{m}}\delta b_{j\bar{n}})\right] \sqrt{g}\mathrm{d}^6 x \,. \tag{2.4.47}$$

This decomposition is also natural from the IIA supergravity[33] effective action, which has complex moduli for Kähler deformations with an additional special Kähler structure, while each complex deformation lives in a complex subspace of a quaternionic space. The complex subspace is in the image of special Kähler projection, while the additional fields which double the degree of freedom to make

[32] At the boundary of the Kähler also a divisor may collapse. In this case $\tilde{t}^k$ is still the area of a curve C_k in D.

[33] For the IIB compactification which has even form Ramond-Ramond fields the assignment of the two types of metric and B-field deformations complex and complexified Kähler to the complex and quaternionic moduli space is reversed.

it to the quaternionic space are odd form Ramond-Ramond fields, invisible in the perturbative action above. Moreover it can be argued from the $\mathcal{N} = 2$ supergravity interactions that this metric is at least for smooth manifolds block diagonal. Another argument in favor of the two complex deformations spaces comes from the marginal deformations of the $(\mathcal{N}, \mathcal{N}) = (2, 2)$ world-sheet theory.

We see from (2.4.46), (2.4.47) that it is natural to complexify the parameter t^k by adding the integrals of the second term in (2.4.46), the anti-symmetric tensor Neveu-Schwarz field b over C_k to the volumes and normalize to define the *complexified Kähler parameters* t^k as

$$t^k = \frac{1}{2\pi i}\int_{C_k}(ib - \omega) = \frac{1}{2\pi i}(b^k - v^k). \tag{2.4.48}$$

It is very important to note that the second term in (2.4.46) merely counts a winding number, describing how often the string winds C_k. Since the action is unchanged by an integer phase shift of $2\pi i n$ with $n \in \mathbb{Z}$ a shift of b^k by $2\pi i$ corresponds to an exact symmetry of the large volume expansion, which is hence invariant under

$$t^k \to t^k + 1\,. \tag{2.4.49}$$

The mirror symmetry conjecture asserts that Calabi-Yau manifolds occur in mirror pairs (M, W) so that the complex moduli space $\mathcal{M}_{cs}(M)$ of M is identified with the complexified Kähler moduli $\mathcal{M}_{cks}(W)$ space of W and vice versa. This requires that all moduli spaces are at least one complex dimensional, because both M and W are Kähler and have to have at least one Kähler modulus. This excludes the so called rigid Calabi-Yau spaces which have no complex structure deformations. However even in these cases the complexified Kähler moduli space $\mathcal{M}_{cks}(M)$ can be mapped to the complex deformation space of a mirror geometry, which can either be described as a higher dimensional manifold [62] or one with anticommuting coordinates [285, 292]. Mirror symmetry also gives a natural choice of the complex parametrization of the complexified Kähler moduli space $\mathcal{M}_{cks}$, simply the complex structure parameters of the mirror t_k^m.[34]

Elliptic Curve

The simplest example of a local and global Torelli theorem and the stringy moduli space is the complex moduli space of an elliptic curve $\mathcal{E}$ that is locally parametrized by periods of the holomorphic (1, 0)- form $\omega := \Omega_1$ over a symplectic basis (A, B)

[34] As a corollary all singularities of $\mathcal{M}_K$ occur at complex codimension one and the cone structure disappears completely.

of $H_1(\mathcal{E}, \mathbb{Z})$ as

$$\tau = \frac{\int_B \omega}{\int_A \omega} \in \mathbb{H}_+ \, . \tag{2.4.50}$$

Here $\mathbb{H}_+$ is the complex upper half plane defined as

$$\mathbb{H}_+ = \{z \in \mathbb{C} | \mathrm{Im}(\tau) > 0\} = SU(1,1)/U(1) \, . \tag{2.4.51}$$

Note that by the definition of elliptic curve $\mathcal{E} = \mathbb{C}/\mathbb{Z}^2(1, \tau)$, i.e. as a parellelogram spanned by the complex numbers $(1, \tau)$ and identified opposite sides, the inequality $\mathrm{Im}(\tau) > 0$ merely means the positivity of the volume of the curve. So the moduli space is globally parametrized by $\tau \in \mathbb{H}_+$ up to the SL$(2, \mathbb{Z})$ monodromy action on Π, which induces a $\Gamma_1 = \mathrm{PSL}(2, \mathbb{Z})$ on τ (A4.3). It can hence be identified with the discrete quotient

$$\mathcal{M}_{cs}(\mathcal{E}) = \Gamma_1 \backslash \mathbb{H}_+ \, , \tag{2.4.52}$$

which is also called the fundamental region $\mathcal{F}$. If we consider a family of Riemann surfaces represented typically by the zero locus of algebraic equation representing the anti canonical class in a Fano 2 fold, which is generically smooth and parametrized by the complex parameter s, the actual monodromy group $\Gamma_{\mathcal{E}(s)} \in \Gamma_1$ is a subgroup of finite index m and $\mathcal{M}_{cs}(\mathcal{E}(s)) = \mathbb{H}_+/\Gamma_{\mathcal{E}(s)}$ so that $\mathrm{Vol}(\mathcal{M}_{cs}(\mathcal{E}(s)))/\mathrm{Vol}(\mathbb{H}_+/\Gamma_1) = m$. According to the theorem of Yau the metric of a Calabi-Yau manifold is fixed once Ω_n and the Kählerform ω is fixed. Specialized to the metric of the torus one has the moduli space

$$\mathcal{M}_g = \mathcal{M}_{cs}(\mathcal{E}) \times \mathbb{R}_+ \, , \tag{2.4.53}$$

where the second factor represents the volume of the two torus. Note however that in string theory or rather in the non-linear sigma model that defines it the volume is complexified by the Neveu-Schwarz B-field as explained in greater generality after (2.4.46). By mirror symmetry the moduli space of the complexified Kähler structure must be a copy of the fundamental region $\mathcal{F}$. Since string theory on the two torus is solvable with all its dependence on the moduli this can be explicitly shown, see [185] for review. The stringy moduli space becomes indeed

$$\mathcal{M}_{string} = \Gamma_\sigma \backslash \mathcal{F}_{cs} \times \mathcal{F}_{ck} \, , \tag{2.4.54}$$

where Γ_σ is generated by three $\mathbb{Z}_2$ representing mirror symmetry, charge conjugation and world-sheet parity of the non-linear sigma model.

Higher Genus Riemann Surfaces

Higher genus Riemann surfaces with $c_1(T\Sigma_{g>1}) < 0$ are of course not Calabi-Yau manifolds in any sense. Nevertheless they also have a moduli space of complex structures and similar considerations apply to the variation of their periods. The difference is that the relevant space defined by the periods is the Siegel upper half space. That is because genus g Riemann surfaces have unlike Calabi-Yau spaces g holomorphic one forms Ω_1^k, $k = 1, \ldots, g$. E.g. if the genus g Riemann-surface is given in hyperelliptic[35] form, i.e. as

$$y^2 = \prod_{i=1}^{2g+2} (x - e_i) , \tag{2.4.55}$$

then they can be written as $\Omega_1^k = \frac{x^{k-1}\mathrm{d}x}{y}$, $k = 1, \ldots, g$. Using them all one gets a period matrix

$$\Pi_{g\times 2g} = (A, B) = \begin{pmatrix} \int_{A^1} \Omega_1^1, & \ldots, & \int_{A^1} \Omega_1^g & \int_{B^1} \Omega_1^1, & \ldots, & \int_{B^1} \Omega_1^g \\ \vdots & \vdots & \vdots & \vdots & \vdots & \vdots \\ \int_{A^g} \Omega_1^1, & \ldots, & \int_{A^g} \Omega_1^g & \int_{B^g} \Omega_1^1, & \ldots, & \int_{B^g} \Omega_1^g \end{pmatrix} . \tag{2.4.56}$$

that defines $\tau_{g\times g} = A^{-1}B$ a symmetric matrix. The Riemann bilinear relations require $\mathrm{Im}(\tau_{g\times g}) > 0$ to be positive. Matrices with this property take values in the Siegel upper half space. Again one has to divide by the monodromy group a finite index subgroup of $\mathrm{Sp}(2g, \mathbb{Z})$.

K3 Surfaces

For K3 surfaces and algebraic K3 surfaces one has likewise global Torelli theorems. To state them we recall that by the Hirzebruch signature formula (2.3.37), (2.3.39) for (Calabi-Yau) 2-folds the difference between positive and negative eigenvalues in the intersection form is $\sigma = \int_{K_3} \frac{1}{3}(c_1^2 - 2c_2) = -\frac{2}{3}\chi(K_3) = -16$. The 22 dimensional lattice $H_2(K_3, \mathbb{Z})$ has hence signature (3, 19). By Poincaré duality it is selfdual. Since $c_1(T_M) = 0 \bmod 2$, a K_3 is spin and one gets by the formula of Wu that the intersection form is even[36] [251, 296]. By the classification of even, self

[35] Note that three points can be brought to prescribed values by an $\mathrm{SL}(2, \mathbb{C})$ transformation acting on the x-plane. Often one brings one of them to complex infinity and rescales y so that the r.h.s. of (2.4.55) is of degree $2g + 1$ as in (2.4.24). However only for $g = 1, 2$ all $3g - 3$ moduli of a genus g Riemann surface can be parametrized by the positions of the remaining $2g - 1$ points, i.e. for $g > 2$ one cannot write the most general Riemann surface in hyperelliptic form.

[36] See [327] for a similar application to CY 4 folds.

dual Lorentzian lattices the intersection form on the K3 has to be

$$\Gamma_{3,19} = \Gamma_{E_8} \oplus \Gamma_{E_8} \oplus \Gamma_{1,1} \oplus \Gamma_{1,1} \oplus \Gamma_{1,1} \,. \tag{2.4.57}$$

where Γ_{E_8} is the negative of the E_8 Cartan matrix and $\Gamma_{1,1} = \begin{pmatrix} 0 & 1 \\ 1 & 0 \end{pmatrix}$. If we parametrize the holomorphic two form $\Omega_2 = x + iy$ with $x, y \in H_2(K3, \mathbb{R})$ then it follows from (2.3.41) that $\langle x, x\rangle = \langle y, y\rangle$ and $\langle x, y\rangle = 0$ and from (2.3.42) $\langle x, x\rangle + \langle y, y\rangle \geq 0$, hence Ω_2 lies in the space like two plane in $\mathbb{R}^{3,19}$ with a fixed orientation, which is reversed by complex conjugation. Hence the moduli space of general complex structures is given by the choice of that two plane. The latter can be parametrized by $O^+(3, 19)$ rotations relative to $\Gamma_{3,19}$ and modulo rotations inside and orthogonal to the two pane. I.e. by the Grassmannian

$$\mathrm{Gr}^+(\Omega_2, \mathbb{R}^{3,16}) = \frac{(O^+(3, 19)}{(O(2) \times O(1, 19))^+} \,, \tag{2.4.58}$$

so that the analog of (2.4.52) is the complex space with 20 complex dimension

$$\mathcal{M}_{cs}(K3) = O^+(\Gamma_{3,19}) \backslash \mathrm{Gr}_+(\Omega_2, \mathbb{R}^{3,19}) \,. \tag{2.4.59}$$

If we construct the K3 by a family of hypersurfaces given as the vanishing locus of an algebraic equation representing the anti canonical class in a Fano 3-fold, the complex moduli space is of different dimension. The reason is that the curve classes that decent from the ambient space stay holomorphic in all complex structures accessible by the family of hypersurfaces and are dual to $(1, 1)$ forms, which represent the Picard group $\mathrm{Pic}(K_3) = H^2(K3, \mathbb{Z}) \cap H^{1,1}(K3)$ of rank ρ. Exactly one of those $(1, 1)$ forms represent the Kähler class and is by (2.3.5) also space like, so that the signature of the Picard lattice is $(1, \rho - 1)$. Therefore one can choose the two-plane representing Ω only within the space $\mathbb{R}^{2,20-\rho}$, which implies that the moduli space is the space with $20 - \rho$ complex dimensions

$$\mathcal{M}_{cs}(K3^{alg}) = O(\Lambda_{\mathrm{Im}(r)^T}) \backslash O(2, 20 - \rho)/(O(2) \times O(20 - \rho)) \,. \tag{2.4.60}$$

Here $\Lambda_{\mathrm{Im}(r)^T}$ is the orthogonal complement of image of the embedding map $r : \mathrm{Pic} \to \Gamma_{3,19}$ in $\Gamma^{3,19}$ and O are as before the lattice automorphisms of the corresponding lattices. The moduli space of the metric of K3 is according to the theorem of Yau given by the choice of Ω_2 and ω, which span the space like subspace. However the choice of ω within this three dimensional subspace turns out to be irrelevant for the metric of the $K3$. The reason is that a it can be undone by rotation representing a choice of complex structures parametrized by an S^2, which does not affect the metric. This is a special feature of hyperkähler manifolds such as $K3$. Hence the metric moduli space does only depend on how we chose the space like three dimensional subspace itself relative to the lattice and together with the volume is given by the 58 real dimensional space

$$\mathcal{M}_g = O(\Gamma_{3,19})\backslash O(3,19)/(O(3)\times O(19))\times \mathbb{R}_+ \,. \tag{2.4.61}$$

Mirror symmetry exchanges in the 20 dimensional cohomology of $(1,1)$ forms of the $K3$, the Picard lattice with its complement. The 22 Neveu-Schwarz B-fields in the non-linear sigma model, see (2.4.46), enhance the moduli space of the metric into an 40 dimensional complex moduli space of string theory on K3 back grounds, see [23] for a review, which is given by,

$$\mathcal{M}_{string} = O(\Gamma_{4,20})\backslash O(4,20)/(O(4)\times O(20))\,. \tag{2.4.62}$$

Calabi-Yau 3-Folds

The moduli space of the metrics of Calabi-Yau 3-folds is not a homogenous space like in the examples of Riemann surfaces or K3 discussed above. Also there is no global Torelli theorem. It has a structure called special Kähler geometry which will be discussed from the geometrical point of view in Sect. 2.5. Much of its structure follows from the existence of the $(n,0)$ form and Griffiths tranversality which hold in any dimensions. It leads however to very different geometrical structures in even and odd dimensions.

We conclude the section with a description of the deformation spaces of 3-folds from the physics points of view. In the low energy effective action of type II A/B string theory these marginal deformations arise as vacuum expectation of complex scalar fields labeling the vacuum manifold of the N=2 supergravity in 4d. The general structure of this vacuum manifold for abelian gauge groups $U(1)^{\#V}$ and $U(1)^{\#H}$ is that it is *locally* of the form $\mathcal{M}_{2\#V}\times \mathcal{Q}_{4\#H}$, where $\mathcal{M}$ is a complex special Kähler manifold for the scalar fields in the vector multiplets [81, 83, 84, 116] and $\mathcal{Q}$ is a quaternionic manifold [70] for the scalar fields in the hypermultiplets. The subscripts indicate the real dimension of the moduli space. Its relation to the perturbative sector of the II A/B string compactifications on the Calabi-Yau 3 folds M and W is as follows

$$\mathcal{M}^{IIA}_{tot}(M)=\mathcal{M}^{IIA}_{2h^{1,1}(M)}\times \mathcal{Q}^{IIA}_{4(h^{2,1}(M)+1)} \qquad \mathcal{M}^{IIB}_{tot}(W)=\mathcal{M}^{IIB}_{2h^{2,1}(W)}\times \mathcal{Q}^{IIB}_{4(h^{1,1}(W)+1)}\,. \tag{2.4.63}$$

One very far reaching definition of the mirror conjecture is that type IIA and type IIB string compactifications are completely identically if M and W are mirror pairs. This in particular implies $\mathcal{M}^{IIA}_{tot}(M)=\mathcal{M}^{IIB}_{tot}(W)$. The best studied object is $\mathcal{M}^{IIB}_{2h^{2,1}(W)}$ since it is literally the complex moduli space of W. The enhancement of the Calabi-Yau metric moduli space from the complex to the quaternionic space $\mathcal{Q}$ of Kähler multiplets is due to the moduli of Ramond fields. The additional quaternionic dimension in $\mathcal{Q}$ comes from the universal dilation, whose scalar components (S,C) contain in particular the type II dilaton S.

2.5 Special Geometry on the Complex Moduli Space

Special geometry is a differential property on complex moduli space $\mathcal{M}_{cs}$ of Calabi-Yau manifolds. In the case of Calabi-Yau 3-folds it guarantees the existence of a holomorphic prepotential from which the Kähler potential and triple structure constants derives. We will now derive the integrality condition of *special geometry* for 3-folds as a corollary to the consideration of type (2.3.41) and Griffiths transversality (2.4.34) in way that generalizes immediately to the derivation of similar differential and algebraic properties for general complex dimension n. Additional algebraic properties occur if n is even. The structure generalizes in fact to $(2, 2)$ superconformal theories independent of their geometric interpretation. In this generality it is known as $tt*$-structure.

2.5.1 Universal Part of Special Geometry

We start with elements of the theory, which apply to the complex moduli spaces of Calabi-Yau spaces of any dimensions namely the Weil-Petersson metric on the complex moduli space $\mathcal{M}_{cs}$, which determines the metric in front of the kinetic terms of the moduli fields. The latter exist geometrically, because by the Tian-Todorov theorem the moduli space of Calabi-Yau manifolds is unobstructed [301, 303]. Its tangent space is described by $H^1(M_n, TM_n)$, which can be identified with $H^{n-1,1}(M_n)$ by contracting with Ω_n. Calabi-Yau manifolds have an unique nowhere vanishing $(n, 0)$ form Ω_n and the real from (2.3.42) is related to the Kähler potential for the metric on the moduli space as

$$e^{-K} = (-1)^{\frac{n(n-1)}{2}} R(\Omega_n) = i^{n^2} \langle \Omega_n, \bar{\Omega}_n \rangle \tag{2.5.1}$$

The Weil-Petersson metric on the complex moduli space $\mathcal{M}_{cs}$ of the Calabi-Yau n-fold M_n is then given by (2.3.20).

The following derivation of special geometry relations is essentially due to Bryant and Griffiths, which address the differential structure of infinitesimal period variations of CY 3-folds.

Because of the Griffiths transversality (2.4.34) one has

$$\partial_i \Omega_n := \frac{\partial \Omega_n}{\partial_{z_i}} = \alpha_i(z)\Omega_n + \chi_i \in \mathcal{F}^{n-1} = H^{n,0} \oplus H^{n-1,1} \ . \tag{2.5.2}$$

Here we have chosen elements χ_i, $i = 1, \ldots, h^{n-1,1}$ in $H^{n-1,1}(M_n)$ and $\alpha_i(z)$ is an auxiliary moduli dependent factor.

First we want to show that it can be given in terms of the Kähler potential as

$$\alpha_i(z) = -K_i =: -\partial_i K \ . \tag{2.5.3}$$

To this end we take a derivative of (2.5.1), which leads to the first equality below

$$-K_i e^{-K} = i^{n^2} \partial_i \langle \Omega_n, \bar{\Omega}_n \rangle = i^{n^2} \alpha_i \langle \partial_i \Omega_n, \bar{\Omega}_n \rangle = \alpha_i e^{-K} \,. \tag{2.5.4}$$

The second equality follows from the anti-holomorphicity $\partial_i \bar{\Omega}_n = 0$ of $\bar{\Omega}_n$, the third from the consideration of type as in (2.3.41). Hence we conclude that the auxiliary factor $\alpha_i(z)$ is indeed $\alpha_i = -K_i$.

It is therefore convenient to define the elements $\chi_i \in H^{n-1,1}(M_n)$ and the complex conjugated ones in $\bar{\chi}_{\bar{i}} \in H^{1,n-1}(M_n)$ by

$$D_i \Omega_n = (\partial_i + K_i)\Omega_n =: \chi_i \qquad \bar{D}_{\bar{i}} \bar{\Omega}_n = (\partial_{\bar{i}} + K_{\bar{i}})\bar{\Omega}_n =: \bar{\chi}_{\bar{i}} \tag{2.5.5}$$

Because of the Tian-Todorov theorem the χ_i, $i = 1, \ldots, h_{n-1,1}$ span a basis of $H^{n-1,1}(M_n)$ if z_i, $i = 1, \ldots, h^{n-1,1}$ are an independent local coordinate basis of $\mathcal{M}_{cs}$. One can then define a metric[37]

$$g_{i\bar{j}} = -i^{n^2} \langle \chi_i, \bar{\chi}_{\bar{i}} \rangle = e^{-K} G_{i\bar{j}} \tag{2.5.6}$$

The last equality follows from the application of $\partial_i \bar{\partial}_{\bar{j}}$ to (2.5.1) and using (anti-)holomorphicity of Ω_n ($\bar{\Omega}_n$), (2.5.5) and considerations of type (2.3.41).

It is obviously useful to repeat these steps for further derivatives of (2.5.5) and use (2.4.34), (2.3.41) to learn about the differential properties of the derivatives. The outcome is known as *special geometry*, which we discuss below for $n = 3$ and 4. Other cases are left to the reader as exercise.

At this point it is useful to summarize the gauge transformation properties and the corresponding covariant derivative of the objects introduced. The no-where vanishing holomorphic Ω_n form lives in a holomorphic line bundle $\mathcal{L}$ over $\mathcal{M}_{cs}$ and transforms as

$$\Omega(z) \to e^{f(z)} \Omega(z) \,. \tag{2.5.7}$$

The Kähler form transforms then in the Kähler line bundle with Kähler transformations (2.3.21)

$$K(z, \bar{z}) \to K(z, \bar{z}) - f(z) - \bar{f}(\bar{z}),$$

so that e^{-K} is a section of $\mathcal{L} \otimes \bar{\mathcal{L}}$. One has natural covariant derivatives w.r.t. to the Weil-Petersson metric, with a connection whose Christoffel symbols have the index structure given by (2.3.23) and in addition with the Kähler connection. On general

[37] Note the additional minus from the i^{p-q} factor in (2.3.42).

sections say $V_{j\bar{j}} \in T^*_{1,0}\mathcal{M}_{cs} \otimes T^*_{0,1}\mathcal{M}_{cs} \otimes \mathcal{L}^{\otimes n} \otimes \bar{\mathcal{L}}^{\otimes m}$ these connections act as

$$D_i V_{j\bar{j}} = \partial_i V_{j\bar{j}} - \Gamma^l_{ij} V_{l\bar{j}} + nK_i V_{j\bar{j}}, \qquad D_{\bar{\imath}} V_{j\bar{j}} = \partial_{\bar{\imath}} V_{j\bar{j}} - \Gamma^{\bar{l}}_{\bar{\imath}\bar{j}} V_{i\bar{l}} + mK_{\bar{\imath}} V_{j\bar{j}}, \tag{2.5.8}$$

with $K_i = \partial_i K$ and $K_{\bar{\imath}} = \partial_{\bar{\imath}} K$.

Griffiths transversality (2.4.34) implies that any combination $\nabla^{\rm GM}_{i_1} \ldots \nabla^{\rm GM}_{i_r}$ of the application of the Gauss-Manin connection in the direction of complex moduli $a_i, i = 1, \ldots, \dim(\mathcal{M}_{cs})$ to Ω_n has the property

$$\underline{\nabla^{\rm GM}_{\underline{i}}}\Omega_n = \nabla^{\rm GM}_{i_1} \ldots \nabla^{\rm GM}_{i_r}\Omega \in \mathcal{F}^{n-r}, \tag{2.5.9}$$

cff Sect. 2.4.3. All derivatives of Ω generate the primary horizontal subspace $H^n_H(M_n)$. By consideration of Hodge type and by (2.4.34) one gets

$$\langle \underline{\nabla^{\rm GM}_{\underline{i}}}\Omega_n, \Omega\rangle = \int_{M_n} (\nabla^{\rm GM}_{i_1} \ldots \nabla^{\rm GM}_{i_r}\Omega_n) \wedge \Omega_n = \begin{cases} 0 & \text{for } 0 \le r < n \\ C_{i_1 \ldots i_n}(z) & \text{for } r = n \end{cases} \tag{2.5.10}$$

Note that one can replace the covariant derivative $\nabla^{\rm GM}_i$ in the definition of the n-point coupling with ordinary derivatives ∂_i, because by similar argument like after (2.5.2) one concludes that only the terms, which do not involve Christoffel symbols or their derivatives change the type enough to give by (2.3.41) a non-vanishing intersection with Ω_n. Therefore $C_{i_1 \ldots i_n}(z)$ is a purely *holomorphic quantity*.

The n-point coupling $C_{i_1 \ldots i_n}$, first introduced by Bryant and Griffiths, is *purely holomorphic* and transforms as a section in

$$C_{i_1 \ldots i_n}(z) \subset \mathcal{L}^2 \otimes {\rm Sym}^n(T^*_{1,0}(\mathcal{M}_{cs})). \tag{2.5.11}$$

As explained in Sect. 2.9.2 there is always a gauge of Ω so that the $C_{i_1 \ldots i_n}(z)$ are rational function in algebraic coordinates z on $\mathcal{M}_{cs}$ that are determined from the Picard-Fuchs differential ideal. Moreover they are always regular at the maximal unipotent or large radius points and have typically poles at the discriminant of the Picard-Fuchs differential ideal.

Let us discuss one simple application. $\Omega \subset \mathcal{L}$ and $D_i\Omega = \chi_i \subset \mathcal{L} \otimes (T^*_{1,0}(\mathcal{M}_{cs})$ therefore

$$[D_i, D_{\bar{j}}]\Omega_n = -G_{i\bar{j}}\Omega_n \, . \tag{2.5.12}$$

This can also be checked by testing the coefficients $\alpha_{i\bar{j}}$ in $[D_i, D_{\bar{j}}]\Omega_n = \alpha_{i\bar{j}}\Omega$ by calculating $\langle [D_i, D_{\bar{j}}]\Omega_n | \bar{\Omega}_n\rangle$ using (2.5.1), (2.5.2) and (2.5.6). Similarly one has from (2.3.26)

$$[D_i, D_{\bar{j}}]\chi_k = -G_{i\bar{j}}\chi_k + R_{i\bar{j}k}{}^p \chi_p \, . \tag{2.5.13}$$

Let us consider the derivative $D_i \chi_{\bar{\imath}}$, which by Griffiths transversality (2.4.34) is expandable in the basis of $(1, n-1)$ forms $(\chi_{\bar{k}})$ and the $(0, n)$ form $\bar{\Omega}_n$ as

$$D_i \chi_{\bar{\jmath}} = \alpha_{i\bar{\jmath}}^{\bar{k}} \chi_{\bar{k}} + \beta_{i\bar{\jmath}} \bar{\Omega}_n \tag{2.5.14}$$

We have from (2.5.6)

$$g_{i\bar{\jmath}} = -i \langle D_i \Omega, \chi_{\bar{\jmath}} \rangle = -i \langle \partial_i \Omega, \chi_{\bar{\jmath}} \rangle = i \langle \Omega, \partial_i \chi_{\bar{\jmath}} \rangle = i \langle \Omega, D_i \chi_{\bar{\jmath}} \rangle = i \beta_{i\bar{\jmath}} \langle \Omega, \bar{\Omega} \rangle = \beta_{i\bar{\jmath}} e^{-K}$$

by partial integration, consideration of type and (2.5.14) and so

$$\beta_{i\bar{\jmath}} = G_{i\bar{\jmath}} \, . \tag{2.5.15}$$

On the other hand taking the complex conjugated version of (2.5.12) $[D_{\bar{\jmath}}, D_i]\bar{\Omega}_n = -G_{i\bar{\jmath}}\bar{\Omega}_n$ one has from consideration of type

$$0 = \langle \chi_l, D_i \chi_{\bar{\jmath}} \rangle = \alpha_{i\bar{\jmath}}^{\bar{k}} \langle \chi_l, \chi_{\bar{\jmath}} \rangle = i \alpha_{i\bar{\jmath}}^{\bar{k}} G_{l\bar{k}} e^{-K} , \qquad \forall \; l, i, \bar{\jmath} \, .$$

Since the Weil-Petersson metric is not degenerate we conclude that $\alpha_{i\bar{\jmath}}^{\bar{k}} \equiv 0$ and hence

$$D_i \chi_{\bar{\jmath}} = G_{i\bar{\jmath}} \bar{\Omega}_n \, . \tag{2.5.16}$$

Similarly we show that the $\alpha_{i\bar{\jmath}}^{k}$ in the ansatz

$$D_i \chi_j = \alpha_{ij}^{k} \chi_k + \beta_{ij}{}^{\bar{k}} \chi_{\bar{k}} \tag{2.5.17}$$

vanish by $0 = \langle D_i \chi_j, \chi_{\bar{\jmath}} \rangle = i \alpha_{ij}^{k} G_{k\bar{\jmath}} \; \forall i, j, \bar{\jmath}$.

Further determination of D_i on basis elements involve the n-point coupling and its factorisation by the Frobenius algebra and depends on the dimension. We turn to the case $n = 3$ now.

2.5.2 Special Geometry on the Complex Moduli Space $\mathcal{M}$ of CY 3-Folds

Special geometry is the principle differential geometrical structure on the varying Hodge structure over the complex moduli space $\mathcal{M}_{cs}$ of Calabi-Yau 3-folds. It is characterized by the existence of a holomorphic prepotential $\mathcal{F}$ from which the Kähler potential for the Weil-Petersson metric on $\mathcal{M}_{cs}$ and the holomorphic triple couplings derive. From the curve counting perspective all data involved are genus zero data.

What is specific about 3-folds is that the $\beta_{ij}^{\bar{k}}$ terms in (2.5.17) are expressible by the 3-point functions and the simplification relative to higher dimensional Calabi-Yau manifolds is that the 3-point functions are, unlike higher point functions, not factorizable in lower point functions. This is familiar in 2d field theory. The relations between the two descriptions is that the 2d field theories—more precisely the (2, 2) superconformal theories discussed in section 2.3.1, for M_n a Calabi-Yau space—is the σ-model on the string world-sheet of strings propagating in the target M_n.

Since we seen already that $\alpha_{ij}^{k} = 0$ for any n, it remains to evaluate the $\beta_{ij}^{\bar{k}}$ term in (2.5.17) via the bi-linear

$$\beta_{ij}{}^{\bar{p}} e^{-K} G_{\bar{p}k} = -i\langle \chi_i, D_j \chi_k\rangle = -i\langle \partial_i \Omega, D_j(\partial_k + K_k)\Omega\rangle = i\langle \Omega, \partial_i \partial_j \partial_k \Omega\rangle = -iC_{ijk}\,. \tag{2.5.18}$$

Hence $\beta_{ij}{}^{\bar{p}} = -e^K C_{ijk} G^{k\bar{k}}$ and we can summarize this with together with the generic equations derived in the previous section, the Tian-Todorov theorem and the non-degeneracy of the metric and the 3-point functions as

$$\begin{array}{c|c} \text{Appl. } D_i & \text{spans} \\ \Omega & H^{3,0} \\ D_i\Omega = (\partial_i + K_i)\Omega = \chi_i & H^{2,1} \\ D_i\chi_j = -ie^K C_{ijk} G^{k\bar{k}} \chi_{\bar{k}} & H^{1,2} \\ D_i \chi_{\bar{k}} = G_{i\bar{k}}\bar{\Omega} & H^{0,3} \\ D_i\bar{\Omega} = 0 & \emptyset. \end{array} \tag{2.5.19}$$

We can evaluate now easily the left handside of (2.5.13)

$$[D_i, D_{\bar{j}}]\chi_k = G_{k\bar{j}}\chi_i - e^{2K} C_{\bar{j}\bar{m}\bar{n}} G^{m\bar{m}} G^{n\bar{n}} C_{ikm}\chi_n \tag{2.5.20}$$

and conclude from (2.5.13) that

$$[D_i, D_{\bar{j}}]^k{}_l = -R_{i\bar{j}}{}^k{}_l = \partial_{\bar{j}}\Gamma^k_{il} = \delta^k_l G_{\bar{j}i} + \delta^k_i G_{\bar{j}l} - C^{km}_{\bar{j}} C_{ilm}\,. \tag{2.5.21}$$

In (2.5.21) the an-holomorphic $C^{kl}_{\bar{j}}$ are defined as

$$C^{kl}_{\bar{j}} = e^{2K} C_{\bar{j}\bar{k}\bar{l}} G^{k\bar{k}} G^{l\bar{l}}, \tag{2.5.22}$$

with $C_{\bar{j}\bar{k}\bar{l}} = (C_{jkl})^*$ are a section of

$$C^{kl}_{\bar{j}} \in \mathcal{L}^{-2} \otimes T^*_{0,1} \otimes \mathrm{Sym}^2(T^{1,0}) \tag{2.5.23}$$

2.5.3 Special Geometry and Periods

Firstly in the discussion of the complex structure moduli space integrals like (2.3.40) and (2.3.42) play a crucial role, because they are used directly in the definition of the metric on the moduli space (2.5.1) as well as in the vanishing conditions and the structure constants (2.5.10), that play an important role in the additional differential geometric structure on the moduli space. Just as recapitulation in simple notation: on a Calabi-Yau 3-fold the Kähler potential and the triple couplings have the geometrical definition as integrals

$$e^{-K}=i\int_M \Omega\wedge\bar\Omega, \qquad C_{ijk}=\int_M \Omega\wedge\partial_i\partial_j\partial_k\Omega. \tag{2.5.24}$$

Here the ∂_i are derivatives w.r.t. to any complex structure parametrization. We have already seen that Griffiths tranversality implies that no covariant derivatives are needed in (2.5.24) and it follows that C_{ijk} is a section of $\mathrm{Sym}_3(T^*_{1,0}(\mathcal{M}_{cs}))\otimes\mathcal{L}^2$. Using this and (2.5.5) we can also write the metric $G_{i\bar j}=\partial_i\bar\partial_{\bar j}K$ entirely using integrals

$$G_{i\bar j}=i^{-n^2}e^K\int_{M_n} D_i\Omega_n\wedge\bar D_{\bar j}\bar\Omega_n=-\frac{\int_M \chi_i\wedge\bar\chi_{\bar j}}{\int_{M_n}\Omega_n\wedge\bar\Omega_n}=-\frac{\langle\chi_i,\bar\chi_{\bar j}\rangle}{\langle\Omega_n,\bar\Omega_n\rangle}. \tag{2.5.25}$$

Note the minus sign, which together with (2.3.42) implies that the eigenvalues of the non-degenerate metric $G_{i\bar j}$ are all positive. This is physically significant as $G_{i\bar j}$ multiplies the kinetic terms of the moduli fields ϕ_i associated to the complex structure deformations.

To actual calculate these integrals and their complex moduli dependence one picks a fixed topological basis on $H^n(M_n,\mathbb{Z})$ and a dual basis on $H_n(M_n,\mathbb{Z})$. In particular we can give concrete formulas for the Kähler potential and the triple coupling based on the expansion of $\Omega(z)$ in terms of the symplectic basis α_I,β^I $I=0,\ldots,h_{21}=h$ defined in (2.4.37)

$$\Omega=X^I(z)\alpha_I-F_I(z)\beta^I\,. \tag{2.5.26}$$

By (2.5.24), (2.4.37) one gets in terms of the periods

$$e^{-K}=i\int_M\Omega\wedge\bar\Omega=i(\bar X^I F_I-X^I\bar F_I)=i\vec\Pi^\dagger\,\Sigma\,\vec\Pi, \tag{2.5.27}$$

where we defined as specialisation of the bilinear η on the horizontal part of the middle homology for 3 dim Calabi-Yau manifolds the symplectic $2h\times 2h$ matrix Σ as

$$\Sigma=\begin{pmatrix}\mathbf{0} & \mathbf{1}_{h\times h}\\ -\mathbf{1}_{h\times h} & \mathbf{0}\end{pmatrix}. \tag{2.5.28}$$

Similarly the 3 point functions can be expressed in terms of the periods as

$$C_{ijk} = X^I \partial_i \partial_j \partial_k F_I - F_I \partial_i \partial_j \partial_k X^I = -X_i^I X_j^J X_k^K F_{IJK} \,. \tag{2.5.29}$$

The last equality in (2.5.29) requires some calculation.[38] Indeed due to the local Torelli theorem, the X^I can be used as homogenous coordinates on the moduli space, or as coordinates on the big moduli space parametrizing complex structures W together with a nonvanishing Ω. The transversality (2.3.41)

$$\langle \Omega I \partial_K \Omega \rangle = F_K - X^I \partial_K F_I = 0 \tag{2.5.30}$$

implies

$$2F_K = \partial_K (X^I F_I) \,. \tag{2.5.31}$$

Taking the prepotential to be

$$F = \frac{1}{2} \left(X^I F_I \right), \tag{2.5.32}$$

we see that the $F_K = \partial_K F$ are completely determined from F and therefore not independent of each other. The transversality (2.5.30) also leads to the Euler equation $F_K = X^I \partial_K F_I = X^I \partial_I F_K$, from which one concludes that F_I is homogeneous of degree one and F is of degree two, $2F = X_I \partial_I F$. These homogeneities can also inferred from the effect of a constant rescaling of Ω, which induces the same scaling of all the X^I.

2.5.4 The Big Moduli Space

Returning to (2.5.29) and viewing the X^I as coordinates on the big moduli space, we compute

$$C_{IJK} = \int_W \partial_I \partial_J \partial_K \Omega \wedge \Omega = X^L F_{IJKL} = -F_{IJK}, \tag{2.5.33}$$

the last equality holding since F_{IJK} is homogeneous of degree -1. From this (2.5.29) follows immediately.

[38] We use $\partial_i = \frac{\partial}{\partial z_i}$ ($\bar\partial_{\bar i} = \frac{\partial}{\partial \bar z_{\bar i}}$) denotes partial derivatives w.r.t. generic complex structure (small moduli space) coordinates ($i = 1, \ldots, h_{21}$) and $\partial_I = \frac{\partial}{\partial X^I}$ for the derivatives w.r.t. to homogeneous moduli space (big moduli space) coordinates $I = 0, \ldots, h_{21}$ and abbreviate $\partial_I F = F_I$ etc.

We consider the X^I as functions of the local coordinates on $\mathcal{M}$ together with X_0 in writing the partial derivatives X_i^I, etc.

In view of the symplectic pairing on $H_3(W, \mathbb{R}^3)$ one can view X^I as positions and

$$P_I := F_I = \frac{\partial}{\partial X^I} F \tag{2.5.34}$$

as the dual momenta.

Again due to the local Torelli theorem, the $(b_3/2) \times (b_3/2)$ matrix

$$X = (X^I_\alpha) = (D_\alpha X^I) \tag{2.5.35}$$

is generically invertible, where we have formally set $D_0 X^I = X^I$. We denote the components of the inverse matrix by X^α_I. The rigid supergravity metric in the large moduli space derives from Kähler potential

$$\mathcal{K} = \frac{i}{2}\left(\bar{X}^I F_I - X^I \bar{F}_I\right) = \frac{i}{2}\vec{\Pi}^\dagger \,\Sigma\, \vec{\Pi}\,. \tag{2.5.36}$$

as

$$\mathcal{G}_{IJ} = \partial_I \bar{\partial}_{\bar{J}} \mathcal{K} = \frac{i}{2}(F_{IJ} - \overline{F_{IJ}}) = -\mathrm{Im}\,(F_{IJ})\ . \tag{2.5.37}$$

It follows that the Christoffel symbols of the metric connection are

$$\Gamma^K_{IJ} = \frac{i}{2} C^K_{IJ} = \frac{i}{2} C_{IJL}\mathcal{G}^{KL}\ . \tag{2.5.38}$$

We now view the X^I as local coordinates on the big moduli space and define

$$\hat{\Omega}_I = \frac{\partial \Omega}{\partial X^I} = \alpha_I - F_{IJ}\beta^J = \alpha_I - \tau_{IJ}\beta^J \tag{2.5.39}$$

as a new frame for $\mathcal{H}^{3,0} \oplus \mathcal{H}^{2,1}$. In the last equality of (2.5.39) we have used the standard notation for the normalized periods τ_{IJ}. Note that the period matrix is now normalized in this frame so that $\int_{A^J} \hat{\Omega}_I = \delta^J_I$, while $\int_{B_J} \hat{\Omega}_I = \tau_{IJ}$. Also, the $\bar{\hat{\Omega}}_{\bar{I}} = \alpha_I - \bar{\tau}_{IJ}\beta^J$ give a frame for $\mathcal{H}^{2,1} \oplus \mathcal{H}^{0,3}$. We have

$$\langle \hat{\Omega}_I, \overline{\hat{\Omega}_J} \rangle = 2i\,\mathrm{Im}(\tau_{IJ})\ . \tag{2.5.40}$$

From Riemann's inequality (2.3.42) we see that τ has signature $(h^{21}, 1)$, i.e. h^{21} positive eigenvalues and one negative eigenvalue.

In order to get a (negative) definite $(b_3/2) \times (b_3/2)$ bilinear form, one infers from (2.3.42) that one has to construct a frame for $\mathcal{H}^{3,0} \oplus \mathcal{H}^{1,2}$. This is of the general form $\hat{\Theta}_I = \alpha_I - \mathcal{N}_{IJ}\beta^J$ where the non-holomorphic $\mathcal{N}_{IJ}$ is simply determined by the transversality (2.3.41) of $\hat{\Theta}_I$ to $\mathcal{H}^{3,0} \oplus \mathcal{H}^{1,2}$ itself

$$\langle \hat{\Theta}_I, \Omega \rangle = 0, \qquad \langle \hat{\Theta}_I, \Xi_J^{(12)} \rangle = 0\,. \tag{2.5.41}$$

Here the $\Xi_J^{(12)}$ are obtained by projecting the basis $\hat{\bar{\Omega}}_J$ to a redundant set of generators for H^{21}, i.e. $\Xi_J^{(12)} = \hat{\bar{\Omega}}_J - \frac{\langle \hat{\bar{\Omega}}_J, \Omega \rangle}{\langle \bar{\Omega}, \Omega \rangle} \bar{\Omega}$. We expand (2.5.41) as

$$-F_I + X^J \mathcal{N}_{IJ} = 0, \qquad -\overline{\tau_{IJ}} + \mathcal{N}_{IJ} - \frac{\left(-F_J + X^K \overline{\tau_{KJ}}\right)\left(-\overline{F_I} + \overline{X^K} \mathcal{N}_{IK}\right)}{-\bar{X}^K F_K + X_K \overline{F_K}} = 0, \tag{2.5.42}$$

where we have used

$$\langle \hat{\bar{\Omega}}_J, \Omega \rangle = -F_J + X^K \overline{\tau_{KJ}}\,. \tag{2.5.43}$$

Upon writing $-\overline{F_I} + \overline{X^J} \mathcal{N}_{IJ} = \overline{X^J}\,(\mathcal{N}_{IJ} - \overline{\tau_{IJ}})$ and similarly replacing F_J, $\bar{F}_J$ with $X^I \tau_{IJ}$ and $\bar{X}^I \bar{\tau}_{IJ}$ elsewhere, these conditions are easily solved in the form

$$\mathcal{N}_{IJ} = \bar{\tau}_{IJ} + \frac{N_{IL} N_{JK} X^L X^K}{N_{KL} X^K X^L}\,, \tag{2.5.44}$$

where $N_{IJ} = \tau_{IJ} - \overline{\tau_{IJ}}$.

Physically, this definition is related to the positivity of the gauge kinetic terms of $N = 2$ vectors including the graviphoton, which are $-\frac{1}{8}(\mathrm{Im}(\mathcal{N}_{IJ}) F^I_{nm} F^{nmJ} + i\mathrm{Re}(\mathcal{N}_{IJ}) F^I_{nm} F^{*nmJ})$. Mathematically $\mathcal{N}_{IJ}$ describes the period matrix of the *Weil intermediate Jacobian*, while τ_{IJ} describes the period matrix of the *Griffiths intermediate Jacobian*.

2.5.5 Inhomogenous Flat Coordinates on the Small Moduli Space

The small moduli space can be parametrized by the flat inhomogeneous coordinates (2.4.40). Here we study the relation between the large and the small moduli space. Indeed it is an easy exercise to give the analogeneous equations to (2.5.27) and (2.5.29) in these coordinates.

The key point is the homogeneity of the prepotential of degree two in terms of the projective coordinates (2.5.31), which we proved from Griffiths transversality. When we pass to inhomogenous coordinates $t^a = X^0/X^a$ we can express the quantities (2.5.27) and (2.5.29) with[39]

$$\mathcal{F}(t) = F(X)/(X^0)^2 \tag{2.5.45}$$

locally as

$$e^{-K} = i[(t^{\bar{\imath}} - t^i)(\mathcal{F}_i + \mathcal{F}_{\bar{\imath}}) + 2(\mathcal{F} - \bar{\mathcal{F}})], \qquad C_{ijk} = \mathcal{F}_{ijk} . \tag{2.5.46}$$

We recall that the ratios $t^i = X^i/X^0$, $i = 1, \ldots, h$ are defined by A-cycle periods $X^I = \int_{A_I} \Omega$, $I = 0, 1, \ldots h$. Here the A-cycles are defined to be a basis of the symplectic basis of no-intersecting cycles defined in (2.4.37). Ω is the unique nowhere vanishing holomorphic $(n, 0)$ form and we use the notation $\mathcal{F}_i = \partial_i \mathcal{F} = \frac{\partial \mathcal{F}}{\partial t^i}$, $\bar{\mathcal{F}} = \mathcal{F}^*$ and $\mathcal{F}_{\bar{\imath}}$ is $(\mathcal{F}_i)^*$, as the complex conjugates.

In fact we derived Eq. (2.5.21) in the inhomogenous but not flat coordinates. This equation can be viewed as a *necessary integrability condition* for the existence of the holomorphic prepotential $\mathcal{F}(t)$ in inhomogeneous coordinates so that the Kählerpotential and the triple couplings derive from it as in (2.5.46).

It is also useful to express the period vector (2.4.38) in terms of the inhomogenous holomorphic prepotential $\mathcal{F}(t)$ as

$$\vec{\Pi} = \begin{pmatrix} F_I \\ \\ X^I \end{pmatrix} = X^0 \begin{pmatrix} 2\mathcal{F}(t) - t^i \partial_i \mathcal{F}(t) \\ \partial_i \mathcal{F}(t) \\ 1 \\ t^i \end{pmatrix} . \tag{2.5.47}$$

We can easily obtain an holomorphic version of (2.5.19) by noting that the n-form can be expanded in terms of periods and in the gauge in the Kähler line bundle $\mathcal{L}$ $X^0 = 1$, we call it $\hat{\Omega}$, and get according to (2.5.26)

$$\hat{\Omega}_0 = \alpha_0 + t^a \alpha_a - \partial_a \mathcal{F} \beta^a - (2\mathcal{F} - t^a \partial_a \mathcal{F}) \beta^0 . \tag{2.5.48}$$

Now we can complete this to a basis in $H^3(M, \mathbb{C})$ by writing the further forms in this gauge as

$$\begin{aligned} \hat{\chi}_a &= \alpha_a - \partial_a \partial_b \mathcal{F} \beta^b - (\partial_a \mathcal{F} - t^b \partial_b \partial_a \mathcal{F}) \beta^0 \\ \hat{\chi}^a &= -\beta^a + t^a \beta^0 \\ \hat{\Omega}^0 &= \beta^0 , \end{aligned} \tag{2.5.49}$$

[39]Often the homogeneous—$F(X)$ and the inhomogenous prepotential $\mathcal{F}(t)$ carry an index 0 to indicate that they are encoding the genus theory world-sheet instanton contributions of the A-model. This corresponds to the physical gauge, which is discussed in more detail in Sect. 4.3.

and expressing the holomorphic version of (2.5.26) as

$$\partial_a \begin{pmatrix} \hat{\Omega}_0 \\ \hat{\chi}^b \\ \hat{\chi}_b \\ \hat{\Omega}^0 \end{pmatrix} = \begin{pmatrix} 0 & \delta_a^c & 0 & 0 \\ 0 & 0 & C_{abc} & 0 \\ 0 & 0 & 0 & \delta_a^b \\ 0 & 0 & 0 & 0 \end{pmatrix} \begin{pmatrix} \hat{\Omega}_0 \\ \hat{\chi}^c \\ \hat{\chi}_c \\ \hat{\Omega}^0 \end{pmatrix} . \tag{2.5.50}$$

This can be seen as the first order form of the holomorphic Picard-Fuchs operators and the nilpotent matrix C_a on the r.h.s can be viewed as a very special case the nilpotent connection of the Gauss Manin operator ∇^{GM}, which can be written as $\partial_a - C_a$, but it is far from being equivalent as the latter needs neither the unique $(n, 0)$ forms nor special geometry.

The situation becomes more interesting if we consider higher dimensional Calabi-Yau manifolds. We will address this situations after the discussion of the twisted world-sheet $N = (2, 2)$ theory where we can assign a natural charge grading to vectors like $(\hat{\Omega}_0, \hat{\chi}_b, \hat{\chi}_b, \hat{\Omega}^0)$ and use the Frobenius structure of topological $N = (2, 2)$ field theory in Sect. 3.6.2.

2.5.6 The Attractor Equations

Consider a dyonic extremal black hole in $N = 2$ supergravity. It has a general dyonic charge, i.e. eventually non-vanishing electric (q_L) and magnetic charge (p^L), that comes from a three-brane wrapping the corresponding internal real dimensional three-cycle in $H_3(M, \mathbb{Z})$ in the Calabi-Yau complex three-fold M. This charge can be expanded in terms of the periods $\Pi^T = (X^L, F_L)$ over the basis (α_I, β^I), $I = 0, \ldots, h_{2,1}$ of the lattice $H_3(M, \mathbb{Z})$

$$Q = q_L X^L - p^L F_L . \tag{2.5.51}$$

Here we mean the holomorphic periods $\partial_{\bar{\imath}} \Pi_k = 0$ with Kähler weights $(1, 0)$ unlike the covariantly holomorphic periods $\hat{\Pi}_k = e^{K/2} \Pi_k = (L^I, M_I)$ with Kähler weights $(\frac{1}{2}, \frac{1}{2})$ [112], which fulfill $D_{\bar{\imath}} \hat{\Pi}_k = 0$. The mass square of the BPS state, which is Kähler transformation invariant, is then given by the charge Q as

$$M_{BPS}^2 = e^K |Q|^2 . \tag{2.5.52}$$

Despite the fact that the black hole solution depends on the values of the scalar fields at infinity, the property of black holes to have no hair, i.e. the fact that their near horizon geometry is described completely in terms of their charges, which in the BPS case also describes their mass, is realized by the so called attractor mechanism [114]. The statement is, that independent of multiplet scalar the values of scalar vev at infinity a damped attractor equation governs their value at the horizon. The attractor point in the vacuum space of the scalars in the vector multiplets is determined by the charges. This realizes in particular the property that the near horizon geometry is independently of the values of scalar vev at infinity and completely determined by the charges.

The essence of this mechanism is that the scalar moduli minimize M^2_{BPS} at the attractor point under the additional constraint

$$e^{-K} = 1\,. \tag{2.5.53}$$

Since this is a manifestly real condition one expects the Lagrange multiplier λ [A] to have a fixed phase and λ turns out to be purely imaginary. If we vary $M^2_{BPS} = e^K|Q|^2$ w.r.t to X^K or $\bar X^K$, without restriction we get zero. Therefore we introduce a small real parameter ϵ so that $M^2_{BPS}(\epsilon) = e^{(1+\epsilon)}K|Q|^2$ and consider the variational equations $\frac{\partial}{\partial X^k} M^2_{BPS}(\epsilon) = \epsilon\lambda \frac{\partial}{\partial X^k}(e^{-K} - 1)$ as well as its complex conjugate and consider the $\epsilon \to 0$ limit, i.e. the first order in ϵ. The equations for the variation under the constraint become

$$\begin{aligned}
\frac{\partial M^2_{BPS}(\epsilon)}{\partial X^K} &= \frac{\bar Q(q_K - p^L F_{KL})}{i^{1+\epsilon}(\bar X^L F_L - X^L \bar F_L)^{1+\epsilon}} + \frac{(1+\epsilon)|Q|^2(\bar F_k - \bar X^L F_{KL})}{i^{1+\epsilon}(\bar X^L F_L - X^L \bar F_L)^{2+\epsilon}} = \epsilon\lambda(\bar F_k - \bar X^L F_{KL})\\
\frac{\partial M^2_{BPS}(\epsilon)}{\partial \bar X^K} &= \frac{Q(q_K - p^L \bar F_{KL})}{i^{1+\epsilon}(\bar X^L F_L F_L - X^L \bar F_L)^{1+\epsilon}} - \frac{(1+\epsilon)|Q|^2(F_k - X^L \bar F_{KL})}{i^{1+\epsilon}(\bar X^L F_L - X^L \bar F_L)^{2+\epsilon}} = -\epsilon\lambda(F_k - X^L \bar F_{KL})\,,
\end{aligned} \tag{2.5.54}$$

so that λ has to be purely imaginary for the two equations to be compatible. Do get (2.5.54) we used $e^{-K} = i(\bar X^L F_L - X^L \bar F_L)$ and $F_L = \frac{\partial}{\partial X^K} F$, $F_{KL} = \frac{\partial^2}{\partial X^K \partial X^L} F$. Further we note the holomorphic prepotential F is homogenous of degree 2 in X^L hence by the Euler-Equations $2F = X^L \frac{\partial}{\partial X^L} F$ and $F_K = X^L F_{KL}$. Moreover F_{KL} is at generic points invertible to F^{KL} with $F^{KL} F_{LM} = \delta^K_M$, so we also have $X^K = F^{KL} F_L$. Indeed contracting the first equation (2.5.54) with X^L or the second with $\bar X^L$ and using the above homogeneity identities we get in both cases a pure imaginary λ

$$\lambda = i e^K |Q|^2\,. \tag{2.5.55}$$

We can summarize the conditions with $\mu = 2ie^K \bar Q$ as

$$(q_K - p^L \bar F_{KL}) = \frac{\mu}{2}(F_k - X^L \bar F_{KL})\,. \tag{2.5.56}$$

Taking the imaginary part of (2.5.56) and the imaginary part of the contraction of (2.5.56) with F^{LK} we get the attractor equations

$$p^L = \mathrm{Re}(\mu X^L), \qquad q_L = \mathrm{Re}(\mu F_L)\,. \tag{2.5.57}$$

We can give a slightly more geometric interpretation of these equations. Let

$$\gamma = p^I \alpha_I - q_I \beta^I \in H^3(M,\mathbb{Z}), \qquad C = q_I A^I - p^I B_I \in H_3(M,\mathbb{Z}) \tag{2.5.58}$$

the forms which correspond to the internal parts of the 10d anti-selfdual field strength $F \in \Gamma(\Lambda^2(M_4) \otimes H^3(M,\mathbb{Z}))$, whose 4d part is the graviphoton or the vector multiplet field strength in Minkowski 4-space, i.e. in $\Gamma(\Lambda^2(M_4))$ and $C \in H_3(M,\mathbb{Z})$ the dual 3-cycle in the internal space which is wrapped by the source brane. Then fixing these data the complex structure of M adjust itself at the attractor point so that the real form

$$\gamma = \gamma^{3,0} + \gamma^{0,3} \tag{2.5.59}$$

is purely of type $(3,0)$ and $(0,3)$. To see the equivalence we note that (2.5.59) simply means that the $(3,0)$ projection of $\gamma^{3,0}$ is proportional to Ω. We identify the proportionality factor as $\gamma^{3,0} = \frac{\mu}{2i}\Omega$ then

$$\mathrm{Im}(\mu\Omega) = \gamma \tag{2.5.60}$$

and integration of this equation over the basis A^I and B_I gives (2.5.57).

2.6 *Action of the Monodromy* $\Gamma \in \mathrm{Sp}(h_3(W),\mathbb{Z})$

The most important structure in the moduli space that characterizes the concrete expressions that appear in special geometry as introduced so far for a concrete complex family $W(z)$ is its monodromy group. In a rather precise sense one can identify the amplitudes that appear in special geometry and its generalization to higher genus as modular forms of this monodromy group. Let us therefore describe the transformation properties of the quantities introduced above.

2.6.1 General Form of the Monodromy Action

The analysis of the monodromy group is conceptually straightforward, but can be technically involved. Let us assume we have specified a suitable resolution of $\mathcal{M}_{cs}$ so that all critical loci of the Picard-Fuchs system are normal crossing divisors $D_k \in \mathcal{M}_{cs}$. We specify a base point at which we fix periods $\vec{\Pi}$, Π, and an orientation of the paths γ_k that encircle these divisors. Then the period vector $\vec{\Pi}$ and the period matrix Π transform under transport along the path γ_k as[40]

$$\vec{\Pi} \mapsto \vec{\Pi}_{\gamma_k} = \gamma_k \vec{\Pi}, \qquad \Pi \mapsto \Pi_{\gamma_k} = \Gamma_k \Pi\,. \tag{2.6.1}$$

[40]We denote the path and the monodromy matrix action on the period vector $\vec{\Pi}$ by the same character γ. The matrix Γ acts on the period matrix Π.

All monodromy matrices γ leave the bilinear integer pairing invariant, i.e. in terms of the intersection matrix Σ we get the characterisation of the possible monodromy matrices

$$\gamma^T \Sigma \gamma = \Sigma \,. \tag{2.6.2}$$

The monodromy group is redundantly generated by the γ_k, $k = 1, \ldots, \#(D)$. In odd dimensions all monodromies γ_k are symplectic, so that the monodromy group of the family $W(z)$ is a subgroup $\Gamma_W \subset \mathrm{Sp}(N, \mathbb{Z})$.[41] Let us write $\gamma = \begin{pmatrix} \mathfrak{d} & \mathfrak{c} \\ \mathfrak{b} & \mathfrak{a} \end{pmatrix}$. Then $\mathfrak{a}, \ldots, \mathfrak{d}$ are integer valued $N/2 \times N/2$ matrices satisfying

$$\mathfrak{a}^T \mathfrak{d} - \mathfrak{c}^T \mathfrak{b} = \mathbf{1}, \quad \mathfrak{a}^T \mathfrak{c} = \mathfrak{c}^T \mathfrak{a}, \quad \mathfrak{b}^T \mathfrak{d} = \mathfrak{d}^T \mathfrak{b} \,. \tag{2.6.3}$$

The components of the period matrix transform as

$$X_\alpha'^{\,I} = \mathfrak{d}^I_J X^J_\alpha + \mathfrak{c}^{IJ} F_{\alpha J}, \qquad F'_{\alpha I} = \mathfrak{b}_{IJ} X^J_\alpha + \mathfrak{a}_I^J F_{\alpha J} \tag{2.6.4}$$

respectively. We have $N = b_3(W)$ for a family of Calabi-Yau 3-folds $W(z)$ and $N = 2g$ for a family of genus g Riemann surfaces $W(z) = \Sigma_g(z)$. By (2.6.4) the $\tau = (\tau_{IJ})$ matrix transforms as

$$\tau \mapsto \tau_\gamma = (\mathfrak{a}\tau + \mathfrak{b})(\mathfrak{c}\tau + \mathfrak{d})^{-1} \,. \tag{2.6.5}$$

The same transformation rule applies to the τ-matrix of a genus g curve and $\mathcal{N} = (\mathcal{N}_{IJ})$. We can see the latter either by transforming the equations (2.5.44) by (2.6.4) and (2.6.5) and then reading off the new solution, or more conceptually by noting that these transformation laws hold for the periods of $\hat{\Theta}_I$ as they do for the periods of any $(b_3/2)$-dimensional space of normalized three-forms, by a calculation analogous to that for $\hat{\Omega}_I$. The transformation properties of the corresponding metrics follow from $\mathrm{Im}(\tau) = (\mathrm{Im}(\tau)_{IJ}) = -(\mathcal{G}_{IJ})$

$$\mathrm{Im}(\tau) \mapsto \mathrm{Im}(\tau_\gamma) = \left((\mathfrak{c}\bar{\tau} + \mathfrak{d})^T\right)^{-1} \mathrm{Im}(\tau)\,(\mathfrak{c}\tau + \mathfrak{d})^{-1} \tag{2.6.6}$$

hence for $\mathrm{Im}(\tau)^{-1} = (\mathrm{Im}(\tau)^{IJ}) = -(\mathcal{G}^{IJ})$ we get

$$\mathrm{Im}(\tau)^{-1} \mapsto \mathrm{Im}(\tau_\gamma)^{-1} = (\mathfrak{c}\tau + \mathfrak{d})\mathrm{Im}(\tau)^{-1}(\mathfrak{c}\tau + \mathfrak{d})^T - 2i(\mathfrak{c}\tau + \mathfrak{d})\mathfrak{c}^T \,. \tag{2.6.7}$$

[41]In the Calabi-Yau 3 fold case Γ_W can have either finite or infinite index in $\mathrm{Sp}(h_3(W), \mathbb{Z})$.

2.6.2 Monodromy Types and Choice of Local Flat Coordinates

The choice of the homogeneous and inhomogenous coordinates consist of two choices, which depends on the locus in the moduli space, where they are to be defined. The choices are

1. picking A-cycles A_I, $I = 0, \ldots, h_{21}(M)$, i.e. cycles in $H_3(M)$ which do not intersect each other $A_I \cap A_J = 0$.
2. Picking A_0 and hence X^0.

This choice is important near critical divisors D_i, $i = 1, \ldots, k$ in the moduli space, around which the periods develop monodromy. Without restriction of generality we can assume that after a suitable blow up of the moduli space all D_i have only normal crossings. Now focus on one divisor D_i. The principle properties of a monodromy matrix γ_i around a divisor D are captured by the minimal integer $k > 0$ and $p \geq 0$ in the equation

$$(\gamma_i^k - \mathbf{1})^{p+1} = 0 \,. \tag{2.6.8}$$

Here p is the smallest integer so that the r.h.s. is zero. For $p = 0$, $k > 1$ one has an $\mathbb{Z}_k$ orbifold singularity. The cases $k = 1$ are the unipotent cases. The conifold in n odd has $p = 1$ and the most relevant case for mirror symmetry is the maximal unipotent case $p = n$. In particular one can show that

$$p \leq n \,. \tag{2.6.9}$$

This implies that the number highest power of logarithm in the solutions is n. E.g. the maximal degeneration of an elliptic curve case can be characterize it as cusp point in the moduli space.

We will make the choice (2) as follows. If the monodromy is of finite order, we diagonalize it and take for X^0 the eigenvector, which vanishes fastest as we approach the critical divisor. If the monodromy is a shift of infinite order, which will we take for X_0 the period over the vanishing cycle, which multiplies the logarithms that create the shift.

2.6.3 Monodromy and Degeneration of the Calabi-Yau Manifold

The local monodromy and the corresponding degeneration of the Calabi-Yau manifolds are of great interest in mathematics and physics. For example in the $SU(N)$ $\mathcal{N} = 2$ Seiberg-Witten geometry, which can be embedded in Calabi-Yau 3-folds [205, 214] one has in the asymptotic free region of the gauge theory with nearly massless "electrically" charged gauge bosons $k = N$ and $n = 1$ in (2.6.8), in regions, where magnetic monopoles become massless one has $k = n = 1$, while in regions with massless electric and magnetic degrees of freedom signal a conformal theory typically one has orbifold singularities [21].

The degeneration of the Hodge structure is also studied extensively in mathematics. It is still a local analysis and as such a linear analysis. In order to enjoy the full modularity properties of the amplitudes one has to explore the moduli space globally and figure out which subgroup $\Gamma_M \subset \mathrm{Sp}(h_3(M), \mathbb{Z})$ is generated by all monodromies. An intermediate concept is local mirror symmetry, which deals with a sub-monodromy problem of the finite periods in a semi local limit, where part of M_n decompactifies. Such limits are typical for a manifold W whose mirror exhibit a fibration structure, see for example the occurrence of $\mathrm{Sl}(2, \mathbb{Z})$ monodromy in the elliptically fibered Calabi-Yau manifolds in [195].

As a local analysis we recall as a simple example the limiting mixed Hodge structure for the unipotent cases, see [22, 86, 256] for reviews. Let $z_i \sim \Delta_i$ be the coordinates on the ith punctured disk D_i° so that $(D^\circ)^r$ is a local neighborhood of the r dimensional moduli space at a normal crossing point. The mixed case is reduced to unipotent case by changing to multicover variables $z^{\frac{1}{k}} \mapsto z$. Since $N_i = M_i - \mathbf{1}$ is nilpotent one can define the Lie algebra generator $\mathfrak{N}_i = \log(M_i)$ as a finite expansion in N_i. N_i and $\mathfrak{N}_i$ have the same kernel and cokernel. Schmid's nilpotent orbit theorem [293] provides an extension $\overline{\mathcal{F}}^p$ of $\mathcal{F}^p$ from $(D^\circ)^r$ to the product of full disks D^r. In particular Schmid extends the Gauss Manin connection to a map $\overline{\nabla} : \overline{\mathcal{F}}^0 \to \overline{\mathcal{F}}^0 \otimes \Omega^1_{\overline{\mathcal{M}}_{cs}}(\log(\Delta_i))$. Here the sheaf of rational one forms $\Omega^1_{\overline{\mathcal{M}}_{cs}}(\log(\Delta_i))$ is in the case of l unipotent divisors locally generated by $\frac{dz_1}{z_1}, \ldots, \frac{dz_l}{z_l}, dz_{l+1}, \ldots, dz_r$. On D^r he introduces single valued periods

$$\Pi_S(z) = \exp\left(-\frac{1}{2\pi i}\log(z_i)\mathfrak{N}_i\right)\Pi(0)\,, \qquad (2.6.10)$$

extends that to a section of $\overline{\mathcal{F}}^p$ and shows that this is a leading order approximation to the periods on the disk.

An important implication of the work of [87, 293] is the ability to define the *limiting mixed Hodge structure*, which describes how the integral Hodge structure of the singular model sits inside the integral Hodge structure of the smooth compact manifold. In particular at $0 \in (D^\circ)^r$ there is a limiting Hodge filtration $\overline{\mathcal{F}}^\bullet = \mathcal{F}^\bullet_{\mathrm{lim}}$ with $\mathfrak{N}_j(\mathcal{F}^p_{\mathrm{lim}}) \subset \mathcal{F}^{p-1}_{\mathrm{lim}}$. Both $\mathfrak{N}_i$ and the extension $\overline{\nabla}_{\theta_{z_j}}(\mathcal{F}^p_{\mathrm{lim}}) \subset \mathcal{F}^{p-1}_{\mathrm{lim}}$ induce a linear map $\mathcal{F}^p_{\mathrm{lim}}/\mathcal{F}^{p-1}_{\mathrm{lim}} \mapsto \mathcal{F}^{p-1}_{\mathrm{lim}}/\mathcal{F}^p_{\mathrm{lim}}$ and are identified as $\overline{\nabla}_{\theta_{z_j}} = -\frac{1}{2\pi i}\mathfrak{N}_j$. Moreover if $\Pi_S(z)$ is a multivalued flat integer section of $\mathcal{H}_\mathbb{Z}$ then $\Pi_S(0)$ is an integral element over 0. The mixed Hodge structure comes from the *monodromy weight filtration* $W_\bullet$ with $W_0 \subset W_1 \subset \ldots \subset W_{2n} = H^n(M_t, \mathbb{C})$. For any linear combination $\mathfrak{N}$ of the $\mathfrak{N}_i$ with strictly positive coefficients one defines $W_0 = \mathrm{Im}(\mathfrak{N}^n)$, $W_1 = \mathrm{Im}(\mathfrak{N}^{n-1}) \cap \mathrm{Ker}(\mathfrak{N})$, $W_2 = \mathrm{Im}(\mathfrak{N}^{n-2}) \cap \mathrm{Ker}(\mathfrak{N}) + \mathrm{Im}(\mathfrak{N}^{n-1}) \cap \mathrm{Ker}(\mathfrak{N}^2), \ldots, W_{2n-1} = \mathrm{Ker}(\mathfrak{N}^n)$. Let $\mathrm{Gr}_k = W_k/W_{k-1}$. It is easy to see that $\mathfrak{N}(W_k) \in W_{k-2}$ and it follows from the Jacobson-Morozov Lemma that $\mathfrak{N}^l : \mathrm{Gr}_{k+l} \sim \mathrm{Gr}_{k-l}$. Much more non-trivially $\mathcal{F}^\bullet_{\mathrm{lim}}$ is a Hodge structure of weight k on Gr_k, which means that $(\mathcal{F}^\bullet_{\mathrm{lim}}, W_\bullet)$ is a mixed Hodge structure. It can be shown that $\mathfrak{N}$ is the lowering operator of a $SL(2, \mathbb{C})$

action on the LMHS [293]. A consequence of the theory is that the maximal unipotent monodromy occurs for $p = n$ and no higher p occurs.

The easiest example is the nodal degeneration of a genus g Riemann surface Σ_g. One can chose a symplectic basis A^i, B_i $i = 0, \ldots, g-1$ such that A^0 degenerates. By the Lefschetz formula (2.6.11) the only cycle which is not monodromy invariant is $M : B_0 \mapsto A^0 + B_0$. $N = \mathfrak{N}$ is nilpotent $\mathfrak{N}^2 = 0$. One has $\mathfrak{N} : B_0 \mapsto A^0$ while all others cycles are annihilated. $W_{-1} := 0$, so $\mathrm{Gr}_0 = \mathbb{Q} \cdot A^0$ of grade 0 must be of type (0, 0), $\mathrm{Gr}_1 = \mathrm{span}_{\mathbb{Q}}\{A^1, \ldots, A^{g-1}, B_1, \ldots, B_{g-1}\} = H_n(\Sigma_{g-1}, \mathbb{Q})$ and $\mathrm{Gr}_2 = \mathbb{Q} \cdot B_0$ of grade 2 must be of type (1, 1). I.e. over $\mathbb{Q}$ the cohomology of Σ_g splits on $H_n(\Sigma_{g-1}, \mathbb{Q})$ and one has a closed sub-monodromy problem on the latter. The situation is very similar for the conifold transition in Calabi-Yau 3-folds, up to the fact that in an actual transition m S^3 with k relations shrink [145].

Conifold Monodromy

According to the Lefschetz formula with some sign corrections by Lamotke [250], an n-cycle Γ transforms along a path in the moduli space encircling the conifold divisor $\epsilon = 0$, where the n-sphere $S^n =: \nu$ vanishes, with positive orientation with the monodromy action on Γ that is either a *symplectic*—for n-odd or an *euclidean* reflection w for n-even, i.e.

$$w(\Gamma) = \Gamma + (-1)^{(n+2)(n+1)/2} \langle \Gamma, \nu \rangle \nu \,. \tag{2.6.11}$$

The self intersection of the n-sphere itself is given by

$$\langle \nu, \nu \rangle = \begin{cases} 0 & , \ n \text{ odd} \\ (-1)^{n/2} \cdot 2 & , \ n \text{ even} \,. \end{cases} \tag{2.6.12}$$

Let us now discuss the two cases in turn:

- n odd: Due to the non-degenerate symplectic pairing the vanishing cycle ν intersects a dual cycle Γ and in order to realize (2.6.11) the periods over these cycles degenerate in the local parametrization $\delta_c = 0$ for n odd like

$$\begin{aligned} X_\nu &= \int_\nu \Omega = \delta_c^{\frac{n-1}{2}} \sum_{k=1}^{\infty} c_k(\check{z}) \delta_c^n \,, \\ F_\Gamma &= \int_\Gamma \Omega = \frac{(-1)^{(n+2)(n+1)/2}}{2\pi i} X_\nu \log(X_\nu) + \sum_{k=0}^{\infty} b_k(\check{z}) \delta_c^k . \end{aligned} \tag{2.6.13}$$

  Here we denote by $\check{z}$ the remaining variables which are, eventually after suitable blow ups in the complex moduli space, transversal to the conifold divisor $\delta_c = 0$.

Usually one cannot determine the integral and cycles directly and rather solves the Picard-Fuchs equations near $\epsilon = 0$, but the Lefschetz formula fixes the relative normalization of the solutions so that $T = \begin{pmatrix} 1 & 0 \\ 1 & 1 \end{pmatrix}$ is the unipotent monodromy for a path that encircles $\delta_c = 0$ counter clockwise. It has been observed quite generally for Calabi-Yau 3 folds in toric varieties that at the generic conifold discriminant, see Sect. 2.9.1 an S^3 shinks to zero, which corresponds in the type IIA side to the $D6$ brane, see Sect. 2.10 for the principal example. This is more generally true and the monodromy corresponds to the Seidel-Thomas twist on the derived category side (2.6.27).

- n even: The S^n intersects with itself and so the typical local behaviour for the $K3$, 4-folds etc is $w(\nu) = -\nu$, for a single S^n and hence

$$X_\nu = \delta_c^{\frac{n-1}{2}} \sum_{k=1}^{\infty} d_k(\check{z}) \delta_c^k \,, \tag{2.6.14}$$

e.g. for 4-folds one finds $X_\nu = \delta_c^{\frac{3}{2}}$ etc. This leads to a $\mathbb{Z}_2$ monodromy. For fourfolds the D-brane class on the IIA for the generic conifold locus on derived category side has been calculated in [46].

2.6.4 Integral Symplectic Basis of Periods

It is a prediction of mirror symmetry that the complex moduli of a Calabi-Yau manifold should $\mathcal{M}_{cs}$ have at least a point of maximal unipotent monodromy, i.e. a point (2.6.8) with $k = 1$ and $p = n$. This follows from the mirror map (2.4.40) and for 3– folds from of the periods in terms of the inhomogenous prepotential (2.5.47) and the leading from of the prepotential for the A model in the large radius region.

For Calabi-Yau spaces in arbitrary dimensions mirror symmetry is expected to hold and using the mirror map (2.4.40) and the form of the central charges of central charge of A branes in terms of the mirror map in the large radius region we will explain in general how to determine an *integral basis* of period integrals, which is symplectic in odd and symmetric in even dimensions.

This method is very useful as the easiest way to determine the periods is not by performing explicit integrals over a choice of an integral (symplectic for n odd) basis in $H_n(M, \mathbb{Z})$—even though depending on the construction some of these integrals can be performed—but rather in finding to solutions the Picard-Fuchs differential equations in terms of the complex moduli. These equations are linear and lead to an a priori arbitrary basis of such solutions. The problem we discuss in this section is how to find the linear combinations of these solutions by mirror symmetry, which do correspond to integrals ($\int_{\Gamma^I} \Omega$) w.r.t. to an integral basis Γ^I of $H_n(M, \mathbb{Z})$.

One can in principle compute the analytic continuation to all those divisors around which the monodromies generate the monodromy group $\Gamma(M)$ and make sure that it respects the even intersection form, e.g. is a subgroup of $\mathrm{Sp}(h_3, \mathbb{Z})$ for 3-folds. Up to a normalization and integer conjugation this will in general determine the periods w.r.t. the integer symplectic basis, but it is especially for multi moduli spaces a very complicated analysis.

Lessons Form B-Field Shift and Griffiths Transversality

Mirror symmetry exchanges the complex structure and the complexified Kähler structure. At generic values of the complexified Kähler structure moduli space the n-point amplitudes $C_{i_1,\ldots,i_n}(t)$ of the A-model are very complicated functions of the Kähler moduli, because these functions are world sheet instanton corrected. However at large volumes $\mathrm{Im}(t^i) = v^i \to \infty$ the latter corrections are exponentially suppressed as $Q_k = \exp(2\pi i t_k)$ and $C_{i_1,\ldots,i_n}(t)$ are dominated by the degree zero maps. For example the classical n-point intersection

$$C^{(cl)}_{i_1,\ldots,i_n} = \int_{M_n} \omega_{i_1} \wedge \ldots \wedge \omega_{i_n} = D_{i_1} \cdot \cdots \cdots \cdot D_{i_n} \tag{2.6.15}$$

get promoted to instanton corrected intersection numbers[42]

$$C_{i_1,\ldots,i_n}(t) = C^{(cl)}_{i_1,\ldots,i_n} + \mathcal{O}(Q)\,. \tag{2.6.16}$$

As we discussed in the non-linear σ model, the shift (2.4.49)

$$t^i \to t^i + 1 \tag{2.6.17}$$

is an exact symmetry and does not change the evaluation of amplitudes such as (2.6.16) in the large radius region, where t^k are physical coordinates. Since (2.6.16) is holomorphic as explained cff. Eq. (2.5.11), the holomorphic amplitudes are forced to be a function of the invariant $\exp(2\pi i t_k)$. Non-holomorphic quantities such as the metric in the moduli space derived from the monodromy invariants Kähler potential (2.5.1) can and do depend for example on the real part $(t^k - \bar{t}^k)$ of t^k.

The main point is that already closed string mirror symmetry predicts that the B field shift (2.6.17) of the A model is realized as a monodromy operation on the period, which represent those coordinates of $\mathcal{M}_{cs}$, which are identified with the complexified Kähler coordinates. Let us denote by z_k the coordinates defining the normal crossing divisors $\{z_k = 0\}$ in $\mathcal{M}_{cs}$ around which the continuation of the periods generate the monodromies $T_k\colon \pi \to T_k\Pi$, where T_k preserves the integral intersection and generate in particular the shift (2.6.17). The large volume point is assumed to correspond to $z_1 = \ldots = z_h = 0$ in the complex moduli space M_{cs} with $h = h_{21}(W) = h_{11}(M)$. Because of the form of the inhomogenous coordinates (2.4.40) this implies that at this point in the complex moduli M_{cs} at least h solutions of the Picard Fuchs system $X^i(z), i = 1, \ldots, h$ have to develop single logarithmic singularities, while at least one—$X^0(z)$—must be holomorphic

[42]In mathematics these are the inter section numbers of the so called *quantum cohomology*.

and monodromy invariant. This is because the only ratio solutions which can exhibit the shifts as monodromies are of the form

$$t^k(z) = \frac{X^k}{X^0} = \frac{\frac{1}{2\pi i} X^0 \log(z_k) + \sigma_k^{(1)}(\underline{z})}{X^0(\underline{z})} = \frac{1}{2\pi i}\log(z_k) + \sigma_k^{\prime(1)}(\underline{z}) \qquad (2.6.18)$$

where $\sigma_k^{(1)}(\underline{z})$ $\sigma_k^{\prime(1)}(\underline{z})$ are regular holomorphic series without poles or branch behaviour.

Let us complete the argument for a point of maximal unipotent monodromy using special Kähler geometry for CY 3-folds, which by mirror symmetry must also be present on the complexified Kähler moduli space $\mathcal{M}_{cks}$. Since special Kähler geometry derives as we have seen from Griffiths transversality, the analysis of the latter allows similar conclusion for higher dimensional Calabi-Yau spaces.

Special geometry implies that there is a prepotential (2.12.21) for the A-model, whose leading term have to go like $\mathcal{F} = -\frac{c^{(0)}_{ijk}}{3!} t^i t^j t^k + \mathcal{O}(t^2) + \mathcal{O}(Q_n)$ in order to reproduce (2.6.16) by (2.6.30). In view of (2.6.30) and (2.6.18) this also implies that there must be h double logarithmic periods and one triple logarithmic period. It is easy to see that this is the maximal possible degeneration of the periods, which corresponds to $k = 1$ and $p = 3$ in (2.6.8).

We notice that by the theory of degenerations sketched in the last section, the leading logarithmic terms of the solutions allow us to determine combinations of them which correspond to actual cycles in the integral homology of W. For example X_0 correspond, as period over the unique vanishing cycle relative to the other, to an element in the integral basis. The single logarithmic solutions must correspond likewise to actual integrals over cycles in an integral basis up to possible additions of X^0, while adding higher logarithmic will lead to different basis elements. Further constraints come from the fact that the monodromies T_k have to preserve the integral intersection form. i.e for 3-folds they have to be in $\mathrm{Sp}(2h+2,\mathbb{Z})$. In explicit cases one can find a conifold locus where the triple logarithmic period vanishes. Since this vanishing cycle is unique and part of the integral basis one can them determine the subleading terms of this period. This allows to fix $\mathcal{F}$ up to quadratic terms in t^i which drop out of $2\mathcal{F} - t^i \partial_{t^i}\mathcal{F}$. However the constraints from the monodromies T_k allow to fix the integral basis for 3- and 4-folds up to conjugation and it has been observed that these subleading terms have an universal topological interpretation in the A-model that we discuss next.

The $\hat{\Gamma}$ Classes and Homological Mirror Symmetry

In multi moduli cases the techniques discussed above are complicated to apply in general and we will invoke in this section basic facts of homological mirror symmetry to find a better way to fix the integral bases and explain the occurrence of universal topological terms mentioned above.

The basic idea is that the period vectors in the integral basis have the interpretation of a basis of the central charge lattice Λ_Q in which the integral charge $Z_{D-brane}(Q)$ takes values. The latter determine the *masses* for the supersymmetric

A-branes, boundaries for the open string A-model, supported on special Lagrangian cycles. In addition this lattice comes with an integral pairing on the middle homology of M and a monodromy group that leaves the pairing invariant. By mirror symmetry completely analogous structures must exist for the supersymmetric B-branes, which are boundaries for the open string B-model and are specified as coherent sheafs on holomorphic sub-manifolds. In odd complex dimension the A-model branes are odd and the B-model brane are even dimensional and on both sides the pairing is symplectic, while in even dimensions for both types of branes the pairing is symmetric. The monodromy of the B model becomes an auto equivalence of the derived category of coherent sheafs in the A-model, respecting the bilinear form and their charges become K-theory charges.

The physical picture behind the *masses* of the branes is quite easy. They are given by the 'volume' of the supporting cycle times universal constants. This is because the branes have a tension and their energy will be proportional to this volume. However since they have also an $U(1)$ connection the mass turns out to be proportional to the absolute value of the complexified volume. For example for A branes on 3-folds the relation between mass, central charge and charges Q is given by the formula

$$M_{D-brane}(Q) = g e^{\frac{K}{2}} |Z_{D-brane}(Q)| = g e^{\frac{K}{2}} \left| Q^I \int_{\Gamma_I} \Omega \right| , \quad \text{with } I = 1, \ldots, h_3(W) , \tag{2.6.19}$$

referring to special Lagrangian cycles $\Gamma_I \in H_3(W, \mathbb{Z})$. Note that the factor $e^{\frac{K}{2}}$ is necessary for the mass to be invariant under Kähler gauge transformations.[43] The key point is that by mirror symmetry a similar expression for the B-branes of the A model must exist and has a simple grading by the volume of the highest dimensional holomorphic submanifold $H_{2i} \in H_{ii}(M, \mathbb{Z})$ of support. By the mirror map this corresponds to a grading in powers of t^k defined in (2.6.18). Given the corresponding expression allows to identify the leading logarithms and hence the corresponding basis of solutions.

Such an expression has been suggested in [183, 199, 217, 240], in term of the $\hat{\Gamma}$-classes. Let us recall the following multiplicative characteristic classes, some of which are also discussed in Appendix 2. The well known Chern class the, A-roof genus, the Todd class and the Γ-roof class introduced in [217, 240]

$$\mathrm{ch}(x) = e^x, \qquad \hat{A}(x) = \frac{x/2}{\sinh(x/2)}, \qquad \mathrm{td}(x) = e^{x/2} \hat{A}(x), \qquad \hat{\Gamma}(x) = \Gamma\left(1 - \frac{x}{2\pi i}\right) . \tag{2.6.20}$$

[43] Sometimes the central charge $Z_{D\text{-}brane}$ is defined as $(Q^I \int_{\Gamma_I} \Omega) e^{N/2}$. Q denotes the string coupling.

For a vector bundle V we may expand the $\hat{\Gamma}$ class as[44]

$$\begin{aligned}\hat{\Gamma}(V) &= \exp\left(\gamma \frac{\mathrm{ch}_1(V)}{2\pi i} + \sum_{k\geq 2} \zeta(k)(k-1)! \frac{\mathrm{ck}_k(V)}{(2\pi i)^k}\right) \\ &= 1 - \frac{i c_1 \gamma}{2\pi} - \frac{1}{48}\left(c_1^2\left(\frac{6\gamma^2}{\pi^2}+1\right) - 2c_2\right) \\ &\quad + \frac{1}{96\pi^3}\left(i(2c_1^3\gamma^3 + \pi^2(c_1^2 - 2c_2)c_1\gamma + 4(c_1^3 - 3c_2c_1 + 3c_3)\zeta(3))\right) + O(4)\,.\end{aligned} \tag{2.6.21}$$

For the tangent bundle of a Calabi-Yau manifold one has hence

$$\hat{\Gamma}(TW) = 1 + \frac{1}{24}c_2 + \frac{i c_3 \zeta(3)}{8\pi^3} + \frac{\left(7c_2^2 - 4c_4\right)}{5760} + \frac{i\left(\pi^2 c_2 c_3 \zeta(3) + 6(c_2c_3 - c_5)\zeta(5)\right)}{192\pi^5} + \mathcal{O}(6)\,. \tag{2.6.22}$$

As has been explained in [199, 217, 240] due to the property

$$\hat{A}(V) = e^{c_1(V)/2}\mathrm{td}(V) = \hat{\Gamma}(V)\hat{\Gamma}(V)\,, \tag{2.6.23}$$

the $\hat{\Gamma}$ class can be viewed as an alternative definition of the square root of the A-roof genus of the tangent bundle $V = TW$ of the mirror in Mukai's modified Chern-Character map, which turns out to be the correct one to get integral auto equivalences (twists) on the K-theory classes of the objects in the A brane category and the central charges that determine Bridgeland stability [52]. This fact makes the notion very useful for the comparison with the integral basis obtained in the last subsection.

Modifying the classical Chern character map from topological K-theory to even cohomology $\mathrm{ch} : K(W) \rightarrow H^{2*}(W, \mathbb{Q})$ in a compatible way with the Hirzebruch Riemann Roch index theorem that defines the bilinear

$$\begin{aligned}\eta_{\alpha\beta} &= Q(\Gamma_\alpha, \Gamma_\beta) = \chi(\mathcal{E}_\alpha, \mathcal{E}_\beta) = .\int_{W_n} \mathrm{ch}(\mathcal{E}_\alpha^*) \wedge \mathrm{ch}(\mathcal{E}_\beta) \wedge \mathrm{td}(TW_n) \\ &= \sum_{p=0}^{n} \mathrm{dimExt}^p_{\mathcal{O}_{W_n}}(\mathcal{E}_\alpha, \mathcal{E}_\beta)(-1)^p\,,\end{aligned} \tag{2.6.24}$$

Mukai defined an alternative Chern-Character map as $\mu(\mathcal{E}) = \mathrm{ch}(\mathcal{E})\sqrt{\mathrm{td(TW)}}$. However as observed later by [199, 217, 240] in order to get the right integer auto equivalences on the K-theory classes one must replace $\sqrt{\mathrm{Td(TW)}}$ by $\hat{\Gamma}(TW)$. The subsequent modification of [199, 240] leads to the central charge formulas of the

[44]Here γ is the Euler-Mascheroni constant. The term drops out in the application below since $\mathrm{ch}_1(TW) = c_1(TW) = 0$.

A-branes given by

$$Z(\mathcal{O}_{H_{2i}}) = \int_{M_n} e^{t\omega} \wedge \hat{\Gamma}(TM_n) \wedge \mathrm{ch}(\mathcal{O}_{H_{2i}}), \qquad i = 0, \ldots, n\,. \tag{2.6.25}$$

Here H_{2i} are suitable holomorphic sub-varieties of complex dimension i. The reason for the second modification is that while the twist $\mathcal{E} \to \mathcal{E} \times \mathcal{O}(1)$ comes out correctly as integer with both definitions

$$\chi(\mathcal{E} \otimes \mathcal{O}(1), \mathcal{F} \otimes \mathcal{O}(1)) = \chi(\mathcal{E}, \mathcal{F})\,, \tag{2.6.26}$$

the Seidel-Thomas twist [289]

$$\chi(\Phi_{\mathcal{O}}(\mathcal{E}), \Phi_{\mathcal{O}}(\mathcal{F})) = \chi(\mathcal{E}, \mathcal{F})\,, \tag{2.6.27}$$

which corresponds to the conifold monodromy, requires the use of the $\hat{\Gamma}$ class and in fact the modified Chern character map $\mu_{\hat{\Gamma}}(\mathcal{E}) = \mathrm{ch}(\mathcal{E})\hat{\Gamma}(TM_n)$ to be integer in the basis at infinity. In particular for $i = n$ the K-theory charge of the top dimensional brane is obtained from (2.6.25) as

$$\begin{aligned} Z(\mathcal{O}_M) = \int_{M_n} e^{t\omega} \wedge \hat{\Gamma}(TM_n) = \int_{M_n} \Bigg(1+\omega t+\left(\frac{J^2t^2}{2}+\frac{c_2}{24}\right)+\left(\frac{\omega^3t^3}{6}+\frac{1}{24}tJc_2+\frac{ic_3\zeta(3)}{8\pi^3}\right) \\ +\left(\frac{\omega^4t^4}{4!}+\frac{1}{48}\omega^2t^2c_2+\frac{7c_2^2-4c_4}{5760}+\frac{i\omega tc_3\zeta(3)}{8\pi^3}\right)+\mathcal{O}(5)\Bigg)+\mathcal{O}(e^{2\pi it})\,, \end{aligned} \tag{2.6.28}$$

where we understand that we restrict to the term of order n for Calabi-Yau n-folds and integrate the class ω, which is dual to the Kähler class of t, and its wedge products with the Chern classes over M_n. All terms are corrected by world-sheet instantons effects of order $\mathcal{O}(Q)$, which can be calculated in the B-model using the exact expressions of the period integrals on the mirror W_n near a point of maximal unipotent monodromy. For multi parameter models $e^{2\pi it^a}$ are exponentially suppressed if all large areas are large $\mathrm{Im}(t^a) \to \infty$.

Similar one gets for the $D0$ and $D2$ brane charges universal formulas

$$\begin{aligned} Z(\mathcal{O}_{pt}) &= 1 + \mathcal{O}(e^{2\pi it})\,, \\ Z(\mathcal{O}_{H_2^{(i)}}[-1]) &= t^i + \mathcal{O}(e^{2\pi it}) \\ Z(\mathcal{O}_{H_4^{(i)}}) &= -\tfrac{1}{2}C^{cl}_{ijk}t^it^j + A_{ij}t^j + c_i + \mathcal{O}(e^{2\pi it})\,. \end{aligned} \tag{2.6.29}$$

Note that $Z(\mathcal{O}_{H_{2k}})$, $k > 1$ have to be chosen compatible so that the auto equivalence acts integral and in accordance with monodromy calculation on the complex structure side.

Because of the requirement that (2.6.27) and (2.6.26) are integer in the same basis, the appearance of the coefficients of the $\hat{\Gamma}$ class, like $\zeta(k)$, in the central charge formula that follows also from the analytic continuation of the period X_ν over vanishing cycle ν at the conifold point to the large complex structure point and the fact that this period maps under homological mirror symmetry to the structure sheaf on W_n. This was used in [61] and [153] to determine these coefficients for threefolds and fourfolds respectively. It is possible to prove (2.6.25) using the hemisphere partition function obtained by localization [181]. As was pointed out in [165] the $\hat{\Gamma}$ class is also compatible with the interpretation of the sphere partition function of [36, 102] as e^{-K} [201].

It is simple to identify the central charges of the A- model in an integer basis with the central charges of the B-model, which are simply the periods on the mirror. For Calabi-Yau 3 folds the comparison reads

$$\begin{pmatrix} Z(\mathcal{O}_M) \\ Z(\mathcal{O}_{H_4}) \\ Z(\mathcal{O}_{pt}) \\ Z(\mathcal{O}_{H_2}([-1])) \end{pmatrix} = X^0 \begin{pmatrix} 2\mathcal{F}(t) - t^i \frac{\partial}{\partial t^i}\mathcal{F}(t) \\ \frac{\partial}{\partial t^i}\mathcal{F}(t) \\ 1 \\ t^i \end{pmatrix} . \tag{2.6.30}$$

Note the grading by areas and higher dimensional volumes captured by powers of t. This allows to compare the t powers or according to (2.6.18) equivalently the leading logarithm in z in order to determine the period vector in inhomogenous coordinates (2.5.47) in an integral monodromy basis, that is given by the l.h.s. of (2.6.30) using (2.6.28), (2.6.29). Using (2.4.43) or rather its complexified version (2.4.48) we can read off the holomorphic prepotential $\mathcal{F}$ form, compare (2.9.58). It is given by

$$\mathcal{F} = -\frac{C^{cl}_{abc}}{6} t^a t^b t^c + \frac{A_{ab}}{2} t^a t^b + \frac{c_2 \cdot D_a}{24} t^a + \frac{\zeta(3)\chi}{2(2\pi i)^3} + \mathcal{F}_{inst}(e^{2\pi i t^1}, \ldots, e^{2\pi t^r}) . \tag{2.6.31}$$

One of the main applications of mirror symmetry is that in (2.9.58) the period vector is reconstructed using the leading logarithmic degenerations of the full solutions to the Picard-Fuchs equations on μ. That is if the leading logarithms match, the full solutions give the sub leading terms at the point of maximal unipotent monodromy, determine all the world sheet instanton corrections in (2.6.31) and predict of all Gromov-Witten invariants on the mirror Calabi-Yau 3-fold μ.

We simplified the notation for the classical intersections by using the equivalence of integrals over forms with the intersection of the dual divisors, in particular

$$C^{cl}_{abc} = D_a \cdot D_b \cdot D_c = \int_M \omega^{(a)} \wedge \omega^{(b)} \wedge \omega^{(c)}, \tag{2.6.32}$$

$$c_a = c_2 \cdot D_a = \int_M c_2(TM) \wedge \omega^{(a)} \tag{2.6.33}$$

and

$$\chi = \int_M c_3(TM)\,. \tag{2.6.34}$$

$\zeta(3)$ is the Riemann ζ-function at argument 3. It comes technically by expanding the Γ class from the 3rd derivative of Γ. More conceptually it has been related to a 4-loop calculation of in the non-linear sigma model [59, 117, 118] or to a degree zero map or $D0$ brane contribution. The quantity A_{ab} is the only one which cannot read off from the $D6$ brane charge $Z(\mathcal{O}_M)$ since it drops out from the combination $2\mathcal{F} - t^i \partial_{t^i} \mathcal{F}$. It requires a proper definition of the $Z(\mathcal{O}_{H_4})$ brane charge that is compatible with integrality of the monodromy in the B-model or the auto-equivalence in the derived category of coherent sheaves in the A-model.

It is satisfying to see that the data that go into (2.6.31) are exactly the topological data that classify Calabi-Yau up to topological type according to the theorem of C.T.C Wall [310] that we review in Sect. 2.7. For fourfolds the basis has been determined along similar lines in [46].

2.7 Examples for Calabi-Yau Mirror Pairs

As we mentioned in the introduction most examples of mirror pairs come from Batyrevs construction of hypersurfaces in toric ambient spaces given by reflexive polyhedra and Batyrevs and Borisovs construction of complete intersections given by reflexive polyhedra with Neff partitions. We first recall the definition of the toric spaces $\mathbb{P}_\Delta$ related to lattice polyhedra Δ. These give rise to Fano, semi-Fano or almost Fano varieties which serve as ambient spaces for the Calabi-Yau hypersurfaces given by the anti-canonical section, more general by complete intersection constraints and even more general by determinantal or smooth non-complete intersection Calabi-Yau spaces. Note that the above ambient spaces don't exhaust the mentioned even the smooth Fano varieties which come in finite families in any dimensions. Smooth Fano manifolds have been classified in dimension one, where the only example is $\mathbb{P}^1$, in dimension two, where the only examples are the del Pezzo surfaces, given either by the blow up of $\mathbb{P}^2$ in $1, \ldots, 8$ points or $\mathbb{P}^1 \times \mathbb{P}^1$. Smooth Fano threefolds have been classified by [200] and [264]. In dimension two and three only a subset of the classified ones correspond to a toric Fano variety given by a $\mathbb{P}_\Delta$, with some restrictions on the points on the faces. For example in Fig. 1 the polyhedra $1, 2, 3, 5, 7$, which have no points on the interior of edges are a smooth Fano surfaces.

Compact Calabi-Yau spaces come in finite numbers in any known construction. The most investigated case are Calabi-Yau 3 folds.

In the present construction this is due to finiteness of reflexive polyhedra in any dimensions, like the 16 two dimensional ones depicted in Fig. 1. In three dimensions there are 4319 while in four dimensions in there are 473800776 [245] reflexive

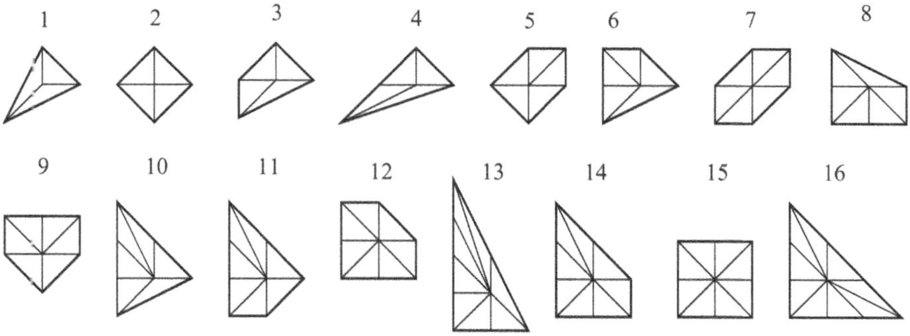

Fig. 1 These are the 16 reflexive polyhedra Δ in two dimensions, which build 11 dual pairs $(\Delta, \hat{\Delta})$. Polyhedron k is dual to polyhedron $17 - k$ for $k = 1, \ldots, 6$. The polyhedra $7, \ldots, 10$ are self-dual. The origin has label 0, the points right to it 1 and the other points are labelled counter clockwise

Polyhedra, while the number of higher dimensional is finite but the number is not exactly known. In principal higher and higher dimensional toric varieties could allow by complete intersection to an arbitrary number of higher co-dimensions Calabi-Yau manifolds, say 3-folds to be definite, but the example of products of projective spaces, a special case of toric varieties, shows again within that this class the topological different Calabi-Yau 3 folds is finite.

The question whether Calabi-Yau 3-folds are topological inequivalent is answered within the more general theorem of C.T.C. Wall for the classification of six manifolds up to homeomorphism, which specializes for Calabi-Yau 3-folds to the following Lemma. Two Calabi-Yau 3-folds in our definition are topological equivalent if

- Their fundamental group is the same. Note that the fundamental group for hypersurfaces in toric ambient spaces is trivial. However discrete fundamental groups can be achieved, by madding w.r.t. a freely acting discrete symmetry.
- The Cohomology groups are the same, which means $H^{2,1}(M)$ and $H^{1,1}(M)$ are the same
- There is choice of basis $\omega^{(i)}$, $i = 1, \ldots, h_{11}$ of $H^{1,1}(M)$ so that the classical triple intersections

$$C^{cl}_{ijk} = \int_M \omega^{(i)} \wedge \omega^{(j)} \wedge \omega^{(k)} = D_a \cdot D_b \cdot D_c$$

  as well as the intersection

$$c_i = \int_M c_2(TM) \wedge \omega^{(i)} = [c_2(TM)] \cdot D_i$$

  are the same.

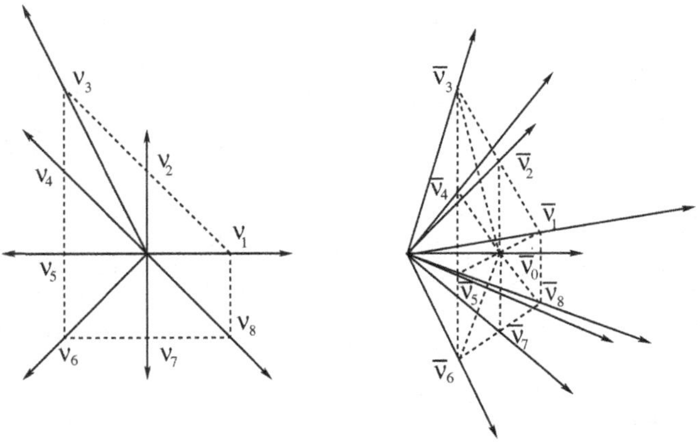

Fig. 2 Here we show the complete fan that yields the compact almost del Pezzo surface $\mathbb{P}_{\Delta^{(14)}}$ associated to the fourteenth polyhedron $\Delta^{(14)}$ from Fig. 1 on the right. On the left we show the non-complete fan $\overline{\Delta^{(14)}}$ that yields $\mathcal{O}(-K_{\mathbb{P}_{\overline{\Delta}^{(14)}}}) \to \mathbb{P}_{\overline{\Delta}^{(14)}}$

On the other hand the toric ambient spaces can also be made to non-compact toric Calabi-Yau spaces given by the anti-canonical line bundle fibred over $\mathbb{P}_\Delta$. Both construction come with explicit mirrors.

But in the non-compact case the construction can be generalized to an infinite number of Gorenstein fans of the type described in Fig. 2 that lead to non-compact Calabi-Yau spaces. For example it is known that $\mathcal{N} = 2$ four dimensional gauge theories can be geometrically engineered for arbitrary gauge groups $SU(N)$ (and $SO(N)$ by an orientifold action) from local Calabi-Yau geometries [214].

2.7.1 Toric Ambient Spaces and Non-compact Calabi-Yau Spaces

The d-dimensional toric[45] almost or semi Fano varieties are most easily classified by d-dimensional reflexive polyhedra. In particular toric almost del Pezzo surfaces are given by reflexive polyhedra in two dimensions, which are depicted in Fig. 1, where also the reflexive pairs $(\Delta, \hat{\Delta})$ are indicated. The anti-canonical class is only semi-positive if there is a point on one edge of the toric diagram, otherwise positive and ample. In particular the polyhedra 1,2,3,5,6 are Fano and correspond to toric del Pezzo surfaces, by the construction explained below. However as in the three dimensional case there are more, namely 9 Fano surfaces constructed as $\mathbb{P}^2$ blown up in up to eight points and $\mathbb{P}^1 \times \mathbb{P}^1$.

We fix the following conventions in arbitrary dimensions. If the dimension d of Δ is important we indicate it as a subscript. Δ is a lattice polyhedron in the lattice Γ

[45] We refer to [125, 270] for a general background in toric geometry.

(whose real completion is denoted by $\Gamma_{\mathbb{R}}$), if it is the convex hull of points $\nu^{(i)} \in \Gamma$ that contain the origin $\nu^{(0)}$ and span Γ. Analogous conventions are made for the dual polyhedron $\hat{\Delta}$, where the above data are all marked with a hat. We denote by $\langle \nu, \hat{\nu} \rangle \in \mathbb{Z}$ the pairing between Γ and the dual lattice $\hat{\Gamma}$. The dual polyhedron $\hat{\Delta}$ is defined by [28]

$$\hat{\Delta} = \{y \in \hat{\Gamma}_{\mathbb{R}} | \langle y, x \rangle \geq -1, \ \forall x \in \Delta\} . \tag{2.7.1}$$

Note that $\widehat{(\hat{\Delta})} = \Delta$ and Δ contains only $\nu^{(0)}$ as inner point. A pair is called reflexive if both Δ and $\hat{\Delta}$ are lattice polyhedra.

Δ together with a triangulation defines a *complete toric fan*[46] Σ. It is spanned by all rays $\Sigma(1)$ from the origin $\nu^{(0)}$ through the points ν_i, $i \neq 0$ as on the l.h.s of Fig. 2. The fan Σ describes for a reflexive polyhedron in real dimension d an (almost) Fano variety $\mathbb{P}_\Sigma$ of complex dimension d, explicitly given in (2.7.3). For simplicity we denote $\mathbb{P}_\Sigma = \mathbb{P}_\Delta$, if Σ comes from a reflexive polyhedron Δ. E.g. in the two-dimensional case $\mathbb{P}_{\Delta_2}$ is a toric (almost) del Pezzo surface S. In this construction a point in Δ different from the origin specifies a ray in the fan Σ_Δ.

We now give the general description of a toric variety starting from a fan Σ. A toric variety X of complex dimension d is by definition a normal variety that contains the algebraic torus $\mathbb{T}_d = (\mathbb{C}^*)^d$ as dense subset together with an action $\mathbb{T}_d \times X \rightarrow X$ that extends the natural action of $\mathbb{T}_d$ on itself [125]. It follows immediately from this definition that a toric variety $\mathbb{P}_\Sigma$ can given most explicitly from a fan Σ as follows [77]. Let ν_i be all the points lying in an integer lattice Γ and spanning the rays $\Sigma(1)$ of Σ in $\Gamma_{\mathbb{R}}$. Denote the vectors[47] $\bar{l}^{(k)}$, $k = 1, \ldots, h$ that specify a basis of linear relations among these points[48]

$$\sum_{i=1}^{s} \bar{l}_i^{(k)} \nu^{(i)} = 0 . \tag{2.7.2}$$

Now $\mathbb{P}_\Sigma$ is given as[49]

$$\mathbb{P}_\Sigma = \frac{\mathbb{C}^{|\Sigma(1)|}[x_1, \ldots, x_{|\Sigma(1)|}] \setminus Z_\Sigma}{\mathrm{Hom}(A_{d-1}(\mathbb{P}_\Sigma), \mathbb{C}^*)} . \tag{2.7.3}$$

[46] A complete toric fan in $\mathbb{R}^d$ covers all $\mathbb{R}^d$.

[47] In $\bar{l}^{(a)}$ the l stands for *l*inear relation. We extend the $\bar{l}^{(a)}$ latter, see Eq. (2.8.18), in way that makes useful to encode data of the periods, drop the bar and call them $l^{(a)}$.

[48] If the fan comes from a polytope $h := l(\Delta_d) - d - 1$, where $l(\Delta)$ are the number of integer points in Δ_d. For general fans Σ, which not necessarily come from Δ_d we have $h := |\Sigma(1)| - d$, where $s = |\Sigma(1)|$ is the number of points spanning the fan Σ. In many cases h can be identified with the number of Kähler parameter.

[49] If Σ comes from a polytope Δ_d in the two ways described above, we also use the notation $\mathbb{P}_{\Delta_d}$ or $\mathbb{P}_{\hat{\Delta}_d}$.

Here $G_\mu := \mathrm{Hom}(A_{d-1}(\mathbb{P}_\Sigma), \mathbb{C}^*) = (\mathbb{C}^*)^{\mathrm{rank} A_{d-1}(\mathbb{P}_\Delta)} \times A_{d-1}(\mathbb{P}_\Sigma)_{\mathrm{tors}}$ and the $\mathbb{C}^*$-action is specified by the $\bar{l}^{(k)}$ as

$$x_i \mapsto x_i(\mu^{(k)})^{\bar{l}_i^{(k)}} \quad \forall i, \tag{2.7.4}$$

with $\mu^{(k)} \in \mathbb{C}^*$. The rays $\Sigma(1)$ of a fan Σ correspond to the toric divisors D_i in the Chow group $A_{d-1}(P_\Sigma)$ of the d-dimensional toric variety $\mathbb{P}_\Sigma$ and we can assign a coordinate x_i, whose vanishing $x_i = 0$ specifies the divisor D_i. We abbreviate by $s = |\Sigma(1)| = \mathrm{rank} A_{d-1}$ the number of one dimensional fans or rays defined by s lattice points spanning the fan Σ. For example in Fig. 2, the right picture shows a *complete fan* that corresponds to a compact geometry, while the left picture shows a *non-complete fan* that defines a related non compact geometry. Z_Σ is the Stanley-Reisner ideal. Its subtraction guarantees well-defined orbits under the torus action. It is determined from a triangulation of Σ and consists of all loci in the intersection of divisors $D_{i_1} \cap \ldots \cap D_{i_r}$ for which the set of corresponding points $\{\nu_{i_1}, \ldots, \nu_{i_k}\}$ are not on a common cone of dimension d. A d-dimensional cone in the left picture in Fig. 2 contains for example all points $C_{12} = \{a_1\nu_1 + a_2\nu_2 | a_1, a_2 \in \mathbb{R}_+\}$ or in the right picture in Fig. 2 all points $C_{012} = \{a_0\bar{\nu}_0 + a_1\bar{\nu}_1 + a_2\bar{\nu}_2 | a_0, a_1, a_2 \in \mathbb{R}_+\}$. The triangulation determines also the generators $l_i^{(k)}$ of the Mori cone, which is dual to the Kähler cone, i.e. to each $l_i^{(k)}$ there is a dual curve whose volume vanishes at the boundary of the Kähler cone. Calculating the triangulations and the Mori cones is an combinatorial problem, which is described in [79] and has been implemented partly in SAGE packages. For the fans coming from the two dimensional reflexive polyhedra the solution can be found in [235].

Each toric variety comes with a natural symplectic structure, which is given by the real 2-form in coordinates $x_k = |x_k|e^{i\theta_k}$ of the algebraic torus $\mathbb{T}_d$ as

$$\omega = \frac{i}{2}\sum_{k=1}^{d} \mathrm{d}x_k \wedge \mathrm{d}\bar{x}_k = \frac{1}{2}\sum_{k=1}^{d} \mathrm{d}|x_k|^2 \wedge \mathrm{d}\theta_k \tag{2.7.5}$$

and extends by the definition of (2.7.3) and (2.7.4) to $\mathbb{P}_\Sigma$.

Non-compact toric Calabi-Yau space $M_{nc} = M_{\bar{\Delta}}$ are canonically obtained from Δ_d the following construction: In a (d+1)-dimensional lattice $\bar{\Gamma}$ spanned by the points with coordinates $\bar{\nu}^{(i)} = (1; \nu^{(i)})$, Δ_d is canonically embedded in the hyperplane at distance one from the origin $O = (0; 0, \ldots, 0) \in \bar{\Gamma}$ as the convex hull $\bar{\Delta}$ of the points $\bar{\nu}^{(i)}$. From O one can span *a non-complete fan* $\bar{\Sigma}$ through $\bar{\Delta}$. I.e. the rays $\Sigma(1)$ that define $\bar{\Sigma}$ go from the origin $\mathcal{O}$ through all points $\bar{\nu}_i, i = 0, \ldots, l(\Delta_d)$, as on the r.h.s of Fig. 2. The non-complete fan $\bar{\Sigma}$ defines $M_{\bar{\Delta}} = \mathbb{P}_{\bar{\Sigma}}$ as a non-compact toric variety with trivial canonical bundle as in (2.7.3). In this construction Δ_d is called the *trace* of the fan $\bar{\Sigma}$. In particular a reflexive polyhedron Δ_d defines a (d+1)-dimensional non-compact toric Calabi-Yau variety given as the total space $M_{\bar{\Delta}_d}$ of the anticanonical line bundle over $\mathbb{P}_{\Delta_d}$, i.e.

$$M_{\bar{\Delta}_d} = \mathcal{O}(-K_{\mathbb{P}_{\Delta_d}}) \to \mathbb{P}_{\Delta_d} \,. \tag{2.7.6}$$

The coordinate ring of $M_{\bar{\Delta}_d}$ is defined also by (2.7.3), with Δ replaced with $\bar{\Delta}$ in the definition of (2.7.2). Note that in this case $\sum_i l_i^{(k)} = 0$, as the points lie in a plane.[50] It is easy to see from (2.7.4) that this condition ensures the existence of a globally defined $(d+1,0)$-form, hence $M_{nc} = M_{\bar{\Delta}_d}$ is a non-compact CY $(d+1)$-fold.

Since we add the new point O in the construction for the non-complete fan the non-compact Calabi-Yau threefold does not require Δ_d to be reflexive. Any maximally triangulated convex polyhedron Δ_d, for which the rays of every $d+1$ dimensional cone, span the lattice $\bar{\Gamma}$, will lead to a smooth non-compact CY $(d+1)$-fold, otherwise to a singular one, which can be crepantly resolved to a smooth $d+1$ fold, by adding lattice points so that the generators of the new cones span $\bar{\Gamma}$. In general the space will be a CY $(d+1)$-fold if the spanning the incomplete fan lie on the hyperplane of distance one to the origin. In particular Δ_d can have an arbitrary number of inner points. For example if $d = 2$ each inner point corresponds to a compact surface inside $M_{nc} = M_{\bar{\Delta}}$. In the case that Δ_2 has no inner points, the compact parts of $M_{nc} = M_{\bar{\Delta}}$ are just curves, as for e.g. for the resolved conifold geometry $\mathcal{O}(-1) \oplus \mathcal{O}(-1) \rightarrow \mathbb{P}^1$.

The latter is also the simplest example for a situation with two different triangulations: We label the points of the $\mathcal{O}(-1) \oplus \mathcal{O}(-1) \rightarrow \mathbb{P}^1$ polyhedron in Fig. 3 second and third picture counter-clockwise starting from the right lower corner, the first triangulation corresponds to a Mori vector $l^{(1)} = (-1, 1, -1, 1)$. The coordinates $(x_2 : x_4)$ are homogeneous coordinates of the compact $\mathbb{P}^1$, whose positive volume is the Kähler cone and $x_2 = x_4 = 0$ is the Stanley-Reisner ideal. x_1 and x_3 are the line bundle coordinates. The coordinates of the flopped $\mathbb{P}^1$ with $l^{(1)} = (1, -1, 1, -1)$ are correspondingly given by $(x_1 : x_3)$ with $x_1 = x_3$ the Stanley-Reisner ideal and x_2 and x_4 coordinates of the $\mathcal{O}(-1) \oplus \mathcal{O}(-1)$ line bundles.

2.7.2 Gauge Linear σ-Model or Symplectic Quotient Perspective

Let us outline an equivalent description of toric geometry as the vacuum manifold of an gauged $(\mathcal{N}, \mathcal{N}) = (2, 2)$ supersymmetric two dimensional linear sigma model [322], with abelian gauge group. Mathematically this corresponds to the symplectic quotient construction of toric varieties, in which one divides or "mods out" the $(\mathbb{C}^*)^h = U(1)^h \times (\mathbb{R}_+)^h$ group action in two steps: In one we identify the abelian gauge group $U(1)^h$, where h was the number of relations in the construction of $\mathbb{P}_\Sigma$, in another we impose a constraint fixing the values $r_i \geq 0$, $i = 1, \ldots, h$. One advantage is that the symplectic structure is more obvious and in particular the Kähler moduli appear explicitly as the $r_i \geq 0$ and can be complexified in supersymmetric language by a so called Fayet-Illiopolous term.

The physical idea, summarized in [322], is that the two dimensional action of the string propagating on a Kähler manifold M can be described by a gauged linear

[50] In the equivalent description by an abelian 2d gauged linear σ-model it ensures the cancellation of the axial anomaly.

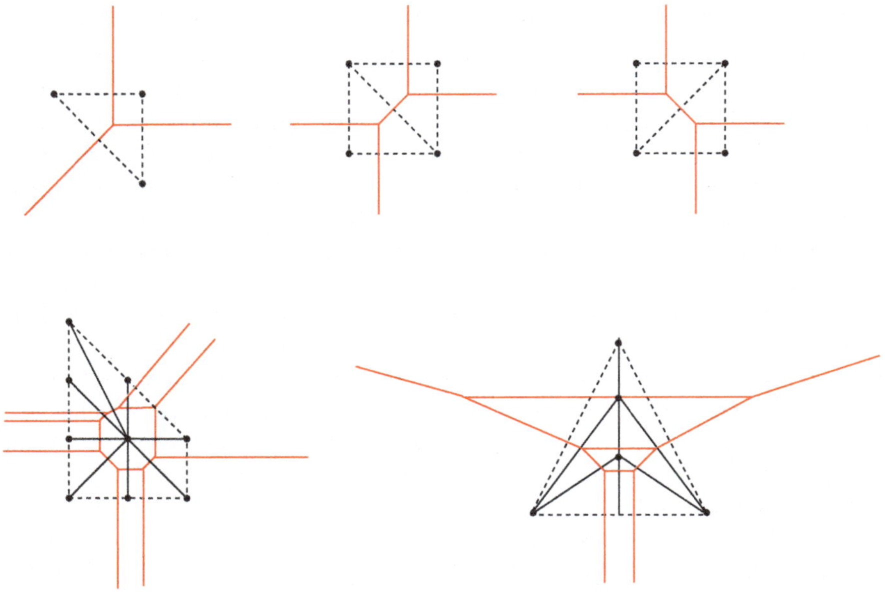

Fig. 3 Here we show the graph of the toric diagram in light black and dashes black and the dual graph in solid red. The latter is interpreted as web of $[p, q]$ 5-branes of the type IIB string. The geometries are $\mathbb{C}^3$, $\mathcal{O}(-1) \times \mathcal{O}(-1) \to \mathbb{P}^1$, with the $\mathbb{P}^1$ flopped, $\mathcal{O}(-K_{P_{\Delta(14)}}) \to P_{\Delta(14)}$, and an A_2 fibration over $\mathbb{P}^1$

σ-model, with a quite simple action with in particular $d + s$ chiral super fields with integral charges under each of the $U(1)$'s, say the k's one that can be grouped into charge vectors $q^{(k)} = \{q_1^{(k)}, \ldots, q_{s+d}^{(k)}\}$, $k = 1, \ldots, h$, that are simply identified with the vectors $l^{(k)}$ encoding the linear relations.

If M is a Calabi-Yau manifold the $(2, 2)$ supersymmetric model must also become conformal. It is argued that the gauged linear σ model indeed has in this case a conformal fix point. The full conformal theory is hard to describe in this approach, but it can be argued the vacuum manifold describes the Calabi-Yau target space and depends only on the charge vectors and a charge invariant superpotential. There is a very nice relationship between geometrical and physical properties. In particular non anomalous R-symmetry of the field theory is equivalent to triviality of the canonical bundle of M and implies

$$\sum_{i=1}^{s} q_i^{(k)} = 0 \, . \qquad (2.7.7)$$

Phases in the field theory correspond partly to geometrical phases in the Kählercone, but there are new phases in the field theory that are only understandable geometrically if one includes all world-sheet instantons corrections to the couplings among the chiral fields, a notion that is know as *quantum geometry*.

To specify the space of classical vacua we need to analyze the zero locus of the scalar potential U. The latter is semi positive and contains the absolute values of $F-$ and D terms. The F terms come from the superpotential $W(X)$ as $F_i = \partial_{X_i} W(X)$ and the D-terms from the gauge theory. The superpotential leads to the hypersurfaces and complete intersections like the one described in Sect. 2.7.3.

If we restrict ourself to the description of the ambient space or local Calabi-Yau spaces we have only the D-terms

$$U = \sum_{k=1}^{h} \frac{1}{2e_k^2} D_k^2, \quad \text{with } D_k = -e_k^2 \left(\sum q_i^{(k)} |x_i| - r_k\right), \tag{2.7.8}$$

where the e_i the coupling constants of the i's gauge group and r_k are identified with the Kähler parameters of our d dimensional variety. In the mathematical literature the D-terms are known the moment maps with respect to the symplectic structure (2.7.5). The toric variety is defined by the field space subject to the vanishing of the D-terms divided by the gauge group

$$\mathbb{P}_\Sigma = \left(\bigcap_{k=1}^{|\Sigma(1)|} D_k^{-1}(0) \right) / G \,. \tag{2.7.9}$$

To illustrate this we consider a theory with $U(1) \times U(1)$ gauge group and charges $q^{(1)} = (1, 1, -2, 0, 0)$, $q^{(2)} = (0, 0, 1, -2, 1)$. This corresponds to the non-compact toric Calabi-Yau space $M_{\overline{\Delta}}$ in (2.7.6), where the two dimensional polyhedron Δ is the fourth in Fig. 1. We labelled its points $\nu_1 = (0, 1)$, $\nu_2 = (-1, -2)$, $\nu_3 = (-1, 0)$, $\nu_4 = (0, 0)$ and $\nu_5 = (1, 0)$. The scalar potential reads then

$$U = \frac{e_1^2}{2}(|x_1|^2 + |x_2|^2 - 2|x_3|^2 - r_1)^2 + \frac{e_2^2}{2}(|x_3|^2 - 2|x_4|^2 + |x_5|^2 - r_2)^2$$

In the phase where r_1 and r_2 are both positive,[51] we see that for our vacua we cannot have $x_1 = x_2 = 0$ or $x_3 = x_5 = 0$. This is how the Stanley-Reisner ideal is encoded in the present approach.

The space M contains a surface S defined by $x_4 = 0$, which is in fact the Hirzebruch surface $F_2 = \mathbb{P}_\Delta$, again with Δ given by the fourth polyhedron in Fig. 1, that is a ruled surface over $\mathbf{P}^1$ with fiber $\mathbf{P}^1$. The non-compact Calabi-Yau manifold $M_{\bar{\Delta}}$ has a compact part which we identify with the Hirzebruch surface F_2 and the total space is identified with F_2 together with the normal bundle on it as

[51]The $U(1)$ generators were chosen so that the phase we are interested in corresponds to requiring r_1 and r_2 to be positive—this is tantamount to choosing a basis for the Mori cone, as we will see.

in (2.7.6). The latter is identified with the canonical line bundle for F_2 given here by the x_4 direction. Our model resolves a curve of A_1 singularities parameterized by $\mathbf{P}^1$; the fibers of F_2 are the vanishing cycles. The well-known cohomology of F_2 is generated by the class s of a section with $s^2 = -2$ and the class f of a fiber. The other intersection numbers are $s \cdot f = 1$ and $f^2 = 0$. There is another section $H = s + 2f$ which is disjoint from s. The section s itself is defined by $x_3 = x_4 = 0$. The curves $x_1 = x_4 = 0$ and $x_2 = x_4 = 0$ are fibers f of the Hirzebruch surface. The section H is identified with the locus $x_4 = x_5 = 0$. The divisors $x_i = 0$ for $i = 1, 2, 3, 5$ are noncompact divisors in X which intersect S in the respective curves f, f, s, H. For this reason, we will sometimes refer to these divisors as f, f, s, H. The divisor S restricts to the surface S itself as the canonical class $K = K_S$ by the adjunction formula.

2.7.3 Global Mirror Symmetry

Hypersurfaces

One key idea in Batyrev's mirror construction [28] is that we can view each polyhedron Δ in two ways. Firstly as defining a projective toric variety $\mathbb{P}_\Delta = \mathbb{P}_{\Sigma_\Delta}$ as explained in Sect. 2.7.1 and secondly as defining the Newton polyhedron for a polynomial $P_\Delta(x)$, where the coordinates of the points determine the exponents of the x_i. These x_i are the coordinates of the projective toric variety defined by the dual polyhedron $\hat{\Delta}$ and $P_\Delta(x) = 0$ defines a Calabi-Yau hypersurface in $\mathbb{P}_{\hat{\Delta}}$. Mirror symmetry then simply exchanges the rôles of Δ and $\hat{\Delta}$. We describe the construction outlining the difference between compact and non-compact toric mirror symmetry.

In the vanishing locus of compact case the Calabi-Yau M, the anti-canonical divisor in $\mathbb{P}_{\hat{\Delta}}$—is defined as the vanishing locus of an ample section of the anti-canonical bundle in $\mathbb{P}_{\hat{\Delta}}$

$$P_\Delta = \sum_{i=0}^{l(\Delta)-1} a_i Y_i = \sum_{\nu^{(i)} \in \Delta} a_i \prod_{\hat{\nu}^{(k)} \in \hat{\Delta}} x_k^{\langle \nu^{(i)}, \hat{\nu}^{(k)} \rangle + 1} = 0 \tag{2.7.10}$$

in the homogeneous coordinate ring x_k of $\mathbb{P}_{\hat{\Delta}}$, defined analogously to the one below (2.7.4), where the fan Σ is defined now by $\hat{\Delta}$.

Here the coefficients a_i parametrize (redundantly) the complex structure of M. In compact mirror symmetry points $\nu^{(i)}$ inside codimension one faces of Δ can be excluded from the sum (products) above, because the corresponding monomials Y_i can be removed by the automorphism group acting on $\mathbb{P}_{\hat{\Delta}}$ as will be discussed in more detail below. Points $\hat{\nu}^{(k)}$ in $\hat{\Delta}$ on codimensions one faces of $\hat{\Delta}$ can also be excluded from the product because the corresponding variables x_k describe (exceptional) divisors $x_k = 0$ (in the resolution of singularities) that lie outside of the hypersurface[52] $P_\Delta = 0$.

In the definition of (2.7.10) we used the fact that any lattice polyhedron $\hat\Delta \in \hat\Gamma$, which contains the origin, comes with a very ample divisor kD_Δ in $\mathbb{P}_{\hat\Delta}$. It is given by its support function $\phi : \hat\Gamma^*_{\mathbb{R}} \to \mathbb{R}$ defined $\phi_{k\Delta}(v) = \min_{\nu \in k\hat\Delta}\langle m, v\rangle$. For reflexive polyhedra $k = 1$ and $P_\Delta = 0$ corresponds by to a generically smooth hypersurface in $\mathbb{P}_{\hat\Delta}$.

To understand the definition of $P_\Delta = 0$ in the Y_i coordinates note that $\mathbb{P}_{\Delta_d}$ can then be embedded as a singular variety in a projective space as follows. Take the points $\{\nu^{(0)}, \ldots, \nu^{(l(\Delta)-1)}\} = \Delta \cap \Gamma$ and map $\mathbb{T}_d \to \mathbb{P}^{l(\Delta)-1}$ by sending $Y \in \mathbb{T}_d$ to $(Y^{\nu^{(0)}}, \ldots, Y^{\nu^{(l(\Delta)-1)}})$, where $Y^{\nu^{(i)}} = \prod_{j=1}^d Y_j^{\nu_j^{(i)}}$ and Y_j are a coordinate basis for $\mathbb{T}_d$. $\mathbb{P}_{\Delta_d}$ is the completion of the image of this map in $\mathbb{P}^{l(\Delta)-1}$. This defines a vector space of Laurent polynomials

$$L(\Delta \cap \Gamma) = \{P_\Delta : P_\Delta = \sum_{\nu^{(i)} \in \Delta\cap\Gamma} a_i Y^{\nu^{(i)}}, a_i \in \mathbb{C}\} \tag{2.7.11}$$

and the geometry M is given by $P_\Delta = 0$. The Laurent polynomials are a good starting point to discuss the complex moduli space, see below. Alternatively we can say that $\mathbb{P}_\Delta$ is be embedded as a singular variety in $\mathbb{P}^{l(\Delta)-1}$ by the constraints[53]

$$\prod_i Y_i^{l_i^{(k)}} = 1 \qquad \forall k \tag{2.7.12}$$

and M is then described by the additional constraint

$$P_\Delta = \sum_{i=0}^{l(\Delta)-1} a_i Y_i = 0 \tag{2.7.13}$$

in $\mathbb{P}^{l(\Delta)-1}$.

Similarly the mirror to M called W is defined as a hypersurface, i.e. as an ample section of the anti-canonical bundle of $\mathbb{P}_\Delta$

$$P_{\hat\Delta} = \sum_{i=0}^{l(\hat\Delta)-1} \hat a_i X_i = \sum_{\hat\nu^{(i)} \in \hat\Delta} \hat a_i \prod_{\nu^{(k)} \in \Delta} y_k^{\langle \hat\nu^{(i)}, \nu^{(k)}\rangle + 1} = 0 \tag{2.7.14}$$

[52] Readers who want to read in parallel the simplest example could look at the example of the quintic in $\mathbb{P}^4$ in Sect. 2.10 or consider the simple elliptic Calabi-Yau with two parameters which is discussed after the data of the polyhedra in (2.7.35).

[53] By construction the Y_i in (2.7.10) viewed as a function of the z_k fulfill this constraint.

as the Newton polynomial of $\hat{\Delta}$ in the coordinate ring of $\mathbb{P}_\Delta$, defined by (2.7.4) and the Stanley Reisner ideal of its coordinate ring (2.7.3).

The quotient construction of mirror symmetry can be realized, if there is an embedding map $\Phi : (\hat{\Delta}, \hat{\Gamma}) \rightarrow (\Delta, \Gamma)$. This defines the êtale map from the X_i coordinates to the Y_i coordinates. For the example of the quintic the relevant charge vector (2.7.2) is $l^{(1)} = (-5, 1, 1, 1, 1, 1)$ and the êtale map is

$$\phi : (Y_0 : Y_1 : \ldots : Y_5) \mapsto (\prod_{i=1}^{5} y_i : y_5^5 : \ldots : y_5^5) = (X_0 : X_1 : \ldots : X_5) , \tag{2.7.15}$$

which is many to one viewed as map x_k to y_k and is made unique by identifying the x_k under the action of the mirror quotient group G. I.e. kern$(\Phi) = G$ and the order of G is the degree of Φ.

For example the pairs $(\Delta_2, \hat{\Delta}_2)$ define one-dimensional compact Calabi-Yau hypersurfaces $P_{\Delta_2}(x) = 0$ in (almost) del Pezzo surfaces $\mathbb{P}_{\hat{\Delta}_2}$, i.e. elliptic curves and all a_i up to one can be set to 0 or 1 by the automorphism group of $\mathbb{P}_{\hat{\Delta}_2}$ and rescalings of x_i in the coordinate ring of $\mathbb{P}_{\hat{\Delta}}$. Let us describe the moduli space of the hypersurfaces more generally. Clearly as for the simple example the a_i defined in (2.7.10) are redundant parameters, from which first of all the action of the automorphism group of $\mathbb{P}_{\hat{\Delta}}$ has to be divided. These automorphisms come in three types [78].

(A1) By the definition of a toric variety $\mathbb{P}_{\hat{\Delta}_2}$ contains the algebraic torus $\mathbb{T}_d$ as an open dense subset. The natural $(\mathbb{C}^*)^{l|\Delta_2|-1}$ action on $\mathbb{C}^{(l|\Delta_2|-1)}[Y_i]$ is reduced by the identifications (2.7.4), as expressed by the exact sequence

$$1 \rightarrow G_\mu \rightarrow (\mathbb{C}^*)^{\Sigma(1)} \rightarrow \mathbb{T}_d \rightarrow 1 , \tag{2.7.16}$$

to an action of $\mathbb{T}_d$ on $\mathbb{P}_{\hat{\Delta}}$ that extends the natural action of $\mathbb{T}_d$ on $\mathbb{T}_d \subset \mathbb{P}_{\hat{\Delta}}$.

(A2) The second type of automorphism are the *weighted homogeneous coordinate transformations*

$$Y_i \mapsto b_0^{(i)} Y_i + \sum_k b_k^{(i)} m_k^{(i)}(\underline{Y}) , \tag{2.7.17}$$

of $\mathbb{P}_{\hat{\Delta}}$. Here $b_l^{(i)} \in \mathbb{C}$ and the monomials $m_k^{(i)}(\underline{Y})$ on the r.h.s do not contain the one Y_i on the l.h.s. They such that both sides of (2.7.17) transform equal under (2.7.4) and P_Δ stays well defined under (2.7.17). Pairs $(Y^i, b_k^{(i)})$ are called roots.

(A3) Further there can be symmetries of the toric polyhedron, which according to [78] have to be identified.

These actions A1–A3 do not leave the general P_Δ invariant unless they are compensated by actions on the a_i. This is the action that has to be divided out to get a first model for the complex moduli space.

To construct this quotient [78] start with the *Laurent polynomial* in d variables as defined in (2.7.11), which we can think of also as dividing from the Y_i in (2.7.10) the analog of the relations (2.7.12), because the coordinate ring Y_i captures all the blow up coordinates, but as far as the complex structure deformations of M_n go, the d variables Y_j that yield a basis for $\mathbb{T}_d$ are sufficient to characterize P_Δ.[54] The statement about the moduli space can now be phrased as

$$\mathcal{M}_{P_\Delta} = \mathbb{P}(L(\Delta \cap \Gamma))/Aut(\mathbb{P}_{\Delta_2})\,. \tag{2.7.18}$$

It has been shown by Batyrev that any complex structure deformation has one representative under the (gauge) orbits (2.7.17), which corresponds to the restricted Newton polynomial of Δ in which only such monomials in (2.7.10) are considered that correspond to points in $\nu^{(i)}$ not inside co-dimension one faces of Δ. We call Δ without those points Δ_0. This can be viewed as a gauge fixing and leads to the definition

$$\mathcal{M}^{simp}_{P_\Delta} = \mathbb{P}(L(\Delta_0 \cap \Gamma))/\mathbb{T}_d. \tag{2.7.19}$$

One can show that the map $\phi : \mathcal{M}^{simp}_{W_\Delta} \to \mathcal{M}_{P_\Delta}$ is at most a *finite cover*. Note that not all symmetries of M_n might be manifest in a chosen gauge.

As it turns out the most interesting points in the analysis below are precisely related to the nature of the *finite covers*. In order to get the right description for $\mathcal{M}_{cs}$ we have to divide $Aut(\mathbb{P}_{\Delta_2})$ by this discrete group G.

It is explained in [28][55] how to calculate the Euler number and $h_{n-1,1}$ and $h_{1,1}$ from the polyhedra, which leads to the beautiful combinatorial formulas

$$\begin{aligned} h^{n-1,1}(M_n) &= h^{1,1}(W_n) \\ &= l(\Delta_{n+1}) - (n+2) - \sum_{\dim\hat\theta = n} l'(\hat\theta) + \sum_{\text{codim}\,\hat\theta_i} l'(\hat\theta_i)l'(\theta_i)\,, \end{aligned} \tag{2.7.20}$$

$$\begin{aligned} h^{1,1}(M_n) &= h^{n-1,1}(W_n) \\ &= l(\hat\Delta_{n+1}) - (n+2) - \sum_{\dim\theta = n} l'(\theta) + \sum_{\text{codim}\,\theta_i} l'(\theta_i)l'(\hat\theta_i)\,. \end{aligned} \tag{2.7.21}$$

[54] Physically this independence of complex parameters from the blow ups moduli reflects, e.g. the decoupling of vector- and hypermultiplets in type IIb compactifications on M_3 to 4d at generic loci in the moduli space.

[55] See Cor. 4.5.1, Cor 4.5.2, Thm 4.5.3.

In this expression θ ($\hat{\theta}$) denote faces of Δ ($\hat{\Delta}$), while the sum is over pairs $(\theta_i, \hat{\theta}_i)$ of dual faces. The $l(\theta)$ and $l'(\theta)$ count the total number of integral points of a face θ and the number of integral points inside the face θ, respectively. Finally, $l(\Delta)$ is the total number of integral points in the polyhedron Δ.

This determines all Hodge numbers for $n \leq 3$ and for $n = 4$ the other Hodge numbers follow from the Hirzebruch-Riemann-Roch theorem, see Appendix 2. The Euler number and $h_{q,1}$ for $0 \leq q \leq d - r$ can also be directly calculated by the formulas[56] given for $r > 0$ in [29]. The above mentioned formulas relate toric divisors and intersections thereof, as well as deformations of (2.7.10) to representatives in the homology groups, while the E-polynomial [30] yields for more general homology groups only information about the dimensions.

Complete Intersections

For the *complete intersections case* [34] one needs r semi-ample Cartier divisors corresponding to r upper convex piecewise linear support functions ϕ_l, which define a nef-partition $E = E_1 \cup \cdots \cup E_r$ of the vertices ρ^* of $\hat{\Delta} := \hat{\Delta}_{n+r}$ into disjoint subsets $E_1, \ldots, E_r$ as

$$\phi_l(\rho^*) = \begin{cases} 1 & \text{if } \rho^* \in E_l, \\ 0 & \text{otherwise.} \end{cases} \tag{2.7.22}$$

Each ϕ_l defines a semi-ample Cartier divisor $D_{0,l} = \sum_{\rho^* \in E_l} D_{\rho^*}$ on $\mathbb{P}_{\hat{\Delta}}$, where D_{ρ^*} is the divisor corresponding to the vertex $\rho^* \in E_l$. The family of Calabi-Yau manifolds M_n is given as a complete intersection $M_n = D_{0,1} \cap \cdots \cap D_{0,r}$ of codimension r in $\mathbb{P}_{\hat{\Delta}}$, where in accordance with the notation in the hypersurface case we define a generic section of $\mathcal{O}(D_{0,l})$ as

$$P_{\Delta_l} = 0, \quad l = 1, \ldots, r \tag{2.7.23}$$

Each ϕ_l defines the lattice polyhedron Δ_l as $\Delta_l = \{x \in \Gamma_{\mathbb{R}} : (x, y) \geq -\phi_l(y) \; \forall \; y \in \hat{\Gamma}_{\mathbb{R}}\}$. These Δ_l support global sections of the semi-ample invertible sheaf $\mathcal{O}(D_{0,l})$, whose explicit form is given by (2.7.10) with Δ replaced by Δ_l. Note that $\sum_{l=1}^{r} \phi_l = \phi$ yields the support function of K and that the Minkowski sum is $\Delta_1 + \cdots + \Delta_r = \Delta$. Moreover giving a partition $\Pi(\Delta) = \{\Delta_1, \ldots, \Delta_r\}$ of a supporting polyhedra Δ_l is equivalent to give $E_1, \ldots, E_r$ and is therefore also called nef-partition. $\nabla_l = \langle \{0\} \cup E_l \rangle \subset \hat{\Gamma}_{\mathbb{R}}$ defines also a nef-partition

[56] This gives again the full Hodge diamond for $n \leq 4$, which is implemented in the software package PALP.

$\Pi^*(\nabla) = \{\nabla_1, \dots, \nabla_r\}$ and $\nabla = \nabla_1 + \dots + \nabla_r$ is also reflexive polyhedron with the following duality relations

$$\begin{array}{ccc} \Gamma_{\mathbb{R}} & & \hat{\Gamma}_{\mathbb{R}} \\ \Delta = \Delta_1 + \ldots + \Delta_r & & \hat{\Delta} = \langle \nabla_1, \ldots, \nabla_r \rangle \\ & (\Delta_l, \nabla_{l'}) \geq -\delta_{ll'} & \\ \hat{\nabla} = \langle \Delta_1, \ldots, \Delta_r \rangle & & \nabla = \nabla_1 + \ldots + \nabla_r \end{array} \quad (2.7.24)$$

where the angle-brackets denote the convex hull of the inscribed polyhedra. By a conjecture due to [29] this construction leads mirror pairs (M_n, W_n) of families of complete intersections Calabi-Yau varieties where M_n is embedded into $\mathbb{P}_{\hat{\Delta}}$ as complete intersection of the sections P_{Δ_l} of the line bundles associated to $D_{0,l}$ specified by Δ_l, while W_n is embedded into $\mathbb{P}_\nabla$ as complete intersection of the sections of $\mathcal{O}(D^*_{0,l})$ specified by ∇_l.

2.7.4 Local Mirror Symmetry

In Sect. 2.7.1 we constructed $n + 1$ dimensional non-compact Calabi-Yau spaces $M_{nc} = M_{\hat{\Delta}_d}$ by a fan defined by the trace of a toric polyhedron in co-dimension a distance one from the origin as in Fig. 2. This situation is the realised for any co-dimension two face of a $d + 2$ dimensional reflexive polyhedron $\hat{\Delta}_{d+2}$. If the compact Calabi-Yau space M, defined as hypersurface in $\mathbb{P}_{\hat{\Delta}_{d+2}}$, is a fibration of a $d - 1$ dimensional Calabi-Yau manifold as fibre, that projects to $\mathbb{P}_{\hat{\Delta}_d}$ one can decompactify this fibre and focus on $M_{nc} = M_{\hat{\Delta}_d}$ and in particular obtain its mirror from a limit of the global mirror W obtained by the Batyrev construction.

This construction was first described in [214] in the context of geometric engineering of gauge theories and basically restricts Batryrevs êtale construction to the coordinate ring defining M_{nc}. More precisely one observes that in the decompactification limit the mirror polynom $P_{\hat{\Delta}}$ restricts to monomials that correspond to the points in $\hat{\Delta}_d$, whose zero locus describes a $d - 1$ dimensional variety, i.e. a curve for the most relevant case that M_{nc} is a CY 3-fold, plus a quadric term which corresponds to the normal direction and carries no complex structure deformation. The mirror description is therefore[57]

$$P_{\Delta_d} = \sum_{i=0}^{l(\Delta_d)-1} a_i X_i = 0\,, \quad (2.7.25)$$

where Δ_d is any d-dimensional polyhedron, not necessarily reflexive. In contrast to the compact mirror symmetry discussed above there are no automorphisms in $M_{\hat{\Delta}_d}$ to remove monomials in (2.7.25), hence the sum runs over all points in Δ_d. The independent deformations are hence as follows: Let $\bar{l}^{(i)}$ generate a basis of

[57] In the following sections we drop the ˆ for notational convenience.

linear relations $\sum_i \bar{l}_i^{(i)} \bar{\nu}^{(i)} = 0$ among the points of $\bar{\Delta}_d$, which define the analog of (2.7.12) in the X_i coordinates. These relations restrict the possibility to undo deformations parametrised by the a_i using rescalings of X_i, leaving $l(\Delta_d) - d - 1$ independent deformations of the B-model. A convenient way to introduce these in the curve in a manifest scaling invariant way is to set first all $a_i = 1$ and modify (2.7.12) to

$$\prod_i X_i^{\bar{l}_i^{(k)}} = z_k \qquad \forall k\ . \tag{2.7.26}$$

Here we use Batyrev's coordinates for the complex structure deformations

$$z_k = \prod_i a_i^{\bar{l}_i^{(k)}} \tag{2.7.27}$$

so that $z_k = 0$ is the large complex structure point.

In this description (2.7.25) with $a_i = 1$, (2.7.26) and a $\mathbb{C}^*$-identification $X_i \sim \mu X_i$ with $\mu \in \mathbb{C}^*$ define the mirror geometry. It can be written as a $(d+1)$-dimensional affine variety by adding to the singular constraint $P_{\Delta_d} = 0$ the trivial non-compact normal directions as quadratic coordinates. E.g. for Δ_2 it is

$$H(x, p, \underline{z}) := P_{\Delta_2}(x, p, \underline{z}) = uv\ . \tag{2.7.28}$$

Note that in order to solve (2.7.26) in favor of two variables say x, p we have to view the X_i as $\mathbb{C}^*$-variables. $P_\Delta(x, p) = 0$ becomes in general a Laurant polynomial in $\mathbb{C}^*$-variables defining a genus g Riemann surface Σ_g with h punctures. Here g is the number of inner points in Δ_2 and $h = l(\Delta_2) - g$.

As an example we take the first polyhedron in Fig. 1. Here $\mathbb{P}_\Delta = \mathbb{P}^2$ and the local Calabi-Yau 3-fold $M_{\bar{\Delta}}$ is the total space of $\mathcal{O}(-3) \to \mathbb{P}^2$. This can can also been constructed as the blow up $\widehat{\mathbb{C}^3/\mathbb{Z}_3}$ of the singular orbifold $\mathbb{C}^3/\mathbb{Z}_3$, with the $\mathbb{Z}_3$-group acting like $x_i \to e^{\frac{2\pi i}{3}} x_i$, $i = 1, 2, 3$ on the $\mathbb{C}^3$ coordinates. The $\mathbb{P}^2$ emerges as the exceptional divisor in the blow up of the singularity at $z_i = 0$, $i = 1, 2, 3$. $\bar{\Delta}$ defining the non-complete fan for this geometry has the points

$$\begin{array}{r|rrr|r|} Div. & & \bar{\nu}_i & & \bar{l} \\ D_{X_0} & 1 & 0 & 0 & -3 \\ D_{X_1} = L & 1 & 1 & 0 & 1 \\ D_{X_2} = L & 1 & 0 & 1 & 1 \\ D_{X_3} = L & 1 & -1 & -1 & 1 \end{array}\ . \tag{2.7.29}$$

with one relation $\bar{l}$. From (2.7.26) we get

$$X_1 X_2 X_3 = z X_0^3\ . \tag{2.7.30}$$

We can use the $\mathbb{C}^*$ action on the X_i to set $X_0 = 1$ and identify $X_1 = x$ and $X_2 = p$. Also we can set $a_i = 1$ for $i = 1, 2, 3$. Then (2.7.28) defining the mirror geometry reads

$$H(x, p; z) = 1 + x + p + \frac{z}{xp} = uv\ . \tag{2.7.31}$$

Here $z = \frac{1}{a_0^3}$ is the only complex structure deformation of $H(x, p; z) = 0$. We will discuss the associated soutions to the Picard-Fuchs equations and the genus zero instantons of this geometry and in fact all local Calabi-Yau geometries associated to the polyhdra in Fig. 1 in Sect. 2.12.

In general the coefficients a_i of the *inner points* deform the complex structure of Σ_g, while the punctures, whose number is given by the independent deformations $l(\Delta) - g - 3$ correspond to the independent residue values of the meromorphic one form λ, which in turn are given by the coefficients of the *non inner points*. In view of their rôle in geometric engineering they are referred to as masses m_i, $i = 1, \ldots, l(\Delta) - g - 3$.

In the case of a del Pezzo basis there is only one inner point whose coefficient a_0 is identified with the complex structure of the elliptic curve $P_{\hat{\Delta}}(X, Y) = 0$, physically related to the gauge coupling of the $U(1)$ theory on the Coulomb branch while $l(\hat{\Delta}_2) - 4$ of the a_i are identified with mass parameters.

There is a physical interpretation for the dual graph associated to a general triangulated polyhedron [2]. It can be viewed as a web of $[p, q]$ five branes for the type IIB string. These 5-branes fill the $0, \ldots, 5$ directions of the five-dimensional space-time. The red lines in Fig. 3 corresponds to the $(5, 6)$-plane, where the 5-branes extend as lines, whose slope is given by the $SL(2, \mathbb{Z})$ charge $[p, q]$.

2.7.5 Fibration Structures and Global Embeddings of the Local Geometries

Let us now discuss the global embeddings of local geometries in compact Calabi-Yau spaces M. In the simplest case those are related to elliptic fibrations. The del Pezzo surface appears as the base and all $(1, 1)$-classes of the del Pezzo surface are $(1, 1)$-classes in M.

The global embeddings of local Calabi-Yau can be studied in via embeddings of the reflexive polyhedra $(\Delta_n^B, \hat{\Delta}_n^B)$ into a pair of reflexive polyhedra $(\Delta_{n+2}, \hat{\Delta}_{n+2})$, so that the anti-canonical hypersurface in $\mathbb{P}_{\hat{\Delta}_{n+2}}$ gives rise to an elliptically fibred Calabi-Yau $(n + 1)$-fold over the toric base $\mathbb{P}_{\hat{\Delta}_n^B}$ with an interesting structure of global sections.

For Calabi-Yau manifolds defined in toric ambient spaces, as above, the fibration structure descends from a toric morphism from the ambient space. Denote[58] by Σ

[58] We again drop the ^.

the fan in Γ generated from Δ and by Σ_B the fan defined from Δ_B in the lattice Γ_B (generated by Δ_B) and identify $\mathbb{P}_\Sigma$ with $\mathbb{P}_\Delta$ etc. Here are the two conditions for a fibration map $\tilde{\phi}$ from the ambient space $\mathbb{P}_\Delta$ to $\mathbb{P}_{\Delta_B}$ with fibre $\mathbb{P}_{\Delta_F}$ [125][59]

(F1) There exist a lattice morphism $\phi : \Gamma \to \Gamma_B$. This is the case if Δ_F is a reflexive lattice sub-polyhedron of the lattice polyhedron Δ and both share the unique inner point. The lattice Γ_F is then in the kernel of ϕ, i.e. one has

$$0 \to \Gamma_F \to \Gamma \to \Gamma_B \to 0\,.$$

(F2) There exists a triangulation of Σ so that every cone $\sigma \in \Sigma$ is mapped under ϕ to a cone $\sigma_B \in \Sigma_B$. In this case there is an $\mathbb{T}_d$-equivariant morphism $\tilde{\phi} : \mathbb{P}_\Delta \to \mathbb{P}_{\Delta_B}$.

For notational simplicity we outline only the embedding of a two-dimensional polyhedra in a four-dimensional polyhedron, which gives rise to an elliptically fibred threefold over a toric (almost) del Pezzo surface $\mathbb{P}_{\hat{\Delta}_2^B}$, specified by $\hat{\Delta}_2^B \in \hat{\Delta}_4$. However everything in this section, except for (2.7.34),[60] generalized trivially to arbitrary dimension.

The reflexive pair $(\hat{\Delta}_4, \Delta_4)$ is the convex hull of the following points

$$\begin{array}{cc}
\hat{\nu}_i \in \hat{\Delta}_4 & \nu_j \in \Delta_4 \\
\left|\begin{array}{cc} & \hat{\nu}_i^F \\ \hat{\Delta}_2^B & \vdots \\ & \hat{\nu}_i^F \\ 0\ldots 0 & \\ \vdots & \hat{\Delta}_2^F \\ 0\ldots 0 & \end{array}\right| &
\left|\begin{array}{cc} & \nu_j^F \\ s_{ij}\Delta_2^B & \vdots \\ & \nu_j^F \\ 0\ldots 0 & \\ \vdots & \Delta_2^F \\ 0\ldots 0 & \end{array}\right|
\end{array}\,. \qquad (2.7.32)$$

Here we consider all points $\hat{\nu}_j^F \in \hat{\Delta}_F$ and define

$$s_{ij} = \langle \nu_i^F, \hat{\nu}_j^F \rangle + 1 \in \mathbb{N}\,. \qquad (2.7.33)$$

Note that we scaled $\Delta_2^B \to s_{ij}\Delta_2^B$. This means to scale the coordinates of the points of Δ_2^B by s_{ij} while keeping the original lattice basis, i.e. $s_{ij}\Delta_2^B$, contains in general more lattice points. Note that the vertices of $\hat{\Delta}$ (Δ) are given by the vertices of the polyhedra $\hat{\Delta}_2^F$ ($\hat{\Delta}_2^F$) and $\hat{\Delta}_2^B$ ($s_{ij}\Delta_2^B$) respectively.

It is obvious that both polyhedra $(\Delta, \hat{\Delta})$ fulfill the condition (F1), but only for $\hat{\Delta}$ it is easy to establish that one can pick a triangulation that also fulfills (F2), see [235] for details. As a consequence M given by $P_{\Delta_4} = 0$ in $\mathbb{P}_{\hat{\Delta}}$ is a smooth and flat

[59]See exercise p.49, where the statements are made at the level of the fans.

[60]For which aspects of the generalization have been discussed in [235].

elliptic fibration. All C.T.C Wall data of M depend in a simple way on the base and the type of the fibration [235]. For example the the Euler number and the Hodge number are given as

$$\chi(M) = -a_F \int_B c_1^2, \qquad h_{11}(M) = l(\hat{\Delta}_B) - 3 + s\,. \tag{2.7.34}$$

where a_F, and the number of sections s depend only on the fibration type, which is in turn specifed by $\Delta^{\hat{F}}$ and $\nu^{\hat{F}}$. The problem in establishing (F2) for W is the scaling of Δ_2^B. W given by $P_{\hat{\Delta}_4} = 0$ in $\mathbb{P}_{\hat{\Delta}}$ is in general only a non-flat elliptic fibration.

Let us give as the most elementary example the smooth elliptic fibration over $\mathbb{P}^2$. In this case[61] we pick for the base $\mathbb{P}^2$, whose toric polyhedron is the convex hull of the points $\hat{\Delta}^B = \text{conv}((1,0),(0,1),(-1,-1))$, for the fibre polynomial $\hat{\Delta}^F = \text{conv}((1,0),(0,1),(-2,-3))$ and for $\hat{\nu}_3^F = (-2,-3)$. Then $\nu_3^F = (-1,-1)$ and $s_{33} = 6$. For this fibre one has one section $s = 1$ and $a_F = 60$.

We list the points which give rise to the coordinate ring of $\mathbb{P} = \mathbb{P}_\Delta/(\mathbb{Z}_{18}\times\mathbb{Z}_6)$, all points $\hat{\nu}_i \in \hat{\Delta}$ and the two vectors of linear relations among them, which correspond to the Mori cone of $\mathbb{P}_{\hat{\Delta}}$, as well the toric divisors $D_{x_i} = \{x_i = 0\}$

$Div.$	ν_i				$\hat{\nu}_i$				$l^{(E)}$	$l^{(B)}$
D_{x_0}	0	0	0	0	0	0	0	0	−6	0
$D_{x_1} = L$	12	−6	−1	−1	1	0	−2	−3	0	1
$D_{x_2} = L$	−6	12	−1	−1	0	1	−2	−3	0	1
$D_{x_3} = L$	−6	−6	−1	−1	−1	−1	−2	−3	0	1
$D_z = E$	0	0	−1	−1	0	0	−2	−3	1	−3
$D_x = 2H$	0	0	2	−1	0	0	1	0	2	0
$D_y = 3H$	0	0	−1	1	0	0	0	1	3	0

(2.7.35)

The classical topological data of the 3-fold M are easily calculable from the toric construction. The Euler is $\chi(M) = -540$, the two independent Hodge numbers are $h^{1,1}(M) = 2$, $h^{2,1}(M) = 272$, the classical triple intersection numbers are given by[62]

$$C^0_{111} = H^3 = 9, \quad C^0_{112} = H^2 \cdot L = 3, \quad C^0_{122} = H \cdot L^2 = 1, \quad C^0_{222} = L^3 = 0 \tag{2.7.36}$$

[61] It can be also written as the zero locus of a degree 18 polynomial in the weighted projective space $\mathbb{P}^4(1,1,1,6,9)$ called $X_{18}(1,1,1,6,9)$.

[62] In the notation of [186] these intersections are encoded in the ring $\mathcal{R} = 9J_E^3 + 3J_E^2 J_B + J_E J_B^2$. H, L are the notations for the divisors used in [64].

where H and L are the divisors dual to the curves defined by the Mori vectors $l^{(E)}$ and $l^{(B)}$ and the Kähler classes J_E and J_B. The intersections with the second Chern class c_2 of M are

$$\int_M c_2 \wedge J_E = [c_2] \cdot H = 36, \qquad \int_M c_2 \wedge J_B = [c_2] \cdot L = 102\,. \tag{2.7.37}$$

2.7.6 Elliptic Fibrations, Flops and Transitions

In the second elliptic fibration that we discuss, we take as base the Hirzebruch surface $\mathbb{F}_1$, which can be seen as blow up of $\mathbb{P}^2$ and corresponds to the polyhedra in 3 Fig. 1. We take the same fibre type. i.e. $a_F = 60$ and $s = 1$ in (2.7.34). The polyhedra are hence specified by

$$\begin{array}{c|ccccc|ccc|ccc}
 & & & \bar{\hat{\nu}}_i & & & l^{(E)} & l^{(f)} & l^{(s)} & \tilde{l}^{(E)} & \tilde{l}^{(f)} = l^{(f)} + l^{(s)} & \tilde{l}^{(s)} \\
D_0 & 1 & 0 & 0 & 0 & 0 & -6 & 0 & 0 & -6 & 0 & 0 \\
D_1 & 1 & 1 & 0 & -2 & -3 & 0 & 0 & 1 & 1 & 1 & -1 \\
D_2 & 1 & 0 & 1 & -2 & -3 & 0 & 1 & 0 & 0 & 1 & 0 \\
D_3 & 1 & -1 & -1 & -2 & -3 & 0 & 0 & 1 & 1 & 1 & -1 \\
D_4 & 1 & 0 & -1 & -2 & -3 & 0 & 1 & -1 & -1 & 0 & 1 \\
D_z & 1 & 0 & 0 & -2 & -3 & 1 & -2 & -1 & 0 & -3 & 1 \\
D_x & 1 & 0 & 0 & 1 & 0 & 2 & 0 & 0 & 2 & 0 & 0 \\
D_y & 1 & 0 & 0 & 0 & 1 & 3 & 0 & 0 & 3 & 0 & 0
\end{array}\,. \tag{2.7.38}$$

Here $\tilde{l}^{(E)} = l^{(E)} + l^{(s)}$, $\tilde{l}^{(f)} = l^{(f)} + l^{(s)}$ and tilde $l^{(s)} = -l^{(s)}$ are the Mori generators of the flopped phase and we denote with E again the elliptic fibre and with f the fibre and s the section of the Hirzebruch surface. This example shows that there are two Calabi-Yau phases possible in the elliptic fibration over $\mathbb{F}_1$, which are related by flopping s a $\mathbb{P}^1$, representing a curve with self-intersection (-1) in the base that is represented by $l^{(s)}$. The geometries of the two phases for, which the triangulation of $\hat{\Delta}$ is induced from the picture in the l.h.s of Fig. 4, is that in the first phase we have an elliptic fibration and a K_3 fibration. Moreover in this phase there is a local geometry over the (-1) curve in the base, which is a rational elliptic fibration known as half K3. We specify the intersections $C^{(0)}_{ijk}$ by the coefficients of an intersection ring and get in the first phase

$$\mathcal{R} = 8J_E^3 + 3J_E^2 J_f + J_E J_f^2 + 2J_E^2 J_s + J_f J_s J_E\,. \tag{2.7.39}$$

Moreover $\int_M c_2 J_E = 92$, $\int_M c_2 J_f = 36$ and $\int_M c_2 J_s = 24$.

For the second phase we flop the $\mathbb{P}^1$ that corresponds to the Mori cone element $l^{(s)}$ and get the triangulation of $\hat{\Delta}$ that is induced from the middle picture in Fig. 4.

Generally if we flop the curve C this changes the triple intersection of the divisors $J_i J_j J_k$ by [326]

$$\Delta_{ijk} = -(C \cdot J_i)(C \cdot J_j)(C \cdot J_k) . \tag{2.7.40}$$

Here the intersection of the curves C_i which correspond to the mori cone vector $l^{(i)}$ with the toric divisors D_k is given by $(C_i \cdot D_k) = l_k^{(i)}$. On the other hand the J_k are combinations of the D_k restricted to the hypersurface so that $(J^k \cdot C_i) = \delta_i^k$.

In addition one has to change the basis in order to maintain positive intersection numbers[63] $\tilde{l}^{(E)} = l^{(E)} + l^{(s)}$, $\tilde{l}^{(f)} = l^{(f)} + l^{(s)}$ and $\tilde{l}^{(s)} = -l^{(s)}$. For the J_i, which transform dual to the curves, we get then the intersection ring in the new basis of the Kähler cone

$$\mathcal{R} = 8\tilde{J}_E^3 + 3\tilde{J}_E^2\tilde{J}_f + \tilde{J}_E\tilde{J}_f^2 + 9\tilde{J}_E^2\tilde{J}_s + 3\tilde{J}_E\tilde{J}_f\tilde{J}_s + \tilde{J}_f^2\tilde{J}_s + 9\tilde{J}_E\tilde{J}_s^2 + 3\tilde{J}_f\tilde{J}_s^2 + 9\tilde{J}_s^3 . \tag{2.7.41}$$

The intersections with c_2 are not affected by the flop, only the basis change has to be taken into account. In this phase a del Pezzo surface of degree one, which is the blow up of $\mathbb{P}^2$ in eight points, can be shrunken to get to the elliptic fibration over $\mathbb{P}^2$.

2.8 *Representations of the (n,0) Form Ω and Special Integrals of Them*

Beside the Kähler form ω an unique nowhere vanishing $(n, 0)$-form $\Omega(\underline{z})$ is the characteristic form on a Calabi-Yau n-fold. It is also the fundamental object to define the period integrals and the variation of Hodge structures with the change of complex structure parameters here symbolically indicated by $\underline{z}$, if we refer to complex structure parameters in which the point of maximal unipotent is at $\underline{z} = 0$, otherwise by $\underline{a}$. We will first study various representations of Ω_n for the compact and non-compact Calabi-Yau n-folds discussed in Sect. 2.7, as well as the period over a distinguished

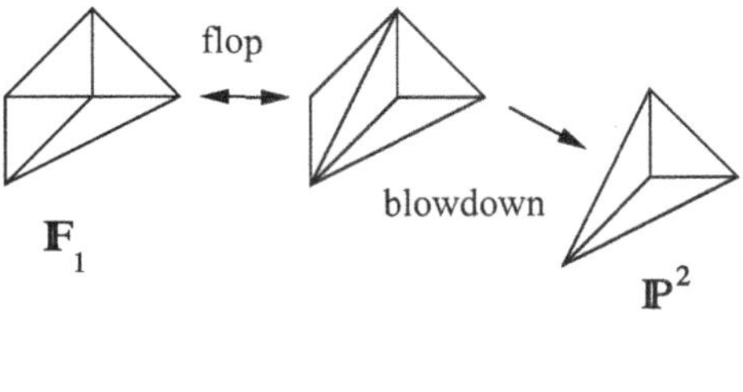

Fig. 4 The base triangulation for the flop in second example and the blowdown of an degree 1 del Pezzo surface

[63] This is one criterion that holds in a simplicial Kähler cone. The full specification is that $\int_C J > 0$, $\int_\mathcal{D} J \wedge J > 0$ and $\int_M J \wedge J \wedge J > 0$ for J in the Kähler cone and C, $\mathcal{D}$ curves and divisors. E.g. if the latter is simplicial and generated by J_i then $J = \sum d_i J_i$ with $d_i > 0$.

cycle, which has the topology of a real n-torus and is generically present in toric embeddings. In the next section we use this information to obtain the complete D-module of Picard-Fuchs operators. The completeness of the latter can be checked with the methods to determine the n-point functions, described in Sect. 2.9.2.

2.8.1 Representations of the Holomorphic (n,0) Form

Let us first explain the different expressions to represent the holomorphic (n,0) forms for the compact Calabi-Yau manifolds given as hypersurfaces or complete intersection that we studied in Sect. 2.7.3.

We assume the reader to be familiar with the $(1, 0)$ for the elliptic curve (2.4.25). However there are equivalent and more symmetric forms given by residuum expressions, which after performing the residuum integral take a form that specialised to (2.4.25) as we explain at the beginning of the next subsection.

Residuum Forms

In particular the nowhere vanishing holomorphic $(n, 0)$-form can be defined in a coordinate patch of the $(n+1)$-dimensional toric ambient space by a contour integral or a residuum at $P_\Delta = 0$

$$\Omega_n = \frac{a_0}{(2\pi i)} \oint_{P_\Delta=0} \frac{1}{P_\Delta} \wedge_{j=1}^{n+1} \frac{dY_j}{Y_j} \,. \tag{2.8.1}$$

Here we used the relations (2.7.12) to eliminate $l(\Delta_{n+1})-n-2$ coordinates, so that P_Δ becomes a Laurant monomial. In particular we scaled $Y_0 = 1$. As in (2.7.11) we have to make sure that the points corresponding to the remaining coordinates are linear independent. One advantage of the representation is that the integral over a real n-cycle $\mathbb{T}^n$ that is induced from the toric ambient space and is defined by the locus $|Y_i| = 0$, $i = 1, \ldots, n+1$ can be readily performed as residuum

$$X^0 = \int_{\mathbb{T}^n} \Omega_n = \frac{a_0}{(2\pi i)^{n+1}} \int_{|Y_i|=0} \frac{1}{P_\Delta} \wedge_{j=1}^{n+1} \frac{dY_j}{Y_j} \,. \tag{2.8.2}$$

In particular in the large volume limit this integral can be performed by identifing $P_\Delta = a_0(1 + R)$ and expanding in small R. This is a valid expansion with finite values as $a_0 \to \infty$ majorises in the large volume limit all a_i, $i \neq 0$ and the factored a_0 cancels in the expression of Ω_n. Using the multinomial formulas to rewrite the powers of R, se get sums restricted by the residuums, see Sect. 2.8.2 for the explicit results. In particular if we use the generalised Mori cones $l^{(k)}$ (2.8.18) to organize these sums we get immediately universal expressions in terms of the complex structure $z_i = 0$ defined in (2.7.27).

This formalism extends to complete intersection where the three forms is defined as

$$\Omega_n = \frac{\prod_{i=1}^r a_{0,i}}{(2\pi i)^r} \oint_{P_{\Delta_1}=0} \cdots \oint_{P_{\Delta_r}=0} \frac{1}{\prod_i P_{\Delta_i}} \wedge_{j=1}^{n+r} \frac{dY_j}{Y_j} \,. \tag{2.8.3}$$

Here the $a_{0,i}$ are coefficient of the coordinate of the inner point in Δ_i. The corresponding Y_* have been scaled to one and the Y_j appearing in (2.8.3) have correspond to linear independent points. The analogous period integral to (2.8.2)

$$X^0 = \frac{\prod_{i=1}^r a_{0,i}}{(2\pi i)^{n+r+1}} \int_{|Y_i|=0} \frac{1}{\prod_i P_{\Delta_i}} \wedge_{j=1}^{n+r} \frac{dY_j}{Y_j} \,. \tag{2.8.4}$$

can be performed in the same manner as (2.8.2) by expanding each of the polynomials as $P_{\Delta_i} = a_{0,i}(1 + R_i)$ and performing the residuum integral using generalised Mori cone vectors (2.8.18) see sectionrefgeneralizedMoricone and [31, 187].

We note that the results $X^0 = 1 + \mathcal{O}(z)$ is always a normalized period in the integral symplectic basis (2.6.30). It is identified with the $D0$ charge at the large volume point.

We can also express (2.8.1) in homogeneous coordinates. In this case we use $l(\Delta_{n+1}) - n - 3$ scaling relations to set all but $n+2$ coordinates to 1. The remaining scaling relation of the remaining $l^{(1)}$ should act non-trivial on the coordinates x_i with weight w_i in analogy to a weighted projective space $x_i \mapsto x_i \lambda^{w_i}$, cfor $i = 1, \ldots, n+2$ and $\lambda \in \mathbb{C}^*$. Then one can check that the following expression is well defined under the remaining scaling relation.

$$\Omega_n(\underline{a}) = \oint_\gamma \frac{\mu_{n+1}}{P_\Delta(x, \underline{a})}, \tag{2.8.5}$$

where γ is a path in $\mathbb{P}_{\hat{\Delta}}$ around $P_\Delta = 0$ and[64] and μ is given as

$$\mu_{n+1} = \sum_{k=1}^{n+2} (-1)^k w_k x_k \mathrm{d}x_1 \wedge \ldots \wedge \widehat{\mathrm{d}x_k} \wedge \ldots \wedge \mathrm{d}x_{n+2} \,. \tag{2.8.6}$$

This is the form the form appears in the work of [149], known as the Griffiths residuum form.

[64]The x_i are obtained from the X_i by setting all but $n+1$ suitable ones to 1. The choice is canonical if $\mathbb{P}_{\hat{\Delta}}$ is the resolution of a weighted projective space $\mathbb{P}^{n+1}(w_1, \ldots, w_{n+2})$. Then the x_i are its coordinates.

Patchwise Form

Here we perform the integral over the small circle γ say in the patch U_k, i.e. $x_k = 1$ to bring the expression of the $(n, 0)$ form to one which is familiar from the study of Riemann surfaces. To simplify notation we call $P_\Delta(x, a)$ just P below.

Note first that in each coordinate patch U_l, $x_l = 1$ and $\mathrm{d}x_l = 0$ so the sum (2.8.6) collapses to a single term. The w_k makes (2.8.6) immediatly applicable to hypersurfaces in weighted projective space [98] $\mathbb{P}^n[w_0, \ldots, w_n]$, which are generalizations of $\mathbb{P}^n$, in which the equivalence class under the $\mathbb{C}^*$ action is defined by

$$[x_0, \ldots, x_n] \sim [\lambda^{w_0} x_0, \ldots, \lambda^{w_n} x_n] \qquad (2.8.7)$$

with $\lambda \in \mathbb{C}^*$ and $w_i \in \mathbb{N}_+$. The Stanley Reisner ideal is $x_0 = \ldots = x_n = 0$. These spaces have $\mathbb{Z}_m$ singularitities normal to the codim k strata given by $x_{i_1} = \ldots = x_{i_k} = 0$ if $m > 1$ is a common factor of the k weights $w_{i_1}, \ldots, w_{i_k}$. The resolutions of these $\mathbb{Z}_m$ singularities do not affect the fact that the complex structure deformations of a hypersurface given by the vanishing of degree d polynomial $P(\lambda^{w_0} x_0, \ldots \lambda^{w_n} x_n) = \lambda^d P(x_0, \ldots, x_n)$, can be studied using the periods defined by (2.8.5) if the hypersurface is a Calabi-Yau manifold, which is equivalent to $d = \sum_{i=0}^n w_i$.

In order to reduce now one integration over $\mathrm{d}x_i$ to the residuum integration $\int \frac{\mathrm{d}p}{p} = 2\pi i$ we perform a coordinate transformation from $(x_1 \ldots \widehat{x_k} \ldots x_{n+2})$ to $(x_1 \ldots \widehat{x_k} \ldots \widehat{x_i} \ldots x_{n+2}, P)$ under which the measure $\mathrm{d}x_1 \wedge \ldots \widehat{\mathrm{d}x_k} \ldots \wedge \mathrm{d}x_{n+1}$ transforms to $\left(\frac{\partial P}{\partial x_i}\right)^{-1} \mathrm{d}x_1 \wedge \ldots \widehat{\mathrm{d}x_k} \ldots \widehat{\mathrm{d}x_i} \ldots \wedge \mathrm{d}x_{n+2} \wedge \mathrm{d}P$. Because of transversality $\mathrm{d}P = 0$ has no common solution with $P = 0$ and we can always pick an k and i so that $\left(\frac{\partial P}{\partial x_i}\right) \neq 0$ for $P = 0$. Therefore the integrand will have a single pole at $\frac{1}{P}$ and integration leads to

$$\Omega_n(\underline{z}) = \frac{a_0 w_k x_k \mathrm{d}x_1 \wedge \ldots \widehat{\mathrm{d}x_k} \ldots \widehat{\mathrm{d}x_i} \ldots \wedge \mathrm{d}x_{n+2}}{\frac{\partial P}{\partial x_i}} =: \frac{a_0 \nu_i^{(n)}}{\frac{\partial P}{\partial x_{i_1}}} . \qquad (2.8.8)$$

As we mentioned at the begining this $(n, 0)$ form is analogous specializes to the wellknown $(1, 0)$ form$\Omega \sim \frac{\mathrm{d}x}{y}$ in the case of an elliptic curve realised as cubic in $\mathbb{P}^2$ with the inhomogeneous equation in the $z = 1$ patch given in the Weierstrass form $y^2 = 4x^3 - g_2 x - g_3$.

Analogously we can express (2.8.3) in homogenous coordinates as

$$\Omega_n = \oint_{\gamma_1} \ldots \oint_{\gamma_r} \frac{\left(\prod_{i=1}^r a_{0,i}\right) \mu_{n+r}}{P_{\Delta_1} \ldots P_{\Delta_r}}, \qquad (2.8.9)$$

where the γ_i, $i = 1, \ldots, r$ are small circles in the ambient space around each constraint P_{Δ_i} and perform the residue integral to arrive at

$$\Omega_n = \frac{\left(\prod_{i=1}^r a_{0,i}\right) \nu^{(n)}_{i_1,\ldots,i_r}}{\frac{\partial P_{\Delta_1}}{\partial x_{i_1}} \cdots \frac{\partial P_{\Delta_r}}{\partial x_{i_r}}}, \tag{2.8.10}$$

where $\mu^{(n+r)}$ and $\nu^{(n)}_{i_1,\ldots,i_r}$ are straightforward generalisations of the forms introduced for the hypersurfaces in (2.8.5) and (2.8.8). The forms (2.8.8) and (2.8.10) are particular useful to perform the partial integration—or Griffiths reduction methods in order to derive the Picard-Fuchs differential modul. With the measure (2.8.6) this form applies to complete intersection of complex dimension n given by $P_1 = \ldots, P_r = 0$ of degree $d_1, \ldots, d_r$ in weigted projective space $\mathbb{P}(w_0, \ldots, w_{n+r})$ if the later are Calabi-Yau, which is equivalent to $\sum_{i=1}^r d_i = \sum_{i=0}^{n+r} w_i$.

Oscillatory Integral Form

In the context of singularities it is particular useful to express periods integrals as oscillatory integrals [22]. For example for homogenous constraints $P = 0$ in weighted projective $\mathbb{P}^4[w_1, \ldots, w_4]$ discussed above one can write the residue form as

$$\Pi_i = \int_{\Gamma_i} \Omega_n = \int_{\gamma_i} \frac{a_0 dx_1 \cdots dx_5}{P}. \tag{2.8.11}$$

Here $\Gamma_i \in H_3(M)$ while $\gamma_i \in H_5(\mathbb{C}^5 \setminus P(x) = 0)$. The latter expression can be transformed into an oscillatory integral of the form

$$\Pi_i = \int_{\gamma_i} \frac{a_0 dx_1 \cdots dx_5}{P} = \int_{\gamma_i^\pm} a_0 e^{\mp P_0(x)} dx_1 \cdots dx_5, \tag{2.8.12}$$

where P_0 is a non-singular from of the constraint, where the deformation parameter a_i are set to zero and $\gamma_i^\pm \in H_5(\mathbb{C}^5, \mathrm{Re}(P_0(x) = \pm\infty)$ and the map from γ_i to $\gamma_i^\pm$ is given by a contour deformation [22, 35].

Meromorphic $(n - k - 1, 0)$ Forms in Local Mirror Symmetry

Here we perform the reduction of the holomorphic $(n, 0)$-form to the meromophic $(n-k-1, 0)$ forms by restricting the coordinates to those living on an $n-k$ plane of $\hat{\Delta}_{n+1}$ for $k \geq 1$ as discussed in Sect. 2.7.4 following in [214]. For example (2.8.1) restricts as

$$\lambda_{n-k-1} = \frac{1}{(2\pi i)} \oint_{P_{\hat{\Delta}_{n-k}}=0} \log(P_{\hat{\Delta}_{n-k}}) \wedge_{j=1}^{n-k} \frac{dX_j}{X_j}. \tag{2.8.13}$$

Let us assume we have a Calabi-Yau threefold i.e. $n = 3$. The interesting case[65] is $k = 1$ in which case λ_1 is a meromorphic one form on a Riemann surface Σ_g. There will be g inner points in $\hat{\Delta}_{n-k}$. This is for example the case if one engineers $\mathcal{N} = 2$ supersymmetric $SU(g+1)$ gauge theory [214]. The Seiberg-Witten curve has g parameters corresponding to the independent Casimirs $u_2, \ldots, u_{g+1}$ of the gauge group. An important consistency condition from geometric engineering is that λ becomes the Seiberg-Witten differential with the property that the $\partial_{u_k}\lambda$ become the g holomorphic one forms. i.e. $\partial_{u_k}\lambda = \omega_{k-1}$ for $k = 1, \ldots, g$. Now in the double scaling limit introduced in [214] to obtain the gauge theory, derivatives w.r.t. to u_k correspond to logarithmic derivatives w.r.t. to the $\tilde{a}_i$ which correspond to inner points in Δ_{n-k}. For example if $g = 1$ we get $\partial_{u_2} = \tilde{a}_0\partial_{\tilde{a}_0}$ where $\tilde{a}_0$ is the unique inner point in Δ_{n-k}. Indeed taking $\tilde{a}_0\partial_{\tilde{a}_0}\lambda_1 = \Omega_1$ as in (2.8.1). Moreover we can perform the integral

$$\begin{aligned}\lambda_1 &= \frac{1}{(2\pi i)}\oint_{P_{\hat{\Delta}_2}=0} \log(P_{\Delta_2})\frac{dx}{x}\frac{dp}{p}\\ &= \frac{1}{(2\pi i)}\oint_{P_{\hat{\Delta}_2}=0} \frac{dP}{P}\log(x)\frac{dp}{p}\\ &= \log(x)\frac{dp}{p}\end{aligned} \tag{2.8.14}$$

This latter form of the differential and the form of the local mirror

$$H(x, p, \underline{z}) := P_{\hat{\Delta}_2}(x, p, \underline{z}) = uv\ . \tag{2.8.15}$$

play a key role when we study local mirror symmetry by the variation of mixed Hodge structures of the compact Calabi-Yau in the non-compact limit.

Moreover we should point out that there is a generalisation to local varieties that are not defined by a single constraint. For example we could consider the quadric in $\mathbb{P}^3$, which has positive first Chern class and construct over it the bundle of the anti-canonical line bundle this yields in general non-compact non-toric Calabi-Yau space and the meromorphic form can be generalized to

$$\lambda_{n-r-1} = \frac{1}{(2\pi i)}\oint_{P_{\Delta_1}=0}\cdots\frac{1}{(2\pi i)}\oint_{P_{\Delta_r}=0}\log(P_{\nabla_1})\ldots\log(P_{\nabla_r})\wedge_{j=1}^{n-1}\frac{dX_j}{X_j}\ . \tag{2.8.16}$$

Likewise one can perform the integral over the algebraic torus $\mathbb{T}_{n-1}$ in the compact part of the ambient space to get the fundamental period.

The occurrence of the logarithm is familiar in the study of mixed Hodge structure associated to singularities, see e.g. [283] or [173] for a review. The variation of the mixed Hodge structure for log Calabi-Yau spaces (X, D) with D a divisor in

[65] $k = 2$ is also possible but gives a $g = 0$ mirror geometry, which is a bit more trivial.

particular the isomorphism

$$\phi : H^3(X \setminus D) \to \bigoplus_{p+q=3} H^q(X, \Omega^p_{\bar{X}}(\log(D)) \qquad (2.8.17)$$

to the log cohomology has been used to calculate superpotentials in [152].

2.8.2 The Generalized Mori Cone and the $\mathbb{T}_{n+r}$ Integral

For Calabi-Yau hypersurfaces or complete intersection M in a general toric ambient space one can often determine the generators of the Mori cone of M from the one of the ambient space. The Mori vectors $\bar{l}^{(a)}$ that span this cone represent curves $C^{(a)}, a = 1, \ldots, h_{11}(M)$ in the Calabi-Yau space ambient space that are dual to the Kähler cone of the ambient space and descend to the Calabi-Yau space M. Often one gets in this way only a sub cone of the Kähler cone of M, whose generators nevertheless define a well defined A-model curve counting problem with a well defined B-model description. The generalized Mori vectors are obtained by the $\bar{l}^{(a)}$ by extending them by the degrees of the complete intersection as follows

$$l^{(a)} = (l^{(a)}_{0,1}, \ldots, l^{(a)}_{0,r}; \bar{l}^{(a)}_1, \ldots, \bar{l}^{(a)}_s), \quad \text{for } a = 1, \ldots, h_{1,1}(M) = h_{2,1}(W). \qquad (2.8.18)$$

Here the negative of their first entries, i.e. $-l^{(a)}_{0,1}, \ldots, -l^{(a)}_{0,r}$, are the *positive* (multi)degree(s) of the algebraic constraints $P_1 = 0, \ldots, P_r = 0$ defining the complete intersection $r > 1$ or hypersurface $r = 1$ Calabi-Yau manifold. If they are non zero the dual divisors correspond to curves in the Kähler cone that cannot be blown down. For the other curves the first entries will be zero. For all vectors the second set of entries $l^{(a)}_1, \ldots, l^{(a)}_s$ are the intersections of the curve $C^{(a)}$ with the toric divisors of the ambient space. These curves and the intersection numbers can be determined purely combinatorial from the toric polyhedra $\hat{\Delta}$ and suitable triangulation that defines a semi ample ambient space. The suitable one are the complete star triangulations, see e.g. [187] for an early application and [79] for an extensive review of toric techniques. There exist now powerful computer programs under the umbrella of SAGE that can find all these triangulation using e.g. the program Topcom, calculate the toric intersection numbers and determine the Kähler cone and the generators of the dual Mori cone, i.e, the $(l^{(a)}_1, \ldots, l^{(a)}_s)$, $a = 1, \ldots, h_{11}(M) = h_{21}(W)$. Once this information is given the topological data that go in the theorem of C.T.C. Wall and the $\hat{\Gamma}$ class of the Calabi-Yau manifold M can be calculated with some care for the hypersurfaces and with more care for the complete intersections. The problem how the Mori cone of $\mathbb{P}_{\hat{\Delta}}$ restricts to the one of M has few subtleties that are discussed in [37]. In general co-dimension one points do not restrict to divisors of M and can therefore not lead to elements that correspond to the Kähler cone on M. In many examples one can therefore just consider singular triangulations of $\hat{\Delta}$ that do not involve these points. However this might lead to a Mori cone that is to small as some curves in the $\mathbb{P}_{\hat{\Delta}}$ still do not descend to M

[37]. For the complete intersections the situation is expected to be much complicated and the problem of restricting the Mori cone has not been discussed in general.

One crucial aspect of the generalized Mori cone in Batyrev's construction is that calculated for M and applied to the complex structure deformation problem of the mirror W its vectors determine the large radius point of W. The same is of course true with M and W exchanged.[66] The large radius point is given by $z_k = 0$ for all k where

$$z_k = (-1)^{\sum_a l^{(b)}_{0,a}} \prod_{i=1}^{n} a_i^{l_i^{(b)}} \qquad k = 1, \ldots, h = h^{21}(W) = h^{11}(M) \tag{2.8.19}$$

in terms of a_i, the coefficients in the polynomial constraints of the complete intersection in the torus variables (2.7.10).

Let us give the simplest examples. For the quintic in $\mathbb{P}^4$ one has a single generalized Mori vector $l = (-5; 1, 1, 1, 1, 1)$. The last entries $\bar{l}^{(a)}_1, \ldots, \bar{l}^{(a)}_s$ parametrize the linear relations among the points v_i, $i = 1, \ldots s$ spanning the rays of $\Sigma(1)$ of $\hat{\Delta}$, while the entries $l^{(a)}_1, \ldots, l^{(a)}_s$ can be understood also a linear relations including the inner point, when $\hat{\Delta}$ is put in a hyperplane at distance one from the origin so that the coordinates become $\bar{v}_i = (1, v_i)$, $i = 0, \ldots, s$, where the origin is shared by the sub polyhedra Δ_k defining the splitting of coordinates into those occuring in the polynomials P_{Δ_k}, $k = 1, \ldots, r$. For this reason all $l^{(a)}$ have the property that $\sum_{k=1}^r l^{(a)}_{0,k} + \sum_{k=1}^s \bar{l}^{(a)}_k = 0$. For example the complete intersection of two cubics in $\mathbb{P}^5$ has a single Mori vector $l = (-3, -3; 1, 1, 1, 1, 1, 1)$.

Let us as a first exercise calculate the $\mathbb{T}_3$ integral (2.8.2) for the quintic generalized in $\mathbb{P}^4$, which is a $\mathbb{T}_4$ integral in the ambient space. We parametrize the dense algebraic torus like in (2.10.6) and by performing the torus integral (2.8.2) we get

$$\begin{aligned} X^0 &= \int_{|Y_1|=1} \cdots \int_{|Y_4|=1} \frac{1}{1 + \sum_{i=1}^4 \frac{a_i}{a_0} Y_i + \frac{a_5}{a_0 Y_1 Y_1 Y_3 Y_4}} \wedge_{i-1}^4 \frac{dY_i}{Y_i} \\ &= \int_{|Y_i|=1} \sum_{l=0}^{\infty} (-1)^l \sum_{\substack{l=0,\ldots,\infty \\ \sum_{i=1}^5 k_i = l}} \left(\frac{l!}{k_1! \ldots k_5!} \right) \left(\frac{a_1}{a_0} Y_1 \right)^{k_1} \cdots \left(\frac{a_4}{a_0} Y_4 \right)^{k_4} \left(\frac{a_5}{a_0} \frac{1}{Y_1 \cdots Y_4} \right)^{k_5} \wedge_{i=1}^4 \frac{dY_i}{Y_i} \\ &= \sum_{l=0}^{\infty} \frac{(5l)!}{(l!)^5} z^l , \end{aligned} \tag{2.8.20}$$

[66] If the classical intersection refer to M, then the complex structure deformation a_i and the $l^{(k)}$ should be hatted as they refer to $\hat{\Delta}$ and (2.7.14). For notational convenience we omit the ^ below.

where $z = (-1)^{l_0} \prod_{i=0}^{5} a_i^{l_i}$. Similarly for the complete intersection of two cubics in $\mathbb{P}^5$ we get in the variables of the algebraic torus $\mathbb{T}_5$

$$\begin{aligned} P_{\nabla_1} &= a_0 Y_0 + \sum_{i=1}^{3} a_i Y_i = a_0 + a_1 Y_1 + a_2 Y_2 + a_3 Y_3 \\ P_{\nabla_2} &= a_0 Y_0 + \sum_{i=4}^{6} a_i Y_i = a_0 + a_4 Y_4 + a_5 Y_5 + a_6 \frac{1}{Y_1 \cdots Y_5} \end{aligned} \tag{2.8.21}$$

and by performing the integral (2.8.4) we get after a very similar calculation as in (2.8.20)

$$X^0 = \sum_{l=0}^{\infty} \frac{(3l)!(3l)!}{(l!)^6} z^l \tag{2.8.22}$$

with $z = \prod_{i=1}^{2} (-a_0)^{l_{0,i}} \prod_{i=1}^{6} a_i^{l_i} = \frac{\prod_{i=1}^{6} a_i}{a_0^6}$. This generalises very naturally for all cases discussed in Sect. 2.7.3. Using the generalised Mori vectors, we can summarize the calculation for the global cases

$$X^0 = \sum_{n_1=0}^{\infty} \cdots \sum_{n_h=0}^{\infty} \frac{\prod_{j=1}^{r} \left(-\sum_{\alpha=1}^{h} l_{0,j}^{(\alpha)} n_\alpha\right)!}{\prod_{j=1}^{|\Sigma(1)|} \left(\sum_{\alpha=1}^{h} l_j^{(\alpha)} n_\alpha\right)!} \prod_{\alpha=1}^{h} z_\alpha^{n_\alpha}, \tag{2.8.23}$$

where h is either $h_{11}(M)$ or $h_{21}(W)$ and the result depends only on the generalized Mori vectors. We note that while the entries $-l_{0,i}^{(\alpha)}$ are positive integers the entries $l_i^{(\alpha)}$ can be positive or negative. We replace $n!$ by $\Gamma(n+1)$ and analyze the pole behaviour of the coefficients in X^0 to define (2.8.23) in case of negative $l_i^{(\alpha)}$.

Let us consider also the local case and evaluate the integral for the local $\mathcal{O}(-3) \to \mathbb{P}^2$, which has the Mori vector[67] $l = (-3, 1, 1, 1)$. We integrate the algebraic two torus and get

$$\begin{aligned} t &= \int_{|Y_1|=0} \int_{|Y_2|=0} \log a_0^{-4} \left(a_0 + a_1 Y_1 + a_2 Y_2 + \frac{a_3}{Y_1 Y_2} \right) \frac{dY_1}{Y_1} \wedge \frac{dY_2}{Y_2} = -3 \log(a_0) \\ &+ \sum_{k=1}^{\infty} \frac{(-1)^k}{k} \int_{|Y_i|=0} \sum_{\sum_{i=1}^{3} k_i = k} \left(\frac{k!}{k_1! k_2! k_3!} \right) \left(\frac{a_1 Y_1}{a_0} \right)^{k_1} \left(\frac{a_2 Y_2}{a_0} \right)^{k_2} \left(\frac{a_3}{a_0 Y_1 Y_2} \right)^{k_3} \frac{dY_1}{Y_1} \wedge \frac{dY_2}{Y_2} \\ &= const + \log(z) + \sum_{k=1}^{\infty} \frac{(-1)^k (3k)!}{k (k!)^3} z^k, \end{aligned} \tag{2.8.24}$$

[67]Note that in this notation we the semicolon in the first position as there is no constraints with a degree.

where $z = \frac{a_1 a_2 a_3}{a_0^3}$. Note that a_1, a_2, a_3 can be set to arbitrary values by the torus automorphism that acts on the local geometry. Further the constant is natural, as λ_1 is a meromorphic form whose residue has to be normalized. If we restrict from the global model discussed in Sect. 2.7.5 to the local model then the global period X^0 becomes 1 in the local limit, while $t = X_{\mathbb{P}^2}/X^0$ represents the complexified volume of the $\mathbb{P}^2$, which since $X^0 = 1$ in the local model is represented as a period, which is the one over the algebraic two torus evaluated in (2.8.24). Let us note further that the elliptic curve C given as a cubic in $\mathbb{P}^2$ has a fundamental period (2.8.23) that corresponds to the generalised Mori vector $l = (-3; 1, 1, 1)$ with $\tilde{z} = -\frac{a_1 a_2 a_3}{a_0^3}$ and $\tilde{z}\frac{d}{d\tilde{z}}t = X_C^0(\tilde{z})$.

The generalised Mori vectors $l^{(a)}$ are technical core data of mirror symmetry for toric complete intersections. Let us end the section by summarizing the multitude of information they contain

1. They contain the degrees of the constraints and the $\mathbb{C}^*$ actions of the toric variety of ambient space and fix thereby M.
2. Equivalently they can be viewed as $U(1)$ charges vectors for the fields in the linear σ model [322].
3. They span the Mori cone of M, which is dual the Kähler cone of M.
4. They specify the point of *maximal unipotent monodromy* in the moduli space of W namely $z^{(a)} = 0$, where the $z^{(a)} = 0$ of (2.8.19) are good local coordinates near this points and all monodromies T^a around $z^{(a)} = 0$, $a = 1, \ldots, h_{21}(M)$ satisfy (2.6.8) with $N = 1$ and $k = \dim_C(W)$.
5. The periods of M are generalized hypergeometric functions with symplectic basis at $z^{(a)} = 0$ given by (2.9.58) and the generalized $l^{(a)}$ provide for those functions the information that the constants a, b, c provide for ordinary hypergeometric functions ${}_2F_1(a, b, c; z)$ (2.9.49).

2.9 *The D-Module of Picard-Fuchs Operators and Their Solutions*

In this section we study the systems of Picard equations for the periods of Calabi-Yau manifolds, their special properties and their solutions.

2.9.1 The General Form of the Picard-Fuchs Equation and Their Singularities

Here we discuss general properties of the different ideal $\mathcal{I}_{PF}$, which is generated from a linear system of operators that annihilate the periods. We denote such a system of generators by

$$\mathcal{L}_k^{(n_k)}\left(z_1,\ldots,z_h;\frac{\partial}{\partial z_1},\ldots\frac{\partial}{\partial z_h}\right),\qquad k=1,\ldots L\,. \tag{2.9.1}$$

Here $\mathcal{L}_k$ is a polynomial in the derivatives w.r.t. to the complex structure parameters and some meromorphic functions in z_i which have at most rational branch cuts. The index n_k indicates the degree of the differential operator, i.e. the highest total order in the derivatives. The differential ideal $\mathcal{I}_{PF}$ is obtained by acting of the left with other differential operators. We do not assume in this sections that z are large complex structure coordinates.

A priori is seems rather difficult to find solutions to such a system, as naively one might think that these are coupled partial differential equations. However the most important fact is that since the solutions are only the finite number of periods, the system is of rather special type. We expect $h^n_{horizontal}(M)$, i.e. for Calabi-Yau 3 folds $h^3(M)$ solutions, since all of $h^3(M)$ is horizontal. For even dimensional manifolds like e.g. Calabi-Yau fourfolds the distinction is important. Is has been found that the in special cases one has to include some of the vertical part of the cohomology of $h^{2,2}$, but in any case the number of solutions will be finite. For the threefold case and all other cases, where only the horizontal cohomology is relevant, one can expects from the Griffith transversality that the maximal degree of the operators will be $n+1$. On the other hand at the maximal unipotent point one has exactly h logarithmic solutions. This excludes the possibility that there are linear differential operators. So $\mathcal{I}_{PF}$ will be generated by degree $2,\ldots,4$ operators for threefolds.

According to analysis of the degenerations of periods we expect only regular singular points as explained in Sects. 2.6.2 and 2.6.3. In a sense the solutions can be viewed as generalizations of hypergeometric functions and in the toric hypersurface case they are exactly that.

To find the singularities of $\mathcal{I}_{PF}$ we will first recall how to find the singularities of M. For hypersurfaces they are given by solving the equations

$$P(x,\underline{z})=0,\qquad dP=\sum_{i=1}^{n+1}\frac{\partial}{\partial x_i}P(x,\underline{z})dx^i=0\,. \tag{2.9.2}$$

This is only possible if we restrict the complex moduli $\underline{z}$ to a complex co-dimension one locus S_M in in the moduli space $\mathcal{M}_{cs}$, which is in general singular itself. Direct solutions of (2.9.2) is difficult except for the simplest examples as the equations are non linear. A way to find S_M is to calculate the resolvent of all the equations in (2.9.2) where the x_i are set at their base locus, i.e. generic x_i, hypersurfaces $x_i=0$ and their intersections. This yields in general many components, which might have points of tangencies, singular points and points with both problems. The *generic component* of discriminant, i.e. when all $x_i\neq 0$ will be a conifold locus. Application of general theorems about desingularizations of Hironaki [175] guarantees that all singularities can always be resolved so that $\hat{S}_M$ are specified by smooth divisors with *normal crossing*, i.e. with no tangencies. This latter fact is true for all representations of M, but the conditions (2.9.2) have to be generalized. E.g. for complete intersections the conditions are

$$\begin{aligned} P_l(x,z)=0, \qquad l=1,\ldots r \quad \text{and} \quad dP_1\wedge\ldots dP_r=0 \\ \leftrightarrow \text{rank}\left(\frac{\partial}{\partial x_i}P_k(x,z)\right)<r, \quad i=1,\ldots,h, \quad k=1,\ldots,r\,, \end{aligned} \tag{2.9.3}$$

where the second line is a restatement of the second condition is the first line.

Certainly we expect the singularities of the $\mathcal{I}_{PF}$ to include the above singularities. The analysis of the latter can be done as follows. Introduce the so called symbols of the $\mathcal{L}_k^{(n_k)}$, by identifying the derivatives $\frac{\partial}{\partial z_k}=\xi_k$ with formal variables ξ_k. Now take the homogenous degree n_k part of $\mathcal{L}_l^{(n_k)}$ including its z-dependent coefficients and consider

$$s^{(n_k)}(z,\xi)=0\,. \tag{2.9.4}$$

The singularities are at the solution of (2.9.4) in which the ξ are fixed to their base locus. This again is a complex co dimension one locus in $\mathcal{M}_{cs}$, called S_{PF} In general we expect $S_M\subset S_{PF}$. S_{PF} can have additional locii, which are orbifold locii, which have monodromy and so called *apparent singularities* around which the solutions of the $\mathcal{I}_{PF}$ have no monodromy. Let us call $\mathcal{I}^m_{PF}$ the locii around which there is monodromy. The theorem of Hironaka [175] guarantees that there is a smooth resolution $\hat{S}_{PF}$ specified by smooth divisors with *normal crossing*.

2.9.2 Griffith Transversality and the *n*-Point Coupling

Crucial properties of the Picard-Fuchs differential ideal $\mathcal{I}_{PF}$ follow from to Griffith transversality and the fact that the n-point function can be chosen to be rational functions in a natural parametrization of the complex moduli space, given for the torically embedded examples by (2.8.19) and are more generally determined by the form (2.9.26). One can use these properties, the existence of a point of maximal unipotent monodromy and integrality of the mirror map and the genus zero instanton expansion to classify Calabi-Yau differential ideals $\mathcal{I}_{PF}$. The idea has been implemented for one parameter cases. It is a powerful idea for classifying Calabi-Yau spaces, since it makes no reference to the actual construction of the variety. An overview of this approach for one parameter families can be found in [308].

Here we want to explore in a pragmatic way the concept of Griffith transversality and the $n-$ point coupling, by explaining in some detail how the latter can be calculated from the Picard-Fuchs equations, with some emphasis on the properties that the latter must have for that program to work.

Let M be a Calabi-Yau n-fold with a $h = h_{n-1,1}$ dimensional complex moduli space, which we parametrize by z_i, $i = 1, \ldots, h$, and Ω its complex $(n, 0)$ form. We define

$$\Xi^{(k_1,\ldots,k_h)} = \int_M \Omega \wedge \partial_{z_1}^{k_1} \cdots \partial_{z_h}^{k_h} \Omega =: \int_M \Omega \wedge \underline{\partial}^{\underline{k}} \Omega =: \Pi(z) \eta \underline{\partial}^{\underline{k}} \Pi(z) , \qquad (2.9.5)$$

with $n_i \in \mathbb{N}_0$. Here $\Pi(z)$ is the period vector over the horizontal middle homology $H_n^{hor}(M)$ and η is the sympletic- or symmetric form on $H_n^{hor}(M)$ for n odd or even respectively. We introduce the abbreviations

$$\underline{k} := \{k_1, \ldots, k_h\}, \quad |\underline{k}| = \sum_{i=1}^{h} k_i, \quad \underline{\partial}^{\underline{k}} := \partial_{z_1}^{k_1} \cdots \partial_{z_h}^{k_h} . \qquad (2.9.6)$$

Due to Griffiths transversality and the definition of the n-point $C_{i_1,\ldots,i_n}$ coupling we assign

$$\Xi^{(k_1,\ldots,k_h)} = \begin{cases} W^{(k_1,\ldots,k_h)} & \text{if } \sum_{i=1}^{h} k_i > n , \\ C^{(k_1,\ldots,k_h)} & \text{if } \sum_{i=1}^{h} k_i = n , \\ 0 & \text{if } \sum_{i=1}^{h} k_i < n , \end{cases} \qquad (2.9.7)$$

Here the $C^{(k_1,\ldots,k_h)}$ are related[68] to the $C_{i_1,\ldots,i_n} \in \mathrm{Sym}_n(T^{1,0*}) \oplus \mathcal{L}^{-2}, i_p = 1, \ldots, h$ for $p = 1, \ldots, n$ in the obvious way, e.g $C^{(n,0,\ldots,0)} = C_{1,\ldots,1}$, with n 1's, $C^{(n-1,1,\ldots,0)} = C_{1,\ldots,1,2}$ with $n-1$ 1's etc.

We assume from now on that the $W^{(\underline{k})}$ obey $\sum_{i=1}^{h} k_i = n + 1$, even though the higher $W^{(\underline{k})}$ still carry interesting information about $\mathcal{I}_{PF}$. It is an elementary exercise to show by taking and combining derivatives of $C^{(k_1,\ldots,k_h)}$ and using the Griffith transversality that

$$W^{(\underline{k})} = \frac{1}{2} \sum_{\substack{\underline{m}^{(l)} \text{ with } m_i^{(l)} \in \mathbb{N}_0 \\ (m_1^{(l)},\ldots,m_l^{(l)}+1,\ldots,m_h^{(l)}) = \\ (k_1,\ldots,k_h)}} (m_k^{(l)} + 1) \partial_{z_l} C^{(\underline{m}^{(l)})} . \qquad (2.9.8)$$

[68] If the coordinate system say z_i is important, we might also denote the $C_{i_1,\ldots,i_n}$ by $C_{z_{i_1},\ldots,z_{i_n}}$, see e.g. (2.9.17).

Here the sum is over all non-negative integral h-tuples $\underline{m}^{(l)} = (m_1^{(l)}, \dots, m_h^{(l)})$, which fulfill the indicated property. E.g. one gets

$$W^{(n+1,0,\dots,0)} = \frac{n+1}{2}\partial_{z_1} C^{(n,0,\dots,0)}, \quad W^{(n,1,0\dots,0)} = \frac{1}{2}(n\partial_{z_1} C^{(n-1,1,0,\dots,0)} + \partial_{z_2} C^{(n,0,\dots,0)}), \ etc. \tag{2.9.9}$$

So far our results depend only mildly on the dimensions not at all on the specific h-parameter family we are considering, since we have used nothing else then Griffiths transversality. For the convenience of the reader we spell out (2.9.8) for the relevant case of the threefolds

$$\begin{aligned}
W^{(4,0,0,0)} &= 2\partial_{z_1} C^{(3,0,0,0)} \\
W^{(3,1,0,0)} &= \tfrac{3}{2}\partial_{z_1} C^{(2,1,0,0)} + \tfrac{1}{2}\partial_{z_2} C^{(3,0,0,0)} \\
W^{(2,2,0,0)} &= \partial_{z_1} W^{(1,2,0,0)} + \partial_{z_2} C^{(2,1,0,0)} \\
W^{(2,1,1,0)} &= \partial_{z_1} W^{(1,1,1,0)} + \tfrac{1}{2}\partial_{z_2} W^{(2,0,1,0)} + \tfrac{1}{2}\partial_{z_3} W^{(2,1,0,0)} \\
W^{(1,1,1,1)} &= \tfrac{1}{2}(\partial_{z_1} C^{(0,1,1,1)} + \partial_{z_2} C^{(1,0,1,1)} + \partial_{z_3} C^{(1,1,0,1)} + \partial_{z_4} C^{(1,1,1,0)}) \ .
\end{aligned} \tag{2.9.10}$$

The actual dependence on the Calabi-Yau manifold comes in via the Picard-Fuchs differentially left graded ideal $\mathcal{I}_{PF}$, which is generated by

$$\mathcal{L}_\alpha = \sum_{\underline{k}} A_\alpha^{(\underline{k})}(z)\underline{\partial}^{\underline{k}}. \quad , \tag{2.9.11}$$

Since $\mathcal{L}_\alpha | \prod(z) = 0$ and using Griffiths duality we immediately obtain the relation

$$\sum_{\underline{k}} A_\alpha^{(\underline{k})}(z)\Xi^{(\underline{k})} = \sum_{|\underline{k}|=n+1} A_\alpha^{(\underline{k})}(z)W^{(\underline{k})} + \sum_{|\underline{m}|=n} A_\alpha^{(\underline{m})}(z)C^{(\underline{m})} = 0. \tag{2.9.12}$$

Here we understand that we have constructed all elements $\mathcal{L}_\alpha \in \mathcal{I}_{PF}$ with differential degree $|\mathcal{L}_\alpha| = n+1$ by taking suitable derivatives of the generators of $\mathcal{I}_{PF}$ if the latter have lower degrees. Note that the ideal $\mathcal{I}_{PF}$ is complete iff

- The complete set of elements $\mathcal{L}_\alpha \in \mathcal{I}_{PF}$ with $|\mathcal{L}_\alpha| = n$ used in (2.9.12) allows to express the

$$p_n(h) = \frac{1}{n!}\prod_{i=0}^{n-1}(h+i)$$

  different n-point functions $C^{(\underline{m})}$ with $|\underline{m}| = n$ in terms in terms of one, say $C(z) = C^{(n,0\dots,0)}(z)$.

- The complete set of elements $\mathcal{L}_\alpha \in \mathcal{I}_{PF}$ with $|\mathcal{L}_\alpha| = n+1$ used in (2.9.12) allows to express the $p_{n+1}(h)$ different n-point functions $W^{(\underline{k})}$ with $|\underline{k}| = n+1$ in terms of these $C^{(\underline{m})}$ with $|\underline{m}| = n$ so eventually in terms of $C(z)$.
- Together with the equations (2.9.8) one can obtain a complete system of differential equations of the form

$$\frac{\partial_{z_k} C(z)}{C(z)} = \frac{p_k(z)}{q_k(z)}, \quad k = 1, \ldots, h, \tag{2.9.13}$$

which can be integrated for $C(z)$ up to a constant c that can be fixed by the classical intersection number of the A-model.

Exercise

1) Prove the relations (2.9.9), (2.9.10). Show that from the Picard-Fuchs equation for the quintic (2.10.13) one gets $A^{(4)} = z^3(5^5 z - 1)$ and[69] $A^{(3)} = 2z(2^2 \cdot 5^5 z - 3)$. Using (2.9.12) and from (2.9.10) $W^{(4)} = 2\partial_z W^{(3)}$ we can integrate

$$C_{zzz} = \exp\left(-\frac{1}{2}\int_c^z \mathrm{d}z' \frac{A^{(3)}}{A^{(4)}}\right) = \frac{5}{z^3(1-5^5 z)}, \tag{2.9.14}$$

where we fixed c to match the A-model normalization $C_{ttt} = 5 + \mathcal{O}(q)$.
2) For the system (2.11.6) we consider first $\mathcal{L}_1, \partial_{z_1}\mathcal{L}_2, \partial_{z_2}\mathcal{L}_2$ in (2.9.12) to express e.g. $W^{(3,0)} = C_{z_1,z_1,z_1}$ in terms of $C_{z_1 z_1 z_2}$, $C_{z_1 z_2 z_2}$ and $C_{z_2 z_2 z_2}$. Using $\partial_{z_1}\mathcal{L}_1, \partial_{z_2}\mathcal{L}_1, \partial_{z_1}^2\mathcal{L}_2, \partial_{z_1}\partial_{z_2}\mathcal{L}_2 = \partial_{z_2}^2\mathcal{L}_2$ in (2.9.12) we may express $W^{(4,0)}$ in terms of $W^{(3,0)}$ and integrate[70] w.r.t. z_1. Proceeding this way we get after rescaling of $a = 1728 z_1$ and $b = 4z_2$ the triple couplings

$$\begin{aligned} C_{aaa} &= \tfrac{4}{a^3\Delta_1}, \quad & C_{aab} &= \tfrac{2(1-a)}{a^2 b\Delta_1}, \\ C_{abb} &= \tfrac{(2a-1)}{ab\Delta_1\Delta_2}, \quad & C_{bbb} &= \tfrac{1+b-a(1+3b)}{2b^2\Delta_1\Delta^2}, \end{aligned} \tag{2.9.15}$$

where we defined the components of the discriminant as

$$\Delta_1 = 1 - 2a - a^2(1-b), \qquad \Delta_2 = (1-b). \tag{2.9.16}$$

The 3 point couplings (2.9.60) with their instanton expansion interpretation in $Q_a = e^{2\pi i t^a}$ in the A-model can now be recovered using the mirror map (2.9.53) in the special gauge at the point of maximal unipotent monodromy $\int_{A_0}\Omega = 1$ of the Kähler line bundle $\mathcal{L}^{-2}$, in the flat coordinates t_a as

$$C_{t_a t_b t_c}(Q) = \frac{1}{X_0^2}\sum_{ijk} \frac{\partial z_i}{\partial t_a}\frac{\partial z_j}{\partial t_b}\frac{\partial z_k}{\partial t_c} C_{z_i z_j z_k}(z(Q)). \tag{2.9.17}$$

[69] For reference we note also $A^{(2)} = z(2^2 \cdot 3^2 \cdot 5^4 z - 7)$, $A^{(1)} = z(2^3 \cdot 35^4 - 1)$ and $A^{(0)} = 120$.

[70] To fix the function $c(z_2)$ in the z_1 integration, we have to calculate $W^{(3,1)}$ and $W^{(2,1)}$ in a similar fashion.

In this section we derived the n-couplings from the differential ideal $\mathcal{I}_{PF}$. It is clear that given the n-points couplings, and even the factorized form in terms of more fundamental three points couplings if $n > 3$, see Sect. 3.6.2, one cannot immediately reconstruct the differential ideal. The reason is that the lower derivatives terms in the differentials operators do not go actually enter the determination of the couplings. However if we in addition put in the information of the homogeneous flat coordinates see Sect. 2.5.5, which do require the knowledge of the A-periods, we can reconstruct the differential equations in the form (2.5.50), whose existence is a consequence of Griffiths transversality.

For example in the special Kähler gauge in $\mathcal{L}$, $X^0 = 1$ in the inhomogeneous flat coordinates see Sect. 2.5.5 one gets from (2.5.50) immediately a quite universal form of the fourth order differential operator for the one parameter models

$$\mathcal{L}^{(4)} = \left(t\frac{d}{dt}\right)^2 \frac{1}{C_{ttt}} \left(t\frac{d}{dt}\right)^2 \tag{2.9.18}$$

As we have shown in Sect. 2.5 this is a consequence of the existence of the unique $(n, 0)$ form, the equation for the Kähler potential (2.5.1) and (2.5.10). It is however possible to conclude from the existence of (2.9.18) directly an invariants condition on the coefficients on the general four order differential operators of one parameter families of Calabi-Yau spaces. Normalizing the coefficient of the fourth order derivative of the single generator of $\mathcal{I}_{PF}$ to one we write this operator

$$\mathcal{L}^{(4)} = \partial^4 + \sum_{k=0}^{3} A^{(k)}(z)\partial^k\,. \tag{2.9.19}$$

Then one can prove that one can bring the operator in flat coordinates into the form (2.9.18) iff the coefficients fulfill the condition

$$W_4 = \frac{1}{2}\partial^2 A^{(3)} + \frac{3}{4}A^{(3)}\partial A^{(3)} + \frac{1}{8}\left(A^{(3)}\right)^3 - \partial A^{(2)} - \frac{1}{2}A^{(2)}A^{(3)} + A^{(1)} = 0\,, \tag{2.9.20}$$

as was pointed out by [13]. This condition is reminiscent to conditions on classical W-algebra of Drinfeld Sokolov Hierarchies [89].

However for the general Calabi-Yau differential ideals $\mathcal{I}_{PF}$ the analogous conditions to (2.9.20) are not known and the one would need the A-periods to get (2.5.50). We therefore put the existence of the unique $(n, 0)$ form Ω, the Griffiths transversality and the real bilinear relations, that is the structures that lead to special geometry, in the focus of the analysis of the general differential properties of $\mathcal{M}_{cs}$. The specific Gauss-Manin connection, its associated differential ideal $\mathcal{I}_{PF}$ and the n-points couplings that derive from it due to Griffiths transversality are in the center of the analysis of the data of the individual family under consideration. It would be interesting to know the analogous conditions to (2.9.20) for simple types of Calabi-Yau ideals $\mathcal{I}_{PF}$.

2.9.3 Indicials and Normal Crossing Variables

The differential equations that arise as Picard-Fuchs equations in Calabi-Yau spaces are of general Fuchsian type. A very useful account of these systems can be found in [334].

In general one would try to first to find a powerseries solutions of $\mathcal{I}_{PF}$ by making and ansatz

$$s(z,c,\alpha) = \sum_{n_1,\ldots,n_h=0}^{\infty} c_{n_1,\ldots,n_h} z_1^{\alpha_1+n_1} \ldots z_h^{\alpha_h+n_h} . \qquad (2.9.21)$$

One would then act with the $\mathcal{L}_k^{(n_k)}$ on this ansatz and fix first from the lowest orders in z the indicials α_j and then attempt to construct the $h^3(M)$ solutions, by solving recursions. For example if the solutions differ in one α_i by a rational number on gets two linear independent solutions with these different values of the indicials as series with different rational powers of one or more z_i must be independent.[71] However when the indicials are degenerate one has to use modified ansätze that involve log's, as we will explain in the next section.

Most interesting in this approach are the singular loci of S_{PF}, because the radii of convergence in a h-dimensional poly disk is bounded by the smallest distance to S_{PF}, so it is most efficient to solve the equations at S_{PF} or rather the normal crossing divisors of $\hat{S}_{PF}$ and try to cover $\mathcal{M}_{cs}$ by patches in which one has convergent solutions along cylindrical neighbourhoods around the normal crossing divisors. At generic points the solutions are analytic and in this sense boring, however they are important, when one wants to find analytic continuation matrices from the basis near one singular locus to another. In order that this program works one has first to construct the resolution $\mathcal{M}_{cs}$ that Hironaka's theorem ensures to exist. What this means in practice that one has to introduced blow up coordinate patches defined by blow up coordinates, which certain ratios of expressions in the old variabels, in which the normal crossing divisors of $\hat{S}_{PF}$ are obtained. Without that new coordinates one cannot construct basis of independent solutions. Examples of such resolutions can be found in [334] and as an example for some physical double scaling limit in [221].

Of course in the one parameter cases many of the difficulties mentioned above evaporate. In particular the divisor are points and the notation of tangency and

[71] There is a nice theory about these indicials as summarized in the Riemann symbol for one parameter families [334].

normal crossing are void and one chose as variables simple the distance to the singular points the cylindrical neighborhoods are just discs etc.

2.9.4 The Basis Dependent Intersection Form and the Transition Matrices

Once one has constructed a good model for the complex moduli space where all special divisors are described locally by the vanishing of normal crossing coordinates $w_1 = 0, \ldots w_h = 0$, it is an easy matter to construct a local basis of solutions. Using the fact about the possible local monodromies in Sect. 2.6.2 it is clear that most general ansatz involves rational branch cuts as well as logarithms up to degree n for Calabi-Yau n-folds, i.e. it must be of the form

$$l(m, \{c^{\underline{k}}\}, \alpha) = \sum_{|\underline{k}|=0}^{m} \log(w_1)^{k_1} \ldots \log(w_h)^{k_h} s(w, c^{\underline{k}}, \alpha) , \tag{2.9.22}$$

with $m \leq n$ and $\alpha_i \in \mathbb{Q}$. It is always possible to construct a basis of linear independent solution for the period vector Π_w of this type around any point $p_w \in \mathcal{M}_{cs}$. Because Griffith transversality holds in any basis and the n-point coupling are rational globally defined functions on $\mathcal{M}_{cs}$ they can be simply expressed in the $\underline{w}$ coordinates as $C^{k_1\ldots k_h}(w) = C_{w_{i_1}\ldots w_{i_n}} = \frac{\partial z_{i_1}}{\partial w_{i_1}} \cdots \frac{\partial z_{i_n}}{\partial w_{i_n}} C_{z_{i_1}\ldots z_{i_n}}$. Then it is always possible to reconstruct the local intersection form η_w so that

$$\Xi^{(k_1,\ldots,k_h)}(w) = \Pi^t_w \eta_w \underline{\partial}^{\underline{k}} \Pi_w \tag{2.9.23}$$

fulfills (2.9.7). It is only necessary to use the equations for $|\underline{k}| \leq n$ and one can show that η_w in a local basis that exhibits locally integer monodromy has rational coefficients.

One significance of this is that it gives Legendrian type of constrains on the *transition matrices*. The latter are defined to connect the basis of period solution, most significantly in an integer (symplectic) basis, between two different (singular) points say $\Pi_z(z)$ at z defined by $\underline{z} = 0$ and $\Pi_w(w)$ at w defined by $\underline{w} = 0$ in $\mathcal{M}_{cs}$. That is the $h \times h$ matrix T_{zw} with the property

$$\Pi_z(z) = T_{zw} \Pi_w(w) . \tag{2.9.24}$$

The complete set of all transition matrices between centers of polydiscs whose radius of convergence covers $\mathcal{M}_{cs}$ defines the periods everywhere. It is essential to study global properties of the periods, finding the attractor points and to understand resurgence relations that yield asymptotic properties of the instanton expansion. Moreover the transition matrices determine the values of the periods in an integral basis at all critical points. The latter have interesting number theoretic meaning and are governed by the motivic Galois group.

The fact that η_w can be constructed by (2.9.23) and (2.9.7) gives explicit quadratic relations, called Legendre relations on the entries of the transition matrices

$$(2\pi i)^{\Delta}\eta_z = T^t_{zw}\eta_w T_{zw}, \tag{2.9.25}$$

where $\Delta \in \mathbb{Z}$ is given by the difference in unipotency of the monodromies around at the points $\underline{z} = 0$ and $\underline{w} = 0$.

2.9.5 The Differential Ideal at the Large Complex Structure Point

The fact that the explicit solutions of the Picard-Fuchs equations can be specified by (2.8.23) or (2.8.24) is helpful in finding the Picard-Fuchs equations. Moreover the Batyrev construction of the complex structure variables z_a from the Mori vectors (2.7.27) ensures that the $z_a = 0$ are normal crossing variables $\forall a$. Further by construction the normal crossing divisor $z_a = 0$ intersect at a point of maximal unipotent monodromy, which is highly degenerate. That means that the indicials α_i in the ansatz (2.9.21), with $h = h_{12}(W)$, for the solutions are all forced to be zero $\alpha_i = 0$ with an $h_3(W)$ fold degeneration.

More explicitly the existence of a point of maximal unipotent monodromy—a term that we use synonymously with large complex structure point—in a suitable complex structure parametrization at $z_a = 0$ implies that the Picard-Fuchs differential ideal $\mathcal{I}_{PF}$ is generated by operators of the form

$$\mathcal{L}_k^{(\underline{n}_k)} = p_k(\theta_1 \ldots \theta_h) + \mathcal{O}(z) q_k(\theta_1, \ldots, \theta_h, z_1, \ldots, z_h)\,, \tag{2.9.26}$$

where p_k and q_k are polynomials in their arguments, i.e. the complex variables z_i as well as in the logarithmic derivatives $\theta_k = z_k \frac{d}{dz_k}$. The multi-degree $\underline{n}_k$ refers to the degrees of the θ_i in p_k and gives the order of the operator $\mathcal{L}^{(\underline{n}_k)}$. The maximal unipotency at $z_k = 0$ implies now that the system $p_k(\theta) = 0$ has an $h_3(W)$ fold degenerate unique solution at $\theta_i = 0$. This restricts the highest powers of θ_i in the p_k, to be larger or equal then the highest power of the θ_i in the q_k. One can show that the degree of p_k in θ_i has to be strictly larger then one. This follows from the existence of h logarithmic solutions at the maximal impotent monodromy point.

Let us view the θ_i as formal variables and not as differential operators. From the $p_k(\theta)$ of a complete ideal of Picard-Fuchs operators $\mathcal{I}_{PF}$ of W one can determine a finite dimensional algebraic ring

$$\mathcal{R}^{coh} = \mathbb{C}[\theta_1, \cdots, \theta_h]/\{p_k(\theta), k = 1, \ldots \#\mathcal{L}\}\,. \tag{2.9.27}$$

It is a corollary to the mirror symmetry hypothesis that the unique highest degree element $n = \dim_{\mathbb{C}}(W)$

$$\mathcal{R}_n^{coh} = \mu \sum_{i_1,\ldots i_n} C^{cl}_{i_1,\cdots i_n} \theta_{i_1} \cdots \theta_{i_n} \tag{2.9.28}$$

determines the classical intersection numbers of the mirror M to W, up to a normalisation μ. Further the number $r(m)$ of classes of elements $\mathcal{R}_m^{coh}$ of degree $m = 0, \cdots, n$ are the same as the dimension of the even vertical cohomology of M, i.e. $r(m) = h^{m,m}(M)$, $m = 0, \cdots, n$. The ring $\mathcal{R}^{coh}$ corresponds to the intersection ring of that cohomology of M in the region of the Kähler cone, which corresponds to the particular large complex structure point in the complex moduli space $\mathcal{M}_{cs}(W)$ of W. Note that there can be more then one point of maximal unipotent monodromy in $\mathcal{M}_{cs}(W)$, which correspond to topological different mirror manifolds M of W. Mirror symmetry suggest of course that these different manifolds M, should be unified in the complexified Kähler moduli space, which includes the $B-$field. We can identify the number of classes of degree m operators also as the dimension of the horizontal cohomology groups $r(m) = h^{n-m,m}(W)$ of W.

We note that the left differential ideal generated by applications of θ_i on the $\mathcal{L}_k^{(n_k)}$ generate differential operators again of the form (2.9.26). We consider the ones whose pure parts in the θ_i, the $p_k(\theta_i, \ldots, \theta_h)$ in (2.9.26) are determined by $\mathcal{R}^{coh}$. The maximal degenerations of the indicials at the maximal unipotent monodromy points and the structure of $\mathcal{R}^{coh}$ implies that there is exactly $r(m)$ solutions of the form (2.9.22) with logarithmic degree m. It is straightforward to determine the coefficients of these logarithmic solutions by simply requiring that they are annihilated by all operators that correspond to elements in $\mathcal{R}^{coh}$ up to an overall constant for each solution.

In principal one can reconstruct much information about $\mathcal{I}_{PF}$ from one solution, such as X^0. Unfortunately it is difficult to find in general the highest power in the z_j in q_k. However in many cases it is possible to find the $\mathcal{L}_k$ that generate the minimal left differential ideal $\mathcal{I}_{PF}$ with the following properties

- Each operator in $\mathcal{I}_{PF}$ annihilates (2.8.23) or (2.8.24) respectively
- The p_k of the $\mathcal{L}_k$ generate a classical intersection ring (2.9.27) .
- Stronger then the former point is the requirement that the operators of degree n and $n+1$ are integrable in the sense that they determine the n-point couplings by the procedure outlined in Sect. 2.9.2 .

We will describe more systematic ways of finding the differential ideal $\mathcal{I}_{PF}$ in the next sections.

2.9.6 Gelfand-Graev-Kapranov-Zelevinsky Method

Here we describe how to obtain the differential ideal for hypersurfaces or complete intersection in toric ambient spaces. In order to do this we consider for the compact case a modified differential $\tilde{\Omega}_n$ starting with the expressions (2.8.1) or (2.8.3)

$$\tilde{\Omega}_n = \frac{\Omega_n}{a_0}, \qquad \tilde{\Omega}_n = \frac{\Omega_n}{\prod_{i=1}^r a_{0,i}}. \tag{2.9.29}$$

After dividing by a_0 or $\prod_{i=1}^r a_{0,i}$ these expression violate the scaling symmetries. However we can write differential operators that annihilate the $\tilde{\Omega}$. This is the so called restricted set of GKZ differential operators of [130, 131] appropriate for the resonant case

$$\mathcal{D}_{l^{(k)}} = \prod_{l_i^{(k)}>0} \left(\frac{\partial}{\partial a_i}\right)^{l_i^{(k)}} - \prod_{l_i^{(k)}<0} \left(\frac{\partial}{\partial a_i}\right)^{-l_i^{(k)}}, \qquad \mathcal{Z}_i = \sum_{j=0}^{r} \bar{\nu}_{j,i}\vartheta_j - \beta_i\,, \tag{2.9.30}$$

where $1 \le k \le |\Delta_0| - d - 1$, $1 \le i \le n+1$, $r = |\Delta_0|$ and $\beta = (1; 0, \dots, 0)$ and $\vartheta_i = a_i \frac{\partial}{\partial a_i}$. The first set of operators $k = 1, \dots, h_{11}(W)$ follows from the linear relation between the points in $\bar{\Delta}$, which are encoded in $\{l^{(i)}\}$. The second set of operators $\mathcal{Z}_i$, $i = 0, \dots, d$ eliminates the $\mathbb{C}^*$ scaling symmetry of P_Δ and the $\mathbb{T}^d$ redundancy A1) in the parametrization of $\mathcal{M}_{cs}$ in terms of the a_i. Note that for the $\mathcal{Z}_0$ operator the $\beta_0 = 1$ term vanishes if it acts on Ω_n instead of $\tilde{\Omega}_n$. This observation is important to eliminate the ϑ differentials in favor of the θ differentials in (2.9.32). We discuss the origin of the $\mathcal{Z}_i$ in for the simplest example in Sect. 2.10.

Let us denote the logarithmic derivatives w.r.t to the a_i by $\vartheta_i = a_i \frac{\partial}{\partial a_i}$ and the one w.r.t. to the Batyrev coordinates z_i by $\theta_k = z_k \frac{\partial}{\partial z_k}$. For both kind we have the trivial commutator

$$[\theta_i, z_i^l] = l z_i^l, \qquad [\vartheta_i, a_i^l] = l a_i\,. \tag{2.9.31}$$

We know that $n + r$ of the a_i are identified by the $\mathbb{T}^{n+r}$ action and are hence redundant. It is easy to see that the $(n + 1 + h)$ a_i can be expressed in terms of the h z_k, $k = 1, \dots h$ by the relation

$$\vartheta_i = l_i^{(k)} \theta_k\,. \tag{2.9.32}$$

Now we consider the first relations in (2.9.30) and convert the ∂_{a_i} derivatives into logarithmic derivatives. In this process the commutators are useful to bring the a_i to the left. Moreover we commute the factor $1/a_0$, $(1/\prod_{i=1}^r a_{0,i})$ out from $\mathcal{D}_{l^{(k)}}\tilde{\Omega}_n = 0$ to get a differential equation on Ω_n. Because of the violation of the scale invariance, the latter will be only vanish up to exact terms, when the differential operator acts on it. That is the latter yields a differential operator for the integrated Ω_n over closed cycles and annihilates the periods. It is further easy to see that the free a_i can, by definition of the z_k variables through the $l^{(k)}$, all be absorbed into z_k. Moreover the ϑ_i can be replaced using (2.9.32), so that the $\mathcal{D}_{l^{(k)}}$ become differential operators that are naturally expressible in the z_i coordinates and will be called $\tilde{\mathcal{L}}_k$. The latter do not generate the differential ideal yet. First we can consider the same procedure for all positive integer linear combinations $\hat{l}_k = \sum_i n_i l^{(i)}$, with $n_i \in \mathbb{N}+$. These will give additional linear operators. While these operators annihilate the periods they annihilate also other functions and don't uniquely fix the $h^3(M)$ periods. However one can factorise these operators eventually after adding them with z and θ dependent coefficients to obtain the minimal ideal in many cases. The simplest case of this factorisation is explained for the quintic after Eq. (2.10.11).

More generally it was observed in [186] that the equations (2.9.30) are sufficient to solve model of type I and that for type II models the infinitesimal version of the invariance of (2.8.5) under the transformation corresponding to the roots (2.7.17) leads to first order differential operators $\mathcal{Z}'_k$, which supplement (2.9.30) and that for general type III models one needs further differential equations $\mathcal{Z}''_k$ coming from relations between monomials corresponding to points at height greater one modulo $\partial_i P_\Delta$. Such relation can be found algorithmically using the Groebner basis for the Jacobian ideal J, see below.

2.9.7 Dwork-Griffiths Reduction Method

From the formal definition of the period $\Pi(z) = \int_{\Gamma_i} \Omega(z)$, with Ω given in (2.10.7) we can alternatively derive a fourth order differential equation for the period in terms of the moduli z by the Dwork-Griffiths reduction method. We mention this methods, because in general the symmetries of the ambient space are not sufficient to find the full set of Picard-Fuchs equations. The key observation for this algorithm can be explained as follows. Consider on the ambient space $\mathbb{P}^{m-1}(w_1, \ldots, w_m)$ the $(m-2)$-form

$$\Phi = \frac{a_0}{P^r} \sum_{i<j} (-1)^{i+j} (w_j x_j A_i - w_i x_i A_j) \mathrm{d}x_1 \wedge \ldots \wedge \widehat{\mathrm{d}x_i} \wedge \ldots \wedge \widehat{\mathrm{d}x_j} \wedge \ldots \wedge \mathrm{d}x_n .$$

Here $P = 0$ is the hyper surface constraint and the $A_i(x)$ are homogeneous of degree d_i in x, i.e. $\sum_{k=1}^m x_k w_k \frac{\partial}{\partial_k} A_i = d_i A_i$. We further assume that $c_1(M) = 0 \leftrightarrow \sum_{i=1}^m w_i = d$, where d is the homogeneous degree of P, $\sum_{k=1}^m x_k w_k \frac{\partial}{\partial_k} P = Pd$. With these assumptions the total derivative of Φ simplifies

$$\begin{aligned} \mathrm{d}\Phi = & \sum_{k=1}^m \left(\frac{a_0 r}{P^{r+1}} A_k \partial_k P - \frac{a_0}{P^r} \partial_k A_k \right) \mu \\ & + \frac{a_0}{P^r} \sum_{j=1}^m (d(1-r) - w_i + d_i) A_i (-1)^j \mathrm{d}x_1 \wedge \ldots \wedge \widehat{\mathrm{d}x_j} \wedge \ldots \wedge \mathrm{d}x_n . \end{aligned}$$

If we choose now the A_j so that $A_j = 0$ for $j \neq k$ and $d_k = d(r-1) + w_k$ so that for $Q(x) := A_k(x)$ the second term vanishes. In other words if

$$\partial_k \left(\frac{Q(x,a)a_0}{P^r} \mu \right) := \frac{\partial}{\partial x_k} \left(\frac{Q(x,a)a_0}{P^r} \mu \right) \tag{2.9.33}$$

is homogeneous of degree 0 w.r.t. the coordinate weights w_i then

$$\frac{a_0 r Q(x,a)\partial_k P}{P^{r+1}} \mu = \frac{a_0 \partial_k Q(x,a)}{P^r} \mu \tag{2.9.34}$$

holds under the integration sign.

Similarly for Calabi-Yau manifolds defined by a transversal complete intersections of s polynomials, i.e. as the zero set $P_1 = \ldots = P_s = 0$ with degree $d_1, \ldots, d_s$ in a weighted projective space $\mathbb{P}^{n+s}[w_1, \ldots, w_{n+s}]$, $c_1(M) = 0 \leftrightarrow \sum_{i=1}^{m} w_i = \sum_{j=1}^{s} d_j$, the analog of (2.10.7) is

$$\Omega = \int_{\gamma_a} \cdots \int_{\gamma_s} \prod_{k=1}^{s} \frac{a_0^{(k)}}{P_s} \mu, \tag{2.9.35}$$

where γ_i are circles around the $P_i=0$ and similar as before $\frac{\partial}{\partial_k}\left(Q(x,a)\prod_{k=1}^{s} \frac{a_0^{(k)}}{P_k}\mu\right)$ is exact iff it is of total degree zero. This leads to the partial integration rule [147]

$$\frac{Q\partial_i P_j}{\prod_{l=1}^{s} P_l^{n_l}} \mu = \sum_{k \neq j} \frac{n_k}{n_j - 1} \frac{P_j}{P_k} \frac{1}{n_j - 1} \frac{P_j \partial_i Q}{\prod_{l=1}^{s} P_l^{n_l}} \mu - \frac{Q\partial_i P_k}{\prod_{l=1}^{s} P_l^{n_l}} \mu \,, \tag{2.9.36}$$

where we omitted the factor $\prod_{k=1}^{s} a_0^{(k)}$, which is however relevant for the correct scaling in (2.10.11) we see in the example below we will commute this factor in after the Griffith reduction calculation.

Let us outline the general idea and exemplify it in simple cases. We restricts to the hypersurface case and consider the graded ring defined by the Jacobian ideal $J = \{\partial_{x_i} P\}$ generated by the partial derivatives of the weighted homogeneous polynomial $P(x)$ of degree $d = \sum_i w_i$ in the weighted projective space $\mathbb{P}^{n+1}[w_1, \ldots, w_{n+2}]$ as

$$\mathcal{R} = \frac{\mathbb{C}[x_1, \ldots, x_{n+2}]}{\{\partial_{x_i} P(x)\}}\,, \tag{2.9.37}$$

with elements $\phi^{i_k}_{dk}(x)$, of weighted degree dk with $k = 0, 1, \ldots, n$ and $i_k = 1, \ldots, h^{hor}_{n-k,k}$. Indeed these ring elements span the horizontal cohomology[72] $H^{hor}_{n-k,k}(M_n, \mathbb{C})$ by the Griffiths residuum formula

$$\chi_{i_k,n} = \int_\gamma \frac{\phi^{i_k}_{dk} a_0 \mu}{P(x)^{k+1}}. \tag{2.9.38}$$

More precisely they yield a basis of rational cohomology. It is easy to calculate the number of $\phi^{i_k}_{dk}(x)$ for a given degree by the Poincaré polynomials, which were used to count the chiral (and anti-chiral) fields in Calabi-Yau/ Landau Ginzburg models [304, 305]. In fact the E-polynomials [30] are based on the same idea.

The idea is take derivatives of the period $\Pi(a)$ w.r.t. the complex structure moduli parametrized by the $a's$ until the numerator contains elements that are reducible w.r.t. to the ideal J. We need to consider only the relevant perturbations, i.e. the ones that do not correspond to co-dimension one points of the Newton polyhedron, and we also use (2.7.19) to set $n+2$ of the a_i to one $a_i = 1$. For example if Δ is simplicial one can set the $a_i = 1$ for the monomials that corresponding to the corners.

Then we rewrite the emerging expression in the numerator using Buchbinders algorithm for a Groeber basis described in more detail below, in a form that enables us to use the partial integration rules (2.9.34) or (2.9.36) w.r.t. x_i. The emerging expressions will have lower powers of P in the denominator and lower homogeneous degree polynomials in x in the numerators.

Eventually all emergent terms can be manipulated into the form of moduli dependent rational functions times lower derivatives of $\Pi(z)$ w.r.t. to the moduli a. The relation derived in this way is one Picard-Fuchs operator.

Let us take as example the cubic in $\mathbb{P}^2$ given by the zero locus of

$$P = x_1^3 + x_2^3 + x_3^3 - 3a x_1 x_2 x_3, \tag{2.9.39}$$

where the factor three in the perturbation has been chosen so that the discriminant is at $a^3 = 1$. Now we consider $\frac{\Pi(a)}{a} = \tilde\Pi(a) = \int_\gamma \frac{\mu}{P}$ and get be taking two derivatives w.r.t. a

$$\begin{aligned}\frac{\partial^2}{\partial^2 a}\tilde\Pi(a) &= \int_\gamma \frac{18(x_1x_2x_3)^2\mu}{P^3} = \int_\gamma \frac{6(x_1^2x_2^2\partial_3 P + a x_1^3 x_2 \partial_2 P + a^2 x_1^2 x_2 x_3 \partial_1 P)\mu}{(1-a^3)P^3}\\ &= \int_\gamma \frac{3a x_1^3\mu}{(1-a^3)P^2} + \frac{6a^2 x_1x_2x_3\mu}{(1-a^3)P^2} = \int_\gamma \frac{a x_1 \partial_1 P\mu}{(1-a^3)P^2} + \frac{3a^2}{(1-a^3)}\frac{\partial}{\partial a}\tilde\Pi(a)\\ &= \frac{a}{(1-a^3)}\tilde\Pi(a) + \frac{3a^2}{(1-a^3)}\frac{\partial}{\partial a}\tilde\Pi(a)\,.\end{aligned} \tag{2.9.40}$$

[72]Throughout the section we assume that all deformations of $P_\Delta(x)$ are polynomial, i.e. no "twisted fields" in other places indicated by ($h^{twisted}_{pq}$).

Hence we derived the operator

$$[(1-a^2)\partial_a^2 - 3a^3\partial_a - a]\tilde{\Pi}(a) = 0 \tag{2.9.41}$$

or after using $\tilde{\Pi} = \frac{1}{a}\Pi(a)$

$$[a^2(1-a^2)\partial_a^2 - a(2+a^3) + 2]\Pi(a) = 0\,. \tag{2.9.42}$$

Using $z = -\frac{1}{27a^3}$ we can rewrite this with $\theta_z = z\partial_z$ as

$$[\theta_z^2 + 3z(3\theta_z + 1)(3\theta_z + 2)]\Pi(z) = 0\,. \tag{2.9.43}$$

The GGKZ method starting with (2.9.30) will lead much quicker to the a third order system

$$\theta_z[\theta_z^2 + 3z(3\theta_z + 1)(3\theta_z + 2)]\Pi(z) = 0\,. \tag{2.9.44}$$

Clearly the actual two periods of the holomorphic differential satisfy this system, but there is one more solution that is obviously not associated with such periods, which can be discarded by dropping in the factorised form the θ_z on the left. The discarded solution has an interpretation in terms of chain integrals, instead of cycle integrals and does play a role in open mirror symmetry.

It is a good exercise to perform the same for the quintic. Here we start with four derivatives of $\tilde{\Pi}(z)$ and the emerging relation is of course the same 4th order generalized hypergeometric differential equation as in (2.10.13).

In the multi moduli examples one has to consider various derivatives of $\Pi(z)$ w.r.t. to different combinations z as starting point and the calculation becomes quite tedious. Nevertheless one can give criteria when the left ideal of differential relations is sufficient to determine $\Pi(z)$ and it is necessary to systematise the calculations somewhat using a Groebner basis for the ring of monomials in the x [104, 186, 187, 211].

Using Buchbergers algorithm one can chose a Groebner basis for the Jacobian ideal J. The properties of the Groebner basis allow to decompose any monomial $m_{kd}^{(i)}(x)$ of degree kd uniquely as follows

$$m_{kd}^{(i)} = q_j^{(i)}(a)\tilde{M}_j(x) + \sum_j Q_j^{(i)}(a,x)\partial_{x_j} P\,. \tag{2.9.45}$$

Here $\tilde{M}_j(x)$ are degree kd monomials in the multiplicative ring $\mathcal{MR}(M_j(x))$ generated by the $M_j(x)$ in (2.7.10), $q_j^{(i)}(a)$ is an unique rational function and the $Q_j^{(i)}(a,x)$ are likewise uniquely determined. In practice one uses (2.9.45), (2.9.34) as follows: One takes $n+1$ derivatives of (2.8.5) w.r.t. to the relevant a_i in (2.7.10). This produces an integrand $\tilde{M}_i(x)/P^{n+2}$, whose numerator is completely reducible by (2.9.45) to the last term on the r.h.s. The first term on the r.h.s of (2.9.45) is zero as the ring $\mathcal{R}$ is empty at this degree. Using (2.9.34) one can reduce the integrand, up to exact terms which do not affect the period integral, to sums of $m^{(i)}_{nd}/P^{n+1}$. Repeating the procedure reduces the $n+1$ th derivative to lower derivatives of (2.8.5) with rational coefficients.

This produces an differential ideal, which is complete if the n-point couplings and its factorized forms can be derived from it. In this case it can be also written in h first order matrix equations like in (3.6.41) in terms of these couplings.

The use of (2.9.36) is explained in an simple example the elliptic curve realised as the complete intersections of two quadrics $P_1 = \frac{1}{2}(x_1^2 + x_2^2 - 2ax_3x_4)$ and $P_2 = \frac{1}{2}(x_3^2 + x_4^2 - 2ax_1x_2)$ in $\mathbb{P}^3$. It is an instructive exercise, with solution in [227] , to derive the Picard-Fuchs operator

$$[a(1-a^4)\partial_a^2 + (1-4a^4)\partial_a - 4a^3]\Pi(a) = 0\,. \tag{2.9.46}$$

As was pointed out in [227] one can use the generalized mori vector $l = (-2, -2; 1, 1, 1, 1,)$ in (2.9.30). This fact is obvious from the relations of the monomials. One can further use scaling relations to eliminate the redundant parameters $a_i, i = 1, \ldots, 4$ and factorize a degree 4 systems to derive

$$\theta_z^2[\theta_z^2 - z(\theta_z + \frac{1}{2})^2]\Pi(z) = 0\,. \tag{2.9.47}$$

whose nontrivial part $\theta_z^2 + z(\theta_z + \frac{1}{2})^2$ is equivalent to (2.9.46) after identifying $z = 1/a^4$.

2.9.8 Determining the Integral Symplectic Basis at the Point of Maximal Unipotent Monodromy

If one knows a complete set of solutions at a large complex structure point, which is characterised by its maximal unipotent monodromy, one can determine an *integral symplectic* basis as a linear combination of these solutions as follows. Classical intersection data of the mirror M determine it's $\hat{\Gamma}$ class and hence the charges of the even branes in an integral symplectic basis on the A model. The precise linear combinations in the B-model are then determined at the large complex structure point by comparing the t^a powers in the solutions of the Picard-Fuchs equations on the r.h.s. in expression (2.6.30) with the one of the l.h.s, which are given even brane masses as calculated from $\hat{\Gamma}$ class.

Part of the classical intersection data for 3-folds are the classical 3-point intersection numbers C^{cl}. More generally the latter can be determined up to a normalisation also on the B-model side from $\mathcal{I}_{PF}$ at a point of maximal unipotent monodromy as discussed in (2.9.27), (2.9.28) for any Calabi-Yau n-fold. The ring

$\mathcal{R}^{coh}$ determines the logarithmic structure of a basis of solutions at the large complex structure point.

We demonstrate the construction of the solutions at the point of maximal unipotent monodromy as well as of the integral basis for the case of so called *hypergeometric Picard-Fuchs systems.* A Picard Fuchs system whose solutions are determined by the data of the generalized Mori cone is a hypergeometric system. Given also the classical intersection ring one can immediately write down a local expansion of a basis of solutions of the Picard Fuchs equations convergent near the large complex structure point. Adding the complete topological data of C.T.C Wall of M as well as the choice of the even brane classes one can take linear combinations of this basis, which correspond to the periods in an *integral symplectic basis.*

We review in the following the essentials and refer to [187] for further details. The particular set of local coordinates z_a on the complex structure moduli space on W are the Batyrevs coordinates defined in (2.8.19). The point of *maximal unipotent monodromy* is then always at $z_k = 0$ for all $k = 1, \ldots, h_{21}(W) = h$.

Let $\varpi_{a_1,\ldots,a_s}$ be obtained by the Frobenius method[73] from the coefficients of the holomorphic function $\varpi(\vec{z}, \vec{\rho})$ defined by

$$\varpi(z_1, \ldots, z_h, \rho_1, \ldots, \rho_h) = \sum_{\{n_k\}} c(n_1 \ldots n_h, \rho_1 \ldots \rho_h) \prod_{k=1}^{h} z_k^{n_k+\rho_k} \qquad \text{with}$$
$$c(n_1, \ldots, n_h, \rho_1, \ldots, \rho_h) = \frac{\prod_{m=1}^{r} \Gamma(1 - \sum_{k=1}^{h} l_{0,m}^{(k)}(n_k + \rho_k))}{\prod_{i=1}^{n} \Gamma(1 + \sum_{k=1}^{h} l_i^{(k)}(n_k + \rho_k))} . \qquad (2.9.48)$$

As the derivatives of (2.9.48)

$$\varpi_{k_1,\ldots,k_s}(z_1, \ldots, z_h) = \left(\frac{1}{2\pi i}\right)^s \partial_{\rho_{k_1}} \ldots \partial_{\rho_{k_s}} \varpi(z_1, \ldots, z_h, \rho_1, \ldots, \rho_h)|_{\{\rho_k=0\}} . \qquad (2.9.49)$$

with respect to the ρ_k, i.e. $\partial_k = \frac{\partial}{\partial \rho_k}$. Note that if it hits the $z_k^{n_k+\rho_k}$ it yields a logarithm $\log(z_k)$ and the factor of $\left(\frac{1}{2\pi i}\right)^s$ avoids shift factors of $2\pi i$ in the monodromy around $z_k = 0$.

We define also $X^0(\underline{z}) = \varpi(z_1, \ldots, z_h, \rho_1, \ldots, \rho_h)|_{\rho_k=0} = \varphi(\underline{z})$ which agrees of course with the (2.8.23) from the direct evaluation of the algebraic torus integral. It is also useful to define the quantities

$$\sigma_{k_1,\ldots,k_s} = (\varpi_{k_1,\ldots,k_s}(z_1, \ldots, z_h)|_{\log(z_k)=0}) , \qquad (2.9.50)$$

[73] The holomorphic period $\varpi(z_1, \ldots, z_h)$ can also be directly integrated using a residuum expression for the holomorphic (3, 0) form [187].

which are together with $\sigma_{k_1,\ldots,k_s}/X^0$ are analytic functions, since $X^0 = 1 + \mathcal{O}(z)$, with a finite radius of convergence near $z_a = 0$. The latter is determined by nearest singularity in the discriminant of the D-moduli of linear Picard-Fuchs operators, which is usually a conifold point. In order to evaluate $\sigma_{k_1,\ldots,k_h}$ from (2.9.49) a relatively intriguing pole cancelation of the Γ functions and their derivatives has to be considered, which is made explicit in [187].

The point of the definition of $\frac{\partial}{\partial \rho_k}$ is that it commutes weakly with the differential operators $\mathcal{L}_j$ defining the D-module, i.e.

$$\left[\frac{\partial}{\partial \rho_k}, \mathcal{L}_j\right] \text{ “ = “ } 0\,, \qquad \forall k, j\,. \tag{2.9.51}$$

In other words we expect that the $\varpi_{k_1,\ldots,k_s}(z_1,\ldots,z_h)$ are closely related to solutions to the Picard-Fuchs equations. Of course this cannot generate an arbitrary number of solutions as there are only $h^3(M)$ periods and therefore only $h^3(M)$ independent solutions. The structure of differential equations at the maximal unipotent monodromy point (2.9.26) implies that the condition for $\varpi_{k_1,\ldots,k_s}(z_1,\ldots,z_h)$ being a solution can be completely determined from the criteria that the $p_k(\theta)$ in (2.9.26) annihilate all logarithms in the solutions. In particular since the degree of $p_k(\theta)$ is strictly greater then one, all the single logarithmic

$$X^a = \varpi_a = \frac{1}{2\pi i}\log(z_a) + \sigma_a \qquad a = 1,\ldots h \tag{2.9.52}$$

are solutions. At the large complex structure point these solutions indeed define the mirror map which are the natural flat coordinates on the Kähler moduli space of the original manifold M as

$$t^a = \frac{X^a}{X^0} = \frac{1}{2\pi i}\log(z_a) + \frac{\sigma_a}{X^0}\,, \qquad a = 1,\ldots,h\,, \tag{2.9.53}$$

and parametrize complexified areas of curves C_a embedded in M, so that the t^a are identified with the complexified Kähler parameters (2.4.48). The logarithmic shift monodromy of the t^a

$$t^a \to t^a + 1, \tag{2.9.54}$$

when z^a is analytically continued counterclockwise around $z^a = 0$ corresponds to the shift of the Neveu-Schwarz b field by an flux quantum (2.4.49). Shift invariants variables are

$$Q^a = \exp\left(2\pi i t^a\right)\,. \tag{2.9.55}$$

The Q^a go exponentially to zero when the area of the curve $\int_{C_a}\omega$ goes to infinite, which happens when z^a approaches the maximal unipotent monodromy point

$z^a = 0$. The latter point is therefore often called the *large area or large volume point*. The latter name is justified as inside the Kähler cone (2.4.45) the volumes of all even cycles goes to infinity, when the areas of the curves go to infinity. The key significance of the Q^a expansion is that their individual terms correspond to worlds-sheet instanton effects, naturally suppressed at large volume, but of increasing importance at smaller volumes. The B-model allows in particular sum up these effects. A key technical problem[74] in the calculation is to invert the exponentiated mirror map (2.9.53) to obtain $z_i(\underline{t})$.

From mirror symmetry one can argue that there will be no solutions in which the order of the derivatives s in (2.4.49) is more then the complex dimension of M, i.e $s > n$. In fact this fits the analysis of the maximal degeneration of periods by W. Schmidt [293].

In [187] it was shown that the $p_k(\theta)$ determine the classical intersection ring and that for Calabi-Yau 3-folds the double and triple logarithmic solutions are given by

$$w_a^{(2)} = \frac{1}{2} \sum_{b,c=1}^{h} C^{cl}_{abc} \varpi_{bc}(z_1, \ldots, z_h), \quad a = 1, \ldots, h. \tag{2.9.56}$$

$$w^{(3)} = \frac{1}{6} \sum_{a,b,c=1}^{h} C^{cl}_{abc} \varpi_{abc}(z_1, \ldots, z_h), \tag{2.9.57}$$

where $C^{cl}_{abc} = D_a \cap D_b \cap D_c$ is the classical intersection ring.

To find the integral symplectic basis it suffices as mentioned to write linear combinations of the periods obtained above to match the complexified area powers t^a in (2.6.30) specified by the prepotential (2.6.31). An integral symplectic basis for the periods is hence given by

$$\Pi = X^0 \begin{pmatrix} 2\mathcal{F}^{(0)} - t^a \partial_{t^a} \mathcal{F}^{(0)} \\ \partial_{t^a} \mathcal{F}^{(0)} \\ 1 \\ t^a \end{pmatrix} = X^0 \begin{pmatrix} \frac{C^{cl}_{abc} t^a t^b t^c}{3!} + c_a t^a - i\chi \frac{\zeta(3)}{(2\pi)^3} + 2f(Q) - t^a \partial_{t^a} f(Q) \\ -\frac{C^{cl}_{abc} t^b t^c}{2} + A_{ab} t^b + c_a + \partial_{t^a} f(Q) \\ 1 \\ t^a \end{pmatrix} \tag{2.9.58}$$

The real coefficients A_{ab} are not completely fixed. They are unphysical in the sense that $K(t, \bar{t})$ and $C_{abc}(Q)$ do not depend on them. The A_{ab} are further restricted by the requirement that the Peccei-Quinn symmetries $t^a \to t^a + 1$ act as integral $Sp(2h^{11} + 2, \mathbb{Z})$ transformations on Π. Note that the prepotential (2.6.31) can be read off from the periods. We put an index (0) on the prepotential $\mathcal{F}^{(0)} = \mathcal{F}$ to stress that it counts the genus zero correction to the classical intersection. In fact the

[74] We wrote an improved code for that (http://hep.itp.tuwien.ac.at/~kreuzer/CY/hep-th/yymmnnn.html).

instanton part in (2.6.31) can be written as

$$\mathcal{F}^{(0)}_{inst} = \frac{1}{(2\pi i)^3} \sum_{\beta \in H_2(M,\mathbb{Z})>0} r_0^\beta Q^\beta = \frac{1}{(2\pi i)^3} \sum_{\beta>0} n_0^\beta \mathrm{Li}_3(Q^\beta)\,. \qquad (2.9.59)$$

Here we defined $Q^\beta = \prod_a e^{2\pi i \beta_a t^a}$ where the tuple $(\beta_1, \ldots, \beta_h)$ specifies a class β in $H_2(M, \mathbb{Z})$. Since t^a are flat coordinates, we have

$$C_{abc}(Q) = \partial_{t^a}\partial_{t^b}\partial_{t^c}\mathcal{F}^{(0)} = C^{cl}_{abc} + \sum_{\beta>0} n_0^\beta \frac{\beta_a\beta_b\beta_c Q^\beta}{1-Q^\beta}\,. \qquad (2.9.60)$$

The sum counts as the last sum in (2.9.59) the genus zero contribution BPS numbers $n_0^\beta \in \mathbb{Z}$. The first sum in (2.9.59) counts the genus zero Gromov-Witten invariants $r_0^\beta \in \mathbb{Q}$.

Note that the term $\frac{\chi\zeta(3)}{2(2\pi i)^3}$ in (2.6.31) can be also view as contributions of the degree zero BPS invariant of $D0$ branes if we extend sum in (2.9.59) to $\beta = 0$ and identify $n_0^{\beta=0} = \frac{\chi(M)}{2}$ as the virtual integral of the degree zero maps and use the fact that $Li_3(1) = \zeta(3)$.

For the quintic the expansion predicts the first line in Table 10. The higher genus predictions will be discussed in Sect. 4.1.1.

2.9.9 The Mellin-Barnes Integrals from Supersymmetric Localisation

One new outcome of the supersymmetric localization [36, 102] are the formulas for partition function of the (2, 2) gauged linear σ model [322] on the hemi-sphere with A- and B-type boundary conditions [181]. These expressions are not valid at large complex structure but rather at the origin of the Coulomb branch typically with enhanced symmetry.

Even though the final formulas for the hemisphere partition function for the σ-model with abelian gauge groups are close to the familiar period expressions in terms of Barnes integrals [61], the formalism yields new conceptual insights in the problem of finding a rational basis $H^n(M_n, \mathbb{Q})$ due to its relation to homological mirror symmetry. It is in addition very general and for non abelian gauge groups it describes Calabi-Yau manifolds embedded in Grassmannians, flag-varieties and those determinantal and more general non-complete intersection embeddings of Calabi-Yau manifolds that are e.g. needed to get F-theory with more than 4 global $U(1)$ factors.

The General Data of the Hemisphere Partition Function

Let[75] G be the rank l_G gauge group of the 2d $(2,2)$ linear σ model and $T \subset G$ its maximal torus. The complex matter fields ϕ_i transform in a G representation space V and carry a charge with respect to T called Q_i. V is also acted on by an R symmetry representation End(V) with charge R_i, i.e. $V|_{\mathbb{C}_R \times T} = \oplus_i \mathbb{C}(R_i, Q_i)$. One has an embedding $e^{\pi i R} =: J \in G$. The superpotential W is a gauge invariant function of the matter fields and homogeneous of degree two w.r.t. the R charge, i.e. $\mathcal{W}(\lambda^R \phi) = \lambda^2 \mathcal{W}(\phi)$. We need to consider only the situations where the twisted superpotential is linear and contains just the term $\tilde{\mathcal{W}} = -\frac{1}{2\pi} t(s)$, where $t = \zeta - i\theta$ is a complex combination of the Fayet-Iliopoulus parameter, identified with the Kähler parameter in the CY phase, and the theta parameter in $\mathbb{R}/\mathbb{Z}^{l_G}$, identified with the periodic B field in the CY phase.

The boundary data for the hemi-sphere are specified by $\mathcal{R} = (M, \mathcal{Q}, \rho, r^*)$, where M is the $\mathbb{Z}_2$ graded vector space of Chan-Paton factors $M = M^{ev} \oplus M^{od}$ over $\mathbb{C}$ and ρ and r^* are representations of G and R on M, i.e. $M|_{\mathbb{C}^R \times T} = \oplus C(r_i, q_i)$. It has the charge integrality property

$$e^{\pi i r^*} \rho(J) = \begin{cases} 1 & \text{on } M^{ev} \\ -1 & \text{on } M^{od} \end{cases} . \tag{2.9.61}$$

$\mathcal{Q}$ is an End$^{od}(M)$ valued holomorphic function in V, which is gauge invariant $\rho^{-1}(g)\mathcal{Q}(g\phi)\rho(g) = \mathcal{Q}(\phi)$, homogeneous $\lambda^{r^*}\mathcal{Q}(\lambda^R\phi)\lambda^{-r^*} = \lambda\mathcal{Q}$ and has the matrix factorization property of W

$$\mathcal{Q}(\phi)^2 = \mathcal{W}(\phi)\mathrm{id}_M . \tag{2.9.62}$$

The general form for the hemi-sphere partition is calculated by supersymmetric localization and reads for the boundary data $\mathcal{R}$ [181]

$$Z_{D^2}(\mathcal{R}) = c(\Lambda, r) \int_\gamma \mathrm{d}^{l_G} s \prod_{\alpha > 0} \alpha(s) \sinh(\pi\alpha(s)) \prod_i \Gamma\left(i Q_i(s) + \frac{R_i}{2}\right) e^{t(s)} \mathrm{Tr}_M\left(e^{\pi i r^*} e^{2\pi\rho(s)}\right) , \tag{2.9.63}$$

where α are the roots of G. The contour γ is chosen in a multidimensional generalization of the contour used in (2.10.23) (rotated by $\frac{\pi}{2}$ to the left), so that it is a deformation if the real locus $it \subset t_{\mathbb{C}} \sim \mathbb{C}^{l_G}$ so that

(C1) the integral is convergent and
(C2) the deformations does not cross poles of the integrand.

[75]Our notation follows [181] and lectures of K. Hori at the University of Michigan in Ann Arbor.

We specialize to the Calabi-Yau n-fold case where the axial $U(1)_A$ anomaly is cancelled, t is not renormalized, $\hat{c} = \mathrm{tr}_V(1-R) - \dim G = n$, the dependence on the scale Λ and radius of the hemi-sphere r disappears and $c(\Lambda, r)$ becomes a normalization constant. We focus the attention to A-branes on W_n, which correspond to twisted line bundles $\mathcal{O}(q^1, \ldots, q^{l_G})$, with $l_G = h_{1,1}(W_n) = h$ with Chern character

$$\mathrm{ch}(\mathcal{O}(q^1, \ldots, q^h)) = \exp\left(\sum_{\alpha=1}^{h} q^\alpha J_\alpha\right) . \tag{2.9.64}$$

The range of independent $\{q_i\}$ is restricted by the relations in the Chow ring of W_n and the corresponding branes for a $\mathbb{Q}$-basis in the K-theory classes of the derived category of coherent sheaves. In particular with standard intersection calculations we can determine (2.6.24) in this basis.

The Case of Abelian Gauge Linear σ-Models

In order not to clutter the notations, we consider abelian gauge groups $G = U(1)^h$ and only matter fields representations and superpotentials that lead to complete intersections. In this case the first product in (2.9.63) is trivial and yields one and we get

$$Z_{D^2}(\mathcal{O}(q^1, \ldots, q^h)) = \frac{2^r i}{(2\pi i)^{n+r}} \int_\gamma \mathrm{d}^h s \prod_{j=1}^{r} \Gamma\left(i l_{0j}^{(\alpha)} s_\alpha + 1\right)$$
$$\prod_{j=1}^{k} \Gamma(i l_j^{(\alpha)} s_\alpha) e^{2\pi(t^\alpha + q^\alpha) s_\alpha} \prod_{j=1}^{r} \sinh(\pi l_{0j}^{(\alpha)} s_\alpha) , \tag{2.9.65}$$

where the sum over α is implicit and we restrict the arguments of $e^{\pi t_\alpha}$ analogously to (2.10.23) to be in the ranges $0 \leq \arg(e^{t_\alpha}) < \min\left(\frac{2\pi}{-l^\alpha_{0,j}}\right)$.

It is not hard to show that the formulas (2.9.65) are indeed a Mellin-Barnes integral representation of (2.9.49) and certain combinations of its $\partial^r_{\underline{\rho}}|_{\underline{\rho}}$ derivatives. We will use similar Barnes integral representations in Sect. 2.10.2. In particular we can rewrite (2.10.23) and its transforms $\varpi_0(\beta^k a)$ using the identity $\Gamma(1-s)\Gamma(s) = \frac{\pi}{\sin(\pi s)}$ in the form (2.9.65) and use contour deformations to analytically continue the Barnes integrals from the orbifold point to the maximal unipotent monodromy point. Two moduli examples where treated in [63, 64]. Elements of a general theory are outlined in [277, 335].

Deriving the Picard-Fuchs Equations

It is useful and at least in the $U(1)^h$ cases straightforward, to derive from (2.9.65) the full system of Picard-Fuchs equations $\mathcal{L}_i$, $i = 1, \ldots, s$. The reason for the usefulness of the remark is that it is hard to read from (2.9.65) all components of the critical locus of the periods, which follow on other hand immediately from the resultant of the symbols of the $\mathcal{L}_i$, $i = 1, \ldots, s$. Moreover the expression (2.9.65) is not useful to analytically continue to most of these components, e.g. to the conifold loci. To derive the $\mathcal{L}_i$, $i = 1, \ldots, s$ that generate the ideal $\mathcal{I}_{PF}$, see Sect. 2.4.3, we follow the observation made in [186] that the classical intersection ring $\mathcal{R}(\underline{\xi}) = c_{i_1,\ldots,i_n}\xi^{i_1}\cdots\xi^{i_n}$ at a large radius point in coordinates[76] at $z_\alpha = e^{-t_\alpha}$ is given by

$$\mathcal{R}(\underline{\xi}) = \mathbb{C}[\xi_1, \ldots, \xi_h]/\mathrm{Id}(\mathcal{S}_i(\underline{\xi}), i = 1, \ldots, s), \tag{2.9.66}$$

where $\mathcal{S}_i$ is obtained as $\mathcal{S}_i = \lim_{t_\alpha \to \infty} \mathcal{L}_i(\partial_{t_\alpha} = \xi_\alpha)$ and Id denotes the multiplicative ideal and the ξ are the *symbols* of the differential ideal $\mathcal{I}_{PF}$. Consider now the monomials $D_{i_1} \cdot D_{i_{|I*|}}$ representing the Stanley Reisner ideal for a given triangulation. Pick a basis K_i of the Chow ring and express the $D_{i_1} \cdot D_{i_{|I*|}}$ in terms of the K_j, $j = 1, \ldots, h$. This yields polynomials $\mathcal{S}_i(K_i = \xi_i)$, $i = 1, \ldots, s - \delta$ which generate part of the ideal Id. The full ideal can be obtained by completing the $\mathcal{S}_i(K_i = \xi_i)$, $i = 1, \ldots, s - \delta$ minimally so that (2.9.66) holds. Now we can act with the $\mathcal{S}_j(\xi_i = \partial_{t_i})$, $j = 1, \ldots, s$ on any period say the one corresponding to the structure sheaf $Z(\mathcal{O}_W)$. This brings down s_i monomials in the integrand, whose exponents can be lowered by the relations $x\Gamma(x) = \Gamma(x + 1)$, $\Gamma(x)\Gamma(1 - x) = \frac{\pi}{\sin(\pi x)}$ and a redefinition of the integration variables. This yields a relation between maximal order derivatives and lower order derivatives of $Z(\mathcal{O})_W$ with polynomial coefficients in the $e^{2\pi i t_\alpha}$ and constitutes a linear differential operator annihilating all $Z(\mathcal{O})_W(q^1, \ldots, q^h)$, i.e. a Picard-Fuchs operator $\mathcal{L}_i$. The differential ideal of the latter completely determines these periods, if (2.9.66) holds. The latter point should also hold in the case of non-abelian gauged linear σ-models, which leads not to differential systems of generalized hypergeometric type, but rather to the Apery type.[77] The $\alpha(s)$ factor in (2.9.63) makes it slightly more non-trivial to lower the powers and rewrite the integral in the standard form described above.

[76] Note that in this section t_α denotes not the quantum corrected Kähler parameter, which is defined in (2.9.53).

[77] Only the one moduli cases of Apery type, like the Grassmannians for which higher genus invariants have been calculated in [161], have conifold loci in different distance from the large complex structure point $\lim_{t \to \infty}$. This is a necessary condition for fast enough convergence of the analytic continuation from the conifold to the large complex structure point, that would be needed to prove the irrationality of $\zeta(2m + 1)$ occurring in the periods of CY $2m + 1$-folds for $m > 1$ at infinity due to (2.6.21) in (2.6.25). We thank Sergei Galkin to point this fact out.

Some Remarks on the Local Cases

We can study the period system on local Calabi-Yau spaces, by replacing the sections $W_{\Delta_l} = 0$ of $D_{0,l}$ by the total space of the bundles $\oplus_{l=1}^{r}\mathcal{O}(-D_{0,l})$ over $\mathbb{P}_{\Delta^*}$. This is the obvious generalization of taking instead of the elliptic curve (quintic hypersurface) defined as a section of the canonical bundle K in $\mathbb{P}^2$ ($\mathbb{P}^4$) the noncompact Calabi-Yau 3(5)-manifold defined as the total space of the anti-canonical line bundle $\mathcal{O}(-3) \to \mathbb{P}^2$ ($\mathcal{O}(-5) \to \mathbb{P}^4$). This is implemented by the following change in the integrand of (2.9.65)

$$\sinh(\pi l_{0j}^{(\alpha)} s_\alpha) \to \frac{\sinh(\pi l_{0j}^{(\alpha)} s_\alpha)}{s_\alpha} . \tag{2.9.67}$$

This process can be done successively leading to an increasing number of non-compact directions.

2.10 *The Quintic and Other One Parameter Families*

Let us now discuss an explicit simple example of such a mirror symmetry computation. The principle example is the quintic in the projective space $\mathbb{P}^4$, which is discussed in great detail in the paper [61]. We follow the notation and the presentation in [191]. The quintic is defined as the zero locus of a special homogeneous polynomial of degree 5 in x_i, which we give here in three parametrisations commonly used in the literature[78]

$$P = \sum_{i=1}^{5} a_i x_i^5 + a_0 \prod_{i=1}^{5} x_i = \sum_{i=1}^{5} x_i^5 - 5\psi \prod_{i=1}^{5} x_i = \sum_{i=1}^{5} x_i^5 - z^{-\frac{1}{5}} \prod_{i=1}^{5} x_i = 0 \tag{2.10.1}$$

The z appears here as one of the 101 possible complex structure deformations of the full family of quintics. A deformation is given by perturbing $P_0 = \sum_{i=1}^{5} x_i^5$ with a parameter multiplying a monomial of degree 5. We count (5) x_i^5, (20) $x_i^4 x_j$, (20) $x_i^3 x_j^2$, (30) $x_i^2 x_j^2 x_k$, (30) $x_i x_j x_k^3$, (20) $x_i x_j x_k x_l^2$, (1) $\prod_{i=1}^{5} x_i$, with $i, j, k, l = 1, \ldots 5$ hence 126 monomials. Not all of those lead to independent complex structure deformations, because the complex linear transformations of the coordinates x_i of $\mathbb{P}^4$ leads to completely equivalent forms of the constraint. The group of those has dimension $5^2 - 1$. Finally there is one relation by $P = 0$ leading to 101 independent

[78]The first is the generic parametrization corresponding to (2.7.10) the third one is the one in the Batyrev parametrization (2.8.19), while in the middle the parameter ψ is as in the original paper of [61].

complex structure deformations. The symmetric deformation in (2.10.4) is chosen with hindsight, because we can see it as the unique complex structure deformation on the mirror manifold of the quintic W. The mirror is constructed as $\mathbb{Z}_5^3$ orbifold of the original quintic M. The orbifold is generated by phase rotations on the homogeneous coordinates $\mathbb{P}^4$

$$x_i \to \exp(2\pi i g_i^{(\alpha)}/5)x_i, \quad \alpha = 1,2,3, \quad i = 1,\ldots,5\,, \tag{2.10.2}$$

with $g^{(1)} = (1,4,0,0,0)$, $g^{(2)} = (1,0,4,0,0)$ and $g^{(3)} = (1,0,0,4,0)$. It leaves precisely the single perturbing monomial $\prod_{i=1}^5 x_i$ in (2.10.1) invariant. This one deformation parameter z can be identified with the one Kähler deformation t of the original quintic M which has Hodge numbers $h^{1,1} = 1$ and $h^{2,1} = 101$. The one element in $H^{1,1}(M)$ comes from the restriction of the unique Kähler form of $\mathbb{P}^5$ to the hyper surface. The 101 elements of $H^1(M,TM)$ we counted above and explained their relation to $H^{2,1}(M)$ in (2.4.15).

The quintic and its mirror is also the simplest example for Batyrev's construction, where one starts with the reflexive polyhedra $(\Delta,\hat\Delta)$ given by

$$\begin{array}{c|cccc|cccc|c} Div. & & \nu_i & & & & \hat\nu_i & & & l^{(1)} \\ D_{x_0} & 0 & 0 & 0 & 0 & 0 & 0 & 0 & 0 & -5 \\ D_{x_1}=H & 4 & -1 & -1 & -1 & 1 & 0 & 0 & 0 & 1 \\ D_{x_2}=H & -1 & 4 & -1 & -1 & 0 & 1 & 0 & 0 & 1 \\ D_{x_3}=H & -1 & -1 & -4 & -1 & 0 & 0 & 0 & 1 & 1 \\ D_{x_4}=H & -1 & -1 & -1 & 4 & 0 & 0 & 0 & 1 & 1 \\ D_{x_4}=H & -1 & -1 & -1 & -1 & -1 & -1 & -1 & -1 & 1 \end{array}\,. \tag{2.10.3}$$

Here we listed the divisors in the toric description of $\mathbb{P}_{\hat\Delta} = \mathbb{P}^4$, which correspond all to the hyperplane class in $\mathbb{P}^4$ and the basis of linear relation between the points in $\hat\Delta$, put in the hyperplane at distance one of the origin in $\mathbb{R}^5$. The polynom (2.7.14) becomes

$$P_{\hat\Delta} = \sum_{i=1}^5 \hat a_i X_i + \hat a_0 X_0 = \sum_{i=1}^5 \hat a_i y_i^5 + \hat a_0 \prod_{i=1}^5 y_i\,. \tag{2.10.4}$$

where we used only the coordinates that correspond to the points in Δ listed in (2.10.3). In particular we set to one the coordinates that correspond to the inner point $y_0 = 1$ and the ones that correspond to points on edges and two faces of Δ, which describe the solutions of the $\mathbb{Z}_5$ fixed curves and the $\mathbb{Z}_5 \times \mathbb{Z}_5$ fixpoints of the $\mathbb{Z}_5^3$ orbifold action on the quintic described above.[79] The remaining coordinates are enough to understand the complex structure of the mirror.

[79]The ones on co-dimension one faces correspond to $\mathbb{Z}_5^3$ fixpoints in $\mathbb{P}^4$ which do not lie on the quintic. The corresponding exceptional divisors do not meet the quintic.

We note that (2.10.4) could also be viewed as a one parameter deformation of P_Δ

$$P_\Delta = \sum_{i=1}^{5} a_i Y_i + a_0 Y_0 = \sum_{i=1}^{5} a_i x_i^5 + a_0 \prod_{i=1}^{5} x_i \ , \tag{2.10.5}$$

only that in (2.10.4) the geometry has been divided by a $\mathbb{Z}_5^3$, so that the overall normalizations of the periods is divided by 5^3. It is customary to use variables in (2.10.5), i.e. (Y_i, x_i, a_i) whenever the complex structure deformations of a geometry is considered and also to write P for P_Δ or $P_{\hat{\Delta}}$. We note that to introduce the variables on the dense algebraic torus $\mathbb{T}_4$ of ambient space, we use (2.7.12). Here this implies $\prod_{i=1}^{5} Y_i = Y_0^5$, we set $Y_0 = 1$ and eliminate Y_5 to arrive at

$$P = a_0 + \sum_{i=1}^{4} a_i Y_i + \frac{a_5}{Y_1 Y_2 Y_3 Y_4} \ . \tag{2.10.6}$$

The holomorphic (3, 0) form can written explicit in every patch U_l of $\mathbb{P}^4$ as a residuum expression [147]

$$\Omega(z) = \int_\gamma \frac{a_0 \mu}{P} \ , \tag{2.10.7}$$

where the contour surrounds the single pole at $P = 0$ inside $\mathbb{P}^4$ and the measure is the specialisation of (2.8.6).

We now discuss the action of $\mathcal{Z}_i$ operators in (2.9.30) on the period Ω and how to use these operators to get the Picard-Fuchs operator in the example of the mirror quintic.

An important consistency condition for Ω is its invariance under the $\mathbb{C}^*$ action $x_i \to \lambda x_i$. Let us consider the parametrization of the complex structure by the parameters a_i, $i = 0, \ldots, 5$ in $P = \sum_{i=1}^{5} a_i x_i^5 + a_0 \prod_{i=1}^{5} x_i$. Theses are redundant parameters and can be "gauged" by the $G_{\mathbb{P}^4} = PGL(N, \mathbb{C}) \times \mathbb{C}^*$ transformation on the homogeneous parameters $(x_1 : \ldots : x_5)$ of $\mathbb{P}^4$ to one parameter. Let us summarize the "gauge invariances" of $\Omega(\underline{a})$, which are obvious from (2.10.7) and (2.8.6).

- It is invariant under the change $a_i \to \rho a_i$ with $\rho \in \mathbb{C}^*$. Defining the logarithmic derivative $\vartheta_i = a_i \frac{\partial}{\partial a_i}$, this homogeneity of degree 0 is expressed as

$$\sum_{i=0}^{5} \vartheta_i \Omega(\underline{a}) = 0 \ . \tag{2.10.8}$$

- It is invariant under the $\mathbb{C}^*$ actions $(a_i, a_j) \to (\rho^{-5} a_i, \rho^5 a_j)$, $i, j = 1, \ldots, 5$ with $\rho \in \mathbb{C}^*$. These are compensated on P by $G_{\mathbb{P}^4}$ transformations $(x_i, x_j) \to (\rho x_i, \rho^{-1} x_j)$, which leave the form μ invariant. As differential relations one has

$$(\vartheta_i - \vartheta_5)\Omega(\underline{a}) = 0, \quad i = 1, \ldots, 4 . \tag{2.10.9}$$

These two equations mean that $\Omega(\underline{a}) = \Omega(z)$ does depend only on the combination $z = -\frac{a_1 a_2 a_3 a_4 a_5}{a_0^5}$, where we chose the sign for latter convenience. Instead of fixing the gauge immediately we first notice the obvious differential relations

$$\left(\frac{\partial}{\partial a_0}\right)^5 \frac{\Omega(\underline{a})}{a_0} = \left(\prod_{i=1}^{n} \frac{\partial}{\partial a_i}\right) \frac{\Omega(\underline{a})}{a_0} , \tag{2.10.10}$$

which as explained after (2.9.30) just follow from the linear relations between the points in the Newton polytop $\bar{\Delta}$ of P.

With $\vartheta_i = a_i \frac{\partial}{\partial a_i}$, $\theta = z \frac{\mathrm{d}}{\mathrm{d}z}$, the commutator $[\vartheta_i, a_i^x] = x a_i$ and $\vartheta_0 = -5\theta$ as well as $\vartheta_i = \theta$ for $i = 1, \ldots, 5$ we rewrite

$$\begin{aligned}
\left(\frac{\vartheta_0}{a_0}\right)^5 \frac{\Omega(\underline{a})}{a_0} &= \frac{1}{a_1 a_2 a_3 a_4 a_5} \left(\prod_{i=1}^{5} \vartheta_i\right) \frac{\Omega(\underline{a})}{a_0} \\
\frac{a_1 a_2 a_3 a_4 a_5}{a_0^5} \left(\prod_{k=1}^{5} (\vartheta_0 - k)\right) \Omega(\underline{a}) &= \left(\prod_{i=1}^{5} \vartheta_i\right) \Omega(\underline{a}) \\
z \prod_{k=1} (5\theta + k) \Omega(z) &= \theta^5 \Omega(z)
\end{aligned} \tag{2.10.11}$$

The last line means that the factorizing differential operator $\mathcal{D} = \theta \mathcal{L} = \theta[\theta^4 - 5z \prod_{i=1}^4 (\theta + i)]$ annihilates $\Omega(z)$ and it also annihilates the periods

$$\Pi_i(z) = \int_{\Gamma_i} \Omega(z) \tag{2.10.12}$$

with $\Gamma_i \in H^3(W)$. One checks that $\mathcal{L}\Omega(z)$ is already exact, i.e. $\int_{\Gamma_i} \mathcal{L}\Omega(z) = 0$ so that the periods $\Pi_i(z) = \int_{\Gamma_i} \Omega(z)$, which correspond to the four independent cycles $\Gamma_i \in H_3(W)$ are determined by the four solutions of differential equation

$$[\theta^4 - 5z \prod_{i=1}^{4} (\theta + i)] \, \Pi(z) = 0 . \tag{2.10.13}$$

Note that the mirror has $h^{2,1} = 1$ and hence 4 elements in the middle cohomology $H^3(M, \mathbb{Z}) = H^{3,0} \oplus H^{21} \oplus H^{12} \oplus H^{03}$. The four period integrals over the dual four homology 3-cycles, which are invariant under the $\mathbb{Z}_5^3$ group correspond to four independent solutions of Eq. (2.10.13). The 3-cycles are in a fixed topological basis of $H^3(M,\mathbb{Z})$. This basis is independent of the complex structure. The trick in the derivation of the differential equation was to fix the gauge symmetry at the very end (last line of (2.10.11)). This results in a considerable simplification in the derivation of the period equations compared with the Dwork-Griffiths reduction method discussed in Sect. 2.9.7. The method is adjusted to derive the systems of Picard-Fuchs operators of multi parameter Calabi-Yau hypersurfaces and complete intersections in toric ambient spaces, which have the corresponding $\mathbb{C}^*$ actions, see [187, 233]. It will give in general as above differential operators allowing for too many solutions, which need to be reduced to lower order differential operators. In the simplest case this is accomplished by factorization.

We will now discuss the solution of the Picard-Fuchs equation at all critical points in the moduli space, with the aim to get everywhere convergent expression for an integral symplectic basis of the periods.

2.10.1 Integral Basis at the Large Complex Structure of Maximal Unipotent Monodromy

Here we have the most general description, which applies immediately to all hypersurfaces and complete intersections in toric varieties, for the moduli described by the generalized l-vectors. Either by solving the Picard-Fuchs equations with up to triple logarithmic ansäzte or by applying the Frobenius method with the data

$$\omega(z,\rho):=\sum_{n=0}^{\infty}\frac{\Gamma(5(n+\rho)+1)}{\Gamma^5(n+\rho+1)}z^{n+\rho}\qquad D^k_\rho\omega:=\frac{1}{(2\pi i)^k k!}\frac{\partial^k}{\partial^k\rho}\omega\Big|_{\rho=0} \tag{2.10.14}$$

as discussed in Sect. 2.9.8 we obtain the following solutions.

$$\begin{aligned}
\omega_0&=\omega(z,0)=\textstyle\sum_{n=0}^{\infty}\frac{(5n)!}{(n!)^5}z^n\\
\omega_1&=D_\rho\omega(z,0)=\tfrac{1}{2\pi i}\left(\omega_0\log(z)+\sigma_1\right)\\
\omega_2&=C^0D^2_\rho\omega(z,\rho)+c\omega_0=\tfrac{C^0}{2\cdot(2\pi i)^2}\left(\omega_0\log^2(z)+2\sigma_1\log(z)+\sigma_2\right)\\
\omega_3&=C^0D^3_\rho\omega(z,\rho)+c\omega_1+e\omega_0=\tfrac{C^0}{6\cdot(2\pi i)^3}\left(\omega_0\log^3(z)+3\sigma_1\log^2(z)+3\sigma_2\log(z)+\sigma_3\right)
\end{aligned} \tag{2.10.15}$$

The constants $C^0=\int_M\omega^3=5$, $a=\frac{1}{2}$, $c=\frac{1}{24}\int_M c_2\wedge\omega=\frac{25}{12}$ and $e=\frac{\zeta(3)}{2(2\pi i)^3}\int_M c_3=-200\frac{\zeta(3)}{2(2\pi i)^3}$ are given by (2.6.30) and (2.12.21).

The σ_k are determined directly from (2.10.13) or the Frobenius method. To the first few orders we have

$$\begin{aligned}\omega_0 &= 1 + 120z + 113400z^2 + \mathcal{O}(z^3),\\ \sigma_1 &= 770z + 810225z^2 + \mathcal{O}(z^3)\,,\\ \sigma_2 &= 1150z + \tfrac{4208175}{2}z^2 + \mathcal{O}(z^3)\,,\\ \sigma_3 &= -6900z - \tfrac{9895125}{2}z^2 + \mathcal{O}(z^3)\,.\end{aligned} \tag{2.10.16}$$

The solutions (2.10.15) can be combined into the period vector $\vec{\Pi}$ with respect to an integer symplectic basis[80] (A^i, B_j) of $H^3(W, \mathbb{Z})$ as follows [61]:

$$\vec{\Pi} = \begin{pmatrix} \int_{B_1} \Omega \\ \int_{B_2} \Omega \\ \int_{A^1} \Omega \\ \int_{A^2} \Omega \end{pmatrix} = \begin{pmatrix} F_0 \\ F_1 \\ X_0 \\ X_1 \end{pmatrix} = \omega_0 \begin{pmatrix} 2\mathcal{F}^{(0)} - t\partial_t \mathcal{F}^{(0)} \\ \partial_t \mathcal{F}^{(0)} \\ 1 \\ t \end{pmatrix} = \begin{pmatrix} \omega_3 + c\,\omega_1 + e\,\omega_0 \\ -\omega_2 - a\,\omega_1 + c\,\omega_0 \\ \omega_0 \\ \omega_1 \end{pmatrix}. \tag{2.10.17}$$

The mirror map t which is identified with the complexified area of a degree one curve is given either by

$$t(z) = \int_C (iJ + B) = \frac{\omega_1}{\omega_0} = \frac{1}{2\pi i}\left(\log(z) + 770z + 717825z^2 + ..\right) \tag{2.10.18}$$

or by

$$\frac{1}{z} = \frac{1}{Q} + 770 + 421375\,Q + 274007500\,Q^2 + \dots\,. \tag{2.10.19}$$

In (2.10.19), we inverted (2.10.18) with $Q = e^{2\pi i t}$. The genus one prepotential reads

$$\mathcal{F}^{(0)} = -\frac{5}{3!}t^3 - \frac{(1/2)}{2}t^2 + \frac{50}{24}t + \frac{-200}{2(2\pi i)^3} + \frac{1}{(2\pi i)^3}\sum_{\beta=1}^{\infty} n_0^\beta \mathrm{Li}_3(Q^\beta)\,, \tag{2.10.20}$$

where the instanton expansion

$$F_{inst}(Q) = \sum_{\beta=0} r_0^\beta Q^\beta = \sum_{\beta=0}^{\infty} n_0^\beta \mathrm{Li}_3(Q^\beta) = \frac{\int_M c_3 \zeta(3)}{2} + 2875q + \frac{4876875}{2}q^2 + \frac{8564575000}{27} + \dots \tag{2.10.21}$$

with $Q = \exp(2\pi i t)$ vanishes in the large radius limit $\mathrm{Im}(t) \to \infty$ exponentially $Q \to 0$. The $r_0^\beta \in \mathbb{Q}$ are the Gromov-Witten invariants for rational curves in the

[80] With $A^i \cap B_j = \delta^i_j$ and zero intersections otherwise.

class $\beta \in H_2(M, \mathbb{Z})$ of the quintic M in $\mathbb{P}^4$, while the $n_0^\beta \in \mathbb{N}$ count[81] the rational curves of degree β on the quintic or the BPS invariants of the topological string on the quintic, which are in the first degree

$$n_0^0 = \frac{\int_M c_3}{2} = \frac{-200}{2}, \quad n_0^1 = 2875, \quad n_0^1 = 609250, \quad n_0^1 = 317206375, \ldots . \tag{2.10.22}$$

Here we also included the degree zero imbedding, because as it turns out the BPS invariants are determined by the fundamental virtual class, which is in this case the Euler density, of its moduli space, which is in this case the Calabi-Yau M itself. The virtual fundamental classes of the moduli spaces all degree maps are is zero, so morally one gets an 'orbifold' point counting problem for the r_0^β. The numbers n_0^β can be interpreted as the number of lines, degenerate genus 0 quadrics, cubics etc.

2.10.2 Expansions Around the Orbifold Point $z = \infty$

The solutions of the Picard-Fuchs equation around the orbifold point $w = \frac{1}{5^5 z} = 0$ are four power series solutions with the indices $\frac{1}{5}, \frac{2}{5}, \frac{3}{5}, \frac{4}{5}$

$$\begin{aligned}\omega_k^{\rm orb} &= w^{\frac{k}{5}} \sum_{n=0}^{\infty} \frac{\left([\frac{k}{5}]_n\right)^5}{[k]_{5n}} \left(5^5 w\right)^n \\ &= -\frac{\Gamma(k)}{\Gamma^5\left(\frac{k}{5}\right)} \int_{C_0} \frac{\mathrm{d}s}{e^{2\pi i s} - 1} \frac{\Gamma^5\left(s + \frac{k}{5}\right)}{\Gamma(5s + k)} \left(5^5 w\right)^{s + \frac{k}{5}}, \qquad k = 1, \ldots, 4 .\end{aligned} \tag{2.10.23}$$

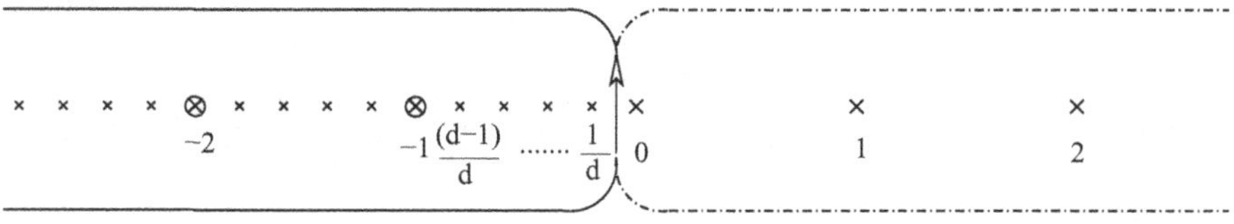

C_∞ for $|w| > 1$ $\qquad$ C_0 for $|w| < 1$.

The Pochhammer symbols are defined as $[a]_n := \frac{\Gamma(a+n)}{\Gamma(a)}$ and we normalized the first coefficient in $\omega_k^{\rm orb} = w^{\frac{k}{5}} + \mathcal{O}(w^{\frac{5+k}{5}})$ to one. The expression in the first line is recovered from the integral representation by noting that the only poles inside C_0, for which the integral converges for $|w| < 0$, are from $g(s) = \frac{1}{\exp(2\pi i s) - 1}$, which behaves at $s_{-n}^\epsilon = n - \epsilon, n \in \mathbb{N}$ as $g(s_{-n}^\epsilon) \sim -\frac{1}{2\pi i \epsilon}$.

[81] In fact in general, i.e. for general classes $\beta \in H_2(M, \mathbb{Z})$ in more complicated manifolds M, $n_0^\beta \in \mathbb{Z}$ and the correct interpretation of the n_0^β is an index in the cohomology of stable pairs.

Up to normalization this basis of solutions is canonically distinguished, as it diagonalizes the $\mathbb{Z}_5$ monodromy at $w = 0$. Similar as for the $\mathbb{C}^3/\mathbb{Z}_3$ orbifold [6], it can be viewed as a twist field basis. Here this basis is induced from the twist field basis of $\mathbb{C}^5/\mathbb{Z}_5$. As it was argued in [6] for $\mathbb{C}^3/\mathbb{Z}_3$, this twist field basis provides the natural coordinates in which the $F^{(g)}$ near the orbifold point can be interpreted as generating functions for orbifold Gromov-Witten invariants. Following up on foundational work on orbifold Gromov-Witten theory [73] and examples given in two complex dimensions [56] this prediction has been checked by direct computation of orbifold Gromov-Witten invariants [75] at genus zero. This provides a beautiful check for the global picture of mirror symmetry.

We will now study the transformation from the basis (2.10.17) to the basis (2.10.23) to make B-model prediction along the lines of [6] with the additional Kähler transformation. Since the symplectic form ω on the moduli space is invariant under monodromy and ω_k^{orb} diagonalizes the $\mathbb{Z}_5$ monodromy, we must have in accordance with the expectation from the orbifold cohomology $H^*(\mathbb{C}^5/\mathbb{Z}_5) = \mathbb{C}1_0 \oplus \mathbb{C}1_1 \oplus \mathbb{C}1_2 \oplus \mathbb{C}1_3 \oplus \mathbb{C}1_4$

$$\omega = \mathrm{d}F_k \wedge \mathrm{d}X_k = -\frac{5^4 s_1}{6}\mathrm{d}\omega_4^{orb} \wedge \mathrm{d}\omega_1^{orb} + \frac{5^4 s_2}{2}\mathrm{d}\omega_3^{orb} \wedge \mathrm{d}\omega_2^{orb} \,. \tag{2.10.24}$$

This is equivalent to determining the symplectic form $\eta = \Sigma_w$ at the orbifold according to the description in Sect. 2.9.4.

The rational factors above have been chosen to match constraints from special geometric discussed below. Similarly the monodromy invariant Kähler potential must have the form

$$e^{-K} = \sum_{k=1}^{4} r_k \omega_k^{\mathrm{orb}} \overline{\omega_k^{\mathrm{orb}}} \,. \tag{2.10.25}$$

To obtain the s_i, r_i by analytic continuation to the basis (2.10.17) we follow [61] for the quintic and the generalisation in [226] for other cases and note that the integral converges for $|w| > 1$ due to the asymptotics of the $f(s) = \frac{\Gamma^5(s+k/5)}{\Gamma(5s+k)}$ term, when the integral is closed along $\mathcal{C}_\infty$ [61]. At $s_n^\epsilon = -n - \epsilon$ the $g(s_n^\epsilon)$ pole is compensated by the $f(s_n^\epsilon)$ zero and at $s_{n,k}^\epsilon = -n - k/5 - \epsilon$ we note the expansions

$$\begin{aligned}
g(s_{n,k}^\epsilon) &= \frac{\alpha^k}{1-\alpha^k} + \frac{2\pi i\alpha^k}{(1-\alpha^k)^2}\epsilon + \frac{(2\pi i)^2\alpha^k(1+\alpha^k)}{2(1-\alpha^k)^3}\epsilon^2 + \frac{(2\pi i)^3\alpha^k(1+4\alpha^k+\alpha^{2k})}{6(1-\alpha^k)^4}\epsilon^3 + \mathcal{O}(\epsilon^4) \\
\tilde{f}(s_{n,k}^\epsilon) &= \frac{C^0\omega_0(n)}{\epsilon^4} + \frac{\kappa\sigma_1(n)}{\epsilon^3} + \frac{1}{\epsilon^2}\left(\frac{\kappa\sigma_2(n)}{2} + \frac{(2\pi i)^2 c_{2J}\omega_0(n)}{24}\right) + \\
&\quad \frac{1}{\epsilon}\left(\frac{\kappa\sigma_3(n)}{6} + \frac{(2\pi i)^2 c_{2J}\sigma_1(n)}{24} + \chi\zeta(3)\omega_0(n)\right) + \mathcal{O}(\epsilon^0) \\
\left(5^5 w\right)^{s_n^\epsilon} &= z^n(1 + \log(z)\epsilon + \tfrac{1}{2}\log(z)^2\epsilon^2 + \tfrac{1}{6}\log(z)^3\epsilon^3 + \mathcal{O}(\epsilon^4))
\end{aligned} \tag{2.10.26}$$

Here $\alpha = \exp(2\pi i/5)$. $C^0 = \int_M J^3$, $c_{2J} = \int_M c_2 J$, and $\chi = \int_M c_3$ are the classical intersection calculated at large volume. The $\omega_0(n)$, $\sigma_i(n)$ are coefficients of the series we encountered in sec. 2.10.1. Performing the residue integration and comparing with (2.10.15), (2.10.17) we get

$$\omega_k^{orb} = \frac{(2\pi i)^4\Gamma(k)}{\Gamma^5\left(\frac{k}{5}\right)}\left(\frac{\alpha^k F_0}{1-\alpha^k} - \frac{\alpha^k F_1}{(1-\alpha^k)^2} + \frac{5\alpha^k(\alpha^{2k}-\alpha^k+1)X_0}{(1-\alpha^k)^4} + \frac{\alpha^k(8\alpha^k-3)X_1}{(1-\alpha^k)^3}\right) \tag{2.10.27}$$

It follows with $r_i = \frac{\Gamma^{10}\left(\frac{k}{5}\right)}{\Gamma^2(k)} c_i$ that

$$\begin{aligned} c_1 &= -c_4 = \alpha^2(1-\alpha)(2+\alpha^2+\alpha^3), \quad c_3 = -c_2 = \alpha(2+\alpha-\alpha^2-2\alpha^3) \\ s_1 &= s_2 = -\frac{1}{5^5(2\pi i)^3} \end{aligned} \tag{2.10.28}$$

and so

$$\begin{pmatrix} F_0 \\ F_1 \\ X_0 \\ X_1 \end{pmatrix} = w^{1/5}\frac{\alpha\Gamma^5\left(\frac{1}{5}\right)}{(2\pi i)^4}\begin{pmatrix} (1-\alpha)(\alpha-1-\alpha^2) \\ \frac{1}{5}(8-3\alpha)(1-\alpha)^2 \\ (1-\alpha+\alpha^2) \\ \frac{1}{5}(1-\alpha)^3 \end{pmatrix} + \mathcal{O}(w^{2/5})\,. \tag{2.10.29}$$

Equation (2.10.28) implies that up to a rational rescaling of the orbifold periods the transformation of the wave function $\mathbb{Z} = \Psi$ [321] from infinity to the orbifold is given by a metaplectic transformation with the same rescaling of the string coupling as for the $\mathbb{C}^3/\mathbb{Z}_3$ case in [4]. The BPS mass formula $m_\Pi = e^{\frac{K/2}{|\Pi|}}$, for Π a period in the Sp(4$\mathbb{Z}$) symplectic basis (2.10.29), implies that there are no massless BPS states at the orbifold. This means that there is no massless RR state in the K-theory charge lattice at the orbifold point. We note further that after rescaling of the orbifold periods the transformation (2.10.27) can be chosen to lie in Sp(4, $\mathbb{Z}[\alpha, \frac{1}{5}]$).

We can define the analogue of mirror map at the orbifold point,

$$s = \frac{\omega_2^{orb}}{\omega_1^{orb}} = w^{\frac{1}{5}}\left(1 + \frac{13w}{360} + \frac{110069w^2}{9979200} + \mathcal{O}(w^3)\right) \tag{2.10.30}$$

where we use the notation s, as in [6], to avoid confusion with the mirror map in the large volume limit. We next calculate the genus zero prepotential at the orbifold point. For convenience let us rescale our periods $\hat{\omega}_{k-1} = 5^{3/2}\omega_k^{orb}$. The Yukawa-Coupling or 3-point function is transformed to the s variables as

$$C_{sss} = \frac{1}{\hat{\omega}_0^2}\frac{5}{w^2(1-w)}\left(\frac{\partial w}{\partial s}\right)^3 = 5 + \frac{5}{3}s^5 + \frac{5975}{6048}s^{10} + \frac{34521785}{54486432}s^{15} + \mathcal{O}(s^{20})\,. \tag{2.10.31}$$

A trivial consistency check of special geometry is that the genus zero prepotential $F^{(0)} = \int ds \int ds \int ds \; C_{sss}$ appears in the periods $\hat{\Pi}^{orb} = (\hat{\omega}_0, \hat{\omega}_1, \frac{5}{2!}\hat{\omega}_2, -\frac{5}{3!}\hat{\omega}_3)^T$ as

$$\hat{\Pi}^{orb} = \hat{\omega}_0 \begin{pmatrix} 1 \\ s \\ \partial_s F^{(0)}_{\text{A-orbf.}} \\ 2F^{(0)}_{\text{A-orbf.}} - s\partial_s F^{(0)}_{\text{A-orbf.}} \end{pmatrix} . \tag{2.10.32}$$

This can be viewed also as a simple check on the lowest order meta-plectic transformation of Ψ which is just the Legendre transformation. Note that the Yukawa coupling is invariant under the $\mathbb{Z}_5$ which acts as $s \mapsto \alpha s$. $\mathbb{Z}_5$ implies further that there can be no integration constants, when passing from C_{sss} to F_0 and the coupling λ must transform with $\lambda \mapsto \alpha^{\frac{3}{2}}\lambda$ to render $F(\lambda, s, \bar{s})$ invariant.

The holomorphic limit $\bar{w} \to 0$ of Kähler potential and metric follows from (2.10.25) by extracting the leading anti-holomorphic behaviour. Denoting[82] by a_k the leading powers of ω_k^{orb} we find

$$\lim_{\bar{w}\to 0} e^{-K} = r_1 \bar{w}^{a_1} \omega_1^{orb}, \qquad \lim_{\bar{w}\to 0} G_{w\bar{w}} = \bar{w}^{a_2-a_1-1} \frac{r_2}{r_1}\left(\frac{a_2}{a_1} - 1\right)\frac{\partial s}{\partial w} . \tag{2.10.33}$$

2.10.3 Expansions Around the Conifold Point

A basis of solutions of the Picard-Fuchs equation around the conifold point $(1 - 5^5 z) = \delta_c \sim 0$ is the following

$$\vec{\Pi}_c = \begin{pmatrix} \omega_0^c \\ \omega_1^c \\ \omega_2^c \\ \omega_3^c \end{pmatrix} = \begin{pmatrix} 1 + \frac{2\delta_c^3}{625} + \frac{97\delta_c^4}{18750} + \mathcal{O}(\delta_c^5) \\ \delta_c + \frac{7\delta_c^2}{10} + \frac{41\delta_c^3}{25} + \frac{1133\delta_c^4}{2500} + \mathcal{O}(\delta_c^5) \\ \delta_c^2 + \frac{37\delta_c^3}{30} + \frac{2309\delta_c^4}{1800} + \mathcal{O}(\delta_c^5) \\ \omega_1^c \log(\delta_c) - \frac{23\,\delta_c^3}{360} - \frac{6397\,\delta_c^4}{60000} + \mathcal{O}(\delta_c^5) \end{pmatrix} \tag{2.10.34}$$

Here we use the superscript "c" in the periods to denote them as solutions around the conifold point. We see that one of the solutions ω_1^c is singled out as it multiplies the log in the solution ω_3^c. By a Lefschetz argument as in section "Conifold Monodromy" it corresponds to the integral over the vanishing S^3 cycle B_1 and moreover the solution containing the log is the integral over dual cycle A_1. Comparing with (2.10.17), (2.6.30) and (2.10.35) shows in the Type IIA interpretation that the $D6$ brane becomes massless. To determine the periods in

[82] This is to make contact with the other one modulus cases. Of course if $a_1 = a_2$ a log singularity appears and the formula does not apply.

the symplectic basis at the conifold we analytically continue the solutions (2.10.17) from $z = 0$ to $\delta_c = 0$ and get

$$\begin{pmatrix} F_0 \\ F_1 \\ X_0 \\ X_1 \end{pmatrix} = \begin{pmatrix} 0 & \frac{\sqrt{5}}{2\pi i} & 0 & 0 \\ a - \frac{i}{2}g & b - \frac{i}{2}h & c - \frac{i}{2}r & 0 \\ d & e & f & -\frac{\sqrt{5}}{(2\pi i)^2} \\ ig & ih & ir & 0 \end{pmatrix} \begin{pmatrix} \omega_0^c \\ \omega_1^c \\ \omega_2^c \\ \omega_3^c \end{pmatrix} = T_{z\delta} \begin{pmatrix} \omega_0^c \\ \omega_1^c \\ \omega_2^c \\ \omega_3^c \end{pmatrix} \tag{2.10.35}$$

Six of the real numbers $a, \ldots, r$ are only known numerically.[83] According to the prescription outlined in section (2.9.4) we can calculate the new intersection form $\eta_\delta = \Sigma_\delta$ as

$$\Sigma_\delta = \begin{pmatrix} 0 & 0 & \frac{5}{2} & 0 \\ 0 & 0 & -\frac{9}{4} & -5 \\ -\frac{5}{2} & \frac{9}{4} & 0 & 0 \\ 0 & 5 & 0 & 0 \end{pmatrix} = (2\pi i)^3 T_{z\delta}^T \Sigma T_{z\delta} \, . \tag{2.10.36}$$

Here Σ is the standard symplectic pairing in the symplectic basis, as determined in our case at the point of maximal unipotent monodromy. This can be read as an constraint that special geometry imposes on the entries on $T_{z\delta}$ and leads to the Legendre relations

$$\begin{aligned} \sqrt{5}d &= 2\pi(bd - qh) \\ 5 &= \frac{16}{5}\pi^3(ra - cg) \\ \frac{9}{4} &= 8\pi^3(\frac{\sqrt{5}}{2\pi}f + ch - br) \, , \end{aligned} \tag{2.10.37}$$

which can be solved for example to yield

$$h = \frac{2\pi bg - \sqrt{5}d}{2\pi a}, \qquad r = \frac{16\pi^3 cg + 5}{16\pi^3 a} \qquad f = \frac{9a + 10b + 16\sqrt{5}\pi^2 cd}{16\sqrt{5}\pi^2 a}. \tag{2.10.38}$$

The new mirror map at $\delta = 0$ should be invariant under the conifold monodromy and vanishing at the conifold. The vanishing period has $D6$ brane charge and is singled out to appear in the numerator of the mirror map. The numerator is not fixed

[83] $a = 6.19501627714957\ldots$, $b = 1.016604716702582\ldots$, $c = -0.140889979448831\ldots$, $d = 1.07072586843016\ldots$, $e = -0.0247076138044847\ldots$, $g = 1.29357398450411\ldots$

up to the fact that ω_3^c should not appear. The simplest mirror map compatible with symplectic form (2.10.36) is

$$t_D(\delta) := \frac{\omega_1^c}{\omega_0^c} = \delta - \frac{3\delta^2}{10} + \frac{11\delta^3}{75} - \frac{9\delta^4}{100} + \frac{5839 t_D^5}{93750} + \mathcal{O}(t_D^6) \quad (2.10.39)$$

$$\delta(t_D) = t_D + \frac{3t_D^2}{10} + \frac{t_D^3}{30} + \frac{t_D^4}{200} + \frac{169 t_D^5}{375000} + \mathcal{O}(t_D^6) \quad (2.10.40)$$

We call this new mirror map the dual mirror map and denote it t_D to distinguish from the large complex structure modulus mirror map. Note that here we changed the parametrization of the discriminant locus in way to make contact to the notation in [191] to $\delta_c = \frac{\delta}{1+\delta}$.

In the holomorphic limit $\bar{\delta} \to 0$, the Kahler potential and metric should behave as $e^{-K} \sim \omega_0^c$ and $G_{\delta\bar{\delta}} \sim \partial_\delta t_D$. Remarkably it turns out that shifts $\omega_0^c \to \omega_0^c + b_1\omega_1^c + b_2\omega_2^c$ does not affect the structure we are interested in. The fact that the b_1 shifts do not affect the amplitudes is reminiscent of the SL(2, $\mathbb{C}$) orbit theorem [293]. It is therefore reasonable to state the results in the more general polarization and define $\hat{\omega}_0^c = \omega_0^c + b_1\omega_1^c + b_2\omega_2^c$. We first determine the genus 0 prepotential checking consistency of the solutions with special geometry. Defining $\hat{t}_D = \frac{\omega_1^c}{\hat{\omega}_0^c}$ and $\Pi_{con} = (\hat{\omega}_0^c, \omega_1^c, \frac{5}{2}\omega_2^c, -5\omega_3^c)^T$ we get

$$\hat{\Pi}^{conif} = \hat{\omega}_0 \begin{pmatrix} 1 \\ \hat{t}_D \\ 2F^{(0)}_{\text{conif.}} - \hat{t}_D \partial_{\hat{t}_D} F^{(0)}_{\text{conif.}} \\ \partial_{\hat{t}_D} F^{(0)}_{\text{conif.}} \end{pmatrix}. \quad (2.10.41)$$

with

$$F^{(0)}_{\text{conif.}} = -\frac{5}{2}\log(\hat{t}_D)\hat{t}_D^2 + \frac{5}{12}(1-6b_1)\hat{t}_D^3 + \left(\frac{5}{12}(b_1-3b_2) - \frac{89}{1440} - \frac{5}{4}b_1^2\right)\hat{t}_D^4 + \mathcal{O}(\hat{t}_D^5). \quad (2.10.42)$$

The Monodromies of the Quintic

Let us shortly discuss the monodromies of the quintic. Originally in the paper by [61] the construction of the integral basis has been performed by calculating local bases at large radius—the orbifold—and the conifold point and using analytic continuation as well as the fact that the monodromy has to be a subgroup of Sp(4, $\mathbb{Z}$) to find the integral basis. In this process it was realized that the vanishing cycle at the conifold correspond to the $D6$ brane i.e. the first entry in the period in (2 6.30). Therefore by the Picard-Lefshetz monodromy (2.6.13) the monodromy $M_{\delta_c=0}$ around $\delta_c = 0$ in the integral basis is determined. Like wise the monodromy

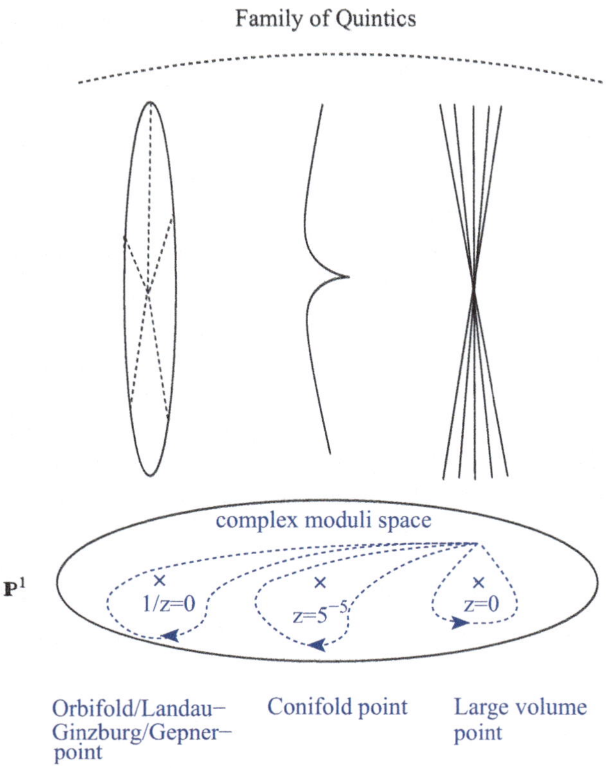

Fig. 5 The family of quintics over the moduli space $\mathbb{P}^1 \setminus \{z = 0, z = 5^{-5}, z = \infty\}$

around $z = 0$ is just given by the B-field shift i.e. $t \to t - 1$ due to the logarithms in the periods in (2.10.15)[84] and therefore trivial to determined from the classical intersection data. Finally since the moduli space is an $\mathbb{P}^1$ without the three points in Fig. 5, we get immediately that the orbifold or Gepner point monodromy is given by $M_{w=0} = M^{-1}_{z=0} M_{\delta_c=0}$, i.e. we can summarise the monodromy matrices

$$M_{\delta_c=0} = \begin{pmatrix} 1 & 0 & 0 & 0 \\ 0 & 1 & 0 & 0 \\ 1 & 0 & 1 & 0 \\ 0 & 0 & 0 & 1 \end{pmatrix}, \quad M^{-1}_{z=0} = \begin{pmatrix} 1 & 1 & -5 & 2 \\ 0 & 1 & -3 & 5 \\ 0 & 0 & 1 & 0 \\ 0 & 0 & -1 & 1 \end{pmatrix}, \quad M_{w=0} = \begin{pmatrix} -4 & 1 & -5 & 2 \\ -3 & 1 & -3 & 5 \\ 1 & 0 & 1 & 0 \\ -1 & 0 & -1 & 1 \end{pmatrix}, \tag{2.10.43}$$

[84] We define it in the negative mathematical sense, i.e. we report $M^{-1}_{z=0}$, see the contours in Fig. 5.

with relations

$$M^5_{w=0} = \mathbf{1}, \qquad M_{w=0} = M^{-1}_{z=0} M_{\delta_c=0} . \tag{2.10.44}$$

2.10.4 Further Hypergeometric One Parameter Families

There are thirteen one parameter models which are realised as smooth hypersurfaces or complete intersection in weighted projective spaces, which have been analysed, also at higher genus in [191] and the fourteens one which is not generically smooth. All these example have hypergeometric Picard-Fuchs operators of the form

$$\mathcal{L} = \theta^4 - \mu z \prod_{i=1} (\theta + a_i) . \tag{2.10.45}$$

The factor λ is given in the Table 2 and so that the discriminant whose vanishing indicates a conifold singularity is given by $\delta_c = (1 - \mu z)$. Using (2.9.14) and the triple intersection given also in Table 2 one gets the triple intersection as

$$C_{zzz} = \frac{D^3}{z^3(1-\mu z)} . \tag{2.10.46}$$

With the classical data in Table 2 we can calculate the monodromy at the maximal unipotent point $z = 0$ from (2.6.31), (2.6.30) as for the quintic. The hypergeometric systems have three regular singular points at $z = 0, z = \frac{1}{\mu}, w = \frac{1}{z} = 0$ and the one at the conifold $z = \frac{1}{\mu}$ has the form of the Seidel-Thomas twist, i,e, is always of the form as $M_{\delta_c=0}$ in (2.10.43). The one at $w = \frac{1}{z} = 0$ follows from $M_{w=0} = M^{-1}_{z=0} M_{\delta_c=0}$. In other words the generators of the monodromy follow immediately from the data in Table 2. The one at $w = \frac{1}{z} = 0$ follows from $M_{w=0} = M^{-1}_{z=0} M_{\delta_c=0}$. The $*$ indicate families whose monodromy group Γ inside $Sp(4, \mathbb{Z})$ is not arithmetic, but thin according to [294, 295]. If no * is indicated the monodromy group Γ is arithmetic inside $Sp(4, \mathbb{Z})$, further properties of these arithmetic subgroups were investigated in [179].

Likewise the genus zero instantons follow from the generalized $l^{(1)}$ vectors

$$l^{(1)} = (-d_1, \ldots, -d_s; w_0, \ldots, w_{4+s}) \tag{2.10.47}$$

by (2.9.58) and (2.9.59). The latter algorithm has been implemented in the Mathematica program INSTANTON.m, which is available on request from the author.

As mentioned special geometry implies strong constraints on the transitions matrices. For example for the one parameter families in [236] the following Lemma that can be generalised to the transition matrices of arbitrary Calabi-Yau 3-folds has been proven.

Table 2 Data of one parameter models with hypergeometric Picard-Fuchs operator

N	Name	a_1, a_2, a_3, a_4	μ	D^3	$c_2.D$	a	χ	d_4	$f_{4,N}$
8	$X^*_{2,2,2,2}(1^8)$	$\frac{1}{2}, \frac{1}{2}, \frac{1}{2}, \frac{1}{2}$	2^8	16	64	0	−128	1	$q - 4q^3 - 2q^5 + O(q^7)$
9	$X_{4,3}(1^5 2)$	$\frac{1}{4}, \frac{1}{3}, \frac{2}{3}, \frac{3}{4}$	$2^6 3^3$	6	48	0	−165	1	$q - 8q^4 + 20q^7 + O(q^9)$
16	$X^*_{4,2}(1^6)$	$\frac{1}{4}, \frac{1}{2}, \frac{1}{2}, \frac{3}{4}$	2^{10}	8	56	0	−176	1	$q + 4q^3 - 2q^5 - 24q^7 + O(q^9)$
25	$X^*_5(1^5)$	$\frac{1}{5}, \frac{2}{5}, \frac{3}{5}, \frac{4}{5}$	5^5	5	50	$\frac{1}{2}$	−200	3	$q + q^2 + 7q^3 - 7q^4 + O(q^7)$
27	$X^*_{3,3}(1^6)$	$\frac{1}{3}, \frac{1}{3}, \frac{2}{3}, \frac{2}{3}$	3^6	9	54	$\frac{1}{2}$	−144	4	$q - 3q^2 + q^4 - 15q^5 + O(q^7)$
32	$X_{4,4}(1^4 2^2)$	$\frac{1}{4}, \frac{1}{4}, \frac{3}{4}, \frac{3}{4}$	2^{12}	4	40	0	−144	3	$q - 8q^3 - 10q^5 - 16q^7 + O(q^9)$
36	$X_{3,2,2}(1^7)$	$\frac{1}{3}, \frac{1}{2}, \frac{1}{2}, \frac{2}{3}$	$2^4 3^3$	12	60	0	−144	1	$q + 18q^5 + 8q^7 + O(q^9)$
72	$X^*_{6,2}(1^5 3)$	$\frac{1}{6}, \frac{1}{2}, \frac{1}{2}, \frac{5}{6}$	$2^8 3^3$	4	52	0	−256	4	$q - 14q^5 - 24q^7 + O(q^9)$
108	$X_6(1^4 2)$	$\frac{1}{6}, \frac{1}{3}, \frac{2}{3}, \frac{5}{6}$	$2^4 3^6$	3	42	$\frac{1}{2}$	−204	4	$q - 9q^5 - q^7 - 63q^{11} + O(q^{13})$
128	$X^*_8(1^4 4)$	$\frac{1}{8}, \frac{3}{8}, \frac{5}{8}, \frac{7}{8}$	2^{16}	2	44	0	−296	12	$q - 2q^3 + 6q^5 - 20q^7 - 14q^{11}$
144	$X_{6,4}(1^3 2^2 3)$	$\frac{1}{6}, \frac{1}{4}, \frac{3}{4}, \frac{5}{6}$	$2^{10} 3^3$	12	32	0	−156	7	$q + 16q^5 + 12q^7 - 64q^{11}$
200	$X_{10}(1^3 25)$	$\frac{1}{10}, \frac{3}{10}, \frac{7}{10}, \frac{9}{10}$	$2^8 5^5$	1	34	$\frac{1}{2}$	−288	14	$q + q^3 - 6q^7 - 19q^{11}$
216	$X_{6,6}(1^2 2^2 3^2)$	$\frac{1}{6}, \frac{1}{6}, \frac{5}{6}, \frac{5}{6}$	$2^8 3^6$	1	22	$\frac{1}{2}$	−120	12	$q + q^5 - 9q^7 - 17q^{11}$
864	$X^*_{2,12}(1^4 46)$	$\frac{1}{12}, \frac{5}{12}, \frac{5}{12}, \frac{11}{12}$	$2^{12} 3^6$	1	46	$\frac{1}{2}$	−484	48	$q - 19q^5 - 13q^7 - 65q^{11}$

$X_{d_1,\ldots d_s}(w_0,\ldots,w_{4+s})$ denotes the zero set of $P_1 = \ldots = P_s = 0$ in the weighted projective space $\mathbb{P}^{4+s}[w_0,\ldots,w_{4+s}]$. The table also contains the classical intersection data as well as the specifications of the weight 4, level N Hecke eigenform $f_{4,N} \in S(\Gamma_0(N))$ that counts solution of the manifold over finite Fields and the first Fourier coefficients at the cusp at $\mathrm{Im}(\tau) \to \infty$ and d_4 is the dimension of the Newforms according to the L-function and Modular Form Data Base (LMFDB) [248]

Lemma 1 *For any CY 3-fold hypergeometric system as in Table 2 let* $\delta = 1-\mu z$ *the conifold variable and* $(\tilde{\Pi}^c)^T = (1+O(\delta)^3, \nu = \delta + O(\delta^2), \delta^2 + O(\delta^3), \nu \log(\delta) + O(\delta^3))$ *a normalization of the periods at the conifold point. Then the transition matrix* $T_{z\delta}$ *between the integral symplectic basis* Π *(2.10.17) and the periods at the conifold* $\Pi = T_{z\delta}\tilde{\Pi}^c$ *(2.10.35) fulfills the following quadratic relation*

$$\frac{1}{(2\pi i)^3}\begin{pmatrix} 0 & 0 & \frac{\kappa}{2} & 0 \\ 0 & 0 & -\alpha\kappa & -\kappa \\ -\frac{\kappa}{2} & \alpha\kappa & 0 & 0 \\ 0 & \kappa & 0 & 0 \end{pmatrix} = T_{z\delta}^T \begin{pmatrix} 0 & 0 & 1 & 0 \\ 0 & 0 & 0 & 1 \\ -1 & 0 & 0 & 0 \\ 0 & -1 & 0 & 0 \end{pmatrix} T_{z\delta} \tag{2.10.48}$$

with

$$\alpha = \frac{3}{4}\left(\sum_{i=1}^{4} a_i - \sum_{i<j}^{4} a_i a_j\right) . \tag{2.10.49}$$

2.10.5 One Parameter Calabi-Yau Families with Non Hypergeometric Calabi-Yau Operators

There are Calabi-Yau manifolds embedded as hypersurfaces, complete intersections, determinantal or even more general embeddings into Grassmannians and Flag varieties that have more general and in particular non-hypergeometric order four Picard Fuchs operators then we saw in the last section. For examples of geometric realisations one has complete intersections in Grassmannians $\mathbb{G}(k,n) = (U(k) \times U(n-k))$ with $U(n)$ are the unitary groups, and denotes the complete intersection as

$$(\mathbb{G}(k,n)|d_1,\ldots,d_l) \ . \tag{2.10.50}$$

There are five well studied examples, whose data we list in Table 3. The differential operators of the mirror were found in [32], by a three step chain of transitions over an intermediate torically embedded Calabi-Yau space in which the mirror operation explained in Sect. 2.7.3 could be performed. The connections between the last two was discovered by Rodland [282]. He realised that the one parameter family, which starts out as the complete intersections in a Grassmannians has a second large volume phase in a Paffian Phase with different C.T.C. Wall data as shown in the Table 3. The higher genus invariants have been calculated for the Rodland case in [184] and for all cases in [161]. Let us report on the Yukawa couplings for the models whose numerator also contain the discriminants

$$C_{zzz}^{\mathbb{G}(2,5|1,2^2)} = \frac{15}{z^3(1-176z-256z^2)}\,, \qquad C_{zzz}^{\mathbb{G}(2,5|1^2,3)} = \frac{20}{z^3(1-297z-729z^2)}\,,$$
$$C_{zzz}^{\mathbb{G}(2,6|1^4,2)} = \frac{28}{z^3(1+4z)(1-108z)}\,, \qquad C_{zzz}^{\mathbb{G}(3,6|1^6)}\frac{42}{z^3(1-z)(1-64z)} \tag{2.10.51}$$

Table 3 Data of five one parameter families of complete intersections in Grassmannians with non-hypergeometric Picard-Fuchs operator

#	Name	D^3	$c_2.D$	a	χ	$\mathcal{L}$
1	$\mathbb{G}(2,5\vert 1^2,3)$	15	66	$\frac{1}{2}$	-150	$\theta^4-3z(3\theta+2)(3\theta+1)(11\theta^2+11\theta+3)-$ $9z^2(3\theta+5)(3\theta+2)(3\theta+4)(3\theta+1)$
2	$\mathbb{G}(2,5\vert 1,2^2)$	20	68	0	-120	$\theta^4-4z(11\theta^2+11\theta+3)(1+2\theta)^2-$ $16z^2(2\theta+3)^2(1+2\theta)^2$
3	$\mathbb{G}(2,6\vert 1^4,2)$	28	76	0	-116	$\theta^4-2z(4+13\theta+13\theta^2)(1+2\theta)^2-$ $12z^2(3\theta+2)(2\theta+3)(1+2\theta)(3\theta+4)$
4	$\mathbb{G}(3,6\vert 1^6)$	42	84	0	-96	$\theta^4-z(6+40\theta+105\theta^2+130\theta^3+65\theta^4)+$ $4z^2(4\theta+5)(4\theta+3)(\theta+1)^2$
5	$\mathbb{G}(2,7\vert 1^7)$	42	84	0	-98	$9\theta^4-z((45+2166z-12z^2+26z^3-z^4)+2(153+4773z-675z^2+$ $87z^3-2z^4)\theta+2(408+7597z-2353z^2+239z^3-3z^4)\theta^2+4(3-z)$ $(85+867z-149z^2+z^3)\theta^3+(519+2258z-1686z^2+295z^3-z^4)\theta^4)$
5’	Pfaff Ph	14	56	0	-98	$\mathcal{L}_{\mathbb{G}(2,7\vert\vert 1^7)}(\theta\to-(\theta+1)$

The model $\mathbb{G}(3, 6|1^6)$ has the interesting feature that at $(1-z)$ a Lense space $L(2,1) = S^3/\mathbb{Z}^2$ shrinks all other points are conifolds points where an S^3 shrinks.

$$C_{zzz}^{\mathbb{G}(2,7|1^7)} = \frac{14(3-z)}{z^3\left(z^3 - 289z^2 - 57z + 1\right)}, \qquad C_{zzz}^{Paff.Ph.)} \frac{14(1-3z)}{z^3\left(z^3 - 57z^2 - 289z + 1\right)} \tag{2.10.52}$$

The moduli space are $\mathbb{P}^1 \setminus \{P_1 \ldots, P_s\}$ where in the examples we have $s \le 6$ points. The Riemann symbol can be used to distinguish actual from apparent singularities. One finds moreover that the conifold closest to $z = 0$ corresponds to the vanishing of the $D6$ brane and determines the growth of BPS states, while at the other points spheres in different homology classes vanish [161].

While these non-hypergeometric families have actually known Calabi-Yau representations [14] generate a list of 403 so called Calabi-Yau differential operators, with a maximal unipotent point at $z = 0$ which implies the form (2.9.26) and condition (2.9.20), which as explained is equivalent to the flat coordinate form (2.9.18). Further it is required that the coefficients of X^0 at this points and the genus zero instantons are integer. There are more complicated conditions at $z = \infty$. Namely that the indicials $\alpha_1^{(1)} \le \alpha_1^{(2)} \le \alpha_1^{(2)} \le \alpha_1^{(4)}$ at $z = \infty$ are rational numbers satisfying $\alpha_1^{(1)} + \alpha_1^{(4)} = \alpha_1^{(2)} + \alpha_1^{(3)} = s \in \mathbb{Q}$ and the eigenvalues $\exp(2\pi i \alpha_k)$ of the monodromy are a product of cyclotomic polynomials.

2.11 Two Parameter Examples

According to our discussion of the fibration structure discussed in Sect. 2.7.5 there are two possibilities for a fibration in a Calab-Yau 3 fold: either a K3 fibration over $\mathbb{P}^1$ or an elliptic fibration over a surface. The K3 fibrations have been much studied in the context of the duality of the heterotic string on $K3 \times T2$ and the type IIA string on K3 fibred Calabi-Yau spaces while the elliptic fibration were studied in the context of F theory compactifications to six dimensions. We give therefore the simplest example of each kind.

2.11.1 The $X_{18=2}(1,1,2,2,6)$ 3-Fold, an $K3$ Fibration Over $\mathbb{P}^1$

As one example of this type consider the hypersurface of degree 12 in $\mathbb{P}(1,1,2,2,6)$, which has $h^{1,1}(M) = 2$ and $h^{2,1}(M) = 128$. We mod M out by an $\mathbb{Z}_{12} \times \mathbb{Z}_6 \times \mathbb{Z}_6$ acting as

$$x_i \to \exp(2\pi i g_i^{(\alpha)}/12)x_i, \quad \alpha = 1,2,3, \quad i = 1,\ldots,5\,, \tag{2.11.1}$$

with $g^{(1)} = (1, 11, 0, 0, 0)$, $g^{(2)} = (2, 0, 10, 0, 0)$ and $g^{(3)} = (2, 0, 0, 10, 0)$. The invariant constraint, which we interpret as mirror admits two complex structure deformations $h^{2,1}(W) = 2$

$$P = a_1 x_1^{12} + a_2 x_2^{12} + a_3 x_3^6 + a_4 x_4^6 + a_5 x_5^2 + a_0 \prod_{i=1}^{5} x_i + a_6 (x_1 x_2)^6 \qquad (2.11.2)$$

It is convenient to express the multiplicative relation between the monomials in (2.11.2) in vectors[85]

$$l^{(1)} = (-6; 0, 0, 1, 1, 3, 1) \qquad l^{(2)} = (0; 1, 1, 0, 0, 0, -2) \qquad (2.11.3)$$

such that equations corresponding to (2.12.17) are now written as

$$\prod_{l_i^{(b)}<0} \left(\frac{\partial}{\partial a_i}\right)^{-l_i^{(b)}} \frac{\Omega(\underline{a})}{a_0} = \prod_{l_i^{(b)}<0} \left(\frac{\partial}{\partial a_i}\right)^{l_i^{(b)}} \frac{\Omega(\underline{a})}{a_0} \qquad b = 1, 2\,. \qquad (2.11.4)$$

Similar symmetry considerations as above lead to the conclusion that $\Pi(\underline{z})$ depends only on

$$z_b = (-1)^{l_0^{(b)}} \prod_i a_i^{l_i^{(b)}}, \qquad b = 1, 2 \qquad (2.11.5)$$

and the reduction of (2.11.4) leads after factorization to the differential operators $\theta_i = z_i \frac{\mathrm{d}}{\mathrm{d}z_i}$

$$\begin{aligned} \mathcal{L}_1 &= \theta_1^2(\theta_1^2 - 2\theta_2) - \prod_{i=0}^{2}(6\theta_1 - (2i+1))z_1 \\ \mathcal{L}_2 &= \theta_2^2 - \prod_{i=1}^{2}(2\theta_2 - \theta_1 - i)z_2\,. \end{aligned} \qquad (2.11.6)$$

The periods in the integral basis of this model can be immediately obtained from (2.9.58) and the monodromies of this example have been calculated in [63, 205]. Further information about this model is given in Sect. 5.1.1.

[85] They will identified with the generators of the Mori cone in section "The Monodromies of the Quintic".

2.11.2 The $X_{18}(1, 1, 1, 6, 9)$ 3-Fold, an Elliptic Fibration over $\mathbb{P}^2$

Here we complete the main example for elliptic fibration whose polyhedron was given in (2.7.35) by supplementing the B-model data. This example has been discussed in [186] and in greater detail in [64].

The mirror manifold is given by the zero locus

$$F_{\hat{\Delta}} = x_0(z^6(x_1^{18} + x_2^{18} + x_3^{18} - b(x_1x_2x_3)^6) - 2^{\frac{1}{3}}\sqrt{3}azx_1x_2x_3x_4x_5 + x^3 + y^2) = 0 \tag{2.11.7}$$

in the space $\mathbb{P}_\Delta$. Note that $z := x_4, x := x_5 =$ and $y := x_6$ are the conventional names of variables in the Weierstrass form of the elliptic fibre. In $\mathbb{P}_\Delta$ there are toric $\mathbb{C}^*$ actions on the coordinates $x_i, i = 1, \ldots, 6$, which can be used to eliminate all a_i, but the two complex structure variables (a, b) of W. This is because two $\mathbb{C}^*$ actions

$$x_i \to \mu_r^{l_i^{(r)}} x_i, \quad \text{with } \mu_r \in \mathbb{C}^*, \tag{2.11.8}$$

are divided out from the coordinate ring of $\mathbb{P}_\Delta$. One can introduce manifestly $\mathbb{C}^*$ invariant combinations the a_i as complex structure variables of W, namely

$$z_i = (-1)^{l_0^{(i)}} \prod_{k=1} a_k^{l_k^{(i)}}, \quad i = 1, \ldots, h_{21}(W_3) = h_{11}(M_3)\,. \tag{2.11.9}$$

In the case at hand $z_1 := z_E = \frac{a_4a_5^2a_6^3}{a_0^6}$ and the corresponding mirror map $t_E = \frac{\log(z_E)}{2\pi i X^0} + \mathcal{O}(z)$ corresponds to the elliptic fibre and similarly $z_2 := z_B = \frac{a_1a_2a_3}{a_4^3}$ and t_B to the base class. Using the $\mathbb{C}^*$ actions given by the $l^{(i)}$ vectors in (2.7.35) on the period integrals $\Pi(z) = \int_{\gamma_3} \Omega$ with $(a = a_0)$

$$\Omega = \oint_{\gamma_\epsilon} \frac{a\mu}{P_{\hat{\Delta}}}, \tag{2.11.10}$$

given by a residuum integral around $P_{\hat{\Delta}}$ with the measure $\mu = \sum_{i=1}^5 (-1)^i w_i dx_1 \wedge \ldots \widehat{dx_i} \ldots \wedge dx_5$, one can derive two Picard-Fuchs (PF) differential equations by reducing (2.11.4) to the z_1, z_2 variables and get

$$\begin{aligned}\mathcal{L}_1 &= \theta_1(\theta_1 - 3\theta_2) - 12z_1(6\theta_1 + 1)(6\theta_1 + 5),\\ \mathcal{L}_2 &= \theta_2^3 + z_2\prod_{i=0}^{2}(3\theta_2 - \theta_1 + i)\end{aligned} \tag{2.11.11}$$

determining the periods from $\mathcal{L}_i \Pi(z) = 0$, $i = 1, 2$. Here $\theta_i = z_i \frac{\partial}{\partial z_i}$. The discriminants of the operators are

$$\Delta_1 = (1 - 432 z_1)^3 - 27 z_2 (432 z_1)^3,$$
$$\Delta_2 = 1 + 27 z_2 \qquad (2.11.12)$$

The 3-point couplings can be computed from the PF operators as in Sect. 2.9.2 or in [186]

$$C_{111} = \frac{9}{z_1^3 \Delta_1}, \quad C_{112} = C_{121} = C_{211} = \frac{3\Delta_3}{z_1^2 z_2 \Delta_1},$$
$$C_{122} = C_{212} = C_{221} = \frac{\Delta_3^2}{z_1 z_2^2 \Delta_1}, \quad C_{222} = \frac{9(\Delta_3^3 + (432 z_1)^3)}{z_2^2 \Delta_1 \Delta_2}, \qquad (2.11.13)$$

where for convenience we can define the factor $\Delta_3 = 1 - 432 z_1$. Again the periods in the integral basis are obtained from (2.9.58). The monodromy group contains as a subgroup $Sl(2, \mathbb{Z})$ [64]. This has the consequence that the amplitudes have modular properties w.r.t. the fibre modulus. The strongest result in this direction so far has been obtained in [195], where it was shown that the all genera results for given base degree are meromorphic Jacobi-Forms.

2.11.3 Integrality of the Mirror Map

While the integrality of instanton expansion of the $\mathcal{F}^{(g)}$ has found, at least physically, a completely satisfactory explanation as counting of BPS states, see Sect. 4.3.3, the integral expansion of all known mirror maps at the point of maximal unipotent monodromy remains physically more mysterious.

We exponentiate (2.9.53) for the quintic, invert it and expand $z(q)$ in $Q = e^{2\pi i t}$. Call $j_q = \frac{1}{z(q)}$ in analogy with the normalized $j_e(q)$ $Sl(2, \mathbb{Z})$ invariant function of the elliptic curve. Both expansions have positive integral coefficients

$$j_e = \frac{1}{q} + 744 + 196884\, q + 21493760\, q^2 + 864299970\, q^3 + 20245856256\, q^4 + \ldots$$
$$j_q = \frac{1}{q} + 770 + 421375\, q + 274007500\, q^2 + 236982309375\, q^3 + 251719793608904\, q^4 + \ldots \qquad (2.11.14)$$

The integrality of j_q should be related to monodromy group $\Gamma \in \mathrm{Sp}(4, \mathbb{Z})$ generated by M_0 and M_1, but it is unknown what the integer coefficients are counting. For the

K3 fibred example discussed in we get for each of the two functions $j_1 = \frac{1}{z_1}$ and $j_2 = \frac{1}{z_2}$ an integral two parameter expansion

$$\begin{aligned} j_1 &= \tfrac{1}{q_1} + 744 + 196884\, q_1 + 21493760\, q_1^2 + 864299970\, q_1^3 + \ldots \\ & q_2 \left(-\tfrac{1}{q_1} + 480 + 1403748\, q_1 + 1203172608\, q_1^2 + \ldots\right) \\ & \vdots \\ j_2 &= \tfrac{1}{q_2} + 2 + q_2 + q_1 \left(\tfrac{1}{q_2} 240 - 240 - 240\, q + 240\, q^2\right) + \\ & q_1^2 \left(\tfrac{1}{q_2} 70920 - 57600 - 26640\, q_2 - 57600\, q_2^2 + 70920\, q_2^3\right) \ldots \end{aligned} \tag{2.11.15}$$

The occurrence of the j-function [63] at $j_e = j_1|_{q_2=0}$ has been related to string duality between type II on to the heterotic string on $K3 \times T2$ [204, 205], see [221] for a review. These primitive observations may point towards number theoretic applications of topological string theory. Intriguing observations for Calabi-Yau manifolds over finite fields have been made in [65].

2.12 *Mirror Geometries Related to Elliptic Curves*

In this section we want to describe the local mirror symmetries related to elliptic curves. In particular these are the local mirrors to local Del Pezzo Calabi-Yau spaces, which we describe in the next subsection. Before we recall a couple of elementary fact about elliptic curves.

2.12.1 Elliptic Curves, Modular Functions and Differential Equations

Every family of elliptic curves can be written in the Weierstrass form, which reads in affine complex coordinates x, y as

$$y^2 - 4x^3 + f(s)x + g(s) = 0, \tag{2.12.1}$$

where $\Delta = f^3 - 27g^2$ is the discriminant of the curve. The complex structure of the elliptic curve $\tau \in \mathbb{H}_+$ is related to f and g by[86]

$$j(\tau) = 12^3 \frac{f^3}{\Delta}. \tag{2.12.2}$$

[86] See section "Modular Forms of $\Gamma_1 = \mathrm{Sl}(2, \mathbb{Z})$" in Appendix 4 for notations.

Defining $q = \exp(2\pi i\tau)$ we have as central quantity the modular invariant j function

$$j(\tau) = \frac{(12E_4)^3}{E_4^3 - E_6^2} = \frac{1}{q} + 744 + 196884q + \dots \,. \tag{2.12.3}$$

This is invariant under the PSL(2, $\mathbb{Z}$) action defined in (A4.3) and is therefore well defined on the fundamental region of complex structures of the elliptic curve

$$\mathcal{F} = \mathrm{PSL}(2, \mathbb{Z})/\mathbb{H}_+ \,. \tag{2.12.4}$$

For $k \in 2\mathbb{N}_+$ we defined with

$$E_k(\tau) = \frac{1}{2\zeta(k)} \sum_{\substack{n,m\in\mathbb{Z}\\(n,m)\neq(0,0)}} \frac{1}{(m\tau+n)^k} = 1 + \frac{(2\pi i)^k}{(k-1)!\zeta(k)} \sum_{n=1}^{\infty} \sigma_{k-1}(n) q^n \,, \tag{2.12.5}$$

the normalized (and regularized by the second equal sign for $k = 2$) Eisenstein series, with $\sigma_k(n)$ the sum of the k-th power of the positive divisors of n and

$$\zeta(k) = \sum_{r\geq 0} \frac{1}{r^k} = -\frac{(2\pi i)^k B_k}{(2k(k-1)!)} \tag{2.12.6}$$

for $k \in 2\mathbb{N}_+$. Here the Bernoulli numbers B_k can be defined by the generating series

$$\sum_{k=0}^{\infty} \frac{B_k x^k}{k!} := \frac{x}{e^x - 1} \,. \tag{2.12.7}$$

The Eisenstein series are modular forms of weight k whose properties we shortly explain in section "Modular Forms of $\Gamma_1 = \mathrm{Sl}(2, \mathbb{Z})$" in Appendix 4.

For a family of elliptic curves f and g depend on one complex modulus say s. For instance for instance the Legendre curve discussed in Sect. 2.4.3 in Eq. (2.4.24) is a projective embedding of a cubic in $\mathbb{P}^2$

$$P = zy^2 - x^3 + zx^2(1+s) - xz^2 s = 0 \tag{2.12.8}$$

into $\mathbb{P}^2$. Now by the automorphism of $\mathbb{P}^2$ namely $\mathrm{Aut}(\mathbb{P}^2) = \mathrm{PGL}(3)$ transformations of $\mathbb{P}^2$ we can bring any cubic as e.g. the Legendre cubic (2.12.8) into the Weierstrassform (2.12.1). This problem arises for any embedding of an elliptic curve into an ambient variety and has been solved systematically for all hypersurfaces in toric ambient spaces, for which the automorphism have been described generally in (2.7.16), (2.7.17). One just has to look at the Newton polynomial of the hypersurfaces in Fig. 6, keep track of the coefficients, most of which are zero for (2.12.8), and then the $f = g_2^*$ and the $g = g_3^*$ are given in (A1.1) of Appendix 1, which yields

$$f = \frac{1}{3}(1 - s + s^2) \qquad g = \frac{1}{54}(2 - 3s - 3s^2 + 2s^3), \qquad j(z) = \frac{4}{27}\frac{1 - s + s^2}{(1 - s^2)^2 s^2}. \tag{2.12.9}$$

By (2.4.50) τ gets related to ratios of periods and by (2.12.2), (2.12.3) to the modular forms given in the case of $\mathrm{Sl}(2, \mathbb{Z})$ by the Eisenstein series. One can make this very concrete and express the vanishing period in terms of the $f = g_2(s)$ $g = g_3(s)$ as well as E_4 and E_6 as in (2.12.16). Moreover periods as we know fulfill differential equations in the family parameter s of the elliptic curve, which is in turn very simply related to the j-function by (2.12.9) and (2.12.3).

How is that differential equation related to modular forms, when expressed in $j(\tau)$? The answer is a proposition, quite simple to proof [54], which Zagier calls in [4] the 'perhaps single most important source of applications of modular forms in other branches of mathematics' and which states the following

Proposition 2 *Let $f(\tau)$ be a holomorphic or meromorphic modular form of positive weight (see section "Modular Forms of $\Gamma_1 = \mathrm{Sl}(2, \mathbb{Z})$" in Appendix 4) on some modular group Γ and $s(\tau)$ a modular function on Γ. Express $f(\tau)$ locally as $\Phi(s(\tau))$ then the function $\Phi(s)$ satisfies a linear differential equation of order $k + 1$ with algebraic coefficients, or with polynomial coefficients if $\Gamma/\mathbb{H}_+$ has genus zero and $s(\tau)$ generates the field of modular functions on Γ.*

A simple example which illustrates many aspects of this is that the meromorphic weight $k = 1$ function is given by

$$\sqrt[4]{E_4(\tau)} = {}_2F_1\left(\frac{1}{12}, \frac{5}{12}; 1; s(\tau)\right), \tag{2.12.10}$$

i.e. the standard hypergeometric function, solving a second order differential equation in $s(\tau) = 12^3/j(\tau)$. We see that (2.12.16) is another application of the proposition and that the Picard-Fuchs equations that $\int_A \omega$ fulfills can be seen as consequence of modularity of the elliptic curve.

2.12.2 The Mirror Geometry of Local Toric del Pezzo Surfaces

In this section we describe as an example for the powerful tools that exists for mirror geometries related to elliptic curves the B-model geometry that arises for local del Pezzo Calabi-Yau spaces. Local del Pezzo Calabi-Yau spaces are defined as the total space $\check{M}$ of the anti-canonical line bundle over del Pezzo surfaces S, which we denote as

$$\begin{array}{ccc} \mathcal{O}(K_S) & \longrightarrow & \check{M} \\ & & \downarrow \pi \\ & & S \end{array} \tag{2.12.11}$$

We saw in Sects. 2.7.4 and 2.7.5 examples mirror of a local toric Pezzo surface and concluded that they are elliptic curves. More precisely the toric del Pezzo surfaces can be described in terms of two dimensional reflexive pairs of polyhedra, which in turn can be all embedded in the polyhedra depicted in Fig. 6. In this case $H(X, Y, \underline{z})$ in (2.8.15) is the Newton polynomial of the reflexive polyhedron and $H(X, Y, \underline{z}) = 0$ is a family ε of genus one curves. We homogenize the Newton polynom as indicated in Fig. 6. For example the most general cubic in $\mathbb{P}^2$ is given by the vanishing locus

$$P = m_1x^3+m_2x^2y+m_3xy^2+m_4y^3+m_5y^2z+m_6yz^2+m_7z^3+m_8xz^2+m_9x^2z+uxyz\,. \tag{2.12.12}$$

If we would be interested in the variation of the complex structure deformations of the cubic as encoded in the periods over the holomorphic $(1, 0)$ forms ω, we could use all automorphism of $\mathbb{P}^2$ to set the parameters except for u to constant values. But as the local mirror description for the anti-canonical line bundle over the del Pezzo given by the global embedding of curve in a non-compact 3 fold leads to two data $[\mathcal{E}, \lambda = \lim_{loc} \Omega]$ that are only invariant under an restricted automorphism we are only allowed to fix 3 parameters to non-zero constants, while the others describe the residue of the meromorphic differential of the third kind λ.

As was explained in [193] it is straightforward to get periods in a suitable symplectic basis $(A, B) \in H_1(\mathcal{E}, \mathbb{Z})$

$$\vec{\pi} = \begin{pmatrix} a(u, \underline{m}) \\ a_D(u, \underline{m}) \end{pmatrix} = \begin{pmatrix} \int_A \lambda \\ \int_B \lambda \end{pmatrix} \tag{2.12.13}$$

w.r.t. the holomorphic differential of the third kind λ directly from the Weierstrass form of the curve

$$y^2 = 4x^3 - xg_2(u, \underline{m}) - g_3(u, \underline{m})\,. \tag{2.12.14}$$

The latter is obtained for all forms of the elliptic curves corresponding to the Newton polynomials that derive from the polyhedra in fig 6 using Nagell's algorithm. The resulting $g_2(u, \underline{m})$ and $g_3(u, \underline{m})$ are given for the three choices for convenience in Appendix 1.

The important point is to note that the interior point is special in the A- as well as in the B-model. In the toric description of the A-model it corresponds to the canonical class of the del Pezzo surface and in the B-model it corresponds to the only dynamical deformation of special geometry. The latter fact implies that

$$\partial_u \int_A \lambda = \partial_u t = \int_A \omega, \tag{2.12.15}$$

where ω is the holomorphic differential of the family $\mathcal{E}_{u,\underline{m}}$. By the theory of families of elliptic curves an integral of the holomorphic $(1, 0)$ form over a vanishing cycle

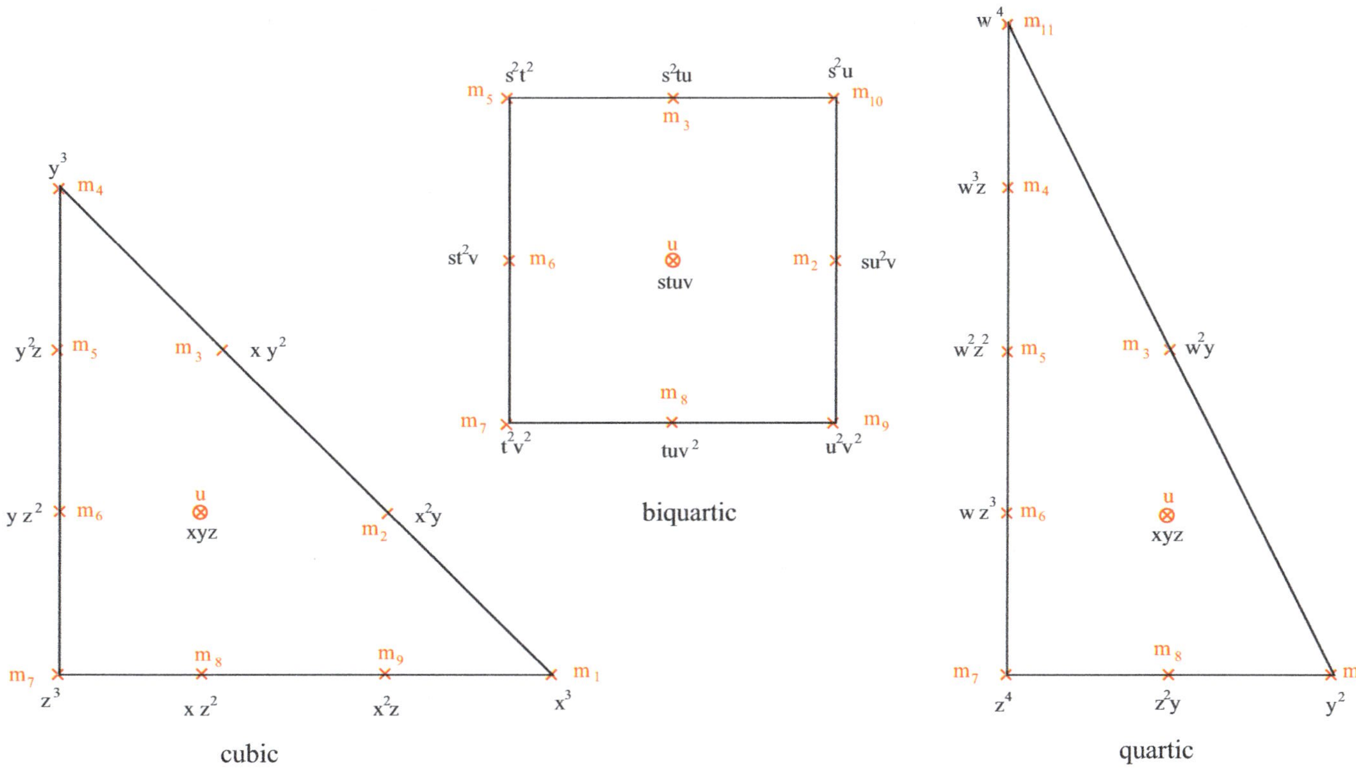

Fig. 6 Two dimensional reflexive polyhedra. From left to right they are the Newton polyhedra for the following families of elliptic curves: The cubic in $\mathbb{P}^2$, the biquadric in $\mathbb{P}^1 \times \mathbb{P}^1$ and the quartic in weighted $\mathbb{P}^2(1, 1, 2)$. All other two dimensional reflexive toric polyhedra corresponding to restricted families of elliptic curves can be embedded into these polyhedra

near any cusp is given by

$$\partial_u t = \sqrt{\frac{E_6(\tau) g_2(u, \underline{m})}{E_4(\tau) g_3(u, \underline{m})}} \,. \tag{2.12.16}$$

The function g_2 and g_3 can be rescaled by

$$\begin{aligned} g_2(u, \underline{m}) &\to g_2(u, \underline{m}) f^3(u, \underline{m}) \\ g_3(u, \underline{m}) &\to g_3(u, \underline{m}) f^2(u, \underline{m}) \,. \end{aligned} \tag{2.12.17}$$

However we can fix the normalization by requiring that $\Delta = g_2^3(u, \underline{m}) - 27 g_3^2(u, \underline{m})$ contains only the geometric components of the discriminant and $\partial_u t \sim \frac{1}{2\pi i} u^{-1}$ at the cusps. The latter normalisation implies that we can express the mirror map by

$$u(t, \underline{m}) = Q_t + \ldots, \tag{2.12.18}$$

with $Q = \exp(2\pi i t)$. This is the leading behaviour of the mirror map, which is fixed to all orders by (2.12.16). In (2.12.16) $\tau(u, \underline{m})$ is a function of u and the m's, which is simply derived by equating the algebraic and the modular definition of the j invariant

$$\frac{1728 g_2^3(u, \underline{m})}{g_2^3(u, \underline{m}) - 27 g_3^2(u, \underline{m})} = \frac{1}{q} + 744 + 192688 q + \ldots \tag{2.12.19}$$

where $q = \exp(2\pi i \tau)$, and solving for $\tau(u, \underline{m})$.

The second period can be derived since special geometry predicts that the period over dual cycle to A fulfills the relation

$$\frac{\partial^2}{\partial^2 t} F = \partial_t t_D = \partial_t \int_B \lambda = -\frac{1}{2\pi i} \tau(t, \underline{m}) \,. \tag{2.12.20}$$

This relation also allows to obtain the holomorphic prepotential and therefor the instanton numbers up to constants w.r.t the integration in t.

$$F(\underline{Q}) = -\frac{c_{ijk}}{3!} + \frac{c_{ij}}{2} t^i t^j + c_i t^i + c + \sum_{\beta \in H_2(M, \mathbb{Z})} n_0^\beta \mathrm{Li}_3(Q^\beta) \tag{2.12.21}$$

The latter can be inferred from the A-model using the $\hat{\Gamma}$ class after choosing an appropriate basis for the Kähler class parameters t_i in the Kähler cone. Here we denote the special class defined in (2.12.16), which is related to the canonical class by $t = t_0$ and the ones related to the masses m_i by t_i. The actual construction of the Kähler cone was done in [235] and allows in general for different phases that involve linear changes in the basis of the t_i, corresponding to the different cones in the secondary fan of the nonlocal toric Calabi-Yau. From the point of view of Gromov-Witten theory classes that do not involve the canonical class count curves in the classes that come from the edges of the toric polyhedra. This lead indeed to a Gromov-Witten theory problem in two complex dimensions, which is trivially solved.

As it is pointed out in [193] also the B-model for higher genus amplitudes for the elliptic families of curves ε with the differential of the third kind λ can be solved very explicitly using the modular expression introduced in this section.

3 The World-Sheet Point of View

The worldsheet point follows simply from the elementary definition of string theory as the map

$$X : \Sigma \rightarrow M\,. \tag{3.0.1}$$

Here we focus on those maps, which embed the worldsheet into the non-trivial part of the space time, i.e. the internal space M, as the maps into the flat space have no interesting non-trivial topology and the worldsheet theory is that of free fields, which is solvable and discussed in early chapters of standard string books. This map comes with an action, which is known as non-linear σ-model action and a curved manifold M is only then a stable background if the non-linear σ-model is conformally invariant. This means in particular that the scale dependence of all couplings to the metric and its derivatives must vanish.

The special holomony target manifolds play an important role again because they lead to extended supersymmetry on the worldsheet. In particular for Riemann manifolds one can have $\mathcal{N} = (1, 1)$ worldsheet supersymmetry, for Kähler manifolds one can have $\mathcal{N} = (2, 2)$ worldsheet supersymmetry.

However general Kähler metrics do not lead to conformally invariant worldsheet theories. For Calabi-Yau 3-folds, it has been established by direct calculations, that there is a choice of metric with $c_1(TM) = 0$ for which the worldsheet theory of the type II superstring is an $\mathcal{N} = (2, 2)$ superconformal theory.

For the heteroric string on CY 3-folds the worldsheet supersymmetry is $\mathcal{N} = (2, 0)$ and additional consistency conditions have to be met for the choices of the gauge bundles to make the theory conformal. In the $E_8 \times E_8$ heterotic string theory is a *standard embedding* of the tangent bundle into one E_8 so that the $SU(3)$ holomomy breaks the gauge group to E_6. This heterotic theory is stable and has some of the simplifying features of the $\mathcal{N} = (2, 2)$ compactifications.

It has been also argued that the question of stability is related to spacetime supersymmetry. As was explained in Sect. 1.1.6 one gets $\mathcal{N} = 2$ and $\mathcal{N} = 1$ space time supersymmetry for Calabi-Yau 3-folds in four dimensions and no enhanced supersymmetry for general Kähler manifolds.

3.1 The $\mathcal{N} = 2$ Worldsheet Superconformal Theories

More precisely the compactifications of the type II superstring on Calabi-Yau threefolds have a global $\mathcal{N} = (2, 2)$ worldsheet susy. An $\mathcal{N} = (1, 1)$ subsymmetry is gauged in the superstring. We assume some familiarity with two dimensional confirmed field theory, see e.g. [90].

3.1.1 The $\mathcal{N} = 2$ Superconformal Algebra

Let us discuss the *chiral* $\mathcal{N} = 2$ part of the superconformal algebra on the worldsheet, which has the following chiral currents with the indicated mode expansions:

- The chiral component of energy momentum tensor[87]

$$T(z) = \sum_{n \in Z} \frac{L_n}{z^{n+2}} \tag{3.1.1}$$

  with conformal dimension h and $U(1)$ charge Q given by $(h, Q) = (2, 0)$.
- An $U(1)$ current

$$J(z) = \sum_{n \in Z} \frac{J_n}{z^{n+1}} \tag{3.1.2}$$

  with $(h, Q) = (1, 0)$
- Two super currents

$$G^{\pm}(z) = \sum_{r \in Z \pm \nu} \frac{G_r^{\pm}}{z^{r+\frac{3}{2}}} \tag{3.1.3}$$

  with $(h, Q) = (\frac{3}{2}, \pm 1)$. Hence

$$G^{\pm}(e^{2\pi i} z) = -e^{\mp 2\pi i \nu} G^{\pm}(z) , \tag{3.1.4}$$

  where $0 \leq \nu < 1$. Since $G^{\pm}(z)$ is fermionic, $\nu = 0$ Ramond- (anti-periodic on the z-plane)[88] and $\nu = \frac{1}{2}$ Neveu-Schwarz boundary conditions (periodic on the z-plane) are particularly natural. However the $\mathcal{N} = 2$ current algebra admits an

---

[87]The standard notation in CFT is quite different then the one common in the discussion of σ models that we used in Sect. 3.2. One uses in CFT $z = \sigma^1 + i\sigma^2$ and $\bar{z} = \sigma^1 + i\sigma^2$ where $\sigma^2 = i\sigma^0$ is the euclidean time. Correspondingly one indicates the left moving sector which carried a $+$ index in Sect. 3.2 by quantities without bar and the right moving carrying before $-$ with quantities with bar. Moreover the unbarred or barred super charges are now distinguished by $-$ and $+$ respectively, e.g. $Q_+ \leftrightarrow G_0^-$, $\bar{Q}_+ \leftrightarrow G_0^+$, $Q_- \leftrightarrow \bar{G}_0^-$ and $\bar{Q}_- \leftrightarrow \bar{G}^+$.

[88]Or more naturally periodic on the cylinder.

$U(1)$ rotation on the super currents as it is seen below. One can therefore consider the more general boundary condition (3.1.4), which leads to the spectral flow on the representations.

The short distance operator expansion reads

$$\begin{aligned}
T(z)T(w) &= \frac{c}{2(z-w)^4} + \frac{2}{(z-w)^2}T(w) + \frac{1}{z-w}\partial_w T(w) + \text{reg},\\
T(z)G^{\pm}(w) &= \frac{3}{2(z-w)^2}G^{\pm}(w) + \frac{1}{z-w}\partial_w G^{\pm}(w) + \text{reg},\\
T(z)J(w) &= \frac{1}{(z-w)^2}J(w) + \frac{1}{z-w}\partial_w J(w) + \text{reg},\\
G^{\pm}(z)G^{\mp}(w) &= \frac{2c}{3(z-w)^3} \pm \frac{2}{(z-w)^2}J(w) + \frac{2T(w) \pm \partial_w J(w)}{z-w} + \text{reg},\\
G^{+}(z)G^{+}(w) &= G^{-}(z)G^{-}(w) = 0 + \text{reg},\\
J(z)G^{\pm}(w) &= \pm\frac{1}{z-w}G^{\pm}(w) + \text{reg},\\
J(z)J(w) &= \frac{c}{3(z-w)^2} + \text{reg},
\end{aligned} \tag{3.1.5}$$

where $C \in R$ is the central charge.

From this algebra it is clear that there is a continuous $O(2)$ rotation defined in the canonical way on a real basis of supercharges $G_x = G^+ + G^-$ and $G_y = i(G^+ - G^-)$. Parametrized by the angle ν its acts like the internal $U(1)$ on the $G^{\pm}$ used in (3.1.4) to define general twisted boundary conditions. In addition there is an exterior $\mathbb{Z}_2 : G_y \to -G_y$ action.

Let us recapitulate the standard procedure in 2d QFT which recovers the algebra of charge operators from an operator algebra such as (3.1.5). To the operator $A(z)$ we assign charge operators $A_\xi = \oint_{C_0} \mathrm{d}z\, \xi(z)A(z)$, where C_0 is a contour around the origin 0 and $\oint_{C_0} \mathrm{d}z := \int_{C_0} \frac{\mathrm{d}z}{2\pi i}$. In particular for $\xi(z) = z^{n+h(A)-1}$ the charges are the modes A_n of $A(z)$. The transformation of the operator $B(w)$ under (δ_{A_ξ}) is generated by the commutator with A_ξ. In radial time ordering the commutator is given by the following contour integrals

$$\begin{aligned}
(\delta_{A_\xi})B(w) = [A_\xi, B(w)]_{\pm} &= \oint_{\substack{C_0\\ |z|>|w|}} \mathrm{d}z\, \xi(z)A(z)B(w) \pm \oint_{\substack{C_0\\ |z|<|w|}} \mathrm{d}z\, \xi(z)A(z)B(w)\\
&= \oint_{C_w} \mathrm{d}z\, \xi(z)A(z)B(w)\,,
\end{aligned} \tag{3.1.6}$$

see figure. Here $[\cdot,\cdot]_{\pm}$ stands for the commutator $(-)$ $[\cdot,\cdot]$ and anticommutator $(+)$ $\{\cdot,\cdot\}$.

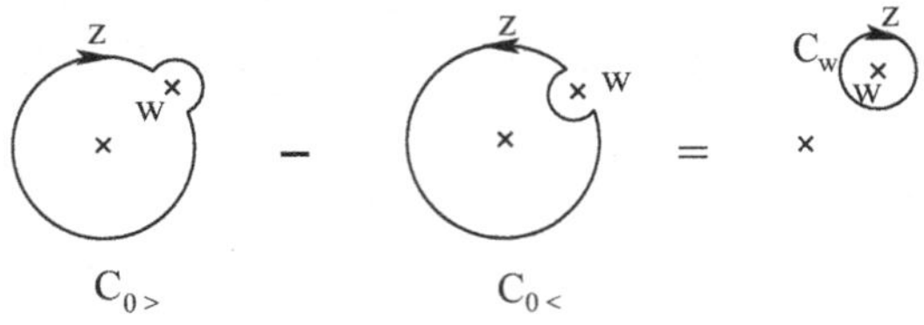

The spatial transformations δ_ξ corresponding to conformal transformations[89] $z \to z + \xi(z)$ are generated by $T(z)$, i.e. $\delta_\xi = \delta_{T_\xi}$. One can integrate (3.1.6) with $\oint_{C'_{w=0}} dw\, z^{m+h(B)-1}$ to recover as residuum the mode algebra from (3.1.5)

$$
\begin{aligned}
[L_m, L_n] &= (m-n)L_{m+n} + \tfrac{c}{12}m(m^2-1)\delta_{m,-n},\\
[L_m, G_r^\pm] &= \left(\tfrac{m}{2} - r\right) G^\pm_{m+r},\\
[L_m, J_n] &= -nJ_{m+n},\\
\{G_r^+, G_s^-\} &= 2L_{r+s} + (r-s)J_{r+s} + \tfrac{c}{3}\left(r^2 - \tfrac{1}{4}\right)\delta_{r,-s},\\
\{G_r^+, G_s^+\} &= \{G_r^-, G_s^-\} = 0,\\
[J_n, G_r^\pm] &= \pm G^\pm_{r+n},\\
[J_m, J_n] &= \tfrac{c}{3}\delta_{m,-n}\,.
\end{aligned}
\tag{3.1.7}
$$

In case that the $\mathcal{N} = (2, 2)$ CFT theory is the internal part of an 10d superstring compactification it must have $c = \bar{c} = 9$ to cancel the Weyl anomaly. This contributions comes from the worldsheet fields which represent coordinates on the internal manifold M. In particular we can identify $n = \dim_{\mathbb{C}}(M) = \frac{c}{3}$.

3.1.2 The Spectral Flow and the Chiral Ring

At this point we review some elementary representation theory of conformal field theories. The Hilbert space of states of a conformal field theory can be organized into highest weight representations. To define them one fixes a highest weight state $|\psi\rangle$ which fulfills

$$
L_n|\psi\rangle = 0, \quad J_m|\psi\rangle = 0, \quad G_r^\pm|\psi\rangle = 0 \qquad \forall m, n, r > 0\,. \tag{3.1.8}
$$

The representations is then generated by applying negative modes of all operators in the algebra. In general there will be null states $|\chi\rangle$, whose norm $\langle\phi, \chi\rangle = 0$ with

[89] These are functions only of z, which can at most be meromorphic in 2d.

all other states $\langle\phi| = |\phi\rangle^\dagger$ in the Hilbert space generated in this way, which will be discarded from the physical Hilbert space. Note that

$$L_n^\dagger = L_{-n}, \qquad J_n^\dagger = J_{-n} \quad \text{and} \quad (G_r^\pm)^\dagger = G_{-r}^\mp . \tag{3.1.9}$$

Any state in the highest weight space can then be classified by it eigenvalues, with respect to the maximal commuting sub algebra. In the case at hand that is generated by the conformal weight operator L_0 and the charge operator J_0, so that a state $|\phi\rangle$ will have

$$L_0|\phi\rangle = h_\phi|\phi\rangle, \qquad J_0|\phi\rangle = q_\phi|\phi\rangle \tag{3.1.10}$$

a conformal weight h_ϕ and an $U(1)$ charge q_ϕ. The operator state correspondence allows to associate these quantum numbers also to the currents or to general fields. Any 2d conformal field theory has a vacuum state $|\text{vac}\rangle$ with $h_{\text{vac}} = 0$. The norm for the vacuum state and the other highest weight states is $\langle\psi, \psi\rangle = 1$. In an unitary theory $\langle\phi, \phi\rangle > 0$ for all fields, this implies $c > 0$ and all h_ϕ values have to fulfill $h_\phi \geq 0$. Note that the modes $G_0^\pm$ do not raise the conformal weight. This means in particular that the Ramond ground states will be degenerate. One has

$$G_0^{\pm'} G_0^{\pm'} = 0, \quad \{G_0^+, G_0^-\} = 2\left(L_0 - \frac{c}{24}\right) . \tag{3.1.11}$$

Unitarity then implies that $h_\phi \geq \frac{c}{24}$ in the Ramond sector. When $h_\phi = \frac{c}{24}$, then $G_0^\pm|\phi\rangle$ are a null states and one has one groundstate $|\phi\rangle$, called a Ramond groundstate. When $h_\phi > \frac{c}{24}$ the $G_0^\pm$ generate four states that are degenerate in the conformal weight. In topological considerations an important role is played by the fermion operator.

$$(-1)^F, \quad \text{where} \quad F|\phi\rangle = \begin{cases} |\phi\rangle \text{ on fermions} \\ 0 \text{ on bosons} \end{cases} \tag{3.1.12}$$

Note that the trace of $(-1)^F$ receives only contributions form the Ramond ground states as the contribution of states with $h_\phi > \frac{c}{24}$ cancels due to the different fermion numbers of the four states.

One can write explicitly the algebra isomorphism associated to the $U(1)$ rotation on the super currents on the modes as [286]

$$\begin{aligned} L_n \to L_n' &= L_n + \nu J_n + \tfrac{1}{6}\nu^2 c\delta_{n,0} \\ J_n \to J_n' &= J_n + \tfrac{1}{3}\nu c\delta_{n,0} \\ G_r^\pm \to (G_r^\pm)' &= G_{r\mp\nu}^\pm . \end{aligned} \tag{3.1.13}$$

In particular the algebra isomorphism allows for a continuous flow from $\nu = 0$ to $\nu = \frac{1}{2}$ that exchanges Neveu-Schwarz (NS) and Ramond (R) boundary conditions on one chiral half and commutes with the GSO projection. Since space time bosons and space fermions have a different association of the R and the N sector on one chiral half the above flow can be identified with the supersymmetry generator on the spacetime spectrum. Note that the outer $\mathbb{Z}_2$ automorphism allows to consider $0 \le |\nu| < 1$ in (3.1.13).

In fact any operator $\mathcal{O}(z)$ with $U(1)$ charge q can be decomposed into a part $\hat{\mathcal{O}}(z)$ which is neutral under the $U(1)$ current and a charge carrying part, i.e.

$$\mathcal{O}_q(z) = \hat{\mathcal{O}}(z) e^{iq\sqrt{\frac{3}{c}}\phi(z)} . \tag{3.1.14}$$

Here we bosonized the $U(1)$ current as $J(z) = \sqrt{\frac{c}{3}}\partial\phi(z)$. Hence using the vertex algebra formalism one can construct an actual *spectral flow operator* $S_\nu \sim e^{i\sqrt{\frac{c}{3}}\nu\phi(z)}$, which shifts the $U(1)$ charge and the conformal dimension according to (3.1.13). Such an operator $S_{\frac{1}{2}}$ is used in the construction of the actual super symmetry operator on the world sheet see [252, 286].

If we consider (3.1.13) with $\nu = \pm'\frac{1}{2}$ we get on the spectrum the shift

$$h \to h' = h \pm' \frac{1}{2} q + \frac{3}{8}, \qquad q \to q' = q \pm' \frac{3}{2} . \tag{3.1.15}$$

In particular the Ramond-Ramond ground states which fulfill [252]

$$G_0^{\pm}|\psi\rangle = 0. \tag{3.1.16}$$

with (3.1.7)

$$h = \frac{c}{24} = \frac{3}{8}, \tag{3.1.17}$$

due to (3.1.7), flow into states in the NS sector into primary fields, which fulfill

$$G^{\pm'}_{-\frac{1}{2}}|\phi\rangle = 0 , \tag{3.1.18}$$

and by (3.1.15)

$$h = \pm' \frac{q}{2}. \tag{3.1.19}$$

States with the plus in (3.1.18) and hence $h = \frac{q}{2}$ are called *chiral states* and such with the minus in (3.1.18) and hence $h = -\frac{q}{2}$ are called *anti-chiral states*.

In a conformal field theory the leading short distance singularity of an operator product expansion is determined by the conformal dimension of the fields on the right and the left $\mathcal{O}_a(z)\mathcal{O}_b(w) = (z-w)^{h_c-h_a-h_b} C^c_{ab}\mathcal{O}_c$ + less sing. Since one has charge conservations and the charge is correlated for chiral with the conformal dimension the chiral and anti-chiral operators form a regular ring under the operator product expansion

$$\mathcal{O}_a\mathcal{O}_b = C^c_{ab}\mathcal{O}_c + \text{reg}. \tag{3.1.20}$$

The operator product expansion contains regular terms, i.e. more information then the (anti-) chiral rings. One has to establish a consistent projection of it to the chiral ring states. First one notice that the positivity of

$$||G^{\pm'}_{-\frac{1}{2}}\phi\rangle|^2 = \langle\phi|G^{\mp}_{\frac{1}{2}}G^{\pm'}_{-\frac{1}{2}}|\phi\rangle = \langle\phi|2L_0 \pm' J_0|\phi\rangle \tag{3.1.21}$$

implies

$$h_\phi \geq \frac{1}{2}|q_\phi|, \quad \forall\phi \tag{3.1.22}$$

Here the equality holds for the (anti-)chiral states, while the regular terms in (3.1.20) represent states with do not satisfy the bound. One can show that each NS state has a chiral primary representative in the sense that there exists an unique decomposition

$$|\phi\rangle = |\phi_0\rangle + G^+_{-\frac{1}{2}}|\phi_1\rangle + G^-_{\frac{1}{2}}|\phi_2\rangle \tag{3.1.23}$$

with $|\phi_0\rangle$ a chiral primary. This the analog of the Hodge decomposition (2.3.33) and the (anti-)chiral states are the analogs of harmonic forms. The decomposition has been argued in [252] in fact using the relation between the zero modes of the super currents and the exterior derivatives on a target space

$$G^+_0 \sim \partial, \quad \bar{G}^+_0 \sim \bar{\partial} \tag{3.1.24}$$

in the Ramond sector and the spectral flow arguments. It is obvious that in correlation functions involving only (anti-)chiral fields the terms in the image of $G^+_{-\frac{1}{2}}$ or $G^-_{+\frac{1}{2}}$ do not matter, just like the choice of the representative of a cohomology class does not matter in an integral of forms over a closed cycle. In this sense (3.1.20) can be viewed as defining a cohomology ring.

It is further easy to show that the charge in the (anti-)chiral ring is bounded by the positivity of

$$||G^{\pm'}_{-\frac{3}{2}}\phi\rangle|^2 = \langle\phi|G^{\mp}_{\frac{3}{2}}G^{\pm'}_{-\frac{3}{2}}|\phi\rangle = \langle\phi|2L_0 \pm' 3J_0 + \frac{2}{3}c|\phi\rangle \tag{3.1.25}$$

which implies

$$|q| \leq \frac{c}{3} = d, \quad \forall \text{ (anti}-\text{)chiral states} . \tag{3.1.26}$$

That truncates the (anti-) chiral ring to a finite ring if the changes are discrete. In fact the (anti-)chiral rings have the important property of defining a Frobenius algebra. The grading is of course the charge grading and all axioms of the Frobenius algebra follow from the axioms of conformal field theory and its consistent truncation.

Massless space-time scalars have $(Q, \bar{Q}) = (\pm 1, \pm 1)$. The states in the chiral- and anti chiral rings with this property are related to the cohomology of M. The (c, c) ring corresponds to the $H^{d-1,1}(M)$ cohomology and the (c, a) ring corresponds[90] to the $H^{1,1}(M)$ cohomology.

In order to define a topological version of the $\mathcal{N} = 2$ algebra one needs to define nilpotent globally define super charge operators Q with $Q^2 = 0$. Of course one wants to define them using the fermionic super charges $G^{\pm}$, which have the nilpotency property (3.1.5). However being fermions they are not globally defined on general worldsheets of genus $g \neq 1$. The reason is that Riemann surfaces are just Kähler and there are no covariant constant spinors unless the surfaces are also Calabi-Yau manifolds, which leaves in one complex dimension just the torus with $g = 1$. For general Riemann surfaces one can still define a topological theory using the fact that there is the $U(1)$ current algebra generated by the currents $J(z)$. The euclidian Lorentz group in two dimensions is just a $SO(2) \cong U(1)$ rotation. For this reason it is possible to combine the $U(1)$ spinor connection from the Lorentz group with the $U(1)$ symmetry connection to define a scalar super charge. This has to be discussed on two chiral halfs of the theory.

For preparation of this we consider first the $(+, -)$ twisting on one chiral half[91] [92, 105]

$$\hat{T}(z) = T(z) \pm' \frac{1}{2} \partial J(z) \qquad \rightarrow \qquad \hat{L}_0 = L_0 \pm' \frac{1}{2} J_0 \tag{3.1.27}$$

then the modifications of (3.1.5) occur in the following short distance expansions

$$\begin{aligned}
\hat{T}(z)\hat{T}(w) &\sim \tfrac{2}{(z-w)^2}\hat{T}(w) + \tfrac{1}{z-w}\partial_w \hat{T}(w) \\
\hat{T}(z)G^{\pm}(w) &\sim \tfrac{3\pm' 1}{2(z-w)^2} G^{\pm}(w) + \tfrac{1}{z-w}\partial_w G^{\pm}(w) \\
\hat{T}(z)J(w) &\sim \tfrac{1}{(z-w)^2} J(w) + \tfrac{1}{z-w}\partial_w J(w) \mp' \tfrac{c}{3(z-w)^3}, \\
G^{+}(z)G^{-}(w) &\sim \tfrac{2c}{3(z-w)^3} + \tfrac{2}{(z-w)^2} J(w) + \tfrac{2}{z-w}\hat{T}(w) + \tfrac{1\mp' 1}{z-w}\partial_w J(w).
\end{aligned} \tag{3.1.28}$$

[90] The (a, a) and (a, c) rings correspond to conjugated fields and contain no independent information.

[91] $\pm'$ marked by a prime are correlated in (3.3.4), (3.1.28).

Let us point out the salient features of the operator product expansions in (3.1.28)

- Since the central term in the first OPE vanishes, no ghost system is required to quantize the world sheet theory.
- By the second OPE either G^+ (+-twist) or G^- (−-twist) become a spin one currents, so either $Q = G_0^+ = \oint G^+$ or $G_0^- = \oint G^-$ becomes conformal, i.e. scalars that are defined on every genus world sheet. The opposite super currents G^- (+-twist) or G^+ (−-twist), become spin 2 fields.
- The above conformal zero modes are recognized as building blocks for nilpotent operators $Q_{A/B}$. $Q_A = G_0^+ + \bar{G}_0^-$ in the case of the $(+,-)$ twist defining the (c,a) twisted chiral ring as cohomology. $Q_B = G_0^+ + \bar{G}_0^+$ for the $(+,+)$ twist defining the (c,c) chiral ring. The relation to geometry of M is[92] for the A-model $Q_A \leftrightarrow \mathrm{d}$ and the for the B-model $Q_B \leftrightarrow \bar\partial$ as discussed in more detail in Sects. 3.4.1, 3.5.1.
- The third OPE shows that $J(z)$ has an anomalous transformation. By arguments familiar from the BRST quantization of the bosonic string this gives rise to an anomaly in the divergence of the current, see [122–124] for a derivation, which can be covariantly written as

$$\int \nabla^\mu J_\mu = -\int \frac{d}{2\pi}\sqrt{h}R = -d\int c_1(\Sigma_g) = d(2g-2). \tag{3.1.29}$$

For $d = \frac{c}{3} = 3$ this comes precisely with the same anomalous coefficient -3 as the ghost current in the BRST quantization of the bosonic string $j_g = -:bc:$, see [279]. Integration of the anomaly in the divergence of the current leads to a $U(1)$-charge violation of $d(2g-2)$ units on a genus g Riemann surface.
- The last OPE finally is like the one between the BRST current and the b ghost. Integration around a contour to isolate G_0^+, yields for the + twist

$$\{Q, G^-(z)\} = T(z)\,, \tag{3.1.30}$$

which echos the main equation $\{Q_{BRST}, b(z)\} = T^{g+m}(z)$ in the BRST quantization of the bosonic string. We have seen already that G^- has $(h,Q) = (2,-1)$, which are precisely the conformal dimension and ghost charges of the $b(z)$ ghost.

To summarize we have for the $(+,+)$ twist [40] exactly the same structure as in the bosonic string if we identify

$$(G^+(z), J(z), T(z), G^-(z)) \leftrightarrow (J_{BRST}(z), j_g = -:bc:(z), T^{m+g}(z), b(z)) \tag{3.1.31}$$

[92]For Calabi-Yau manifolds this identifications can be viewed as convention and is reversed in [40].

and similar for the anti chiral half. This implies also $Q_B \leftrightarrow Q_{BRST}$ and the ghost number becomes $U(1)_A$ charge.

Let us finally discuss a concrete realisation of this structure proposed by Gepner, which provides exact conformal field theory descriptions at special moduli values of Calabi-Yau-families. These values are called Gepner points. The main point in Gepners construction is to identify the internal $c = \bar{c} = 9$ theory with an orbifold of a tensor product of minimal (2, 2) superconformal field theories. The factor theories are constructed as cosets of supersymmetric, WZW models, see [220] for a general discussion. WZW models and cosets are an important source of rational CFT's beyond $c > 1$. In the simplest case based on a $(SU(2) \times U(1))/U(1)$ coset the central charge is

$$c_k = \frac{3k}{k+3}, \qquad k \in N\,. \tag{3.1.32}$$

Primary states $|l, q, s\rangle$ of the algebra (3.1.5) are labeled in the minimal models by integers which have the following *standard range*[93]

$$\begin{aligned}
&0 \leq l \leq k, \\
&0 \leq |q - s| \leq l \\
&s = \left\{ \begin{array}{ll} 0, 2 & \text{Neveu} - \text{Schwarz} - \text{sector} \\ \pm 1 & \text{Ramond} - \text{sector} \end{array} \right\} \\
&l + q + s = 0 \bmod 2
\end{aligned} \tag{3.1.33}$$

and have conformal dimension and charge

$$h = \frac{l(l+2)}{4(k+2) - q^2} + \frac{s^2}{8}, \qquad Q = -\frac{q}{k+2} + \frac{s}{2}. \tag{3.1.34}$$

Above we discussed only the right moving part of the theory. There is a remarkable $A - D - E$ classification behind the question how to combine the $\chi_{l,q,s}$ and $\chi_{\bar{l},q,s}$ characters to a modular invariant one loop partition function [66]. Note that we consider only $l \neq \bar{l}$ in the left and right combination of characters. That is because all possible shifts of q, s w.r.t. $\bar{q}, \bar{s}$ are obtainable in a separate step by orbifold constructions w.r.t. to simple current symmetries. The simplest way to get a modular invariant theory is to start with a left right symmetric theory with states $|l, q, s; l, q, s\rangle$, this corresponds to the A-series. Considering only this series there are 145 possibilities to build a tensor product theory with $\bar{c} = c = \sum_{i=1}^{5} c_{k_i} = 9$. Note that at most one k_j is allowed to be zero, because of the $c = 9$ condition. The 145 is the same number as the one of $c_1(T_M) = 0$ Fermat hypersurfaces in WCP^4, i.e. with $\sum_{i=1} w_i = d$, see Appendix 3. In fact identifying $m_i = d/w_i = k_i + 2$ it

[93] For the orbifold procedure the following equivalences are important $q \sim q \bmod 2(k+2)$, $s = s \bmod 4$ and $|l, q, s; \bar{l}, \bar{q}, \bar{s}\rangle \sim |k - l, q, s; k - \bar{l}, \bar{q} + k + 2, \bar{s} + 2\rangle$.

is easy to see that the second enumeration lead to the same diophantic problem and yields Fermat type polynomial constraints in WCP^4

$$P = \sum_{i=1}^{5} a_i x_i^{m_i} = 0 , \tag{3.1.35}$$

which define CY 3-folds. The simplest possibility is $k_i = 3$ for $i = 1, \ldots, 5$. This leads to $d = 5$, $w_i = 1$, $i = 1, \ldots 5$, the in $\mathbb{P}^4$. Gepners orbifold construction divides the symmetric tensor product by a symmetry group which is generically the subgroup $G = \mathbb{Z}_{least\ com.\ mult.\{k_i\}} \times (\mathbb{Z}_2)^{r+1}$ among the group generated by the simple currents and constructs a modular invariant orbifold. The effect is that the factor theories and the space-time part are either all in the NS-NS sector or all in the R-R sector and that the charges in the internal NS-NS sector become odd integers [132, 133]. It is then easy to see that states in (c, c) ring from the invariant sector[94] of the orbifold are of the form $\bigotimes_i |l_i, l_i, 0; l_i, l_i, 0\rangle$. For the tensor product model that corresponds to the quintic this leads in view of (3.1.33) to 101 elements. The counting is the same that leads to the 101 independent complex structure deformations under Eq. (2.10.4), which are identified with elements in $H^{2,1}(M)$. All states in the (a, c) ring are from the twisted sector. They are more complicated to count but one checks that they yield the number of independent elements in $H^{1,1}(M)$. It is also straightforward to identify the orbifold action, like e.g. (2.10.2), (2.11.1), that leads to the mirrors W of the manifolds M in (3.1.35) in the conformal field theory context and to check that it indeed exchanges the (c, c) with (c, a) ring [121, 144]. A fascinating idea has been to use Cardy states [280] to classify D-branes as boundary conditions in the rational CFT at the Gepner-point and compare with geometric pictures of D-branes [55] in particular the triangulated category of coherent sheaves over M for the B-branes or the category of special Lagrangian submanifolds of M for the A-branes respectively.

3.2 Supersymmetric Nonlinear σ-Models

A $1d$ (supersymmetric) σ-model is simply a $1d$ field theory associated to a manifold M such that the fields are coordinates (and supercoordinates) of M, which depend only on one variable. It is natural to think this one variable as the time and the whole setup as (supersymmetric) quantum mechanics on M. In $2d$ dimensional σ models, the case relevant to string theory, the coordinates (and supercoordinates) of M depend on two variables the WS coordinates of the string and σ-model fields can be viewed as a map $x : \Sigma \to M$ from the worldsheet Σ to the target space M.

94 In general there might be (c, c) states in the twisted sectors but for the smooth hypersurfaces, such as the quintic, there are none.

We search in these models for field configurations which are fixpoints under some super symmetry transformation. The super symmetry generators become nilpotent operators Q on the Hilbert space of the field theory. The cohomology of Q is a natural structure to extract topological invariants of the classical bosonic configuration space. In more interesting situations indices can occur, which are invariant under some deformations, but are *family indices* w.r.t. others. Physically the family indices can be particular correlation functions. Their dependence on certain geometrical deformation parameters, e.g. of the target space metric, can often be exactly calculated e.g. in an all genera string loop expansion. This is the main physical benefit from topological theories. Apart form this more interesting geometry there is only one *new conceptual issue* in the 2d case and that are potential *anomalies* of the $2d$ quantum field theory on the WS.

The original references for the following are [247, 317] and especially [319]. We have adopted the conventions from the review [182]. There is a well known dictionary between properties of the worldsheet theory and properties of M. In particular if M is a Kähler manifold the σ-model will have $(2, 2)$ worldsheet supersymmetry [336]. The inverse statement is not quite true, i.e. one can construct more general geometric backgrounds that allow for $(2, 2)$ worldsheet supersymmetric σ-models [128].

In order to have superconformal invariance M has to be a Calabi-Yau manifold. A Calabi-Yau manifold is Kähler manifold with vanishing first Chern class of its tangent bundle $c_1(TM) = 0$. As we have seen in Sect. 2.4.1 this is equivalent to the statement that there exists a hermitian metric g for which the Ricci curvature vanishes $R_{i\bar{j}} = 0$. This in turn is equivalent to the statement that the holomomy group of M is contained in $SU(3)$. We call a Calabi-Yau threefold a manifold where the holonomy is the full $SU(3)$ (or a least $SU(2) \times Z_2$), which implies that there are exactly two covariant constant spinors on M. This leads to $N = 2$ supergravity theories in 4d for the compactification of type II on M. We will start the discussion of the symmetries of the actions at the classical level and comment then on the potential anomalies and their cancellation.

3.2.1 $N = (1, 1)$ Nonlinear σ-Model

Let us first treat the $N = (1, 1)$ case. For this case the target space needs to have just a Riemannian metric. We parametrize the map $x : \Sigma \to M$ by x^I, where $I \ldots, d$ where d is the real dimension of M. The worldsheet is parametrized by $z, \bar{z}$, hence x is given in local coordinates as $x^I(z, \bar{z})$ The fields of the σ model have the following transformation properties under worldsheet and target space reparametrizations. With K and $\bar{K}$ the canonical and anti-canonical bundle of Σ and TM the complexified tangent bundle of M one has WS-fermions which transform as

$\psi_+^I \in \Gamma(\bar{K}^{\frac{1}{2}} \otimes x^*(TM))$ and $\psi_-^I \in \Gamma(K^{\frac{1}{2}} \otimes x^*(TM))$, where Γ denotes sections of the indicated bundles. The Lagrangian of the non-linear 2d σ-model is then given by

$$L = 2t \int_\Sigma \mathrm{d}^2 z \left(\frac{1}{2} g_{IJ}(x) \partial_z x^I \partial_{\bar{z}} x^J + \frac{i}{2} g_{IJ} \psi_-^I D_z \psi_-^J + \frac{i}{2} g_{IJ} \psi_+^I D_{\bar{z}} \psi_+^J + \frac{1}{4} R_{IJKL} \psi_+^I \psi_+^J \psi_-^K \psi_-^L \right) . \tag{3.2.1}$$

The covariant derivatives $D_{\bar{z}}$ (D_z) are obtained using the pullback of the Levi-Civita connection from M as

$$D_{\bar{z}} \psi_+^I = \frac{\partial}{\partial \bar{z}} \psi_+^I + \frac{\partial x^J}{\partial \bar{z}} \Gamma^I_{JK} \psi_+^K \tag{3.2.2}$$

and R_{IJKL} is the Riemann-Tensor of M.[95] Here we assumed a flat world-sheet or a local trivialization of $K^{\frac{1}{2}}$, so that no spin connection appears in (3.2.2). Soon global properties of $K^{\frac{1}{2}}$ and $\bar{K}^{\frac{1}{2}}$ become all important.

With Grassmann valued supersymmetry parameters $\epsilon_- \in \Gamma(K^{-\frac{1}{2}})$ and $\epsilon_+ \in \Gamma(\bar{K}^{-\frac{1}{2}})$ one checks at the classical level that (3.2.1) is invariant under the following supersymmetry transformations

$$\begin{aligned} \delta x^I &= -\epsilon_- \psi_+^I + \epsilon_+ \psi_-^I \\ \delta \psi_+^I &= i\epsilon_- \partial x^I + \epsilon_+ \psi_-^K \Gamma^I_{KM} \psi_+^M \\ \delta \psi_-^I &= -i\epsilon_+ \partial x^I + \epsilon_- \psi_+^K \Gamma^I_{KM} \psi_-^M . \end{aligned} \tag{3.2.3}$$

From these Eq. (3.2.3) we would like to define nilpotent operators. The obstruction is that there are no global trivial sections of $K^{-\frac{1}{2}}$ or $\bar{K}^{-\frac{1}{2}}$ unless $g = 1$. This means that there no global supersymmetry transformations on the worldsheet unless[96] $g = 1$.

In the case of the worldsheet being a torus one can chose globally defined sections $\epsilon_- \in \Gamma(K^{-\frac{1}{2}})$ and $\epsilon_+ \in \Gamma(\bar{K}^{-\frac{1}{2}})$ to obtain globally defined supersymmetry generators $Q_-^2 = 0$ and $Q_+^2 = 0$ on the Hilbert space $\mathcal{H}$. E.g. we can chose $\epsilon_\pm$ both to be in trivial sections of $K^{-\frac{1}{2}}$ and $\bar{K}^{-\frac{1}{2}}$ respectively. In view of (3.2.3) we have to chose corresponding trivializations for $\psi_+^I \in \Gamma(\bar{K}^{\frac{1}{2}} \otimes x^*(TM))$ and $\psi_-^I \in \Gamma(K^{\frac{1}{2}} \otimes x^*(TM))$ and this simply means that the fermions will have periodic boundary conditions on T^2. These boundary conditions are called *twisted* boundary conditions. Q_- and Q_+ are globally defined and $Q_+|\Psi\rangle = Q_-|\Psi\rangle = 0$ for $\Psi \in \mathcal{H}$

[95]By $x^*(TM)$ we denoted the pullback of TM to Σ.

[96]The quest for covariant constant spinors is familiar on the target space in order to obtain space-time supersymmetric compactifications as discussed in Sect. 1. It requires restricted holonomies, see Sect. 1.1.5, which is equivalent to the familiar $c_1(TM) = 0$ condition for $N = 2$ ($N = 1$) II (heterotic) compactifications 6d internal manifolds.

forces the cohomological states to be in the $E = 0$ super symmetric ground state of the Hamiltonian [315]

$$H = \frac{1}{2}\{Q_+, Q_-\} = \frac{1}{2}(\mathrm{dd}^* + \mathrm{d}^*\mathrm{d})\,. \tag{3.2.4}$$

Generically the non-trivial information in the double twisted model is the *Witten index*. It is simplest written in the operator formalism

$$\chi(M) = \mathrm{Tr}(-1)^F q^{H_+}\bar{q}^{H_-} = \mathrm{Tr}(-1)^F, \tag{3.2.5}$$

where $F = F_+ + F_-$ and F_+/F_- count the left/right moving fermion numbers so that $\{(-)^{F_\pm}, Q_\pm\} = 0$ while $[(-)^{F_\mp}, Q_\pm] = 0$. The σ model cohomology is equivalent to the cohomology of M, much in the same way as we will made explicit in Sects. 3.4.1 and 3.5.1 for the $N = (2,2)$ case. Since (3.2.4) is the Laplacian and the fermion number, measured by $(-1)^F$, corresponds to the form degree and the Witten index is equal to the Euler number $\chi(M)$ of M [315]. The insertion of $(-1)^F$ kills the information about the time evolution and spatial excitation of the string. The latter fact reduces the model to constant maps, i.e. supersymmetric quantum mechanics on M, i.e. the index can also be obtained starting with a $1d$ supersymmetric σ model on M. The consideration that leads to the index is referred to as *quantizing the zero mode sector*. If further global quantum numbers are present one can get finer information then just the Euler number, by inserting the corresponding charge operator in the trace. These ideas play a rôle in extracting BPS numbers for instance associated to branes see Sect. 4.3.4.

Much more detailed information survives for example in the string context if one chooses only ϵ_+ to be in a trivial section. The corresponding index is called the *elliptic genus*[97]

$$\mathcal{E}(M) = \mathrm{Tr}(-1)^{F_+} q^{H_+}\bar{q}^{H_-} = \mathrm{Tr}(-1)^{F_+}\bar{q}^{H_-}\,. \tag{3.2.6}$$

Here only the left moving states are forced in the left moving groundstate. The trace over the right moving states explores information which goes beyond cohomological information of M. It can be defined for 2d supersymmetric field theories and is conformally invariant even if the underlying field theory is not [323]. It requires $(-)^{F_+}$ not to be anomalous, which is essentially equivalent to M being spin [322]. It carries information, which is robust under certain deformations. In the case of the σ model on M, $\mathcal{E}(M)$ is the Dirac index of the loop space of M [314, 316]. This index varies with the volume parameters of M, but is independent of the complex structure of M and is the first example of the promised family indices. There are further simple refinements possible, if as below in the $N = (2,2)$ theories F_- comes

[97]Unfortunately there many notations common to distinguish the left- and right moving sectors in this context unbarred/barred for euclidean worldsheets, R/L, $+/-$ and without tilde/with tilde are maybe most often used.

from an $U(1)_L$ current $F_- = \oint J_L$. If the latter is not anomalous one can insert $(-1)^{\theta F_-}$ in the trace in (3.2.6) and even if the $U(1)_L$ is broken to Z_K Eq. (3.2.6) with $\exp(\frac{i\pi}{k}F_-)$ inserted is still an index. A theme of the lecture is to explore more sophisticated family indices mainly in the $N = (2, 2)$ context and even at genus one there are further refinements such as (4.1.20).

3.2.2 Compactifications with $N = (2, 2)$ World Sheet Supersymmetry

The additional structure that allows to define more general family indices for the $(2, 2)$ worldsheet theories are right and left $U(1)_{R/L}$ symmetries, so called R-symmetries. Since the nilpotent Q operators are derived from the supersymmetry transformations and since there are no covariant constant spinors for world sheets of genus $g \neq 1$ there will be no well defined supersymmetry operators on general Σ_g without further modifications. For the topological theory to make sense at all genera g we "change" the transformation properties of the fields, so that the supersymmetry transformation becomes a scalar operator on the world sheet. This modification is implemented by twisting the world sheet Lorentz group either by the vector $U(1)_V = U(1)_L + U(1)_R$ or the axial $U(1)_A = U(1)_L - U(1)_R$ symmetry. To do this we first gauge the R-symmetries. Then we combine the $U(1)$ gauge connection with the spin connection to a twisted world sheet spin connection. Contrary to the $U(1)_V$, the $U(1)_A$ current develops an quantum anomaly proportional to $\int_\Sigma x^*(c_1(TM))$. Therefore the B model, which is obtained by twisting with the $U(1)_A$ connection, is only well defined on Calabi-Yau manifolds ($c_1(TM) = 0$), while the A model, which is obtained by twisting with the $U(1)_V$ connection can be considered on any Kähler manifold.

3.2.3 The (2, 2) Non-linear σ-Model

Let us now study this twisting mechanism in the Kähler case, which has at the classical level a $N = (2, 2)$ supersymmetry and hence the necessary $U(1)$ symmetries. The action is given by

$$S = 2t \int_\Sigma \mathrm{d}^2 z \Big(- g_{i\bar{j}} \partial_\mu x^i \partial^\mu x^{\bar{j}} + i g_{\bar{i}i} \psi_-^{\bar{i}} D_z \psi_-^i + i g_{\bar{i}i} \psi_+^{\bar{i}} D_{\bar{z}} \psi_+^i + R_{i\bar{i}j\bar{j}} \psi_+^i \psi_+^{\bar{i}} \psi_-^j \psi_-^{\bar{j}} \Big) . \tag{3.2.7}$$

Here we have split the index I into i and $\bar{i}$ according to the Kähler decomposition of the CY metric. Such a metric can locally be written as $g_{i\bar{j}} = \partial_i \partial_{\bar{j}} K(x^i, x^{\bar{i}})$ and its Levi-Civita connection in Kähler geometry is pure in the indices $\Gamma^i_{jk} = g^{i\bar{j}} \partial_j g_{k\bar{j}}$ as discussed in more detail in Sect. 2.3.1. On a non-flat Riemann surface Σ one has the connection

$$\begin{aligned} D_{\bar{z}} \psi_+^i &= \partial_{\bar{z}} \psi_+^i + \tfrac{i}{2} \omega_{\bar{z}} \psi_+^i + \Gamma^i_{kl} \partial_{\bar{z}} x^k \psi_+^l \\ D_z \psi_-^i &= \partial_z \psi_-^i - \tfrac{i}{2} \omega_z \psi_+^i + \Gamma^i_{kl} \partial_z x^k \psi_-^l \, , \end{aligned} \tag{3.2.8}$$

where ω_z and $\omega_{\bar{z}}$ are the components of the spin connection of Σ.

In superfield formalism one can write $L = 2t \int d\theta^4 K(\mathbf{X}^i, \bar{\mathbf{X}}^{\bar{i}})$, where the chiral field $\mathbf{X}^i$ has components $x^i, \psi^i_\pm, F^i$. F^i is an auxiliary field that has has no kinetic terms and can be eliminated from the action by its equation of motion $F = \Gamma^i_{ij}\psi^j_+\psi^k_-$. This offshell superfield formalism is particularly useful when one couples a holomorphic superpotential $W(x^i)$ to the action, which only possible for non-compact target spaces M. This formalism is worked out in detail including the off-shell supersymmetry transformations in [247] and reviewed in [182]. For notational brevity we restrict ourselves to the onshell formalism.

Classically there are now twice as many super symmetries, one set for the holomorphic and one set for the antiholomorphic space time indices. They are generated by $\epsilon_+ \in \Gamma(K^{\frac{1}{2}})$, $\epsilon_- \in \Gamma(\bar{K}^{\frac{1}{2}})$ and $\bar{\epsilon}_\pm$. The latter are sections of the same bundles but have opposite charges under $U(1)_A$ and $U(1)_V$. The super symmetry transformations are

$$
\begin{aligned}
\delta x^i &= -\epsilon_-\psi^i_+ + \epsilon_+\psi^i_- \\
\delta x^{\bar{i}} &= \bar{\epsilon}_-\psi^{\bar{i}}_+ - \bar{\epsilon}_+\psi^{\bar{i}}_- \\
\delta\psi^i_+ &= 2i\bar{\epsilon}_-\partial_+ x^i + \epsilon_+\psi^j_+\Gamma^i_{jm}\psi^m_- \\
\delta\psi^{\bar{i}}_+ &= -2i\epsilon_-\partial_+ x^{\bar{i}} + \bar{\epsilon}_+\psi^{\bar{j}}_-\Gamma^{\bar{i}}_{\bar{j}\bar{m}}\psi^{\bar{m}}_+ \\
\delta\psi^i_- &= -2i\bar{\epsilon}_+\partial_- x^i + \epsilon_-\psi^j_+\Gamma^i_{jm}\psi^m_- \\
\delta\psi^{\bar{i}}_- &= 2i\epsilon_+\partial_- x^{\bar{i}} + \bar{\epsilon}_-\psi^{\bar{j}}_-\Gamma^{\bar{i}}_{\bar{j}\bar{m}}\psi^{\bar{m}}_+ \, .
\end{aligned}
\tag{3.2.9}
$$

The relation between the existence of two supersymmetries and the decomposition of the exterior derivative on Kähler manifolds into a holomorphic and antiholomorphic derivative $d = \bar{\partial} + \partial$, which gives rise to the Hodge decomposition of cohomology groups into $H^{p,q}(M)$, has been discussed first by [336]. The fields x^i, $x^{\bar{i}}$, $\psi^i_\pm$ and $\psi^{\bar{i}}_\pm$ transform as before under WS transformations. W.r.t. the spacetime transformations one has now simply a splitting of $TM_{\mathbb{C}}$ into $T^{1,0}M \oplus T^{0,1}M$ with i referring to $T^{1,0}M$ and $\bar{i}$ referring to $T^{0,1}M$, so e.g. $\psi^i_+ \in \Gamma(\bar{K}^{\frac{1}{2}} \otimes x^*(T^{1,0}M))$ e.t.c. All transformation properties are summarized in Table 4.

The action of the $U(1)_V$ and $U(1)_A$ symmetries are conveniently formulated in superfield formalism, i.e. we expand any field in Grassmann valued $\theta^+, \theta^-, \bar{\theta}^+, \bar{\theta}^-$ complex fermionic spinor coordinates on which complex conjugation is given by $(\theta^\pm)^* = \bar{\theta}^\pm$. The WS Lorentz transformation acts on $t = x^0$ and $s = x^1$ (with $(1,1)$ signature) and on spinors as

$$
\begin{aligned}
\begin{pmatrix} x^0 \\ x^1 \end{pmatrix} &\to \begin{pmatrix} \cosh\gamma & \sinh\gamma \\ \sinh\gamma & \cosh\gamma \end{pmatrix} \begin{pmatrix} x^0 \\ x^1 \end{pmatrix} \\
\theta^\pm &\to e^{\pm\frac{\gamma}{2}}\theta^\pm \\
\bar{\theta}^\pm &\to e^{\pm\frac{\gamma}{2}}\bar{\theta}^\pm
\end{aligned}
\tag{3.2.10}
$$

Table 4 Space time transformation of the non linear σ-model fields after + and − twist

	Section before twisting	Section (+) twist	Section (−) twist
x	$x^*(TM)$	$x^*(TM)$	$x^*(TM)$
ψ_-^i	$x^*(T^{1,0}) \otimes K^{\frac{1}{2}}$	$x^*(T^{1,0})$	$x^*(T^{1,0}) \otimes K$
$\bar{\psi}_-^{\bar{i}}$	$x^*(T^{0,1}) \otimes K^{\frac{1}{2}}$	$x^*(T^{0,1}) \otimes K$	$x^*(T^{0,1})$
ψ_+^i	$x^*(T^{1,0}) \otimes \bar{K}^{\frac{1}{2}}$	$x^*(T^{1,0})$	$x^*(T^{1,0}) \otimes \bar{K}$
$\bar{\psi}_+^{\bar{i}}$	$x^*(T^{0,1}) \otimes \bar{K}^{\frac{1}{2}}$	$x^*(T^{0,1}) \otimes \bar{K}$	$x^*(T^{0,1})$

Classically and in non-anomalous theories one can chose the twisting on the left movers ψ_-^i, $\psi_-^{\bar{i}}$ and the right movers ψ_+^i, $\psi_+^{\bar{i}}$ independently

Since the fermionic variables anticommute w.r.t. to each other the Taylor expansion in them contains only 2^4 terms

$$\Phi(x, \theta^\pm, \bar{\theta}^\pm) = x(t,s) + \theta^+\psi_+(t,s) + \theta^-\psi_-(t,s) + \bar{\theta}^+\bar{\psi}_+(t,s) + \bar{\theta}^-\bar{\psi}_-(t,s) + \theta^+\theta^- A_{+-}s, t + \ldots \tag{3.2.11}$$

In this sense one can think superspace as a thin space in the fermionic directions, which contains no second order derivative information in any fermionic direction. The relation to calculus with differential forms is very obvious. The action of the vector $U(1)_V$ and axial $U(1)_A$ symmetries on all component fields is induced from

$$\begin{aligned} e^{i\alpha F_V} &: \Phi(x, \theta^\pm, \bar{\theta}^\pm) \mapsto e^{i\alpha q_V}\Phi(x, e^{-i\alpha}\theta^\pm, e^{i\alpha}\bar{\theta}^\pm) \\ e^{i\beta F_A} &: \Phi(x, \theta^\pm, \bar{\theta}^\pm) \mapsto e^{i\beta q_A}\Phi(x, e^{\mp i\beta}\theta^\pm, e^{\pm i\beta}\bar{\theta}^\pm) . \end{aligned} \tag{3.2.12}$$

Let us denote now the four supersymmetry operators corresponding to $\epsilon^\pm$ and $\bar{\epsilon}^\pm$ transformations by $Q_\mp$ and $\bar{Q}_\mp$ respectively. A general supersymmetry transformation is then generated by the operator

$$\hat{\delta} = i\epsilon_+ Q_- - i\epsilon_- Q_+ - i\bar{\epsilon}_-\bar{Q}_- + i\bar{\epsilon}_+\bar{Q}_+ , \tag{3.2.13}$$

where $(Q^\pm)^\dagger = \bar{Q}_\pm$ and $\hat{\delta}^\dagger = -\hat{\delta}$.

More generally for any infinitesimal field transformation $\delta_Q\phi$ we will denote the infinitesimal transformation on the field operator $\delta\mathcal{O}_\phi$ by $\delta_Q\mathcal{O}_\phi = [Q, \mathcal{O}_\phi]_\pm$, where Q is the corresponding generating operator. Let M be the generator of two dimensional Lorentz rotations $SO(1,1)$. It is convenient to make the Wick rotation $x^0 = -ix^2$ and we call $M_E = iM$ the generator of the compact Euclidean rotation group $U(1)_E$. Beside the supersymmetry generators one has on the WS: H the generator of (euclidean) time translations, P generator of translations. Furthermore there are the R-charge operators associated to the $U(1)_V$ and $U(1)_A$ currents called F_V and F_A. These generators fulfill the algebra

$$\begin{aligned}
&Q_+^2 = Q_-^2 = \bar{Q}_+^2 = \bar{Q}_-^2 = 0, \\
&\{Q_\pm, \bar{Q}_\pm\} = H \pm P, \qquad \{\bar{Q}_+, \bar{Q}_-\} = \{Q_+, Q_-\} = \{Q_-, \bar{Q}_+\} = \{Q_+, \bar{Q}_-\} = 0, \\
&[M_E, Q_\mp] = \mp Q_\pm, \qquad [M_E, \bar{Q}_\pm] = \mp \bar{Q}_\pm, \\
&[F_V, Q_\pm] = -Q_\pm, \qquad [F_V, \bar{Q}_\pm] = \bar{Q}_\pm, \\
&[F_A, Q_\pm] = \mp Q_\pm, \quad [F_V, \bar{Q}_\pm] = \pm \bar{Q}_\pm,
\end{aligned} \tag{3.2.14}$$

It becomes soon important that $Q_\pm$ and $\bar{Q}_\pm$ have opposite charges under the R symmetry groups. As already stated F_A is present at the quantum level only for Calabi-Yau manifolds, the conformal case, while F_V is generically present. See [247] for a further discussion of this algebra.

3.3 Twisting the $N = (2, 2)$ Theories and Cohomological Field Theories

Twisting amounts to a modification the Euclidean rotation group $U(1)_E$ by a generator of the global $U(1)$ R-symmetry groups and defines the new generator of the Euclidean rotation group $U(1)_{E'}$ as $M'_E = M_E + R$. As explained our goal is to make some of fermionic Q operators scalar w.r.t. M'_E, so that they are well defined on all genera world-sheets. These "scalar" operators can then be used to define a cohomological theory on an arbitrary Riemann surface. The term twisting is familiar in the orbifold context, where it means to modify the boundary conditions of a field along cycles of the worldsheet by an element g of a global symmetry group G, e.g. for the torus with a A cycle of length 2π a field is periodically identified by $\phi(x + 2\pi) = g\phi(x)$. The analogy is appropriate since also in the above case we change the boundary conditions of some fermionic fields to become periodic. We encountered such twisting already in the discussion of Witten index and the elliptic genus.

Here the twisting is implemented by *gauging* the $U(1)$-R symmetry group and adding the corresponding gauge connection A^R_μ to the spin connection, so that the transformation property of the spinor fields depend now on their R charge. An important consequence of gauging the $U(1)$-R symmetry is that the gauge field modifies the energy momentum tensor, see (3.3.4). Since we are dealing with a 2d quantum field theory this program of gauging the R symmetry might be obstructed by anomalies. The potentially dangerous terms in the action are the fermion kinetic terms $ig_{\bar{i}i}\psi^{\bar{i}}_- D_z \psi^i_- + ig_{\bar{i}i}\psi^{\bar{i}}_+ D_{\bar{z}}\psi^i_+$ in (3.2.7). As explained in reviews on anomalies such as [43] the vector $U(1)_V$ will never be anomalous. The anomaly density for the axial current is calculated also in [43] and from (3.2.8) we see that we have a Dirac operator on Σ_g coupled to a connection of a bundle, which is the pullback by x of the holomorphic tangent bundle to Σ written as $x^*(T^{0,1}M)$. The Atiyah-Singer index theorem [26] for the *twisted* spin complex, see [267] for a review, gives us then the answer that the axial $U(1)_A$ current violation is

$$\int_\Sigma \partial_\mu j^\mu_A = 2\int_\Sigma c_1(x^*(T^{1,0}M)) = 2\int_\Sigma x^*(c_1(T^{1,0}M)) = 2[C]\cdot c_1(TM)\,. \tag{3.3.1}$$

This breaks the $U(1)_A$ symmetry generically to a Z_2. For a discussion of the $U(1)_A$ anomaly in the linear σ-model context see [322]. By $[C]$ we denote the curve class to which Σ maps.

The most important consequence of the above result is that on a Calabi-Yau manifold where $c_1(TM)=0$ we can twist by the $U(1)_A$ and the $U(1)_V$ symmetry as both are *anomaly free*. In the (2, 2) theory we have therefore two fundamentally different possibilities to twist

$$\begin{array}{ll} \mathrm{A-Twist:} & M_{E'} = M_E + F_V \\ \mathrm{B-Twist:} & M_{E'} = M_E + F_A\,. \end{array} \tag{3.3.2}$$

The tables below record how the twisting changes the WS transformation properties of the fields. We do this first for the the so + and the − twist first. In the above notation of Table 4 the A twist corresponds to a $(-,+)$ twist, i.e. to a combination of the $(-)$ twist on ψ_-, $\bar\psi_-$ and the $(+)$-twist on ψ_+, $\bar\psi_+$, while the B twist is $(+,+)$ twist, i.e. a combination of the $(+)$ twist on ψ_-, $\bar\psi_-$ and the $(+)$-twist on ψ_+, $\bar\psi_+$. There are the possibilities of an $(+,-)$ twist and an $(-,-)$ twist making $\bar Q_A$ and $\bar Q_B$ nilpotent operators. They lead to the definition of conjugated cohomological sectors and considered for themselves not to new theories. However as explained in Sect. 3.6.1 the combined geometry of the sectors conjugated to each other leads to an interesting geometry, the so called tt^* geometry.

The effects of the twisting on the fields and the supersymmetry transformation can be summarized in the Tables 5 and 6 respectively.

As it is clear from Table 6 and (3.2.14) the following combinations

$$\begin{aligned} Q_A &= Q_- + \bar Q_+ \\ Q_B &= \bar Q_- + \bar Q_+ \end{aligned} \tag{3.3.3}$$

are now scalar, nilpotent operators which can be used to define two different cohomological theories, the topological A- and the topological B-model respectively. Mirror symmetry exchanges the − twist with the + twist on the ψ_-, $\bar\psi_-$ side. Even before twisting Q_A and Q_B define cohomological theories on the plane and the torus, where covariantly constant spinors exist. One can also choose to twist only the say ψ_-, $\bar\psi_-$ side. The indices of so called half-twisted models are the closest analogs of the elliptic genus (3.2.6) at higher genus [319, 329]. This indices are

Table 5 Space time transformation of the non linear σ-model fields and charges after A and B twist

			Before twisting		A twist $(-,+)$			B twist $(+,+)$		
	$U(1)_V$	$U(1)_A$	$U(1)_E$	spin		$U(1)'_E$	spin		$U(1)'_E$	spin
x	0	0	0	$\mathbf{1}_C$	x	0	$\mathbf{1}_C$	x	0	$\mathbf{1}_C$
ψ^i_-	-1	1	1	$K^{\frac{1}{2}}$	χ^i	0	$\mathbf{1}_C$	ρ^i_z	2	K
$\bar\psi^{\bar i}_+$	1	1	-1	$\bar K^{\frac{1}{2}}$	$\chi^{\bar i}$	0	$\mathbf{1}_C$	$-\frac{1}{2}(\theta^{\bar i}+\eta^{\bar i})$	0	$\mathbf{1}_C$
$\bar\psi^{\bar i}_-$	1	-1	1	$K^{\frac{1}{2}}$	$\rho^{\bar i}_z$	2	K	$\frac{1}{2}(\theta^{\bar i}-\eta^{\bar i})$	0	$\mathbf{1}_C$
ψ^i_+	-1	-1	-1	$\bar K^{\frac{1}{2}}$	$\rho^i_{\bar z}$	-2	$\bar K$	$\rho^i_{\bar z}$	-2	$\bar K$

We also indicate the names of the fields in the A and B model

Table 6 Space time transformation of the supersymmetry generators after the A and B twist

			Before Twisting		A-twist		B-twist	
	$U(1)_V$	$U(1)_A$	$U(1)_E$	spin	$U(1)'_E$	spin	$U(1)'_E$	spin
Q_-	-1	1	1	$K^{\frac{1}{2}}$	0	$\mathbf{1}_C$	2	K
$\bar Q_+$	1	1	-1	$\bar K^{\frac{1}{2}}$	0	$\mathbf{1}_C$	0	$\mathbf{1}_C$
$\bar Q_-$	1	-1	1	$K^{\frac{1}{2}}$	2	K	0	$\mathbf{1}_C$
Q_+	-1	-1	-1	$\bar K^{\frac{1}{2}}$	-2	$\bar K$	-2	$\bar K$

shared between the A and the B model and contain information about the couplings of $\mathbf{1272\bar{7}}$ in the heterotic string with standard embedding.

We denote the gauge current, which corresponds to the gauge variations δA^R_μ by J^R_μ. The twisting modifies the energy momentum tensor to

$$\hat T_{\mu\nu} = T_{\mu\nu} + \frac{1}{4}\left(\epsilon^\lambda_\mu \partial_\lambda J^R_\nu + \epsilon^\lambda_\nu \partial_\lambda J^R_\mu\right) . \tag{3.3.4}$$

In the action of the gauged theory of covariant theory the world sheet emerges a coupling

$$\Delta S = \int_\Sigma J^\mu \omega_\mu = \frac{1}{2}\int_\Sigma J\bar\omega + \bar J\omega = \frac{1}{2}\int_\Sigma R\phi + \text{total der.} , \tag{3.3.5}$$

to the spin connection ω. In the third equality we bosonized the $U(1)_R$ current $\partial\phi = J$ and integrated partially. Contact terms of operators with the this expression will play a rôle in determining properties of the correlation functions.

3.3.1 Generalities on the Physical Observables

One calls an operator *a chiral* operator or (c, c) operator ϕ if

$$[Q_B, \phi] = 0 . \tag{3.3.6}$$

Chiral and twisted chiral superfields play an important rôle in formulating the general $(2, 2)$ worldsheet theory, see [322]. The lowest component ϕ of chiral superfield Φ obeys $[\bar{Q}_{\pm}, \phi] = 0$ and is a hence a chiral operator. An operator ϕ is called *twisted chiral* or (a, c) if

$$[Q_A, \phi] = 0 \,. \tag{3.3.7}$$

The lowest component v of a twisted chiral superfield Σ obeys $[\bar{Q}_+, v] = [Q_-, v] = 0$ and is hence a twisted chiral operator. $[\bar{Q}_-, \phi_-] = 0$ and $[Q_-, \phi_-] = 0$ define left chiral- and antichiral operators while $[\bar{Q}_+, \phi_+] = 0$ and $[Q_+, \phi_+] = 0$ define right chiral- and antichiral operators.

The key concept is now to define a cohomological theory whose observables are the equivalence classes $[\phi]$ of Q *closed* operators. To be closed the operators have to fulfill $[Q, \phi] = 0$ and the equivalence relation is as usually up to *exact* operators $\mathcal{E} = [Q, \Lambda]_{\pm}$, i.e.

$$\phi \sim \phi + [Q, \Lambda]_{\pm} \,. \tag{3.3.8}$$

If the vacuum is annihilated by Q, which is the case if Q comes from a unbroken symmetry as above, then the correlation function of the Q closed operators does not depend on the representative of the class

$$\begin{aligned} \langle \phi_1 \ldots (\phi_k + \{Q, \Lambda\}) \ldots \phi_n \rangle &= \langle \phi_1 \ldots \phi_n \rangle \pm \langle 0 | \phi_1, \ldots \phi_{k-1} \Lambda \phi_{k+1} \ldots \phi_n Q | 0 \rangle \\ &\quad \pm \langle 0 | Q \phi_1, \ldots \phi_{k-1} \Lambda \phi_{k+1} \ldots \phi_n | 0 \rangle \\ &= \langle \phi_1 \ldots \phi_n \rangle \end{aligned} \tag{3.3.9}$$

Above the $\pm$ signs are uncorrelated and the two terms vanish independently if the vacuum is Q invariant. The analogy of the definition of topological correlators with cohomological intersections $\int_M \omega_1 \wedge \ldots \wedge (\omega_k + \mathrm{d}\lambda) \wedge \ldots \wedge \omega_n = \int_M \omega_1 \wedge \ldots \wedge \omega_k \wedge \ldots \wedge \omega_n$ is not just formal in the case of the $(2, 2)$-sigma model as we will see.

An important property of these operators is that they form position independent rings. Using the algebra (3.2.14), the properties of the twisted chiral operators and $[\{A, B\}, C] = \{[A, C], B\} + \{A, [B, C]\}$ it is easy to see that e.g.

$$\begin{aligned} \frac{i}{2}\left(\frac{\partial}{\partial x^0} + \frac{\partial}{\partial x^1}\right)\phi &= [(H + P), \phi] = [\{Q_+, \bar{Q}_+\}, \phi] = \ldots = \{Q_B, [Q_+, \phi]\} \\ \frac{i}{2}\left(\frac{\partial}{\partial x^0} - \frac{\partial}{\partial x^1}\right)\phi &= [(H - P), \phi] = [\{Q_-, \bar{Q}_-\}, \phi] = \ldots = \{Q_B, [Q_-, \phi]\} \end{aligned} \tag{3.3.10}$$

and similar for the A model. Combining (3.3.9) and (3.3.10) one sees that the correlation functions of the twisted chiral operators do not depend on the position of the insertions of the operators, which is also true for the chiral operators. The ring structure comes from the operator product expansion see (3.1.20). It is obvious respects the symmetry that the OPE of two (twisted) chiral fields is (twisted) chiral again and by (3.3.10) position independent. One defines the structure constants of the ring in a basis of the ring ϕ_k as

$$\phi_i\phi_j = C_{ij}^k\phi_k + [Q,\Lambda]_\pm\,, \tag{3.3.11}$$

i.e. identifying an element on the right hand side up to exacts term. The ring satisfies the usual associativity $C_{jl}^m C_{ik}^l = C_{lk}^m C_{ij}^l$. The unit $\phi_0 = 1$ is always (twisted) chiral, so $C_{0j}^k = C_{j0}^k = \delta_j^k$.

The position independence (3.3.10) and its realization on p-form operators can be formulated in a covariant way as the so called *descend equations*, see [92] for a review. If $\mathcal{O}^{(0)} = \phi$ is a Q closed position independent 0-form operator, one can define the following non-local n-form operators

$$\begin{aligned} 0 &= [Q,\mathcal{O}^{(0)}] \\ \mathrm{d}\mathcal{O}^{(0)} &= \{Q,\mathcal{O}^{(1)}\} \\ \mathrm{d}\mathcal{O}^{(1)} &= [Q,\mathcal{O}^{(2)}] \\ \mathrm{d}\mathcal{O}^{(2)} &= 0\,. \end{aligned} \tag{3.3.12}$$

Using (3.3.10) and the corresponding relation for the A-model one can find the descend operators for the (a,c) and the (c,c) ring states explicitly noting that $Q_-\mathrm{d}z$ $(\bar{Q}_-\mathrm{d}z)$ and $Q_+\mathrm{d}\bar{z}$ $(\bar{Q}_+\mathrm{d}\bar{z})$ are covariant combinations

$$\begin{aligned} &\mathrm{A-mod.}\ \mathcal{O}_A^{(1)} = i\mathrm{d}z[\bar{Q}_-,\mathcal{O}_A^{(0)}] - i\mathrm{d}\bar{z}[Q_+,\mathcal{O}_A^{(0)}],\ \mathcal{O}_A^{(2)} = \mathrm{d}z\mathrm{d}\bar{z}\{Q_+,[\bar{Q}_-,\mathcal{O}_A^{(0)}]\}, \\ &\mathrm{B-mod.}\ \mathcal{O}_B^{(1)} = i\mathrm{d}z[Q_-,\mathcal{O}^{(0)}] - i\mathrm{d}\bar{z}[Q_+,\mathcal{O}_B^{(0)}],\ \mathcal{O}_B^{(2)} = \mathrm{d}z\mathrm{d}\bar{z}\{Q_+,[Q_-,\mathcal{O}_B^{(0)}]\}\,. \end{aligned} \tag{3.3.13}$$

The descend equations truncate, because of the anti symmetrization in the world-sheet indices. The $\bar{Q}_B$ and $\bar{Q}_A$ operators define the (a,a) and (c,a) ring states which we call $\bar{\mathcal{O}}_B^{(0)}$ and $\bar{\mathcal{O}}_A^{(0)}$ respectively. Their descendants $\bar{\mathcal{O}}_B^{(1,2)}$ and $\bar{\mathcal{O}}_A^{(1,2)}$ are defined as in (3.3.13) with the barred and unbarred Q operators exchanged. As an easy exercise one checks that $\mathcal{O}_B^{(2)}$ $(\bar{\mathcal{O}}_B^{(2)})$ and $\mathcal{O}_A^{(2)}$ $\bar{\mathcal{O}}_A^{(2)}$ are $\bar{Q}_B$ (Q_B) and $\bar{Q}_A$ (Q_A) exact.

The significance of the descend p-form operators is that one can integrate them over closed p-cycles C_p of the WS (or more general the topological field theory space-time) to obtain *non-local operators* $\mathcal{O}(C_p) = \int_{C_p}\mathcal{O}^{(p)}$, which are automatically Q closed, because of Stokes theorem $[Q,\mathcal{O}(C_p)]_\pm = \int_{C_p}[Q,\mathcal{O}^{(p)}]_\pm = \int_{C_p}\mathrm{d}\mathcal{O}^{(p-1)} = \int_{\partial C_p}\mathcal{O}^{(p-1)} = 0$. Reversed use of Stokes theorem shows that the topological equivalence class of $\mathcal{O}(C_p)$ depends only the homology class of C_p. For a $p-1$ chain S with $C_p - C'_p = \partial S$ the difference $\mathcal{O}(C_p) - \mathcal{O}(C'_p) = \int_{\partial S}\mathcal{O}^{(p)} = \int_S \mathrm{d}\mathcal{O}^{(p)} = [Q,\int_S\mathcal{O}^{(p+1)}]_\pm$ is Q exact.

What is of importance is that integrals of the two form operators $\int_\Sigma \mathcal{O}_i^{(2)}$ defined above can be added to the topological action as deformations preserving the (2, 2) world-sheet supersymmetry formally as

$$S = \int_\Sigma \mathrm{d}z^2 \mathcal{L}_0 + \sum_{i=1}^{r} t^i \int_\Sigma \mathcal{O}_i^{(2)} . \qquad (3.3.14)$$

After the A twist we can define zero form operators $\mathcal{O}^{(0)}_{w_{i\bar{j}}} = w_{i\bar{j}}\chi^i\chi^{\bar{j}}$, which have $(U(1)_V, U(1)_A)$ charges (0, 2), see Table 5. This charge is offset by $Q_+, \bar{Q}_-$ in (3.3.13), as seen from Table 6 so that $\mathcal{O}^{(2)}_{w_{i\bar{j}}}$ is neutral. As we shall see these operators are associated to elements in $H^{1,1}(M)$ (3.4.3), (3.4.4). Similarly the operators associated to elements in $A \in H^1(M, TM)$ (3.5.9) in the B-model $\mathcal{O}^{(0)}_A = w^i_{\bar{j}}\eta^{\bar{j}}\theta_i$ have $(U(1)_V, U(1)_A)$ charge (2, 0) which is offset by Q_+, Q_- so that $\mathcal{O}^{(2)}_A$ in (3.3.13) is neutral. Derivatives w.r.t. to t^i bring down such operators in the correlation functions and neutrality implies that arbitrary derivatives do no violate any selection rule. Generically this extends the theory to a family of theories. In the above discussion we omitted the consideration of $w_{ij}\chi^i\chi^j \leftrightarrow H^{2,0}(M)$ in the A-model and bi-vectors $w^{ij}\theta_i\theta_j \leftrightarrow H^0(M, \Lambda^2 TM)$ as these cohomology groups are trivial on manifolds with strict $SU(3)$ holonomy.[98] Perturbations w.r.t. the full set of operators have been considered in [27, 319].

3.3.2 A First Look at the Metric (In)dependence and Topological String Theory

In a topological theory the correlation functions are not only formally position independent, but decouple formally from variations of the worldsheet metric $h^{\mu\nu}$. Classically the energy momentum tensor $T_{\mu\nu} = \frac{1}{\sqrt{h}}\frac{\delta S}{\delta h^{\mu\nu}}$ is the generator of those variations. From the first order variation of the weight factor e^S one gets a dependence of a correlation function on metric variations $\delta h^{\mu\nu}$

$$\delta_h \langle \mathcal{O} \rangle_g = \langle \mathcal{O} \int_{\Sigma_g} \sqrt{h} \mathrm{d}^2\sigma \delta h^{\mu\nu} T_{\mu\nu} \rangle_g . \qquad (3.3.15)$$

In a topological theory $\delta_h \langle \mathcal{O} \rangle_g = 0$ does not require that $T_{\mu\nu} = 0$, but in virtue of (3.3.9) that it is exact

$$T_{\mu\nu} = \{Q, G_{\mu\nu}\} . \qquad (3.3.16)$$

[98] A slight modification of the twisting procedure makes the descend operators to these fields neutral [210].

This structure ensures general covariance or *topological invariance*. It plays a key role in covariant quantization of string theory, where $Q^2 = 0$ is the BRST operator and the part of $G_{\mu\nu}$ is played by the antighost field $b_{\mu\nu}$. It is also is the starting point of closed string field theory formulations [324]. One can have topological invariance independently of conformal invariance and also independently of the decoupling between ghost and matter sector [324]. For instance the A model relies on this structure and can be defined on Kähler manifolds on which the σ model is not conformally invariant.

In string theory we integrate the world-sheet metric h of Σ_g over all possible choices $\mathcal{H}_g$. Some review references for the following short account of the metric dependence are [91, 119, 279] from the physical and [10] from the mathematical perspective. Classically the integral over h is invariant under diffeomorphism and Weyl- and conformal transformations of the metric

$$\tilde{h}_{ab}(\tilde{\sigma}) = \exp[2\omega(\sigma)]\frac{\partial\sigma^c}{\partial\tilde{\sigma}_a}\frac{\partial\sigma^d}{\partial\tilde{\sigma}_b}h_{cd} \,. \tag{3.3.17}$$

These "gauge" invariances are present at quantum level in critical string theory, which requires an anomaly cancellation for the latter. The integral over the metric hence contains huge gauge orbits over the diffeomorphism- and the Weyl group, which we divide from the path integral measure and consider

$$\mathcal{M}_g = \mathrm{LGT}\backslash\mathcal{H}_g/(\mathrm{diff}_0 \times \mathrm{Weyl})_g = \mathrm{LGT}\backslash\mathcal{T}_g \,. \tag{3.3.18}$$

Large gauge transformations (LGT) refer to discrete diffeomorphism of Σ_g not connected to the identity the so called *mapping class group* $\mathrm{LGT} = \frac{\mathrm{diff}}{\mathrm{diff}_0}$, which does not affect the dimension or other local properties in the interior of $\mathcal{M}_g$. Focussing on the latter means considering the Teichmüller space $\mathcal{T}_g = \mathcal{H}_g/(\mathrm{diff}_0 \times \mathrm{Weyl})$. Locally near a reference metric h^0_{ab} we can linearize the problem and once this is done it is easy to see the key property that this moduli space is *finite dimensional*. Infinitesimal Weyl and diffeomorphism transformations are read of from (3.3.17)

$$\begin{aligned}\tilde{\delta}h_{ab} &= 2\delta\omega h_{ab} - \nabla_a\delta\sigma_b - \nabla_b\delta\sigma_a \\ &= (2\delta\omega - \nabla_c\delta\sigma^c)h_{ab} - 2(P_1\delta\sigma)_{ab}\end{aligned} \tag{3.3.19}$$

with $(P_1\delta\sigma)_{ab} = \frac{1}{2}(\nabla_a\delta\sigma_b + \nabla_b\delta\sigma_a - h_{ab}\nabla_c\delta\sigma^c)$. The scalar product for the linearized metric deformations $\delta^i h_{ab}$ near $h^{(0)}_{ab}$ is

$$G^{ij} = \langle\delta^i h_{ab}|\delta^j h_{ab}\rangle = \int_\Sigma \mathrm{d}^2\sigma\sqrt{h}\,\delta^i h_{ab}\,\delta^j h^{ab} \,, \tag{3.3.20}$$

where $\delta^i h^{ab} := h^{(0)ac}h^{(0)bd}\delta^i h_{cd}$ is compatible with the first order approximation. It has a straightforward generalization for other tensors on Σ transforming in $(\otimes^q_{i=1}T\Sigma)\otimes(\otimes^p_{i=1}T^*\Sigma)$ and allows us to define the adjoint of linear operators such

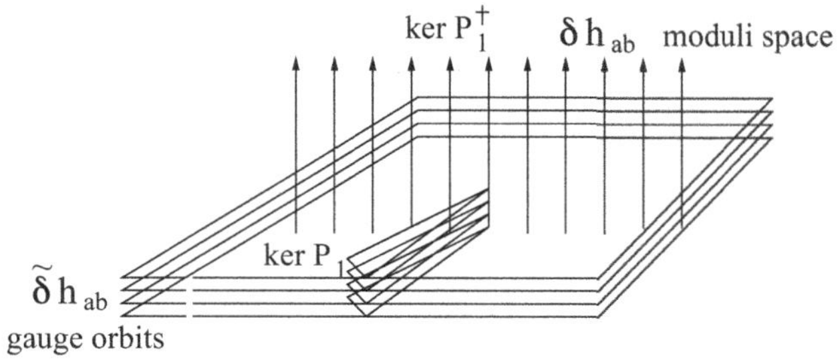

Fig. 7 Schematic of the objects in the linearisation of the metric variations

as P_1, see [257]. Locally $\mathcal{T}_g$ is parametrized by the linear changes δh_{ab} of the metric, which are orthogonal to $\tilde{\delta} h_{ab}$ of (3.3.19), i.e. $0 = \langle \delta h_{ab} | \tilde{\delta} h_{ab} \rangle = \langle \delta h_{ab} | (2\delta\omega - \nabla \cdot \delta\sigma) h_{ab} \rangle - 2\langle \delta h_{ab} | (P_1 \delta\sigma)_{ab} \rangle = \langle h^{ab} \delta h_{ab} | (2\delta\omega - \nabla \cdot \delta)\sigma \rangle - 2\langle (P_1^\dagger \delta h)_b | \delta\sigma_a \rangle$. Up to a small subtlety (dependence), which we discuss below, the free variation of $\delta\sigma_a$ and $(2\delta\omega - \nabla \cdot \delta\sigma)$ span $T^*\Sigma$ and the space of functions on Σ so that the required orthogonality enforces the conditions

$$h^{ab} \delta h_{ab} = 0, \qquad (P_1^\dagger \delta h)_b = 0 \, . \tag{3.3.21}$$

The first is tracelessness of δh_{ab} and in a hermitian gauge choice $h^0_{z\bar{z}}$ we see in [257] that the second means holomorphicity of δh_{ab}. I.e. $\delta h_{zz}(z) = \phi(z)_{zz}$ are components of *holomorphic quadratic differentials*. Holomorphicity of a quadratic differentials in one complex dimension is equivalent to harmonicity and the spectrum of the Laplacian is finite on compact Σ, which establishes this *key property*.

It is easy to connect this to the discussion in section "First Order Complex Structure Deformations". If we pick a metric $h^0_{z\bar{z}}$ we can define from ϕ^* the components of the so called Beltrami differentials $\mu_{\bar{z}}^{z} = h^{\bar{z}z} \phi^*_{\bar{z}\bar{z}}$. Holomorphicity of ϕ implies that $\mu_{\bar{z}}^{z} \mathrm{d}\bar{z} \frac{\partial}{\partial z} \in H^1(T\Sigma)$ is a harmonic representatives. Section "First Order Complex Structure Deformations" uses Čech-cohomology to ignore trivial changes of the metric by complex reparametrizations, which relates by (2.3.34) to the gauge condition $(P_1^\dagger \delta h)_b = 0$. To summarize can span the tangent space $T\mathcal{M}_g$ of the complex moduli space by $\mu_{\bar{z}}^{k\; z}(z) \mathrm{d}\bar{z} \frac{\partial}{\partial z}$ and the cotantgent space $T^*\mathcal{M}_g$ by $\phi^{(k)}_{zz} \mathrm{d}z\mathrm{d}z$ with $k = 1, \ldots, h^1(T\Sigma)$. For the a hermitian choice $h_{z\bar{z}}$ of the metric the pairing (3.3.20) becomes a Kähler metric $G^{i\bar{j}} = \int_\Sigma \mathrm{d}^2 z (h^{z\bar{z}})^2 \phi^i \phi^{*\,\bar{j}}$ called the *Weil-Peterson metric*. We depict the objects used in the definition of $\mathcal{M}_g$ in Fig. 7.

Let us come to the small subtlety mentioned above. If $\delta\sigma^a$ is in the kernel of P_1, i.e. $(P_1 \delta\sigma)_{ab} = 0$ we may pick a $\delta\omega$ so that $\langle \delta h_{ab} | \tilde{\delta} h_{ab} \rangle = 0$, without restricting δh_{ab}. Such vector fields $\delta\sigma_a$ in the kernel of P_1 are elements of $H^0(T\Sigma)$, appropriately called *conformal Killing* fields, as they don't change the conformal class of h_{ab}. So apart from restricting changes of the metric to complex structure changes only, which is the main effect of the division by the gauge group, we have to subtract these null vectors because they appear in the numerator of (3.3.18).

Hence the *expected dimension* of $\mathcal{M}_g$ is $h^1(T\Sigma) - h^0(T\Sigma)$, which we calculate in Appendix 2 by *Hirzebruch-Riemann-Roch* (A2.11) to be $3g - 3$.

To avoid the peculiarities of $h^0(T\Sigma) \neq 0$ (3 and 1 for $g = 0$ and $g = 1$) consider $g > 1$ and let $z^a =: m^a$, $a = 1, \ldots, 3g - 3$ the complex structure variables of Σ. We can describe then a first order deformation of the metric modulo Weyl and diffeomorphisms as $\int_\Sigma \mathrm{d}^2\sigma\sqrt{h}\tilde{\delta} h^{ab} T_{ab} = \int_\Sigma \mathrm{d}^2 z \mu^{(a)\,z}_{\bar{z}} \delta m^a T_{zz} + \bar{\mu}^{a\,\bar{z}}_{z} \delta\bar{m}^a \bar{T}_{\bar{z}\bar{z}}$ and if we insert that in (3.3.15) we conclude that

$$\frac{\partial}{\partial m^a}\langle \mathcal{O}\rangle_g = \langle \mathcal{O}\int_\Sigma \mathrm{d}^2 z \mu^{a\,z}_{\bar{z}} T_{zz}\rangle_g =: \langle \mathcal{O} T^a\rangle_g \tag{3.3.22}$$

and similarly $\frac{\partial}{\partial \bar{m}^a} = \langle \mathcal{O}\bar{T}^a\rangle_g$. Equation (3.3.16) is strictly true, so the argument that cohomological states and the vacuum are Q closed would make *topological string theory* completely metric independent and therefore *trivial*! However the argument involving the invariance of the vacuum fails, because the measure on the moduli space of higher genus Riemann surfaces, which is part of the vacuum definition is not Q closed. It is a real $6g - 6$ form μ_g for surfaces of $g > 1$ and the argument fails in a very specific way. If we act with Q on it, it gives an exact form, as we will see in detail in Sect. 4.1. This is like a descend equation, but with the exterior derivative acting in the moduli space direction. By Stokes or rather Dolbeaults theorem the contribution to the integral can then only come from the *boundary of* $\mathcal{M}_g$, which represents degenerate Riemann surfaces. If the vacuum is not Q closed we cannot trust the argument about position independence either. In the moduli space $\mathcal{M}_{g,n}$ with insertion of n operators the codimension one locus, where two operators coincide is part of the boundary components. Its contribution has to be taken into account by so called *contact terms*. Most of what topological string theory is about is the organizing of the contributions of these boundaries. The questions which boundaries do give contributions leads to the stable compactifications on $\overline{\mathcal{M}}_{g,n}$ in which only the boundary components are included, which are in complex codimension one. These facts will govern the coupling of the A and the B-model to WS gravity as discussed in Sect. 4.1.

This section sketched the leap that one can take in topological string theory from a hopeless looking path integral over $\mathcal{D}h$ to essentially a combinatorial problem. The linear approximations to the moduli space of Σ_g scratched the surface of this subject by one ϵ to be exact. We have not established global properties including existence. We will say more about that for Calabi-Yau manifold in Sect. 2.4.2 and leave the reader in the case of Riemann surfaces with the literature [10].

3.4 The Topological A-Model

As mentioned above the gauged $U(1)_V$ symmetry becomes never anomalous and this topological model can be defined on any Kähler manifold.

3.4.1 *A Model Without Worldsheet Gravity*

In this section we want to describe the operators and correlation functions of topological A topological and their relation to the geometry of the target space M. We denote the anticommuting scalars from Table 5 $\chi^i := \psi^i_-$ and $\chi^{\bar\imath} := \bar\psi^{\bar\imath}_+$ and the one forms i.e. sections of K and $\bar K$ by $\rho^{\bar\imath}_z = \bar\psi^{\bar\imath}_-$ and $\rho_{\bar z} := \psi^i_+$. The action becomes then

$$L = 2t \int d^2 z \Big(g_{i\bar\jmath} \partial_\nu x^i \partial^\nu x^{\bar\jmath} + i\epsilon^{\mu\nu} b_{i\bar\jmath} \partial_\mu x^i \partial_\nu x^{\bar\jmath} - i g_{i\bar\jmath} \rho^{\bar\jmath}_z D_{\bar z} \chi^i \\ + i g_{i\bar\jmath} \rho^i_{\bar z} D_z \chi^{\bar\jmath} - \frac{1}{2} R_{i\bar k j\bar l} \rho^i_{\bar z} \chi^j \rho^{\bar k}_z \chi^{\bar l} \Big) , \quad (3.4.1)$$

where we added the term involving the antisymmetric 2-form $b_{i\bar\jmath} \in H_2(M, Z)$, which plays an important rôle in the bosonic sector of the topological A model. Supersymmetry $\delta = \bar\epsilon_- \bar Q_+ + \epsilon_+ Q_-$ acts by

$$\begin{aligned} &\delta x^i = \epsilon_+ \chi^i, && \delta x^{\bar\imath} = \bar\epsilon_- \chi^{\bar\imath} \\ &\delta\rho^i_{\bar z} = 2i\bar\epsilon_- \partial_{\bar z} x^i + \epsilon_+ \Gamma^i_{jk} \rho^j_{\bar z} \chi^k, && \partial\chi^{\bar\imath} = 0 \\ &\delta\chi^i = 0, && \delta\rho^{\bar\imath}_z = -2i\bar\epsilon_+ \partial_z x^{\bar\imath} + \bar\epsilon_- \Gamma^{\bar\imath}_{\bar\jmath\bar k} \rho^{\bar k}_z \chi^{\bar\jmath} \end{aligned} \quad (3.4.2)$$

with $\delta^2 = 0$. There is a fixpoint of δ on the fermionic zero mode configuration with x^i a holomorphic map $x : \Sigma_g \to M$, i.e. $\partial_z \bar x^{\bar\jmath} = \partial_{\bar z} x^i = 0$, on which the path integral will localize by the fermionic zero mode integration, so that the bosonic integration reduced to a integration over the moduli space $\mathcal{M}$ of such holomorphic maps.[99] This moduli space will be labeled by the following topological data: the genus g of Σ_g and the homology class $\beta = [x_*(\Sigma_g)] \in H_2(M, Z)$ of the image of Σ_g in M. The 0-form correlation observables are combinations of x^i, $x^{\bar\imath}$ and χ^i, $\chi^{\bar\imath}$ the latter are anticommutating operators and can be identified with the forms on M, i.e $\chi^i \leftrightarrow dx^i$ and $\chi^{\bar\imath} \leftrightarrow dx^{\bar\imath}$ One checks that under this correspondence Q_- and $\bar Q_+$ are identified with the exterior derivatives of Dolbeault cohomology ∂ and $\bar\partial$. Since then $Q = Q_- + \bar Q_+$ is identified with the deRham operator $d = \partial + \bar\partial$ one can summarize the correspondence between the BRST cohomology of the Q_A and the deRham cohomology of M as follows. For each form

$$W = w_{I_1,\ldots,I_n}(x) dx^{I_1} \wedge \ldots \wedge dx^{I_n} \quad (3.4.3)$$

on M there is a topological operator

$$\mathcal{O}^{(0)}_{W(P)} = w_{I_1,\ldots,I_n}(x) \chi^{I_1} \ldots \chi^{I_n}(P) \quad (3.4.4)$$

[99] In considering only $Q_A = \bar Q_+ + Q_-$, i.e. setting $\epsilon_+ = \bar\epsilon_-$ one neglects structure, which would give information about the individual cohomology groups of $\mathcal{M}$.

of the A-model and the operation of Q_A is identified with the exterior derivative

$$\{Q_A, \mathcal{O}_W\} = -\mathcal{O}_{\mathrm{d}W}\,, \tag{3.4.5}$$

where the form degree n of W is identified with the ghost number of $\mathcal{O}_W$, since χ has ghost number $+1$.

The action can be written as

$$S = it\int_\Sigma \mathrm{d}^2 z\{Q, V\} + t\int_\Sigma x^*(\omega), \qquad \text{with} \qquad V = g_{i\bar{j}}\left(\rho_z^{\bar{i}}\partial_{\bar{z}}x^j + \partial_z x^{\bar{i}}\rho_{\bar{z}}^j\right) \tag{3.4.6}$$

and

$$\int_\Sigma x^*(\omega) = \int_\Sigma \mathrm{d}^2 z\left(\partial_z x^i \partial_{\bar{z}} x^{\bar{j}} g_{i\bar{j}} - \partial_{\bar{z}} x^i \partial_z x^{\bar{j}} g_{i\bar{j}}\right) = \omega\cdot\beta \ge 0, \tag{3.4.7}$$

where ω is the Kähler form $\omega = i g_{i\bar{j}}\mathrm{d}z^i \mathrm{d}z^{\bar{j}}$ and β is the cohomology class $[x_*(\Sigma)]$ of the image of Σ and the positivity holds if ω is in the Kähler cone. If the antisymmetric tensor field is B is non-zero we replace ω by a complexified Kähler form $\omega_c = iB - \omega = i(b_{i\bar{j}} - g_{i\bar{j}})\mathrm{d}z^i\mathrm{d}z^{\bar{j}}$.

The correlation function of the physical operators

$$\langle\prod_{i=1}^n \mathcal{O}_i\rangle_\beta = e^{-it\beta\cdot\omega}\int_{\mathcal{M}_\beta} \mathcal{D}x\mathcal{D}\chi\mathcal{L}\rho e^{-it\{Q,\int V\}}\prod_{i=1}^n \mathcal{O}_i \tag{3.4.8}$$

depends on the metric of M only via the Kähler class ω (or on the complexified Kähler class $\omega_{\mathbb{C}}$). Other metric dependence in particular on the complex structure of M as well as on Σ_g appears in V. However this dependence appears only as a Q exact expression in (3.4.8) and decouples by (3.3.9) from the topological correlation function. Moreover taking the derivative w.r.t. t implies by (3.3.9) that the second factor is independent of t and the correlation can be calculated for ω in the Kählercone for Re $t > 0$ in limit of infinite t i.e. at the classical minimum of the action. This is another way to understand the supersymmetric localization to the fixpoints of the Grassmann symmetries. If we write

$$\begin{aligned} S_B &= \int_\Sigma g_{i\bar{j}}\left(\partial_z x^i \partial_{\bar{z}} x^{\bar{j}} + \partial_{\bar{z}} x^i \partial_z x^{\bar{j}}\right) \\ &= 2\int_\Sigma g_{i\bar{j}}\partial_{\bar{z}} x^i \partial_z x^{\bar{j}} + \int_\Sigma x^*(\omega) \end{aligned} \tag{3.4.9}$$

it is obvious that this minimum is taken at holomorphic maps $\partial_{\bar{z}} x^i = \partial_z x^{\bar{j}} = 0$. This equation requires to specify a holomorphic structure j on Σ_g and one J on M. For fixed J and fixed j there will no maps for $g > 0$. Only if we couple the theory to gravity and integrate over j we have a chance to get contributions from finite dimensional integrals over an infinite series of components of moduli spaces of holomorphic maps, which are labeled by g and the class $\beta \in H^2(M, Z)$. I.e. the path integral collapses to these integrals. It should be stressed that the A model does not require an integrable complex structure J. A symplectic structure and a compatible almost complex structure are sufficient for the above arguments.

Let us discuss the selection rules for $g = 0$ correlators $\langle \prod_{k=1}^n \mathcal{O}_{W_k} \rangle_\beta$. We note from Table 5 and the identification of χ^i and $\chi^{\bar{i}}$ that χ^i has charge $q_l = -1$ and $q_r = 0$ under the left and right $U(1)_{l/r}$ respectively, while $\chi^{\bar{i}}$ has $q_l = 0$ and $q_r = 1$. Because of the splitting of the tangent bundle of $M = T^{(1,0)} \oplus T^{(0,1)}$ we can associate to $\mathcal{O}_{W_k}$ an element in the Dolbeault cohomology group $H^{(p_k,q_k)}$. Since the vector $U(1)_V$ is unbroken in the quantum theory we get a charge conservation constraint $q_V = \sum_{k=1}^n p_k - \sum_{k=1}^n q_k = 0$ for the correlator to respect vector charge conservation. For the classical axial charge we would get naively $q_A = \sum_{k=1}^n p_k + \sum_{k=1}^n q_k = 0$. However the $U(1)_A$ is anomalous. Looking at the kinetic terms of ρ and χ we see that its anomaly is given by the index of the *twisted Dolbeault complex* on Σ (A2.27) which is calculated by the Hirzebruch-Riemann-Roch theorem as explained at the end of Appendix 2 to be

$$\begin{aligned} q_A &= \#(\chi 0 \text{ modes}) - \#(\rho\ 0 \text{ modes}) = 2(h^0(x^*(TM)) - h^1(x^*(TM))) \\ &= 2\int_\Sigma \mathrm{ch}(x^*(TM^{(1,0)}))\mathrm{td}(T\Sigma) = 2(c_1(TM)\cdot\beta + \dim_C M\ (1-g))\,. \end{aligned} \tag{3.4.10}$$

Combining the constraints we get

$$\sum_{k=1}^n q_k = \sum_{k=1}^n p_k = c_1(TM)\cdot\beta + \dim_C M(1-g)\,. \tag{3.4.11}$$

In particular for $g = 0$ we can have a non-vanishing coupling $\langle \mathcal{O}_{W_i}\mathcal{O}_{W_j}\mathcal{O}_{W_k}\rangle$, where all the W_l are $(1,1)$-forms.

With two non-degenerate pairings we can associate a divisor $D_k \in H_4(M)$ to each $W^{(k)}_{(1,1)}$. One can pick a (2, 2) form $W_{(l)\ (2,2)}$ so that $\int_M W^{(k)}_{(1,1)} \wedge W_{(l)\ (2,2)} = \delta^k_l$ as well as $\int_{D_i} W_{(j)\ (2,2)} = \delta^i_j$. If β denotes the cohomology class of the image C of the worldsheet in M then we can write the product $\beta \cdot \omega = 2\pi \sum_{k=1}^{h^{1,1}} t_k d_k$, where $d_k = C \cap D_k$ is the number of intersections of C with D_k or the degree of C w.r.t. D_k. The degree 0 map with $d_k = 0$ for all k is special. It maps the three punctured sphere $\Sigma_{0,3}$ to a point in M. One can always find a representative of $W^{(k)}_{(1,1)}$ that has δ-function support on D_k. This implies that the point in $\mathcal{O}_{W^k}(P_k)$ maps to D_k. If $\Sigma_{0,3}$ maps to a point in M the path integral collapses hence to the intersection

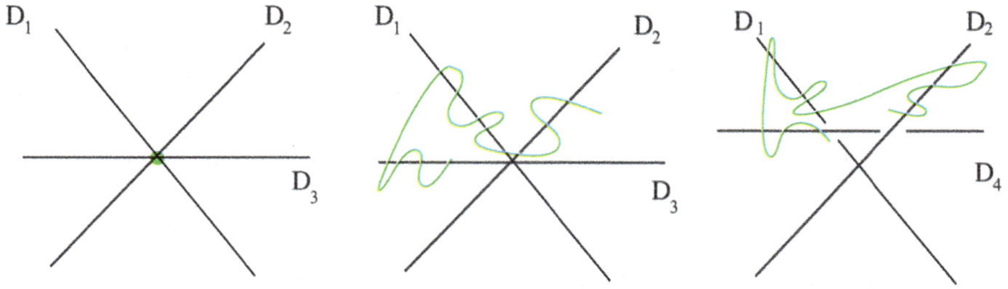

Fig. 8 This figure shows instanton corrections to the coupling C_{123} with $D_1 \cap D_2 \cap D_3 = O(1)$ and C_{124} with $D_1 \cap D_2 \cap D_4 = 0$. From the left to the right we pictured an instanton of degree 0 contributing of $O(1)$ to C_{123}, an instanton of degree $d_1 = 5, d_2 = 3, d_3 = 4$ contributing $\sim q_1^5 q_2^3 q_3^4$ to C_{123} and an instanton of degree $d_1 = 5, d_2 = 4, d_4 = 3$ contributing $\sim q_1^5 q_2^4 q_4^3$ to C_{124}. Roughly speaking for large radii second the coupling C_{124} is expected to be exponentially suppressed against the first C_{123}. The precise statement depends on the growth of $r^{g=0}_{\{d_i\}}(D_i, D_j, D_k)$. Such collective effects of the intantons can be analyzed best in the B-model

number of $D_i \cap D_j \cap D_k$. We define $Q_k = e^{-2\pi i t_k}$, then the correlation function[100] is

$$C_{ijk}(t) = \langle \mathcal{O}_{W_i} \mathcal{O}_{W_j} \mathcal{O}_{W_k} \rangle = D_i \cap D_j \cap D_k + \sum_{\{d_i\} \neq \{0\}} r^{g=0}_{\{d_i\}}(D_i, D_j, D_k) \prod_{i=1}^{h^{1,1}} Q_i^{d_i} \,. \tag{3.4.12}$$

This deformed intersection is piece of the structure known as *quantum cohomology ring* of M. It is a deformation of the classical cohomology ring on M by the parameters Q_k. One needs in general the deformations of all pairings $[m] : H^{\otimes n} \to H$ indexed by $m \in H^*(\mathcal{M}_{0,n+1})$, see [258] and [78] for a review, which we can be provided on the mirror side. Note that the relation to classical intersections in the limit picks a natural normalization of the operators $\mathcal{O}_W$ and of their two-point functions, see Fig. 8.

One collective effect of the instantons corrections is that the correlation functions $C_{ijk}(t)$ behaves smoothly at singularities in codimension two in M as for instance through flop transitions [25, 322].

We note from table 2 and 3 and from (3.3.13) that the $U(1)_V$ as well as the $U(1)_A$ charge of the operator $\mathcal{O}^{(2)}_{W_j}$ vanishes. In view of (3.3.14) this means that non-vanishing derivatives of $C_{jkl}(t)$ such as

$$\frac{\partial}{\partial t^i} \langle \mathcal{O}_{w^j} \mathcal{O}_{w^k} \mathcal{O}_{w^l} \rangle \Big|_{t^i=0} = \langle \mathcal{O}_{w^j} \mathcal{O}_{w^k} \mathcal{O}_{w^l} \int_\Sigma \mathcal{O}^{(2)}_{w^i} \rangle \tag{3.4.13}$$

[100] We abbreviate $\prod_{i=1}^{h^{1,1}} q_i^{d_i} = q^\beta$ in the following.

do exist according to the selection rules. This non-vanishing correlators signal that a non-trivial deformation family exist, but do not contain new information once $c_{jkl}(t)$ is known as function after summing up all intantons or easier from a B-model calculation. By $SL(2, \mathbb{C})$ invariance on S^2 there is a symmetry between fixing any three of the $\{i, j, k, l\}$ points and integrating over the fourth. This implies that

$$\partial_i C_{jkl}(t) = \partial_j C_{ikl}(t) \tag{3.4.14}$$

which is the integrability condition for the existence of a function $\mathcal{F}^{(0)}(t)$ with the property that

$$C_{ijk}(\underline{t}) = \partial_i \partial_j \partial_k \mathcal{F}^{(0)}(t) \ , \tag{3.4.15}$$

where we defined $\partial_i = \frac{\partial}{\partial t^i}$. This is in perfect accordance with facts concerning $\mathcal{F}(t)$ from the analysis of the vector moduli space of $N = 2$ supergravity in 4d, which is identified in type IIA compactifications with complexified Kähler moduli space. This facts can also be established in the complex structure deformation space, see Sect. 2.10.17, which again is identified by mirror symmetry with the complexified Kähler moduli space of the A-model. We should finally note that Eqs. (3.4.13)–(3.4.15) are not written covariantly, but rather in special flat coordinates. Covariant derivatives are discussed in the B-model section.

3.4.2 The A-Model Coupled to Gravity

We will not say much of the A-model coupled to world-sheet gravity except that we explain some of the index theoretical calculations in Appendix 2. In particular the important formula for the virtual dimension of the moduli spaces of higher genus maps in to the class $\beta \in H_2(M, \mathbb{Z})$ has been motivated there as

$$\dim \text{vir}\, \overline{\mathcal{M}}_{g,n}(M, \beta) = \int_\beta c_1(TM) + (\dim M - 3)(1 - g) + n \ . \tag{3.4.16}$$

Physically on Calabi-Yau n-folds there are no insertion operators within the (2, 2) super conformal world sheet theories on higher genus world sheets.[101] The integrated operators like in (3.4.13) do of course not kill any automorphism in (A2.15) and hence do not contribute to the n in (3.4.16), but rather describe the moduli dependence of the corresponding amplitude. Hence $n = 0$ for the topological string on Calabi-Yau manifolds. This highlights the important role that Calabi-Yau manifolds play in this theory, as for them $\dim \text{vir}\, \overline{\mathcal{M}}_{g,n}(M, \beta) = 0$. This

[101] It does make sense to deform the theory by descend operators away from the (2, 2) super conformality. This is used e.g. in the calculation of relative Gromov-Witten invariants in the mathematical context.

implies that in generic situations one has a zero dimensional moduli space of maps, which means that one literally counts maps, with finite orders of automorphism of automorphism divided out, which can make the result of the "counting" rational. In particular non-zero results can be expected in this case for all classes and genera. The generic situation in which the actual dimension is the virtual dimension, i.e. zero, is rarely realised, but there is a virtual obstruction theory which guarantees that the actual dimension is always positive and that there is always a virtual fundamental class against which on can integrate on this positive dimensional space to get a number.

3.5 The Topological B-Model

Since the axial $U(1)_A$, whose gauge connection is added to the spin connection to define the B-model, develops an anomaly of its current proportional to $\int_\Sigma \partial_\mu j_A^\mu \sim \int_\Sigma x^*(c_1(TM))$ the twisted B-model is only consistent for Kähler manifold with vanishing first Chern class, i.e. Calabi-Yau manifolds.

3.5.1 The Topological B Without Worldsheet Gravity

The scalar BRST operator is in this case,

$$Q_B = \bar{Q}_- + \bar{Q}_+ \,, \tag{3.5.1}$$

see Table 6. The scalar fields on the worldsheet are conveniently chosen as

$$\eta^{\bar{i}} := -(\psi_-^{\bar{i}} + \psi_+^{\bar{i}}), \qquad \theta_j := g_{j\bar{i}}(\psi_+^{\bar{i}} - \psi_-^{\bar{i}}) \,, \tag{3.5.2}$$

while the one form fields are

$$\rho_z^i := \psi_-^i \quad \text{of type } (1,0), \qquad\qquad \rho_{\bar{z}}^i := \psi_+^i \quad \text{of type } (0,1). \tag{3.5.3}$$

The supersymmetry transformation $\delta = \bar{\epsilon}\bar{Q}_+ + \bar{\epsilon}\bar{Q}_-$ is obtained by setting $\bar{\epsilon}_+ = -\bar{\epsilon}_- = \bar{\epsilon}$ and $\epsilon_\pm = 0$ in (3.2.9) and using the above field identifications

$$\begin{aligned} \delta x_i &= 0, & \delta x^{\bar{i}} &= \bar{\epsilon}\eta^{\bar{i}} \\ \delta\theta_i &= 0, & \delta\eta^{\bar{i}} &= 0 \\ \delta\rho^i_\mu &= \pm i\bar{\epsilon}\partial_\mu x^i \,. \end{aligned} \tag{3.5.4}$$

The zero form observables $\mathcal{O}^{(0)}$ are now related to forms in $\Omega^{(0,p)}(M, \Lambda^q T^{0,1} M)$ with the identification of the scalar Grassmann fields on the worldsheet to forms and vectors on M $\eta^{\bar{\imath}} \leftrightarrow \mathrm{d}x^{\bar{\imath}}$ and $\theta_i \leftrightarrow \frac{\partial}{\partial x^i}$. I.e. to each form on M of type

$$W = \omega_{\bar{\imath}_1 \ldots \bar{\imath}_p}^{j_1 \ldots j_q} \mathrm{d}x^{\bar{\imath}_1} \wedge \ldots \wedge \mathrm{d}x^{\bar{\imath}_p} \frac{\partial}{\partial x^{j_1}} \wedge \ldots \wedge \frac{\partial}{\partial x^{j_q}} \tag{3.5.5}$$

we associate a 0-form operator on Σ

$$\mathcal{O}_W^{(0)} = \omega_{\bar{\imath}_1 \ldots \bar{\imath}_p}^{j_1 \ldots j_q} \eta^{\bar{\imath}_1} \ldots \eta^{\bar{\imath}_p} \theta_{j_1} \ldots \theta_{j_q} . \tag{3.5.6}$$

One checks that the Q_B operator is identified with the Dolbeault operator $\bar{\partial}$ which increases the anti holomorphic form degree

$$0 \stackrel{\bar{\partial}}{\to} \Omega^{00}(M, \Lambda^q T^{1,0} M) \stackrel{\bar{\partial}}{\to} \Omega^{01}(M, \Lambda^q T^{1,0} M) \stackrel{\bar{\partial}}{\to} \ldots \stackrel{\bar{\partial}}{\to} \Omega^{0d}(M, \Lambda^q T^{1,0} M) \stackrel{\bar{\partial}}{\to} 0 . \tag{3.5.7}$$

and one has with $\{Q_B, \mathcal{O}_W^{(1)}\} = -\mathcal{O}_{\bar{\partial} W}^{(0)}$ the identification

$$H^*_{Q_B} = \frac{\mathrm{Ker}\ Q_B}{\mathrm{Im}\ Q_B} = \bigoplus_{p,q=0}^{d} H^{0,p}(M, \Lambda^q T^{1,0} M) . \tag{3.5.8}$$

The selection rules from the R-symmetries are as before $\sum_i p_i = \sum_i q_i = d(1-g)$. It follows that for $g = 0$ we have again only one possibility of a non-vanishing three point function $\langle \mathcal{O}_{A^{(i)}} \mathcal{O}_{A^{(j)}} \mathcal{O}_{A^{(k)}} \rangle$, if we consider three local operators $\mathcal{O}_{A^{(k)}}$ associated to

$$A^{(k)} = \omega_{\bar{j}}^{(k)\ i} \mathrm{d}x^{\bar{j}} \frac{\partial}{\partial x^i} \in H^1(M, T^{1,0} M) . \tag{3.5.9}$$

Equation (3.5.4) shows that there is a fixpoint of the fermionic symmetry at the constant maps

$$\partial_\mu x^i = 0 . \tag{3.5.10}$$

We expect therefore that all contributions to the path integral are localized to constant maps. *This is the main simplification in the B-model.* For constant maps Σ_g is mapped to a point in M. These maps are of course much easier to control then the holomorphic maps of the A-model and in particular they are not affected by the sizes, i.e. Kähler parameter of M. The B-model without worldsheet gravity is rather like a Kaluza-Klein reduction of a point particle theory. By writing the action in the form

$$S = it \int_\sigma \{Q_B, V\} + tW \tag{3.5.11}$$

with

$$V = g_{i\bar{j}}(\rho_z \partial_{\bar{z}} x^{\bar{j}} + \rho^i_{\bar{z}} \partial_z x^{\bar{j}}) \tag{3.5.12}$$

and

$$W = \int_{\Sigma_g} (-\theta_i D\rho^i - \frac{i}{2} R_{i\bar{i}j\bar{j}} \rho^i \wedge \rho^j \eta^{\bar{i}} \theta_k g^{\bar{j}k}) \tag{3.5.13}$$

one can conclude the following: W does not depend on the complex structure of Σ, which decouples from the B-model at genus 0. The Kähler variations of W are Q_B exact and decouple likewise. It is also t independent as t can be absorbed in a field redefinition in W. For more details see [319]. In the off shell formulation of [247, 249] one can simply write the complete action as Q commutator $S = \{Q_B, \tilde{V}\}$ which makes the above points more obvious.

Since the fixpoints of the fermionic maps of the B-model are constant maps, mapping all Σ to a point in the Calabi-Yau manifold M, their moduli space contains M and in the special case of the three punctured sphere, i.e. in the case of the three point function it is actually M, since these three points can be fixed on S^2 by an $SL(2, \mathbb{C})$ transformation and the sphere itself has no complex deformations. For this reason all we have to find is a canonical measure on M, which we integrate over M to get the three point function. Using Kaluza Klein reduction methods this measure has been found in [298]

$$C_{ijk}(z) = \langle \mathcal{O}^{(0)}_{A^i} \mathcal{O}^{(0)}_{A^j} \mathcal{O}^{(0)}_{A^k} \rangle = \int_M \Omega \wedge A^{(i)\ i_1}_{\bar{j}_1} A^{(j)\ i_2}_{\bar{j}_2} A^{(k)\ i_3}_{\bar{j}_3} \Omega_{i_1 i_2 i_3} \mathrm{d}x^{\bar{j}_1} \wedge \mathrm{d}x^{\bar{j}_2} \wedge \mathrm{d}x^{\bar{j}_3} \,. \tag{3.5.14}$$

Here $\Omega(z)$ is the unique non-vanishing holomorphic (3, 0) form, which exists on every Calabi-Yau, see Sect. 2.4.1. Using the isomorphism (2.4.15) $A^{(j)} \mapsto \hat{A}^{(j)}$ we can define a non-holomorphic two point function

$$i g_{i\bar{j}} = \int_M \hat{A}^{(i)} \wedge \overline{\hat{A}}^{(\bar{j})} \,. \tag{3.5.15}$$

The definitions (3.5.14) and (3.5.15) are a special case of (2.5.10) and (2.5.6) respectively.

3.6 The tt^* Structure

In this section we want to discuss the tt^* structure, which combines the geometrical structure of special geometry of Sect. 2.5 with the structure of the $\mathcal{N} = (2, 2)$ world-sheet theories, which is more general, and prepares for the structures we need to understand the properties of higher genus world-sheet theories that lead to the holomorphic anomaly equations in Sect. 4.

3.6.1 tt^* Equations, Special Geometry and Contact Terms

The tt^* equations describe the geometry of the ground states of $\mathcal{N} = (2,2)$ two dimensional theories. The construction does not require necessarily conformal invariance, but rather the following structure. A nilpotent operator Q and its adjoint $Q^\dagger$

$$\{Q, Q^\dagger\} = H \tag{3.6.1}$$

and a conserved fermion number. Q and its adjoint $Q^\dagger$ define rings of cohomological operators $\mathcal{R}$ and $\mathcal{R}^*$ respectively. The advantage of the approach is that it derives the relevant geometry with minimal assumptions. E.g. special Kähler geometry follows just with an additional requirement on integral charge conservation for the A-model the B-model and even the more exotic cases introduced in [128] follow from the construction. To make contact with the previous sections this can be realized as

$$Q = \begin{cases} Q_A = Q_- + \bar{Q}_+, & \mathcal{R} = (a,c) \\ Q_B = \bar{Q}_- + \bar{Q}_+, & \mathcal{R} = (c,c) \end{cases} \qquad Q^\dagger = \begin{cases} Q_A^\dagger = \bar{Q}_- + Q_+, & \mathcal{R}^* = (c,a) \\ Q_B^\dagger = Q_- + Q_+, & \mathcal{R}^* = (a,a) \end{cases} \tag{3.6.2}$$

As explained we have to twist the theories by identifying the corresponding A^R gauge connection with the spin connection. Since only the fermion number must be conserved [68] one needs only a Z_2 anomaly free subgroup of the $U(1)_R$-currents. The tt^* geometry is applicable to $N = (2,2)$ 2d field theories with marginal (conformal) but also relevant (non-conformal) deformations. While these theories might not have a geometrical target space realization, it is still[102] useful to think of a formal correspondence to the deRham (Dolbeault) cohomology on a manifold M with $(Q, Q^\dagger, H) \sim (\mathrm{d}, \mathrm{d}^*, \Delta)$

The Ramond *vacuum states*, compare (3.1.16), are defined by

$$Q|\alpha\rangle = 0, \qquad |\alpha\rangle \sim |\alpha\rangle + Q|\lambda\rangle. \tag{3.6.3}$$

Such states play the rôle of harmonic forms. We call the space of these vacua $\mathcal{H}$. The *operator state correspondence* of 2d QFT associates to every operator $\phi \in \mathcal{R}$ acting on a any vacuum state α a state $|\phi\rangle_\alpha = \phi|\alpha\rangle$. In order to avoid too many indices we call the zero-form operators $\mathcal{O}^{(0)} = \phi$ and the two form operators $\mathcal{O}^{(2)} = \mathcal{O}$. If

[102]For σ model on M this formal correspondence becomes an actual correspondence.

$Q^+|\alpha\rangle = 0$ then $|\phi\rangle_\alpha = \phi|\alpha\rangle$ is closed, the Hodge decomposition (2.3.33) applies $|\phi\rangle_\alpha = |\phi_0\rangle_\alpha + Q|\phi_-\rangle_\alpha + Q^\dagger|\phi_+\rangle_\alpha$ and by that we get a map

$$\Pi_h : |\phi\rangle_\alpha \mapsto |\phi_0\rangle_\alpha \tag{3.6.4}$$

from $\mathcal{R}$ to $\mathcal{H}$. If α is fixed and as will soon see there is *preferred choice*, we can find a canonical map from the ring $\mathcal{R}$ to the Ramond-Ramond groundstates. Moreover every $\phi \in \mathcal{R}$ induces a map

$$\phi : |\alpha\rangle \mapsto |\phi_0\rangle_\alpha \tag{3.6.5}$$

from $\mathcal{H}$ to $\mathcal{H}$. Everything we said from Eq. (3.6.3) on, could have been said verbatim for the conjugated sector defined by $Q^\dagger$. In particular we get for the same choice of α a second basis of $\mathcal{H}$, which we call $|\bar{\imath}\rangle$, $\bar{\jmath} = 1, \ldots, r$. If one has unbroken $U(1)_{R/L}$ symmetries as in Sect. 3.1 one could single out $|\alpha\rangle$ as the lowest charge state in the Ramond-Ramond groundstate.

The following path integral argument requires only conserved fermion number. In the operator approach [17, 92] to 2d field theory one defines a state in the Hilbert space H of 2d theory by the path integral over a half sphere HS^2 bounding an S^1. Parametrize the S^1 by θ and denote the fields generically by $\phi(\theta)$. The path integral is a functional of the boundary field configuration $\phi(\theta) \in L^2$ on the S^1 and defines a state $|\phi\rangle$ in H as in (3.6.7). Anti periodic boundary conditions for fermionic states on contractible loops as S^1 on HS^2 are the natural boundary conditions in the path integral so that (3.6.7) does not yield periodic Ramond-Ramond states in H. However the connection A^R_μ of the gauged $U(1)$ R-symmetry couples to the fermion number with charge $\frac{1}{2}$, i.e. acts like a spin connection ω_μ. When one transports the fermion along the S^1, the connection is integrated to a Wilson loop phase rotation acting on the fermionic state as

$$e^{\pi i \oint_{S^1} \omega \mathrm{d}x} = e^{\pi i \int_{HS^2} \mathrm{d}\omega} = e^{\pi i \int_{HS^2} \frac{R}{2\pi i}\sqrt{h}} = e^{\pi i \int_{HS^2} c_1(T)} = -1, \tag{3.6.6}$$

which rectifies the periodicity. A projection to the Ramond-Ramond groundstates at the boundary can now be achieved by attaching a cylinder of length T to HS^2, see Fig. 9. Call the combined surface $H_T S^2$. The "evolution" of a state $|\phi\rangle$ defined by the original boundary S^1 of HS^2 to the far boundary is described by $e^{-HT}|\phi\rangle$. If the length T of the cylinder goes to infinity only the groundstates in $\mathcal{H}$ survive, because they have 0 as energy eigenvalue of H, cff (3.1.17).

After this preparation we can define the path integral version of a projector (3.6.4)

$$|i\rangle = \lim_{T \to \infty} \int \mathcal{D}\phi e^{-\int_{H_T S^2} L(\phi)} \phi_i = \Pi_p(\phi_i)\,. \tag{3.6.7}$$

The $T \to \infty$ limit makes the projector only sensitive to cohomological information of ring states $\phi \in \mathcal{R}$ or $\bar{\phi} \in \mathcal{R}^*$. Exact pieces have non-zero energy and are

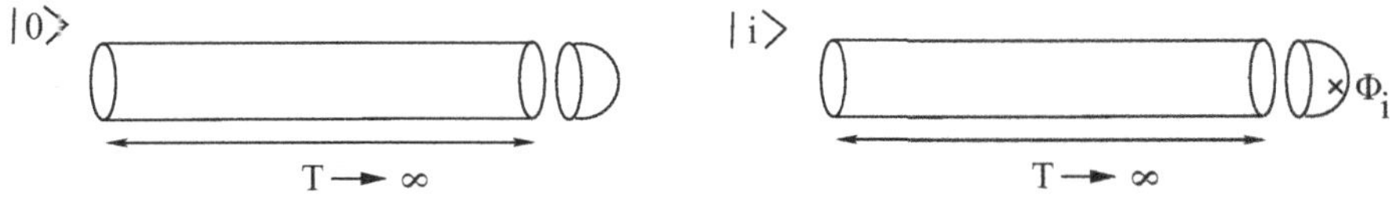

Fig. 9 Path integral projectors to the Ramond-Ramond ground states $\mathcal{H}$

completely suppressed. Note that $\Pi(\mathbf{1}) = |0\rangle$ defines a *preferred vacuum* state. We call the image of a basis $\phi_i \in \mathcal{R}$, $i = 0, \ldots, r$ with $\Phi_0 = \mathbf{1}$ in $\mathcal{H}$ the *topological basis* $|i\rangle = \Pi_p(\phi_i)$. By the operator state correspondence we can also represent the rings (3.3.11) on the vacuum states

$$\phi_i |j\rangle = C_{ij}^k |k\rangle \tag{3.6.8}$$

The path integral (3.6.7) with insertions of $\bar{\phi}_i \in \mathcal{R}^*$ defines the *anti-topological basis* $|\bar{\imath}\rangle = \Pi_p(\bar{\phi}_i)$. The two basis of $\mathcal{H}$ namely $|i\rangle$ and $|\bar{\imath}\rangle$ must be related by a linear transformation, the real structure,

$$|i\rangle = M_i^{\bar{\imath}} |\bar{\imath}\rangle \ . \tag{3.6.9}$$

The CPT theorem of the 2d field theory states that the effect of complex conjugating all expressions in (3.6.7) sends $|i\rangle \to |\bar{\imath}\rangle$, i.e. $|\bar{\imath}\rangle = M_{\bar{\imath}}^j |j\rangle$ which implies $MM^* = 1$. One has a topological bilinear pairing

$$\langle i|j\rangle = \eta_{ij} \tag{3.6.10}$$

and an hermitian bilinear pairing called the tt^* metric

$$\langle \bar{\imath}|j\rangle = g_{\bar{\imath}j} \ , \tag{3.6.11}$$

which are in an obvious way related by the real structure

$$g^{\bar{l}i}\eta_{ij} = M_j^{\bar{l}} \ . \tag{3.6.12}$$

Note that $\langle i| \neq (|i\rangle)^\dagger$. Both bilinear pairings can be defined by the path integral as in Fig. 10. These objects are topological to different extend. Changing the representative of the Q cohomology class $|i\rangle \mapsto |i\rangle + Q|\lambda\rangle$ or $\langle j| \mapsto \langle j| + \langle\lambda|Q$ will do nothing in $\langle i|j\rangle$ as $|j\rangle$ and $\langle i|$ are Q closed. Due to (3.3.10) the pairing η_{ij} is independent of the position. That is true for all length/diameter ratios of the cylinder, i.e. the cylinder is not needed at all in the definition. For the pairing $g_{\bar{\imath}j}$ with $\langle\bar{\imath} \mapsto \langle\bar{\imath}| + \langle\lambda|Q^\dagger$ and $|i\rangle \mapsto |i\rangle + Q|\lambda\rangle$ the argument does not apply as $|j\rangle$ is not $Q^\dagger$ and $\langle\bar{\imath}|$ not Q closed. However from (3.6.1) and $Q|\lambda\rangle \neq 0$ ($\langle\lambda|Q^\dagger \neq 0$) follows that these exact states have positive energy. The only states with zero energy are R-R vacua. I.e. in the case of $g_{\bar{\imath}j}$ we need the $T \to \infty$ limit to define a topological quantity.

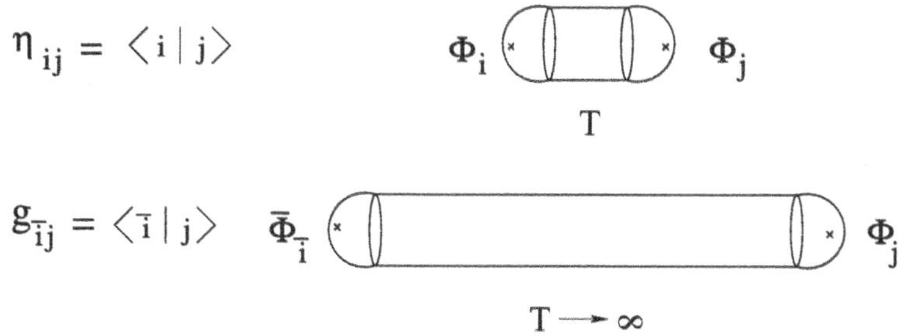

Fig. 10 Path integral representation of the topological pairing η_{ij} and the topological-antitopological pairing $g_{\bar{i}j}$

Locally the tangent space of the (t, t^*) moduli space is spanned by elements from $\mathcal{R}(t)$ and $\mathcal{R}^*(t^*)$. It is clear that the pairing η_{ij} depends only on the t moduli. Moreover one shows that as metric it is completely flat, i.e. all components of the curvature tensor vanish similar as in $d < 1$ strings [95]. One can therefore find coordinates which make the metric η_{ij} constant. This defines the moduli dependent basis of $\mathcal{R}$. As it is clear from the construction of the basis $|i\rangle$ and $|\bar{i}\rangle$ via the projection of moduli dependent elements in the rings $\mathcal{R}$ and $\mathcal{R}^*$ they will depend on the moduli $\underline{m} = (\underline{t}, \underline{t}^*)$. In the Landau-Ginzburg approach [305] η_{ij} is explicitly defined in terms of the Landau Ginzburg superpotential as $\eta_{ij} = \mathrm{Res}_W[\phi_i \phi_j]$ with

$$\mathrm{Res}_W[\Phi] = \frac{1}{(2\pi i)^n} \int_\Gamma \frac{\Phi(X)\mathrm{d}X^1 \wedge \ldots \wedge \mathrm{d}X^n}{\partial_1 W \ldots \partial_n W} = \sum_{\mathrm{d}W} \Phi(X)\det{}^{-1}[\partial_i \partial_j W]\,. \tag{3.6.13}$$

Another approach to define η_{ij} is via the supersymmetric Schroedinger equation [67]. We will not dwell deeper into the derivation of (3.6.13), except for remarking that it is a zero dimensional analog of the Griffith residuum expressions (2.8.3), (2.8.8) used in Sect. 2.8.1 to define the periods, with the identification $W = P$.

The tt^* equations describe how the vacuum states in $\mathcal{H}$ vary over the moduli space or deformation space of the theory parametrized by $\underline{m}$. One calls the corresponding bundle also $\mathcal{H}$. Let e_γ be a basis, i.e. a section in $\mathcal{H}$, and denote its connection

$$A^\alpha_{\beta\gamma} = g^{\alpha\kappa} \langle e_\kappa | \partial_\beta | e_\gamma \rangle. \tag{3.6.14}$$

If the basis of $\mathcal{H}$ changes by a "gauge" transformation $|e_\gamma\rangle \mapsto |e'_\gamma\rangle = \Lambda_{\gamma\delta}|e_\delta\rangle$ then the connection undergoes a gauge transformation $A \mapsto \Lambda^{-1} A \Lambda + \Lambda^{-1}\mathrm{d}\Lambda$. Let us consider the perturbation

$$S = \int_\Sigma \mathrm{d}^2 z \mathcal{L}_0 + \sum_i t^i \int_\Sigma \mathrm{d}^2 z \mathcal{O}_i + \sum_{\bar{i}} t^{*\bar{i}} \int_\Sigma \mathrm{d}^2 z \bar{\mathcal{O}}_{\bar{i}}\,, \tag{3.6.15}$$

where the two-form descendants are called $\mathcal{O}_i := \mathcal{O}_i^{(2)}$. It is easy to show that the following mixed indices of this connection vanish in the holomorphic basis. Consider e.g. $A^i_{\bar{i}j}$ using (3.6.42) we can write $A^i_{\bar{i}j} = g^{i\bar{k}}\langle\bar{k}|\partial_{\bar{i}}|j\rangle = \eta^{ik}\langle k|\partial_{\bar{i}}|j\rangle$. By (3.3.12) we can write $\int_\Sigma \bar{\mathcal{O}}_{\bar{i}} = [Q, \Lambda]$ and since ϕ_j is Q closed we can write $\partial_{\bar{i}}|j\rangle = \Pi_h([Q,\Lambda]\phi_j) = Q\Pi_h(\Lambda\phi_j) = Q(\Lambda|j\rangle)$. Since $\langle k|Q = 0$ is closed this expression vanishes

$$A^i_{\bar{i}j} = 0\,. \tag{3.6.16}$$

Similarly one shows that $A^i_{k\bar{j}} = \eta^{il}\langle l|\partial_k|\bar{j}\rangle = 0$.

The metric connection is characterized by

$$0 = D_k g_{i\bar{j}} = \partial_k g_{i\bar{j}} - (\partial_k\langle i|)|\bar{j}\rangle - \langle i|\partial_{\bar{k}}|\bar{j}\rangle = (\partial_k\langle i|)|\bar{j}\rangle\,. \tag{3.6.17}$$

From this and the $\bar{D}_{\bar{k}}$ derivative, we get formulas for A^j_{km} and $A^{\bar{j}}_{\bar{k}\bar{m}}$

$$A^j_{km} = g^{j\bar{j}}\partial_k g_{m\bar{j}}, \qquad A^{\bar{j}}_{\bar{k}\bar{m}} = g^{m\bar{j}}\partial_{\bar{k}} g_{m\bar{m}}\,. \tag{3.6.18}$$

as hermitian connection of g. Indeed the topological basis $|i\rangle$ and the anti-topological basis $|\bar{i}\rangle$ form holomorphic and antiholomorphic sections of the vacuum bundle over the moduli $\underline{m}$ and one gets the vanishing of the following components of the curvature

$$[D_i, D_j] = [\bar{D}_{\bar{i}}, \bar{D}_{\bar{j}}] = 0\,. \tag{3.6.19}$$

The most important relation comes from analyzing the $[D_i, \bar{D}_{\bar{i}}]$ curvature term. Let us do this for definiteness for the B model. Since the twisting (3.1.28) is so that $\bar{Q}_+(z) \sim G^+(z)$ and $\bar{Q}_-(z) \sim \bar{G}^+(z)$ have dimension one, we can define

$$\bar{Q}_+ = \oint \mathrm{d}z G^+(z), \qquad \bar{Q}_- = \oint \mathrm{d}z \bar{G}^+(z)\,. \tag{3.6.20}$$

Here we adopt the notation to use the CFT conventions for the twisted currents. The commutators and anticommutators in the definition of the descendants (3.3.13) can be represented by (3.1.6) as

$$\begin{aligned} \mathcal{O}_i := \mathcal{O}_i^{(2)} &= \{Q_+, [Q_-, \phi_i(u)]\} \sim \oint_{C_u} \mathrm{d}z G^-(z) \oint_{C'_u} \mathrm{d}w \bar{G}^-(w)\phi_{\bar{i}}(u)\,, \\ \bar{\mathcal{O}}_{\bar{i}} := \bar{\mathcal{O}}_{\bar{i}}^{(2)} &= \{\bar{Q}_+, [\bar{Q}_-, \bar{\phi}_{\bar{i}}(u)]\} \sim \oint_{C_u} \mathrm{d}z G^+(z) \oint_{C'_u} \mathrm{d}w \bar{G}^+(w)\bar{\phi}_{\bar{i}}(u) \end{aligned} \tag{3.6.21}$$

We calculate $[D_i, \bar{D}_{\bar{i}}]$ in the topological $|l\rangle$ basis i.e.

$$\begin{aligned}
[D_i, \bar{D}_{\bar{j}}]^l_k &= \partial_i A^l_{\bar{j}\ k} - \partial_{\bar{j}} A^l_{i\ k} = \eta^{lp}[(\partial_i\langle p|)\bar{\partial}_{\bar{j}}|k\rangle - (\bar{\partial}_{\bar{j}}\langle p|)\partial_i|k\rangle] \\
&= \eta^{lp}\Pi\left(\phi_p \int_{HS^2_L}\{Q_+,[Q_-,\phi_i]\}\right)\Pi\left(\int_{HS^2_R}\{\bar{Q}_+,[\bar{Q}_-,\bar{\phi}_{\bar{j}}]\}\phi_k\right) \\
&\quad -\eta^{lp}\Pi\left(\phi_p \int_{HS^2_L}\{\bar{Q}_+,[\bar{Q}_-,\bar{\phi}_{\bar{j}}]\}\right)\Pi\left(\int_{HS^2_R}\{Q_+,[Q_-,\phi_i]\}\phi_k\right) \\
&= \eta^{lp}\left[\Pi\left(\phi_p \int_{HS^2_L}\partial\bar{\partial}\phi_i\right)\Pi\left(\int_{HS^2_R}\bar{\phi}_{\bar{j}}\phi_k\right) - \Pi\left(\phi_p\int_{HS^2_L}\bar{\phi}_{\bar{j}}\right)\Pi\left(\left(\int_{HS^2_R}\partial\bar{\partial}\phi_i\right)\phi_k\right)\right] \\
&= \eta^{lp}\left[\Pi\left(\phi_p \int_{HS^2_L}\bar{\phi}_{\bar{j}}\right)\Pi\left(\int_{C_R}(\partial_{\tau_2}\phi_i)\phi_k\right) - \Pi\left(\phi_p\oint_{C_L}\partial_{\tau_2}\phi_i\right)\Pi\left(\left(\int_{HS^2_R}\bar{\phi}_{\bar{j}}\right)\phi_k\right)\right] \\
&= \eta^{lp}\left[\Pi\left(\phi_p \int_{HS^2_L}\bar{\phi}_{\bar{j}}\right)\Pi\left(\left(\oint_\Gamma H(z)\oint_{C_R}\phi_i\right)\phi_k\right)\right. \\
&\quad \left. - \Pi\left(\phi_p\oint_\Gamma H(z)\int_{C_L}\phi_i\right)\Pi\left(\left(\int_{HS^2_R}\bar{\phi}_{\bar{j}}\right)\phi_k\right)\right],
\end{aligned} \tag{3.6.22}$$

where the contours of the $G^-(z)$, $\bar{G}^-(z)$ $G^+(z)$, $\bar{G}^+(z)$ integration are as in Fig. 11. Moreover we consider operators ϕ in the (c, c) and $\bar{\phi}$ in the (a, a) ring, e.g. ϕ is $\bar{Q}_+$ and $\bar{Q}_-$ closed. In the language of current algebras that means that the short distance expansion of $\phi(v)$ with $\bar{Q}_+(z) \sim G^+(z)$ and $\bar{Q}_-(w) \sim \bar{G}^+(z)$ has no pole and $\phi(v)$ can be ignored when deforming Γ_z and Γ_w. The contours e.g. of the term in the third line can be deformed as in Fig. 11 and the contours of $G^-(z)$, $\bar{G}^-(z)$ encircling $G^+(z)$, $\bar{G}^+(z)$ give the L_{-1} and $\bar{L}_{-1}$ acting as ∂ and $\bar{\partial}$ derivatives on ϕ_i by (3.3.10). Similar manipulations apply to the term in the second line of (3.6.22). Applying Gauss's law in both terms gives the integral over the normal derivative $-\partial_{\tau_2}$. The minus sign is due to the orientation of τ_2. The normal direction is "time" evolution by H, i.e. $\partial_{\tau_2} = \partial_n\phi_i = [H, \phi_i]$, which is used in the last line of (3.6.22), where $H(z)$ is integrated around ϕ_i From now on we exploit the topological nature of the theory and take ordered limits of Σ

$$\text{first}: \quad T_R, T_L \to \infty, \qquad \text{second}: \quad T \to \infty \tag{3.6.23}$$

as depicted Fig. 12. The tubes are all normalized to have perimeter 1. Elongation T_R and T_L projects ϕ_p and ϕ_k to the Ramond-Ramond vacuum state $\langle p|$ and $|k\rangle$ respectively. The procedure of the limits is a prescription how to deal with short distance singularities and the only such issue in topological field theory are *contact terms* see (3.6.31).

The action of H on these states yields zero. The two terms in the last line of (3.6.22) are transformed into each other by exchanging the left- and right infinity. We discuss the $-\Pi\left(\phi_p \int_{HS^2_L}\bar{\phi}_{\bar{j}}\right)\Pi\left(\left(\oint_\Gamma H(z)\oint_{C_R}\phi_i\right)\phi_k\right)$ explicitly. Vanishing of $H|k\rangle$ means that H may considered as acting on the full state $\Pi\left(\left(\oint_{C_R}\phi_i\right)\phi_k\right)$. In Hilbert space notation is denoted as $H|(\oint_{C_R}\phi_i)|k\rangle$ and similar $\Pi\left(\phi_p\int_{HS^2_L}\bar{\phi}_{\bar{j}}\right)$ as

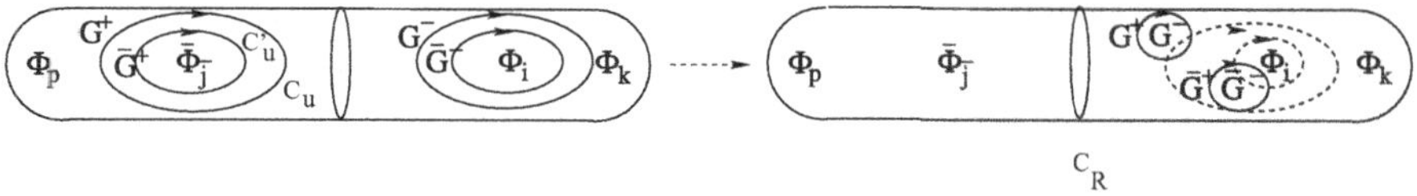

Fig. 11 Contour manipulation on Σ in the evaluation of $[D_i, \bar{D}_{\bar{j}}]^l_k$

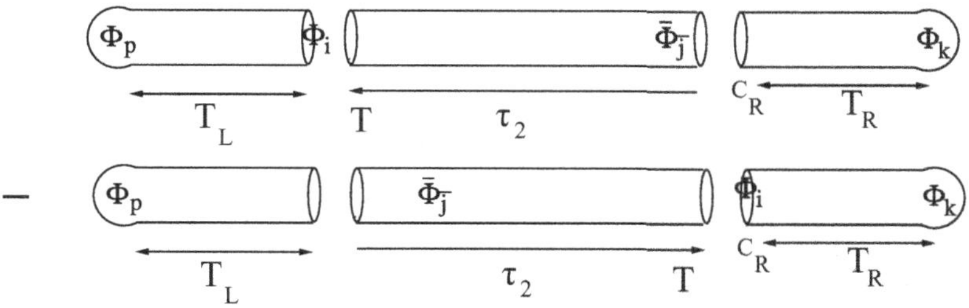

Fig. 12 Limits taking in the evaluation of $[D_i, \bar{D}_{\bar{j}}]^l_k$

$\langle p| \int_{HS_L^2} \bar{\phi}_{\bar{j}}|$. We can move the H integral to the left and since ϕ_p is projected to the groundstate the non-vanishing contribution comes from its action on $\int_{HS_L^2} \bar{\phi}_{\bar{j}}$. If the insertion of $\bar{\phi}_{\bar{j}}$ is on the most left part in Fig. 12 it will also be projected to the groundstate in the $T \to \infty$ limit and annihilated by H. Therefore is remains to consider the contribution from integral over the middle tubus whose length is parametrized by T. This integral is $\int_{tube} \bar{\phi}_{\bar{j}} = \int_0^T \mathrm{d}\tau_2 \oint_{C_L} \mathrm{d}\theta \bar{\phi}_{\bar{j}}$. H creates τ_2 translations, so $[H, \bar{\phi}_{\bar{j}}] = -\partial_{\tau_2} \bar{\phi}_{\bar{j}}$ and the integration over τ_2 becomes trivial. Note that only the lower boundary $\tau_2 = 0$ contributes. The upper boundaries, where $\bar{\phi}_{\bar{j}}$ is near ϕ_i in both contributions see Fig. 12, cancels. Therefore

$$
\begin{aligned}
[D_i, \bar{D}_{\bar{j}}]^l_k &= \eta^{lp} \lim_{T_{L/R}\to\infty} \Big[\Pi\left(\phi_p \int_{HS_L^2} \bar{\phi}_{\bar{j}}\right) \Pi\left((\oint_\Gamma H \oint_{C_R} \phi_i)\phi_k\right) \\
&\quad - \Pi\left(\phi_p \oint_\Gamma H \int_{C_L} \phi_i\right) \Pi\left(\int_{HS_R^2} \bar{\phi}_{\bar{j}}\phi_k\right)\Big] \\
&= \eta^{lp}[\langle p| \left(\int_{Tu} \bar{\phi}_{\bar{j}}\right) H \left(\oint_{C_R} \phi_i\right) |k\rangle - \langle p| \left(\int_{C_L} \phi_i\right) H \left(\int_{Tu} \bar{\phi}_{\bar{j}}\right) |k\rangle] \\
&= \eta^{lp} \lim_{T\to\infty} [\langle p| \left(\oint_{C_L} \bar{\phi}_{\bar{j}}\right) e^{-HT} \left(\oint_{C_R} \phi_i\right) |k\rangle \\
&\quad - \langle p| \left(\int_{C_L}\right) \phi_i e^{-HT} \left(\oint_{C_R} \bar{\phi}_{\bar{j}}\right) |k\rangle] \\
&= (\bar{C}_{\bar{j}} C_i)^l_k - (C_i \bar{C}_{\bar{j}})^l_k = -[C_i, \bar{C}_{\bar{j}}]^l_k
\end{aligned}
\tag{3.6.24}
$$

This is the main identity within the tt^* equations. The others are easier to derive and all are summarised below in the topological basis as

$$\begin{aligned} [D_i, \bar{D}_{\bar{j}}] &= -[C_i, \bar{C}_{\bar{j}}] \\ [D_i, D_j] = [\bar{D}_{\bar{i}}, \bar{D}_{\bar{j}}] &= [D_i, \bar{C}_{\bar{j}}] = [\bar{D}_{\bar{i}}, C_j] = 0 \\ D_i C_j = D_j C_i &\qquad \bar{D}_{\bar{i}} \bar{C}_{\bar{j}} = \bar{D}_{\bar{j}} \bar{C}_{\bar{i}} \end{aligned} \tag{3.6.25}$$

We can now define a *flat* $[\nabla_i, \nabla_j] = [\nabla_i, \bar{\nabla}_{\bar{j}}] = [\bar{\nabla}_{\bar{i}}, \bar{\nabla}_{\bar{j}}] = 0$ *connection*

$$\nabla_i = D_i + \alpha C_i, \qquad \bar{\nabla}_{\bar{j}} = \bar{D}_{\bar{j}} + \alpha^{-1} \bar{C}_{\bar{j}} . \tag{3.6.26}$$

The sections of the vacuum bundle are identified with the periods in the Calabi-Yau σ model context. The above *flat connection* can be identified with the *Gauss Manin connection*, see Sect. 2.10.17.

Since it is flat it seems that the theory is *trivial*! However flat connections can still have *monodromies*, over non simply connected manifolds, see Fig. 5, which are the essential data of our theories. Where do these monodromies come from? The key is that (2.4.63), which is based on a local consideration of the tangent spaces of metric deformations at a generic point of the moduli space fails at singular degenerations of the space time Calabi-Yau manifold. At these loci charged Ramond-Ramond states become light, the simplest example is the charged *black hole* at the conifold [297], which sits in a hyper multiplet. In the presence of massless charged states the supergravity argument for the factorization (2.4.63) into hyper- and vector multiplets does not apply either. In fact the logarithm in third period that produces the monodromy M_1 in (2.10.43) can be interpreted as the one loop correction of the vector multiplet gauge coupling due to the massless hypermultiplet. An intriguing experimentally verifiable occurrence of monodromies of flat connections is the *Berry Phase* in quantum mechanics [38] see [267] for a review.

The tt^* equations describe the essence of the WS super symmetry constraints on the topological correlators. These equations have in general to be supplemented with information about the structure constants C_{ij}^l and boundary conditions. But already with one $U(1)$ i.e. R symmetry charge constraints they become powerful. E.g. for $d < 1$ (3.1.22) implies $|Q| < 1$ moreover these theories are rational and have finitely many chiral primaries in this charge range. We assign to the t^i of say the (c, c) ring (3.6.15) the weight $w_i = (1 - Q_i) > 0$. The last equation (3.6.25) called *associativity* guarantees the existence of a potential $\mathcal{F}$ with $C_{ijk} = D_i D_j D_k \mathcal{F}$. As discussed one can chose flat coordinates, which we call for convenience also t^i such that $C_{ijk} = \partial_i \partial_j \partial_k \mathcal{F}$. Charge conservation implies that $\mathcal{F}$ is homogeneous of degree 2 in the weights w_i of the t^i, i.e. a finite polynomial and *associativity* determines its coefficients up to an overall normalization. These constraints imply indeed that there is a completely solvable discrete infinite set of $d < 1$ $N = (2, 2)$ theories with an ADE classification, that can be identified with the classification of Kleinian singularities. For $d > 1$ there are zero and negative weight t^i and this simple way of approaching the problem loses its grip. The bordering case $d = 1$ are elliptic singularities.

However if $d \in \mathbb{Z}$ and the R charges are also integer, we expect from Sect. 3.1 that beside world-sheet super symmetry also space-time super symmetry constrains the correlators. Let us show that (3.6.25) implies for the Calabi-Yau σ models on threefolds $d = 3$ and odd integer R charges *special Kähler geometry*. In the holomorphic basis we use (3.6.16) to write $[D_i, \bar{D}_{\bar{j}}]^k_l = -\bar{\partial}_{\bar{j}} A^k_{il} = -[C_i, \bar{C}_{\bar{j}}]$. With $(C^k_{il})^\dagger = \bar{C}^{\bar{l}}_{\bar{i}\bar{k}}$ and hence $C^k_{\bar{j}m} = g^{k\bar{k}} C^{\bar{m}}_{\bar{j}\bar{k}} g_{\bar{m}m}$ we write

$$\bar{\partial}_{\bar{j}} A^k_{il} = [C_i, \bar{C}_{\bar{j}}]^k_l = [C_i, g^{-1} C^\dagger_j g]^k_l \,. \tag{3.6.27}$$

In the case of Calabi-Yau σ model the R charge conservation law forbids many correlators, see Sects. 3.4.1 and 3.5.1. In particular $g_{0\bar{k}} = g^{0\bar{k}} = 0$ for $\bar{k} \neq \bar{0}$ and $C^k_{i0} = \delta^k_i$ and $C^{\bar{k}}_{\bar{i}\bar{0}} = \delta^{\bar{k}}_{\bar{i}}$. If we specialize (3.6.27) to $k = l = 0$ we can write

$$\begin{aligned}
\bar{\partial}_{\bar{j}} A^0_{i0} = \bar{\partial}_{\bar{j}}(g^{0\bar{k}} \partial_i g_{0,\bar{k}}) &= [C_i, g^{-1}(C_j)^\dagger g]^0_0 \\
\bar{\partial}_{\bar{j}} \partial_i \log(g_{0\bar{0}}) &= -g^{0\bar{0}} C^{\bar{k}}_{\bar{j}\bar{0}} g_{\bar{k}i} \\
&= -\frac{g_{\bar{j}i}}{g_{0\bar{0}}} \,.
\end{aligned} \tag{3.6.28}$$

As follows from the identification (3.5.5), (3.5.6) in the B-model and (2.4.16) or Serre duality (2.4.7) the vacuum states $|0\rangle$ and $|\bar{0}\rangle$ are associated to the holomorphic $(n, 0)$ and anti-holomorphic $(0, n)$ forms. In particular

$$e^{-K} = i \int_M \Omega \wedge \bar{\Omega} = \langle \bar{0} | 0 \rangle \tag{3.6.29}$$

and comparing (2.5.25), (2.5.24) with (3.6.28), (3.6.29) we identify the Weil-Peterson metric with a sub-block of the tt^* metric

$$G_{i\bar{j}} = g_{i\bar{j}} e^K \,. \tag{3.6.30}$$

In (3.6.24) we have related the curvature of $g_{i\bar{j}}$ to a bilinear in the 3-point functions and with (3.6.30) this becomes the special geometry relation (2.5.21). In other words tt^* in genus 0 implies special Kähler geometry, but the main virtue of the formalism is that it generalized special Kähler geometry to higher genus. This will become essential to solve the B-model.

It is worth mentioning the closely related *contact term approach* to the definition of the connection (3.6.17), see e.g. [246] for a short introduction. It does use conformal invariance and restricts the analysis to exactly marginal ring operators. If the operators are exactly marginal for all values of $t = (t, \bar{t})$ of marginal perturbation parameters as (3.6.15) then the most general short distance expansion in the basis e_γ of them is

$$\mathcal{O}_\alpha(z)\mathcal{O}_\beta(0) \sim \frac{G_{\alpha\beta}}{|z|^4} + \Gamma^\gamma_{\alpha\beta}\delta^2(z)\mathcal{O}_\gamma(0) \,. \tag{3.6.31}$$

Clearly this expansion is compatible with dimensional analysis as $\delta^2(z) = \frac{\partial}{\partial z}\frac{1}{\bar{z}}$. Marginality implies in first order in t that $\int d^2z\langle\mathcal{O}_\alpha(z)\mathcal{O}_\beta(1)\mathcal{O}_\gamma(0)\rangle$ gets only contributions from $z = 1$ and $z = 0$, which explains that only the δ-function appears on the right of (3.6.31) in this order. Exact marginality means that scale independence, i.e. vanishing β functions, are maintained to all orders in t. To next order follows the closing on exactly marginal operators, as opposed to arbitrary $(1, 1)$ operators, on the right in (3.6.31). The *Zamolodchikov metric* is defined as the sphere correlator

$$G_{\alpha\beta} = \langle\mathcal{O}_\alpha(1)\mathcal{O}_\alpha(0)\rangle \tag{3.6.32}$$

and because of conformal invariance it does not require a limit as in the tt^* case. Taking the derivatives with respect to perturbations one gets

$$\frac{\partial G_{\alpha\beta}}{\partial t_\gamma} = \int d^2z\langle\mathcal{O}_\alpha(z)\mathcal{O}_\beta(1)\mathcal{O}_\beta(0)\rangle = \Gamma^\delta_{\alpha\gamma}G_{\delta\beta} + \Gamma^\delta_{\gamma\beta}G_{\delta\alpha} \,, \tag{3.6.33}$$

which establishes $\Gamma^\delta_{\alpha\gamma}$ as connection of the *Zamolodchikov metric*. So far the discussion of the contact terms has been about a general ansatz and in particular all $\Gamma^\delta_{\alpha\gamma}$ could have been zero. However [141] observed first that in order to ensure marginality in superconformal theories with non trivial triple couplings C^k_{ik} the contact terms have to be present, which is of course required to get (3.6.25). The virtue of the tt^* equations is to generalize this analysis to all ring states replacing $\Gamma^\delta_{\alpha\beta}$ with $A^\delta_{\alpha\beta}$ and non-conformal theories.

As an exercise one may derive the special geometry relation in $N = (2, 2)$ SCFT using the contact term approach as a specialisation of the derivation of the tt^* equations. The decomposition of α, β into $j\bar{j}$ comes from the possibility of picking the holomorphic basis in $N = (2, 2)$ WS theories. Of course the real challenge is to understand the occurrence of the monodromies, which we identified as the data of the theory, which however requires to understand the spacetime Ramond-Ramond states.

Mathematically the tt^*-structure is related to the TERP-structure, see review of Hertling and Sabbah [174], and the non-commutative Hodge structure [217]. It was in the latter context that the $\hat{\Gamma}$ classes were found in [217].

3.6.2 The Frobenius Algebra, tt^* Structure and Gauss Manin Connection

In this section we discuss the Frobenius structure, which is associated naturally to any topologically twisted (2, 2) supersymmetric world-sheet theory. In view of the relation of the cohomology ring of the topological field theory and the cohomology[103] of the target space M, this chapter further links the two dimensional description given above with the target space descriptions of the horizontal[104] cohomology that was discussed in Sect. 2.4.

A Frobenius algebra has the following elements. It is a graded vector space $\mathcal{A} = \oplus \mathcal{A}^{(i)}$, $i \geq 0$ with a symmetric non-degenerate bilinear form η and a cubic form

$$C^{(i,j,k)} : \mathcal{A}^{(i)} \otimes \mathcal{A}^{(j)} \otimes \mathcal{A}^{(j)} \to \mathbb{C}. \tag{3.6.34}$$

Here is the list of defining properties:

(FAs) Symmetry: $C^{(i,j,k)}_{abc} = C^{(\sigma(i,j,k))}_{\sigma(abc)}$ under any permutation of indices.
(FAd) Degree:[105] $C^{(i,j,k)} = 0$ unless $i + j + k = n$.
(FAu) Unit: $C^{(0,i)}_{1ab} = \eta^{(i)}_{ab}$.
(FAnd) Non-degeneracy: $C^{(1,j)}$ is non-degenerate in the second slot.
(FAa) Associativity: $C^{(i,j)}_{abp}\eta^{pq}_{(n-i-j)}C^{(i+j,k)}_{qcd} = C^{(i,k)}_{acq}\eta^{qp}_{(n-i-k)}C^{(i+k,j)}_{pbd}$.

Here the latin indices $a, b, c, \ldots$ refer to a choice of basis $\mathcal{A}^{(i)}_a$, $a = 1, \ldots, \dim(\mathcal{A}^{(i)})$ of $\mathcal{A}^{(i)}$. The above defines a commutative algebra as follows

$$\mathcal{A}^{(i)}_a \cdot \mathcal{A}^{(j)}_b = C^{(i,j)}_{abq}\eta^{qp}_{(i+j)}\mathcal{A}^{(i+j)}_p = C^{(i,j)p}_{ab}\mathcal{A}^{(i+j)}_p . \tag{3.6.35}$$

These structures are intrinsic to the (c, c)- and (a, c)-rings of the worldsheet (2, 2) superconformal theory. The charges of the $U(1)_A$ and $U(1)_V$ currents in the (2, 2) supersymmetry algebra define the grading in the (c, c)- and the (a, c) rings respectively and the rest of the Frobenius structure follows from the axioms of conformal field theory and the closing of the rings up to Q_B, Q_A exact terms, see Sect. 3.3.1 and [182]. Note that in families of Frobenius algebras $\eta^{(p)}_{ab}$ is a constant topological pairing, while the general $C^{(i,j)}_{abc}(\underline{t})$ varies with the deformation parameter.

In the B-model the (c, c) ring is identified for fixed complex structure with the elements $B^{(p,q)}_a$ in $H^p(M, \wedge^q TM)$. We consider only those elements which are mapped by contraction with Ω to elements $\mathcal{B}^{(p)} = \Omega(B)$ in $H^n_{\rm prim}(M_n)$, i.e. in particular for $p = q$. The Hodge type for this complex structure, which can be taken at the point of maximal unipotent monodromy, defines the grading, so that

[103] Quantum cohomology in the A-model.

[104] Vertical in the A model.

[105] Because of this property the last index indicating the degree dropped in the following.

one gets, compare (3.5.14) and (2.5.10) with $\langle x, y\rangle = Q(y, x)$,

$$C^{(p,q)}_{abc} = Q(\Omega(B^{(p)}_a \wedge B^{(q)}_b \wedge B^{(n-p-q)}_c), \Omega)\,, \tag{3.6.36}$$

which depends also on the Kählergauge of Ω. For complex families, i.e. deformations w.r.t. elements with $(p = q = 1)$, the grade of $\mathcal{B}^{(p)}$ is encoded in the Hodge filtration parameter of $\mathcal{F}^{n-p}$ See (2.4.32) and we give a covariant definition of (3.6.36) below.

The (a, c) ring is mapped to the quantum cohomology extension of H^*_{deRham}. On the vertical cohomology in $H^{p,p}(M_n)$ the grading of the A-model is simply identified with the form degree. We allow again only the complexified Kähler deformation family w.r.t. to elements with $p = 1$. These deformation families of rings are pairwise identified by mirror symmetry on M and W and at the point of maximal unipotent monodromy the gradings can be matched using the monodromy weight filtration. This gives important information about the basis of the cohomology and homology groups in $H^n_{\text{prim}}(M_n)$.

One can chose a basis $A^I_{(p)}$ of the homology $H_n(M_n, \mathbb{Z})$ and a dual one $\alpha^{(p)}_I$ for cohomology of the primary horizontal subspace so that

$$\int_{A^I_{(p)}} \alpha^{(q)}_J = \delta^I_J \delta^q_p, \qquad \int_{M_n} \alpha^{(q)}_I \wedge \alpha^{(p)}_J = \begin{cases} 0 & \text{if } p+q > n, \\ \eta^{(q)}_{IJ} & \text{if } p+q = n\,. \end{cases} \tag{3.6.37}$$

Here $p = 0, \ldots, n$ denotes a grading which can be related to the Hodge type given a point in the moduli space. As mentioned above the most useful one is the large complex structure point.

The information in the Gauss-Manin connection (2.4.33) and in the Picard Fuchs ideal $\mathcal{I}_{PF}$ is equivalent. Combined with Griffiths transversality this information determines the Frobenius structure. From $\mathcal{I}_{PF}$ and the differential and algebraic relations that follow from Griffiths transversality one can calculate the Frobenius structure constants explicitly, see Sect. 2.9.2 as well as [61] for the quintic and [186] for any Calabi-Yau manifold. More abstractly the Frobenius structure can be identified on the B-model cohomology ring by choosing appropriate basis vectors $\mathcal{B}^{(p)}_a$ in $\mathcal{F}^{n-p}$ with the properties

$$\eta^{(p)}_{ab} = Q(\mathcal{B}^{(p)}_a, \mathcal{B}^{(n-p)}_b), \qquad C^{(1,p)}_{abc} = Q(\nabla_a \mathcal{B}^{(p)}_b, \mathcal{B}^{(n-p-1)}_c)\,. \tag{3.6.38}$$

The last equation of the two equations defining the ring homomorphism may be by comparing with (3.6.36) stated shortly as

$$\mathcal{A}^{(1)}_\alpha \cdot \quad \leftrightarrow \quad \nabla_\alpha \cdot\ . \tag{3.6.39}$$

FAs) is fulfilled because the Gauss-Manin connection is flat $[\nabla_a, \nabla_b] = 0$, FAnd) because of the Tian-Todorov Lemma and the rest of the axioms follow from Griffiths transversality. Note that associativity determines all possible couplings and that $\mathcal{B}_a^{(p)}$ can be readily expanded in the $\alpha_I^{(p)}$ basis with the following upper triangular property

$$\int_{A^I{}_{(p)}} \mathcal{B}_J^{(q)} = \begin{cases} 0 & \text{if } p < q \\ \delta_J^I & \text{if } p = q \end{cases} . \tag{3.6.40}$$

We can restate the Gauss-Manin connection in an easy form. Using the operator state correspondence in 2d field theory we write ($\mathcal{B}^{(0)} = \Omega_n, \mathcal{B}_{\alpha_1}^{(1)}, \mathcal{B}_{\alpha_2}^{(2)}, \ldots, \mathcal{B}_{\alpha_{n-2}}^{(n-1)}, \mathcal{B}^{(n)}$) as ($|0\rangle$, $|\alpha_1\rangle$, $|\alpha_2\rangle, \ldots$, $|\alpha_{n-2}\rangle$, $|n\rangle$). Since the the Gauss-Manin connection becomes the ordinary derivative in flat coordinates, which are given by a ratio of $t^\kappa = X^\kappa / X^0$ of the projective complex coordinates X^I see (2.5.26) for 3-folds and (2.4.40) in general, as e.g. the mirror map at large complex structure. Using the Griffiths-Frobenius structure on the B-model one can write the Gauss Manin connection as

$$\partial_{t_\kappa} \begin{pmatrix} |0\rangle \\ |\alpha_1\rangle \\ |\alpha_2\rangle \\ \vdots \\ |\alpha_{n-3}\rangle \\ |\alpha_{n-2}\rangle \\ |n\rangle \end{pmatrix} = \begin{pmatrix} 0 & \delta_{\kappa,\alpha_1} & 0 & 0 & \ldots & 0 & 0 \\ 0 & 0 & C^{(1,1)\ \alpha_2}_{\kappa,\alpha_1} & 0 & \ldots & 0 & 0 \\ 0 & 0 & 0 & C^{(1,2)\ \alpha_3}_{\kappa,\alpha_2} & \ldots & 0 & 0 \\ & & & & & & \\ 0 & 0 & 0 & 0 & \ldots & C^{(1,n-2)\ \alpha_{n-2}}_{\kappa,\alpha_{n-3}} & 0 \\ 0 & 0 & 0 & 0 & \ldots & 0 & \delta_{\kappa,\alpha_{n-2}} \\ 0 & 0 & 0 & 0 & \ldots & 0 & 0 \end{pmatrix} \begin{pmatrix} |0\rangle \\ |\alpha_1\rangle \\ |\alpha_2\rangle \\ \vdots \\ |\alpha_{n-2}\rangle \\ |\alpha_{n-2}\rangle \\ |n\rangle \end{pmatrix}, \tag{3.6.41}$$

for $\kappa = 1, \ldots, h_{n-1,1}(M_n)$. This specializes to the 3-fold case given in Sect. 2.5.5 and [88], see also [78] section 5.6 and [71] for a physics derivation from $\mathcal{N} = 2$ special geometry.

The *real structure*

$$g^{(q)}_{\alpha\bar{\beta}} = \langle\bar{\beta}, q|n-q, \alpha\rangle = R(\mathcal{B}_\alpha^{(n-q)}, \mathcal{B}_\beta^{(q)}) , \tag{3.6.42}$$

and the worldsheet parity operation

$$\langle\bar{\alpha}, q| = \langle\beta| M_{\bar{\alpha}}^{(q)\,\beta} , \tag{3.6.43}$$

which fulfills the worldsheet CPT constraints $MM^* = 1$, extend the Griffiths-Frobenius package on the mixed Hodge structure to the tt^*- structure [40].

In particular one can chose the basis $\mathcal{B}_\alpha^{(p)}$ compatible with the real structure, i.e.

$$\mathcal{B}_\alpha^{(p)} = \overline{\mathcal{B}^{(n-p)}}_{\bar{\alpha}} . \tag{3.6.44}$$

The degree one elements are the well known tangents vectors to the complex deformation space of Tian and Todorov

$$\mathcal{B}^{(1)}_\alpha = D_\alpha \Omega, \qquad \overline{\mathcal{B}}^{(1)}{}_{\bar\alpha} = \bar D_{\bar\alpha}\bar\Omega\,. \tag{3.6.45}$$

Note that $g^{(q)}_{\alpha\bar\beta}$ is the Zamolodichkov metric, $g_{i\bar j}$ is related to the Weil-Petersson metric $G_{i\bar j}$ by $g^{(1)}_{i\bar j} = e^{-K}G_{i\bar j} = \langle\bar 0|0\rangle G_{i\bar j}$ and has blockform w.r.t. to the grading. The higher degree operators are given for $p = 2, \ldots, \lfloor\frac{n}{2}\rfloor$ by

$$\mathcal{B}^p_\alpha = D^{(p)}_\alpha\Omega, \qquad \overline{\mathcal{B}}^{(p)}{}_{\bar\alpha} = \overline{D}^{(p)}_{\bar\alpha}\bar\Omega, \tag{3.6.46}$$

with $D^{(p)}_\alpha = \frac{1}{p!}\kappa_{\alpha;i_1,\ldots,i_p}D_{i_1}\ldots D_{i_p}$. These operators are closely related to the *Frobenius operator* $D^{(p)}(\underline{\rho})$ and as the latter determined at the point of maximal unipotent monodromy from the *symbol* of the Picard-Fuchs differential ideal $\mathcal{I}_{PF}$ on M_n as in (2.9.66) or equivalently from the information in the Chow ring of W_n. In order to fix them completely cases one has to construct the integer basis, which we do in section "The $\hat\Gamma$ Classes and Homological Mirror Symmetry" and Sect. 2.9.8. The formalism of the $\hat\Gamma$ class generalized to arbitrary CY n-folds, see e.g. for 4-folds [46].

3.7 *Kodaira-Spencer Gravity as Space-Time Action for the B-Model*

There are three space time actions known, which reproduce as classical equations of motion the unobstructedness of complex structures on the Calabi-Yau: Kodaira-Spencer gravity [40], Hitchins three-form action [177] and Hitchins general threeform action [178]. The first [40] and the last [134, 278] reproduce the B-model also at one loop. But even Einstein's gravity poses no problem up to one loop [299]. While it is not clear how the suggested spacetime descriptions make sense as full quantum theory, the worldsheet B-model approach makes remarkable predictions at higher loops.

Kodaira-Spencer theory of gravity is theory on M which couples exclusively to the complex moduli of M. Its tree level result reproduces the B-model without the coupling to worldsheet gravity, i.e. its genus zero sector [40]. It is a space time gravity theory in the sense that is does couple to the Calabi-Yau metric as far as complex structure dependence is concerned. It reproduces the Eq. (2.4.13) in the form $\bar\partial A(z) + \frac{1}{2}\partial(\widehat{A(z)\wedge A(z)}) = 0$ as its equation of motion and its Feynman graph expansion corresponds to the iterative solution to that equation exactly in the form as given Sect. 2.4.2. In fact by the ∂, $\bar\partial$-Lemma we have shown e.g. in the second induction step that one has an ψ_1 with $\partial\bar\partial\psi_1 = \widehat{[A_1, A_1]}$, hence $\hat A_2 = \frac{1}{2}\partial\psi_1$.

By (2.4.19) the first statement means also $\bar\partial\psi = (\widehat{A_1 \wedge A_1})$. Combining the two facts one gets a solution for $\hat{A}_2$ in the form

$$\hat{A}_2 = -\frac{1}{2\bar\partial}\partial(\widehat{A_1 \wedge A_1}) = \mathcal{P}(\widehat{A_1 \wedge A_1})\,. \tag{3.7.1}$$

We have used a "gauge" $\partial \hat{A}_k = 0$ and it is easy to see that the recursive solution comes with the freedom $\hat{A}_k + \bar\partial\lambda$, which one can fix be requiring $\bar\partial^* A_k = 0$. We can then define the "propagator" as $\mathcal{P} = -\frac{1}{2\bar\partial}\partial = -\bar\partial^*\frac{1}{2\Delta_{\bar\partial}}\partial$. With this "propagator" one can recursively write the solutions to $\hat{A}_k$. E.g. $\hat{A}_3 = 2\mathcal{P}(A_1 \wedge (\mathcal{P}(\widehat{A_1 \wedge A_1}))^\vee)^\wedge$. It follows from the construction of A_k that only $\hat{A}_1$ fulfills the Laplace equation, while A_k for $k > 1$ correspond to "massive modes" (Fig. 13).

It is not hard to see [40], that the Kodaira-Spencer action

$$\lambda^2 S(\hat{A}_1, \hat{A}_m, z_0) = \int_M \frac{1}{2}\hat{A}_m\frac{1}{\partial}\bar\partial\hat{A}_m + \frac{1}{6}((A_1 + A_m) \wedge (A_1 + A_m))^\wedge \wedge (A_1 + A_m)^\wedge \tag{3.7.2}$$

has $\bar\partial(\hat{A}_1 + \hat{A}_m) + \frac{1}{2}\partial((A_1 + A_m) \wedge (A_1 + A_m))^\wedge = 0$ as e.o.m. and reproduces the Feynman graph expansion above. Here we have defined as A_m the massive part of $A(z)$ and z_0 is background value of the complex structure. It has further be shown that (3.7.2) is the reduction of closed string field theory to the topological modes and it has been argued that its path integral defines the generating function for all correlators of the topological B-model coupled to worldsheet gravity. However the action has not been made sense of as quantum theory. So its solution is indirect by means of the holomorphic anomaly equation of the topological B-model. Nevertheless the divergent factors in the graph expansion of (3.7.2) lead to an analysis of the leading behavior at the boundaries of the complex moduli space of the Calabi-Yau space once the ones of the three point couplings are known. For one modulus t one gets [40]

$$F_g \sim \frac{[\partial_t^3 C_{ttt}]^{2g-2}}{[\partial_t C_{ttt}]}$$

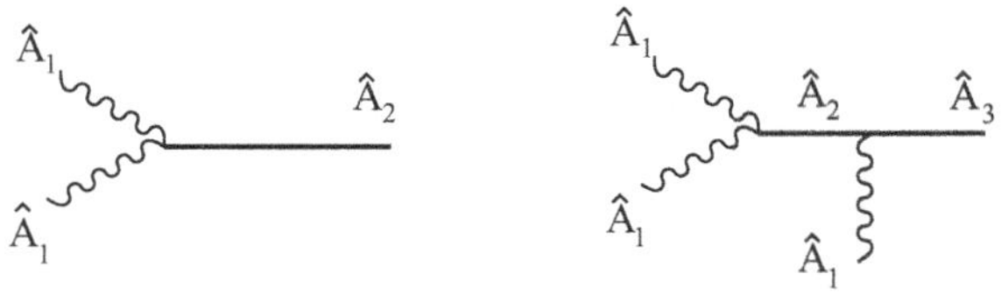

Fig. 13 Perturbative solution of the Kodaira-Spence equation in Tians form $\bar\partial A(z) + \frac{1}{2}\partial(\widehat{A(z) \wedge A(z)}) = 0$ by Feynmann graphs with massless fields (weavy lines) and massive fields (solid lines)

This result will be useful to fix the holomorphic ambiguity as it predicts the leading behaviour at the conifold, see Sect. 4.5.

4 The B-model at Higher Genus

In this chapter we use the world-sheet formalism on higher genus Riemann surfaces to couple the B-model to topological gravity. From the boundary strata in the moduli space of these Riemann surfaces comes the holomorphic anomaly, that we derive next following [40]. Note that in this derivation the formalism of special geometry, discussed in Sect. 2.5, which like the closely related tt^* formalism is a genus zero story, gets extended to higher genus. The holomorphic anomaly equations are recursive equations for the higher genus $F^{(g)}$, roughly the higher genus generalisations of the prepotential[106] $\mathcal{F}^{(0)} = F^{(0)}$, that need as starting data the genus zero amplitude $\mathcal{F}^{(0)}$, all the special geometry data associated to it and the genus one amplitude $F^{(1)}$. The holomorphic anomaly equations can be solved up to kernel, the holomorphic or modular ambiguity, once the propagators are defined, either using repetitively the special geometry algebra (2.5.21) by a simple integration-by-parts procedure [40] or by the Feynman rules of an auxiliary action [40]. Both methods lead expression for the $F^{(g)}$ in which the number of terms grow exponentially. We review these methods shortly but focus on a third method which is known as *direct integration*. This uses the idea hat all holomorphic dependence is in certain almost modular generators of a ring of automorphic forms that close under a holomorphic derivative, known as *Serre derivative*, and that the $F^{(g)}$ are generated by the generators a ring of meromorphic and an-holomorphic building blocks, which closes under the full covariant derivatives (4.2.33). In this formalism which uses more efficiently the modular invariances of the $F^{(g)}$ the terms to be calculated grow only polynomial. Strictly speaking this formalism can only proved in the local cases where the rings are rings of quasi modular forms. We exemplify it therefore first for the local $\mathbb{P}^2$ in Sect. 4.2.4.

The holomorphic anomaly equations determine the $F^{(g)}$ only up to meromorphic sections of $\mathcal{L}^{2-2g}$ over the complex moduli space $\mathcal{M}_{cs}$ of the Calabi-Yau 3-fold called $f_g(z)$. We discuss the important question of how to fix these sections in Sect. 4.5.

4.1 *Coupling the B Model to Topological Gravity*

We consider again the moduli space introduced in Sect. 3.3.2

$$\mathcal{M}_g = \text{large gauge transf.}\backslash\mathcal{H}_g/(\text{diff}\times\text{Weyl})_g \ .$$

[106]Holomorphic quantities are in the following denote by calligraphic characters.

with expected dimension $3g-3$ (A2.12). In the covariant quantization of string theory the metric independence of the theory, up to this finite dimensional space (3.3.18) we presently discuss, is expressed by a nilpotent BRST operator just like in (3.3.16). Conformal invariance is maintained for σ models on Calabi-Yau spaces. To take advantage of this extra bonus of the B-model note that in a conformal fields theory $T^{\mu}_{\mu}=0$ and (3.3.16) splits in the following two components corresponding to $T_{zz}=T(z)$ and $T_{\bar{z}\bar{z}}=\bar{T}(z)$. Now we can borrow literally the treatment of the measure from the critical bosonic string. In the case of the bosonic string the situation is exactly as in the topological B-model on a Calabi-Yau 3 fold (3.1.31), where the ghost number is identified with the $U(1)$ axial charge of the B-model. The geometrical reason for this equivalence is that (A2.13) and (A2.14) give the same anomaly if $\dim_{\mathbb{C}}(M)=3$ and $c_1(TM)=0$. As we saw in Sect. 3.1.2 the $b(z)$ and the Q_{BRST} operator have ghost number -1 and 1 respectively and there is a ghost number anomaly of $6g-6=-3\chi(\Sigma_g)$ on a higher genus worldsheet, which corresponds to the axial current anomaly $6g-6=-3\chi(\Sigma_g)$.We can use therefore the same measure over the complex moduli space as in the bosonic string. From the Beltrami-Differentials $\mu^k=\mu^{k\,z}_{\bar{z}}\mathrm{d}\bar{z}\partial_z$, $k=1,\ldots,3g-3$ in $H^1(T\Sigma_g)$, which represent tangent directions of $\mathcal{M}_g$, we define

$$B^k := \int_{\Sigma_g}\sqrt{h}h^{\alpha\gamma}h^{\beta\delta}\delta^{(k)}h_{\alpha\beta}G_{\gamma\delta}=\int_{\Sigma_g}\mathrm{d}^2z(G_{zz}\mu^{k\,z}_{\bar{z}}+G_{\bar{z}\bar{z}}\bar{\mu}^{k\,\bar{z}}_{z})=\beta^k+\bar{\beta}^k. \tag{4.1.1}$$

The definition of $B^{(k)}$ in itself does not require conformal invariance but just (3.3.16). We used after the second equality the standard metric in a conformal gauge and the expressions for the Beltrami-Differentials. In the last equality we used $(2,2)$ supersymmetry and the fact that after the B-twist the G^-, $\bar{G}^-$ are $h=2$ fields to define

$$\beta^k=\int_{\Sigma_g}\mathrm{d}^2z\, G^-\mu^k, \qquad \bar{\beta}^k=\int_{\Sigma_g}\mathrm{d}^2z\,\bar{G}^-\bar{\mu}^k\,. \tag{4.1.2}$$

Because of the antisymmetry of G and the Kähler structure on the moduli space $\mathcal{M}_g$ the quantity

$$\mu_g=\langle\prod_{k=1}^{6g-6}B^k\rangle\cdot[\mathrm{d}M]=\left\langle\prod_{k=1}^{3g-3}\beta^k\bar{\beta}^k\right\rangle\cdot[\mathrm{d}m\wedge\mathrm{d}\bar{m}] \tag{4.1.3}$$

is a top-form on $\mathcal{M}_g$. Here $\cdot[\mathrm{d}M]$ or $\cdot[\mathrm{d}m\wedge\mathrm{d}\bar{m}]$ means contraction with $\mathrm{d}M_{i_1}\wedge\ldots\wedge\mathrm{d}M_{i_{6g-g}}$ or $\mathrm{d}m_{i_1}\wedge\mathrm{d}\bar{m}_{i_1}\wedge\ldots\wedge\mathrm{d}m_{i_{3g-3}}\wedge\mathrm{d}\bar{m}_{i_{3g-3}}$ and suitable normalization. That is we inserted $6g-6$ times $\beta^{(k)}$ to compensate the ghost or axial anomaly, which is by the index theorems (cff Appendix 2) identified with the dimension of $\mathcal{M}_g$. The integral

$$F^{(g)} = \int_{\mathcal{M}_g} \mu_g \tag{4.1.4}$$

is the central observable of the topological B model.

How does this discussion of the dimension of the moduli space relate to dimension of the moduli space in the A-model? In the A-model one can compute the geometrical virtual dimension of the moduli space of non-trivial holomorphic maps using the Riemann Roch theorem and finds that the expected or virtual complex dimension of the moduli of stable maps is [182]

$$\begin{aligned} \mathrm{vdim}_C \overline{\mathcal{M}}_{g,n}(M,\beta) &= h^0(x^*(TM)) - h^1(x^*(TM)) + \dim\ \mathrm{Def}(\Sigma,\underline{p}) - \dim\ \mathrm{Aut}(\Sigma,\underline{p}) \\ &= c_1(TM)\cdot\beta + (\dim_C M - 3)(1-g) + n\,, \end{aligned} \tag{4.1.5}$$

where one calculates the first two terms by (A2.14) and the last two by (A2.13) with addition of moduli for marked points. The upshot is that the deformations of the metric $\mathcal{M}_g$ are offset by the obstructions of having a nontrivial holomorphic map to M, so that the virtual dimension of the moduli space of maps is zero. In the B-model we kill the deformation space of $\mathcal{M}_g$ by viewing the B-model fields as ghost system from which we construct a top form to integrate over $\mathcal{M}_g$. The topological B-model is one of those examples of string theories, where general covariance (3.3.16) is maintained by an Q_{BRST} operator, whose charge violation measure the dimension of the moduli space, but the decoupling of ghost and matter sector is not imposed [324].

As part of the prerequisite for coupling topological theories to gravity [318] the measure μ_g must be closed $\mathrm{d}\mu_g = 0$. To see that consider

$$0 = \langle\{Q, \prod_{k=1}^{6g-5} B^k\}\rangle = \sum_{j=1}^{6g-5}(-1)^{j-1}\langle B^1\ldots\{Q,B^j\},\ldots B^{5g-5}\rangle \tag{4.1.6}$$

and use the fact that $\{Q, B^i\}$ yields the $T^i = \int_{\Sigma_g} \mathrm{d}^2 z T\mu^i$, whose insertions can be interpreted as derivative on $\mathcal{M}_g$ according to (3.3.22). A second prerequisite is that μ_g is basic, i.e. that it vanishes for all variations of the metric induced by infinitesimal diffeomorphism. These correspond to the last two terms in (3.3.19) and the property is easily checked. We will show below explicitly by manipulations similar to the one that lead to (4.1.6) that the Q commutator of the measure is exact. The metric dependence comes hence from the boundaries of $\mathcal{M}_g$. Combinatorial the calculation is similar non-topological higher string loop calculations, apart from the fact that the latter involved much more sophisticated integrals, whose nature in the case of the super moduli spaces have became only clear recently [99]. The compactifications of $\mathcal{M}_{g,n}$ is identical to the one discussed in topological gravity [320]. Its boundary components come from pairwise collision of inserted points and nodes. In $2d$ gravity one gets from these boundaries the topological recursion relations. In the case of the B-model there is an interesting modification namely that the boundary components contribute only in anti-holomorphic derivatives of $\mathcal{F}_g$, which gives rise to recursion relations involving anti-holomorphic derivatives. Since without boundary component contributions the $\mathcal{F}^{(g)}$ would be holomorphic one calls these recursions the *holomorphic anomaly equations*. They are no more anomalous then the topological recursion relations.

4.1.1 The Holomorphic Anomaly

We want to consider in this section perturbations of a more general form then in Sect. 3.3.1 namely

$$S = \int_\Sigma \mathrm{d}^2 z \mathcal{L}_0 + \sum_i t^i \int_\Sigma \mathcal{O}_i + \sum_i \bar{t}^{\bar{\imath}} \int_\Sigma \bar{\mathcal{O}}_{\bar{\imath}} \, . \tag{4.1.7}$$

Here the WS two-form field $\mathcal{O} = \mathcal{O}^{(2)}$ is the B-model field (3.3.13) which comes from a $\phi = \mathcal{O}^{(0)}$ in the (c, c) ring. We will use here the CFT notation introduced in Sect. 3.3.1, i.e. $\mathcal{O}_i := \{Q_+, [Q_-, \phi_i]\} \sim \{G_0^-, [\bar{G}_0^-, \phi_i]\}$ and $\bar{\mathcal{O}}_{\bar{\imath}} := \{\bar{Q}_+, [\bar{Q}_-, \bar{\phi}_{\bar{\imath}}]\} \sim \{G_0^+, [\bar{G}_0^+, \bar{\phi}_{\bar{\imath}}]\}$. In an unitary theory $\bar{t}^{\bar{\imath}} = (t^i)^*$, but it will be important in the following to view $\bar{t}^{\bar{\imath}}$ as an independent parameter. As explained in Sect. 3.3.1 the WS two-form fields in (4.1.7) are neutral. Therefore we can expect that arbitrary $n - point$ functions like for $g > 1$

$$C^{(g)}_{i_1,\ldots,i_n} = \int_{\mathcal{M}_g} \langle \int \mathcal{O}_{i_1} \ldots \int \mathcal{O}_{i_n} \prod_{k=1}^{3g-3} \beta^k \bar{\beta}^k \rangle \tag{4.1.8}$$

do not vanish. Is it stands (4.1.8) is not well defined. We first have to specify how to deal with the contact terms, which are necessarily present in an interacting supersymmetric theory, see (3.6.24) or (3.6.31). Now in the case $g = 0$ there are the three $PSL(2, \mathbb{C})$ conformal Killing fields. The zero mode integral of their superpartners compensates for three descendant operations and with the $PSL(2, \mathbb{C})$ symmetry we set three points to $0, 1, \infty$. The generic genus zero correlation is then

$$C^{(0)}_{i_1,\ldots,i_n} = \int_{\mathcal{M}_0} \langle \phi_{i_1}(0) \phi_{i_2}(1) \phi_{i_3}(\infty) \int \mathcal{O}_{i_4} \ldots \int \mathcal{O}_{i_n} \rangle \tag{4.1.9}$$

This has no contact interaction among the first 3 fields. It is natural to make this function symmetric in its indices. Therefore we exclude *all* contact interactions from the regions of the integrations. This the regularization we adopt for general g.

In view of (4.1.7) we can insert $\int_\Sigma \mathcal{O}_i$ operators by taking t^i derivatives ∂_i of $C^{(g)}_{i_1,\ldots,i_n}$ in an attempt to obtain $C^{(g)}_{i,i_1,\ldots,i_n}$. In order to achieve our short distance regularization we have to subtract the would be contact terms in the integration over Σ. This is *very naturally* achieved by taking covariant derivatives w.r.t. the Weil-Peterson metric, i.e. $\partial_i \to \partial_i - \Gamma_i$. In the tt^* formalism we can isolate the contact term as the difference between $\partial_i(Q_+Q_-|j\rangle) - \mathcal{O}_j\partial_i|0\rangle = [(A_i)^k_j\mathcal{O}_k - (A_i)^0_0\mathcal{O}_j]|0\rangle$. The logic is that in the term $\partial_i(Q_+Q_-|j\rangle)$ the field $\mathcal{O}_i$ in the integral $\int_\Sigma \mathcal{O}_i$ explores the region near $\mathcal{O}_j$ in (3.6.7), while in the second it does not. The Q_+Q_- generate the descendant field from ϕ_j in (3.6.7) that are needed in order to compare the two terms. In particular applying this to $|j\rangle = |0\rangle$ and using (3.6.28), (3.6.29) we get a contact term with the $\mathbf{1}$ operator $(A_i)^0_0 \cdot \mathbf{1} = -\partial K \cdot \mathbf{1}$. Roughly speaking this non triviality of the vacuum comes from the coupling of ϕ_j to the $U(1)_R$ current (3.3.5). One can argue that the above contact term is proportional to the integral of R integrated over the Riemann surface. The above consideration for the half sphere (3.6.7) , fixes the normalization and in general gives the Euler number χ of Σ. Subtracting both contact terms one concludes that the insertion of $\int_\Sigma \mathcal{O}^{(2)}_i$ into on a genus g correlation function with the right short distance prescription is given by the covariant derivative of $C^{(g)}_{i_1,\ldots,i_n}$

$$D_i = \partial_i - \Gamma_i - (2-2g)\partial_i K, \tag{4.1.10}$$

This implies the fact that $C^{(g)}_{i_1,\ldots,i_n}$ is a section of a tensor bundle over the complex moduli space $\mathcal{M}$ of the Calabi-Yau manifold $\mathcal{M}$ transforming in

$$C^{(g)}_{i_1\ldots i_n}(z) \subset \mathcal{L}^{2-2g} \otimes \mathrm{Sym}^n(T^*_{1,0}(\mathcal{M}_{cs})). \tag{4.1.11}$$

in as a generalisation of the genus zero discussion in Sect. 2.5 see (2.5.11). First note that the contact algebra analysis yields that all correlators can be obtained from the vacuum correlators F^g as

$$C^{(g)}_{i_1,\ldots,i_n} = D_{i_1}\ldots D_{i_n}F^{(g)}\,. \tag{4.1.12}$$

In particular they are symmetric, because of the vanishing of the corresponding curvature terms in Kähler connections. The line bundle factor $\mathcal{L}^{2-2g}$ in (4.1.11) can be understood by building the higher genus Riemann surface Σ_g by sewing it from spheres. This involves $g+1$ of "propagator" insertions $S^{ij} \in \mathcal{L}^{-2}$. The insertion of the propagator follows from the $\bar\partial_{\bar k}$ integrated version of (4.1.17) and (4.2.7) and the transformation w.r.t. to the Kähler line bundle form (4.2.7) or the relations in Section in (3.6.1). Further we have seen from Griffiths transversality that $F^{(0)}(\underline{X})$ is homogenous of degree 2 in X^a or said differently a section $F^{(0)}(\underline{X}) \in \mathcal{L}^2$. By the rules of sewing rules of topological field theory it follows that the genus g amplitudes $F^{(g)}(\underline{X})$ are of homogeneous of degree $2-2g$ i.e. $F^{(g)}(\underline{X}) \in \mathcal{L}^{2-2g}$. Concretely this can be seen as follows: the n-point $C_{i_1,\ldots,i_{g+1}} = D_{i_1}\ldots, D_{i_n}\mathcal{F} \in$

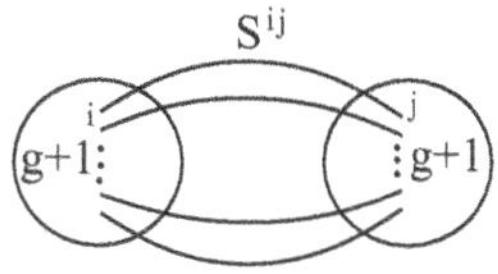

Fig. 14 A degenerate genus g curve made from two degree 2 genus zero $g+1$ function and $g+1$ propagators

$\mathrm{Sym}^n(T^{*(1,0)}) \otimes \mathcal{L}^2$ and the simplest degenerate genus g curve can be build by Fig. 14, which involves $(g+1)$ propagators and hence the $F^{(g)}(\underline{X})$ and their derivatives transforms as claimed.

Let us therefore investigate similarly as in Sect. 3.6.22 the derivative w.r.t. $\bar{t}_{\bar{i}}$ of the correlator

$$\begin{aligned}\frac{\partial}{\partial \bar{t}_{\bar{i}}}\mathcal{F}^g &= \int_{\mathcal{M}_g}\left\langle \oint_{C_w} G^+ \oint_{C'_w} \bar{G}^+ \bar{\phi}_{\bar{i}}(w) \prod_{k,\bar{k}=1}^{3g-3} \beta^k \beta^{\bar{k}} \right\rangle \cdot [\mathrm{d}m \wedge \mathrm{d}\bar{m}] \\ &= \int_{\mathcal{M}_g} 4 \sum_{i\bar{i}=1}^{3g-3} \frac{\partial^2}{\partial m_i \partial \bar{m}_{\bar{i}}} \left\langle \phi_{\bar{i}}(w) \prod_{k\neq i} \beta^k \prod_{\bar{k}\neq \bar{i}} \beta^{\bar{k}} \right\rangle \cdot [\mathrm{d}m \wedge \mathrm{d}\bar{m}] \\ &= \int_{\mathcal{M}_g} \partial\bar{\partial}\lambda^{6g-8} = \int_{\partial\mathcal{M}_g} \lambda^{6g-8}\end{aligned} \tag{4.1.13}$$

The contour of G^+, $\bar{G}^+$ are originally as in Fig. 11 encircling $\bar{\phi}(w)$. The deformation and splitting of the contour yields a sum of terms in which the G^+ and $\bar{G}^+$ encircle one $\oint_{C_u} \mathrm{d}w G^+(w) G^-(u)\mu^k = 2T(u)\mu^k$ and one $\oint_{C_u} \mathrm{d}w \bar{G}^+(w)\bar{G}^-(u)\mu^{\bar{k}} = 2\bar{T}(u)\mu^{\bar{k}}$ in each summand. Together with the integral in the definition of the β^k and $\bar{\beta}^k$ and the charges $\bar{Q}_+(Q_+)$ and $\bar{Q}_-(Q_-)$ associated to $G^+(z)(G^-(z))$ and $\bar{G}^+(z)(\bar{G}^-(z))$ we can write the result of the contour deformation as

$$\begin{aligned}\{\bar{Q}_-, \beta^k\} &= \int_{\Sigma_g} \mathrm{d}^2 z T \mu^k =: T^k \\ \{\bar{Q}_+, \bar{\beta}^k\} &= \int_{\Sigma_g} \mathrm{d}^2 z \bar{T} \bar{\mu}^k =: \bar{T}^k .\end{aligned} \tag{4.1.14}$$

In Sect. 3.6.1 where the $G^-(u)$, $\bar{G}^-(u)$ are integrated over a contour we got the L_{-1} mode of the T, which corresponds to derivative of a insertion position. Here we get the T^k and $\bar{T}^k$, which convert according to (3.3.22) into a derivative in the moduli space. Both effects are related and lead to exact forms on $\mathcal{M}_g$ and $\mathcal{M}_{g,n}$. The boundary components $\partial\mathcal{M}_g$, where the integral in the last line of (4.1.13) contributes according to Cauchy's theorem are in real codimension two as indicated by the form degree of λ. They are the standard stable degenerations [182]. The whole point specific of the B-model is to now figure out what boundary contributions are. This turns out to be easier then in the $2d$ gravity case. It is a bosonic string higher loop sewing consideration [279] with simplifications. There will be no new information in the contributions from colliding points above what we summarized in (4.1.12).

It remains to analyze the A and B degeneration depicted in Figs. 15 and 16 respectively. Near the boundary component in the moduli space corresponding to the degenerate surface in the Figs. 15 and 16, the normal direction to the boundary can be parametrized by the length of the tube τ_2. The moduli space of the boundary components consist of the $3g-6$ dimensional moduli space of the irreducible curves of genus $g-1$ in case A or h and $g-h$ in case B respectively with measure $[\mathrm{d}\hat{m} \wedge \mathrm{d}\hat{\bar{m}}]$. That is we loose three complex dimensions in the moduli space of the irreducible components and hence three $\beta\bar{\beta}$. As we make the tube infinitely long or equivalently infinitesimal thin the data remembered about the shape are merely the two insertion points w and u, the length and the twist of the tube. In particular two $\beta\bar{\beta}$ are replaced by $(\oint_{C_x} G^- \oint_{C'_x} G^- \phi_X(x))$ with $x = u, w$ and since we want calculate a string amplitude we have to insert a complete set of states for the ϕ_X. The contribution of the boundary is hence

$$\int_{\partial M_g} [\hat{\mathrm{d}}m \wedge \hat{\mathrm{d}}\bar{m}][dw][du]\frac{\partial}{\partial \tau_2}\left\langle \int \bar{\phi}_{\bar{j}} (\oint_{C_u} G^- \oint_{C'_u} G^- \phi_i)\eta^{ij} (\oint_{C_w} G^- \oint_{C'_w} G^- \phi_j) \prod_{a=1}^{3g-6} \hat{\beta}^a \hat{\bar{\beta}}^a \right\rangle \tag{4.1.15}$$

The integration over $[du]$ and $[dw]$ is over the fibre Σ_g of the universal curve. We can hence convert, e.g. the $\oint_{C_u} G^- \oint_{C'_u} G^- \phi_i$ insertions in a descendant field $\mathcal{O}^{(2)}_j$ integrated over Σ_g. Only if the $\int \bar{\phi}_{\bar{j}}$ integral extends over the tube one gets a contribution proportional to τ_2 which does not cancel under the derivative in (4.1.15) and one can focus on this integration domain. The correlation function factorizes

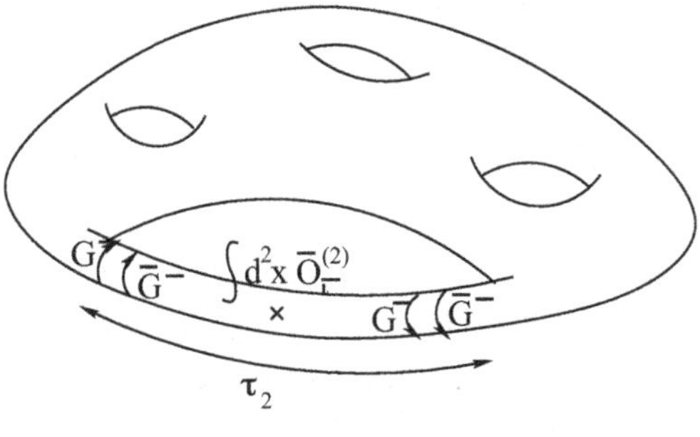

Fig. 15 A-type sewing

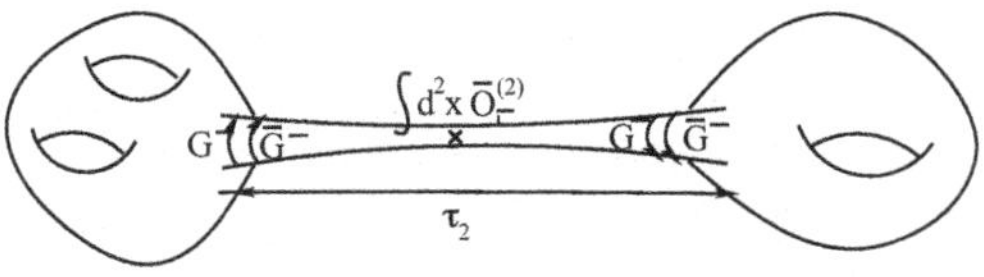

Fig. 16 B-type sewing

upon complete insertion of states in operator approach, which gives

$$\int_{\partial M_g}[d\hat{m}\wedge d\hat{\bar{m}}]\frac{\partial}{\partial\tau_2}\langle k|\int_{tube}\bar{\phi}_{\bar{j}}|l\rangle\eta^{ik}\eta^{lj}\left\langle\left(\int_\Sigma\mathcal{O}_i\right)\left(\int_\Sigma\mathcal{O}_j\right)\prod_{a=1}^{3g-6}\hat{\beta}^a\hat{\bar{\beta}}^a\right\rangle. \tag{4.1.16}$$

Here we also used the fact that propagation on the tube projects on the groundstate. With the manipulations from Sect. 3.6.1 and the normalizing the perimeter of the tube to one we get

$$\begin{aligned}\langle k|\int_{tube}\bar{\phi}_{\bar{j}}|l\rangle\eta^{ik}\eta^{lj} &= \langle\bar{k}|\int_{tube}\bar{\phi}_{\bar{j}}|\bar{l}\rangle M^{\bar{k}}_{k}\eta^{ik}M^{\bar{l}}_{l}\eta^{lj}\\ &= \tau_2\langle\bar{k}|\bar{\phi}_{\bar{j}}|\bar{l}\rangle e^{2K}G^{i\bar{k}}G^{j\bar{l}} = \tau_2\bar{C}_{\bar{k}\bar{j}\bar{l}}e^{2K}G^{i\bar{k}}G^{j\bar{l}} =: \tau_2\bar{C}^{ij}_{\bar{k}}\end{aligned} \tag{4.1.17}$$

Using this result in the boundary contribution of the A or B type degeneration and (4.1.12) one gets the holomorphic anomaly equation [40]

$$\bar{\partial}_{\bar{k}}F^{(g)} = \frac{1}{2}\bar{C}^{ij}_{\bar{k}}\left(D_iD_jF^{(g-1)}+\sum_{r=1}^{g-1}D_iF^{(r)}D_jF^{(g-r)}\right) \tag{4.1.18}$$

The factor $\frac{1}{2}$ comes the fact that we over count the integration over $\mathcal{O}_i$ and $\mathcal{O}_j$ in (4.1.16) by two in the A degeneration, as the $\mathcal{O}_i \leftrightarrow \mathcal{O}_j$ does not change the complex structure and in the B degeneration we doubled the non symmetric terms.

For $g = 1$ the situation is more tricky and interesting. Because of $h^0(T^2) = 1$ we have to kill the infinite automorphism by the insertion of one operator to start with a stable curve. Hence we have to consider $\bar{\partial}_{\bar{k}}\partial_m\mathcal{F}^{(1)}$. That leads in addition to the A degeneration to a contact term between $\mathcal{O}_i\mathcal{O}_{\bar{j}}$

$$\bar{\partial}_{\bar{k}}\partial_m F^{(1)} = \frac{1}{2}\bar{C}^{ij}_{\bar{k}}C_{mij}+\left(\frac{\chi}{24}-1\right)G_{\bar{k}m}\,. \tag{4.1.19}$$

The first term above is from the A type degeneration. The second contact term sees global properties of the Calabi-Yau which is quite interesting. There are two ways to normalize this contact term. Compare with the operator

$$F_1(t,\bar{t}) = \frac{1}{2}\int\frac{d^2}{\tau_2}\mathrm{Tr}(-1)^F F_L F_R q^H\bar{q}^{\bar{H}}\,. \tag{4.1.20}$$

formulation $\mathcal{F}^{(1)}$ [39] and calculate the $t\bar{t}$ term as in [69]. Or relate it to the Ray-Singer torsion[107] [40] and use the family index theorem of [44, 45].

---

[107] See [278] for an application to Hitchins action.

4.1.2 Wave Function and Background Independence

The topological string partition function Z is given by the exponential

$$Z = \exp(F) \tag{4.1.21}$$

of the all genera free energy

$$F(\lambda, z) = \sum_{g=0}^{\infty} \lambda^{2g-2} F^{(g)}(z) \,. \tag{4.1.22}$$

Here it must be assumed that the topological string coupling $\lambda \in \mathcal{L}^{2g-2}$ so that the exponential is invariant under Kähler gauge transformations. Witten [321] reinterprets the holomorphic anomaly equations of (4.1.18) and (4.1.19) as a special realisation of background independence in string theory. In this formulation Z is interpreted as a wave function $Z = \Psi$ on the complex moduli space $\mathcal{M}$ of M which arises from the quantization of the middle cohomology $H^3(M, \mathbb{R})$. The latter space is equipped with the necessary symplectic form via its intersection pairing. The holomorphic anomaly equations are interpreted as an infinitesimal consequence of the freedom to choose a polarisation on $H_3(M, \mathbb{R})$, i.e.to fix a choice of holomorphic coordinates or in the symplectic language to fix a choice of coordinates as positions and dual momenta. The form in which this infinitesimal change of polarization is realized is according to [321] in form of heat equation type of operator acting on the wave function

$$\left(\frac{\partial}{\partial \bar{z}^{\bar{a}}} + \frac{\lambda^2}{2} C^{bc}_{\bar{a}} \frac{D}{Dz^b} \frac{D}{Dz^c} \right) \Psi = 0 \,. \tag{4.1.23}$$

In the (4.1.23) the z^a, $a = 1, \ldots, h_{21}(W)$ are generic complex coordinates on $\mathcal{M}$ and the derivatives $D_b = \frac{D}{Dz^b}$ are meant to be taken covariantly in the inner derivative of the exponential. A derivative D_b is in general covariant w.r.t. to the Weil-Petersson metric, which derives from the Kähler potential (2.5.24) as well as to connection the Kähler line bundle $\mathcal{L}$, see (4.1.10). The coupling $C^{bc}_{\bar{a}} = e^{2K} \bar{C}_{\bar{a}\bar{b}\bar{c}} G^{\bar{a}a} G^{\bar{b}b}$ is a section of $\mathcal{L}^{-2} \otimes \mathrm{Sym}^2(T^{(1,0)}_{\mathcal{M}}) \otimes T^{*(0,1)}_{\mathcal{M}}$. Equation (4.1.23) almost but unfortunately not quite reproduces the holomorphic anomaly equations. Changing the range of the genus index and allowing for additional terms to reproduce the genus one holomorphic anomaly in a fairly obvious way [40] write down a closely related *master equation*, which does reproduce (4.1.18) and (4.1.19), but whose relation to background independence is not quite clear. One problem are the projective coordinates for the global Calabi-Yau spaces that force one to interpret λ as a non-trivial section of $\mathcal{L}^{2g-2}$. This problem evaporates in the local cases. A concrete application of the idea that Z is fulfils a wave function equation in the local case has been found in [5]. It was shown in [6] likewise in the local case that the wave function transformation property for the change of polarisation and analytic continuation allows to expand the topological string partition function at various loci in the moduli space that require different coordinates choices t_* compare Eq. (4.3.1). In the Nekrasov-Shatashvili limit see Sect. 4.4 the quantum mechanical interpretation has been studied in [7] and many subsequent works that can be found in [115]. Because of the simplification in the rigid supergravity formalism in the large moduli space that resembles the local case the authors [96, 151, 309] set up the formalism in the large moduli space. In [96, 160] the heat equation is studied in global Calabi-Yau cases in which the moduli is a symmetric space, which is likewise similar to the local case.

4.2 Integration of the Holomorphic Anomaly Equation

The integration of the holomorphic anomaly equation up to holomorphic terms is a relatively straightforward application of the special geometry relations that we derived in Sect. 2.5.2. The key object is an an-holomorphic potential S for the an-holomorphic triple couplings $\bar{C}_{\bar{a}\bar{b}\bar{c}}$, w.r.t. to the covariant derivatives $\bar{D}_{\bar{a}}$ defined as

$$\bar{D}_{\bar{a}}\bar{D}_{\bar{b}}\bar{D}_{\bar{c}}S = e^{2K}\bar{C}_{\bar{a}\bar{b}\bar{c}}\,, \qquad \bar{D}_{\bar{a}}\bar{D}_{\bar{b}}S = S_{\bar{a}\bar{b}}\,, \qquad \bar{D}_{\bar{a}}S = S_{\bar{a}}\,. \tag{4.2.1}$$

4.2.1 Propagators and Non-holomorphic Prepotential

The close relation between the propagators and the metric in the large moduli space has been pointed out in [151]. In order to solve the holomorphic anomaly equation in the big moduli discussed in Sect. 2.5.4 space one introduces propagators Δ^{IJ} that has to satisfy the equation

$$\bar{\partial}_K\Delta^{IJ} = \frac{i}{4}\bar{C}_K^{IJ} = \frac{i}{4}\bar{C}_{KLM}\mathcal{G}^{IL}\mathcal{G}^{JM} \tag{4.2.2}$$

This equation has a very simply solution

$$\Delta^{IJ} = -\frac{1}{2}\mathcal{G}^{IJ} + \mathcal{A}^{IJ} = \hat{\Delta}^{IJ} + \mathcal{A}^{IJ}\,, \tag{4.2.3}$$

where $\mathcal{A}^{IJ} = \mathcal{A}^{JI}$ is a holomorphic ambiguity, i.e. holomorphic functions, which do not affect the defining property (4.2.2). Note that the non-holomorphic part of the propagator $\hat{\Delta}^{IJ} = -\frac{1}{2}\mathcal{G}^{IJ}$ is proportional to the metric in the big moduli space and the property (4.2.2) is a simple consequence of the covariance of the latter $\bar{D}_K\mathcal{G}^{IJ} = 0$ and (2.5.38). Another consequence of its covariance $D_K\mathcal{G}^{IJ} = 0$ and (2.5.38) is

$$\partial_K\hat{\Delta}^{IJ} = -iC_{KLM}\hat{\Delta}^{IL}\hat{\Delta}^{JM} \tag{4.2.4}$$

The latter equation can be viewed as the generalisation of the an-holomorphic piece of Ramanujans identity for the weight two almost holomorphic modular form $\hat{E}_2 =$ of SL(2, $\mathbb{Z}$), cff (A4.13). Indeed we note, due to $\hat{\Delta}^{IJ} = -\frac{1}{2}(\text{Im}(\tau)^{-1})^{IJ}$ and (2.6.7), that the propagator $\hat{\Delta}^{IJ}$ transforms almost but not quite as a second rank tensor with two holomorphic indices under modular transformations. The shift in (2.6.7) should be cancelled by an appropriate choice of $\mathcal{A}^{IJ}$, which transforms inhomogeneous under modular transformations as E_2 see (A4.9), so that (4.2.6) yields a modular invariant definition of $S^{\alpha\beta}$.

The metric in the homogeneous coordinates is easily related to the expressions in the inhomogeneous coordinates using Griffiths tranversality and (2.5.35) one has

$$(G_{\alpha\bar{\beta}}) = \begin{pmatrix} -1 & 0 \\ 0 & G_{a\bar{b}} \end{pmatrix} = \left(-\frac{\langle \Omega_\alpha, \bar{\Omega}_{\bar{\beta}} \rangle}{\langle \Omega, \bar{\Omega} \rangle} \right) = (2e^K X^I_\alpha \mathcal{G}_{IJ} \bar{X}^J_{\bar{\beta}}) \tag{4.2.5}$$

and equivalently $G^{\alpha\bar{\beta}} = \frac{e^{-K}}{2} X^\alpha_I \mathcal{G}^{IJ} \bar{X}^{\bar{\beta}}_J$. The negative eigenvalue corresponds to the entry $G_{0\bar{0}} = G^{0\bar{0}} = -1$, while $G_{a\bar{b}}$ is the positive definite metric for the complex structure moduli with signature $+h_{12}(W)$. Similarly it was observed in [151] that

$$(S^{\alpha\beta}) = \begin{pmatrix} 2S & -S^b \\ -S^a & S^{ab} \end{pmatrix} = (i X^\alpha_I \Delta^{IJ} X^\beta_J) \,. \tag{4.2.6}$$

This relation been established in [151] by using (4.2.5) and (2.5.19) to show that the propagators S, S^a, S^{ab} fulfill the defining relations of the propagators. To see this note that

$$\partial_{\bar{c}} S = G_{a\bar{c}} S^a, \qquad \partial_{\bar{c}} S^a = G_{b\bar{c}} S^{ab}, \qquad \partial_{\bar{c}} S^{ab} = C^{ab}_{\bar{c}} := e^{2K} \bar{C}_{\bar{a}\bar{b}\bar{c}} G^{\bar{a}a} G^{\bar{b}b}, \tag{4.2.7}$$

is equivalent to

$$\bar{D}_{\bar{a}} S = S_{\bar{a}}, \qquad \bar{D}_{\bar{a}} \bar{D}_{\bar{b}} S = S_{\bar{a}\bar{b}}, \qquad \bar{D}_{\bar{a}} \bar{D}_{\bar{b}} \bar{D}_{\bar{c}} S = e^{2K} \bar{C}_{\bar{a}\bar{b}\bar{c}} \,, \tag{4.2.8}$$

where $S_{\bar{a}} = G_{a\bar{a}} S^a$ etc. We can focus on the an-holomorphic part $\hat{\Delta}^{IJ}$ in (4.2.6) and use (4.2.5) to lower the indices to get

$$(S_{\bar{\alpha}\bar{\beta}}) = \begin{pmatrix} 2S & S_{\bar{b}} \\ S_{\bar{a}} & S_{\bar{a}\bar{b}} \end{pmatrix} = (4ie^{2K} \bar{X}^I_{\bar{\alpha}} \hat{\Delta}_{IJ} \bar{X}^J_{\bar{\beta}}) = (e^{2K} \bar{X}^I_{\bar{\alpha}} (\bar{F}_{IJ} - F_{IJ}) \bar{X}^J_{\bar{\beta}}) \tag{4.2.9}$$

The first equation in (4.2.8) is immediate, because $\bar{F}_{IJK} \bar{X}^K = 0$ and when $\bar{D}_{\bar{a}}$ acts on $\bar{X}^I$ the covariant derivative in the complex conjugate of (2.5.19) under the A^I integral that defines $\bar{X}^I$ gives $\bar{X}_{\bar{a}}$ which produces as result $S_{\bar{a}}$. The second covariant derivate $\bar{D}_{\bar{a}}$ in the second equation produces likewise zero, when acting on $\bar{F}_{IJ}$. It can also act on $\bar{X}^J_{\bar{b}}$. Then it produces by the conjugate of (2.5.19) a term propotional to X_k^J which vanishes however due to the zeros in (4.2.5). The covariant derivative $\bar{D}_{\bar{a}}$ in the third equation of (4.2.8) gives by the conjugate of (2.5.29) a term proportional to $\bar{C}_{\bar{a}\bar{b}\bar{c}}$ when acting on $\bar{F}_{IJ}$ and also by the conjugate of (2.5.19) and (4.2.5) when acting on $\bar{X}^I_{\bar{b}}, \bar{X}^I_{\bar{c}}$. Together this gives (4.2.8), which proves (4.2.6), (4.2.9). In the same way we can consider holomorphic covariant derivatives D_k on $S^{\alpha\beta} = G^{\alpha\bar{\alpha}} G^{\beta\bar{\beta}} S_{\bar{\alpha}\bar{\beta}}$ and get by (2.5.19)

$$D_k S^{\alpha\beta} = -C_{ijk} S^{\alpha i} S^{\beta j} + \delta_k^{\beta} S^{\alpha 0} + \delta_k^{\alpha} S^{\beta 0} \tag{4.2.10}$$

We see that from the metric in the large moduli space one gets immediately a solution for the an-holomorphic prepotential $\mathcal{S}$

$$\mathcal{S} = -e^{-2K} \bar{S} = \frac{i}{2} \bar{X}^0_I \hat{\Delta}^{IJ} \bar{X}^0_J = -\frac{1}{2} X^I (F_{IJ} - \bar{F}_{IJ}) X^J \, . \tag{4.2.11}$$

$\mathcal{S}$ transforms as section of $\mathcal{L}^2$ and by the complex conjugate of the last equation in (4.2.8) it can be viewed as an an-holomorphic potential for the triple intersection w.r.t. to the covariant derivatives

$$C_{abc} = -D_a D_b D_c \mathcal{S} \, . \tag{4.2.12}$$

However since (4.2.6), (4.2.9) are contracted with two holomorphic or two anti-holomorphic period matrices, the propagator matrices just based on $\hat{\Delta}^{IJ}$, $\hat{\Delta}_{IJ}$ are not modular invariant. This is unlike the $G_{\alpha\bar{\beta}}$ (or its inverse), which by (2.6.6) are modular invariant. We should therefore construct Δ^{IJ} as a tensor that transforms as under modular transformation by finding a modular invariant quantity that depends on $\mathrm{Im}(\tau_{IJ})$ and whose derivative w.r.t τ_{IJ} produces $\hat{\Delta}^{IJ}$ and something holomorphic. The resolution to this problem comes from the consideration of the genus one amplitude or Ray-Singer torsion.

4.2.2 Propagators and Ray-Singer Torsion

The genus one amplitude $F^{(1)}$ can be integrated [39] from its holomorphic anomaly equation (4.1.19) with $C^{kl}_{\bar{j}} := e^{2K} C_{ikl} C_{\bar{j}\bar{k}\bar{l}} G^{l\bar{l}} G^{k\bar{k}}$ and $G_{i\bar{j}} = \partial_i \bar{\partial}_{\bar{j}} K$

$$\partial_i \bar{\partial}_{\bar{j}} F^{(1)} = \frac{1}{2} C_{ikl} C^{kl}_{\bar{j}} - \left(\frac{\chi}{24} - 1\right) G_{i\bar{j}} \tag{4.2.13}$$

using the general formula for the Ricci-curvature $R_{i\bar{j}} = \partial_i \bar{\partial}_{\bar{j}} \log \det(G_{a\bar{b}})$ and the special geometry relation (2.5.21). This implies $R_{i\bar{j}} = R_{i\bar{j}l\bar{l}} G^{l\bar{l}} = G_{i\bar{j}}(h_{11}+1) - C_{ikl} C^{lk}_{\bar{j}}$ hence $\frac{1}{2} C_{ikl} C^{kl}_{\bar{j}} = \frac{1}{2} \partial_i \bar{\partial}_{\bar{j}} \big(\frac{1}{2}(h_{11}+1)K - \log\det(G_{a\bar{b}})\big)$ from which one gets

$$F^{(1)} = -\frac{1}{2} \log\det(G_{a\bar{b}}) + \left(\frac{1}{2}(h_{11}+1) - \frac{\chi}{24} + 1\right) K + \log(f(z)) + \log(f(\bar{z})) \, . \tag{4.2.14}$$

Here the topological Euler number $\chi = \chi(M) = -\chi(W)$ and $h_{11} = h_{11}(M) = h_{21}(W)$ are the ones of the original manifold M. The $f(z)$ are meromorphic functions in the modular invariant parameters z and hence by (2.5.27), (4.2.5) we see that $F^{(1)}$ is modular invariant. In [40] relations of (4.2.14) to other objects have been pointed out. It is shown to be the weighted sum of logarithms $F^{(1)} = \frac{1}{2}\sum_{q=0}^{\dim(W)} I(\wedge^q T_W^*)$ of the Ray-Singer torsions $I(V)$ of the vector bundle V, where all vector bundles over taken to be ones of the universal family of Calabi-Yau spaces over their complex structure moduli space [44, 45]. It has also been argued in [40] that it is the one loop free energy of Kodaira spencer gravity whose action was identified with the Hitchin functional. This statement has been made more precise in the sense that $F^{(1)}$ is the one loop free energy the generalized Hitchin functional [278].

To lift expressions like (4.2.13), (4.2.14) from the small to the big moduli space one notes that on tensors on the big moduli space one has [151]

$$D_i = X_i^I D_I, \qquad \bar{D}_{\bar{j}} = \bar{X}_{\bar{i}}^I \bar{D}_I . \tag{4.2.15}$$

Useful for this purpose is also the identity

$$X_l^L X_m^M C_{\bar{i}}^{lm} = \frac{1}{4} \bar{X}_{\bar{i}}^I \bar{C}_I^{LM} . \tag{4.2.16}$$

So (4.2.13) lifts to

$$\partial_{X^I} \bar{\partial}_{\bar{X}^J} F^{(1)} = \frac{1}{8} C_{ILM} \bar{C}_J^{LM} - \left(\frac{\chi}{24} - 1\right) \partial_{X^I} \bar{\partial}_{\bar{X}^J} K \tag{4.2.17}$$

Using the special geometry relation in the big moduli space $R_{IJ} = \partial_{X^I}\bar{\partial}_{\bar{X}^J} \log\det(\mathrm{Im}\tau) = -\frac{1}{4} C_{ILM}\bar{C}_J^{LM}$ one gets

$$F^{(1)} = -\frac{1}{2}\log\left(\det(\mathrm{Im}(\tau))|\Phi(\tau)|^2\right) + \left(1 - \frac{\chi}{24}\right) K + \log\left(|f(z)|^2\right) \tag{4.2.18}$$

Here we dropped possible constant terms, and splitted the holomorphic ambiguity, which is of course not fixed in the integration of (4.2.17) in two terms $f(z)$ and $\Phi(\tau)$. The terms can be identified by comparing with (4.2.14) and noting by (4.2.5) that

$$\det(G_{a\bar{b}}) = -e^{(h_{11}+1)K} \det(2\mathrm{Im}(\tau_{IJ}))\det(X_\alpha^I)\det(\bar{X}_{\bar{\beta}}^I) . \tag{4.2.19}$$

The factor $e^{(h_{11}+1)K}$ absorbs the $(h_{11}+1)$ factor in (4.2.14) and we identify

$$\Phi(\tau) = \det(X_\alpha^I) = (X^0)^{h_{11}+1} \det\left(\frac{\partial}{\partial z^i}\frac{X^j}{X^0}\right) . \tag{4.2.20}$$

The second equal sign implies that Φ is purely holomorphic. The non-holomorphic line bundle connection in X^I_a drops out in the determinant [72].

Note now that all terms in (4.2.18) are individually symplectically invariant. Let us focus on the on the first one $F^{(1)}_{trunc} = -\frac{1}{2}\log\left(\det(\mathrm{Im}(\tau))|\Phi(\tau)|^2\right)$. This comes from integrating $\partial_{X^I}\bar\partial_{\bar X^J}F^{(1)}_{trunc} = \frac{1}{8}C_{ILM}\bar C^{LM}_J$ without integration constant. If we integrate w.r.t. $\bar X^J$ and use $\frac{\partial}{\partial X^I} = C_{ILM}\frac{\partial}{\tau_{LM}}$ we get $2i\frac{\partial}{\partial\tau_{IJ}}F^{(1)}_{trunc} = \Delta^{IJ}$. By applying this to the first term in (4.2.18) we get

$$\Delta^{IJ} = -\frac{1}{2}\mathcal{G}^{IJ} + \mathcal{A}^{IJ}, \tag{4.2.21}$$

with

$$\mathcal{A}^{IJ} = i\frac{\frac{\partial\Phi(\tau)}{\partial\tau_{IJ}}}{\Phi(\tau)}. \tag{4.2.22}$$

Since $F^{(1)}_{trunc}(\tau)$ was modular invariant this $\Delta^{IJ}(\tau)$ transforms as $\Delta^{IJ}(\tau_\gamma) = (\mathfrak{c}\tau+\mathfrak{d})\Delta^{IJ}(\mathfrak{c}\tau+\mathfrak{d})^T$. In view of the transformation property of (2.6.7) we see that $\mathcal{A}^{IJ}$ has to transform like a quasimodular tensor form of weight two, so that the inhomogenous term in the $\mathcal{A}^{IJ}$ transformation cancels the inhomogenous term in (2.6.7). Because of (2.6.4) ,(4.2.6) the $S^{\alpha\beta}$ are now modular invariants. This derivation of the propagator is analogous to the derivation of the propagator on an elliptic curve as almost holomorphic Eisentein series $\hat E_2(\tau)$. From (A4.33) we get

$$S^{tt} = \frac{c^2}{12}\left(E_2 - \frac{3}{\pi\,\mathrm{Im}(\tau)}\right) = \frac{c^2}{12}\hat E_2 = -2c^2 D_\tau F^{(1)}. \tag{4.2.23}$$

I.e. the function $\Phi(\tau)$ is here identified with the $\eta(\tau)$ invariant of the elliptic curve. For genus two mirror curves with $\mathrm{Im}(\tau)_{pq}$, $p,q = 1,2$ one obtains a very similar formular

$$S^{ij} = \frac{1}{2\pi i}\frac{C^i_p C^i_q}{10}\left(\frac{\partial}{\partial\tau_{pq}}\log(\chi_{10}(\tau)) - \frac{5}{2\pi}(\mathrm{Im}(\tau)^{-1})^{pq}\right), \tag{4.2.24}$$

where now χ_{10} is the Igusa cusp form and the constants C^i_p are related to the intersection numbers in the A model but can also be calculated as the χ^α_I in (4.2.6) in the local limit of the B-model.

One can also derive expressions for the propagators and Eq. (4.2.10) using (2.5.21). First we can integrate (2.5.21) w.r.t. $\bar z^{\bar j}$ by noting that all terms can be written as $\bar\partial_{\bar j}$ derivative using $G_{i\bar j} = \partial_i\bar\partial_{\bar j}K$ and the last equation in (4.2.8) as

$$\Gamma^k_{il} = \delta^k_l K_i + \delta^k_i K_l - C_{iln}S^{kn} + a^k_{il} \tag{4.2.25}$$

This equation can be solved for S^{kn} by inverting the triple coupling C_{iln} to $C^{(p)il} = \left((C_{(p)})^{-1}\right)^{il}$, so that $C^{(p)il}C_{(p)lk} = \delta^i_k$

$$S^{kl} = C^{(p)km}\left(\delta^l_m K_{(p)} + \delta^l_{(p)} K_m - \Gamma^l_{(p)m} + a^l_{(p)m}\right) \tag{4.2.26}$$

A priori there is such a determining equation for every p with $p = 1, \ldots, h_{12}(W)$. However the non-degeneracy of the triple coupling guarantees only that there is at least one p in this range. In particular for the elliptically fibred Calabi-Yau spaces, the algebraic triple couplings are not invertible over the index $p = e = 1$ which corresponds to the elliptic fibre. In (4.2.25) $a^k_{il} = a^k_{li}$ is holomorphic and undetermined. However for S^{kl} to become a tensor $\mathrm{Sym}^2\left(T^{(1,0)}\right) \otimes \mathcal{L}^{-2}$ it has to compensate the inhomogeneous transformation of the Christoffel symbol. This property can be guaranteed by the definition (4.2.2), (4.2.6). It is tedious, but straightforward to derive (4.2.10) with the ambiguities a^k_{il} and further ones a^{lm}_k, a^{lk}, a_k and a_{kl} occurring due to further integration of $\bar\partial_k D_k S^{lm}$ $\bar\partial_k D_k S^l$, $\bar\partial_k D_k S$ and $\bar\partial_k D_k K_l$ with respect to $\bar z^{\bar k}$. The number of these ambiguities adds up modulo their obvious symmetries to $h_{21}(h^2_{21} + 2h_{21} + 2)$. They are not independent, but are all related to the symmetric $(h_{21}+1) \times (h_{21}+1)$ tensor of ambiguities (4.2.22), which provides also right modular properties.

In the first step of the rederivation of (4.2.10) one considers $\bar\partial_{\bar j} D_k S^{lm}$, applies (4.1.10) to evaluate the covariant derivative and uses (2.5.21), when $\bar\partial_{\bar j}$ hits the Christoffel symbols. Using also $D_k C^{lm}_{\bar j} = D_k(e^{2K}\bar C_{\bar j\bar m\bar l}G^{l\bar l}G^{m\bar m}) = 0$, because $\bar C_{\bar j\bar m\bar l}$ is anti-holomorphic, the mixed Christoffel symbols vanish and $G^{l\bar l}$ as well as e^{2K} are covariantly constant, $G_{i\bar j} = \partial_i\bar\partial_{\bar j}K = \bar\partial_{\bar j}K_i$, one arrives at $\bar\partial_{\bar j} D_k S^{lm} = \delta^l_k S^m_{\bar j} + \delta^m_k S^l_{\bar j} - C^{lc}_{\bar j} C_{ack} S^{am} - C^{cm}_{\bar j} C_{ack} S^{al}$. Using (4.2.7) and the holomorphicity of C_{abc} the r.h.s can also written as a total derivative w.r.t. $\bar\partial_{\bar j}$, which can be integrated to (4.2.10) plus a holomorphic ambiguity. This and the similar evaluation of $\bar\partial_{\bar j} D_k S^l$, $\bar\partial_{\bar j} D_k S$ and $\bar\partial_{\bar j}\partial_k K_l = G_{a\bar j}\Gamma^a_{kl}$ yields

$$\begin{aligned} D_k S^{lm} &= -C_{abk} S^{la} S^{mb} + \delta^l_k S^m + \delta^m_k S^l + a^{lm}_k\,, \\ D_k S^l &= -C_{kab} S^a S^{bl} + 2\delta^l_k S + K_a a^{al}_k + a^l_k\,, \\ D_k S &= -\tfrac{1}{2} C_{kab} S^a S^b + \tfrac{1}{2} K_a K_b a^{ab}_k + K_a a^a_k + a_k\,, \\ D_k K_l &= -K_k K_l - S^a C_{kla} + C_{kla} S^{ab} K_b + a_{kl}\,. \end{aligned} \tag{4.2.27}$$

4.2.3 Propagators and Direct Integration Formulas

As was reviewed in Sect. 4.1.1, the generalisation of holomorphic anomaly for genus one (4.2.13) to arbitrary genus was found in [40] to be

$$\bar{\partial}_{\bar{j}} F^{(g)} = \frac{1}{2} C^{kl}_{\bar{j}} \left(D_k D_l F^{(g-1)} + \sum_{h=1}^{g-1} D_l F^{(h)} D_l F^{(g-h)} \right) \tag{4.2.28}$$

With (4.2.15), (4.2.16) this can be written in the large moduli space as

$$\bar{\partial}_{\bar{X}^I} F^{(g)} = -\frac{i}{8} C^{KL}_{\bar{I}} \left(D_K \partial_L F^{(g-1)} + \sum_{h=1}^{g-1} \partial_K F^{(h)} \partial_L F^{(g-h)} \right), \tag{4.2.29}$$

Using (4.2.2) and hence $\bar{\partial}_{\bar{I}} F^{(g)} = \frac{i}{4} C^{KL}_{\bar{J}} \frac{\partial F^{(g)}}{\partial \Delta^{KL}}$ this can be written immediately written in a form suitable for direct integration

$$\frac{\partial F^{(g)}}{\partial \Delta^{KL}} = \frac{1}{2} \left(D_K \partial_{\bar{X}^L} F^{(g-1)} + \sum_{h=1}^{g-1} \partial_K F^{(h)} \partial_L F^{(g-h)} \right). \tag{4.2.30}$$

In the small moduli space $\{\underline{S}, K_i\} = \{S^{ij}, S^i, S, K_i\}$ are viewed as a redundant set of non-holomorphic generators similar to the Eisenstein series $\hat{E}_2$ for $\mathrm{Sl}(2,\mathbb{Z})$, with respects to one wishes to integrate in the recursive direct integration step. To implement this direct integration in the small moduli space is convenient redefine the propagators [11, 330] as

$$\tilde{S}^{ij} = S^{ij}, \qquad \tilde{S}^i = S^i - S^{ia} K_a, \qquad \tilde{S} = S - S^a K_a + S^{ab} K_a K_b \tag{4.2.31}$$

The motivation is that their derivatives w.r.t to $\bar{z}^{\bar{c}}$ contain all $C^{ab}_{\bar{c}}$. i.e. $\bar{\partial}_{\bar{c}} \tilde{S}^{ij} = \bar{\partial}_{\bar{c}} S^{ij} = C^{ij}_{\bar{c}}$, $\bar{\partial}_{\bar{c}} \tilde{S}^i = -C^{ia}_{\bar{c}} K_a$, $\bar{\partial}_{\bar{c}} \tilde{S} = \frac{1}{2} C^{ab}_{\bar{c}} K_a K_b$, by (4.2.8) and $G_{a\bar{b}} = \partial_{\bar{a}} \bar{\partial}_{\bar{b}} K = \bar{\partial}_{\bar{b}} K_a$. Hence if one changes the independent generators as in $\tilde{F}^{(g)}(\underline{\tilde{S}}, K_i, \underline{G}) = F^{(g)}(\underline{S}(\underline{\tilde{S}}), K_i, \underline{G})$, where $\underline{G}$ are holomorphic generators similar to the Eisenstein series E_4, E_6 for $\mathrm{Sl}(2,\mathbb{Z})$, and dropping all tildes again one finds

$$\begin{aligned} \bar{\partial}_{\bar{j}} F^{(g)} &= C^{ab}_{\bar{j}} \frac{\partial F^{(g)}}{\partial S^{ab}} + G_{a\bar{j}} \left(\frac{\partial F^{(g)}}{\partial S} S^a + \frac{\partial F^{(g)}}{\partial S^b} S^{ab} + \frac{\partial F^{(g)}}{\partial K_a} \right) \\ &= C^{ab}_{\bar{j}} \left(\frac{\partial F^{(g)}}{\partial S^{ab}} - \frac{\partial F^{(g)}}{\partial S^a} K_b + \frac{\partial F^{(g)}}{\partial S} K_a K_b \right) + G_{a\bar{j}} \frac{\partial F^{(g)}}{\partial K_a} \end{aligned} \tag{4.2.32}$$

In view of (4.2.28) the proportionality to $C^{ab}_{\bar{j}}$ is useful to write (4.2.28) in the same form as (4.2.30), namely as derivative w.r.t. to the propagators. Hence one proves inductively that $\frac{\partial F_g}{\partial K_a} = 0$ as in (4.2.28) there is no term proportional to $G_{a\bar{j}}$ on the r.h.s. Therefore the independent non-holomorphic generators of the F_g are $\{\underline{\tilde{S}}\}$ [11, 330]. To evaluate the r.h.s. one transforms (4.2.27), for convenience to normal derivatives, of the $\{\underline{\tilde{S}}\}$ as in [11]

$$\begin{aligned}
\partial_k \tilde{S}^{lm} &= C_{kab}\tilde{S}^{la}\tilde{S}^{mb} + \delta^l_k \tilde{S}^m + \delta^m_k \tilde{S}^l - a^l_{ka}\tilde{S}^{am} - a^m_{ka}\tilde{S}^{la} + a^{lm}_k, \\
\partial_k \tilde{S}^l &= C_{kab}\tilde{S}^{la}\tilde{S}^b + 2\delta^l_k \tilde{S} - a^l_{ka}\tilde{S}^a - a_{ka}\tilde{S}^{la} + a^l_k, \\
\partial_k \tilde{S} &= \frac{1}{2}C_{kab}\tilde{S}^a\tilde{S}^b - a_{ka}\tilde{S}^a + a_k, \\
\partial_k K_l &= K_k K_l - C_{kla}\tilde{S}^{ab}K_b + a^a_{kl}K_a - C_{kla}\tilde{S}^a + a_{kl}.
\end{aligned} \tag{4.2.33}$$

This equations together with the fact that we can solve for $\tilde{S}^{ij} = S^{ij}$ (4.2.26) implies the formulas [12],

$$\begin{aligned}
\tilde{S}^i &= \tfrac{1}{2}\left(\partial_i \tilde{S}^{ii} - C_{imn}\tilde{S}^{mi}\tilde{S}^{ni} + 2a^i_{im}\tilde{S}^{mi} - a^{ii}_i\right), \\
\tilde{S} &= \tfrac{1}{2}\left(\partial_i \tilde{S}^i - C_{imn}\tilde{S}^m\tilde{S}^{ni} + a^i_{im}\tilde{S}^m + a_{im}\tilde{S}^{mi} - a^i_i\right),
\end{aligned} \tag{4.2.34}$$

where there is no sum over equal indices.

Under the assumption that the propagators are functionally independent and that the $F^{(g)}$ are independent of K_a, which is proven recursively, we can now use (4.2.32) to the l.h.s of (4.2.28) and (4.2.33) to evaluate the r.h.s. of these equations respectively and by comparing then coefficients of $(K_a)^0$, K_a, $K_a K_b$ and arrive at equations for partial derivative, which are

$$\begin{aligned}
\frac{\partial F^{(g)}}{\partial S^{ij}} &= \frac{1}{2}\partial_i(\partial'_j F^{(g-1)}) + \frac{1}{2}(C_{ijl}S^{lk} - s^k_{ij})\partial'_k F^{(g-1)} + \frac{1}{2}(C_{ijk}S^k - h_{ij})c_{g-1} \\
&\quad + \frac{1}{2}\sum_{h=1}^{g-1}\partial'_i F^{(h)}\partial'_j F^{(g-h)}, \\
\frac{\partial F^{(g)}}{\partial S^i} &= (2g-3)\partial'_i F^{(g-1)} + \sum_{h=1}^{g-1} c_h \partial'_i F^{(g-h)}, \\
\frac{\partial F^{(g)}}{\partial S} &= (2g-3)c_{g-1} + \sum_{h=1}^{g-1} c_h c_{g-h},
\end{aligned} \tag{4.2.35}$$

where the c_g is defined as

$$c_g = \begin{cases} \frac{\chi}{24} - 1, & g = 1; \\ (2g-2)F^{(g)}, & g > 1. \end{cases} \tag{4.2.36}$$

We have also used the notation ∂' to denote

$$\partial'_i F^{(g)} = \begin{cases} \partial_i F^{(g)} + (\frac{\chi}{24} - 1)K_i, & g = 1; \\ \partial_i F^{(g)}, & g > 1, \end{cases} \tag{4.2.37}$$

i.e. on the right hand in (4.2.35), we use the integrated version of (4.2.13) w.r.t. $\bar{\partial}_{\bar{j}}$ and the definition of the S^{ij} in (4.2.7) for $\partial_i F^{(1)}$ omitting the $-(\frac{\chi}{24} - 1)K_i$ term. Similar equations in the big moduli space have been derived in [167].

Since the propagators are symmetric $S^{ij} = S^{ji}$, we can choose to use only the S^{ij} with $i \le j$. In the case of $i \ne j$, the right hand side of first equation in (4.2.35) need to be multiplied by an extra factor of 2 to take account of the double contribution.

The advantage of formulating the holomorphic anomaly equation in the form (4.2.35) is that if we assume that the S^{ij}, $S^i = S^{i\varphi}$ and $S = S^{\varphi\varphi}$ are functionally independent modular generators then the equations can be integrated directly recursively to reach the following conjecture

Conjecture 3 Integration of (4.2.36) yields for each $F^{(g)}$ a minimal in-homogenous polynomial of degree $3g - 3$ with the weighted degree $(1, 2, 3)$ for the propagators S^{ij}, S^i and S respectively and whose coefficients are meromorphic functions in the complex moduli z.

The latter two points are already clear, because given the type of degenerations of the world-sheet Riemann surface [40] found an action whose Feynman graphs reproduces precisely those degenerations. The propagators are literally the once derived above and the vertices are the derivatives of the $F^{(g)}$ with the following additional rules $n, m = 0, 1, 2 \ldots$

$$F^{(0)}_{\varphi^n} = 0, \ F^{(0)}_{i\varphi^n} = 0, \ F^{(0)}_{ij\varphi^n} = 0, \ F^{(g)}_{i_1,\ldots i_m,\varphi^{n+1}} = (2g - 2 + n + m) \cdot F^{(g)}_{i_1,\ldots i_m,\varphi^n}$$

The graph contribution is divided by the following symmetry factors: $k!$ for k equal (self)links joining the same vertices, 2 for each selflink $S^{\varphi\varphi}$, S^{ij} times the order of the graph automorphism obtained by permuting the vertices. For example form this actions one gets purely by Feynman graph combinatorics for the genus 2 amplitude those degeneration are captured by the 12 graphs in Fig. 17. As it is clear from Fig. 17, e.g. the first two, the second two the third, e.t.c two graphs give rise to the same contributions in the polynomial for $F^{(g)}$. The advantage of the *direct integration* method, is that the terms in the polynomial grow of course polynomial, while the number of Feynman graphs growth exponentially. In a sense the closing of the rings of propagators (4.2.33) and (4.2.28) induce all Ward identities among the Feynman amplitudes.

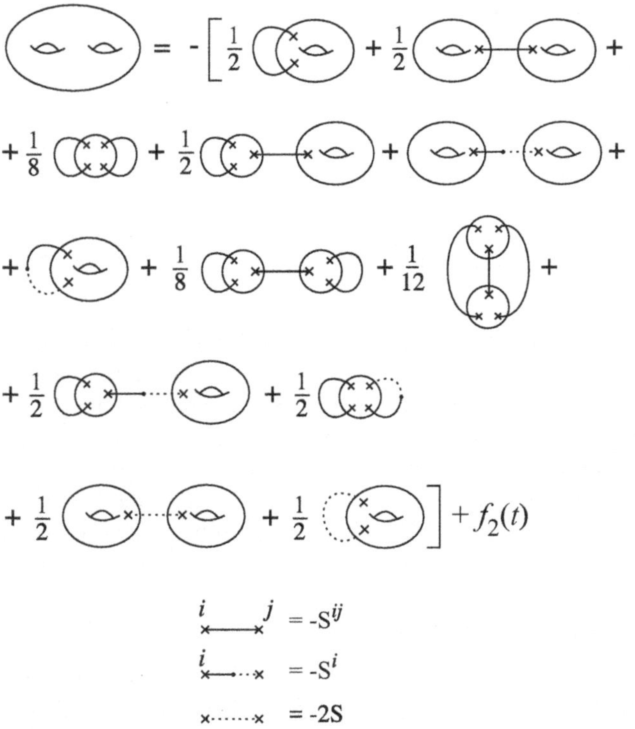

Fig. 17 Contributions of degenerate world-sheets to the genus two amplitude after [40]

The above Feynman term expansion for the $F^{(g)}$ can also be obtained by integration-by-parts method using recursively (4.2.28) and using the commutator $[\bar{\partial}_{\bar{i}}, D_j]$ to pull out a $\bar{\partial}_{\bar{i}}$ on the r.h.s., as described in [40].

Of course the methods of *direct integration* relies on the assumption that there is a ring of modular forms which generate the $F^{(g)}$ and that in particular the anholomorphic elements in this ring are in dependent and behave like (4.2.33) under derivation. That is exactly what happens in the local case for which the ring is the ring of almost holomorphic modular forms and Conjecture 3 can be proven in local examples as we will show in the next Sect. 4.2.4.

4.2.4 Calculation and Checks from the Local Toric Calabi-Yau Cases

Indeed all expressions that are needed in the direct integration of the holomorphic anomaly equations can be written in terms of modular forms of $\Gamma(3)$, which are summarised in section "Modular Forms of $\Gamma(3)$" in Appendix 4 as was first spelled out in [6]. The complex deformation parameter of the family of mirror curves that is equivalent to (2.7.31) with $z = -(1/3\psi)^3$ is

$$\{\sum_{i=1}^{3} x_i^3 - 3\psi \prod_{i=1}^{3} x_i = 0 | (x_1 : x_2 : x_3) \in \mathbb{P}^2\} . \tag{4.2.38}$$

is given as function of the complex structure parameter $\tau \in \mathbb{H}_+$ by

$$\psi(\tau) = \frac{a - c - b}{d}, \tag{4.2.39}$$

where a,b,c and d are defined in section "Modular Forms of $\Gamma(3)$" in Appendix 4.

The flat Kähler parameter of the A-model can be found by integrating

$$\frac{\partial t}{\partial \psi} = -\sqrt{3}\frac{d}{\eta} \tag{4.2.40}$$

and the discriminant can be written as

$$\Delta = 1 - \psi^3 = -3^3 \frac{\eta^{12}}{d^4} .$$

From the expression (4.2.14) in the local limit in which the Kähler-potential K becomes an irrelevant constant $F_1(\tau) = -\log(\sqrt{\mathrm{Im}(\tau)}|\eta(\tau)|^2) + \frac{1}{24}\log(\Delta)$ we get in the holomorphic limit up to an infinite constant

$$\mathcal{F}_1(\tau) = -\log(\eta(\tau)) + \frac{1}{24}\log(\Delta) \; = -\frac{1}{6}\log(d\eta^3),$$

The 3-point function $C_{ttt} = \frac{\partial^3}{\partial t^3}\mathcal{F}^{(0)}$ can be written as

$$C_{ttt} = -\frac{1}{3}\frac{\partial \tau}{\partial t} = -\frac{1}{3}\frac{\partial \psi}{\partial t}\frac{\partial \tau}{\partial \psi}$$

and using the modular expressions for ψ, for $\frac{\partial t}{\partial \psi}$, and the formulae for logarithmic derivatives in section "Modular Forms of $\Gamma(3)$" in Appendix 4, we get

$$C_{ttt} = -\frac{1}{3^{5/2}}\frac{d}{\eta^9} . \tag{4.2.41}$$

Also the $\Gamma(3)$-invariant Yukawa coupling, expressed in terms of the globally defined variable ψ is easily obtained from (2.9.17), where we note that $X^0 = 1$ in the local limit

$$C_{\psi\psi\psi} = \left(\frac{\partial t}{\partial \psi}\right)^3 C_{ttt} = -\frac{9}{\Delta}. \tag{4.2.42}$$

We note that in the local limit the only propagator is given by (4.2.23). Consequently only the first equation in (4.2.35) with the additional simplification in the local limit that $c_g = 0$ and $\partial_i' = \partial_i$ is relevant. Hence the general form of the higher genus amplitudes reads

$$F_g = C^{g-1} \sum_{k=0}^{3(g-1)} \hat{E}_2^k h_g^{(3g-3-k)}(K_2, K_4, K_6), \tag{4.2.43}$$

where we defined the weight -6 object

$$C = \frac{d^2}{2^9 3^4 \eta^{18}} = \frac{1}{1536} C_{ttt}^2$$

and the ring of modular forms of $\Gamma(3)$ generating the weight $2d$ forms $h_g^{(d)}$ is given by

$$K_2 = -\alpha^2 \frac{(a-\alpha c)^2}{\eta^2}, \qquad K_4 = \frac{1}{\alpha^2-1} \frac{ac(a+c)(\alpha^2 a - c)}{\eta^4}, \qquad K_6 = \frac{(ac)^2(a+c)^2}{\eta^6}.$$

The coefficients $h_g^{(k>0)}$ of $\hat{E}_2$ are fixed either by the direct integration of (4.2.35) or by the Feynman graph expansion and we obtain for example for the holomorphic ambiguity

$$\begin{aligned} h_2^{(0)} &= F_2 - X\left(5\hat{E}_2^3 + \hat{E}_2^2 K_2 + \tfrac{1}{3}\hat{E}_2 K_2^2\right), \\ h_3^{(0)} &= F_3 - X^2(180\hat{E}_2^6 + 240\hat{E}_2^5 K_2 + 4\hat{E}_2^4(145K_2^2 - 1008K_4) \\ &\quad + \tfrac{32}{9}\hat{E}_2^3(199K_2^3 - 1908K_2K_4 + 648K_6) + \tfrac{4}{5}\hat{E}_2^2(563K_2^4 - 7936K_2^2K_4 + 26496K_4^2) \\ &\quad + \tfrac{16}{15}\hat{E}_2(149K_2^5 - 2536K_2^3K_4 + 11952K_2K_4^2 - 3456K_4K_6)). \end{aligned} \tag{4.2.44}$$

As was pointed out in [163] the conifold gap condition (4.5.8) and the regularity at the orbifold points are sufficient to fix this ambiguity completely in the local models and in the case at hand we get

$$\begin{aligned} h_2^{(0)} &= \frac{11}{69120} + \frac{1}{34560\Delta} - \frac{1}{7680\Delta^2}, \\ h_3^{(0)} &= \frac{17}{6289280} + \frac{269}{464448640\Delta} - \frac{19393}{278691840\Delta^2} + \frac{337}{2211840\Delta^3} - \frac{373}{4128768\Delta^4}. \end{aligned} \tag{4.2.45}$$

Note that the holomorphic limit is simply given by $\mathrm{Im}(\tau) \to \infty$ which has the effect of replacing the propagator, i.e. the almost holomorphic Eisenstein series $\hat{E}_2$, with the quasi modular Eisenstein series E_2 see (A4.11). As discussed in section "Differentiable Rings of Modular Forms" in Appendix 4 the ring of almost holomorphic forms generated by $(\hat{E}_2, E_4, E_6)$ and almost holomorphic forms (E_2, E_4, E_6) are isomorphic as differential rings when the *Masss derivative* is replaced by the *Serre derivative*. Constructing the an-holomorphic $F^{(g)}$ or the holomorphic $\mathcal{F}^{(g)}$ from (4.2.35) gives formally the same polynomials (4.2.43) in the generators $(\hat{E}_2, K_2, K_4, K_6)$ or (E_2, K_2, K_4, K_6) respectively. It was shown in [151] that the same is true for general Calabi-Yau 3 folds, i.e. the holomorphic limit can always be described by replacing the an-holomorphic propagators, which are transform modular, with their holomorphic limit which transform quasimodular. Note further that the B-model expressions (4.2.43) contain the exact moduli dependence and using the wave function transformation property of Z they can be evaluated everywhere in the moduli space. That was used in [6] to predict orbifold Gromov-Witten invariants and in [163] to fix the ambiguities (4.2.45) to genus 105. As we mentioned above the model is completely solvable by the gap conditions and the orbifold regularity.

Of course the mirror map defined by the modular expression (4.2.39) and (4.2.39) can be also obtained using the Picard-Fuchs operator, which due to (2.9.43) and (2.12.15) $(3\psi = u)$ is given by

$$\mathcal{D} = [\theta_z^2 + 3z(3\theta_z + 1)(3\theta_z + 2)]\theta_z \tag{4.2.46}$$

or since this is hypergeometric from $\bar{l}^{(3)} = (-3, 1, 1, 1)$ from (2.9.53) noting that there are no $l_{0,m}$ in the definition of $c(n, \rho)$ in (2.9.48) and hence no numerators.

4.3 The Integer Structure of the BPS Invariants

To see the integer structure of the BPS invariants requires to make a physically or geometrically motivated gauge choice in the Kähler line bundle and to take a holomorphic limit for the suitable polarisation in $H_3(W, \mathbb{R})$ for the singular point in moduli space under consideration. First we describe these two concepts, and after completing the example of the $\mathcal{O}(-3) \to \mathbb{P}^2$ local geometry, we come to the general physical and mathematical arguments for the BPS integrality structure.

4.3.1 Physical Gauge and Holomorphic Limit

The integer structure and the BPS invariants are obtained from the $F^{(g)}$ calculated from the holomorphic anomaly in a physical gauge and after taking an holomorphic limit. This physical gauge corresponds to a choice of the section $\mathcal{L}^{2-2g}$ and depends on the locus in $\mathcal{M}_{cs}$ in question. For the Gromov-Witten invariants the relevant locus is the maximal unipotent point where one has always a holomorphic solution $X_m^0 = 1 + \mathcal{O}(z)$. The physical gauge is given by

$$F_p^{(g)}(t) = (X_m^0)^{2g-2} F^{(g)}(z(t)) = (X_m^0)^{2g-2} F^{(g)}(X(z(t))) \,, \tag{4.3.1}$$

where we also passed to the in-homogeneous coordinates chosen by the mirror map $t^a = X^a_m/X^0_m$. Here we stressed by the index m that both choices depend on the point in the moduli space under consideration. I.e. there are physical gauge choices at other points in the moduli space, which involve different choices of in-homogenous coordinates. The choice of these coordinates is dictated by the flatness condition and a simple behaviour under the local monodromy. E.g. at the maximal unipotent point there are $h_{11}(M) = h_{21}(W)$ variables t_i which have simple shift monodromies (2.9.54) so that the Q^a defined in (2.9.55) are monodromy invariant. Generally one picks $t^a = X^a_*/X^0_*$, such that they become locally flat coordinates w.r.t. to the Weil-Petersson connection and that they have the simplest possible monodromy transformation, i.e. minimal shift symmetry at the maximal unipotent monodromy point or minimal phase rotation for the eigenvectors of X^0_o, X^a_o for irreducible orbifold monodromy action, as for example in Sect. 2.10.2. The choice for X^a at the conifold is given by the vanishing cycle, but the choice of $X^0 = 1+\mathcal{O}(\delta)$ in Sect. 2.10.3 is ambiguous and also do not change e.g. the gap properties as discussed in [191].

In addition the generating function for the Gromov-Witten invariants are holomorphic functions, so a holomorphic limit at the base point in the t^a coordinate t^a_0 that corresponds to point in the moduli space.

$$\mathcal{F}^{(g)}(t) = \lim_{\bar{t}^a \to \bar{t}^a_0} F^{(g)}_p(t) \,. \qquad (4.3.2)$$

Here $F^{(g)}_p(t)$ is an an-holomorphic and $\mathcal{F}^{(g)}(t)$ is a holomorphic or in general a meromorphic function.[108] The holomorphic limit is defined by the *canonical coordinates* t^a and one can infer further properties for this choice from the existence of a well defined limit. In particular one can show that in the holomorphic limit the leading behaviour the metric data is a follows

$$e^{-K} \sim X^0_*, \qquad G_{j\bar{\imath}} \sim \partial_{z_j} t^i_* C_{i\bar{\imath}}, \qquad (4.3.3)$$

where $C_{i\bar{\imath}}$ is a constant matrix, that can be chosen to be diagonal. Note that from this expressions we get the connection Γ^i_{jk} with the holomorphic indices and by (4.2.26), (4.2.34) we can calculate the holomorphic limit of the propagators at each point $*$. Since the coefficients in the polynomials in Conjecture 3 are globally defined rational functions in z, it is indeed the fastest to evaluate the $\mathcal{F}^{(g)}(t_*)$ by replacing in the $F^{(0)}(S^{ij}, S^i, S, z)$ the propagators by the local holomorphic expressions and perform the physical gauge choice (4.3.1). It was shown in [6] more conceptional that this corresponds to change in the polarisation of the wave function Z.

[108] $\mathcal{F}^{(g)}(t)$ for $g > 1$ only has poles. $\mathcal{F}^{(0)}$ and $\mathcal{F}^{(1)}$ have logarithmic branch points.

4.3.2 The A-Model Results for the Local $\mathbb{P}^1$ and $\mathbb{P}^2$ Geometry

Mathematically Gromov-Witten invariants of higher genus can be calculated by localisation, but only in non-compact toric Calabi-Yau manifolds. For example using special weights and closed expressions for certain classes of Hodge integral [111] prove the following all genera result for the blown up conifold geometry $\mathcal{O}(-1)\oplus\mathcal{O}(-1)\rightarrow\mathbb{P}^1$ as

$$\mathcal{F}(\lambda,t)=\sum_{g=0}^{\infty}\lambda^{2g-2}\mathcal{F}^{(g)}(t)=\sum_{g=0}^{\infty}\sum_{d=1}^{\infty}r_d^g\lambda^{2g-2}Q^d\,, \tag{4.3.4}$$

where the r_d^g are the Gromov-Witten invariants defined physically in (3.4.12), for mathematical definition see [182], $d\in\mathbb{Z}$ specifies the degree in $H_2(M,\mathbb{Z})$, which is generated just by the $\mathbb{P}^1$ and $q:=\exp(2\pi t)$. The result of Faber and Pandharipande [111] gives all r_d^g by the formula

$$\mathcal{F}(\lambda,t)=\sum_{d=1}^{\infty}\frac{Q^d}{d\left(\sin\frac{d\lambda}{2}\right)^2}\,. \tag{4.3.5}$$

In this simple geometry we can understand all contributions as the multicovering of the rigid $\mathbb{P}^1$, which is the only non-trivial holomorphic curve in this geometry, by maps of various degree and genus. This calculation includes a proof of the Aspinwall-Morrison formula [24] and extends it to higher genus for rigid $\mathbb{P}^1$ curves.

In general non-compact toric Calabi-Yau can support holomorphic curves of arbitrary genera in infinitely many classes β where these multi-covering formulas and their generalisations, discussed in the next sections can be tested. E.g. for the closed string amplitudes on $\mathcal{O}(-3)\rightarrow\mathbb{P}^2$ [228] obtain by localisation

$$\begin{aligned}
\mathcal{F}^{(0)}&=-\tfrac{t^3}{18}+3\,Q-\tfrac{45\,Q^2}{8}+\tfrac{244\,Q^3}{9}-\tfrac{12333\,Q^4}{64}+\tfrac{211878\,Q^5}{125}\ldots\\
\mathcal{F}^{(1)}&=-\tfrac{t}{12}+\tfrac{Q}{4}-\tfrac{3\,Q^2}{8}-\tfrac{23\,Q^3}{3}+\tfrac{3437\,Q^4}{16}-\tfrac{43107\,Q^5}{10}\ldots\\
\mathcal{F}^{(2)}&=\tfrac{\chi}{5720}+\tfrac{Q}{80}+\tfrac{3\,Q^3}{20}-\tfrac{514\,Q^4}{5}+\tfrac{43497\,Q^5}{8}\ldots\\
\mathcal{F}^{(3)}&=-\tfrac{\chi}{145120}+\tfrac{Q}{2016}+\tfrac{Q^2}{336}+\tfrac{Q^3}{56}+\tfrac{1480\,Q^4}{63}-\tfrac{1385717\,Q^5}{336}\ldots\\
\mathcal{F}^{(4)}&=\tfrac{\chi}{87091200}+\tfrac{Q}{57600}+\tfrac{Q^2}{1920}+\tfrac{7\,Q^3}{1600}-\tfrac{2491\,Q^4}{900}+\tfrac{3865234\,Q^5}{1920}\ldots\\
\mathcal{F}^{(5)}&=-\tfrac{\chi}{2554675200}+\tfrac{Q}{1774080}+\tfrac{Q^2}{14080}+\tfrac{61\,Q^3}{49280}+\tfrac{4471\,Q^4}{22176}-\tfrac{65308319\,Q^5}{98560}\ldots\,.
\end{aligned} \tag{4.3.6}$$

and by the general multicovering formulas (4.3.18) one extracts the integer BPS numbers $n^{(g)}_{\beta=d}$ reported in Table 7.

The combinatoric of the A-model localisation calculation is involved. E.g. for the genus 5 degree 5 terms one has to sum over $\sim 10^4$ graphs, however the localisation procedure is completely algorithmic an will determine all invariants for all toric Calabi-Yau spaces. Another method to obtain the r_d^g or from then the $n_d^{(g)}$ is the topological vertex [4].

Table 7 The (weighted) number of BPS states n_d^g for the local $\mathbf{P}^2$ case

d	g=0	g=1	g=2	g=3	g=4
1	3	0	0	0	0
2	−6	0	0	0	0
3	27	−10	0	0	0
4	−192	231	−102	15	0
5	1695	−4452	5430	−3672	1386
6	−17064	80948	−194022	290853	−290400
7	188454	−1438086	5784837	−1536990	29056614
8	−2228160	25301295	−155322234	649358826	−2003386626
9	27748899	−443384578	3894455457	−23769907110	109496290149
10	−360012150	7760515332	−93050366010	786400843911	−5094944994204

Due to the non-trivial holomorphic curves in all degrees it is hard to give $\mathcal{F}(\lambda, t)$ in closed form, even though closed expressions for the $\mathcal{F}^{(g)}(t)$ can be given using mirror symmetry and the B-model [228] and described by the B-model analysis in terms of modular forms as explained in Sect. 4.2.43.

4.3.3 Schwingerloop Calculation of BPS Invariants from Branes Wrapping Curves

Many fascinating topological and physical ideas enter the reinterpretation of $\mathcal{F}^{(g)}(t)$ as BPS counting function [140]. The argument splits in a supergravity consideration and a geometrical part

- The $N = 2$ supergravity action contains terms $\sum_{g>0} \int_{M^4} \mathrm{d}^4x \mathcal{F}^{(g)}(t, \bar{t}) T_-^{2g-2} R_- \wedge R_-$, which couple the anti-selfdual part of the curvature R_+ with the anti-selfdual part of the graviphoton field strength T_-. The above terms are part of the component form of $\int_{M^4} \mathrm{d}^4x\mathrm{d}^4\theta \mathcal{F}^{(g)}(t, \bar{t})(W^2)^{g-1}$, where $W^2 = \epsilon_{ij}\epsilon_{kl} W^{ij}_{\mu\nu} W^{kl}_{\mu\nu}$ and $W^{ij}_{\mu\nu} = \epsilon^{ij} T_{\mu\nu} - R_{\mu\nu\eta\delta}\theta^i \sigma^{\eta\delta}\theta^j + \ldots$ is a chiral multiplet. The structure of $N = 2$ supergravity in Type II string on M implies that in the topological limit $\mathcal{F}^{(g)}(t) = \lim_{\bar{t}\to\bar{t}_0} \mathcal{F}^{(g)}(t, \bar{t})$ in the one loop contribution

$$S^{N=2}_{1-loop} = \int \mathrm{d}^4x R_-^2 \mathcal{F}(\lambda, t) , \tag{4.3.7}$$

is identified with the topological string free energy (4.3.18) [18, 40] after a suitable identification of T_- with λ. It depends only on vector multiplets. This statements require like (2.4.63) a certain genericity assumptions. Moreover supergravity puts the following restriction on this amplitude [18]. It is generated at one-loop and at one-loop only, the corresponding graph is shown in Fig. 18. The only particles which can contribute in the loop are BPS states. Their mass is

Fig. 18 BPS saturated one-loop graph contribution to $T_-^{2g-2}R_-^2$

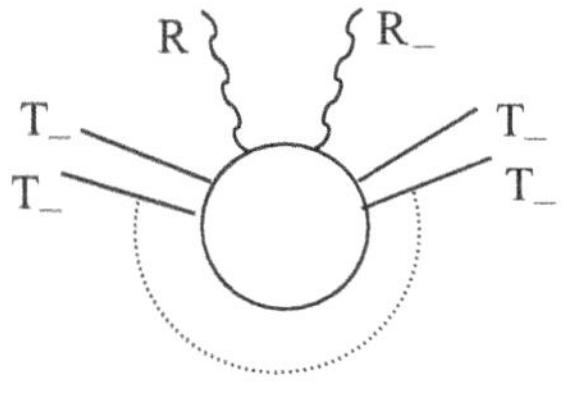

determined by their charge. Once *mass* and *spin* of the BPS particle is known is contribution to $\mathcal{F}^{(g)}(t)$ can be evaluated by a Schwinger-loop calculation.
- In the geometrical consideration one has to identify the *mass* and *spin* of BPS particle with the geometrical properties of the embedded branes. The mass is easy and will be discussed below. The spin part is more complicated and is discussed in Sect. 4.3.4

Because the type II string coupling g_s is in a hyper multiplet and the above decoupling one expects that the strongly coupled M-theory- $g_s \gg 1$ and the weakly coupling IIA description $g_s \ll 1$ are equivalent points of view. The former description involves BPS states as coming from $M2$ branes the latter as coming from $D2 - D0$ bound states. In both cases the extended branes wrapping curves C in M in the class β. The mass is given straightforwardly as

$$m(\beta, k) = \frac{1}{\lambda}2\pi i\beta \cdot t + 2\pi ik = \frac{1}{\lambda}\sum_{i=1}^{h^{1,1}} t_i \int_{C_\beta} \omega_i + 2\pi ik, \qquad \beta \in H^2(M, \mathbf{Z}), \qquad k \in \mathbf{Z}\,, \tag{4.3.8}$$

were the first term is the minimal volume of the curve on which the extended brane wraps. The second can be either viewed as the momentum k of $M2$ on the M theory circle or as the number k of $D0$ branes. The latter form in arbitrary numbers k a bound states with the $D2$ brane.

Consider now an M-theory compactification on M to five dimensions. The space time BPS states fall into representations $\mathcal{R} = [j_-, j_+]$ of the Little Group of the 5D Lorentz group $L = \mathrm{SO}(4) \simeq \mathrm{SU}(2)_- \times \mathrm{SU}(2)_+$ and have a mass m related to their charge by the BPS formula. As mentioned the low energy interpretation of the free energy $\mathcal{F}$ in 4d relates it to the 5d BPS spectrum through a Schwinger one loop calculation of the 4d $\int_{M^4} T_-^{2g-2}R_-^2$ effective terms.[109] Note that these 4d calculations are sensitive to the off shell quantum numbers, i.e. to $\mathrm{SU}(2)_- \times \mathrm{SU}(2)_+$. Only BPS particles annihilated by the supercharges in the $(\mathbf{0}, \frac{1}{2})$ representation contribute to the loop. They couple to the anti-selfdual graviphoton field strength T_- and the anti-selfdual curvature R_- only via their left spin eigenvalues of their representation under L. The right $+$ representation content enters solely via its

[109] A similar one loop calculation corrects the effective gauge coupling $\frac{1}{g^2(G,p^2)}$ through threshold effects in heterotic strings [209].

multiplicity and a sign $(-1)^{2j_+^3}$, in particular any contribution of long multiplets is projected out by these signs. To summarize, the dependence of $\mathcal{F}$ on the BPS spectrum is via a supersymmetric index

$$I(\alpha,\tau)=\mathrm{Tr}_{\mathcal{H}}(-1)^F e^{-\alpha j_-^3-\tau H}\ , \tag{4.3.9}$$

where $F = 2j_-^3 + 2j_+^3$, and all spin information entering $\mathcal{F}$ is carried by $\left[\left(\frac{1}{2}\right)_- + 2(\mathbf{0})_-\right]$ times the following combination

$$\sum_{j_-^3,j_+^3}(-1)^{2j_R^3}(2j_+^3+1)N^\beta_{j_-^3,j_+^3}[j_-]=\sum_{g=0}^{\infty}n_\beta^{(g)}I_g\ . \tag{4.3.10}$$

The multiplicities of the BPS states $N^\beta_{j_-^3,j_+^3}$ enters only via the index like quantity $n_\beta^{(g)}$. Indeed the basis change of the left spin from $[j_-]$ to

$$I_g=\left(2[0]_-+\left[\frac{1}{2}\right]_-\right)^{\otimes g} \tag{4.3.11}$$

relates the left spin $-$ to the genus g of C as explained in Sect. 4.3.4 and defines the integer Gopakumar-Vafa invariants n_β^g associated to a holomorphic curve C of genus g in the class β. It is sometimes convenient to change from the irreducible highest weight representations $\left[\frac{i}{2}\right]$ to the above basis by

$$\left(2[0]+\left[\frac{1}{2}\right]\right)^{\otimes n}=\sum_i\left(\binom{2n}{n-i}-\binom{2n}{n-i-2}\right)\left[\frac{i}{2}\right]\ , \tag{4.3.12}$$

because

$$\mathrm{Tr}_{I^n}(-1)^{2\sigma_3}e^{-2\sigma_3 s}=(-1)^n\left(2\sinh\frac{s}{2}\right)^{2n}\ . \tag{4.3.13}$$

Reduction on the circle leads to a four dimensional theory with $N = 2$ supersymmetry arising from the type IIA reduction on M. The superalgebra and the symmetry acting on it does not change. Only now the 4d mass gets shifted by a Kaluza-Klein momentum on the circle. After this compactification the charge lattice of the BPS states is naturally identified with the K-theory charge of the type IIA D_{2k} branes

$$\Gamma=(q_0,q_A,p^A,p^0)\in\oplus_{i=0}^3H^{2i}(M)\ . \tag{4.3.14}$$

For particles at rest in 4d the eigenvalues of the Hamiltonian H are the BPS masses $M = |\Gamma \cdot \Pi|$. The vector Π is the instanton corrected in the type IIA theory, but it can be mapped by mirror symmetry to the period vector of the holomorphic 3-form of the mirror of M and calculated exactly by the period. The relation between the left spin and the D_0 and brane charge q_0 is [127]

$$q_0 = 2\frac{j_-}{(p^0)^2} \,. \tag{4.3.15}$$

At large radius in the A-model and if we assume there are no or at least no light $D4$ branes the relevant BPS states are D-brane bound states with charge $(Q_6, Q_4, Q_2, Q_0) = (1, 0, \beta, k)$ where $\beta \in H_2(M, \mathbb{Z})$. Their mass is given by (4.3.8). As explained in [140], see [182] for review, the one-loop integral (4.3.7) is calculated in a constant graviphoton background, which depends only on the left spin j_- of particles in the loop. The calculation is very similar to the normal Schwinger-loop calculation. The latter computes the one-loop effective action in an $U(1)$ gauge theory, which comes from integrating out massive particles P coupling to a constant background $U(1)$ gauge field. For a self-dual background field $T_{12} = T_{34} = T$ it leads to the following one-loop determinant evaluation:

$$S^S_{1-loop} = \log \det\left(\nabla + m^2 + 2e\,\sigma_- F\right) = \int_\epsilon^\infty \frac{ds}{s} \frac{\mathrm{Tr}(-1)^f \exp(-sm^2)\exp(-2se\sigma_- F)}{4\sin^2(seF/2)} \,. \tag{4.3.16}$$

Here the $(-1)^f$ takes care of the sign of the log of the determinant depending on whether P is a boson or a fermion, and σ_- is the Cartan element in the left Lorentz representation of P. To apply this calculation to the $N = 2$ supergravity case one notes, that the graviphoton field couples to the mass, i.e., we have to identify $e = m$. The loop has two R_- insertions and an arbitrary, (for the closed string action even) number F_- of graviphoton insertions. It turns out [140] that the only supersymmetric BPS states in the Lorentz representation

$$\left[\left(\frac{\mathbf{1}}{\mathbf{2}}, \mathbf{0}\right) + 2(\mathbf{0}, \mathbf{0})\right] \otimes \mathcal{R} \tag{4.3.17}$$

contribute to the loop. Here $\mathcal{R}$ is an arbitrary Lorentz representation of $SO(4)$. Moreover the two R_- insertions are absorbed by the first factor in the Lorentz representation (4.3.17), and the coupling of the particles in the loop to F_- insertions in the $N = 2$ evaluation works exactly as in the non-supersymmetric Schwinger-loop calculation above for P in the representation $\mathcal{R}$. In performing it for all $m(\beta, k)$

and g we note that I_g is a very convenient basis as $\mathrm{Tr}_{I_g}(-)^F e^{-2i\tau j_-^3\lambda} = \left(\sin\frac{\tau\lambda}{2}\right)^{2g}$ and that the sum over k gives a δ function, which makes the $d\tau$ integration trivial, so that we get quite straightforwardly

$$\begin{aligned}\mathcal{F}(\lambda,t) &= \sum_{g=0}^{\infty}\lambda^{2g-2}\mathcal{F}^{(g)}(t)\\ &= \frac{c(t)}{\lambda^2}+l(t)+\sum_{g=0}^{\infty}\sum_{\beta\in H_2(M,\mathbb{Z})}\sum_{m=1}^{\infty} n_\beta^{(g)}\frac{1}{m}\left(2\sin\frac{m\lambda}{2}\right)^{2g-2} Q^{\beta m}\\ &= \frac{c(t)}{\lambda^2}+l(t)+\sum_{g=0}^{\infty}\sum_{\beta\in H_2(M,\mathbb{Z})}\sum_{m=1}^{\infty} n_\beta^{(g)}(-1)^{g-1}\frac{[m]^{(2g-2)}}{m}Q^{\beta m},\end{aligned} \tag{4.3.18}$$

with

$$Q^\beta = e^{i\sum_{i=1}^{h^{1,1}} t_i \int_{C_\beta}\omega_i}, \qquad [x] := q_\lambda^{\frac{x}{2}} - q_\lambda^{-\frac{x}{2}}, \qquad q_\lambda = e^{i\lambda}.$$

The cubic term $c(t)$ in the Kähler parameters t_i is the classical part of the prepotential $\mathcal{F}^{(0)}$ given in (2.6.31) without the constant term, and $l(t) = \sum_{i=1}^{h}\frac{t_i}{24}\int_M c_2\wedge\omega_i$ is the classical part[110] of $\mathcal{F}^{(1)}$. Using the expansion

$$\frac{1}{m}\frac{1}{\left(2\sin\frac{m\lambda}{2}\right)^2} = \sum_{g=0}\lambda^{2g-2}(-1)^{g+1}\frac{B_{2g}}{2g(2g-2)!}m^{2g-3} \tag{4.3.19}$$

and a $\zeta(x) = \sum_{m=1}^{\infty}\frac{1}{m^x}$ regularization of the sum over m with $\zeta(-n) = -\frac{B_{n+1}}{n+1}$, we see that for $g \ge 2$ the $\beta = 0$ constant map terms from localisation [111]

$$\langle 1\rangle_{g,0}^M = (-1)^g\frac{\chi}{2}\int_{\mathcal{M}_g} c_{g-1}^3 = (-1)^g\frac{\chi}{2}\frac{|B_{2g}B_{2g-2}|}{2g(2g-2)(2g-2)!} \tag{4.3.20}$$

are reproduced if we set $n_0^{(0)} = -\frac{\chi}{2}$. This choice also reproduces the constant term proportional to $\zeta(3)$ in $\mathcal{F}^{(0)}$. In $\mathcal{F}^{(1)}$ there is a $\zeta(1)$ term which requires an additional regularization. More importantly expanding (4.3.18) in λ and comparing

[110] These terms do not follow entirely from the Schwinger-loop calculation and added here for completeness.

with (4.3.4) predicts the *multicovering formulas* at all genera. Specialized to one Kähler class such that β is identified with the degree $d \in \mathbb{Z}$ we get

$$
\begin{aligned}
\mathcal{F}^{(0)} &= \frac{D^3 t^3}{3!} + t \int_M c_2 \wedge \omega - i \frac{\chi}{2(2\pi)^3} \zeta(3) + \sum_{d=1}^{\infty} n_d^{(0)} \mathrm{Li}_3(Q^d)\,, \\
\mathcal{F}^{(1)} &= \frac{t \int_M c_2 \wedge \omega}{24} + \sum_{d=1}^{\infty} \left(\frac{1}{12} n_d^{(0)} + n_d^{(1)} \right) \mathrm{Li}_1(Q^d)\,, \\
\mathcal{F}^{(2)} &= \frac{\chi}{5760} + \sum_{d=1}^{\infty} \left(\frac{1}{240} n_d^{(0)} + n_d^{(2)} \right) \mathrm{Li}_{-1}(Q^d), \\
\mathcal{F}^{(g)} &= \frac{(-1)^g \chi |B_{2g} B_{2g-2}|}{4g(2g-2)!(2g-2)} \\
&\quad + \sum_{d=1}^{\infty} \left(\frac{|B_{2g}| n_d^0}{2g(2g-2)!} + \frac{2(-1)^g n_d^2}{(2g-2)!} \pm \ldots - \frac{g-2}{12} n_d^{g-1} + n_d^g \right) \mathrm{Li}_{3-2g}(Q^d).
\end{aligned}
\tag{4.3.21}
$$

Using resummations like (4.3.19) one checks that the partition function $\mathcal{Z} = \exp(\mathcal{F})$ has the following product form[111]

$$
\mathcal{Z}_{\mathrm{GV}}(M, \lambda, Q) = \prod_{\beta} \left[\left(\prod_{r=1}^{\infty} (1 - q_\lambda^r Q^\beta)^{r n_\beta^{(0)}} \right) \prod_{g=1}^{\infty} \prod_{l=0}^{2g-2} (1 - q_\lambda^{g-l-1} Q^\beta)^{(-1)^{g+r} \binom{2g-2}{l} n_\beta^{(g)}} \right]
\tag{4.3.22}
$$

in terms of the invariants $n_\beta^{(g)}$. This product form resembles the Hilbert scheme of symmetric products written in terms of partition sums over free fermionic and bosonic fields with an integer $U(1)$ charge as well as the closely related product form for the elliptic genus of symmetric products. As it has already been pointed out in [139], it is also reminiscent of the Borcherds product form of automorphic forms of $O(2, n, \mathbb{Z})$, see [48] and [239] for a review. In the Borcherds multiplicative lift the idea is that integrality of the $n_\beta^{(g)}$ is related to the fact that they are Fourier coefficients of other (quasi)automorphic Jacobi form, see also [219].

4.3.4 Geometric Interpretation of the BPS Numbers

As usual in theory of BPS solitons the degeneracy of the BPS states comes from the cohomology of the moduli space of the solitonic solutions, in this case of the brane solution. This moduli space is the vacuum manifold of the brane world volume theory, which is parametrized by the zero modes and the cohomological information is extracted by quantizing this zero mode sector as shortly discussed in Sect.3.2.1.

[111]Here we dropped the $\exp(\frac{c(t)}{\lambda^2} + l(t))$ factor of the classical terms at genus 0, 1.

In the following we will discuss only single wrapped branes. For the M2 brane the eleven dimensional tangent space splits $0 \to N_8 \to T_{11} \to T_{M2} \to 0$. The normal space N_8 is decomposed into $N \times \mathcal{N}$, where $\mathcal{N}$ is the normal direction in the CY M and N are the spacial directions of $5d$ Minkowski space. The CY tangent space splits as well $0 \to \mathcal{N} \to T_M \to T_C \to 0$. The unbroken space-time symmetries $G_{N_8} = SO(4)_N \times U(2)_{\mathcal{N}}$ transversal to the brane become R-symmetries of the fields on the brane-world-volume. For holomorphic curves in n complex dimensional Kähler manifolds the generic structure group of normal bundle $SO(2(n-1))$ restricts because of property (2) in Bergers list, Sect.1.1.5, to $U(n-1)_{\mathcal{N}}$. For Calabi-Yau manifolds it follows from the adjunction formula (A3.8) and the vanishing of the first Chern-class that $c_1(\det(U(n-1))_{\mathcal{N}}) = c_1(T^*C)$, i.e. over C the $U(1)_{\mathcal{N}} \in U(n-1)_{\mathcal{N}}$ can be identified with the $U(1)_{\mathcal{N}}$ connection in the canonical bundle $K_C = T^*C$. This identification of the R-symmetry transformation of the normal bundle with the WS transformations on C leads to a natural twisting of the brane-world-volume theory [41].

Let us describe the transformation properties of theses fields on the brane under $G_T = SO(2,1)$ the Lorentzgroup on the brane and $G_{N_8} = SO(4)_N \times U(1)_{L,\mathcal{N}} \times SU(2)_{R,\mathcal{N}}$ the R-symmetry from the normal direction

- Before twisting the eight fermions[112] $\psi \in [s, \mathbf{8}_s]$ transforms as spinor with helitity $s = \pm\frac{1}{2}$ under G_T and as spinor under $G_N = SU(2)_{L,\,N} \times SU(2)_{R,\,N} \times U(1)_{L,\mathcal{N}} \times SU(2)_{R,\mathcal{N}}$. The $U(1)_{L,\mathcal{N}}$ connection is identified with the connection in K_C. It changes the helicity of fields in $\sqrt{K}$ therefore by $0, \pm\frac{1}{2}$ depending on their $U(1)_{L,\mathcal{N}}$ charge

$$\begin{aligned}
\psi &\in \left[s, [(0,\tfrac{1}{2})_N \otimes (0,\tfrac{1}{2})_{\mathcal{N}}] \oplus [(\tfrac{1}{2},0)_N \otimes (\pm 1, 0)_{\mathcal{N}}]\right] \\
\psi_T &\in \left[\pm\tfrac{1}{2}, (0,\tfrac{1}{2})_N \otimes (\tfrac{1}{2})_{R,\mathcal{N}}\right] \oplus \left[2(0), (\tfrac{1}{2},0)_N \otimes (0)_{R,\mathcal{N}}\right] \oplus \left[\pm 1, (\tfrac{1}{2},0)_N \otimes (0)_{R,\mathcal{N}}\right]
\end{aligned} \tag{4.3.23}$$

  here the $U(1)_{L,\mathcal{N}}$ charge is combined with the helicity in G_T to the first entry $h = \pm\frac{1}{2}, 0, \pm 1$ in the twisted representation ψ_T, which implies that the field is a section of K_C^h.
- For the eight bosons ϕ corresponding to the coordinates of the normal directions

$$\begin{aligned}
\phi &\in \left[0, [(\tfrac{1}{2},\tfrac{1}{2})_N \otimes (0,0)_{\mathcal{N}}] \oplus [(0,0)_N \otimes (\pm 1,\tfrac{1}{2})_{\mathcal{N}}]\right] \\
\phi_T &\in \left[0, (\tfrac{1}{2},\tfrac{1}{2})_N \otimes (0)_{R\mathcal{N}}\right] \oplus \left[\pm\tfrac{1}{2}, (0,0)_N \otimes (\tfrac{1}{2})_{R\mathcal{N}}\right]
\end{aligned} \tag{4.3.24}$$

Clearly the zero modes of the bosons transforming as $\left[\pm\frac{1}{2}, (0,0)_N \otimes (\frac{1}{2})_{R\mathcal{N}}\right]$ and the fermions transforming as $\left[\pm\frac{1}{2}, (0,\frac{1}{2})_N \otimes (\frac{1}{2})_{R\mathcal{N}}\right]$ correspond to deformations

[112] In this section we set denote the left spin index $-$ by L and the right spin index $+$ by R.

(and superdeformations) of C in the CY direction and parametrize the moduli space $\mathcal{M}_C$ of movements of C within M. Fermionic and bosonic zero modes form the field content of a supersymmetric σ model on $\mathcal{M}_C$ and after quantization one gets the cohomology of the moduli space of $\mathcal{M}_C$ weighted in addition with the R quantum number of the fermions modes from their $SU(2)_{N,R}$ transformation. The corresponding representations are identified with the Lefshetz decomposition of the cohomology of the Kähler manifold (2.3.35) $\mathcal{M}_C$. Other fermionic modes in ψ_T transform as 2 scalars the $(0, 0)$ and $(1, 1)$ form and g holomorphic and g antiholomorphic one forms on the genus g curve C if the latter does not degenerate. The corresponding zero modes are then forms on a $(2g + 2)$ dimensional Jacobian torus, which form $SU(2)_{N,\ L}$ representations $[(\frac{1}{2}, 0)+2(0, 0)]^{g+1}$, cff. (2.3.35). By the definition (4.3.10) only the multiplicity $(2j_R^3 + 1)$ and the sign $(-1)^{2j_R^3}$ of the cohomology of $\mathcal{M}_C$ are relevant for the determination of n_β^g. This alternating sum is just the Euler number $(-1)^m\chi(\mathcal{M}_C)$, with $m = \dim_{\mathbb{C}}(\mathcal{M}_C)$. For classes β in M with non degenerate genus g curves we get therefore as coefficient of I_{g+1}

$$n_\beta^g = (-1)^m\chi(\mathcal{M}_C)\ . \tag{4.3.25}$$

An instructive example is a that of a ruled surface (RS) inside M. Familiar ruled surfaces are the Hirzebruch surfaces F_n fibrations of a $\mathbb{P}^1$ bundle over $\mathbb{P}^1$. More generally the base can be a higher genus surface Σ_g. We want to calculate the n_β^0 for β the class of the fibre. The genus zero fibre curve $C = \mathbb{P}^1$ is smoothly embedded and zero is the maximal genus of a curve in the class. Due to the fibration structure of the RS the moduli space $\mathcal{M}_C = \Sigma_g$ is identified with the base. Therefore (4.3.25) applies and gives $n_\beta^g = (-1)^1\chi(\Sigma_g) = 2g - 2$. The embedding of Σ_g is locally described by $\mathcal{O}(r)\otimes\mathcal{O}(s) \to \Sigma_g$ with $r+s = 2g-2$. Unless $g = 0$ $(r = s = -1)$ the curve Σ_g is not rigid in M and for $g > 0$ the curve Σ_g can be deformed to $2(g - 1)$ points in M, as in Fig. 19.

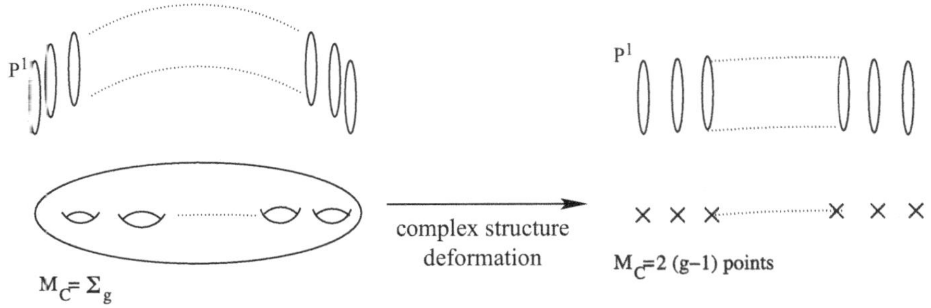

Fig. 19 The index n_β^g of the $D2$-$D0$ moduli space of the fibre in a ruled surface is constant under complex deformations, while the $N^g_{j_L^3,j_R^3}$ jump

The $SU(2)_{N,R}$ content before the deformation is $R = 2g[0] - \left[\frac{1}{2}\right]$ with $\chi(R) = 2g - 2 = -\chi(\Sigma_g)$ and after deformation $R' = (2g-2)[0]$ with $\chi(R') = 2g - 2 = +\chi(2(g-1)\ pts)$. I.e. the total BPS numbers $N^g_{j^3_L, j^3_R}$ change by states with $\left[2[0] - \left[\frac{1}{2}\right]\right]$ right representation content, when the complex structure moduli space of M is deformed. In particular in contrast to the $n^{(g)}_\beta$, the $N^\beta_{j^3_L, j^3_R}$ are not invariant under the change of the complex structure. Notice that the successful microscopic interpretation of the 5d black hole entropy requires deformation invariance and relies on the index-like quantity n^g_β and not on $N^g_{j^3_R, j^3_L}$.

Example Such ruled surfaces appear typically if one embeds the Calabi-Yau in a weighted projective space. E.g. the degree 14 hypersurface in $WCP^4(1, 2, 2, 2, 7)$, see Appendix 3, contains a ruled surface with a genus 15 curve as base.[113] The genus $g = 15$ curve is *semi* stable because the relevant complex deformation moduli are frozen as an artifact of the embedding. For other realization of the same family that is not necessarily the case.

Above situation of a ruled surface is a good example to get a rough idea of some concepts of *virtual intersection theory*. The virtual dimension of the brane moduli space on CY 3-folds is expected to be zero by (4.1.5) or here equivalently by (4.3.30). In this preferred situation the intersection problem is reduced to point counting, but the situation might not be achievable as in the example of the ruled surface above and the dimension of the moduli space remains positive. In this particular case the *excess intersection calculation* amounts to integrate $c_1(T\Sigma_g)$ over Σ_g.

In the type IIA picture one transversal direction parametrized previously by a scalar in $\left[0, (\frac{1}{2}, \frac{1}{2})_N \otimes (0)_{R,\mathcal{N}}\right]$ is dualized on the 3d World-Volume to a $U(1)$ gauge field. The flat $U(1)$ connection has $2g$ zero modes on C exactly as the $\left[\pm 1, (\frac{1}{2}, 0)_N \otimes (0)_{R,\mathcal{N}}\right]$ fermions in ψ_T. Since these zero-modes parametrize the $2g$ dimensional torus Jac(C), called the Jacobian of C see [148] Chap 2.7, one gets a supersymmetric quantum mechanics (SQM) on the moduli space $\mathcal{M}$ with a fibration structure Jac(C) $\to \mathcal{M} \to \mathcal{M}_C$, see Fig. 20. The proposal [140] for the $\mathrm{SU}(2)_{N\ L} \times \mathrm{SU}(2)_{N\ R}$ action on $\mathcal{M}$ is that $H^*(\mathcal{M}) = N^\beta_{j^3_L, j^3_R}[j^3_{L=fibre}, j^3_{R=base}]$. Again one can conclude that the contribution n^g_β of smooth genus curves in the class β is the $(-1)^{2j^3_R}$ weighted sum of the right representations multiplying the non degenerate fibre contribution I_g in the decomposition of the left representation. This is

$$n^g_\beta = (-1)^{dim_C(\mathcal{M}_C)} \chi(\mathcal{M}_C). \tag{4.3.26}$$

[113] Such case have been investigated [213, 223], because there have interesting gauge symmetry enhancements, when the $\mathbb{P}^1$ shrinks.

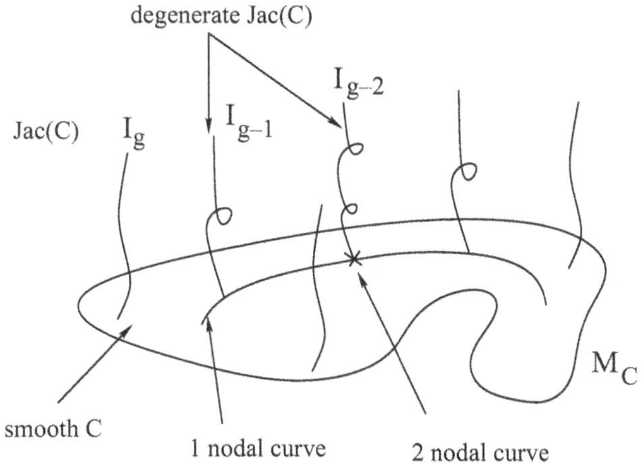

Fig. 20 Moduli space of $D2$-$D0$ brane bound states as a Jacobian fibration over the deformation space $\mathcal{M}_C$

On the other extreme are the curves which are maximally degenerate. They have genus zero and come from genus g curves with g nodes. The Euler number of the fibres with δ nodes is $\chi(I_{g-\delta}) =: \delta_{g,\delta}$. Due to the fibration structure the, Euler number of $\chi(\mathcal{M})$ is calculated as the Euler number of the locus in the base where the completely degenerate fibres sit. This is the $(-1)^{2j^3_R}$ weighted sum of the right representations on the cohomology of this locus and therefore

$$n^0_\beta = (-1)^{\dim_C(\mathcal{M})}\chi(\mathcal{M}) . \qquad (4.3.27)$$

In [215] a calculational scheme for the intermediate cases was given. E.g. if no *reducible fibres* contribute one obtains

$$n^{g-\delta}_\beta = (-1)^{(\dim(\mathcal{M}_C)+\delta)}\sum_{p=0}^{\delta} b_{g-p,\delta-p}\chi(\mathcal{C}^{(p)}), \quad b_{g,k} := \frac{2}{k!}\prod_{i=1}^{k-1}(2g-(k+2)+i), \quad b_{g,0} := 1 . \qquad (4.3.28)$$

Here $\mathcal{C}^{(p)}$ is the moduli space of the curve C with p points, e.g. $\mathcal{C}^{(0)} = \mathcal{M}_C$. In the case that C lies in a surface S in M, one can use similarly as in (5.1.2) formulas for the cohomology of Hilbert scheme to calculate $\chi(\mathcal{C}^{(p)})$, see [215] for examples.

Donaldson-Thomas Invariants

As we saw above we obtain BPS states by wrapping D-branes on supersymmetric cycles in M. More generally we can wrap 6-branes on M itself, 4-branes on divisors and 2-branes on a curves $C \subset M$, possibly bound to some 0-branes. We leave out the

4-branes as we don't know an index yet carrying deformation invariant information. At the level of RR charges a configuration of the other branes can be cast into a short exact sequence of the form

$$0 \longrightarrow \mathcal{I} \longrightarrow \mathcal{O}_M \longrightarrow \mathcal{O}_Z \longrightarrow 0 \tag{4.3.29}$$

where $\mathcal{I}$ is the ideal sheaf describing this configuration of one D6 brane on M, a D2 brane on the curve C and k D0 branes and Z is the subscheme of M consisting of the curve C and the points at which the 0-branes are supported. Counting BPS states therefore leads to the study of the moduli space $I_k(M, \beta)$ of such ideal sheaves $\mathcal{I}$, which has two discrete invariants: the class $\beta = [Z] \in H_2(M, \mathbb{Z})$ and the number of 0-branes $k = \chi(\mathcal{O}_Z)$ plus an integral contribution form C. With the analogue of the Hirzebruch-Riemann-Roch theorem for sheaves, the Grothendieck-Riemann-Roch theorem,[114] one can calculate the virtual dimension of the deformations of ideal sheaves $\mathcal{I}$ inside a threefold M as [261]

$$\dim_{vir} = \dim\ \mathrm{Ext}^1_0(\mathcal{I}, \mathcal{I}) - \dim \mathrm{Ext}^2_0(\mathcal{I}, \mathcal{I}) = c_1 \cdot \beta\ . \tag{4.3.30}$$

This reflects again the special rôle of Calabi–Yau threefolds and one expects that the number of BPS states with these charges is obtained by counting points. As is in the case of Gromov–Witten invariants, these configurations can appear in families, and one has to work with the virtual fundamental class. However the situation for Donaldson–Thomas invariants is considered easier in many respects. For example there is no finite automorphism group acting on $I_k(M, \beta)$ so one expects directly integer BPS numbers as result. This number is called the Donaldson–Thomas invariant $\tilde{n}^{(k)}_\beta$ [101, 300].

Since both invariants, Gopakumar–Vafa and Donaldson–Thomas, keep track of the number of BPS states, they should be related. The relation is in fact a consequence of the S-duality in topological strings [268], and takes the form reported in (4.3.22). The factor on the r.h.s of (4.3.22) comes from the constant maps and gives the McMahon function M $(q_\lambda) = \prod_{n\geq 0} \frac{1}{(1-q_\lambda^n)^n}$ to the power $\frac{\chi}{2}$. This function appears also in Donaldson–Thomas theory [261], calculable on local toric Calabi–Yau spaces e.g. with the vertex [4]. However, in Donaldson–Thomas theory the power of the McMahon function is χ. Note also that if (4.3.18) holds then $\mathcal{F}$ or $\mathcal{Z}$ restricted to a class β is always a finite degree rational function in q_λ symmetric in $q_\lambda \to \frac{1}{q_\lambda}$, since the genus is finite in a given class β. Thanks to this observation one can read from the comparison of the expansion of $\mathcal{Z}$ in terms of Donaldson–Thomas invariants $\tilde{n}^{(m)}_\beta \in \mathbb{Z}$

$$\mathcal{Z}^{hol}_{\mathrm{DT}}(M, q_\lambda, Q) = \sum_{\beta, k \in \mathbb{Z}} \tilde{n}^{(k)}_\beta q_\lambda^k Q^\beta \tag{4.3.31}$$

[114]For Calabi-Yau 3 folds there is an even simpler argument that the difference below vanishes. Serre duality applies the Ext groups and relates Ext^1_0 and Ext^2_0 on three folds with trivial canonical bundle.

with the expansion in terms of Gopakumar–Vafa invariants [261]

$$\mathcal{Z}^{hol}_{\mathrm{GV}}(M, q_\lambda, Q)\mathrm{M}(q_\lambda)^{\frac{\chi(M)}{2}} = \mathcal{Z}^{\mathrm{hol}}_{\mathrm{DT}}(M, -q_\lambda, Q) \tag{4.3.32}$$

the precise relation between $\tilde{n}^{(m)}_\beta$ and $n^{(g)}_\beta$. Equation (4.3.18) and (4.3.22) then relate the two types of invariants to the Gromov–Witten invariants $r^{(g)}_\beta \in \mathbb{Q}$ as in (4.3.4).

Pandharipande-Thomas Invariants

The geometrical interpretation of the BPS invariants introduced by [140, 215] as reviewed above can be used to deal with fairly smooth Jacobian fibrations. It found a rigorous mathematical definition in terms stable pairs due to Pandharipande and Thomas [273, 275]. The basic ideas are similar then in section "Donaldson-Thomas Invariants", but the setup of stable pairs was made specifically to clarify the assertions made in [215] and also in order to the provide mathematical proofs. Let us shortly review the main ingredients and results and start with the definition of stable pairs.

Definition 4 A *stable pair* on a smooth threefold X consists of a sheaf $\mathcal{F}$ on X and a section $s \in H^0(\mathcal{F})$ such that

- $\mathcal{F}$ is pure of dimension 1
- s generates $\mathcal{F}$ outside of a finite set of points

A stable pair is a D6-D2-D0 brane bound state, and can be written as a complex

$$\mathcal{I}^\bullet : \mathcal{O}_M \xrightarrow{s} \mathcal{F}.$$

Let $P_n(\mathrm{M}, \beta)$ denote the moduli space of stable pairs with $\mathrm{ch}_2(\mathcal{F}) = \beta,\ \chi(\mathcal{F}) = n$. Then if X is Calabi-Yau, $P_n(M, \beta)$ supports a *symmetric* obstruction theory. See [34] for the definitions and basic properties of symmetric obstruction theories.

There are only a few things that we need to know about symmetric obstruction theories. The basic idea of a symmetric obstruction theory is that the obstructions are dual to their deformations. For stable pairs, the space of first order deformations is $\mathrm{Ext}^1(\mathcal{I}^\bullet, \mathcal{I}^\bullet)$ and the space of obstructions is $\mathrm{Ext}^2(\mathcal{I}^\bullet, \mathcal{I}^\bullet)$. These are dual by Serre duality.

An important feature of symmetric obstruction theories is that they have virtual dimension 0, since deformations and obstructions have the same dimension.

If $\mathcal{M}$ is the moduli space associated with a symmetric obstruction theory and $\mathcal{M}$ is smooth, then the corresponding virtual number is $(-1)^{\dim(\mathcal{M})}e(\mathcal{M})$, where $e(\mathcal{M})$ is the topological euler characteristic. This is because the bundle describing the deformations is the tangent bundle of $\mathcal{M}$, so the obstruction bundle must be the cotangent bundle of $\mathcal{M}$, and the euler class of the cotangent bundle is $(-1)^{\dim(\mathcal{M})}e(\mathcal{M})$.

In general, the virtual number is a weighted euler characteristic. See [34] for more details.

Now let M be Calabi-Yau and let $P_n(M, \beta)$ be the moduli space of stable pairs with $\mathrm{ch}_2(F) = \beta$ and $\chi(F) = n$, and let $P_{n,\beta}$ be the associated invariant, i.e. the degree of the virtual fundamental class of $P_n(M, \beta)$. These invariants can be arranged in a generating function

$$\mathcal{Z}^{hol}_{pt} = \sum_{n,\beta} P_{n,\beta} q^n Q^\beta . \tag{4.3.33}$$

We let Z_{GW} be the generating function for disconnected Gromov-Witten invariants:

$$\mathcal{Z}^{hol}_{GW} = \exp\left(\mathcal{F}'_{GW}(\lambda, Q)\right), \quad \mathcal{F}'_{GW}(\lambda, Q) = \sum_{\beta\neq 0}\sum_{g} \Gamma^\beta_g \lambda^{2g-2} Q^\beta , \tag{4.3.34}$$

where Γ^β_g is the Gromov-Witten invariant. The fundamental conjecture from which everything will follow is

Conjecture 5 After the change of variables $q = -e^{i\lambda}$, we have $\mathcal{Z}^{hol}_{PT} = \mathcal{Z}^{hol}_{GW}$.

Conjecture 5 is known to be true in the toric case [274].

Remarks on More General Integral Structures on Calabi-Yau n-Folds

We will discuss the refined integral structure in Sect. 4.4 and as we have seen the refined invariants are in general not invariant under complex structure of Calabi-Yau 3-folds as we saw in the example of the ruled surface. Integral structures appear also in the open string sectors. The corresponding formula is given in (5.2.7). In particular the disk amplitude enjoys an integral structure given by (5.2.8). This is the same integral structure then for the three point function of Calabi-Yau fourfolds. In particular for α a fourcycle in the vertical homology (3.6.36) evaluates as analog to (2.9.60) as [46, 146, 232, 262]

$$C^{(2,1)}_{\alpha,b,c} = C^{0\,(2,1)}_{\alpha,b,c} + \sum_{\beta\in H_2(M_4,\mathbb{Z})} \frac{\beta_a \beta_b n^{(0)\,\alpha}_\beta}{1-\prod_{i=1}^{h^{1,1}} Q_i^{d_i}} \prod_{i=1}^{h^{1,1}} Q_i^{d_i} . \tag{4.3.35}$$

This implies that there exist potentials $\mathcal{F}^{(0)\alpha}$ of the form

$$\mathcal{F}^{(0)\alpha} = \frac{1}{2} C^{0\,(2,1)}_{\alpha,b,c} t^a t^b + class + \sum_{\beta\in H_2(M_4,\mathbb{Z})} n^{(0)\,\alpha}_\beta \mathrm{Li}_2(Q^\beta) . \tag{4.3.36}$$

The sub leading classical terms have been determined in [46] and further properties of the $\mathcal{F}^{(0)\alpha}$ in [46, 262].

As the formula (3.4.16) shows the *genus one* contributions are expected to be non-trivial for all Calabi-Yau n-folds irrespectively of their dimensions. The corresponding analog of the Ray-Singer formula and the genus one multi covering contributions that lead to integer genus one invariants have been worked out in [224] together with a proof of (4.3.35) for fourfolds and in [276] analogous formulas for five folds.

4.4 *Refined Topological Invariants*

The BPS interpretation of the amplitude is obtained by computing it via a Schwinger loop integral with BPS states running in the loop [140] as discussed in Sect. 4.3.3.

4.4.1 Refined Integral Structure

At least formally it makes sense to couple the representation $\mathcal{R} = [j_-, j_+]$ to the insertions of the self- *and the* anti-self-dual part of a background gravi-photon field strength $T = \epsilon_1 dx^1 \wedge dx^2 + \epsilon_2 dx^3 \wedge dx^4$ and two insertions of the background curvature 2-form R respectively. Passing to spinor notation for the field T, one gets $\epsilon_-^2 = -\det T_{\alpha\beta}$ and $\epsilon_+^2 = \det T_{\dot\alpha\dot\beta}$, with $\epsilon_\pm = \frac{1}{2}(\epsilon_1 \pm \epsilon_2)$. Then the anti-self-dual and self-dual parts of the field strength couple to the left and right spin j_- and j_+ of the BPS particle respectively. The Schwinger loop calculation for these amplitudes yields, with $q_\pm = e^{-2\epsilon_\pm}$ is given by

$$S^s_{1-loop}(\epsilon_\pm) = -\int_\epsilon^\infty \frac{ds}{s} \frac{\mathrm{Tr}_{\mathcal{R}}(-1)^{\sigma_+^3+\sigma_-^3} e^{-sm} q_-^{s\sigma_-^3} q_+^{s\sigma_+^3}}{4\left(\sinh^2\left(\frac{s\epsilon_-}{2}\right) - \sinh^2\left(\frac{s\epsilon_+}{2}\right)\right)} . \tag{4.4.1}$$

If gravity can be decoupled, i.e. a rigid limit of supergravity exists then there emerges a further $SU(2)_{\mathcal{R}}$ symmetry acting on the algebra as symmetry group. In this case the individual degeneracies $N^\beta_{j_- j_+}$ are protected and it makes sense to define a 5d BPS super trace which keeps track of both j_- and j_+ as

$$Z_{\mathcal{BPS}}(\epsilon_-, \epsilon_+, t) = \mathrm{Tr}_{\mathcal{BPS}}(-1)^{2(J_-+J_+)} e^{-2\epsilon_- J_-} e^{-2\epsilon_+ J_+} e^{-2\epsilon_{\mathcal{R}} J_{\mathcal{R}}} e^{\beta H} . \tag{4.4.2}$$

In [74] it was shown that the presence of an $U(1)_{\mathcal{R}} \in SU(2)_{\mathcal{R}}$ isometry in the non-compact toric Calabi-Yau manifold allows to define and calculate the individual $N^\beta_{j_- j_+}$ by localisation.

The Schwinger-Loop integral (4.4.1) is analogously calculated as in Sect. 4.3.3 and yields

$$\mathcal{F}(\epsilon,t) = \sum_{\substack{2j_-,2j_+=0\\k=1}}^{\infty} \sum_{\beta\in H_2(M,\mathbb{Z})} (-1)^{2(j_-+j_+)} \frac{N^\beta_{j_-j_+}}{k} \frac{\sum_{m_-=-j_-}^{j_-} q_-^{km_-}}{2\sinh\left(\frac{k\epsilon_1}{2}\right)} \frac{\sum_{m_+=-j_+}^{j_+} q_+^{km_+}}{2\sinh\left(\frac{k\epsilon_2}{2}\right)} e^{-k\beta\cdot t}$$

$$= \lim_{E\to E_0} \frac{1}{\epsilon_1\epsilon_2} \sum_{g=0}^{\infty}\sum_{m=0}^{2g} \epsilon_1^{2g-m}\epsilon_2^m F^{(+)}_{m,g} = \sum_{n,g=0}^{\infty} (\epsilon_1+\epsilon_2)^{2n}(\epsilon_1\epsilon_2)^{g-1} F^{(n,g)}(t)\,. \tag{4.4.3}$$

The sum over $m_\pm$ is taken in integral increments both for $j_\pm$ integral and half-integral. The main difference to the unrefined integrality is that the n^g_β are only indices given by (4.3.10) and can hence be in $\mathbb{Z}$, while the physical interpretation of the $N^\beta_{j_-j_+}$ as the number of BPS states implies of course that they are all positive integers and in fact dimensions of vectors spaces. In particular if there is a symmetry group acting on the BPS states the degeneracies $N^\beta_{j_-j_+}$ must fall into dimensions of representations of that group.

Formula (4.4.3) can easily be exponentiated upon expanding the $\sinh(x)$. This yields the following expression for the partition function [198],

$$\mathcal{Z} = \prod_\beta \prod_{2j_\pm=0}^{\infty} \prod_{m_\pm=-j_\pm}^{j_\pm} \prod_{m_1,m_2=1}^{\infty} \left(1-q_-^{m_-}q_+^{m_+}e^{\epsilon_1(m_1-\frac{1}{2})}e^{\epsilon_2(m_2-\frac{1}{2})}e^{-\beta\cdot t}\right)^{(-1)^{2(j_-+j_+)+1}N^\beta_{j_-j_+}}\,. \tag{4.4.4}$$

4.4.2 The Refined Holomorphic Anomaly Equation

The refined holomorphic anomaly equations are obtained for the $F^{(m,g)}$ in [192, 243][115]

$$F(s,t,g_s) = \log Z = \sum_{m,g=0}^{\infty} s^m g_s^{2g-2} F^{(m,g)}(t)\,, \tag{4.4.5}$$

where $s=(\epsilon_1+\epsilon_2)^2$ and $g_s^2=\epsilon_1\epsilon_2$ as

$$\bar\partial_{\bar i} F^{(n,g)} = \frac{1}{2}\bar C^{jk}_{\bar i}\left(D_jD_kF^{(n,g-1)} + {\sum_{m,h}}' D_jF^{(m,h)}D_kF^{(n-m,g-h)}\right), \quad n+g>1\,, \tag{4.4.6}$$

[115]The equations in [243] were derived in the gauge theory context and differ from the one in [192]. The difference was explained in [242].

where the prime denotes omission of $(m,h) = (0,0)$ and $(m,h) = (n,g)$ in the sum and the first term on the right hand side is set to zero if $g = 0$. Beside of $F^{(0,0)} = \mathcal{F}^{(0)}$ and $F^{(0,1)} = F^{(1)}$, the holomorphic anomaly conditions have to be supplemented by

$$\mathcal{F}^{(1,0)} = \frac{1}{24}\log(\Delta_{con}\prod_i z_i^{a_i}) \tag{4.4.7}$$

as initial conditions. Here Δ_{con} is the complete conifold discriminant and the exponents a_i can be determined by vanishing of BPS invariants at the large radius.

Examples of Refined Invariants

For example for the $\mathcal{O}(-3) \rightarrow \mathbb{P}^2$ geometry one can refine the B-model solutions discussed in Sect. 4.2.43 to solve the B-model amplitudes $F^{(m,g)}$ using the refined gap condition discussed in Sect. 4.5.4 and obtain in the holomorphic limit the BPS invariants that refine the ones given in Table 7 to the ones given in Table 8.

This number coincide with the ones defined by localisation [74] or obtained by the refined topological vertex [198]. Using *direct integration techniques* developed in [192–194] the refined theory on all del Pezzo Surfaces and on the $\frac{1}{2}$ K3 surface [158] can be solved.

Refining (1.1.16) and other heterotic/Type II predictions like the KKV formula for $K3 \times T2$ [215] and the all genera predictions of the n^g_β for fiber classes of K3 fibrations [260] is also possible [192, 216] and definitions for the refined invariants in these cases have been suggested in [216] (Table 9).

As an example we exposit the refinement of the Yau and Zaslow predictions for the K3 surface (1.1.16) from [192, 216]. Let $[j]_x := x^{-j} + x^{-j+1} + \cdots + x^{j-1} + x^j$. Then we can obtain the refined invariants from a refinement of the Hilbert-Scheme formula in [215] as

$$\sum_{h=0}^{\infty}\sum_{j_-}\sum_{j_+} \mathsf{N}^h_{j_-,j_+}\,[j_-]_{u^2}[j_+]_{y^2}\, q^h = \prod_{n=1}^{\infty}\frac{1}{(1-u^{-1}y^{-1}q^n)(1-u^{-1}yq^n)(1-q^n)^{20}(1-uy^{-1}q^n)(1-uyq^n)}, \tag{4.4.8}$$

where the sums over j_- and j_+ are both taken over $\frac{1}{2}\mathbb{Z}_{\geq 0}$. The relation above determines the $N^h_{j_-,j_+}$ for $h \geq 0$ listed in Table 9. As an example for the possible detection of a symmetry action by recognising the dimension of its representations in "simple" decomposition of the refined BPS states, we note the following. The

Table 8 Non vanishing BPS numbers $N^d_{j_-,j_+}$ of local $\mathcal{O}(-3) \to \mathbb{P}^2$ up to $d = 6$

d	$j_-\backslash j_+$	0	$\frac{1}{2}$	1	$\frac{3}{2}$	2	$\frac{5}{2}$	3	$\frac{7}{2}$	4	$\frac{9}{2}$	5	$\frac{11}{2}$	6	$\frac{13}{2}$	7	$\frac{15}{2}$	8	$\frac{17}{2}$	9	$\frac{19}{2}$	10	$\frac{21}{2}$	11	$\frac{23}{2}$	12	$\frac{25}{2}$	13	$\frac{27}{2}$
1	0			1																									
2	0						1																						
3	0							1																					
	$\frac{1}{2}$										1																		
4	0						1				1				1														
	$\frac{1}{2}$									1		1		1															
	2												1																
	$\frac{3}{2}$															1													
5	0			1				1		1		2		2		2		1											
	$\frac{1}{2}$						1		1		2		2		3		2		1										
	1									1		1		2		2		2		1									
	$\frac{3}{2}$												1		1		2		1		1								
	2															1		1		1									
	$\frac{5}{2}$																		1										
	3																					1							
6	0		1		1		3		2		6		4		8		5		7		2		2						
	$\frac{1}{2}$			1		2		3		5		6		9		9		10		7		5		1		1			
	1				1		1		3		3		7		7		11		9		9		4		2				
	$\frac{3}{2}$							1		1		3		4		7		7		10		6		4					
	2										1		1		3		4		7		6		6		2		1		
	$\frac{5}{2}$													1		1		3		3		5		3		2			
	3																1		1		3		3		3		1		
	$\frac{7}{2}$																			1		1		2		1		1	
	4																						1		1		1		
	$\frac{9}{2}$																									1			
	5																												1
d	j_-/j_+	0	$\frac{1}{2}$	1	$\frac{3}{2}$	2	$\frac{5}{2}$	3	$\frac{7}{2}$	4	$\frac{9}{2}$	5	$\frac{11}{2}$	6	$\frac{13}{2}$	7	$\frac{15}{2}$	8	$\frac{17}{2}$	9	$\frac{19}{2}$	10	$\frac{21}{2}$	11	$\frac{23}{2}$	12	$\frac{25}{2}$	13	$\frac{27}{2}$

candidate is the Mathieu group $\mathbb{M}_{24}$ acting on the category of coherent sheaves on K3 and whose representations are

$$1, 23, 45, 231, 252, 253, 483, 770, 990, 1035, 1265,$$
$$1771, 2024, 2277, 3312, 3520, 5313, 5544, 5796, 10395\,,$$

where the representations of dimension $45, 231, 770, 990, 1035$ come in complex conjugated pairs and there is an extra real representation of dimension 1035. Note that apart from the 20 all numbers decompose in dimensions of representations in a relatively simple way

$$N^3_{0,0} = 1981 = 2 \cdot 990 + 1, \qquad N^4_{0,0} = 13938 = 2 \cdot 5313 + 3312$$
$$N^4_{\frac{1}{2},\frac{1}{2}} = 2233 = 2 \cdot 990 + 253, \quad N^5_{1,1} = 2254 = 1771 + 483$$
$$N^6_{\frac{3}{2},\frac{3}{2}} = 2255 = 1265 + 990,$$

Table 9 The nonvanishing $R^h_{j_-,j_+}$ for $h \leq 4$ for the K3 surface

$N^0_{\frac{i}{2},\frac{j}{2}}$	i=0
j=0	1

$N^1_{\frac{i}{2},\frac{j}{2}}$	j=0	1
i=0	20	
1		1

$N^2_{\frac{i}{2},\frac{j}{2}}$	j=0	1	2
i=0	231		
1		21	
2			1

$N^3_{\frac{i}{2},\frac{j}{2}}$	j=0	1	2	3
i=0	1981		1	
1		252		
2	1		21	
3				1

$N^4_{\frac{i}{2},\frac{j}{2}}$	j=0	1	2	3	4
i=0	13938		21		
1		2233		1	
2	21		253		
3		1		21	
					1

a fact that has been further analysed in [170].

4.5 Fixing the Holomorphic Ambiguity

As it is evident from (4.1.18) each new $F^{(g)}$ for $g > 1$ is determined only up to a meromorphic section $f_g(z)$ of $\mathcal{L}^{2-2g}$ called the holomorphic ambiguity at genus g. It that depends on the invariant parameters complex structure parameters z. These sections are restricted by the fact that for physical and mathematical reasons the total $F^{(g)}$ have a quite restricted pole and regularity structure at the critical divisors of $\mathcal{M}_{cs}$. Most notably at each conifold divisor there is in the local transversal coordinate t_c a pole of order t_c^{2-2g} and regularity in the sub-leading terms in the $F^{(g)}$. This pole structure can be calculated with a Schwinger loop computation since there is an effective action with massless particle spectrum at the singularity, which is at finite distance from the interior of the moduli space. Generically and even at orbifold divisors the $F^{(g)}_{phys}$ in the physical gauge have to be regular. At the point of maximal unipotent monodromy the $F^{(g)}$ for $g \ge 1$ have to be regular and for $g = 0$ there are only logarithmic singularities given by the intersection numbers (2.6.31) for $g = 0$ and by (4.5.3) for $g = 1$. In addition $F^{(1)}$ has a pre described logarithmic singularity at the conifold by (4.5.4).

These are the types of singularities that occur in the one parameter families discussed in Sects. 2.10.4 and 2.10.5. However in more generic multi parameter models there are other types of singularities. In particular interesting are the singularities in the FHSV model [113] discussed [151, 222] where a fourbrane becomes light and one has a dual integral BPS expansion. If an effective action exists one can figure out the leading behaviour of $F^{(g)}$, but even in some case like the shrinking del Pezzo surfaces there is information as we can solve the latter in their entire moduli space [193]. The canonical threefold singularities are non classified however many things are known for hypersurface [281] and quotient singularities [332]. Generally the expected pole behaviour and the regularity condition leads to an ansatz for the $f_g(z)$ that we discuss in the next section.

4.5.1 The Ansatz for the Holomorphic Ambiguity

The counting function for the GW invariants is obtained as a holomorphic limit of the result of the integration $\mathcal{F}(t) = \lim_{\bar t\to\infty} F^g(t,\bar t)$ of (4.1.19), (4.1.18). It depends strongly on the possible holomorphic ambiguities $f_g(z)$ and gives together with the other boundary divisors general constraints on the latter. In view of the conifold singularities one makes for the holomorphic ambiguities the following ansatz

$$f_g(z) = \sum_{i=1}^{D}\sum_{k=0}^{t(i)} \frac{p_i^{(k)}(z)}{\Delta_i^k}, \tag{4.5.1}$$

where D is the number of components Δ_i of the discriminant and $t(i)$ gives the maximal singularity that one has at the corresponding type of divisor. In particular for the conifold divisors $t(i) = 2g-2$. If in the large complex structure variables the point $z_i \to \infty$ is regular the $p_i^{(k)}(z)$ are polynomials that are generically bounded to have highest degree

$$t(i) \times \mathrm{ord}(\Delta_i) + \sigma t(i).$$

Here the shift σ is due to the fact that the regularity condition holds for the physical $F^{(g)}_{phys}$. It is therefore determined by the vanishing of $e^{-K} \sim X^0_* = \Delta^\sigma_* + \mathcal{O}(\Delta_*^{a>\sigma})$ at the corresponding divisor $*$. In particular if $\sigma > 0$ then the regularity condition is weaker and higher order poles are allowed for the f_g at this divisor $*$ and consequently the $p_i^{(k)}(z)$ are less constrained. Note that the number of conifold divisors in the one moduli cases is the order $\mathrm{ord}(\Delta_c)$ of the unreduced conifold discriminant and from each conifold component we get from the gap condition (4.5.8) $2g - 1$ independent conditions. Hence the condition would be sufficient if the only singular behaviour is from conifolds and $\sigma = 0$ for other divisors. This precisely the case in the local models as $e^{-K} \sim X^0 = c$ becomes globally constant and the Mirror Riemann surface have generically no worse singularities then conifolds, also called nodes. This was used in [163] to argue that the topological string B-model is completely solvable in the local case and in [194] the same was shown for the refined topological B-model.

In general for compact Calabi-Yau 3 folds the discussion is more complicated because there more general types of singularities and the fact that e^{-K} varies in a non-trivial way. For the quintic the only singular behaviour is the gap at the conifold, but at $z \to \infty$ there is the orbifold and X^0 has local vanishing order $\sigma = 1/5$ Using the expansion (4.5.8) and the regularity leaves therefore a moderate grow in the undetermined unknowns in $f_g(z)$ of order

$$\sigma[2g-2] = \left[\frac{2g-2}{5}\right] . \tag{4.5.2}$$

This is supplemented by the vanishing of the BPS invariants $n^g_{\beta=d}$ in good approximation for $d < \sqrt{g}$ due to the adjunction formula. Hence one gets roughly $\sqrt{g}$ additional constraints. The point where the linear growth (4.5.2) majorizes the curve $d = \sqrt{g}$ is the point where one runs out of boundary conditions can be seen in Fig. 21. The conclusion is that the topological string on the quintic is solvable for $g \leq 51$. Using the topological data of the quintic discussed in section "The Monodromies of the Quintic" one obtains the BPS numbers in Table 10.

4.5.2 Boundary Conditions from Light BPS States

Boundaries in the moduli space $\mathcal{M}_{cks}(M)$ correspond to degenerations of the manifold M and general properties of the effective action can be inferred from the physics of the lightest states. More precisely the light states relevant to the $F^{(g)}$ terms in the $N = 2$ actions are the BPS states. Let us first discuss the boundary conditions for $F^{(1)}$ at the singular points in the moduli space.

- At the point of maximal unipotent monodromy in the mirror manifold W, the Kähler areas, four, and six volumes of the original manifold M are all large. Therefore the lightest string states are the constant maps $\Sigma_g \to pt \in M$. For

Table 10 BPS invariants n_d^g on the Quintic hypersurface in $\mathbb{P}^4$ [191]

g	d=1	d=2	d=3	d=4	d=5	d=6
0	2875	609250	317206375	242467530000	229305888887625	248249742118022000
1	0	0	609250	3721431625	12129909700200	31147299733286500
2	0	0	0	534750	75478987900	871708139638250
3	0	0	0	8625	−15663750	3156446162875
4	0	0	0	0	49250	−7529331750
5	0	0	0	0	1100	−3079125
6	0	0	0	0	10	−34500
7	0	0	0	0	0	0

g	d=7	d=8	d=9
0	295091050570845659250	375632160937476603550000	503840510416985243645106250
1	715784060228807617500	1549905417529615684181250	3240644643102795856570087500
2	518546255661726962500	225168410631059177667500	814649217868395665025601250
3	111468926053022750	130346459840858345500 0	952321365916921756899150 0
4	245477430615250	2551750225483422675 0	5077234965144335614982500
5	−1917984531500	46569889619570625	1028074359449310831975 0
6	1300955250	−47185210090950 0	30884164195870217250
7	4874000	2876330661125	−135197508177440750
8	0	−1670397000	193765229097112 5
9	0	−6092500	−12735865055000
10	0	0	18763368375
11	0	0	5502750
12	0	0	60375
13	0	0	0

g	d=10	d=11
0	704288164978454686113488249750	1017913203569692432490203659468875
1	6628637743914140967424065763000	1336442091735463067608016312923750
2	2619106395286732590955451374500	7757206271485037501990496914497500
3	529399661897916624420404068250	2457496729082220699996115276347500
4	564669022311863868292985660 0	448475557200658307168403004753750
5	302653046360802682731297875	469508660948449138653717762000 0
6	694875009474861138496273 0	26778976421684176016869138162 5
7	401795199961582390768 00	735709924295207023870887000 0
8	−2530103276608330315 0	7274265159936800289770125 0
9	1155593062739271425	140965985795732693440000
10	−17976209529424700	722850712031170092000
11	150444095741780	−1899895525748217125 0
12	−454092663150	3536502289027385 00
13	50530375	−4041708780324500
14	−286650	22562306494375
15	−5700	−29938013250
16	−50	−7357125
17	0	−86250
18	0	0

these Kaluza-Klein reduction, i.e. a zero mode analysis of the A-twisted non-linear σ-model is sufficient to calculate the leading behaviour[116] of $F^{(1)}$ as [40]

$$F^{(1)} = \frac{t_i}{24} \int c_2 \wedge J_i + \mathcal{O}(e^{2\pi i t}) . \qquad (4.5.3)$$

Here $2\pi i\, t_i = \frac{X^i}{X^0}$ are the canonical Kähler parameters, c_2 is the second Chern class, and J_i is the basis for the Kähler cone dual to 2-cycles C_i defining the $t_i := \int_{C_i} \hat{J} = \int_{C_i} \sum_i t_i J_i$.

- At the conifold divisor in the moduli space $\mathcal{M}_{cs}(W)$, W develops a nodal singularity, i.e., a collapsing cycle with S^3 topology. As discussed in Sect. 2.10.3 this corresponds to the vanishing of the total volume of M. The leading behaviour at this point is universally [187]

$$F^{(1)} = \frac{1}{12} \log(t_c) + O(t_c) . \qquad (4.5.4)$$

This leading behaviour has been physically explained as the effect of integrating out a non-perturbative hypermultiplet, namely the extremal black hole of [297]. Its mass $\sim t_c$, see (4.5.6), goes to zero at the conifold and it couples to the $U(1)$ vector in the $N = 2$ vector multiplet, whose lowest component is the modulus t_c. The factor $\frac{1}{12}$ comes from the gravitational one-loop β-function, which describes the running of the $U(1)$ coupling [306]. A closely related situation is the one of a shrinking lense space S/G. As explained in [138] one gets in this case several BPS hyper multiplets as the bound states of wrapped D-branes, which modifies the factor $\frac{1}{12} \rightarrow \frac{|G|}{12}$ in the one loop β-function (4.5.4).
- The gravitational β-function argument extends also to non-perturbative spectra arising at more complicated singularities, e.g. with gauge symmetry enhancement and adjoint matter [223].

For the case of the one parameter families the above boundary information and the fact is sufficient to fix the holomorphic ambiguity in $F^{(1)}$.

4.5.3 The Gap Condition

To learn from the effective action point of view about the higher genus boundary behaviour, let us recall that the $F^{(g)}$ as in $\mathcal{F}(\lambda, t) = \sum_{g=1}^{\infty} \lambda^{2g-2} \mathcal{F}^{(g)}(t)$ give rise the one loop term (4.3.7) where R_- is the anti self-dual part of the curvature and we identify λ with T_-, the anti self-dual part of the graviphoton field strength.

What are the microscopic BPS states that run in the loop? They are related to non-perturbative RR states, which are the only charged states in the Type II compactification. They come from branes wrapping cycles in the Calabi-Yau, and

[116]The leading of $F^{(0)}$ at this point is similarly calculated and given in (2.12.21).

as BPS states their masses are proportional to their central charge (2.5.52), (2.6.19). For example, in the large radius in the type IIA string on M, the mass is determined by integrals of complexified volume forms over even cycles. E.g., as we saw in Sect. 4.3.3 the mass of a 2 brane wrapping a holomorphic curve $C_\beta \in H^2(M,\mathbb{Z})$ is given by[117]

$$m_\beta = \frac{1}{g_s}\int_{C_\beta \in H^2(M,\mathbb{Z})} (iJ + B) = \frac{1}{g_s} 2\pi i t\cdot \beta =: \frac{1}{g_s} t_\beta \, . \tag{4.5.5}$$

We note that $H^2(M,\mathbb{Z})$ plays here the role of the charge lattice. In the type IIB picture the charge is given by integrals of the normalized holomorphic (3, 0)-form Ω. In particular the mass of the extremal black hole that vanishes at the conifold is given by

$$m_{BH} = \left| \frac{1}{g_s \int_{A_c} \Omega} \int_{S^3} \Omega \right| =: \left| \frac{1}{g_s} t_c \right| , \tag{4.5.6}$$

where A_c is a suitable non-vanishing cycle at the conifold. It follows from the discussion in the Sect. 4.3.3 that with the identification $e = m$ and after a rescaling $s \to s\lambda/e$ in (4.3.18), as well as absorbing F into λ, one gets a result for (4.3.7)

$$S^s_{1-loop}(\lambda, t) = \int_\epsilon^\infty \frac{\mathrm{d}s}{s} \frac{\mathrm{Tr}(-1)^f \exp(-st)\exp(-2s\sigma_L\lambda)}{4\sin^2(s\lambda/2)} \, . \tag{4.5.7}$$

Here t are the regularized masses, c.f. (4.5.5), (4.5.6) of the light particles P that are integrated out, f is their spins in $\mathcal{R}$, and σ_L is the Cartan element in the representation $\mathcal{R}$.

Let us turn to type IIB compactifications near the conifold. As it was checked with the gravitational β-function in [306] there is precisely one BPS hypermultiplet with the Lorentz representation of the first factor in (4.3.17) becoming massless at the conifold. In this case the Schwinger-Loop calculation (4.5.7) simply becomes

$$F(\lambda, t_c) = \int_\epsilon^\infty \frac{\mathrm{d}s}{s} \frac{\exp(-st_c)}{4\sin^2(s\lambda/2)} + \mathcal{O}(t_c^0) = \sum_{g=2}^\infty \left(\frac{\lambda}{t_c}\right)^{2g-2} \frac{(-1)^{g-1}B_{2g}}{2g(2g-2)} + \mathcal{O}(t_c^0) \, . \tag{4.5.8}$$

Since there are no other light particles, the above Eq. (4.5.8) encodes all singular terms in the effective action. There will be regular terms coming from other massive states. This property of (4.5.8) is called the *gap condition*. The gap condition was first observed in N=2 Seiberg-Witten theory and matrix models [189] where one has additional independent checks on the F_g.

117 Here we choose a phase so that t_β is real.

As mentioned an interesting singularity appears in the FHSV model [113] models discussed [151, 222] where a 4-cycle shrinks in the IIA picture and one has a shortened gap and dual integral BPS expansion.

4.5.4 The Refined Gap Condition

Performing the refined Schwinger loop integral (4.4.1) with a single dyon of vanishing mass $|t_c|$ and expanding it in ϵ_1, ϵ_2, and $\frac{1}{t_c}$ gives us the leading behavior of each $F^{(n,g)}$ near the conifold point from the corresponding coefficients of $(\lambda^2 = (\epsilon_1\epsilon_2)$ and $s = (\epsilon_1 + \epsilon_2)^2)$ [192]

$$\begin{aligned} F(s,\lambda,t_c) &= \int_0^\infty \frac{ds}{s}\frac{\exp(-st_c)}{4\sinh(s\epsilon_1/2)\sinh(s\epsilon_2/2)} + \mathcal{O}(t_c^0) \\ &= \big[-\frac{1}{12}+\frac{1}{24}(\epsilon_1+\epsilon_2)^2(\epsilon_1\epsilon_2)^{-1}\big]\log(t_c) \\ &\quad + \frac{1}{\epsilon_1\epsilon_2}\sum_{g=0}^{\infty}\frac{(2g-3)!}{t_c^{2g-2}}\sum_{m=0}^{g}\hat{B}_{2g}\hat{B}_{2g-2m}\epsilon_1^{2g-2m}\epsilon_2^{2m}+\ldots \\ &= \big[-\frac{1}{12}+\frac{1}{24}sg_s^{-2}\big]\log(t_c)+\big[-\frac{1}{240}g_s^2+\frac{7}{1440}s-\frac{7}{5760}s^2g_s^{-2}\big]\frac{1}{t_c^2} \\ &\quad +\big[\frac{1}{1008}g_s^4-\frac{41}{20160}sg_s^2+\frac{31}{26880}s^2-\frac{31}{161280}s^3g_s^{-2}\big]\frac{1}{t_c^4}+\mathcal{O}(t_c^0) \\ &\quad + \text{contributions to } 2(g+n)-2>4\,. \end{aligned} \tag{4.5.9}$$

Hence, e.g.,

$$F^{(0,2)} = -\frac{1}{240}\frac{1}{t_c^2}+\mathcal{O}(t_c^0),\quad F^{(1,1)}=\frac{7}{1440}\frac{1}{t_c^2}+\mathcal{O}(t_c^0),\quad F^{(2,0)}=-\frac{7}{5760}\frac{1}{t_c^2}+\mathcal{O}(t_c^0)\,. \tag{4.5.10}$$

The leading behavior of (4.5.9) is the same as that of the S^1 compactification of the $c=1$ string,[118] where the same integral appears [154].

The fact that near conifold points in moduli space, the relation

$$F^{(n,g)} = \frac{N^{(n,g)}}{t_c^{2(g+n)-2}}+\mathcal{O}(t_c^0) \tag{4.5.11}$$

[118] As was noted for $\epsilon_1 = -\epsilon_2$ in [135].

holds, i.e. the absence of subleading poles in the t_p expansion, is referred to as the refined gap condition of the $F^{(n,g)}$ at these points [189, 191]. We note that these refined gap conditions are sufficient to fix the refined holomorphic ambiguity also *for the local models*. For the global cases they will be not sufficient and because of the absence of the $U(1)_{\mathcal{R}}$ symmetry it is not clear what the calculation precisely means.

4.5.5 Bounds on the BPS Numbers from Castelnuovo Theory

The basic idea of the Castelnouvo bound for the n^g_β is that in a certain class $\beta \in H_2(M,\mathbb{Z})$ it is generically not possible to have an embedded curve C of arbitrary high genus g in M. For the Gromov-Witten invariants if there was a map at genus g' then there will by multi coverings of this map n^g_β in (4.3.4) at all genera $g > g'$ as the formula (4.3.5) predicts even for the simplest case of an isolated genus zero curve. The formula (4.3.18) is supposed to give the multi coverings for maps with arbitrary genus. If the geometric interpretation of Sect. 4.3.4 or rather the definition of Pandharipande-Thomas invariants reproduces (4.3.18), then in each class $\beta \in H_2(M,\mathbb{Z})$ there will be a g_{max} beyond which $n^{g\geq g_{max}}_\beta = 0$ because there is no curve of higher genus and all contributions from lower genus curves are already subtracted. At least qualitatively it is easy to understand why such a bound is there from the adjunction formula

$$C^2 + K\cdot C = (2g-2)\,. \tag{4.5.12}$$

On the Calabi-Yau manifold M and $K = 0$ and in one class specified by β one gets roughly

$$g(\beta) = \frac{\beta^2+2}{2}\,. \tag{4.5.13}$$

However not all curves are smooth and generic, e.g. in the quintic, so the bound might be violated and in particular not saturated. As explained in [215] curves in projective within the quintic are either plane curves in $\mathbb{P}^2$, curves in $\mathbb{P}^3$, or $\mathbb{P}^4$. In all case one gets from Castelnuovo theory a bound on g, which grows still for large $d = \beta$ like $g(d) \sim d^2/2$. For a detailed exposition of curves in projective space see [167]. Using this information one can determine which curves above are realised and contribute to the BPS numbers. These statements generalize to the hypersurfaces and complete intersections with one Kähler modulus in weighted projective spaces. In fact if there is a smooth curve one cannot only determine the exact bound, but using (4.3.26) or the specialized versions of (4.3.28)

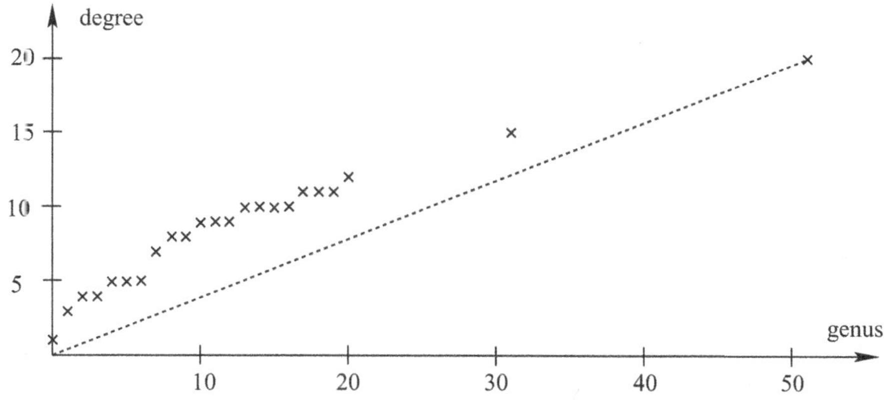

Fig. 21 The points that follow roughly the $g = d^2/2$ curve represent the known Castelnouvo bounds, while the line represent the number of unknowns in the holomorphic ambiguity f_g after imposing the regularity at the orbifold and the gap condition [191]

$$
\begin{aligned}
n_d^{\tilde{g}-1} &= (-1)^{\dim(\mathcal{M})+1}\left(e(\mathcal{C}) + (2\tilde{g}-2)e(\mathcal{M})\right) \\
n_d^{\tilde{g}-2} &= (-1)^{\dim(\mathcal{M})+1}\left(e(\mathcal{C}^{(2)}) + (2\tilde{g}-4)e(\mathcal{C}) + \tfrac{1}{2}(2\tilde{g}-2)(2\tilde{g}-5)e(\mathcal{M})\right) .
\end{aligned} \tag{4.5.14}
$$

one can make nontrivial checks on the curve counts in Table 10. The most detailed information on the bound we have is in Fig. 21. As it turns out there is a smooth curve at genus 51 and for $g \leq 51$ the vanishing of the n_β^g and the bounds that we get from the orbifold regularity and the gap condition are sufficient to solve the topological string on the quintic.

4.5.6 Higher Genus Results for the Quintic

In Table 10 we present the non-vanishing n_d^g for the quintic in $\mathbb{P}^4$ up to degree $d = 11$. Similar results for hypergeometric one parameter Calabi-Yau manifolds can be found in [191]. An analysis of the asymptotic growth of the BPS numbers relevant for the comparison of the mirco state—and the Bekenstein Hawking entropy was performed in [190].

5 Further Higher Genus Techniques

This chapter is much less self contained then the other parts of the lecture. Its intention is to give an overview over further techniques to solve the topological string at high genus. While these techniques give beautiful all genera results, especially when combined with automorphic symmetries that lead to modular or automorphic forms, these techniques are all restricted so far to lower dimensional sub-geometries of the Calabi-Yau manifolds.

The techniques fall basically in two types:

- The *modular techniques*[119] rely on an understanding of the discrete symmetries that govern the theory and in particular its correlation functions. Due to the finiteness of the ring of modular forms, once the weight and other indices of the correlation function is know then, if the latter is sufficiently restricted by physical or geometrical boundary conditions, it can be reconstructed. In a sense the solution of the B-model discussed in Sect. 4 follows the same logic. The problem was in part that the knowledge of the generators of the ring of modular forms is incomplete on a compact Calabi-Yau 3-fold and more severely that the boundary conditions are not sufficient. Both objections as we already saw evaporate in the local case, as the gap is sufficient and the forms become either modular forms or Siegel forms.
- Another good situation is if the global Calabi–Yau has a fibration structure either by K3 or elliptic curves, so that the modular generators at least in this directions restrict to easier automorphic forms or even strictly modular forms. Moreover in the case of K3 fibrations there is duality between Type II and the heterotic string, preserving all discrete symmetries of course. In the heteroric string these symmetries combined with world sheet techniques provide powerful methods to calculate the one-loop amplitude which provides all genera information in the $K3$ fiber. In the case of elliptic fibrations there has been recently a lot of progress inspired also by world-sheet techniques of auxiliary dual string quiver theories in which the topological string partition function is calculated as an *elliptic genus*. Further six and five dimensional non-trivial QFT theories from F-theories on elliptic Calabi-Yau spaces provide further structures that help to determine the topological string partition function.
- The *large N techniques* use 't Hooft's idea of gauge theory/string theory duality. The problem is that for string theory on the compact Calabi-Yau the dual gauge theory is not yet known. However in the local Calabi-Yau cases the gauge theory is often known. In particular for all local toric Calabi-Yau spaces and is either a matrix model or a 3d-Chern Simons theory.

With start our discussion with the modular approach.

[119] We use the word modular loosely not only to refer subgroups of the modular group $SL(2, \mathbb{Z})$ and the corresponding ring of modular forms, but also to other automorphic group and automorphic forms and functions.

5.1 Modularity from Fibration Structures and Dualities

Here we address modular techniques whose modularity originates either from duality with the heterotic string and is associated to an K3 fibred Calabi-Yau 3 fold or to six dimensional super conformal theories associated elliptically fibred Calabi-Yau 3-folds or more general geometries which have a dual quiver gauge linear sigma model description.

5.1.1 K3 Fibrations and Heterotic Type II String Dualities

As the above ideas originate to some extend from the duality between N=2 Type II string and the heterotic string, some of the strongest predictions for the n^g_β invariants on compact Calabi-Yau manifolds can be made if dual pairs of heterotic/type II compactifications are known (1.1.14). The relevant Calabi-Yau manifolds are K3 fibrations over $\mathbb{P}^1$ [204, 229] and under the moduli identification the heterotic weak coupling limit translates to infinite volume limit of the base $\mathbb{P}^1$. The heterotic string prediction relies on a perturbative WS one-loop calculation in the weak coupling limit and makes therefore only predictions for n^g_β if β is a class entirely in the $K3$ fibre. Information about other classes $\hat\beta$ is suppressed, because for them one has $q^{\hat\beta} \to 0$ in the weak coupling/infinite base limit. The one-loop (torus) amplitude is [19]

$$\mathcal{F}^g = \int_{\mathcal{F}} \mathrm{d}\tau\, \tau_2^{2g-3} \frac{1}{|\eta|^4} \sum_{even} \frac{i}{\pi} \partial_\tau \left(\frac{\theta\left[{a \atop b}\right](\tau)}{\eta(\tau)} \right) Z^{int}_g \begin{bmatrix} a \\ b \end{bmatrix}, \qquad Z^{int}_g \begin{bmatrix} a \\ b \end{bmatrix} = \langle : (\partial X)^{2g} : \rangle \,. \tag{5.1.1}$$

The integrand can be understood as an index on the heterotic WS theory very similar to (4.1.20) [169] and the integral over the fundamental region $\mathcal{F}$ of the torus can be calculated using the modular properties of the integrand in an ingeneous way [48, 97, 169, 222, 260]. For the $K3$ fibrations without reducible fibres one finds in the holomorphic limit [233]

$$\mathcal{F}^{\rm hol}(\text{Fibre}_{K_3}, \lambda, q) = \frac{\Theta(q)}{q} \left(\frac{1}{2\sin(\frac{\lambda}{2})} \right)^2 \prod_{n\geq 1} \frac{1}{(1-q_\lambda q^n)^2 (1-q^n)^{20} (1-q_\lambda^{-1} q^n)^2} \,. \tag{5.1.2}$$

Similar as in the case of elliptic Del-Pezzo surfaces S embedded in Calabi-Yau manifolds [188] the product factor can be interpreted as the Goettsches formula for the cohomology of the resolved Hilbert scheme of points on the surface S or the $K3$ respectively. The formula (5.1.2) can also be viewed as an extension of the analysis of [333] to a situation with less supersymmetry.

Example The degree 12 hypersurface in the weighted projective space $WCP(1,1,2,2,6)$, see Appendix 3 is a K3 fibration, which is dual to the ST heterotic string discussed in [204]. In this case $\Theta(q)$ is [218] $\frac{\Theta(q)}{\eta^{24}} = -\frac{2\sigma E_4 F_6}{\eta^{24}} = -\frac{2}{q} + 252 + 2496q^{\frac{1}{4}} + 223752q + \ldots$, where $\sigma(q) = \sum_{n\in\mathbb{Z}} q^{\frac{n^2}{4}}$ and $F_2(q) = \sum_{n\in\mathbf{Z}_{>0},\text{odd}} \sigma_1(n) q^{\frac{n}{4}}$ generate the ring of modular forms for the congruence subgroup $\Gamma^0(4)$, and $F_6 = E_6 - 2F_2(\sigma^4 - 2F_2)(\sigma^4 - 16F_2)$. The embedding of the Picard lattice of the $K3$ into the Calabi-Yau M is specified by the replacement of $\lambda^{2g-2}q^l \to \frac{1}{(2\pi i)^{3-2g}} \sum_{n^2/4=l} \mathrm{Li}_{3-2g}(q^{\beta n})$ in (5.1.2), where β is the single class in the $K3$ fibre. Comparing with (4.3.18) one gets predictions in a closed form for n^g_β for all g and all β. Below the first few are listed

g	$\beta = 1$	2	3	4	...
0	2496	223752	38637504	9100224984	...
1	0	−492	−1465984	−1042943520	...
2	0	−6	7488	50181180	...
3	0	0	0	−902328	...
4	0	0	0	1164	...
5	0	0	0	12	...
⋮	⋮	⋮	⋮	⋮	...

Many of these predictions from string duality have been checked in [233] using geometrical techniques. Using the direct integration technique described in Sect. 4.2.3 for this two parameter model this model has been to large extend solved by B-model techniques confirming the higher genus invariants in the K3 fibre but also supplementing them with such that have non trivial degree on the base [162].

Moreover in a closed related elliptic fibration the STU model, an K3 fibration that is also an elliptic fibration over $\mathbb{P}^1 \times \mathbb{P}^1$ the results for the invariant in the K3 fibre have been mathematically proven and used to prove the SYZ conjecture (1.1.16) for non-primitive classes [234].

Particular strong results have been obtained for the FHSV model [113] in [222]. This is special K3 fibration obtained as an $\mathbb{Z}_2$ orbifold of $K3 \times T_2$ that acts like the Enriques involution on the $K3$ and as hyperelliptic involution on the two torus T_2 and has $\mathbb{Z}_2 \times SU(2)$ holonomy leading for type II to $N = 2$ in four dimensions theory with strong simplifications. In particular the prepotential is purely classical without instanton corrections or equivalently the moduli space is a symmetric group space [222]. However the $g \geq 1$ instanton sectors are non-trivial and in [151, 222] a direct integration formalism has been developed that allows to calculate all higher genus instanton up to base degree 5.

5.1.2 Elliptic Fibrations and Elliptic Genus

Here we report on recent progress on the topological string on elliptically fibered Calabi-Yau spaces, which implies that for given base degree the all genus BPS invariants in all fibre classes are given by certain meromorphic Jacobi Forms. This was first observed in local models based on an elliptic surface, first for smooth fibrations [164], then for singular fibrations [166] and finally also in global models [195]. In the local case in particular in [166] there are easy 2d gauge linear σ-model quiver description and the higher genus invariants are captured by the elliptic genus of the σ model, which makes the occurrence of Jacobi forms very natural. We focus on the global case.

Let us denote by M the elliptically Calabi-Yau manifold, by B its base and by $\mathcal{E}$ its fibre with B as section

$$\begin{array}{ccc} \mathcal{E} \to & M & \\ & \downarrow \pi \, . & \\ & B & \end{array} \tag{5.1.3}$$

We will first assume that the fibration has only I_1 fibres over the base as for example in the model discussed in Sect. 2.11.2 where the base is $\mathbb{P}^2$ or in the elliptic fibration phase in the example in which the flop is described (2.7.38), where the base is the Hirzebruch surface $\mathbb{F}_1$. We start with some features of the classical geometry.

Denote by $\{[\tilde{C}^k]\}$, $k = 1, \ldots, h_{11}(B) = h_{11}(M) - 1$ the generators of the Mori cone of B, and by $\{[D'_k]\}$ the dual basis for the Kähler cone of B. Since B is a section, $[\tilde{C}^k]$ are curve classes on M. Further $[\tilde{C}^e]$ denotes the class of the elliptic fiber. To summarize $\{[\tilde{C}^e], [\tilde{C}^k]\}$ are generators for the Mori cone of M and the dual basis of the Kähler cone of M is generated by

$$[\tilde{D}_k] = \pi^*[D'_k], \qquad [\tilde{D}_e] = [E] + \pi^* c_1(B), \tag{5.1.4}$$

where $[E]$ is the divisor class of the section. The complexified Kähler areas of the curves in the base and the fibre are $\tilde{T}^k = \frac{1}{2\pi i}\int_{\tilde{C}^k} ib - \omega$ and $\tilde{\tau} = \frac{1}{2\pi i}\int_{\tilde{C}^e} ib - \omega$ respectively.

We assume that $c_1(B) = -K_B$ of B is semi-positive.[120] One can then easily calculate the following classical intersections on M [235]

$$\tilde{D}_e^3 = \int_B c_1(B)^2, \qquad \tilde{D}_e^2 \cdot \tilde{D}_k = a_k, \qquad \tilde{D}_e \cdot \tilde{D}_i \cdot \tilde{D}_j = c_{ij} \tag{5.1.5}$$

with

$$a_k = c_1(B) \cdot D'_k, \qquad a^k = c_1(B) \cdot \tilde{C}^k \, . \qquad c_{ij} = D'_i \cdot D'_j. \tag{5.1.6}$$

[120] This is fulfilled for all bases given by the 2d reflexive lattice polytopes shown in Fig. 1.

To exhibit the modular properties it is convenient to transform the complexified Kähler parameter to

$$\tau = \tilde{\tau}, \qquad T^k = \tilde{T}^k + \frac{a^k}{2}\tau \tag{5.1.7}$$

and the curve classes accordingly. We denote the exponentiated variables by $q = \exp(2\pi i \tau)$, and $Q_k = \exp(2\pi i T^k)$. The point is that in this basis there is a subgroup of the monodromy group of the mirror Γ_W, which after identification by the mirror map generates an $PSL(2,\mathbb{Z})$ action on τ and does not act, up to exponentially small terms in Q_k, on the modified base classes T^k. For example the S transformation monodromy in the basis defined from the classical intersection numbers by (2.6.30) is [196]

$$\begin{array}{c} \qquad X^0 \quad X^e \quad \tilde{X}^k \qquad F_0\; F_e \qquad F_k \\ \mathbb{S} = \left(\begin{array}{cccccc} 0 & 1 & 0\ldots & 0 & 0 & \ldots 0 \\ -1 & 0 & 0\ldots & 0 & 0 & \ldots 0 \\ a^k & 0 & 0\ldots & 0 & 0 & -c^{kj} \\ -1 & 0 & 0\ldots & 0 & 1 & -a_j \\ 0 & c_1^2(B)+1 & a_j & -1 & 0 & -a_j \\ a_k & -a_k & c_{kj} & 0 & 0 & \ldots 0 \end{array}\right) \begin{array}{c} X^0 \\ X^e \\ \tilde{X}^k \\ F_0 \\ F_e \\ F_k\,. \end{array} \end{array} \tag{5.1.8}$$

The T transformation is given by the b shift in the τ parameter. We defined a Kähler gauged partition function

$$\mathfrak{Z} = \exp(\mathfrak{F}) = \exp\left(\sum_{g=0}^{\infty} g_s^{2g-2}\mathfrak{F}^{(g)}(\underline{t})\right), \quad \text{with} \quad \mathfrak{F}^{(g)}(\underline{t}) = (X^0)^{2g-2}(\underline{t})F^{(g)}(\underline{X})\,. \tag{5.1.9}$$

and expand the latter in the base classes as

$$\mathfrak{Z}(\tau, \underline{T}, g_s) = \mathfrak{Z}_{\beta=0}\left(1 + \sum_{\substack{\beta\in H_2(B,\mathbb{Z})\\ \beta\neq 0}} \mathfrak{Z}_\beta(\tau, g_s) Q^\beta\right), \tag{5.1.10}$$

then the $\mathfrak{Z}_\beta(\tau, g_s)$ have the following properties[121]:

[121] The definitions for the Jacobi forms are collected in section "Jacobi Forms" in Appendix 4. In the section $g_s = \lambda$ is the topological string coupling.

Property 1 $\mathfrak{Z}_\beta(\tau, g_s)$ is a *meromorphic Jacobi form* of weight zero $k = 0$ and index

$$m_\beta = \frac{1}{2}\beta \cdot (\beta - c_1(B)), \tag{5.1.11}$$

where the topological string coupling $\lambda = g_s$ is identified up to multiple with the elliptic argument z of the Jacobi form

$$g_s = 2\pi z\,. \tag{5.1.12}$$

Property 2

$$\mathfrak{Z}_\beta(\tau, z) = \frac{1}{\eta^{12\beta\cdot c_1(B)}} \frac{\varphi_\beta(\tau, z)}{\prod_{l=1}^{b_2(B)} \prod_{s=1}^{\beta_l} A(\tau, sz)} \tag{5.1.13}$$

where $\varphi_\beta(\tau, z)$ is a *weak Jacobi form* of weight

$$w_\beta = 6\beta \cdot c_1(B) - 2\sum_{l=1}^{b_2(B)} \beta_l \tag{5.1.14}$$

and index

$$m_\beta = \frac{1}{6}\sum_{l=1}^{b_2(B)} \beta_l(1+\beta_l)(1+2\beta_l) + \frac{1}{2}\beta \cdot (\beta - c_1(B))\,. \tag{5.1.15}$$

Property 3 The Castelnouvo bounds that predict the vanishing of BPS indices n_g^β for $g \geq \mathcal{O}(d_e)$ in the classes $\kappa = (d_e, \beta)$ for $\beta = k\tilde{\beta}$ and $\tilde{\beta}a$ primitive class, determine together with the genus zero and one results the weak Jacobi form φ_β for all positive $k \in \mathbb{N}$, if

$$m_\beta \leq 0\,. \tag{5.1.16}$$

This properties can be shown using the holomorphic anomaly (4.1.23) and the expected pole structure in (4.3.18). Property 3 and the finiteness of the ring of *weak Jacobi* forms that generate $\varphi_\beta(\tau, z)$ is strong enough to solve many local models. For example the E-string can be solved completely and very efficiently [196]. The corresponding φ are ($Q = E_4$, $R = E_6$ and the definition of the weak Jacobi forms A and B is given in A4.49)

$$\begin{aligned} \varphi_1 &= -Q, \\ \varphi_2 &= \tfrac{1}{288}(A^2(5R^2 - Q^3) - 8ABQR + 4B^2Q^2), \\ &etc \end{aligned} \tag{5.1.17}$$

yielding the BPS invariants in Tables 11 and 12.

Table 11 The GV invariants $n^g_{(d_E,1)}$, i.e. for the base degree $\beta = 1$

$g\backslash d_E$	0	1	2	3	4	5	6
0	1	252	5130	54760	419895	2587788	13630694
1	0	−2	−510	−11780	−142330	−1212930	−8207894
2	0	0	3	772	19467	257796	2391067
3	0	0	0	−4	−1038	−28200	−403530
4	0	0	0	0	5	1308	37991
5	0	0	0	0	0	−6	−1582
6	0	0	0	0	0	0	7
7	0	0	0	0	0	0	0

Table 12 The GV invariants $n^g_{(d_E,2)}$, i.e. for the base degree $\beta = 2$

$g\backslash d_E$	0	1	2	3	4	5	6
0	0	0	−9252	−673760	−20534040	−389320128	−5398936120
1	0	0	760	205320	11361360	317469648	5863920760
2	0	0	−4	−25604	−3075138	−135430120	−3449998524
3	0	0	0	1296	494144	38004700	1400424188
4	0	0	0	−6	−45172	−7279258	−416416202
5	0	0	0	0	1844	918292	90943340
6	0	0	0	0	−8	−69026	−14214528
7	0	0	0	0	0	2408	1505880
8	0	0	0	0	0	−10	−97272
9	0	0	0	0	0	0	2988
10	0	0	0	0	0	0	−12
11	0	0	0	0	0	0	0

The formalism can be also refined and extended to all classes in the surface within the local geometry using of course more generalised rings of Jacobi Forms. For smooth cases, i.e. the refined E-string and chains of E- and M-strings, this is done in [158]. In the refined case ϵ_1 and ϵ_2 feature as elliptic arguments of the Jacobi-forms, similar as the mass parameters for the E-string, for which the corresponding Weyl-invariant Jacobi-Forms were given in [284]. Singular geometries have been treated in [85], where the results exhibits an affine Weyl-Invariance that corresponds to the affine Weyl Invariance $\hat{\mathfrak{g}}$ in the Kodaira resolution diagram of a singularity at co-dimension one in B that leads to a gauge theory $\mathfrak{g}$ in F-theory.

In the compact $X_{18}(1, 1, 1, 6, 9)$ manifold the condition $m_\beta \leq 0$ is violated for all curve classes in the base. Nevertheless using the Castelnouvo bound, the gap condition and the specific form (5.1.13) of the amplitudes very high genus (all degrees and $g \leq 189$) and base degree (all genera and fibre degrees and base degree $\beta \leq 20$) computations can be performed [195].

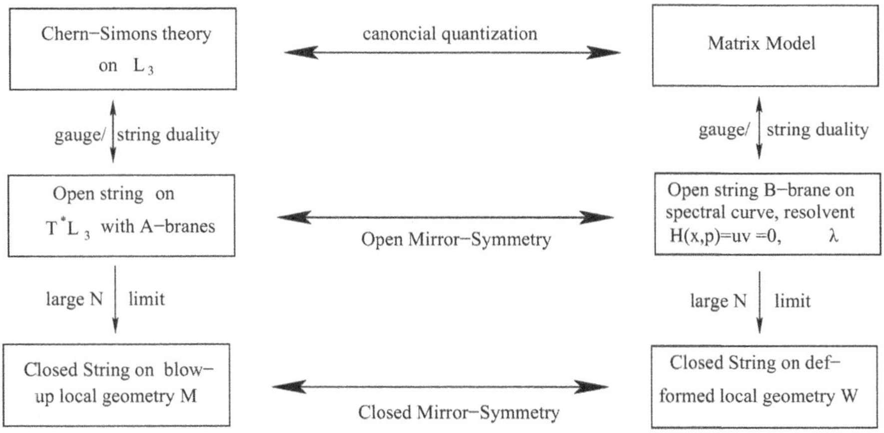

Fig. 22 Scheme of the relations and dualities in large N approaches to the topological string on local Calabi-Yau threefolds

5.2 *Open String Amplitudes and Large N Techniques*

Large N-techniques that rely on the duality between gauge theory with high rank gauge group are extremely powerful tools to get higher genus information for open and closed topological string theory on local Calabi Yau spaces. Let us start then with a small overview over the A- and $B-$ model applications in Fig. 22. In the preparation for the large N techniques in the B model we start then with some review on the disk amplitude.

5.2.1 Chain Integrals and Open String Disk Amplitudes

Formulas for the open string amplitudes can be derived from the super potential W [328]. The latter can be obtained by reducing the holomorphic Chern Simons action [324]

$$W = \int_W \Omega \wedge \mathrm{Tr}[A \wedge \bar{\partial} A + \frac{2}{3} A \wedge A \wedge A] \tag{5.2.1}$$

that extends over the hole Calabi-Yau manifold to lower dimensional branes. The action can be reduced in particular to curves $C \in M$ and evaluated in the geometries discussed in Sect. 2.7.4, as has been described in [1].

The tangent bundle of M split over C as

$$\mathcal{T}(M) = \mathcal{T}(C) \oplus \mathcal{N}(C)\,. \tag{5.2.2}$$

Parameterizing the independent sections in the normal bundle $\mathcal{N}(C)$ with two complex parameters $\phi^{1,2}(z)$ that depend on the coordinate z of C and using the fact that due to triviality of the canonical class of W the determinant bundle on $\mathcal{N}(C)$ is isomorphic to $\mathcal{T}^*(C)$, i.e. each $V_z = \mathcal{T}^*(C)$ can be identified with $V_z = \Omega_{ijz}\phi^i \wedge \phi^j$, one gets the reduction of (5.2.1) to C for a single brane as

$$W(C) = \int_C \Omega_{ijz}\phi^i \bar{\partial}_z \phi^j dz d\bar{z}. \tag{5.2.3}$$

Here one chooses a coordinate system in which Ω_{ijz} is constant. Moreover locally one can write $\Omega = d\mu$, i.e. $\Omega_{ijz} = \partial_z \mu_{ij} + perms$ and rewrite (5.2.3) as [1]

$$W(C) = \int_C \mu, \tag{5.2.4}$$

which as μ is not exact is well defined as long as C has no boundary. This is not directly a meaningful physical quantity in space time. Rather one has to consider two curves C and C^*. The correct physical picture comes from a $D5$ brane wrapping a curve and filling space time. Assume in the left half space of the spacial $\mathbb{R}^3$ it wraps C^* and in the right C. Then along the $2+1$ dimensional subspace that constitute the boundary one has a domain wall in space time whose tension is given as

$$W(C) = \int_C \mu - \int_{C^*} \mu = \int_\Gamma \Omega \tag{5.2.5}$$

by the period integral over a 3-chain Γ whose boundaries are $\partial\Gamma = C - C^*$. C^* is viewed as a reference curve. If W would have been an elliptic curve Σ_1 and $\mathbb{C}/\Lambda$ its Jacobian[122] $J(\Sigma_1)$, the map (5.2.5) would be just the Abel-Jacobi map

$$w : \Sigma_g \to J(\Sigma_1), \quad w(p) = \int_{p^*}^{p} \omega_1 \,. \tag{5.2.6}$$

I.e. the super potential can be viewed as a higher dimensional analog of the Abel-Jacobi map, where the points p^*, p are replaced by curves constituting the boundaries of a chain. If C^* and C lie in a holomorphic family of curves $W(C) = 0$ and in this sense $W(C)$ measured the obstruction to deform C^* holomorphically. The order of the obstruction can be in simple situation calculated by algebra geometry and yields the order of the leading terms in $W(C)$ in the considered perturbation [206, 207]. The corresponding evaluation of the chain integrals is a complicated by in principle mathematically well understood subject, see [265] for a review with physical applications. It leads typically to Picard-Fuchs like equations with non-homogenous terms as for example in the pioneering work [311].

[122] In this case it is also an 2 torus, but in the obvious general generalization to Σ_g using all ω_i, $i = 1, \ldots, g$ holomorphic one forms, it is a $\mathbb{T}^{2g}$ torus.

Mirror symmetry maps on Calabi-Yau 3 folds maps holomorphic sub-manifold to special Lagrangian 3-cycles L and can applied in any situation in which the W and M and the corresponding sub-manifolds are known, i.e. in particular for the Batyrev mirrors. Using the correspondence of Chern-Simons theory as String theory and a similar integral as (4.3.18) [271] find the multi covering formula

$$F(t, Y_i) = i \sum_{m=1}^{\infty} \sum_{\beta,\mathcal{R},g} \frac{N_{\beta,\mathcal{R},s}}{2m\sin(m\lambda/2)} e^{n(t\cdot\beta+ig\lambda)} \mathrm{Tr}_{\mathcal{R}} \prod_{i=1}^{b_1(S)} Y_i^m \tag{5.2.7}$$

where Y parametrises the open string moduli and $\mathcal{R}$ is a representation basis that encodes the boundary topology of the open string. In particular for super potential generated by disk contributions one has

$$W = \sum_{m=0}^{\infty} \sum_{\beta,\vec{w}} \frac{N_{\beta,\vec{w}}}{m^2} Q^{m\beta} y^{m\vec{w}} = \sum_{\beta,\vec{w}} N_{\beta,\vec{w}} \mathrm{Li}_2(Q^\beta y^{\vec{w}}), \tag{5.2.8}$$

where $\vec{w}$ keeps track of the winding[123] in the relative homology $H_2^{rel}(M, L, \mathbb{Z})$.

First concrete evaluations of such disk instanton superpotentials have been performed in the local mirror symmetry context where the mirror geometry (2.7.28) is captured by the mirror curve $H(x, p, z) = 0$ [1, 2]. As the x, p are $\mathbb{C}^*$ variables it is natural to parametrize them as $x = e^u$ and $p = e^v$, with the reduction[124] of $\Omega = du \wedge dv \wedge \frac{dz}{z}$ to $H = 0$ described in (2.8.16) the action (5.2.3) becomes

$$W(C) = \int_C \frac{dz d\bar{z}}{z} u \bar{\partial}_z v \tag{5.2.9}$$

After performing the $dz/z \sim d\theta$ integral one is left with a radial integral in the z plane given by

$$W(v) = \int_{v_*}^{v} \lambda = \int_{v_*}^{v} u dv, \tag{5.2.10}$$

which readily fixes W up to a constant because from (5.2.10)

$$\partial_v W = u(v) \tag{5.2.11}$$

we see that the non compact geometry the chain integral (5.2.5) has really been reduced to an Abel-Jacobi map albeit with respect to the meromorphic differential

[123] The transitions between winding and representation basis is discussed in [4].

[124] z in (2.7.28) is a closed string modulus and not a coordinate of C is in this section.

$\lambda = udv$ on a punctured Riemann surface, which parametrizes the open string moduli space.

The question how to define the open/close mirror map, i.e $v(Q, y)$ and $z(Q)$ for non-trivial geometries was answered in [2]. The closed string map is always given by (2.9.53) see also the discussion at the end of Sect. 4.2.43, but the open string mirror is generally defined in form

$$\log(y) = v + \sum_{\alpha=1}^{k} r_v^\alpha (\log(z^\alpha) - t^\alpha)\,, \tag{5.2.12}$$

where the $r_v^\alpha \in \mathbb{Q}$ are determined in [2]. This information allows to make disk instanton prediction for the Harvey-Lawson special Lagrangian [168] in all toric local geometries as we show in our running example $\mathcal{O}(-3) \to \mathbb{P}^2$. Simultaneously localisation calculation by mathematicians of disk instanton invariants [212] confirmed the results but also observed that the invariants depend on an integer f parametrizing the weight choice of the toric $\mathbb{C}^*$ action used in the localisation. Given the fact that after a conifold transition with branes, described [271], the A-model geometry with Harvey Lawson Branes is described in the simplest case by a real Chern-Simons theory on the S^3 in the T^*S^3 geometry with branes ending on a knot (the unknot) in S^3, the integer ambiguity $f \in \mathbb{Z}$ was interpreted as the framing choice of that knot [2].

To be explicit the mirror curve Σ given by the affine equation in $\mathbb{C}^*$ variables as $H(x, p) = 0$ one has the reparameterisation group G_Σ

$$G_\Sigma = \mathrm{SL}(2, \mathbb{C}) \times \begin{pmatrix} 0 & 1 \\ 1 & 0 \end{pmatrix}, \tag{5.2.13}$$

acting on the coordinates (x, p) by

$$(x, p) \mapsto (x^a p^b, x^c p^d), \qquad \begin{pmatrix} a & b \\ c & d \end{pmatrix} \in G_\Sigma. \tag{5.2.14}$$

These transformations preserve the symplectic form

$$\left| \frac{dx}{x} \wedge \frac{dp}{p} \right| \tag{5.2.15}$$

on $\mathbb{C}^* \times \mathbb{C}^*$. In particular the shift subgroup The action of $G_\S$ is given by

$$(x, p) \mapsto (xp^f, p), \qquad f \in \mathbb{Z} \tag{5.2.16}$$

corresponds to the discrete framing choice.

Disks on Harvey-Lawson Branes in the $\mathcal{O}(-3) \to \mathbf{P}^2$ Geometry

The $\mathcal{O}(-3) \to \mathbf{P}^2$ geometry is defined the $\mathbb{C}^*$ action $l^{(1)} = (-3, 1, 1, 1)$ according to the gauge linear sigma model description compare (2.7.8) the F-term potential for this case reads

$$U = |x^1|^2 + |x^2|^2 + |x^3|^2 - 3|x^0|^2 = r \tag{5.2.17}$$

and defines the three dimensional subspace in $\mathbb{R}^3$ above the four 2-faces shown in Fig. 23. This this subspace is the base of the 3-torus fibration defining the toric Calabi-Yau space.

The charges defining the Harvey-Lawson special Lagrangian L are $q_1 = (1, 0, -1, 0)$, $q_2 = (0, 1, -0, -1)$. The corresponding half line

$$|x^1|^2 - |x^0|^2 = c^1, \quad |x^2|^2 - |x^0|^2 = c^2$$

defines one dimension of L, the other two are given by two circles in the torus fibration selected by the slope of the half line, so that L is special Lagrangian and $h_1(L, \mathbb{Z}) = 1$. The latter fact implies that there is non-trivial circle in L on which the disks can end with winding w, see [2, 168] for more detailed explanations.

There are three phases for the special Harvey-Lawson Lagrangian in the $\mathcal{O}(-3) \to \mathbf{P}^2$ geometry depending on the classical open string parameters $c^1, c^2 \in \mathbb{R}$ depicted in Fig. 23[125]

$$\begin{array}{rl} \text{Phase I:} & r_t > c^1 > 0, \quad c^2 = 0 \\ \text{Phase II:} & c^1 = 0, \quad r_t > c^2 > 0 \\ \text{Phase III:} & c^1 = c^2, \quad 0 < c^1 \end{array}$$

Using the form of the mirror curve (2.7.31) as well as (5.2.10) and (2.9.53), (5.2.12) and (5.2.8) it is straight forward to evaluate the $N_{\beta,w}$ as in Tables 13 and 14.

5.2.2 Remodelling the B-Model by Matrix Model Techniques

As it is visible in Fig. 22 there are two large N approaches to the open topological string theory on non-compact Calabi-Yau geometries. The left its large N three dimensional Chern-Simons theory and the right are large N matrix model techniques. These approaches are related by mirror symmetry [3].

While the former approach leads via conifold like transition to the topological vertex [4] that solves the open/closed topological A-model on all local toric Calabi-Yau manifolds in the large radius limit, the second one can by understood as an

[125] The complex parameter $\hat{v}$ in $y = e^{\hat{v}}$ in (5.2.8)comes due to a complexification by the A field on L.

Fig. 23 Three phases of the A–brane on $O(-3) \to \mathbf{P}^2$. Phases I and II are related by $\mathbf{Z}_3$ symmetry of the $\mathbf{P}^2$ from [2]

$$
\begin{aligned}
&\text{Phase I}: && r_t > c^1 > 0, \quad c^2 = 0\\
&\text{Phase II}: && c^1 = 0, \quad r_t > c^2 > 0\\
&\text{Phase III}: && c^1 = c^2, \quad 0 < c^1
\end{aligned}
$$

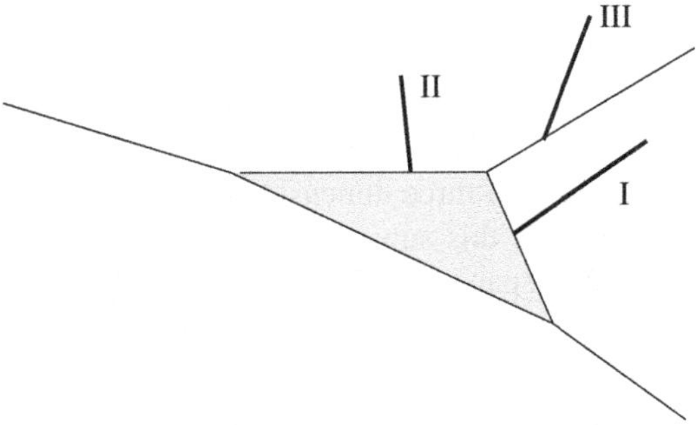

Table 13 Integer disk invariants $N_{\beta,w}$ for brane I or II in the $O(K) \to \mathbf{P}^2$ geometry in the canonical framing

w	$\beta = 0$	1	2	3	4	5	6	7	8
−5	0	0	0	0	0	5	−84	1200	−16854
−4	0	0	0	0	−2	28	−344	4360	−57760
−3	0	0	0	1	−10	102	−1160	14274	−185988
−2	0	0	−1	4	−32	326	−3708	45722	−598088
−1	0	1	−2	12	−104	1085	−12660	159208	−2112456
1	1	−1	5	−40	399	−4524	55771	−729256	9961800
2	0	−1	7	−61	648	−7661	97024	−1293185	17921632
3	0	−1	9	−93	1070	−13257	173601	−2371088	33470172
4	0	−1	12	−140	1750	−22955	312704	−4396779	63460184
5	0	−1	15	−206	2821	−39315	559787	−8136830	120497011

extension of the B-model techniques to the open string, likewise only for local B-model Calabi-Yau geometries specified by the Riemann surface $H(x, p, z) = 0$ in

Table 14 Integer disk invariants $N_{\beta,w}$ for brane III in the $O(K) \to \mathbf{P}^2$ geometry in the canonical framing

w	$\beta = 0$	1	2	3	4	5	6	7	8
1	−1	2	−5	32	−286	3038	−35870	454880	−6073311
2	0	1	−4	21	−180	1885	−21952	275481	−3650196
3	0	1	−3	18	−153	1560	−17910	222588	−2926959
4	0	1	−4	20	−160	1595	−17976	220371	−2869120
5	0	1	−5	26	−196	1875	−20644	249120	−3205528
6	0	1	−7	36	−260	2403	−25812	306095	−3889116
7	0	1	−9	52	−365	3254	−34089	397194	−4981102
8	0	1	−12	76	−528	4578	−46812	535639	−6627840

the B-model geometry. According to the theme of the lecture we outline here shortly the second approach.

As in Sect. 4 these B-model techniques give answers for the amplitudes that are analytic in the moduli and hence valid everywhere in the moduli space. For example using them it is possible to evaluate open higher genus amplitudes at orbifold points [3, 51].

The question is simply formulated: given formal expansion of the open string free energy (5.2.7)

$$F(Q, y, \lambda) = \sum_{\substack{\beta \in H_2^{rel}(X,L) \\ g,w}} N_{\beta,w,g} Q^\beta y^w \lambda^{g-1} = \sum_{g,w} F_{g,w}(Q) y^w \lambda^{g-1} \tag{5.2.18}$$

with $w = (w_1, \ldots, w_h)$. Can we find an analytic expression for the amplitude

$$A_h^{(g)}(z, \underline{p}) = \sum_w F_{g,w}(z) p_1^{w_1} \ldots p_h^{w_h} \quad ? \tag{5.2.19}$$

Here the B-model closed string moduli z and open string variable $p_i = \log(v_i)$ are related to the Q and y_i by the open/closed string mirror map (5.2.12). Of course one of the most important amplitude is the disk amplitude that we calculated in the last section

$$W = A_1^{(0)}(z, p) = \int_{v_*}^{v} u dv = \int_{p^*}^{p} \log(x) \frac{dp}{p} =: \int_{p^*}^{p} \lambda \tag{5.2.20}$$

The relation between matrix models and local geometries in particular N=2 and N=1 Seiberg-Witten geometries has been pointed out in [3, 93, 94]. These works started with a concrete matrix model and set up the geometry as its spectral curve, but it is has been long realised that at least the perturbative large $1/N$ or equivalently large genus expansion can be also reconstructed from the spectral curve Σ, the disk amplitude or more precisely the corresponding meromorphic differential and the annulus amplitude or more precisely the Bergmann Kernel [9, 20, 110].

It was first suggested in [259] and further developed in [50] to use this matrix model recursion to reconstruct the amplitudes (5.2.19). The main difference in the normal set up considered usually in the matrix model literature [9, 20, 110] is that the matrix model spectral curve is normally an affine curve in $\mathbb{C}^2$ while the mirror curve is defined in $(\mathbb{C}^*)^2$ and that therefore the meromorphic differential $\lambda = \log(x)\frac{dp}{p}$ is different then the one $\phi = ydp$ used in usual matrix model application. For completeness we give the starting data for the curve in $(\mathbb{C}^*)^2$ and the recursion relations as set up in [110]:

The starting data are

- the ramification points on the spectral curve Σ $r_i \in \Sigma$ of the projection map $\Sigma \to \mathbb{C}^*$ onto the x-axis, i.e. , the points $q_i \in \Sigma$ such that $\frac{\partial H}{\partial p}(r_i) = 0$. Near a ramification point, there is again two points $r, \bar{r} \in \Sigma$ with the same projection $x(r) = x(\bar{r})$;

- the meromorphic differential

$$\lambda(s, z) = \log p(s) \frac{dx(s)}{x(s)} \tag{5.2.21}$$

on Σ, which descends from the symplectic form

$$\frac{dx}{x} \wedge \frac{dp}{p} \tag{5.2.22}$$

on $\mathbb{C}^* \times \mathbb{C}^*$.

- the Bergmann kernel $B(r, s)$ on C, which is the unique meromorphic differential with a double pole at $r = s$ with no residue and no other pole, and normalized such that

$$\oint_{a_I} B(r, s) = 0, \tag{5.2.23}$$

where (a_I, b^I) is a canonical basis of cycles for C. The Bergmann kernel is related to the prime form $E(p, q)$ by

$$B(r, s) = \partial_r \partial_s E(r, s). \tag{5.2.24}$$

We will also need the closely related one-form

$$dE_s(r) = \frac{1}{2} \int_r^{\bar{r}} B(s, \xi), \tag{5.2.25}$$

which is defined locally near a ramification point r_i.

The recursion [110] defines an infinite sequence of meromorphic differentials on Σ called $W_h^{(g)}(s_1, \ldots, s_h)$, $g, h \in \mathbb{Z}^+$, $h \geq 1$. By definition

$$W_1^{(0)}(r_1) = 0, \qquad W_2^{(0)}(r_1, r_2) = B(r_1, r_2), \tag{5.2.26}$$

while the remaining differentials are generated recursively by taking residues at the ramification points:

$$\begin{aligned} W_{h+1}^{(g)}(r, r_1 \ldots, r_h) = \sum_{q_i} \operatorname*{Res}_{r=r_i} \frac{dE_r(s)}{\Phi(r) - \Phi(\bar{r})} \Big(& W_{h+2}^{(g-1)}(r, \bar{r}, s_1, \ldots, s_h) \\ & + \sum_{l=0}^{g} \sum_{J \subset H} W_{|J|+1}^{(g-l)}(r, s_J) W_{|H|-|J|+1}^{(l)}(\bar{r}, s_{H \backslash J}) \Big). \end{aligned} \tag{5.2.27}$$

Here we denoted the set $H = \{1, \cdots, h\}$, and given any subset $J = \{i_1, \cdots, i_j\} \subset H$ we defined $s_J = \{s_{i_1}, \cdots, s_{i_j}\}$. The root $\phi(r)$ is defined as $\phi(r) = \int_0^r ydx$.

The annulus amplitude, $(g, k) = (0, 2)$, is given by removing the double pole from the Bergmann kernel:

$$A_2^{(0)} = \int \left(B(s_1, s_2) - \frac{ds_1 ds_2}{(s_1 - s_2)^2} \right). \tag{5.2.28}$$

The closed string amplitudes $F^{(g)}$, except for the prepotential can be obtained as follows: Let $\phi(r)$ be an arbitrary anti-derivative of $\lambda(r)$, i.e. $d\phi(r) = \lambda(r)$. Then the $F^{(g)}$, $g \geq 1$

$$F^{(g)} = \frac{1}{2 - 2g} \sum_{r_i} \operatorname*{Res}_{r=r_i} \phi(r) W_1^{(g)}(r). \tag{5.2.29}$$

The above amplitudes are in the B-model coordinates and to extract the $N_{\beta,w,g}$ via (5.2.7), (5.2.18), (5.2.19) the closed (2.9.53) and open (5.2.12)mirror maps have to be used. An important fact is that the closed string amplitudes (5.2.29) are invariant under G defined in (5.2.13). The reason is that the different G choices lead to λ's which differ by exact terms, irrelevant for closed cycle integrals but crucial for open chain integrals, which do change. The latter fact was used e.g. in [53] to use the recursion (5.2.27) to calculate colored HOMFLY polynomials of torus knots. In general one has to use the augmentation variety and construct λ and the Bergman kernel to perform similar calculations of the colored HOMFLY polynomials for arbitrary knots [157]. This example might show how many concrete applications the remodelling of the B-model [50, 259] has.

Acknowledgements These lectures grew out of collaborations with many dear colleagues and friends as well as many students. It is a great pleasure to thank them for sharing their insights and knowledge in scientific and other matters with me. Let me thank Mina Aganagic, Murad Alim, Per Berglund, Vincent Bouchard, Nana Geraldine Cabo Bizet, Jinwon Choi, Cesar Fierro Cota, Gottfried Curio, Ti Ming Chiang, Robbert Dijkgraaf, Fabian Fischbach, Jürgen Fuchs, Jens Erler, Matthias Gaberdiel, Thomas Grimm, Jie Gu, Thaisa Guio, Micheal Hecht, Babak Haghighat, Shinobu Hosono, Kentaro Hori, Tae Won Ha, Min Xin Huang, Hans Jockers, Shamit Kachru, Amir Kian Kashani-Poor, Sheldon Katz, Denis Klevers, Max Kreuzer, Karl Landsteiner, Calin Lazaroiu, Wolgang Lerche, Bong Lian, Guglielmo Lockhart Dieter Lüst, Jan Manschot, Marcos Marino, Davesh Maulik, Peter Mayr, Yaron Oz, Seth Quackenbush, Rahul Pandharipande, Sara Pasquetti, Maximilan Poretschkin, Marco Rauch, Jonas Reuter, Erwin Riegler, Shi-shyr Roan, Yongbin Ruan, Ingo Runkel, Christoph Scheich, Emanuel Scheidegger, Marc Schiereck, Michael G. Schmidt, Thorsten Schimannek, David Shih, Masoud Soroush, Piotr Sulkowski, Stefan Theisen, AlirezaTavanfar, Cumrun Vafa, Daniel Vieira Lopes, Nick Warner, Marlene Weiss, Martin Westerholt-Raum, Shimon Yankielowicz, Hung Yu Yeh, Shing-Tung Yau, Don Zagier, Eric Zaslow and Michele Del Zotto. I am in particular indebted to Yongbin Ruan who invited me to give these Lectures at Michigan University in Ann Arbor and the MSRI in Berkeley were the Lectures were finalised during the Program "Enumerative geometry beyond numbers".

Appendix 1: Weierstrassforms

Here we provide the data for the Weierstrass forms for all local toric del Pezzo surfaces discussed in Sect. 2.12.2.

Weierstrass Form of the Cubic in $\mathbb{P}^2$

$$\begin{aligned}
12g_2^C =\ & 144(m_1m_3m_5m_7 + m_1m_4m_6m_8+m_2m_4m_7m_9)\\
& -16(m_2m_3m_6m_8+m_2m_5m_6m_9+m_3m_5m_8m_9)+16(m_2^2m_6^2+m_3^2m_8^2+m_5^2m_9^2)\\
& -48(m_1m_3m_6^2+m_2^2m_5m_7+m_1m_5^2m_8+m_2m_4m_8^2+m_3^2m_7m_9+m_4m_6m_9^2)\\
& +(24(m_1m_5m_6+m_2m_3m_7+m_2m_5m_8+m_3m_6m_9+m_4m_8m_9)-216m_1m_4m_7)u\\
& -8(m_2m_6+m_3m_8+m_5m_9)u^2+u^4
\end{aligned}$$

$$\begin{aligned}
216g_3^C =\ & 48m_2m_3m_5m_6m_8m_9-5832m_1^2m_4^2m_7^2-1296(m_1m_2m_3m_5m_6m_7+m_1m_2m_4m_5m_7m_8\\
& +m_1m_3m_4m_6m_7m_9+m_1m_4m_5m_6m_8m_9+m_2m_3m_4m_7m_8m_9)\\
& +3888(m_1^2m_4m_5m_6m_7+m_1m_2m_3m_4m_7^2 + m_1m_4^2m_7m_8m_9) - 576(m_1m_3^2m_6^2m_8\\
& +m_1m_3m_5^2m_8^2+m_2^2m_4m_6m_8^2+m_2^2m_5^2m_7m_9+m_2m_4m_6^2m_9^2+m_3^2m_5m_7m_9^2)\\
& +864(m_1m_2m_4m_6^2m_8 + m_1m_3^2m_5m_7m_8+m_1m_3m_4m_6m_8^2+m_1m_3m_5^2m_7m_9\\
& +m_2^2m_4m_6m_7m_9+m_2m_4m_5m_7m_9^2)+96(m_2^2m_3m_6^2m_8+m_2m_3^2m_6m_8^2+m_2^2m_5m_6^2m_9\\
& +m_3^2m_5m_8^2m_9+m_2m_5^2m_6m_9^2+m_3m_5^2m_8m_9^2)-144(m_1m_2m_5^2m_6m_8 + m_2^2m_3m_5m_7m_8\\
& +m_1m_3m_5m_6^2m_9+m_2m_3^2m_6m_7m_9+m_2m_4m_5m_8^2m_9+m_3m_4m_6m_8m_9^2)\\
& +216(m_1^2m_5^2m_6^2+m_2^2m_3^2m_7^2+m_2^2m_5^2m_8^2+m_3^2m_6^2m_9^2+m_4^2m_8^2m_9^2)\\
& -64(m_2^3m_6^3+m_3^3m_8^3+m_5^3m_9^3)+288(m_1m_2m_3m_6^3+m_2^3m_5m_6m_7+m_2m_3m_4m_8^3\\
& +m_1m_5^3m_8m_9+m_3^3m_7m_8m_9 + m_4m_5m_6m_9^3) - 864(m_1^2m_4m_6^3\\
& +m_1^2m_5^3m_7+m_1m_3^3m_7^2+m_2^3m_4m_7^2+m_1m_4^2m_8^3+m_4^2m_7m_9^3)+(720(m_1m_3m_5m_6m_8\\
& +m_2m_3m_5m_7m_9+m_2m_4m_6m_8m_9) - 1296(m_1m_2m_4m_6m_7\\
& +m_1m_3m_4m_7m_8+m_1m_4m_5m_7m_9)+864(m_1m_2m_5^2m_7+m_1m_3^2m_6m_7\\
& +m_2^2m_4m_7m_8+m_1m_4m_5m_8^2+m_1m_4m_6^2m_9+m_3m_4m_7m_9^2)-144(m_1m_2m_5m_6^2\\
& +m_2^2m_3m_6m_7+m_2^2m_5m_6m_8+m_2m_3^2m_7m_8+m_2m_3m_5m_8^2+m_1m_5^2m_6m_9\\
& +m_2m_3m_6^2m_9+m_2m_5^2m_8m_9+m_3^2m_6m_8m_9+m_3m_4m_8^2m_9+m_3m_5m_6m_9^2\\
& +m_4m_5m_8m_9^2))u + (24(m_2m_3m_6m_8+m_2m_5m_6m_9+m_3m_5m_8m_9)\\
& +48(m_2^2m_6^2+m_3^2m_8^2+m_5^2m_9^2) - 648(m_1m_3m_5m_7+m_1m_4m_6m_8+m_2m_4m_7m_9)\\
& -72(m_1m_3m_6^2+m_2^2m_5m_7+m_1m_5^2m_8+m_2m_4m_8^2+m_3^2m_7m_9+m_4m_6m_9^2))u^2\\
& +(540m_1m_4m_7+36(m_1m_5m_6+m_2m_3m_7+m_2m_5m_8+m_3m_6m_9\\
& +m_4m_8m_9))u^3 - 12(m_2m_6+m_3m_8+m_5m_9)u^4+u^6
\end{aligned} \tag{A1.1}$$

Weierstrassform of the Biquadric in $\mathbb{P}^1 \times \mathbb{P}^1$

$$\begin{aligned}
12g_2^B =\ & 224m_{10}m_5m_7m_9-48(m_2^2m_5m_7+m_{10}m_5m_8^2+m_{10}m_6^2m_9+m_3^2m_7m_9)\\
& -16(m_{10}m_2m_6m_7+m_2m_3m_6m_8+m_{10}m_3m_7m_8+m_2m_5m_6m_9+m_3m_5m_8m_9)\\
& +16(m_2^2m_6^2+m_{10}^2m_7^2+m_3^2m_8^2+m_5^2m_9^2)+24(m_2m_3m_7+m_2m_5m_8\\
& +m_{10}m_6m_8+m_3m_6m_9)u-8(m_2m_6+m_{10}m_7+m_3m_8+m_5m_9)u^2+u^4
\end{aligned}$$

$$
\begin{aligned}
216 g_3^B = {} & 48(m_{10}m_2m_3m_6m_7m_8+m_2m_3m_5m_6m_8m_9)-144(m_2^2m_3m_5m_7m_8+m_{10}m_2m_5m_6m_8^2\\
&+m_2m_3^2m_6m_7m_9+m_{10}m_3m_6^2m_8m_9)+288(m_2^3m_5m_6m_7+m_{10}m_3m_5m_8^3+m_{10}m_2m_6^3m_9\\
&+m_3^3m_7m_8m_9)-960(m_{10}m_2m_5m_6m_7m_9+m_{10}m_3m_5m_7m_8m_9)\\
&+216(m_2^2m_3^2m_7^2+m_2^2m_5^2m_8^2+m_{10}^2m_6^2m_8^2+m_3^2m_6^2m_9^2)-576(m_{10}m_2^2m_5m_7^2\\
&+m_{10}^2m_5m_7m_8^2+m_2^2m_5^2m_7m_9+m_{10}^2m_6^2m_7m_9+m_{10}m_3^2m_7^2m_9+m_{10}m_5^2m_8^2m_9\\
&+m_{10}m_5m_6^2m_9^2+m_3^2m_5m_7m_9^2)+2112(m_{10}^2m_5m_7^2m_9+m_{10}m_5^2m_7m_9^2)+96(m_{10}m_2^2m_6^2m_7\\
&+m_{10}^2m_2m_6m_7^2+m_2^2m_3m_6^2m_8+m_{10}^2m_3m_7^2m_8+m_2m_3^2m_6m_8^2+m_{10}m_3^2m_7m_8^2\\
&+m_2^2m_5m_6^2m_9+m_3^2m_5m_8^2m_9+m_2m_5^2m_6m_9^2+m_3m_5^2m_8m_9^2)-64(m_2^3m_6^3+m_{10}^3m_7^3\\
&+m_3^3m_8^3+m_5^3m_9^3)+(720(m_{10}m_2m_5m_7m_8+m_2m_3m_5m_7m_9+m_{10}m_3m_6m_7m_9\\
&+m_{10}m_5m_6m_8m_9)-144(m_2^2m_3m_6m_7+m_{10}m_2m_3m_7^2+m_2^2m_5m_6m_8+m_{10}m_2m_6^2m_8\\
&+m_2m_3^2m_7m_8+m_{10}^2m_6m_7m_8+m_2m_3m_5m_8^2+m_{10}m_3m_6m_8^2+m_2m_3m_6^2m_9\\
&+m_2m_5^2m_8m_9+m_3^2m_6m_8m_9+m_3m_5m_6m_9^2))u+(24(m_{10}m_2m_6m_7+m_2m_3m_6m_8\\
&+m_{10}m_3m_7m_8+m_2m_5m_6m_9+m_3m_5m_8m_9)-480m_{10}m_5m_7m_9-72(m_2^2m_5m_7\\
&+m_{10}m_5m_8^2+m_{10}m_6^2m_9+m_3^2m_7m_9)+48(m_2^2m_6^2+m_{10}^2m_7^2+m_3^2m_8^2\\
&+m_5^2m_9^2))u^2+36(m_2m_3m_7+m_2m_5m_8+m_{10}m_6m_8+m_3m_6m_9)u^3-12(m_2m_6+m_{10}m_7\\
&+m_3m_8+m_5m_9)u^4+u^6
\end{aligned}
\tag{A1.2}
$$

Weierstrassform of the Quartic in $\mathbb{P}^2(1,1,2)$

$$
\begin{aligned}
12 g_2^Q = {} & -48m_4m_6m_9^2+192m_{11}m_7m_9^2+48(m_3^2m_7m_9+m_{11}m_8^2m_9)\\
&+16(m_3^2m_8^2+m_3m_5m_8m_9+m_5^2m_9^2)-24(m_3m_6m_9+m_4m_8m_9)u\\
&+(-8m_3m_8+8m_5m_9)u^2+u^4
\end{aligned}
$$

$$
\begin{aligned}
216 g_3^Q = {} & 64(m_5^3m_9^3-m_3^3m_8^3)+96(m_3m_5^2m_8m_9^2-m_3^2m_5m_8^2m_9)-144m_3m_4m_6m_8m_9^2\\
&-1152m_{11}m_3m_7m_8m_9^2-2304m_{11}m_5m_7m_9^3+216(m_3^2m_6^2m_9^2+m_4^2m_8^2m_9^2)\\
&-576(m_3^2m_5m_7m_9^2+m_{11}m_5m_8^2m_9^2)-288(m_3^3m_7m_8m_9+m_{11}m_3m_8^3m_9+m_4m_5m_6m_9^3)\\
&+864(m_{11}m_6^2m_9^3+m_4^2m_7m_9^3)+(144(m_3^2m_6m_8m_9+m_3m_4m_8^2m_9)\\
&-144(m_3m_5m_6m_9^2+m_4m_5m_8m_9^2)+864(m_3m_4m_7m_9^2+m_{11}m_6m_8m_9^2))u\\
&+(72(m_3^2m_7m_9+m_{11}m_8^2m_9)+48(m_3^2m_8^2+m_5^2m_9^2)-24m_3m_5m_8m_9-72m_4m_6m_9^2\\
&-576m_{11}m_7m_9^2)u^2-36(m_3m_6m_9+m_4m_8m_9)u^3+(12m_5m_9-12m_3m_8+)u^4+u^6
\end{aligned}
\tag{A1.3}
$$

Appendix 2: Characteristic Classes of Holomorphic Vector Bundles

In this section we provide background on holomorphic vector bundles, Chern classes and index theorems.

Index Theorems for Holomorphic Vector Bundles

The holomorphic tangent bundle of M is an example of a holomorphic vector bundle E with a hermitian metric, which we call h_{ab} in the general case. The connection one form

$$A_k = (\partial_k h)h^{-1}, \qquad A_{\bar{k}} = 0 \tag{A2.1}$$

defines the unique affine connection, which is compatible with the hermitian metric, i.e $\nabla h = 0$, and compatible with the complex structure. One defines the curvature two form as $F = \mathrm{d}A + A \wedge A$. The differential geometry approach to Chern classes $c_i(E) \in H^{2i}(M, \mathbb{R})$ of a rank r holomorphic vector bundle is to define them in terms of symmetric function of the eigenvalues of the curvature form as

$$c(E) = \det(1 + \frac{i}{2\pi}F) = 1 + \sum_i c_i(E) = 1 + \frac{i}{2\pi}\mathrm{Tr}F + \ldots \tag{A2.2}$$

and to prove then that they do not depend on the metric [42, 302].

Topologically one can represent the Chern class c_k as the Poincaré dual to the degeneracy cycle

$$D_{r-k+1}(\sigma) = \{x : \sigma_1(x) \wedge \ldots \sigma_{r-k+1}(x) = 0\}, \tag{A2.3}$$

where $r - k + 1$ generic $\mathcal{C}^\infty$-sections σ_i of E become linearly dependent. This is described as Gauss Bonnet formula II in Chap 3.3 of [150], see also [126, 176] for the approach using classifying spaces. The simplest example of the above dual descriptions arise for line bundles $\mathcal{L}$. Let $|\sigma|^2$ be a metric on a line bundle L, where σ is a section of L. Local trivialization of L are $\phi : L|_U \to U \times \mathbb{C}$, where s_U is a holomorphic function and $|\sigma|^2 = h(x)|s_U|^2$ for some function $h(x)$, which is positive if the metric is. The curvature 2-form given by

$$\mathcal{R} = \bar{\partial}\partial \log h(x) \tag{A2.4}$$

defines the Chern-class of L represented by $c_1(L) = \frac{i}{2\pi}[\mathcal{R}] \in H^2(M)$. This class is Poincaré dual to the divisor class $[D]$ which defines L and is uniquely recovered from L as the locus where the generic section vanishes. As a corollary the first Chern class of a holomorphic vector bundle is also the first Chern class of the determinant bundle $L_D = \wedge^r E$

$$c_1(E) = c_1(L_D)\,. \tag{A2.5}$$

For the tangent bundle we identify the curvature 2-form F with $\Theta^j_{\bar{i}} = g^{j\bar{p}} R_{i\bar{p}k\bar{l}}\mathrm{d}x^k \wedge \mathrm{d}x^{\bar{l}}$ and get a representative for $c_1(TM)$ (which we also call $c_1(M)$)

$$c_1(M) = \frac{i}{2\pi}\Theta^i_i = \frac{i}{2\pi}R_{k\bar{l}}\mathrm{d}x^k \wedge \mathrm{d}x^{\bar{l}} = -\frac{i}{2\pi}\partial\bar{\partial}\log\det(g_{k\bar{l}})\,. \tag{A2.6}$$

The canonical line bundle is the determinant line bundle of the holomorphic tangent bundle $K_M = \wedge^n T^{*\,1,0}M$. By (A2.5) and (A2.10) we have therefore

$$-2\pi c_1(K_M) := -2\pi c_1(\wedge^n T^{*\,1,0}M) = -2\pi c_1(T^*M) = 2\pi c_1(TM)\,. \qquad \text{(A2.7)}$$

Let us derive this also using as an explicit representative of the Chern class the curvature 2-form. Given an complex structure and a Kähler metric $g_{i\bar{j}}$ we have a connection on $T^{*\,1,0}M$ described by the holomorphic Christoffel symbols. This connection induces a connection on the line bundle K_M and a straightforward calculation shows on total antisymmetric forms $[\nabla_i, \nabla_{\bar{j}}]\omega_{i_1,\ldots,i_n} = -R_{i\bar{j}}\omega_{i_1,\ldots,i_n}$ Therefore we can identify $h(x)$ of (A2.4) with $\det^{-1}(g_{i\bar{j}})$ and by (A2.4) the first Chern class of K_M is

$$-2\pi c_1(K_M) = -i[\mathcal{R}] = 2\pi c_1(TM)\,. \qquad \text{(A2.8)}$$

If one uses the Poincaré Hopf theorem that the Euler number $\chi(M)$ of a manifold of dim n is given by the sum of indices of zeros of a generic vector field, i.e. a section of the tangent bundle, then by (A2.3) the dual to $c_n(TM)$ is D_1. Counting these zeros leads then to the Gauss-Bonnet formula

$$\chi(M) = D_1 \cap M = \int_M c_n(TM)\,. \qquad \text{(A2.9)}$$

Let us discuss further properties of the Chern classes. By (A2.2) one has $c_0(E) = 1$, $c_{k>r}(E) = 0$ and the Whitney product formula $c(E \oplus F) = c(E)c(F)$ from the properties of the determinant, see [49] for a proof from the topological definition. It is also easy to see [150] that for the dual bundle E^*

$$c_k(E^*) = (-1)^k c_k(E) \qquad \text{(A2.10)}$$

and $c_k(f(E)) = f^* c_k(E)$ for $f : M \to M'$ a differentiable mapping. A further important property is the *splitting principle* [49]. For an exact sequence of holomorphic vector bundles or sheaves one has $0 \to E \to F \to G \to 0$ one has $c(F) = c(E)c(G)$. One considers often classes x_i such that $c(E) = \prod_{i=1}^r (1 + x_i)$ where x_i are Chern classes of line bundles. One reason that this is useful is that the *splitting principle* implies that if one wants to derive polynomial identities among Chern classes of vector bundles, one may replace the vector bundles by direct sums of line bundles. This opens up a calculational machinery with classes, which behave e.g. more natural on direct products as the *Chern character* $\mathrm{Ch}(E) = \sum_{i=1}^r e^{x_i}$. All expression are polynomial, defined by expanding up to degree r in x_i. Obviously $\mathrm{Ch}(E \oplus F) = \mathrm{Ch}(E) + \mathrm{Ch}(F)$ and $\mathrm{Ch}(E \otimes F) = \mathrm{Ch}(E)\mathrm{Ch}(F)$. A little playing with symmetric functions reveals $\mathrm{Ch}(E) = r + c_1 + \frac{1}{2}(c_1^2 - 2c_2) + \frac{1}{6}(c_1^3 - 3c_1c_2 + 3c_3) + \ldots$, where we set $c_k = c_k(E)$. Similar is the Todd genus defined $\mathrm{td}(E) = \prod_{i=1}^r \frac{x_i}{1-e^{-x_i}} = 1 + \frac{1}{2}c_1 + \frac{1}{12}(c_1^2 + c_2) + \frac{1}{24}c_1c_2 + \ldots$. A central theorem is the Hirzebruch-Riemann-Roch formula, which gives the arithmetic genus $\chi(E) = \sum_k (-1)^k h^k(E)$ of a vector bundle over a manifold M, see [176] for the proof

$$\chi(E) = \int_M \mathrm{ch}(E) \wedge \mathrm{td}(TM) \,. \tag{A2.11}$$

A close variant of this index using the $\hat{A}$-class instead of the Todd class has been discussed in (2.6.25).

In Sects. 3.4.1 and 4.1 we needed applications of (A2.11). Namely to count the deformation space (3.3.18) of a Riemann surface[126] Σ_g. As seen in section "First Order Complex Structure Deformations" the complex structure moduli of the metric are given by elements in the Čech cohomology group $H^1(T)$ with $T = T\Sigma$ and for $g > 1$ there are no conformal Killing vectors generating global diffeomorphisms i.e. one has $h^0(T) = 0$. However for $g = 1$ the shift $z \to z + \lambda$ on the torus accounts for $h^0 = 1$ and for $g = 0$ the three generators of $PSL(2, \mathbb{C})$ $z \to \frac{az+c}{cz+d}$ on S^2 account for $h^0 = 3$. For a vector bundle V of rank $r$over the Riemann surface Σ the formula (A2.11) gives

$$\begin{aligned} h^0(\Sigma, V) - h^1(\Sigma, V) &= \int_\Sigma \mathrm{ch}(V) \wedge \mathrm{td}(T) = \int_\Sigma (r + c_1(V))(1 + \frac{1}{2} c_1(T)) \\ &= \int_\Sigma c_1(V) + r(1-g) \,. \end{aligned} \tag{A2.12}$$

The *virtual dimension* of the deformation space is obtained by setting $V = T$ with rank 1

$$\dim \mathcal{M}_g = h^1(T) - h^0(T) = -\int_\Sigma \mathrm{ch}(T) \wedge \mathrm{td}(T) = 3g - 3 \,. \tag{A2.13}$$

In the integral over the metric moduli space in string amplitudes one sacrifices in the $g = 0, 1$ cases $h^0 = 3, 1$ additional parameters, the position of insertion points, to offset the negative contributions to (A2.13) from the conformal Killing fields. Another application leads to the formula (3.4.10) describing the dimension of the deformation space of holomorphic maps $x : \Sigma \to M$. The movement of the curve in M is described infinitesimal by a vector field $x^i \to x^i + \epsilon \xi^i$ on M. The vector field must be holomorphic $\partial_{\bar{z}} \xi = 0$ so that the deformed map stays holomorphic. Also we are not counting vector fields which correspond to reparametrizations of Σ.

[126] This related by the Atiayh-Singer index formula to the index of the Dirac operator and hence to the ghost zero modes. An overview about index formulas for physicist can be found in [106] and the connections to the zero modes is in explained e.g. in [279].

That is we look at elements of $H^0_{\bar\partial}(\Sigma, x^*(TM)) = H^0(x^*(TM))$ and (A2.11) gives us

$$h^0(x^*(TM)) - h^1(x^*(TM)) = \int_\Sigma (\text{dim}M + x^*(c_1(TM)))(1 + \frac{1}{2}c_1(T))$$
$$= c_1(TM)\cdot\beta + \text{dim}M(1-g)\,. \qquad \text{(A2.14)}$$

Generically the movement of the map is unobstructed and $H^1(x^*(TM)) = 0$. In the case the above is also the dimension of the deformation space. In the case of Calabi-Yau three folds we get for genus 0 that the dimension of the deformation space is 3. We can think about this in two ways. Either we don't fix points on S^2, then we have to mod out by the 3 dim automorphism group $PL(2,\mathbb{C})$ of S^2 and the expected dimension of the moduli space is 0. That is the way the corrections in $\mathcal{F}^{(0)}$ are interpreted. Or we kill $PL(2,\mathbb{C})$ by marking three points on the S^2 required to map into three divisors, which put three constraints and yields again a zero dimensional moduli space. Counting genus zero maps with three marked points and zero dimensional moduli space is the interpretation of corrections in $C_{ijk}(t)$ in (3.4.12).

To get the virtual complex dimension of the moduli space $\overline{\mathcal{M}}_{g,n}(M,\beta)$ of arbitrary genus g maps in a class $\beta\in H_2(M,\mathbb{Z})$ one has to add, according to the deformation exact sequence [182] Chapter 24,

$$\begin{aligned}&0\to \text{Aut}(\Sigma, p_1,\ldots,p_n;x)\to\text{Aut}(\Sigma,p_1,\ldots,p_n;x)\to\\&\text{Def}(x)\to\text{Def}(\Sigma,p_1,\ldots,p_n;x)\to\text{Def}(\Sigma,p_1,\ldots,p_n)\to\\&\text{Ob}(x)\to\text{Ob}(\Sigma,p_1,\ldots,p_n;x)\to 0\end{aligned} \qquad \text{(A2.15)}$$

the difference of $\Delta_{\text{dim}}(\text{Def, Aut}) = \text{dim Def}(\Sigma, p_1,\ldots, p_n) - \text{dim Aut}(\Sigma, p_1,\ldots,p_n)$ to (A2.14) and calculate

$$\begin{aligned}\text{dim}\,\overline{\mathcal{M}}_{g,n}(M,\beta) &= h^0(x^*(TM)) - h^1(x^*(TM)) + \Delta_{\text{dim}}(\text{Def, Aut})\\ &\int_\beta c_1(TM) + (\text{dim}\,M-3)(1-g)+n\,.\end{aligned} \qquad \text{(A2.16)}$$

Let us introduce the *Pontrjagin classes* for real vector bundles V as the Chern class of the complexification of $V_{\mathbb{C}}$ of V [176]

$$p_k(V) = (-1)^k c_{2k}(V_{\mathbb{C}}) \qquad \text{(A2.17)}$$

The *Euler class* of the real vector rank r bundle V can now be defined as $e^2(V) = p_{\frac{r}{2}}(V)$. The Gauss-Bonnet formula, e.g. $\int_M e(TM) = \chi(M)$ fixes the sign. The Pontrjagin class of a complex vector bundle E is defined via the Pontrjagin of its *realization* $E_{\mathbb{R}} = E\oplus\bar{E}$ as $p_k(E) = (-i)^k c_{2k}(E_{\mathbb{R}})$. By the splitting principle and Whitneys formula [49] one gets $c_r(E) = e(E_{\mathbb{R}})$. The *A-roof or Dirac genus* is defined as symmetric polynomial in x_i^2 and can therefore be expressed in terms of the Pontrjagin classes $\hat{A}(E) = \prod_{j=1}^{r} \frac{x_j/2}{\sinh(x_j/2)} = 1 - \frac{1}{24}p_1 + \frac{1}{5760}(7p_1^2 - 4p_2) + \ldots$. A useful formula with applications to the Calabi-Yau tangent bundle is that $\mathrm{td}(E) = e^{c_1(E)}\hat{A}(E)$.

Let us explore some consequences. Let V is a vector bundle over X, $\chi(X, W) = \sum_{i=0}^{n}(-1)^i \mathrm{dim} H^i(M, V)$ and $c_0[X], \ldots, c_n[X]$, Chern classes of X and $d_0[W], \ldots, d_r[V]$ Chern classes of W and write (A2.11) as in [176]

$$\chi(M, V) = \kappa_n \left[\sum_{i=1}^{q} e^{\delta_i} \prod_{i=1}^{n} \frac{\gamma_i}{1 - e^{-\gamma_i}} \right]. \tag{A2.18}$$

where $\kappa_n[]$ means taking the coefficient of the n'th homogeneous form degree, the γ_i and δ_i are the formal roots of the total Chern classes: $\sum_{i=0}^{n} c_i[M] = \prod_{i=1}^{n}(1 - \gamma_i)$ and $\sum_{i=0}^{q} d_i[M] = \prod_{i=1}^{q}(1 - \delta_i)$. We want to use the index formula to compute the arithmetic genera $\chi_q = \sum_p (-1)^p \mathrm{dim} H^p(M, \Omega^q)$. First we will evaluate (A2.18) for $V = T_M$. This is done by expressing the formal roots, via symmetric polynomial, in terms of the Chern classes c_i and yields for the two,three and four dimensional cases the following formulas for $\chi_q = \sum_{p=1}^{\mathrm{dim}(M)} (-1)^p h^{p,q}$:

$$\mathrm{dim}(M) = 2: \qquad \chi_0 = \frac{1}{12}\int_M (c_1^2 + c_2), \tag{A2.19}$$

$$\mathrm{dim}(M) = 3: \qquad \begin{array}{ll} a.) & \chi_0 = \frac{1}{24}\int_X (c_1 c_2) \\ b.) & \chi_1 = \frac{1}{24}\int_X (c_1 c_2 - 12 c_3), \end{array} \tag{A2.20}$$

$$\mathrm{dim}(X) = 4: \qquad \begin{array}{ll} a.) & \chi_0 = \frac{1}{720}\int_X (-c_4 + c_1 c_3 + 3c_2^2 + 4c_1^2 c_2 - c_1^4) \\ b.) & \chi_1 = \frac{1}{180}\int_X (-31 c_4 - 14 c_1 c_3 + 3c_2^2 + 4c_1^2 c_2 - c_1^4) \\ c.) & \chi_2 = \frac{1}{120}\int_X (79 c_4 - 19 c_1 c_3 + 3c_2^2 + 4c_1^2 c_2 - c_1^4) \end{array} \tag{A2.21}$$

For Calabi-Yau manifolds we have $c_1 = 0$ and $\chi_0 = 2$. Using this in (A2.21) implies a relation among the Hodge numbers of fourfolds say

$$h^{2,2} = 2(22 + 2h^{1,1} + 2h^{3,1} - h^{2,1}). \tag{A2.22}$$

The Euler number of the fourfold can thus be written as

$$\chi(X) = 6(8 + h^{1,1} + h^{3,1} - h^{2,1}). \tag{A2.23}$$

Atiyah-Singer Index Theorem and Anomalies

One can generalize the proof for (A2.11) in [176] to obtain [26]

$$\text{index}\, D = (-1)^n \int_M \frac{1}{e(TM)} \sum_p (-1)^p \text{ch}(\text{E}^\text{p}) \wedge \text{td}(TM_{\mathbb{C}})\,. \qquad \text{(A2.24)}$$

Examples:

- *De Rham complex:* If $E^i = \Lambda^i T^* M$ on an even $m = 2l$ dimensional manifold and $D = d$ is the exterior derivative, then using the relation of the Euler class to the top Chern class $e(TM) = \prod_{i=1}^{l} x_i(TM_{\mathbb{C}})$, see cff (A2.17), we get

$$\text{index}\, d = \int_M e(M) = \chi(M)\,. \qquad \text{(A2.25)}$$

- *Dolbeault complex:* If $E^i = \Omega^{0,i}$ on a complex m dimensional manifold and $D = \bar\partial$ then

$$\text{index}\, \bar\partial = \sum_{k=1}^{m} (-1)^k h^{0,k} = \int_M \text{td}(TM)\,. \qquad \text{(A2.26)}$$

  is the *arithmetic genus*.
- *Twisted Dolbeault complex:* If $E^i = \Omega^{0,i} \otimes E$ with E a holomorphic vector bundle on a complex m dimensional manifolds and $D = \bar\partial_V$ then

$$\text{index}\, \bar\partial_V = \chi(E) = \sum_k (-1)^k h^k(E) = \int_M \text{ch}(E)\text{td}(TM)\,. \qquad \text{(A2.27)}$$

  is the Hirzebruch-Riemann-Roch formula.
- Spin complex: If E is the 2-complex $E^\pm = P \times S^\pm$ over a $2n$ dimensional manifold, where P is the principal Spin$(2n)$ bundle and $D^+ = P^+ D$, with D is the Dirac operator coupled to the spin connection then

$$\text{index}\, D^+ = \int_M \hat{A}(TM)\,. \qquad \text{(A2.28)}$$

- *Twisted Spin complex:* If $E^\pm_E = E^\pm \times E$, where E is a gauge bundle $D^+_E = P^+ D$, with connection A_μ and D is the Dirac operator coupled to the spin connection and E, i.e. $D = i\gamma^\alpha e^\mu_\alpha(\partial_\mu + \omega_\mu + A_\mu)$

$$\text{index}\, D^+_E = \int_M \hat{A}(TM)\text{ch}(E)\,. \qquad \text{(A2.29)}$$

- *bc system:* The following standard example from bosonic string theory [91, 279] uses techniques of this and the last section. Let $T^n = T^{q-p}$ be a section of $(\otimes_{i=1}^q T\Sigma) \otimes (\otimes_{i=1}^p T^*\Sigma)$ over a Riemann surface and compare [47]

$$\begin{aligned} \nabla_n^z : T^n \to T^{n+1},\ \nabla_n^z = h^{z\bar z}\partial_{\bar z} T, \qquad & (\nabla_n^z)^\dagger = -\nabla_z^{n+1}, \\ \nabla_z^n : T^n \to T^{n-1},\ \nabla_z^n = (h^{z\bar z})^n \partial_z[(h_{z\bar z})^n T],\ & (\nabla_z^n)^\dagger = -\nabla_z^{n-1}, \end{aligned} \tag{A2.30}$$

where the inner product is $\langle T_1, T_2\rangle = \int_\Sigma d^2 z\sqrt{h}(h^{z\bar z})^n T_1^* T_2$. In a conformal theory real traceless symmetric tensors transforming as a subbundle of $S^n = T^n \oplus T^{-n}$ are of special interest and of the form $\Phi = (\phi, (h_{z\bar z})^n \phi^*)$ with $\phi = \phi^{\overbrace{z,\ldots z}^{n}}$. One defines on them

$$\begin{aligned} P_n &= \nabla_n^z \oplus \nabla_z^{-n} : S^n \to S^{n+1} \\ P_n^\dagger &= -(\nabla_z^{n+1} \oplus \nabla_{-n-1}^z) : S^{n+1} \to S^n\ . \end{aligned} \tag{A2.31}$$

where the inner product is= $\langle \Phi_1, \Phi_2\rangle \int_\Sigma d^2 z\sqrt{h}(h^{z\bar z})^n(\phi_1^*\phi_2 + \phi_2^*\phi_1)$. Note that the choice of the metric is $h_{z\bar z} = h_{\bar z z} = \frac{1}{2}e^{2\sigma}$, $h^{z\bar z} = h^{\bar z z} = 2e^{-2\sigma}$ with vanishing pure components. P_1 above is as in (3.3.19). In particular the $b = (b^{zz}, (h_{z\bar z})^2 b^{\bar z\bar z})$, $c = (c^z, h_{z\bar z}c^{\bar z})$ system has the action $S = \frac{1}{\pi}\langle n, P_1 c\rangle = \frac{1}{\pi}\int d^2 z(b_{zz}\partial_{\bar z}c^z + b_{\bar z\bar z}\partial_z c^{\bar z})$. We want to calculate the anomaly density of the $U(1)$ $c^z \to e^{-i\theta_z}c^z$, $c^{\bar z} \to e^{i\theta_{\bar z}}c^{\bar z}$, $b_{\bar z\bar z} \to e^{-i\theta_{\bar z}}b_{\bar z\bar z}$ and $b_{zz} \to e^{i\theta_z}b_{zz}$ *ghost number current*. The Laplacians above become $\Delta_1 = P_1^\dagger P_1$ and $\Delta_2 = P_1 P_1^\dagger$ with $\sigma(\delta_i) : \pi^* S^i \to \pi^* S^i$ is an isomorphism outside the zero section. One expands $c = \sum_n c_n\psi_n$ and $b = \sum_n b_n\phi^n$ as eigenfunctions of $\Delta_{1/2}$ orthonormal w.r.t. the inner product $\langle \Phi_1, \Phi_2\rangle$ applies the Noether procedure as well as the analysis of the transformation of the fermionic measure as in [123] and one finds the anomalies of the ghost currents $j_z = b_{zz}c^z$ and $j_{\bar z} = b_{\bar z\bar z}c^{\bar z}$ is $\partial_{\bar z} j_z = \pi A(z, \bar z)$ and $\partial_z j_{\bar z} = \pi A(z, \bar z)$ with $\int_\Sigma A(z, \bar z) = \sum_n\langle\psi_n, \psi_n\rangle - \sum_m\langle\phi_m, \phi_m\rangle$. Again these sums contribute only if the eigen functions ψ_n of Δ_1 and ϕ_m of Δ_2 are zero modes. E.g. if $\Delta_1\psi_n = \lambda_n\psi_n, \lambda > 0$ then $\lambda_n(P_1\psi_n) = P_1\Delta_1\psi_n = \Delta_2(P_1\psi_n)$ is a eigenfunction of Δ_2, so the corresponding contributions to the sum cancel and the integral over the anomaly density is $\ker\Delta_1 - \ker\Delta_2 = \ker P_1 - \mathrm{coker} P_1 = \mathrm{index P_1} = \mathrm{index}\nabla_1^z + \mathrm{index}\nabla_z^{-1} = \frac{3}{2}\chi(\Sigma) + \frac{3}{2}\chi(\Sigma)$. Here we used in the last step (A2.24) with $\mathrm{index}\nabla_z^{-1} = \mathrm{index}\nabla_1^z = -\int_\Sigma \frac{\mathrm{ch}(T\Sigma)-\mathrm{ch}(T\Sigma\otimes T\Sigma)}{e(\Sigma)}\mathrm{Td}(T\Sigma_C)$ with $e(\Sigma) = c_1(T\Sigma)$ and the expressions of Appendix 2. Hence the anomaly density must be $A(z, \bar z) = \frac{3}{2\pi}\sqrt{h}R$ and the current anomaly in covariant form is

$$\partial_\mu j^\mu = 3\sqrt{h}R \tag{A2.32}$$

A physics approach to proof (A2.24 is to evaluate the anomaly density

$$\partial_\mu j_5^\mu + 2iA(x) = 0, \qquad \text{with} \int d^{2n}x A(x) = \int d^{2n}x \sum_n \psi_n^\dagger\gamma_5\psi_n = \sum_n\langle\psi_n|\gamma_5|\psi_n\rangle\ , \tag{A2.33}$$

by a *heat kernel regularization*, see [124] for a review. Here the quantity $A(x)$ is called the *anomaly density*. For the vector $U(1)_V$ symmetry the contribution of the a_n and b_n cancels. Now since iD is hermitian the eigen spaces spanned by $|\psi_n\rangle$ with $iD|\psi_n\rangle = \lambda_n|\psi_n\rangle$ are orthogonal to each other. On the other hand as $\{iD, \gamma_5\} = 0$, the eigenvalues of the states $|\psi_n\rangle$ and $\gamma_5|\psi_n\rangle$ are negatives of each other. Therefore the sum in $A(x)$ has only contributions from the zero modes $\lambda_l = 0$. With the γ_5 in the trace the total current violation evaluates to

$$\begin{aligned}\text{index } D^+ &= \int \mathrm{d}^{2n}x A(x) = \#(+0 \text{ modes}) - \#(-0 \text{ modes}) \\ &= \text{dim Ker } \partial - \text{dim Ker } \partial^\dagger = \text{dim Ker } \partial - \text{dim coker } \partial \,,\end{aligned} \tag{A2.34}$$

where the last equality used that $\partial = D^+ = P^+D$ ($\partial^\dagger = P^-D$) is a Fredholm operator, i.e. kernel and cokernel are finite dimensional, and linear algebra. Nothing about the above principal setting will change if in addition to the spin connection we couple to a gauge bundle as well and consider $D = i\gamma^\alpha e_\alpha^\mu(\partial_\mu + \omega_\mu + A_\mu)$. For instance the calculation of our last example using the heat kernel i.e. without resorting to the index theorem is an exercise whose solution is found in Appendix B2 of [91]. Interesting are also the proofs by supersymmetric localisation [16, 120].

Appendix 3: Simple Examples of Calabi-Yau Spaces

In this Appendix we discuss concrete examples of CY-hypersurfaces in (weighted) projective spaces.

Toric CY-hypersurfaces

The tool that makes constructing of Calabi-Yau spaces easy is the perfect control over the first Chern class in algebraic geometry. As an application of some statements in Appendix 2 we want to calculate the first Chern class of $\mathbb{P}^n$, following [49]. As every projective space $\mathbb{P}^n$ has a tautological sequence

$$0 \to H^* \to \mathbb{P}^n \times \mathbb{C}^{n+1} \to Q \to 0 \,. \tag{A3.1}$$

$H^* = \{(l, x) \in \mathbb{P}^n \times \mathbb{C}^{n+1} | x \in \hat{l}\}$, where $\hat{l}$ is the line in $\mathbb{C}^{n+1}$, which defines l as point in $\mathbb{P}^n$, and the quotient space Q is defined by (A3.1). H^* is parametrized by the homogeneous variables $[x_1 : \ldots : x_{n+1}]$, which, as maps to $\mathbb{C}$, are section of the dual space H, called the hyperplane bundle. We can write tangent vectors in $T\mathbb{P}^n$ as linear combinations of $(\sum_{k=1}^{n+1} a_k^i x_k)\frac{\partial}{\partial x_i}$, which is scaling invariant under the $\mathbb{C}^*$ action and maps $H^{\oplus(n+1)}$ to $T\mathbb{P}^n$. There is a kernel $\mathbb{C}$ of that map, namely we have $\sum x_i \frac{\partial}{\partial x_i} = 0 \in T\mathbb{P}^n$ as it just generates the scaling action. These facts are expressed in the Euler sequence

$$0 \to \mathbb{C} \to H^{\oplus(n+1)} \to T\mathbb{P}^n \to 0 \,. \tag{A3.2}$$

The Chern class of $\mathbb{C}$ is 1 and the Whitney formula and (trivial) splitting principle gives

$$c(T\mathbb{P}^n) = (1+x)^{n+1} \,, \tag{A3.3}$$

where we denoted $x = c_1(H)$.

A weighted projective space WCP^n is defined similarly as $\mathbb{P}^n$ cff. (2.2.2), only that $\mathbb{C}^*$ acts now by ($\lambda \in \mathbb{C}^*$)

$$(z_1, \ldots, z_{n+1}) \sim (\lambda^{w_1} z_1, \ldots, \lambda^{w_{n+1}} z_{n+1}) \,, \tag{A3.4}$$

where the integral weights w_i contain no common factor. Common factors k in subsets of the weights lead to $\mathbb{Z}_k$ quotient singularities of WCP^n. A similar argument as before shows that [98]

$$c(TWCP^n) = \prod_{i=1}^{n+1} (1 + w_i x) \,, \tag{A3.5}$$

All weights are in $\mathbb{Z}$ and order to be compact $w_i > 0$. This prevents us to define compact WCP with $c_1(TWCP^n) = 0$, but $WCP(-2,1,1)$ is a well known example of a non-compact Calabi-Yau two manifold, better know as $\mathcal{O}(-2)$ line bundle over $\mathbb{P}^1$ called $\mathcal{O}(-2) \to \mathbb{P}^1$. The notion $\mathcal{O}(n) \to \mathbb{P}^1$ means the following. If we introduce local coordinates on $\mathbb{P}^1$, i.e. according to (2.2.3) $z^{(1)} = z_2/z_1$ in $\mathcal{U}^{(1)}$ and $z^{(2)} = z_1/z_2 = 1/z^{(1)}$ in $\mathcal{U}^{(2)}$, we have local coordinates $(l^{(i)}, z^{(i)})$ on $\mathcal{O}(n) \to \mathbb{P}^1$ with the transition function

$$(l^{(2)}, z^{(2)}) = \left(\frac{l^{(1)}}{(z^{(1)})^n}, \frac{1}{z^{(1)}} \right) \,. \tag{A3.6}$$

$\mathcal{O}(-2)$ can be viewed as the cotangent bundle over $\mathbb{P}^1$ parametrized by $l dx$ and $\Omega = dl \wedge dx$ is a non-vanishing $(2,0)$ form. Note that $c_1(\mathcal{O}(n) = nH$.

Compact examples as easily obtained, e.g. as hypersurfaces in the projective spaces above. Let us consider a smooth degree d hypersurface M in $\mathbb{P}^n$. M is defined as zero locus of a degree d polynomial P, which is sufficiently general so that $P = 0$ and $\mathrm{d}P = 0$ has no common solution. It is a section of $H^d = \mathcal{O}_{\mathbb{P}^n}(d)$. Since P is smooth we have a splitting of the tangent bundle $T\mathbb{P}^n$ as follows

$$0 \to TM \to T\mathbb{P}^n|_M \to N_M \to 0 \,, \tag{A3.7}$$

where N_M is the normal bundle to M, which is identified with $\mathcal{O}(d)|_M$ because P is a coordinate of N near M. $\mathrm{Ch}(H^d) = e^{dx} = 1 + c_1(H^d) = 1 + dx$, i.e. $c_1(H^d) = dx$ and the adjunction formula gives

$$c(M) = \frac{(1+x)^{n+1}}{(1+dx)} = 1 + (n+1-d)x + \ldots, \tag{A3.8}$$

i.e. a Calabi-Yau hypersurface in $\mathbb{P}^n$ has to have degree $d = n + 1$. In this case P is a section $\mathcal{O}(K_{\mathbb{P}^n})$ of the canonical line bundle $K = -[c_1(\mathbb{P}^n)]$. This gives in for dimension three one case, the quintic in $\mathbb{P}^4$. For weighted projective spaces one has

$$c(M) = \frac{\prod_{i=1}^{n+1}(1+w_i x)}{(1+dx)} = 1 - (d - \sum_i w_i)x + \ldots, \tag{A3.9}$$

where the degree d of a quasihomogeneous polynomial P is defined by the scaling $P(\lambda_1^w z_1, \ldots, \lambda^{w_{n+1}} z_{n+1}) = \lambda^d P(z_1, \ldots, z_{n+1})$. Together with the transversality condition $\mathrm{d}P = 0$ at $P = 0$ it leads 7555 examples of Calabi-Yau threefolds [225]. This sample contains many mirror pairs.

This in turn has a fairly obvious generalisation to hypersurfaces and complete intersections, which live over coordinate ring of a general toric varieties $\mathbb{P}_\Delta$and enjoy an explicit mirror construction given by Batyrev as discussed in Sect. 2.7.3 . The reflexive polyhedra Δ in four dimensions relevant for the CY threefold case have been classified [244]. This class of Calabi-Yau manifolds exihibits about 30.000 different Hogde numbers. As explained previously h^{11} and h^{21} are the only independent ones and the corresponding distribution for the sample is shown[127] in Fig. 24.

Appendix 4: Modular Forms

In this appendix, we collect some facts regarding the ring of modular forms. Of particular importance for the holomorphic anomaly equation is the theory of differentiation on these rings which leads to the notation quasi modular—and almost holomorphic modular forms, which is connected to the theory of Jacobi forms. A good reference for the modular forms is the book of [54]. For the Jacobi forms we refer to [107] and [82].

[127]Special thanks to Maximillian Kreuzer for sending me this figure.

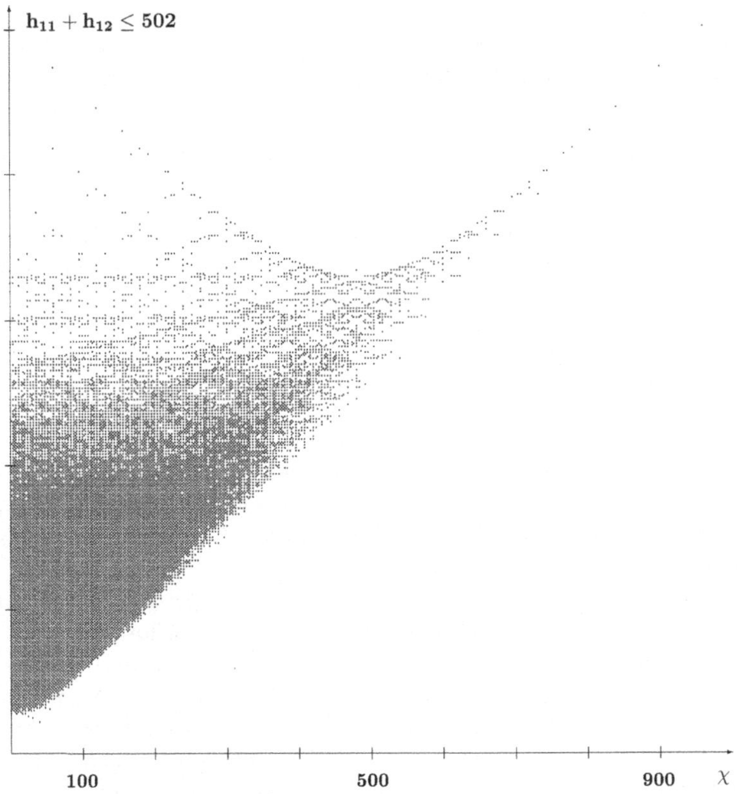

Fig. 24 Hodge numbers of hypersurface in toric varieties

Modular Forms of $\Gamma_1 = \mathrm{Sl}(2, \mathbb{Z})$

In this appendix we give an account of the ring of holomorphic modular forms, as well as derivative structures including quasi modular forms and almost holomorphic modular forms. We also comment on the relation to the topological B-model. Finely we include some useful facts regarding the Hecke operations. Most of our formulas are for $\Gamma_1 = \mathrm{Sl}(2, \mathbb{Z})$ even though qualitatively many statements generalise to finite index subgroups of Γ_1 as we point out occasionally.

The Eisenstein Series and the Ring of Holomorphic Modular Forms

We define $q := e^{2\pi i\tau}$, with

$$\tau \in \mathbb{H}_+ = \{\tau \in \mathbb{C} \,|\, \mathrm{Im}(\tau) = \frac{1}{2i}(\tau - \bar{\tau}) > 0\} \tag{A4.1}$$

and the projective action PSL(2, $\mathbb{Z}$) of

$$\Gamma_1 = \mathrm{SL}(2,\mathbb{Z}) = \left\{ \gamma = \begin{pmatrix} a & b \\ c & d \end{pmatrix} \,\middle|\, ad - bc = 1,\ a,b,c,d \in \mathbb{Z} \right\} \qquad \text{(A4.2)}$$

on $\mathbb{H}_+$ by

$$\tau \mapsto \tau_\gamma = \frac{a\tau + b}{c\tau + d}, \qquad \text{(A4.3)}$$

for $\gamma \in \Gamma_1$. We like to study expression, which have easy transformation properties under the transformation (A4.3), for example geometrically it is desirable to define functions which are well defined on the fundamental region of the complex structure $\mathcal{F} = \mathrm{PSL}(2,\mathbb{Z})/\mathbb{H}_+$, compare (2.12.4). Particular relevant objects are *modular forms* of Γ_1 that transform as

$$f_k(\tau_\gamma) = (c\tau + d)^k f_k(\tau) \qquad \text{(A4.4)}$$

with *weight* $k \in \mathbb{Z}$ for all $\tau \in \mathbb{H}_+$ and $\gamma \in \Gamma_1$, are meromorphic for $\tau \in \mathbb{H}_+$ and grow like $\mathcal{O}(e^{C\mathrm{Im}(\tau)})$ for $\mathrm{Im}(\tau) \to \infty$ and $\mathcal{O}(e^{C/\mathrm{Im}(\tau)})$ for $\mathrm{Im}(\tau) \to 0$ with $C > 0$. A strategy to build modular forms of weight k is to sum over orbits of Γ_1

$$G_k = \frac{1}{2} \sum_{\substack{m,n\in\mathbb{Z} \\ (m,n)\neq(0,0)}} \frac{1}{(m\tau + n)^k}. \qquad \text{(A4.5)}$$

It is easy to see that this expression transforms like (A4.4), converges absolutely for $k > 2$ and vanishes for k odd. One can proof [54] the central fact

Proposition 6 *G_4, G_6, or E_4, E_6 which are just a more convenient normalisation of G_4, G_6 explained below, generate freely the, by k, graded ring of holomorphic modular forms $M(\Gamma_1) = \oplus_k M_k(\Gamma_1)$.*

The decisive step in the proof is to show the dimension of the ring at weight k is bounded in a way that becomes an equality for the polynomials in E_4, E_6 modulo possible relations. In fact holomorphic modular forms can only exist if k even, $k \geq 0$ and the dimension is then given by

$$\dim M_k = \begin{cases} \lceil \frac{k}{12} \rceil + 1 \ \text{if}\ k \neq 2 \bmod 12\,, \\ \lceil \frac{k}{12} \rceil \ \text{if}\ k = 2 \bmod 12\,. \end{cases} \qquad \text{(A4.6)}$$

One writes $M(\Gamma_1)[E_4, E_6]$ to indicate that this ring of holomorphic modular forms is generated by E_4, E_6. The statement is very powerful, for if one knows that a quantity is a modular form of a given weight k, one has only to fix finite coefficient of a polynomial in E_4, E_6 to anchor it.

Still one may spot two shortcomings. Firstly the ring $\mathcal{M}(\Gamma_1)$ does not close under ordinary differentiation. Since $D_\tau = \frac{d}{2\pi i d\tau}$ has weight 2, one needs a 'modular form' of weight 2 to achieve such a closure. We will discuss the closure more in the next subsection and end the section with the construction of a weight 2 object, with a formalism that also leads to general convenient forms for (A4.5) in all cases. The definition of G_2 is achieved by an ϵ regularization in the sum $G_{2,\epsilon} = \frac{1}{2}\sum_{\substack{m,n\in\mathbb{Z}\\(m,n)\neq(0,0)}} \frac{1}{(m\tau+n)^k|m\tau+n|^\epsilon}$ after which it is possible to define $G_2 = \lim_{\epsilon\to 0} G_{2,\epsilon}$. Then all G_k, $k \in 2\mathbb{Z}$, $k \geq 2$ have a Fourier expansion[128] in $q = \exp(2\pi i\tau)$

$$G_k(\tau) = \frac{(2\pi i)^k}{(k-1)!}\left(-\frac{B_k}{2k} + \sum_{n=1}^{\infty} \sigma_{k-1}(n) q^n\right), \tag{A4.7}$$

with $\sigma_k(n) = \sum_{p|n} p^k$ the sum of kth powers of positive divisors of n and $\sum_{k=0}^{\infty} \frac{B_k x^k}{k!} = \frac{x}{e^x-1}$ defining the Bernoulli numbers B_k, e.g. $B_2 = \frac{1}{6}$, $B_4 = -\frac{1}{30}$, $B_6 = \frac{1}{42}$, $B_8 = -\frac{1}{30}$, $B_{10} = \frac{5}{66}$, $B_{12} = -\frac{691}{2730}$, $B_{14} = \frac{7}{6}$ etc.

In the standard definition of the Eisenstein series E_k the sum runs over coprime (m, n), which yields a proportionality $G_k(\tau) = \zeta(k)E_k(\tau)$, where $\zeta(k) = \sum_{n\geq 1}\frac{1}{n^k}$. Therefore the Eisenstein series are just a different normalization of (A4.7) and given for $k = 2m$ as

$$E_{2m}(\tau) = 1 + \frac{2}{\zeta(1-2m)}\sum_{n=1}^{\infty}\frac{n^{2m-1}q^n}{1-q^n} \qquad m \geq 1\,, \tag{A4.8}$$

Very much like in QFT the regularization for $k = 2$ introduces an anomaly in the symmetry transformation so that E_2 transforms

$$E_2(\tau_\gamma) = (c\tau+d)^2 E_2(\tau) - \frac{6ic}{\pi}(c\tau+d) \tag{A4.9}$$

with an inhomogeneous term. This inhomogenoeus transformation behaviour is called *quasi modular*. It follows by an elementary calculation from (A4.3) that

$$\frac{1}{\mathrm{Im}(\tau_\gamma)} = \frac{(c\tau+d)^2}{\mathrm{Im}(\tau)} - 2ic(c\tau+d) = \frac{|c\tau+d|^2}{\mathrm{Im}(\tau)} \tag{A4.10}$$

transforms also quasi modular. Using (A4.10) and (A4.9) we can arrange it that the inhomogeneous terms cancel so that

$$\hat{E}_2(\tau) = E_2(\tau) - \frac{3}{\pi\,\mathrm{Im}(\tau)} \tag{A4.11}$$

transforms like a modular form of weight 2, albeit not a holomorphic one.

[128]Note that the Eisenstein series start with coefficient 1.

Differentiable Rings of Modular Forms

In this section we like to study rings of differential forms which closes under derivatives. Many of the mentioned properties of these can be found in [208] and in the review [54]. Lets us denote the normal derivative by

$$D = D_\tau = \frac{\mathrm{d}}{2\pi i \mathrm{d}\tau} = q\frac{\mathrm{d}}{\mathrm{d}q}. \tag{A4.12}$$

We start with an remarkable observation attributed to Ramanujan.

Proposition 7 *The generators* (E_2, E_4, E_6) *form a ring of quasi modular forms, which by the Ramanujan identities*

$$DE_2 = \frac{1}{12}(E_2^2 - E_4), \quad DE_4 = \frac{1}{3}(E_2E_4 - E_6), \quad DE_6 = \frac{1}{2}(E_2E_6 - E_4^2)\,. \tag{A4.13}$$

closes under differentiation the $D = q\frac{\mathrm{d}}{\mathrm{d}q}$.

To understand the structure of differential rings a more systematic, we look now closer at the properties of the derivatives. For any function f defined on the upper half plane $\mathbb{H}_+$ it is convenient to introduce the notation of a *modular weight k transformation*

$$f|_k\gamma = (c\tau + d)^{-k} f(\tau_\gamma)\,. \tag{A4.14}$$

It does not imply any additional property of f, but if f is a modular function of weight k then

$$f|_k\gamma = f\,. \tag{A4.15}$$

One also uses for $\gamma, \gamma' \in \Gamma$ the notation[129] (A4.14) as

$$f|_k(\gamma + \gamma') = f_k\gamma + f_k\gamma'\,. \tag{A4.16}$$

Since

$$\frac{\mathrm{d}\tau_\gamma}{\mathrm{d}\tau} = \frac{1}{(c\tau + d)^2} \tag{A4.17}$$

[129]This should be obvious as $\gamma\gamma'$ is the group operation and $\gamma + \gamma'$ makes no sense in Γ.

the derivative of a weight zero modular modular form becomes a weight $k = 2$ modular form or $D : M_0(\Gamma) \to M_2(\Gamma)$, but D does not map in general modular forms of weight k into modular forms of weight $k + 2$ as one can see from

$$D_{\tau_\gamma} f_k(\tau_\gamma) = (c\tau + d)^2 D_\tau (c\tau + d)^k f_k(\tau) = (c\tau + d)^{k+2} D_\tau f_k(\tau) + c(c\tau + d)^{k-1} f_k(\tau) \tag{A4.18}$$

or in a shorter notation using (A4.14)

$$Df_k|_{k+2}\gamma = Df_k + \frac{c}{c\tau + d} f_k \, . \tag{A4.19}$$

In order to define a covariant derivative one can use the inhomogeneous terms in (A4.9) or (A4.10) to cancel the inhomogeneous term in (A4.19) and to define covariant derivative operators of weight two. At the same time we have to modify the ring of holomorphic modular forms $M(\Gamma_1)[E_4, E_6]$ either to the one of quasi modular forms $M^!(\Gamma_1)[E_2, E_4, E_6]$ by adding the Eisenstein series E_2 to the generators or to the one of almost holomorphic forms $\hat{M}(\Gamma_1)[\hat{E}_2, E_4, E_6]$ by adding $\hat{E}_2$. This allows to defined covariant operators operators to act on these rings called the

- The *Maass derivative*

$$D^M f_k = \left(D - \frac{k}{4\pi \mathrm{Im}(\tau)} \right) f_k \tag{A4.20}$$

- or the *Serre derivative*

$$D^S f_k = \left(D - \frac{k}{12} E_2 \right) f_k \, . \tag{A4.21}$$

The derivatives have the following properties

$$D : M^!_k(\Gamma_1) \to M^!_{k+2}(\Gamma_1), \quad D^M : \hat{M}_k(\Gamma_1) \to \hat{M}_{k+2}(\Gamma_1), \quad D^S : M_k(\Gamma_1) \to M_{k+2}(\Gamma_1) \, . \tag{A4.22}$$

Note that Eq. (A4.13) hold with D replaced by D^M and $E_2(\tau)$ replaced by $\hat{E}_2(\tau)$. In particular the rings $M^!(\Gamma_1)$ and $\hat{M}(\Gamma)$ are isomorphic and behave identical under the corresponding derivatives. The isomorphic identification is achieved just by replacing $E_2 \leftrightarrow \hat{E}_2$. Of course D^S operators close also on $M^!(\Gamma_1)$ and it naturally restricts to $M(\Gamma)$. We further note that for finite index subgroups Γ of SL(2, $\mathbb{Z}$) one has in similar ring structures for the modular forms but $M(\Gamma)$ has more generators then just E_4 and E_6. Nevertheless $\hat{M}(\Gamma)$ as well as $M^!(\Gamma)$ can be defined as before just by adding $\hat{E}_2$ or E_2 respectively. The formulas (A4.43) are important examples of such generalisations. In view of the importance of the ring of quasi modular forms shorter notions for the generators are common

$$P := E_2, \quad Q := E_4, \quad \text{and} \quad R := E_6 . \tag{A4.23}$$

Something special happens if we correlate the number of derivatives with the weight the transformation. Then even without modifying anything about the derivative or the rings one can show by an elementary calculation *Bol's identity*, which holds on any holomorphic function on the upper half plane $F : \mathbb{H} \to \mathbb{H}$.

Proposition 8 *Bol's identity states that the $k-1$ derivative of the weight $2-k$ modular transfomation of F equals modular wight k transformation of its $k-1$'s derivative, i.e.*

$$D^{k-1}(F|_{2-k}\gamma) = (D^{k-1}F)|_k\gamma . \tag{A4.24}$$

Defining $\mathcal{H}_l$ as the space of holomorphic functions on the the upperhalf plane with the action of SL$(2,\mathbb{R})$ on it as defined in (A4.14) we have hence an Sl$(2,\mathbb{R})$ invariant operation (or intertwinner)

$$D^{k-1} : \mathcal{H}_{2-k} \to \mathcal{H}_k . \tag{A4.25}$$

The Relation to the More General Formalism of the Topological B-Model

From the physical point of view there is a story behind the above well known mathematical facts, which are best understood starting from the genus one torus amplitude of the bosonic string. The latter can be of course calculated using the interesting unfolding technique that uses space time—and world sheet modularity to integrate over the fundamental domain of the world sheet torus [97] and the result is mirror symmetric

$$F^{(1)} = -\log\left(\sqrt{\mathrm{Im}(\tau)}|\eta(\tau)|^2\right) - \log\left(\sqrt{\mathrm{Im}(T)}|\eta(T)|^2\right) . \tag{A4.26}$$

in the complex parameter and the complexified Kähler parameter of the space time torus respectively.

The holomorphic propagator, which can be made proportional to E_2, compare (4.2.23) in the $\mathrm{Im}(\tau) \to \infty$ limit needs some regularization, which breaks T duality. The latter is restored by adding the non-holomorphic term (A4.11). The modular anomaly and the holomorphic anomaly are in this sense counterparts, which cannot both be realized at least perturbatively. T-duality is physically better motivated. Attempts in the literature, e.g. in an interesting paper [109], to define a holomorphic and modular non-perturbative completion by summing over orbits seem to make sense only if absolute convergence in the moduli is established.

$F^{(1)}$ is an index, which is finite for smooth compact spaces. It diverges therefore only from singular configurations, that occur if e.g. the discriminant of the elliptic curve given for the *Weierstrass* form of the elliptic curve as

$$y^2 = 4x^3 - \frac{4}{3}xE_4 - \frac{8}{27}E_6 \tag{A4.27}$$

as

$$\Delta(\tau) = \eta^{24}(\tau) = q\prod_{n=1}^{\infty}(1-q^n)^{24} = \frac{1}{1728}(E_4^3(\tau) - E_6^2(\tau)), \tag{A4.28}$$

vanishes. Here η is the Dedekind η function which can be defined as an infinite product

$$\eta(\tau) = q^{\frac{1}{24}}\prod_{n=1}^{\infty}(1-q^n). \tag{A4.29}$$

Note that the absolute modular invariant, called j-invariant, for the curve (A4.27) is

$$j = 1728\frac{E_4^3}{E_4^3 - E_6^2} = \frac{1}{q} + 744 + 196884q + 21493760q^2 + \mathcal{O}(q^3). \tag{A4.30}$$

The Dedekin η function transforms[130] as

$$\eta(\tau+1) = e^{\frac{\pi i}{12}}\eta(\tau), \qquad \eta\left(-\frac{1}{\tau}\right) = \sqrt{-i\tau}\,\eta(\tau), \tag{A4.31}$$

i.e. with weight $\frac{1}{2}$, but additional phases, which are called a multiplier system. From (A4.13), (A4.28) one calculates

$$d_\tau \log(\eta(\tau)) = \frac{1}{24}E_2(\tau). \tag{A4.32}$$

Further from (A4.10) we see that $\sqrt{\mathrm{Im}(\tau)}|\eta(\tau)|^2$ is an almost holomorphic modular invariant and from (A4.13), (A4.11), (A4.28) that

$$D_\tau \log(\sqrt{\mathrm{Im}(\tau)}|\eta|^2) = \mathrm{d}_\tau \log(\sqrt{\mathrm{Im}(\tau)}|\eta|^2) = \frac{\hat{E}_2}{24}. \tag{A4.33}$$

---

[130] As can be seen from (A4.29) and (A4.41), (A4.38).

Theta Functions and Modular Forms

Theta functions are examples of Jacobi forms, which are more abstractly discussed in the next section. Our conventions for the theta functions associated to the spin structure on the torus are

$$\Theta\begin{bmatrix} a \\ b \end{bmatrix}(\tau, z) = \sum_{n\in\mathbb{Z}} e^{\pi i(n+a)^2\tau + 2\pi i(z+b)(n+a)} . \tag{A4.34}$$

The Jacobi theta functions are then $\theta_1 = -\Theta\begin{bmatrix} \frac{1}{2} \\ \frac{1}{2} \end{bmatrix}$, $\theta_2 = \Theta\begin{bmatrix} \frac{1}{2} \\ 0 \end{bmatrix}$, $\theta_3 = \Theta\begin{bmatrix} 0 \\ 0 \end{bmatrix}$ and $\theta_4 = \Theta\begin{bmatrix} 0 \\ \frac{1}{2} \end{bmatrix}$. In particular, we have

$$\theta_1(\tau, z) = z \cdot \eta(\tau)^3 \exp\left(\sum_{k=1}^{\infty} \frac{B_{2k}}{2k(2k)!}(iz)^{2k} E_{2k}(\tau)\right) . \tag{A4.35}$$

For convenience we give the expansion for the individual Jacobi theta functions

$$\begin{aligned}
\vartheta_1(\tau, z) &= \vartheta[{}^1_1](\tau, z) = i\sum_{n\in\mathbf{Z}}(-1)^n q^{\frac{1}{2}(n+1/2)^2} e^{i\pi(2n+1)z}, \\
\vartheta_2(\tau, z) &= \vartheta[{}^1_0](\tau, z) = \sum_{n\in\mathbf{Z}} q^{\frac{1}{2}(n+1/2)^2} e^{i\pi(2n+1)z}, \\
\vartheta_3(\tau, z) &= \vartheta[{}^0_0](\tau, z) = \sum_{n\in\mathbf{Z}} q^{\frac{1}{2}n^2} e^{i\pi 2nz}, \\
\vartheta_4(\tau, z) &= \vartheta[{}^0_1](\tau, z) = \sum_{n\in\mathbf{Z}}(-1)^n q^{\frac{1}{2}n^2} e^{i\pi 2nz},
\end{aligned} \tag{A4.36}$$

where $q = e^{2\pi i\tau}$. When $z = 0$ we will simply denote $\vartheta_2(\tau) = \vartheta_2(0|\tau)$ (notice that $\vartheta_1(0|\tau) = 0$). The theta functions $\vartheta_2(\tau)$, $\vartheta_3(\tau)$ and $\vartheta_4(\tau)$ have the following product representation:

$$\begin{aligned}
\vartheta_2(\tau) &= 2q^{1/8}\prod_{n=1}^{\infty}(1-q^n)(1+q^n)^2, \\
\vartheta_3(\tau) &= \prod_{n=1}^{\infty}(1-q^n)(1+q^{n-\frac{1}{2}})^2, \\
\vartheta_4(\tau) &= \prod_{n=1}^{\infty}(1-q^n)(1-q^{n-\frac{1}{2}})^2
\end{aligned} \tag{A4.37}$$

and under modular transformations they behave as vector valued modular forrms of weight $\frac{1}{2}$:

$$\begin{aligned} \vartheta_2(-1/\tau) =& \sqrt{\frac{\tau}{i}}\vartheta_4(\tau), & \vartheta_2(\tau+1) =& e^{i\pi/4}\vartheta_2(\tau), \\ \vartheta_3(-1/\tau) =& \sqrt{\frac{\tau}{i}}\vartheta_3(\tau), & \vartheta_3(\tau+1) =& \vartheta_4(\tau), \\ \vartheta_4(-1/\tau) =& \sqrt{\frac{\tau}{i}}\vartheta_2(\tau), & \vartheta_4(\tau+1) =& \vartheta_3(\tau). \end{aligned} \tag{A4.38}$$

The theta function $\vartheta_1(\tau, z)$ has the product representation

$$\vartheta_1(\tau, z) = -2q^{\frac{1}{8}} \sin(\pi z) \prod_{n=1}^{\infty}(1-q^n)(1-2\cos(2\pi z)q^n + q^{2n}). \tag{A4.39}$$

We also have the following useful identities:

$$\vartheta_3^4(\tau) = \vartheta_2^4(\tau) + \vartheta_4^4(\tau), \tag{A4.40}$$

and

$$\vartheta_2(\tau)\vartheta_3(\tau)\vartheta_4(\tau) = 2\,\eta^3(\tau). \tag{A4.41}$$

One has the following doubling formulae,

$$\begin{aligned} \eta(2\tau) =& \sqrt{\frac{\eta(\tau)\vartheta_2(\tau)}{2}}, & \vartheta_2(2\tau) =& \sqrt{\frac{\vartheta_3^2(\tau)-\vartheta_4^2(\tau)}{2}}, \\ \vartheta_3(2\tau) =& \sqrt{\frac{\vartheta_3^2(\tau)+\vartheta_4^2(\tau)}{2}}, & \vartheta_4(2\tau) =& \sqrt{\vartheta_3(\tau)\vartheta_4(\tau)}, \\ \eta(\tau/2) =& \sqrt{\eta(\tau)\vartheta_4(\tau)}. \end{aligned} \tag{A4.42}$$

The formulae for the derivatives of the theta functions are also useful:

$$\begin{aligned} q\frac{d}{dq}\log\vartheta_4 =& \frac{1}{24}\left(E_2 - \vartheta_2^4 - \vartheta_3^4\right), \\ q\frac{d}{dq}\log\vartheta_3 =& \frac{1}{24}\left(E_2 + \vartheta_2^4 - \vartheta_4^4\right), \\ q\frac{d}{dq}\log\vartheta_2 =& \frac{1}{24}\left(E_2 + \vartheta_3^4 + \vartheta_4^4\right), \end{aligned} \tag{A4.43}$$

and from these one finds

$$q\frac{d}{dq}\log\eta = \frac{1}{24}E_2(\tau) \tag{A4.44}$$

Modular Forms of $\Gamma(3)$

Here we summarise the modular forms of $\Gamma(3)$, that are relevant to solve the B-model of $\mathcal{O}(-3) \to \mathbb{P}^2$ [6]. For the congruence subgroup $\Gamma(3)$ of $\mathrm{Sl}(2,\mathbb{Z})$, the relevant theta constants (taking their third powers) are

$$a := \theta^3\begin{bmatrix}\frac{1}{6}\\ \frac{1}{6}\end{bmatrix}(\tau,0), \quad b := \theta^3\begin{bmatrix}\frac{1}{6}\\ \frac{1}{2}\end{bmatrix}(\tau,0), \quad c := \theta^3\begin{bmatrix}\frac{1}{6}\\ \frac{5}{6}\end{bmatrix}(\tau,0), \quad d := \theta^3\begin{bmatrix}\frac{1}{2}\\ \frac{1}{6}\end{bmatrix}(\tau,0),$$

satisfying the identities

$$b = a + c, \qquad d = a + \alpha b,$$

with $\alpha = e^{\frac{2\pi i}{3}}$. Moreover, the Dedekind η-function is given by $\eta^{12} = \frac{i}{3^{3/2}}abcd$.

We need derivative formulae for these theta constants as well. Let us first define the six following modular forms of weight 2:

$$\begin{aligned} t_1 &= \tfrac{ac}{\eta^2}, & t_2 &= \tfrac{ab}{\eta^2}, & t_3 &= \tfrac{bc}{\eta^2}, \\ t_4 &= \tfrac{bd}{\eta^2}, & t_5 &= \tfrac{ad}{\eta^2}, & t_6 &= \tfrac{cd}{\eta^2}. \end{aligned}$$

Then we found the relations:

$$\begin{aligned} 8q\tfrac{d}{dq}\log a &= \tfrac{1}{3}E_2\left(\tfrac{\tau+1}{3}\right) = E_2(\tau) - \tfrac{2}{3}(t_4 + t_6 + \alpha t_3), \\ 8q\tfrac{d}{dq}\log b &= \tfrac{1}{3}E_2\left(\tfrac{\tau}{3}\right) = E_2(\tau) + \tfrac{2}{3}(t_1 - t_5 + t_6), \\ 8q\tfrac{d}{dq}\log c &= \tfrac{1}{3}E_2\left(\tfrac{\tau+2}{3}\right) = E_2(\tau) + \tfrac{2}{3}(t_4 + t_5 - \alpha^2 t_2), \\ 8q\tfrac{d}{dq}\log d &= 3E_2(3\tau) = E_2(\tau) + \tfrac{2}{3}(-t_1 + \alpha^2 t_2 + \alpha t_3)\,. \end{aligned}$$

Note that the second equality in each line are 'triple' analogs of the doubling identities for the Eisenstein series $E_2(\tau)$.

Jacobi Forms

Jacobi forms [107] are functions $\phi : \mathbb{H} \times \mathbb{C} \to \mathbb{C}$ that depend on a modular parameter $\tau \in \mathbb{H}$ and an elliptic parameter $z \in \mathbb{C}$. They transform under the action of the modular group on $\mathbb{H} \times \mathbb{C}$, given by

$$\tau \mapsto \tau_\gamma = \frac{a\tau + b}{c\tau + d}, \quad z \mapsto z_\gamma = \frac{z}{c\tau + d} \quad \text{with} \quad \begin{pmatrix} a & b \\ c & d \end{pmatrix} \in \mathrm{SL}(2; \mathbb{Z}) , \tag{A4.45}$$

as

$$\phi\left(\tau_\gamma, z_\gamma\right) = (c\tau + d)^k e^{\frac{2\pi i m c z^2}{c\tau+d}} \phi(\tau, z) . \tag{A4.46}$$

Furthermore, they enjoy the property of quasi-periodicity,

$$\phi(\tau, z + \lambda\tau + \mu) = e^{-2\pi i m(\lambda^2\tau + 2\lambda z)} \phi(\tau, z) \quad \forall \lambda, \mu \in \mathbb{Z} . \tag{A4.47}$$

$k \in \mathbb{Z}$ is called the *weight* and $m \in \mathbb{N}$ the *index* of the Jacobi form.

Due to the periodicity under $\tau \mapsto \tau + 1, z \mapsto z + 1$, the Jacobi forms enjoy a double Fourier expansion

$$\phi(\tau, z) = \sum_{n,r} c(n, r) q^n y^r , \qquad \text{where } q = e^{2\pi i \tau}, \quad y = e^{2\pi i z} . \tag{A4.48}$$

It is in fact more appropriate to write the coefficients as $c(n, r) = C(4nm - r^2, r)$ as the combination $4nm - r^2$ is invariant under the transformation (A4.47) and $C(4nm - r^2, r)$ has a periodicity of $2m$ in r. Holomorphic Jacobi forms satisfy the constraint $c(n, r) = 0$ unless $4mn \geq r^2$, cusp forms satisfy $c(n, r) = 0$ unless $4mn > r^2$, while for weak Jacobi forms, one imposes the condition $c(n, r) = 0$ unless $n \geq 0$.

According to [107], the ring of weak Jacobi forms of integer index is freely generated over the ring of modular forms by the two generators $\phi_{-2,1}(\tau, z)$ and $\phi_{0,1}(\tau, z)$ of index 1. Introducing the notation

$$A(\tau, z) = \phi_{-2,1}(\tau, z) \quad \text{and} \quad B(\tau, z) = \phi_{0,1}(\tau, z) , \tag{A4.49}$$

we see that the vector space of weak Jacobi forms of weight k and index m is equal to

$$J^{\text{weak}}_{k,m} = \bigoplus_{j=0}^{m} M_{k+2j}(\Gamma_1) A^j B^{m-j} . \tag{A4.50}$$

The generators A and B are of index $m = 1$ and weight -2 and 0 respectively. They can be defined as

$$\begin{aligned} A(\tau, z) &= -\frac{\theta_1(\tau, z)^2}{\eta^6(\tau)} \\ B(\tau, z) &= 4\left(\frac{\theta_2(\tau, z)^2}{\theta_2(\tau, 0)^2} + \frac{\theta_3(\tau, z)^2}{\theta_3(\tau, 0)^2} + \frac{\theta_4(\tau, z)^2}{\theta_4(\tau, 0)^2}\right) . \end{aligned} \tag{A4.51}$$

In order to accommodate our convention for the normalization of ϵ_i, we will also use the notation

$$\mathcal{A}(\tau, 2\pi z) := A(\tau, z)\ , \ \mathcal{B}(\tau, 2\pi z) := B(\tau, z)\ , \ \vartheta_i(\tau, 2\pi z) := \theta_i(\tau, z),\ i = 1, \ldots, 4\ . \tag{A4.52}$$

Using the Jacobi triple product for θ_1 and the notation

$$x_m = (2\sin \pi m z)^2 = -(y^{\frac{m}{2}} - y^{-\frac{m}{2}})^2\ , \quad y = \exp(2\pi i z)\ , \tag{A4.53}$$

one finds that the weak Jacobi form A has a simple product form

$$A(\tau, z) = (y^{\frac{1}{2}} - y^{-\frac{1}{2}})^2 \prod_{n=1}^{\infty} \frac{(1-q^n y)^2 (1-\frac{q^n}{y})^2}{(1-q^n)^4} \tag{A4.54}$$

$$= -x_1 \prod_{n=1}^{\infty} \frac{(1 + x_1 q^n - 2q^n + q^{2n})^2}{(1-q^n)^4}\ . \tag{A4.55}$$

The weight zero index one weak Jacobi form B is one half of the elliptic genus of K3,

$$\chi(\mathrm{K3}; q, y) = 2B(\tau, z) = \left(2y + 20 + \frac{2}{y}\right) + \left(\frac{20}{y^2} - \frac{128}{y} + 216 - 128y + 20y^2\right) q + \mathcal{O}(q^2)\ , \tag{A4.56}$$

and it enjoys the following expansion in x_1

$$B(\tau, z) = -x_1(1 - 10x_1 q + x_1^2 q^2) + \sum_{n=0}^{\infty} q^n g_n(x_1)\ , \tag{A4.57}$$

with $g_n(x_1)$ a polynomial in x_1 of order n. Note that $A(\tau, z)$ vanishes when $z = 0$, while $B(\tau, 0) = 12$, as can be seen from the expansion of these Jacobi forms in z with quasi modular coefficients

$$A(\tau, z) = -z^2 + \frac{E_2}{12} z^4 + \frac{-5E_2^2 + E_4}{1440} z^6 + \frac{35E_2^3 - 21E_2E_4 + 4E_6}{362880} z^8 + \mathcal{O}(z^{10}),$$
$$B(\tau, z) = 12 - E_2 z^2 + \frac{E_2^2 + E_4}{24} z^4 + \frac{-5E_2^3 - 15E_2E_4 + 8E_6}{4320} z^6 + \mathcal{O}(z^8). \tag{A4.58}$$

The real zeros of A coincide with the zeros of x_1. All complex zeros are obtained as $\mathrm{SL}(2, \mathbb{Z})$ images of these zeros.

The weak Jacobi forms $A(\tau, nz)$ of index n^2 will play an important role in string theory on elliptic fibrations. $A(\tau, n_1 z)$ is divisible by $A(\tau, n_2 z)$ if the integer n_1 is divisible by n_2. Based on this observation, it is convenient to define a more primitive weak Jacobi form, due to Zagier,

$$P_d(\tau, z) = \prod_{k|d} A(\tau, kz)^{\mu(d/k)}, \tag{A4.59}$$

where $\mu(n)$ is the Möbius function. The first few P_d are given by

$$P_1 = A_1\,, \quad P_2 = \frac{A_2}{A_1}\,, \quad P_3 = \frac{A_3}{A_1}\,\ldots. \tag{A4.60}$$

For any d, one can show that $P_d(\tau, z)$ has no poles and vanishes only at primitive d-torsion points, i.e. at $z = 2\pi(n_1 + \tau n_2)/d$ for integers n_1, n_2 with the greatest common divisor $\gcd(n_1, n_2, d) = 1$. So $P_d(\tau, z)$ is a weak Jacobi form and can be written as a polynomial in $A(\tau, z)$, $B(\tau, z)$, E_4, E_6. We can express $A(\tau, nz)$ in terms of these more basic building blocks via

$$A(\tau, nz) = \prod_{k|n} P_k(\tau, z). \tag{A4.61}$$

References

1. Aganagic, M., Vafa, C.: Mirror symmetry, D-Branes and counting holomorphic disk (2000). hep-th/0012041
2. Aganagic, M., Klemm, A., Vafa, C.: Mirror Symmetry, D-branes and the duality web. Z. Naturforsch. **A 57**, 1 (2002). arXiv:hep-th/0105045
3. Aganagic, M., Klemm, A., Marino, M., Vafa, C.: Matrix model as a mirror of Chern-Simons theory. J. High Energy Phys. **0402**, 010 (2004). https://doi.org/10.1088/1126-6708/2004/02/010 [hep-th/0211098]
4. Aganagic, M., Klemm, A., Marino, M., Vafa, C.: The topological vertex. Commun. Math. Phys. **254**, 425 (2005). https://doi.org/10.1007/s00220-004-1162-z [hep-th/0305132]
5. Aganagic, M., Dijkgraaf, R., Klemm, A., Marino, M., Vafa, C.: Topological strings and integrable hierarchies. Commun. Math. Phys. **261**, 451 (2006). https://doi.org/10.1007/s00220-005-1448-9 [hep-th/0312085]
6. Aganagic, M., Bouchard, V., Klemm, A.: Topological strings and (almost) modular forms. Commun. Math. Phys. **277**, 771 (2008). https://doi.org/10.1007/s00220-007-0383-3. [hep-th/0607100]
7. Aganagic, M., Cheng, M.C.N., Dijkgraaf, R., Krefl, D., Vafa, C.: Quantum geometry of refined topological strings. J. High Energy Phys. **1211**, 019 (2012). https://doi.org/10.1007/JHEP11(2012)019. [arXiv:1105.0630 [hep-th]]
8. Aharony, O., Bergman, O., Jafferis, D.L., Maldacena, J.: J. High Energy Phys. **0810**, 091 (2008). https://doi.org/10.1088/1126-6708/2008/10/091. [arXiv:0806.1218 [hep-th]]
9. Akemann, G.: Higher genus correlators for the Hermitian matrix model with multiple cuts. Nucl. Phys. B **482**, 403 (1996). https://doi.org/10.1016/S0550-3213(96)00542-1. [hep-th/9606004]

10. Albeverio, S., Jost, J., Paycha, S., Scarlatti, S.: A Mathematical Introduction to String Theory. London Mathematical Society Lecture Notes Series, vol. 225. Cambridge University Press, Cambridge (1997)
11. Alim, M., Lange, J.D.: Polynomial structure of the (open) topological string partition function. J. High Energy Phys. **0710**, 045 (2007). https://doi.org/10.1088/1126-6708/2007/10/045. [arXiv:0708.2886 [hep-th]]
12. Alim, M.: Proc. Sympos. Pure Math. **88**, 197 (2014). https://doi.org/10.1090/pspum/088/01464. [arXiv:1401.5537 [hep-th]]
13. Almkvist, G., Zudilin, W.: Differential equations, mirror maps and zeta values. In: Yui, V.N., Yau, S.-T., Lewis, J.D. (eds.) Mirror Symmetry. AMS/IP Studies in Advanced Mathematics, vol. 38, pp. 481–515. International Press & American Mathematical Society, Providence (2007). arXiv:math/0402386v2 [math.NT]
14. Almkvist, G., van Enckevort, C., van Straten, D., Zudilin, W.: Tables of Calabi–Yau equations (2005). arXiv:math/0507430 [math.AG]
15. Alvarez, O.: Topological quantization and cohomology. Commun. Math. Phys. **100**, 279 (1985)
16. Alvarez-Gaume, L.: Supersymmetry and the Atiyah-Singer index theorem. Commun. Math. Phys. **90**, 161 (1983)
17. Alvarez-Gaume, L., Gomez, C., Moore, G.W., Vafa, C.: Strings in the operator formalism. Nucl. Phys. B **303**, 455 (1988)
18. Antoniadis, I., Gava, E., Narain, K.S., Taylor, T.R.: Topological amplitudes in string theory. Nucl. Phys. B **413**, 162 (1994). [arXiv:hep-th/9307158]
19. Antoniadis, I., Gava, E., Narain, K.S., Taylor, T.R.: N=2 type II heterotic duality and higher derivative F terms. Nucl. Phys. B **455**, 109 (1995). [arXiv:hep-th/9507115]
20. Ambjørn, J., Chekhov, L., Kristjansen, C.F., Makeenko, Y.: Matrix model calculations beyond the spherical limit. Nucl. Phys. B **404**, 127 (1993). Erratum: [Nucl. Phys. B **449**, 681 (1995)]. https://doi.org/10.1016/0550-3213(93)90476-6, 10.1016/0550-3213(95)00391-5 [hep-th/9302014]
21. Argyres, P.C., Douglas, M.R.: New phenomena in SU(3) supersymmetric gauge theory. Nucl. Phys. B **448**, 93 (1995). https://doi.org/10.1016/0550-3213(95)00281-V. [hep-th/9505062]
22. Arnold, V.I., Gusein-Zade, S.M., Varchenko, A.N.: Singularities of Differentiable Maps, vol. 2. Birkhäuser/Springer, New York (2012)
23. Aspinwall, P.S.: K3 surfaces and string duality. In: Yau, S.T. (ed.) Differential Geometry Inspired by String Theory, pp. 1–95. International Press, Vienna (1996) [hep-th/9611137]
24. Aspinwall, P.S., Morrison, D.R.: Topological field theory and rational curves. Commun. Math. Phys. **151**, 245 (1993). https://doi.org/10.1007/BF02096768 [hep-th/9110048]
25. Aspinwall, P.S., Greene, B.R., Morrison, D.R.: Space-time topology change and stringy geometry. J. Math. Phys. **35**, 5321 (1994)
26. Atiyah, M.F., Singer, I.M.: The index of an elliptic operator I& III. Ann. Math. **87**, 484 (1968)
27. Barannikov, S., Kontsevich, M.: Frobenius manifolds and formality of Lie algebras of polyvector fields. Int. Math. Res. Not. **4**, 201–215 (1998). [alg-geom/9710032]
28. Batyrev, V.: Dual polyhedra and mirror symmetry for Calabi-Yau hypersurfaces in toric varieties. J. Algebraic Geom. **3**(3), 493–535 (1994)
29. Batyrev, V.V., Borisov, L.A.: On Calabi-Yau complete intersections in toric varieties (1994). alg-geom/9412017
30. Batyrev, V.V., Dais, D.I.: Preprint - Batyrev, V.V. (rec. October 1994), 39 pp. [alg-geom/9410001]
31. Batyrev, V.V., van Straten, D.: Generalized hypergeometric functions and rational curves on Calabi-Yau complete intersections in toric varieties. Commun. Math. Phys. **168**, 493 (1995). https://doi.org/10.1007/BF02101841 [alg-geom/9307010]
32. Batyrev, V.I., Ciocan-Fontanine, I., Kim, B., Straten, D.v.: Conifold transitions and mirror symmetry for Calabi-Yau complete intersections in Grassmannians (1997). alg-geom/9710022

33. Becker, K., Becker, M., Strominger, A.: Fivebranes, membranes and non-perturbative string theory (1995). hep-th/9507158
34. Behrend, K.: Donaldson-Thomas type invariants via microlocal geometry. Ann. Math. **170**, 1307–1338 (2009)
35. Belavin, A., Gepner, D., Kononov, Y.: Flat coordinates for Saito Frobenius manifolds and String theory. Theor. Math. Phys. **189**(3), 1775 (2016) [Teor. Mat. Fiz. **189**(3), 429 (2016)]. https://doi.org/10.1134/S0040577916120096 [arXiv:1510.06970 [hep-th]]
36. Benini, F., Cremonesi, S.: Partition functions of $\mathcal{N} = (2, 2)$ gauge theories on S^2 and vortices. Commun. Math. Phys. **334**(3), 1483 (2015). https://doi.org/10.1007/s00220-014-2112-z [arXiv:1206.2356 [hep-th]]
37. Berglund, P., Katz, S., Klemm, A.: Mirror symmetry and the moduli space for generic hypersurfaces in toric varieties. Nucl. Phys. B **456**, 153 (1995). [arXiv:hep-th/9506091]
38. Berry, D.W.: Quantal phase factors accompanying adiabatic changes. Proc. R. Soc. Lond. Ser. A Math. Phys. Sci. **392**, 45 (1984)
39. Bershadsky, M., Cecotti, S., Ooguri, H., Vafa, C.: Holomorphic anomalies in topological field theories. Nucl. Phys. B **405**, 279 (1993). [arXiv:hep-th/9302103]
40. Bershadsky, M., Cecotti, S., Ooguri, H., Vafa, C.: Kodaira-Spencer theory of gravity and exact results for quantum string amplitudes. Commun. Math. Phys. **165**, 311 (1994). [arXiv:hep-th/9309140]
41. Bershadsky, M., Vafa, C., Sadov, V.: D-Branes and topological field theories. Nucl. Phys. B **463**, 420 (1996). [arXiv:hep-th/9511222]
42. Besse, A.L.: Einstein Manifolds. Ergebnisse der Mathematik und ihrer Grenzgebiete (3), 510 p. Springer, Berlin (1987)
43. Bilal, A.: Lectures on anomalies (2008). arXiv:0802.0634 [hep-th]
44. Bismut, J.-M., Freed, D.S.: The analysis of elliptic families. II. Dirac operators, eta invariants, and the holonomy theorem. Commun. Math. Phys. **107**, 103–163 (1986)
45. Bismut, J.-M., Gillet, H., Soul, C.é, Analytic torsion and holomorphic determinant bundles. I. Bott-Chern forms and analytic torsion. Commun. Math. Phys. **115**, 49–78 (1988)
46. Bizet, N.C., Klemm, A., Lopes, D.V.: Landscaping with fluxes and the E8 Yukawa Point in F-theory (2014). arXiv:1404.7645 [hep-th]
47. Blumenhagen, R., Lüst, D., Theisen, S.: (2013). https://doi.org/10.1007/978-3-642-29497-6
48. Borcherds, R.: Automorphic forms with singularities on Grassmannians (1996). alg-geom/9609022
49. Bott, R., Tu, L.W.: Differential Forms in Algebraic Topology. Springer, Berlin (1982)
50. Bouchard, V., Klemm, A., Marino, M., Pasquetti, S.: Remodeling the B-model. Commun. Math. Phys. **287**, 117 (2009). https://doi.org/10.1007/s00220-008-0620-4 [arXiv:0709.1453 [hep-th]]
51. Bouchard, V., Klemm, A., Marino, M., Pasquetti, S.: Commun. Math. Phys. **296**, 589 (2010). https://doi.org/10.1007/s00220-010-1020-0 [arXiv:0807.0597 [hep-th]]
52. Bridgeland, T.: Stability conditions on triangulated categories. Ann. Math. (2) **166**, 317–345 (2007)
53. Brini, A., Eynard, B., Marino, M.: Torus knots and mirror symmetry. Annales Henri Poincare **13**, 1873 (2012). https://doi.org/10.1007/s00023-012-0171-2 [arXiv:1105.2012 [hep-th]]
54. Bruinier, J.H., van der Geer, G., Harder, G., Zagier, D.: The 1-2-3 of Modular Forms. Springer Univesitext. Springer, Berlin (2008)
55. Brunner, I., Douglas, M.R., Lawrence, A.E., Romelsberger, C.: J. High Energy Phys. **0008**, 015 (2000) [arXiv:hep-th/9906200]
56. Bryan, J., Graber, T., Pandharipande, R.: The orbifold quantum cohomology of C^2/Z_3 and Hurwitz-Hodge integrals (2005). [arXiv:math.AG/0510335]
57. Candelas, P., del la Ossa, X.: Lectures on complex manifolds. In: Alvarez-Gaume, L., et al. (eds.) Superstrings '87, Proceedings of the Trieste Spring School. World Scientific, Singapore (1987)
58. Candelas, P., de la Ossa, X.: Moduli space Of Calabi-Yau manifolds. Nucl. Phys. B **355**, 455 (1991)

59. Candelas, P., Freeman, M.D., Pope, C.N., Sohnius, M.F., Stelle, K.S.: Higher order corrections to supersymmetry and compactifications of the heterotic string, Phys. Lett. B **177**, 341 (1986). https://doi.org/10.1016/0370-2693(86)90764-1
60. Candelas, P., Green, P.S., Hubsch, T.: Rolling Among Calabi-Yau Vacua. Nucl. Phys. B **330**, 49 (1990)
61. Candelas, P., De La Ossa, X.C., Green, P.S., Parkes, L.: A pair Of Calabi-Yau manifolds as an exactly soluble superconformal theory. Nucl. Phys. B **359**, 21 (1991)
62. Candelas, P., Derrick, E., Parkes, L.: Generalized Calabi-Yau manifolds and the mirror of a rigid manifold. Nucl. Phys. B **407**, 115 (1993). https://doi.org/10.1016/0550-3213(93)90276-U [hep-th/9304045]
63. Candelas, P., De La Ossa, X., Font, A., Katz, S., Morrison, D.R.: Mirror symmetry for two parameter models. I. Nucl. Phys. B **416**, 481 (1994). [arXiv:hep-th/9308083]
64. Candelas, P., Font, A., Katz, S., Morrison, D.R.: Mirror symmetry for two parameter models. 2. Nucl. Phys. **B429**, 626–674 (1994). arXiv:hep-th/9403187
65. Candelas, P., de la Ossa, X., Rodriguez-Villegas, F.: Calabi-Yau manifolds over finite fields. I& II (2000,) arXiv:hep-th/0012233. (2004) arXiv:hep-th/0402133
66. Capelli, A., Itzykson, C., Zuber, J.B.: Modular invariant partition functions in two dimensions. Nucl. Phys. **B280**, 445 [FS 18] (1987); Capelli, A.: Modular invariant partition functions of superconformal theories. Phys. Lett. **B215**, 82 (1987)
67. Cecotti, S.: Geometry of N=2 Landau-Ginzburg families. Nucl. Phys. B **355**, 755 (1991)
68. Cecotti, S., Vafa, C.: Topological anti-topological fusion. Nucl. Phys. B **367**, 369 (1991)
69. Cecotti, S., Vafa, C.: Ising model and N=2 supersymmetric theories. Commun. Math. Phys. **157**, 139 (1993) [arXiv:hep-th/9209085]
70. Cecotti, S., Ferrara, S., Giradello, L.: Goemetry of Type II superstrings and the moduli of superconformal field theories. Int. J. Mod. Phys. **A4**, 2475 (1989)
71. Ceresole, A., D'Auria, R., Ferrara, S., Lerche, W., Louis, J.: Picard-Fuchs equations and special geometry. Int. J. Mod. Phys. A **8**, 79 (1993). https://doi.org/10.1142/S0217751X93000047 [hep-th/9204035]
72. Ceresole, A., D'Auria, R., Ferrara, S.: The Symplectic structure of N=2 supergravity and its central extension. Nucl. Phys. Proc. Suppl. **46**, 67 (1996). [hep-th/9509160]; Andrianopoli, L., Bertolini, M., Ceresole, A., D 'Auria, R., Ferrara, S., Fre, P., Magri, T.: N=2 supergravity and N=2 superYang-Mills theory on general scalar manifolds: dymplectic covariance, gaugings and the momentum map. J. Geom. Phys. **23**, 111 (1997). [hep-th/9605032]
73. Chen, W., Ruan, Y.: Orbifold Gromow-Witten theory. In: Orbifolds in Mathematics and Physics, Madison WI 2001. Contemporary Mathematics, vol. 310, pp. 25–85. American Mathematical Society, Providence (2002)
74. Choi, J., Katz, S., Klemm, A.: The refined BPS index from stable pair invariants. Commun. Math. Phys. **328**, 903 (2014). https://doi.org/10.1007/s00220-014-1978-0 [arXiv:1210.4403 [hep-th]]
75. Coates, T., Corti, A., Iritani, H., Tseng, H.-H.: Wall-crossing in toric Gromow-Witten theory I: crepant examples [arXiv:math.AG/0611550]
76. Cornalba, M., Griffiths, P.A.: Some transcendental aspects of algebraic geometry. In: Algebraic geometry. Proceedings of Symposia in Pure Mathematics, vol. 29, pp. 3–110. American Mathematical Society, Providence (1974)
77. Cox, D.: The Homogeneous coordinate ring of a toric variety. J. Algebraic Geom. **4**, 17 (1995) [alg-geom/9210008]
78. Cox, D.A., Katz, S.: Mirror Symmetry And Algebraic Geometry. Mathematical Surveys and Monographs, vol. 68. American Mathematical Society, Providence (1999)
79. Cox, D.A., Little, J.B., Schenck, H.K.: Toric Varieties. Graduate Studies in Mathematics, vol. 124. American Mathematical Society, Providence (2011)
80. Cremmer, E., Julia, B., Scherk, J.: Supergravity theory in eleven-dimensions. Phys. Lett. B **76**, 409 (1978)
81. Cremmer, E., et al.: Vector multiplets coupled to N=2 supergravity: superhiggs effect, flat potentials and geomtric structures. Nucl. Phys. **250**, 385 (1985)

82. Dabholkar, A., Murthy, S., Zagier, D.: Quantum black holes, wall crossing, and mock modular forms. arXiv:1208.4074 [hep-th]
83. de Wit, B., van Proyen, A.: Potentials and symmetries of general gauged N=2 supergravity-Yang-Mills models. Nucl. Phys. **245**, 89 (1984)
84. de Wit, B., Lauwers, P.G., Van Proyen, A.: Lagrangians of N=2 supergravity matter systems. Nucl. Phys. **B255**, 569 (1980)
85. Del Zotto, M., Gu, J., Huang, M.X., Kashani-Poor, A.K., Klemm, A., Lockhart, G.: J. High Energy Phys. **1803**, 156 (2018) https://doi.org/10.1007/JHEP03(2018)156. [arXiv:1712.07017 [hep-th]]
86. Deligne, P.: Équations différentielles à points singuliers réguliers. Lecture Notes in Mathematics, vol. 163. Springer, Berlin-New York (1970)
87. Deligne, P.: Théorie de Hodge. III. Inst. Hautes Études Sci. Publ. Math. **44**, 5–77 (1974)
88. Deligne, P.: Local Behavior of Hodge Structures at Infinity, pp. 683–699. Mirror Symmetry II. American Mathematical Society, Providence (1997)
89. Di Francesco, P., Itzykson, C., Zuber, J.-B.: Classical W-algebras. Commun. Math. Phys. **140**(3), 543–567 (1991)
90. Di Francesco, P., Mathieu, P., Sénéchal, D.: Conformal Field Theory. Springer, Berlin (1997)
91. D'Hoker, E., Phong, P.H.: The geometry of string perturbation theory. Rev. Mod. Phys. **60**, 917 (1988)
92. Dijkgraaf, R.: Intersection theory, integrable hierarchies and topological field theory (1992). arXiv:hep-th/9201003
93. Dijkgraaf, R., Vafa, C.: Matrix model, topological strings and supersymmetric gauge theories. Nucl. Phys. **B644**, 3 (2002). arXiv:hep-th/0206255
94. Dijkgraaf, R., Vafa, C.: On geometry and matrix models. Nucl. Phys. **B644**, 21 (2002). arXiv:hep-th/0207106
95. Dijkgraaf, R., Verlinde, H., Verlinde, E.: Topological strings in d<1. Nucl. Phys. B **352**, 59 (1991)
96. Dijkgraaf, R., Verlinde, E.P., Vonk, M.: On the partition sum of the NS five-brane (2002). hep-th/0205281
97. Dixon, L.J., Kaplunovsky, V., Louis, J.: Moduli dependence of string loop corrections to gauge coupling constants, Nucl. Phys. B **355**, 649 (1991)
98. Dolgachev, V.: Weighted Projective Spaces. Lectures Notes in Mathematics. Springer, Berlin (1982)
99. Donagi, R., Witten, E.: Supermoduli space is not projected. Proc. Symp. Pure Math. **90**, 19 (2015) [arXiv:1304.7798 [hep-th]]
100. Donaldson, S.: Riemann Surfaces. Oxford Graduate Texts in Mathematics, vol. 22. Oxford University Press, Oxford (2011)
101. Donaldson, S.K., Thomas, R.P.: Gauge theory in higher dimensions, In: The Geometric Universe, pp. 31–47. Oxford University Press, Oxford (1998)
102. Doroud, N., Gomis, J., Le Floch, B., Lee, S.: Exact results in D=2 supersymmetric gauge theories. J. High Energy Phys. **1305**, 093 (2013). https://doi.org/10.1007/JHEP05(2013)093 [arXiv:1206.2606 [hep-th]]
103. Duff, M.J., Nilsson, B.E.W., Pope, C.N.: Kaluza-Klein Supergravity. Phys. Rep. **130**, 1 (1986)
104. Dwork, B.: On the zeta function of a hypersurface. II. Ann. Math. (2) **80**, 227–299 (1964)
105. Eguchi, T., Yang, S.-K.: N=2 superconformal models as topological field theories. Mod. Phys. Lett. **A5** 1693 (1990)
106. Eguchi, T., Gilkey, P.B., Hanson, A.: Gravitation, gauge theory and differential geometry. Phys. Rep. **66**, 213 (1980)
107. Eichler, M., Zagier, D.: The Theory of Jacobi Forms. Progress in Mathematics. Birkhäuser, Boston (1985)
108. Erler, J., Klemm, A.: Comment on the generation number in orbifold compactifications. Commun. Math. Phys. **153**, 579 (1993) [arXiv:hep-th/9207111]
109. Eynard, B., Marino, M.: J. Geom. Phys. **61**, 1181 (2011). https://doi.org/10.1016/j.geomphys.2010.11.012 [arXiv:0810.4273 [hep-th]]

110. Eynard, B., Orantin, N.: Commun. Num. Theor. Phys. **1**, 347 (2007). https://doi.org/10.4310/CNTP.2007.v1.n2.a4 [math-ph/0702045]
111. Faber, C., Pandharipande, R.: Hodge integrals and Gromow-Witten theory. Invent. Math.**139**, 173 (2000). math.AG./9810173
112. Ferrara, S., Kallosh, R., Strominger, A.: N=2 extremal black holes. Phys. Rev. **D52**, 5412 (1995). hep-th/9508072
113. Ferrara, S., Harvey, J.A., Strominger, A., Vafa, C.: Second quantized mirror symmetry. Phys. Lett. B **361**, 59 (1995). [arXiv:hep-th/9505162]
114. Ferrara, S., Gibbons, G.W., Kallosh, R.: Black holes and critical points in moduli space. Nucl. Phys. B **500**, 75 (1997)
115. Fischbach, F., Klemm, A., Nega, C.: (2018). arXiv:1803.11222 [hep-th]
116. Freed, D.S.: Special Kaehler manifolds. Commun. Math. Phys. **203**, 31 (1999) [arXiv:hep-th/9712042]
117. Freeman, M.D., Pope, C.N.: Beta functions and superstring compactifications. Phys. Lett. B **174**, 48 (1986). https://doi.org/10.1016/0370-2693(86)91127-5
118. Freeman, M.D., Pope, C.N., Sohnius, M.F., Stelle, K.S.: Higher order σ model counterterms and the effective action for superstrings. Phys. Lett. B **178**, 199 (1986). https://doi.org/10.1016/0370-2693(86)91495-4
119. Friedan, D.: Introduction to Polykov's String Theory. Les Houches (1982)
120. Friedan, D., Windey, P.: Nucl. Phys. **B235**, 395 (1984)
121. Fuchs, J., Klemm, A., Scheich, C., Schmidt, M.G.: Spectra and symmetries of Gepner models compared to Calabi-Yau compactifications. Ann. Phys. **204**, 1 (1990)
122. Fujikawa, K.: Path integral for gauge theories with fermions. Phys. Rev. **D21**, 2848 (1980)
123. Fujikawa, K.: Path integral for relativistic strings. Phys. Rev. **D25**, 2584 (1981)
124. Fujikawa, K., Suzuki, H.: Path Integrals and Quantum Anomalies. Clarendon, Oxford (2004)
125. Fulton, W.: Introduction to Toric Varieties. Princeton University Press, Princeton (1993)
126. Fulton, W.: Intersection Theory. Springer, Berlin (1998)
127. Gaiotto, D., Strominger, A., Yin, X.: New connections between 4D and 5D black holes. J. High Energy Phys. **0602**, 024 (2006). hep-th/0503217
128. Gates, S.J., Hull, C.M., Rocek, M.: Twisted multiplets and new supersymmetric nonlinear sigma models. Nucl. Phys. B **248**, 157 (1984)
129. Gauntlett, P.J.: Branes, calibrations and supergravity. In: Strings and Geometry. Clay Mathematics Proceedings, vol. 3, pp. 79–126. American Mathematical Society, Providence (2004)
130. Gelfand, I.M., Graev, M.I., Zelevinskiĭ, A.V.: Holonomic systems of equations and series of hypergeometric type. Dokl. Akad. Nauk SSSR **295**(1), 14–19 (1987)
131. Gelfand, I.M., Zelevinskiĭ, A.V., Kapranov, M.M.: Equations of hypergeometric type and Newton polyhedra. Dokl. Akad. Nauk SSSR **300**(3), 529–534 (1988)
132. Gepner, D.: Lectures on N=2 String Theory. Lectures at Spring School on Superstrings, Trieste, April 3–14, 1989
133. Gepner, D.: Spacetime supersymmetries in compactified string theory and superconformal models. Nucl. Phys. **B296**, 757 (1989)
134. Gerasimov, A.A., Shatashvili, S.: Towards integrability of topological strings I: three-forms on Calabi-Yau manifolds. arXiv:hep-th/0409238
135. Ghoshal, D., Vafa, C.: C = 1 string as the topological theory of the conifold. Nucl. Phys. B **453**, 121 (1995) [arXiv:hep-th/9506122]
136. Givental, A.B.: Equivariant Gromov-Witten invariants. Int. Math. Res. Not. **13**, 613–663 (1996)
137. Givental, A.B.: A mirror theorem for toric complete intersections. Kashiwara, M., Matsuo, A., Saito, K., Satake, I.I. (eds.) Topological Field Theory, Primitive Forms and Related Topics. Progress in Mathematics, vol. 160. Birkhäuser, Boston (1998)
138. Gopakumar, R., Vafa, C.: Branes and fundamental groups (1997) [arXiv:hep-th/9712048]
139. Gopakumar, R., Vafa, C.: M-theory and topological strings. I (1998) [arXiv:hep-th/9809187]
140. Gopakumar, R., Vafa, C.: M-theory and topological strings. II (1998) [arXiv:hep-th/9812127]
141. Green, M., Seiberg, N.: Nucl. Phys. **B299**, 559 (1988)

142. Green, M.B., Schwarz, J.H., Witten, E.: Superstring Theory. Loop Amplitudes, Anomalies and Phenomenology, vol. 2, 596 pp. Cambridge Monographs On Mathematical Physics. Cambridge University Press, Cambridge (1987)
143. Greenberg, M.J., Harper, J.R.: Algerbraic geometry: a first course. Benjamin/Cummings, Reading (1981)
144. Greene, B.R., Plesser, M.R.: Duality in Calabi-Yau moduli space. Nucl. Phys. B **338**, 15 (1990)
145. Greene, B.R., Morrison, D.R., Strominger, A.: Black hole condensation and the unification of string vacua. Nucl. Phys. B **451**, 109 (1995). https://doi.org/10.1016/0550-3213(95)00371-X [hep-th/9504145]
146. Greene, B.R., Morrison, D.R., Plesser, M.R.: Mirror manifolds in higher dimension. Commun. Math. Phys. **173**, 559 (1995) [AMS/IP Stud. Adv. Math. **1**, 745 (1996)]. https://doi.org/10.1007/BF02101657 [hep-th/9402119]
147. Griffiths, P.: On the periods of certain rational integrals I and II. Ann. Math. **90**, 460 and 498 (1969)
148. Griffiths, P. (ed.) Topics in Transcendental Algebraic Geometry. Annals of Mathematical Studies. Princeton University Press, Princeton (1984)
149. Griffiths, P.P., Harris, J.K.: Principles of Algebraic Geometry, vol. 197. Wiley Online Library (1978)
150. Griffiths, P., Harris, J.: Principles of Algebraic Geometry. Wiley, New York (1978)
151. Grimm, T.W., Klemm, A., Marino, M., Weiss, M.: Direct integration of the topological string. J. High Energy Phys. **0708**, 058 (2007). https://doi.org/10.1088/1126-6708/2007/08/058 [hep-th/0702187 [HEP-TH]]
152. Grimm, T.W., Ha, T.-W., Klemm, A., Klevers, D.: The D5-brane effective action and superpotential in N=1 compactifications. Nucl. Phys. B **816**, 139 (2009) [arXiv:0811.2996 [hep-th]]
153. Grimm, T.W., Ha, T.W., Klemm, A., Klevers, D.: Computing Brane and flux superpotentials in F-theory compactifications. J. High Energy Phys. **1004**, 015 (2010). https://doi.org/10.1007/JHEP04(2010)015 [arXiv:0909.2025 [hep-th]]
154. Gross, D.J., Klebanov, I.R.: One-dimensional string theory on a circle. Nucl. Phys. B **344**, 475 (1990)
155. Gross, M., Huybrechts, D., Joyce, D.: Calabi-Yau Manifolds and Related Geometries. Springer, Berlin (2003)
156. Grothendieck, A.: Fondements de la Géométrie Algébrique Commentaires. Seminaire Bourbaki **7**, 297–298 (1957–1962)
157. Gu, J., Jockers, H., Klemm, A., Soroush, M.: Knot invariants from topological recursion on augmentation varieties. Commun. Math. Phys. **336**(2), 987 (2015). https://doi.org/10.1007/s00220-014-2238-z [arXiv:1401.5095 [hep-th]]
158. Gu, J., Huang, M.x., Kashani-Poor, A.K., Klemm, A.: Refined BPS invariants of 6d SCFTs from anomalies and modularity. J. High Energy Phys. **1705**, 130 (2017). https://doi.org/10.1007/JHEP05(2017)130 [arXiv:1701.00764 [hep-th]]
159. Guio, T.C.d.C., Jockers, H., Klemm, A., Yeh, H.Y.: Effective action from M-theory on twisted connected sum G_2-manifolds. https://doi.org/10.1007/s00220-017-3045-0. arXiv:1702.05435 [hep-th]
160. Gunaydin, M., Neitzke, A., Pioline, B.: J. High Energy Phys. **0612**, 070 (2006). https://doi.org/10.1088/1126-6708/2006/12/070 [hep-th/0607200]
161. Haghighat, B., Klemm, A.: Topological strings on Grassmannian Calabi-Yau manifolds. J. High Energy Phys. **0901**, 029 (2009). https://doi.org/10.1088/1126-6708/2009/01/029 [arXiv:0802.2908 [hep-th]]
162. Haghighat, B., Klemm, A.: Solving the Topological String on K3 Fibrations. J. High Energy Phys. **1001**, 009 (2010) https://doi.org/10.1007/JHEP01(2010)009. [arXiv:0908.0336 [hep-th]]
163. Haghighat, B., Klemm, A., Rauch, M.: Integrability of the holomorphic anomaly equations. J. High Energy Phys. **0810**, 097 (2008). https://doi.org/10.1088/1126-6708/2008/10/097. [arXiv:0809.1674 [hep-th]]

164. Haghighat, B., Lockhart, G., Vafa, C.: Fusing E-strings to heterotic strings: E+E?H. Phys. Rev. D **90**(12), 126012 (2014). https://doi.org/10.1103/PhysRevD.90.126012. [arXiv:1406.0850 [hep-th]]
165. Halverson, J., Jockers, H., Lapan, J.M., Morrison, D.R.: Perturbative corrections to Kähler moduli spaces. Commun. Math. Phys. **333**(3), 1563 (2015). https://doi.org/10.1007/s00220-014-2157-z [arXiv:1308.2157 [hep-th]]
166. Haghighat, B., Klemm, A., Lockhart, G., Vafa, C.: Strings of minimal 6d SCFTs. Fortsch. Phys. **63**, 294 (2015). https://doi.org/10.1002/prop.201500014 [arXiv:1412.3152 [hep-th]]
167. Harris, J.: Curves in Projective Space. Séminaire de mathématiques supérieures. Les Presses de l'Université de Montréal, Montréal (1982)
168. Harvey, R., Lawson, H.B.: Calibrated geometries. Acta Math,**148**, 47 (1982)
169. Harvey, J.A., Moore, G.W.: Algebras, BPS states, and strings. Nucl. Phys. B **463**, 315 (1996) [arXiv:hep-th/9510182]
170. Harvey, J.A., Moore, G.W.: Conway subgroup symmetric compactifications of heterotic string (2018). arXiv:1712.07986 [hep-th]
171. Heckman, J.J., Morrison, D.R., Vafa, C.: On the classification of 6D SCFTs and generalized ADE orbifolds. J. High Energy Phys. **1405**, 028 (2014); Erratum: [JHEP **1506**, 017 (2015)]. https://doi.org/10.1007/JHEP06(2015)017,10.1007/JHEP05(2014)028 [arXiv:1312.5746 [hep-th]]
172. Heckman, J.J., Morrison, D.R., Rudelius, T., Vafa, C.: Fortsch. Phys. **63**, 468 (2015). https://doi.org/10.1002/prop.201500024 [arXiv:1502.05405 [hep-th]]
173. Hertling, K.: Frobenius Manifolds and Moduli Spaces for Singularities. Cambridge University Press, Cambridge (2002)
174. Hertling, C., Sabbah, C.: Examples of non-commutative Hodge structures. J. Inst. Math. Jussieu. **10**, 635–674 (2011)
175. Hironaka, H.: Resolutions of singularities of an algebraic variety over a field of characteristic zero I, II. Ann. Math. **79**, 109 (1964)
176. Hirzebruch, F.: Topological Methods in Algebraic Topology, 3rd edn. Springer, Berlin (1995)
177. Hitchin, N.: The Geometry of three-forms in six and seven dimensions (2000). arXiv:math.DG/0010054
178. Hitchin, N., Generalized Calabi-Yau manifolds. Q. J. Math. **54**, 281 (2003)
179. Hofmann, J., van Straten, D., Some monodromy groups of finite index in $Sp_4(\mathbb{Z})$. J. Aust. Math. Soc. **99**(1), 48–62 (2015). arXiv:1312.3063
180. Horava, P., Witten, E.: Heterotic and type I string dynamics from eleven-dimensions. Nucl. Phys. B **460**, 506 (1996). https://doi.org/10.1016/0550-3213(95)00621-4. [hep-th/9510209]
181. Hori, K., Romo, M.: Exact results in two-dimensional (2,2) supersymmetric gauge theories with boundary (2013). arXiv:1308.2438 [hep-th]
182. Hori, K., Katz, S., Klemm, A., Pandharipande, R., Thomas, R., Vafa, C., Vakil, R., Zaslow, E.: Mirror Symmetry. American Mathemtical Society, Providence (2003)
183. Hosono, S.: Adv. Theor. Math. Phys. **4**, 335 (2000) [hep-th/0007071]
184. Hosono, S., Konishi, Y.: Higher genus Gromov-Witten invariants of the Grassmannian, and the Pfaffian Calabi-Yau threefolds [arXiv:math.AG/0704.2928]
185. Hosono, S., Klemm, A., Theisen, S.: Lectures on mirror symmetry. Lect. Notes Phys. **436**, 235 (1994) [hep-th/9403096]
186. Hosono, S., Klemm, A., Theisen, S., Yau, S.T.: Mirror symmetry, mirror map and applications to Calabi-Yau hypersurfaces. Commun. Math. Phys. **167**, 301 (1995) [arXiv:hep-th/9308122]
187. Hosono, S., Klemm, A., Theisen, S., Yau, S.T.: Mirror symmetry, mirror map and applications to complete intersection Calabi-Yau spaces. Nucl. Phys. B **433**, 501 (1995) [arXiv:hep-th/9406055]
188. Hosono, S., Saito, M.H., Takahashi, A.: Holomorphic anomaly equation and BPS state counting of rational elliptic surface. Adv. Theor. Math. Phys. **3**, 177 (1999) [arXiv:hep-th/9901151]

189. Huang, M.x., Klemm, A.: Holomorphic anomaly in gauge theories and matrix models. J. High Energy Phys. **0709**, 054 (2007). https://doi.org/10.1088/1126-6708/2007/09/054 [hep-th/0605195]
190. Huang, M.x., Klemm, A., Marino, M., Tavanfar, A.: Black holes and large order quantum geometry. Phys. Rev. D **79**, 066001 (2009) https://doi.org/10.1103/PhysRevD.79.066001 [arXiv:0704.2440 [hep-th]]
191. Huang, M.x., Klemm, A., Quackenbush, S.: Topological string theory on compact Calabi-Yau: Modularity and boundary conditions. Lect. Notes Phys. **757**, 45 (2009). https://doi.org/10.1007/978-3-540-68030-73 [hep-th/0612125]
192. Huang, M.x., Klemm, A.: Direct integration for general Ω backgrounds. Adv. Theor. Math. Phys. **16**(3), 805 (2012). https://doi.org/10.4310/ATMP.2012.v16.n3.a2 [arXiv:1009.1126 [hep-th]]
193. Huang, M.X., Klemm, A., Poretschkin, M.: Refined stable pair invariants for E-, M- and $[p, q]$-strings. J. High Energy Phys. **1311**, 112 (2013). https://doi.org/10.1007/JHEP11(2013)112 [arXiv:1308.0619 [hep-th]]
194. Huang, M.x., Kashani-Poor, A.K., Klemm, A.: The Ω deformed B-model for rigid $\mathcal{N} = 2$ theories. Annales Henri Poincare **14**, 425 (2013). https://doi.org/10.1007/s00023-012-0192-x [arXiv:1109.5728 [hep-th]]
195. Huang, M.x., Katz, S., Klemm, A.: Topological string on elliptic CY 3-folds and the ring of Jacobi forms. J. High Energy Phys. **1510**, 125 (2015). https://doi.org/10.1007/JHEP10(2015)125 [arXiv:1501.04891 [hep-th]]
196. Huang, M.x., Katz, S., Klemm, A.: Elliptifying topological string theory (work in progress)
197. Hull, C.M., Townsend, P.K.: Nucl. Phys. B **438**, 109 (1995). https://doi.org/10.1016/0550-3213(94)00559-W [hep-th/9410167]
198. Iqbal, A., Kozcaz, C., Vafa, C.: The Refined topological vertex. J. High Energy Phys. **0910**, 069 (2009). https://doi.org/10.1088/1126-6708/2009/10/069 [hep-th/0701156]
199. Iritani, H.: Ruan's conjecture and integral structures in quantum cohomology (2008). arXiv/math.AG/0809.2749
200. Iskovskih, V.A.: Fano 3-Folds I. Math. USSR-Izv. **11**, 485–527 (1977); Erratum, Fano 3-Folds II. Math. USSR-Izv. **12**, 469–506 (1978)
201. Jockers, H., Kumar, V., Lapan, J.M., Morrison, D.R., Romo, M.: Two-sphere partition functions and Gromov-Witten invariants. Commun. Math. Phys. **325**, 1139 (2014). https://doi.org/10.1007/s00220-013-1874-z [arXiv:1208.6244 [hep-th]]
202. Joyce, D.D.: Compact Riemannian 7-manifolds with holonomy G_2. I, II. J. Differ. Geom. **43**, 291–328, 329–375 (1996)
203. Joyce, D.D.: Compact Manifolds with Special Holonomy. Oxford University Press, Oxford (2000)
204. Kachru, S., Vafa, C.: Exact results for N=2 compactifications of heterotic strings. Nucl. Phys. B **450**, 69 (1995). [arXiv:hep-th/9505105]
205. Kachru, S., Klemm, A., Lerche, W., Mayr, P., Vafa, C.: Nonperturbative results on the point particle limit of N=2 heterotic string compactifications. Nucl. Phys. B **459**, 537 (1996). [arXiv:hep-th/9508155]
206. Kachru, S., Katz, S.H., Lawrence, A.E., McGreevy, J.: Mirror symmetry for open strings. Phys. Rev. D **62**, 126005 (2000). https://doi.org/10.1103/PhysRevD.62.126005 [hep-th/0006047]
207. Kachru, S., Katz, S.H., Lawrence, A.E., McGreevy, J.: Open string instantons and superpotentials. Phys. Rev. D **62**, 026001 (2000). https://doi.org/10.1103/PhysRevD.62.026001 [hep-th/9912151]
208. Kaneko, M., Zagier, D.: A generalized Jacobi theta function and quasimodular forms. In: The Moduli Space of Curves (Texel Island, 1994). Progress in Mathematics, vol. 129, pp. 165–172. Birkhäuser, Boston (1995)
209. Kaplunovsky, V., Louis, J.: On Gauge couplings in string theory. Nucl. Phys. B **444**, 191 (1995) [arXiv:hep-th/9502077]
210. Kapustin, A., Li, Y.: Topological sigma-models with H-flux and twisted generalized complex manifolds (2004). arXiv:hep-th/0407249

211. Katz, N.M.: On differential equations satisfied by period matrices. Publ. Math. IHES **35**, 71–106 (1968)
212. Katz, S., Liu, C.-C.M.: Enumerative geometry of stable maps with Lagrangian boundary conditions and multiple covers of the disc. Adv. Theor. Math. Phys. **5**, 1–49 (2001)
213. Katz, S., Morrison, D.R., Ronen Plesser, M.: Enhanced gauge symmetry in type II string theory. Nucl. Phys. B **477**, 105 (1996) [arXiv:hep-th/9601108]
214. Katz, S.H., Klemm, A., Vafa, C.: Geometric engineering of quantum field theories. Nucl. Phys. B **497**, 173 (1997). https://doi.org/10.1016/S0550-3213(97)00282-4 [hep-th/9609239]
215. Katz, S., Klemm, A., Vafa, C.: M-theory, topological strings and spinning black holes. Adv. Theor. Math. Phys. **3**, 1445 (1999) [arXiv:hep-th/9910181]
216. Katz, S., Klemm, A., Pandharipande, R.: On the motivic stable pairs invariants of K3 surfaces (2014). arXiv:1407.3181 [math.AG]
217. Katzarkov, L., Kontsevich, M., Pantev, T.: Hodge theoretic aspects of mirror symmetry. In: From Hodge theory to integrability and TQFT tt*-geometry. Proceedings of Symposia in Pure Mathematics, vol. 78, pp. 87–174 (2008)
218. Kawai, T.: String duality and modular forms. Phys. Lett. B **397**, 51 (1997) [arXiv:hep-th/9607078]
219. Kawai, T., Yoshioka, K.: String partition function and infinite products [arXiv:hep-th/0002169]
220. Kazama, Y., Suzuki, H.: Characterization Of N=2 superconformal models generated by coset space method. Phys. Lett. B **216**, 112 (1989); New N=2 superconformal field theories and superstring compactification. Nucl. Phys. B **321**, 232 (1989)
221. Klemm, A.: On the geometry behind N = 2 supersymmetric effective actions in four dimensions (1997). arXiv:hep-th/9705131
222. Klemm, A., Marino, M.: Counting BPS states on the enriques Calabi-Yau. Commun. Math. Phys. **280**, 27 (2008). https://doi.org/10.1007/s00220-007-0407-z [hep-th/0512227]
223. Klemm, A., Mayr, P.: Strong coupling singularities and non-abelian gauge symmetries in $N = 2$ string theory. Nucl. Phys. B **469**, 37 (1996) [arXiv:hep-th/9601014]
224. Klemm, A., Pandharipande, R.: Enumerative geometry of Calabi-Yau 4-folds. Commun. Math. Phys. **281**, 621 (2008). https://doi.org/10.1007/s00220-008-0490-9 [math/0702189]
225. Klemm, A., Schimmrigk, R.: Landau-Ginzburg string vacua. Nucl. Phys. B **411**, 559 (1994). [arXiv:hep-th/9204060]; Kreuzer, M., Skarke, H.: On the classification of quasihomogeneous functions. Commun. Math. Phys. **150**, 137 (1992) [arXiv:hep-th/9202039]
226. Klemm, A., Theisen, S.: Considerations of one modulus Calabi-Yau compactifications: Picard-Fuchs equations, Kahler potentials and mirror maps. Nucl. Phys. B **389**, 153 (1993) [arXiv:hep-th/9205041]
227. Klemm, A., Theisen, S.: Mirror maps and instanton sums for complete intersections in weighted projective space. Mod. Phys. Lett. **A9**, 1807 (1994) [hepth/9304034]
228. Klemm, A., Zaslow, E.: Local mirror symmetry at higher genus. In: Winter School on Mirror Symmetry, Vector Bundles and Lagrangian Submanifolds (Cambridge, MA, 1999). AMS/IP Studies in Advanced Mathematics, vol. 23, pp. 183–207 (2001) [arXiv:hep-th/9906046]
229. Klemm, A., Lerche, W., Mayr, P.: K3 Fibrations and heterotic type II string duality. Phys. Lett. B **357**, 313 (1995) [arXiv:hep-th/9506112]
230. Klemm, A., Mayr, P., Vafa, C.: BPS states of exceptional non-critical strings (1996). [arXiv:hep-th/9607139]
231. Klemm, A., Lerche, W., Mayr, P., Vafa, C., Warner, N.P.: Selfdual strings and N=2 supersymmetric field theory. Nucl. Phys. B **477**, 746 (1996). https://doi.org/10.1016/0550-3213(96)00353-7 [hep-th/9604034]
232. Klemm, A., Lian, B., Roan, S.S., Yau, S.T.: Calabi-Yau fourfolds for M- and F-theory compactifications. Nucl. Phys. B **518**, 515 (1998) [arXiv:hep-th/9701023]
233. Klemm, A., Kreuzer, M., Riegler, E., Scheidegger, E.: Topological string amplitudes, complete intersection Calabi-Yau spaces and threshold corrections. J. High Energy Phys. **0505**, 023 (2005). https://doi.org/10.1088/1126-6708/2005/05/023 [hep-th/0410018]

234. Klemm, A., Maulik, D., Pandharipande, R.R., Scheidegger, E.: Noether-Lefschetz theory and the Yau-Zaslow conjecture. J. Am. Math. Soc. **4**, 1013–1040 (2010)
235. Klemm, A., Manschot, J., Wotschke, T.: Quantum geometry of elliptic Calabi-Yau manifolds (2012). arXiv:1205.1795 [hep-th]
236. Klemm, A., Scheidegger, E., Zagier, D.: Periods and quasiperiods of modular forms and D-brane masses (in preparation)
237. Kodaira, K.: Complex Manifolds and Deformations of Complex Structures. Springer, Berlin (1985)
238. Kodaira, K.: Complex Manifolds and Deformation of Complex Structures. Classics in Mathematics, x+465 p. Springer, Berlin (2015)
239. Kontsevich, M.: Product formulas for modular forms on O(2,n) (after R.Borcherds) (1997) [alg-geom/9709006]
240. Kontsevich, M.: The Gamma genus. Talk at the Arbeitstagung Bonn, May 2013
241. Kovalev, A.: Twisted connected sums and special Riemannian holonomy. J. Reine Angew. Math. **565**, 125–160 (2003)
242. Krefl, D., Walcher, J.: (2010). arXiv:1010.2635 [hep-th]
243. Krefl, D., Walcher, J.: Lett. Math. Phys. **95**, 67 (2011). https://doi.org/10.1007/s11005-010-0432-2 [arXiv:1007.0263 [hep-th]]
244. Kreuzer, M., Skarke, H.: Complete classification of reflexive polyhedra in four dimensions. Adv. Theor. Math. Phys. **4**, 6 (2000) [hep-th/0002240]
245. Kreuzer, M., Skarke, H.: Complete classification of reflexive polyhedra in four-dimensions. Adv. Theor. Math. Phys. **4**, 1209 (2002) [hep-th/0002240]
Please check and correct if necessary.
246. Kutasov, D.: Geometry on the space of conformal field theories and contact terms. Phys. Lett. **B 220**, 153 (1989)
247. Labastida, J.M., Llatas, M.: Topological matter in two dimensions. Nucl. Phys. **B379**, 220 (1992). arXiv/hep-th/9112051
248. L-function and Modular Form Data Base (LMFDB) at http://www.lmfdb.org
249. Labastida, J.M.F., Marino, M.: Type B topological matter, Kodaira Spencer theory and mirror symmetry. Phys. Lett. **B333**, 386 (1994). arXiv/hep-th/9405151
250. Lamotke, K.: The topology of complex projective varieties after S. Lefschetz, Topology **20**(1), 15–51 (1981)
251. Lawson, H.B., Michelson, M.-L.: Spin Geometry. Princeton University Press, Princeton (1989)
252. Lerche, W., Vafa, C., Warner, N.P.: Chiral rings in N=2 superconformal theories. Nucl. Phys. **B324**, 371 (1989)
253. Lian, B.H., Liu, K.F., Yau, S.T.: Mirror principle. 1. Asian J. Math. **1**, 729 (1997)
254. Lian, B.H., Liu, K.F., Yau, S.T.: Mirror principle. 2. Asian J. Math. **3**, 109 (1999)
255. Lian, B.H., Liu, K.F., Yau, S.T.: Mirror principle. 3. Asian J. Math. **3**, 771 (1999)
256. Looijenga, E.: Trento notes on Hodge theory. Rend. Semin. Mat. Univ. Politec. Torino **69**, 113–148 (2011)
257. Lust, D., Theisen, S.: Lect. Notes Phys. **346**, 1 (1989). https://doi.org/10.1007/BFb0113507
258. Manin, Y.I.: Frobenius manifolds, quantum cohomology, and moduli spaces. Colloquium Publications of AMS, vol. 47. American Mathematical Society, Providence (1999)
259. Marino, M.: Open string amplitudes and large order behavior in topological string theory. J. High Energy Phys. **0803**, 060 (2008). https://doi.org/10.1088/1126-6708/2008/03/060 [hep-th/0612127]
260. Marino, M., Moore, G.W.: Counting higher genus curves in a Calabi-Yau manifold. Nucl. Phys. B **543**, 592 (1999) [arXiv:hep-th/9808131]
261. Maulik, D., Nekrasov, N., Okounkov, A., Pandharipande, R.: Gromow-Witten theory and Donaldson-Thomas theory I (2003) [arXiv:math.AG/0312059]
262. Mayr, P.: Mirror symmetry, N=1 superpotentials and tensionless strings on Calabi-Yau four folds. Nucl. Phys. B **494**, 489 (1997). https://doi.org/10.1016/S0550-3213(97)00196-X [hep-th/9610162]

263. Moore, G.W., Seiberg, N.: Polynomial equations for rational conformal field theories. Phys. Lett. B **212**, 451 (1988)
264. Mori, S., Mukai, S.: Classifications of Fano 3-folds with $B_2 \geq 2$? Algebraic and topological theories. Papers from the Symposium Dedicated to the Memory of Dr. Takehiko Miyata (Kinosaki, October 30–November 9, 1984). Kinokuniya Company Ltd., Tokyo (1986)
265. Morrison, D.R., Walcher, J.: D-branes and normal functions. Adv. Theor. Math. Phys. **13**(2), 553 (2009). https://doi.org/10.4310/ATMP.2009.v13.n2.a5 [arXiv:0709.4028 [hep-th]]
266. Nahm, W.: Supersymmetries and their representations. Nucl. Phys. B **135**, 149 (1978)
267. Nakahara, M.: Geometry, Toplogy and Physics. Adam Hilger, Bristol (1990)
268. Nekrasov, N., Ooguri, H., Vafa, C.: S–duality and topological strings (2004). arXiv:hep-th/0403167
269. Newlander, A., Nirenberg, L.: Complex analytic coordinates in almost complex manifolds. Ann. Math. **65**, 351–404 (1957)
270. Oda, T.: Convex bodies and algebraic geometry. Springer, Berlin, Heidelberg (1988)
271. Ooguri, H., Vafa, C.: Knot invariants and topological strings. Nucl. Phys. B **577**, 419 (2000) [arXiv:hep-th/9912123]
272. O'Raifeartaigh, L., Straumann, N.: Early history of gauge theories and Kaluza-Klein theories (1998). hep-ph/9810524
273. Pandharipande, R., Thomas, R.P.: Curve counting via stable pairs in the derived category. Invent. Math. **178**, 407–447 (2009). arXiv:0707.2348 [math.AG]
274. Pandharipande, R., Thomas, R.P.: The 3-fold vertex via stable pairs. Geom. Topol. **13**, 1835–1876 (2009). arXiv:0709.3823 [math.AG]
275. Pandharipande, R., Thomas, R.P: Stable Pairs and BPS Invariants. J. Am. Math. Soc. **23**, 267–297 (2010). arXiv:0711.3899 [math.AG].
276. Pandharipande, R., Zinger, A.: Enumerative geometry of Calabi-Yau 5-folds. Adv. Stud. Pure Math. **59**, 239 (2010)
277. Passare, M., Tsikh, A.K., Cheshel, A.A.: Multiple Mellin-Barnes integrals as periods on Calabi-Yau manifolds with several moduli. Teoret. Mat. Fiz. **109**, 381–394 (1996)
278. Pestun, V., Witten, E.: The Hitchin functionals and the topological B-model at one loop (2005). arXiv:hep-th/0503083
279. Polchinski, J.: String Theory, I & II. Cambridge University Press, Cambridge (1998)
280. Recknagel, A., Schomerus, V.: Nucl. Phys. B **531**, 185 (1998) [arXiv:hep-th/9712186]
281. Reid, M.: Young person's guide to canonical singularities. Proceedings of Symposia in Pure Mathematics, vol. 46, pp. 345–414. American Mathematical Society, Providence (1987)
282. Rodland, E.A.: The Pfaffian Calabi-Yau, its mirror and their link to the Grassmannian $G(2, 7)$. Compos. Math. **122**(2), 135–149 (2000) [arXiv:math.AG/9801092]
283. Saito, K.: Theory of logarithmic differential forms and logarithmic vector fields. J. Fac. Sci. Univ. Tokyo **27**, 265–291 (1980) [Jairo]
284. Sakai, K.: Topological string amplitudes for the local $\frac{1}{2}$K3 surface. Prog. Theor. Exp. Phys. **2017**(3), 033B09 (2017). https://doi.org/10.1093/ptep/ptx027 [arXiv:1111.3967 [hep-th]]
285. Schwarz, A.S.: Sigma models having supermanifolds as target spaces. Lett. Math. Phys. **38**, 91 (1996). https://doi.org/10.1007/BF00398301 [hep-th/9506070]
286. Schwimmer, A., Seiberg, N.: Comments on the N=2,3,4 superconformal Algebras in two dimensions. Phys. Lett. **184B**, 191 (1987)
287. Seiberg, N., Witten, E.: Electric - magnetic duality, monopole condensation, and confinement in N=2 supersymmetric Yang-Mills theory. Nucl. Phys. B **426**, 19 (1994); Erratum: [Nucl. Phys. B **430**, 485 (1994)]. https://doi.org/10.1016/0550-3213(94)90124-4,10.1016/0550-3213(94)00449-8 [hep-th/9407087]
288. Seiberg, N., Witten, E.: Monopoles, duality and chiral symmetry breaking in N=2 supersymmetric QCD. Nucl. Phys. B **431**, 484 (1994). https://doi.org/10.1016/0550-3213(94)90214-3 [hep-th/9408099]
289. Seidel, P., Thomas, R.: Braid group actions on derived categories of coherent sheaves. Duke Math. J. **108**(1), 37–108 (2001)

290. Sen, A.: String string duality conjecture in six-dimensions and charged solitonic strings. Nucl. Phys. B **450**, 103 (1995). https://doi.org/10.1016/0550-3213(95)00320-R [hep-th/9504027]
291. Sen, A.: Unification of string dualities. Nucl. Phys. Proc. Suppl. **58**, 5 (1997). https://doi.org/10.1016/S0920-5632(97)00409-X [hep-th/9609176]
292. Sethi, S.: Supermanifolds, rigid manifolds and mirror symmetry. Nucl. Phys. B **430**, 31 (1994) [AMS/IP Stud. Adv. Math. **1**, 793 (1996)], https://doi.org/10.1016/0550-3213(94)90649-1 [hep-th/9404186]
293. Schmidt, W.: Variation of Hodge structure: the singularities of the period map. Invent. Math. **22**, 211 (1973)
294. Singh, S.: Arithmeticity of the 4 monodromy groups associated to the Calabi-Yau threefolds (2014). arXiv:1308.4039
295. Singh, S., Venkataramana, T.N.: Arithmeticity of certain symplectic hypergeometric groups (2014). arXiv:1208.6460
296. Spanier, E.H.: Algebraic Topology. Springer, Berlin (1966)
297. Strominger, A.: Massless black holes and conifolds in string theory. Nucl. Phys. **B245** 96 (1995)
298. Strominger, A., Witten, E.: Commun. Math. Phys. **101**, 341 (1985)
299. 't Hooft, G.: Perturbative quantum gravity. In: Zichichi, A. (ed.) Proceedings of the International School of Subnuclear Physics, Erice 2002, From Quarks and Gluons to Quantum Gravity. Subnuclear Series, vol. 40, p. 249. World Scientific, Singapore (2002). http://www.phys.uu.nl/~thooft/
300. Thomas, R.P.: A holomorphic Casson invariant for Calabi–Yau 3-folds, and bundles on K3 fibrations. J. Differ. Geom. **54**, 367–438 (2000) [arXiv:math.ag/9806111]
301. Tian, G.: Smoothness of the universal deformation space of compact Calabi-Yau manifolds and its Weil-Peterson metric. In: Yau, S.T. (eds.) Mathematical Aspects of String Theory. World Scientific, Singapore (1987)
302. Tian, G.: Canonical Metrics in Kähler Geoemtry. Birkhäuser, Boston (2002)
303. Todorov, A.: The Weyl-Peterson geometry of the moduli-space of $SU(n \geq 3)$ (Calabi-Yau) manifolds I. Commun. Math. Phys. **126**, 325 (1989)
304. Vafa, C., Superstring Vacua. In: Trieste Proceedings 1989 in High Energy Physics and Cosmology, pp. 145–177 (1989)
305. Vafa, C., Topological Landau-Ginzburg models. Mod. Phys. Lett. A **6**, 337 (1991)
306. Vafa, C.: A Stringy test of the fate of the conifold. Nucl. Phys. B **447**, 252 (1995) [arXiv:hep-th/9505023]
307. Vafa, C.: Evidence for F theory. Nucl. Phys. B **469**, 403 (1996). https://doi.org/10.1016/0550-3213(96)00172-1 [hep-th/9602022]
308. van Straten, D.: Calabi–Yau operators (2017). arXiv:1704.00164 [math.AG]
309. Verlinde, E.P.: (2004). hep-th/0412139
310. Wall, C.C.T.: Classification problems in differential topology. V: On certain 6- manifolds. Invent. Math. **1**, 355–374 (1966). Corrigendum. Ibid. **2**, 306 (1967)
311. Walcher, J.: Opening mirror symmetry on the quintic. Commun. Math. Phys. **276**, 671 (2007). https://doi.org/10.1007/s00220-007-0354-8 [hep-th/0605162]
312. Wang, Y.M.: Parallel spinors and parallel forms. Ann. Global Anal. Geom. **7**, 59–68 (1989)
313. West, P.C.: E(11) and M theory. Class. Quant. Grav. **18**, 4443 (2001). https://doi.org/10.1088/0264-9381/18/21/305 [hep-th/0104081]
314. Witten, E.: The index of the Dirac operator in loop-space. In: Landweber, P.S. (ed.) Elliptic Curves and Modular Forms in Algebraic Topology. Lecture Notes in Mathematics, vol. 1326. Springer, Berlin (1980)
315. Witten, E.: Constraints on supersymmetry breaking. Nucl. Phys. **B202**, 253 (1982)
316. Witten, E.: Elliptic genera and quantum field theory. Commun. Math. Phys. **109**, 525 (1987)
317. Witten, E.: Topological sigma model. Commun. Math. Phys. **118**, 411 (1988)
318. Witten, E.: On the structure of the topological phase of two-dimensional gravity. Nucl. Phys. B **340**, 281 (1990)
319. Witten, E.: Mirror manifolds and topological field theory (1991). arXiv:hep-th/9112056

320. Witten, E.: Two-dimensional gravity and intersection theory on moduli space. Surv. Differ. Geom. **1**, 243 (1991)
321. Witten, E.: Quantum background independence in string theory (1993). arXiv:hep-th/9306122
322. Witten, E.: Phases of N=2 theories in two-dimensions. Nucl. Phys. **B403**, 159–222 (1993). arXiv:hep-th/9301042 [hep-th]
323. Witten, E.: On the Landau description of $N = 2$ minimal models. Int. Mod. Phys. **A9**, 4783 (1994). hep-th/9304026
324. Witten, E.: Chern-Simons gauge theory as a string theory. Prog. Math. **133**, 637 (1995) [arXiv:hep-th/9207094]
325. Witten, E.: String theory dynamics in various dimensions. Nucl. Phys. B **443**, 85 (1995). https://doi.org/10.1016/0550-3213(95)00158-O [hep-th/9503124]
326. Witten, E.: Phase transitions in M theory and F theory. Nucl. Phys. B **471**, 195 (1996) [hep-th/9603150]
327. Witten, E.: On flux quantization in M theory and the effective action. J. Geom. Phys. **22**, 1 (1997). https://doi.org/10.1016/S0393-0440(96)00042-3 [hep-th/9609122]
328. Witten, E.: Branes and the dynamics of QCD. Nucl. Phys. B **507**, 658 (1997). https://doi.org/10.1016/S0550-3213(97)00648-2 [hep-th/9706109]
329. Witten, E.: Two-dimensional models with (0,2) supersymmetry: perturbative aspects (2005). hep-th/0504078
330. Yamaguchi, S., Yau, S.T.: Topological string partition as polynomials. J. High Energy Phys. **0407**, 047 (2004). https://doi.org/10.1088/1126-6708/2004/07/047 [hep-th/0406078]
331. Yau, S.-T.: On the Ricci curvature of a compact Kähler manifold and the complex Monge-Ampére equations I. Commun. Pure Appl. Math.**31**, 339 (1978)
332. Yau, S.S.-T., Yu, Y.: Gorenstein quotient singularities in dimension three. Mem. Am. Math. Soc. **105** viii+88 (1993); 339 (1978)
333. Yau, S.T., Zaslow, E.: BPS states, string duality, and nodal curves on K3. Nucl. Phys. B **471**, 503 (1996) [arXiv:hep-th/9512121]
334. Yoshida, M.: Fuchsian differential equations. With Special Emphasis on the Gauss-Schwarz Theory. Friedr. Vieweg & Sohn, Braunschweig (1987)
335. Zhdanov, O.N., Tsikh, A.K.: Computation of multiple Mellin-Barnes integrals by means of multidimensional residues. Dokl. Akad. Nauk **358**, 154–156 (1998)
336. Zumino, B.: Supersymmetry and Kähler manifolds. Phys. Lett. B **87**, 203 (1979)

Geometric Quantization with Applications to Gromov-Witten Theory

Emily Clader, Nathan Priddis, and Mark Shoemaker

Contents

Abstract In the physics literature, geometric quantization is an operation that arises from an attempt to make sense of the passage from a classical theory to the corresponding quantum theory. In mathematics, on the other hand, the work of Alexander Givental and others has revealed quantization to be a powerful tool for studying Gromov–Witten-type theories in higher genus. For example, if the quantization of a symplectic transformation matches two total descendent potentials, then the original symplectic transformations matches their genus-zero theories, and, at least when a semisimplicity condition is satisfied, the converse is also true. In these notes, we give a mathematically-minded presentation of quantization of symplectic vector spaces, and we illustrate how quantization appears in specific applications to Gromov–Witten theory.

E. Clader (✉)
Department of Mathematics, San Francisco State University, San Francisco, CA, USA
e-mail: eclader@sfsu.edu

N. Priddis
Department of Mathematics, Brigham Young University, Provo, UT, USA
e-mail: priddis@mathematics.byu.edu

M. Shoemaker
Department of Mathematics, Colorado State University, Fort Collins, CO, USA
e-mail: mark.shoemaker@colostate.edu

E. Clader, Y. Ruan (eds.), *B-Model Gromov-Witten Theory*, Trends in Mathematics,
https://doi.org/10.1007/978-3-319-94220-9_3

1 Preface

The following notes were prepared for the "IAS Program on Gromov-Witten Theory and Quantization" held jointly by the Department of Mathematics and the Institute for Advanced Study at the Hong Kong University of Science and Technology in July 2013. Their primary purpose is to introduce the reader to the machinery of geometric quantization with the ultimate goal of computations in Gromov-Witten theory. These notes are expository, and the authors make no claim to originality of any of the material appearing in them.

First appearing in this subject in the work of Alexander Givental and his students, quantization provides a powerful tool for studying Gromov-Witten-type theories in higher genus. For example, if the quantization of a symplectic transformation matches two total descendent potentials, then the original symplectic transformation matches the Lagrangian cones encoding their genus-zero theories; we discuss this statement in detail in Sect. 5.5. Moreover, according to Givental's Conjecture (see Sect. 6.2), the converse is true in the semisimple case. Thus, if one wishes to study a semisimple Gromov-Witten-type theory, it is sufficient to find a symplectic transformation identifying its genus-zero theory with that of a finite collection of points, which is well-understood. The quantization of this transformation will carry all of the information about the higher-genus theory in question.

In addition, quantization is an extremely useful combinatorial device for organizing information. Some of the basic properties of Gromov-Witten theory, for instance, can be succinctly expressed in terms of equations satisfied by quantized operators acting on the total descendent potential. To give another example, even if one is concerned only with genus zero, the combinatorics of expanding the relation between two theories into a statement about their generating functions can be unmanageable, but when it is expressed via quantization this unwieldy problem obtains a clean expression. The relationship between a twisted theory and its untwisted analogue, discussed in Sect. 6.3, is a key instance of this phenomenon.

At present, there are very few references on the subject of quantization as it is used in this mathematical context. While a number of texts exist (such as [4, 10, 26]), these tend to focus on the quantization of finite-dimensional, topologically non-trivial symplectic manifolds. Accordingly, they are largely devoted to explaining the structures of polarization and prequantization that one must impose on a symplectic manifold before quantizing, and less concerned with explicit computations. These precursors to quantization are irrelevant in applications to Gromov-Witten theory, as the symplectic manifolds one must quantize are simply symplectic vector spaces. However, other technical issues arise from the fact that the symplectic vector spaces in Gromov-Witten theory are typically infinite-dimensional. We hope that these notes will fill a gap in the existing literature by focusing on computational formulas and addressing the complications specific to Givental's set-up.

The structure of the notes is as follows. In Sect. 2, we give a brief overview of preliminary material on symplectic geometry and the method of Feynman diagram expansion. We then turn in Sect. 3 to a discussion of quantization of finite-dimensional vector spaces. We obtain formulas for the quantizations of functions on such vector spaces by three main methods: direct computation via the canonical commutation relations (which may also be expressed in terms of quadratic Hamiltonians), Fourier-type integrals, and Feynman diagram expansion. All three methods yield equivalent results, but this diversity of derivations is valuable in pointing to various generalizations and applications. We then include, Sect. 4, an interlude on basic Gromov-Witten theory. Though this material is not strictly necessary until Sect. 6, it provides motivation and context for the material that follows.

Section 5, is devoted to quantization of infinite-dimensional vector spaces, such as arise in applications to Gromov-Witten theory. As in the finite-dimensional case, formulas can be obtained via quadratic Hamiltonians, Fourier integrals, or Feynman diagrams, and for the most part the computations mimic those of Sect. 3. The major difference in the infinite-dimensional setting though, is that issues of convergence arise, which we make an attempt to discuss whenever they come up. Finally, in Sect. 6, we present several of the basic equations of Gromov-Witten theory in the language of quantization, and mention a few of the more significant appearances of quantization in the subject.

2 Preliminaries

Before we begin our study of quantization, we will give a quick overview of some of the prerequisite background material. First, we review the basics of symplectic vector spaces and symplectic manifolds. This material will be familiar to most of our mathematical audience, but we collect it here for reference and to establish notational conventions; for more details, see [8] or [25]. Less likely to be familiar to mathematicians is the material on Feynman diagrams, so we cover this topic in more detail. The section concludes with the statement of Feynman's theorem, which will be used later to express certain integrals as combinatorial summations. The reader who is already experienced in the methods of Feynman diagram expansion is encouraged to skip directly to Sect. 3.

2.1 Basics of Symplectic Geometry

2.1.1 Symplectic Vector Spaces

A **symplectic vector space** (V, ω) is a vector space V together with a nondegenerate skew-symmetric bilinear form ω. We will often denote $\omega(v, w)$ by $\langle v, w\rangle$.

One consequence of the existence of a nondegenerate bilinear form is that V is necessarily even-dimensional. The standard example of a real symplectic vector space is $\mathbb{R}^{2n}$, with the symplectic form defined in a basis $\{e^\alpha, e_\beta\}_{1\leq\alpha,\beta\leq n}$ by

$$\langle e^\alpha, e^\beta\rangle = \langle e_\alpha, e_\beta\rangle = 0, \quad \langle e^\alpha, e_\beta\rangle = \delta^\alpha_\beta. \tag{1}$$

In other words, $\langle\ ,\ \rangle$ is represented by the (skew-symmetric) matrix

$$J = \begin{pmatrix} 0 & I \\ -I & 0 \end{pmatrix}.$$

In fact, any finite-dimensional symplectic vector space admits a basis in which the symplectic form is expressed this way. Such a basis is called a **symplectic basis**, and the corresponding coordinates are known as **Darboux coordinates**.

If $U \subset V$ is a linear subspace of a symplectic vector space (V, ω), then the **symplectic orthogonal** U^ω of U is

$$U^\omega = \{v \in V \mid \omega(u, v) = 0 \text{ for all } u \in U\}.$$

One says U is **isotropic** if $U \subset U^\omega$ and **Lagrangian** if $U = U^\omega$, which implies in particular that, if V is finite-dimensional, $\dim(U) = \frac{1}{2}\dim(V)$.

A **symplectic transformation** between symplectic vector spaces (V, ω) and (V', ω') is a linear map $\sigma : V \to V'$ such that

$$\omega'(\sigma(v), \sigma(w)) = \omega(v, w).$$

In what follows, we will mainly be concerned with the case $V = V'$, and it will be useful to express the symplectic condition on a linear endomorphism $\sigma : V \to V$ in terms of matrix identities in Darboux coordinates. Choose a symplectic basis for V and express σ in this basis via the matrix

$$\sigma = \begin{pmatrix} A & B \\ C & D \end{pmatrix}.$$

Then σ is symplectic if and only if $\sigma^T J \sigma = J$, which in turn holds if and only if the following three identities are satisfied:

$$AB^T = BA^T \tag{2}$$

$$CD^T = DC^T \tag{3}$$

$$AD^T - BC^T = I. \tag{4}$$

Using these facts, one obtains a convenient expression for the inverse:

$$\sigma^{-1} = \begin{pmatrix} D^T & -B^T \\ -C^T & A^T \end{pmatrix}.$$

An invertible matrix satisfying (2)–(4) is known as a **symplectic matrix**, and the group of symplectic matrices is denoted $\mathrm{Sp}(2n, \mathbb{R})$.

2.1.2 Symplectic Manifolds

Symplectic vector spaces are the simplest examples of the more general notion of a **symplectic manifold**, which is a smooth manifold equipped with a closed nondegenerate two-form, called a **symplectic form**.

In particular, such a 2-form makes the tangent space T_pM at any point $p \in M$ into a symplectic vector space. Just as every symplectic vector space is isomorphic to $\mathbb{R}^{2n}$ with its standard symplectic structure, Darboux's Theorem states that every symplectic manifold is locally isomorphic to $\mathbb{R}^{2n} = \{(x_1, \ldots, x_n, y_1, \ldots, y_n)\}$ with the symplectic form $\omega = \sum_{i=1}^{n} dx_i \wedge dy_i$.

Perhaps the most important example of a symplectic manifold is the cotangent bundle. Given any smooth manifold N (not necessarily symplectic), there is a canonical symplectic structure on the total space of T^*N. To define the symplectic form, let $\pi : T^*N \to N$ be the projection map. Then one can define a one-form λ on T^*N by setting

$$\lambda|_{\xi_x} = \pi^*(\xi_x)$$

for any cotangent vector $\xi_x \in T_x^*N \subset T^*N$. This is known as the **tautological one-form**. The **canonical symplectic form** on T^*N is $\omega = -d\lambda$. The choice of sign makes the canonical symplectic structure agree with the standard one in the case where $N = \mathbb{R}^n$, which sits inside of $T^*N \cong \mathbb{R}^{2n}$ as $\mathrm{Span}\{e^1, \ldots, e^n\}$ in the basis notation used previously.

In coordinates, the tautological one-form and canonical two-form appear as follows. Let $q_1, \ldots, q_n$ be local coordinates on N. Then there are local coordinates on T^*N in which a point in the fiber over $(q_1, \ldots, q_n)$ can be expressed as a local system of coordinates $(q_1, \ldots, q_n, p^1, \ldots, p^n)$, and in these coordinates,

$$\lambda = \sum_{\alpha} p^{\alpha} dq_{\alpha}$$

and

$$\omega = \sum_{\alpha} dq_{\alpha} \wedge dp^{\alpha}.$$

The definitions of isotropic and Lagrangian subspaces generalize to symplectic manifolds, as well. A submanifold of a symplectic manifold is **isotropic** if the restriction of the symplectic form to the submanifold is zero. An isotropic submanifold is **Lagrangian** if its dimension is as large as possible—namely, half the dimension of the ambient manifold.

For example, in the case of the symplectic manifold T^*N with its canonical symplectic form, the fibers of the bundle are all Lagrangian submanifolds, as is the zero section. Furthermore, if $\mu : N \to T^*N$ is a closed one-form, the graph of μ is a Lagrangian submanifold. More precisely, for any one-form μ, one can define

$$X_\mu = \{(x, \mu(x)) \mid x \in N\} \subset T^*N,$$

and X_μ is Lagrangian if and only if μ is closed. In case N is simply-connected, this is equivalent to the requirement that $\mu = df$ for some function f, called a **generating function** of the Lagrangian submanifold X_μ.

Finally, the notion of symplectic transformation generalizes in an obvious way. Namely, given symplectic manifolds (M, ω_1) and (N, ω_2) a **symplectomorphism** is a smooth map $f : M \to N$ such that $f^*\omega_2 = \omega_1$.

2.2 *Feynman Diagrams*

The following material is drawn mainly from [12]; another reference on the subject of Feynman diagrams is Chapter 9 of [19].

Consider an integral of the form

$$\hbar^{-\frac{d}{2}} \int_V e^{-S(x)/\hbar} dx, \tag{5}$$

where V is a d-dimensional vector space, $\hbar$ is a formal parameter, and

$$S(x) = \frac{1}{2} B(x, x) + \sum_{m \geq 0} \frac{g_m}{m!} B_m(x, \dots, x) \tag{6}$$

for a bilinear form B and m-multilinear forms B_m on V, where g_m are constants. The integral (5) can be understood as a formal series in the parameters $\hbar$ and g_m.

2.3 *Wick's Theorem*

We begin by addressing the simpler question of computing integrals involving an exponential of a bilinear form without any of the other tensors.

Let V be a vector space of dimension d over $\mathbb{R}$, and let B be a positive-definite bilinear form on V. Wick's theorem will relate integrals of the form

$$\int_V l_1(x) \cdots l_N(x) e^{-B(x,x)/2} dx,$$

in which $l_1, \ldots, l_N$ are linear forms on V, to pairings on the set $[2k] = \{1, 2, \ldots, 2k\}$. By a **pairing** on $[2k]$, we mean a partition of the set into k disjoint subsets, each having two elements. Let Π_{2k} denote the set of pairings on $[2k]$. The size of this set is

$$|\Pi_{2k}| = \frac{(2k)!}{(2!)^k k!},$$

as the reader can check as an exercise.

An element $\sigma \in \Pi_k$ can be viewed as a special kind of permutation on the set $[2k]$, which sends each element to the other member of its pair. Write $[2k]/\sigma$ for the set of orbits under this involution.

Theorem 1 (Wick's Theorem) *Let $l_1, \ldots, l_N \in V^*$. If N is even, then*

$$\int_V l_1(x) \ldots l_N(x) e^{\frac{-B(x,x)}{2}} dx = \frac{(2\pi)^{d/2}}{\sqrt{\det B}} \sum_{\sigma \in \Pi_N} \prod_{i \in [N]/\sigma} B^{-1}(l_i, l_{\sigma(i)}).$$

If N is odd, the integral is zero.

Proof First, apply a change of variables such that B is of the form $B(x, x) = x_1^2 + \cdots + x_d^2$. The reader should check that this change of variables changes both sides of the equation by a factor of $\det(P)$, where P is the change-of-basis matrix, and thus the equality prior to the change of variables is equivalent to the result after the change. Furthermore, since both sides are multilinear in elements of V^* and symmetric in $x_1, \ldots, x_d$, we may assume $l_1 = l_2 = \cdots = l_N = x_1$. The theorem is then reduced to computing

$$\int_V x_1^N e^{\frac{-(x_1^2 + \cdots + x_d^2)}{2}} dx.$$

This integral indeed vanishes when N is odd, since the integrand is an odd function. If N is even, write $N = 2k$. In case $k = 0$, the theorem holds by the well-known fact that

$$\int_{-\infty}^{\infty} e^{\frac{-x^2}{2}} dx = \sqrt{2\pi}.$$

For $k > 0$, we can use this same fact to integrate out the last $d - 1$ variables, by which we see that the claim in the theorem is equivalent to

$$\int_{-\infty}^{\infty} x^{2k} e^{-\frac{-x^2}{2}} dx = \sqrt{2\pi} \frac{(2k)!}{2^k k!}. \tag{7}$$

To prove (7), we first make the substitution $y = \frac{x^2}{2}$. Recall that the gamma function is defined by

$$\Gamma(z) = \int_0^\infty y^{z-1} e^{-y} dy$$

and satisfies $\Gamma(\frac{1}{2}) = \sqrt{\pi}$, as well as

$$\Gamma(z+k) = (z+k-1)(z+k-2)\cdots z \cdot \Gamma(z)$$

for integers k. Thus, we have:

$$\begin{aligned}
\int_{-\infty}^{\infty} x^{2k} e^{\frac{-x^2}{2}}\, dx &= 2\int_0^\infty x^{2k} e^{\frac{-x^2}{2}}\, dx \\
&= 2\int_0^\infty (2y)^{k-1/2} e^{-y}\, dy \\
&= 2^{k+1/2}\Gamma(k+\tfrac{1}{2}) \\
&= 2^{k+1/2}(k-\tfrac{1}{2})(k-\tfrac{3}{2})\ldots(\tfrac{1}{2})\Gamma(\tfrac{1}{2}) \\
&= \sqrt{2\pi}\frac{(2k)!}{2^k k!},
\end{aligned}$$

which proves the claim. □

2.3.1 Feynman's Theorem

Now let us return to the more general integral

$$Z = \hbar^{-\frac{d}{2}} \int_V e^{-S(x)/\hbar} dx,$$

which, for reasons we will mention at the end of the section, is sometimes called a **partition function**. Recall that S has an expansion in terms of multilinear forms given by (6).

Because a pairing can be represented by a graph all of whose vertices are 1-valent, Wick's theorem can be seen as a method for expressing certain integrals as summations over graphs, in which each graph contributes an explicit combinatorial term. The goal of this section is to give a similar graph-sum expression for the partition function Z.

Before we can state the theorem, we require a bit of notation. If $\mathbf{n} = (n_0, n_1, \ldots)$ is a sequence of nonnegative integers, all but finitely many of which are zero, let $G(\mathbf{n})$ denote the set of isomorphism classes of graphs with n_i vertices of valence

i for each $i \geq 0$. Note that the notion of "graph" here is very broad: they may be disconnected, and self-edges and multiple edges are allowed. If Γ is such a graph, let

$$b(\Gamma) = |E(\Gamma)| - |V(\Gamma)|,$$

where $E(\Gamma)$ and $V(\Gamma)$ denote the edge set and vertex set, respectively. An **automorphism** of Γ is a permutation of the vertices and edges that preserves the graph structure, and the set of automorphisms is denoted $\mathrm{Aut}(\Gamma)$.

We will associate a certain number F_Γ to each graph, known as the **Feynman amplitude**. The Feynman amplitude is defined by the following procedure:

1. Put the m-tensor $-B_m$ at each m-valent vertex of Γ.
2. For each edge e of Γ, take the contraction of tensors attached to the vertices of e using the bilinear form B^{-1}. This will produce a number F_{Γ_i} for each connected component Γ_i of Γ.
3. If $\Gamma = \bigsqcup_i \Gamma_i$ is the decomposition of Γ into connected components, define $F_\Gamma = \prod_i F_{\Gamma_i}$.

By convention, we set the Feynman amplitude of the empty graph to be 1.

Theorem 2 (Feynman's Theorem) *One has*

$$Z = \frac{(2\pi)^{d/2}}{\sqrt{\det(B)}} \sum_{\mathbf{n}=(n_0,n_1,\ldots)} \left(\prod_{i=0}^{\infty} g_i^{n_i}\right) \sum_{\Gamma \in G(\mathbf{n})} \frac{\hbar^{b(\Gamma)}}{|Aut(\Gamma)|} F_\Gamma,$$

where the outer summation is over all sequences of nonnegative integers with almost all zero.

Before we prove the theorem, let us compute a few examples of Feynman amplitudes to make the procedure clear.

Example 3 Let Γ be the following graph:

Given $B : V \otimes V \to \mathbb{R}$, we have a corresponding bilinear form $B^{-1} : V^\vee \otimes V^\vee \to \mathbb{R}$. Moreover, $B_1 \in V^\vee$, and so we can write the Feynman amplitude of this graph as

$$F_\Gamma = (B^{-1}(-B_1, -B_1))^2.$$

Example 4 Consider now the graph

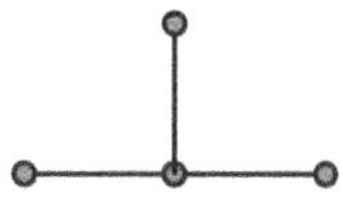

Associated to B is a map $\overline{B} : V \to V^\vee$, and in this notation, the Feynman amplitude of the graph can be expressed as

$$F_\Gamma = -B_3(\overline{B}^{-1}(-B_1), \overline{B}^{-1}(-B_1), \overline{B}^{-1}(-B_1)).$$

Proof of Feynman's Theorem After a bit of combinatorial fiddling, this theorem actually follows directly from Wick's theorem. First, perform the change of variables $y = x/\sqrt{\hbar}$, under which

$$Z = \int_V e^{-B(y,y)/2} e^{\sum_{m\geq 0} g_m (\frac{-\hbar^{\frac{m}{2}-1} B_m(y,\ldots,y)}{m!})} dy.$$

Expanding the exponential as a series gives

$$\begin{aligned} Z &= \int_V e^{-B(y,y)/2} \prod_{i\geq 0} \sum_{n_i \geq 0} \frac{g_i^{n_i}}{(i!)^{n_i} n_i!} \left(-\hbar^{\frac{i}{2}-1} B_i(y,\ldots,y)\right)^{n_i} dy \\ &= \sum_{\mathbf{n}=(n_0,n_1,\ldots)} \left(\prod_{i\geq 0} \frac{g_i^{n_i}}{(i!)^{n_i} n_i!} \hbar^{n_i(\frac{i}{2}-1)} \right) \int_V e^{-B(y,y)/2} \prod_{i\geq 0} \left(-B_i(y,\ldots,y)\right)^{n_i} dy. \end{aligned}$$

Denote

$$Z_{\mathbf{n}} = \int_V e^{-B(y,y)/2} \prod_{i\geq 0} (-B_i(y,\ldots,y))^{n_i} dy.$$

Each of the factors $-B_i(y, \ldots, y)$ in this integral can be expressed as a sum of products of i linear forms on V. After unpacking the expression in this way, $Z_{\mathbf{n}}$ becomes a sum of integrals of the form

$$\int_V e^{-B(y,y)/2} \left(\text{one linear form}\right)^{n_1} \left(\text{product of two linear forms}\right)^{n_2} \cdots,$$

each with an appropriate coefficient, so we will be able to apply Wick's theorem.

Let $N = \sum_i i \cdot n_i$, which is the number of linear forms in the above expression for $Z_{\mathbf{n}}$. We want to express the integral $Z_{\mathbf{n}}$ using graphs. To this end, we draw n_0 vertices with no edges, n_1 vertices with 1 half-edge emanating from them, n_2 vertices with 2 half-edges, n_3 vertices with 3 half-edges, et cetera; these are sometimes called "flowers". We place $-B_i$ at the vertex of each i-valent flower. A pairing $\sigma \in \Pi_N$ can be understood as a way of joining pairs of the half-edges to form full edges, which yields a graph $\Gamma(\sigma)$, and applying Wick's theorem we get a number $F(\sigma)$ from each such pairing σ. One can check that all of the pairings σ giving rise to a particular graph $\Gamma = \Gamma(\sigma)$ combine to contribute the Feynman amplitude $F_{\Gamma(\sigma)}$. In particular, by Wick's theorem, only even N give a nonzero result.

At this point, we have

$$Z = \frac{(2\pi)^{d/2}}{\sqrt{\det(B)}} \sum_{\mathbf{n}} \left(\prod_{i\geq 0} \frac{g_i^{n_i}}{(i!)^{n_i} n_i!} \right) \sum_{\sigma\in\Pi_N} \hbar^{b(\Gamma(\sigma))} F_{\Gamma(\sigma)},$$

where we have used the straightforward observation that the exponent $\sum n_i(\frac{i}{2} - 1) = \frac{N}{2} - \sum n_i$ on $\hbar$ is equal to the number of edges minus the number of vertices of any of the graphs appearing in the summand corresponding to $\mathbf{n}$. All that remains is to account for the fact that many pairings can yield the same graph, and thus we will obtain a factor when we re-express the above as a summation over graphs rather than over pairings.

To compute the factor, fix a graph Γ, and consider the set $P(\Gamma)$ of pairings on $[N]$ yielding the graph Γ. Let H be the set of half-edges of Γ, which are attached as above to a collection of flowers. Let G be the group of permutations of H that preserve flowers; this is generated by permutations of the edges within a single flower, as well as swaps of two entire flowers with the same valence. Using this, it is easy to see that

$$|G| = \prod_{i\geq 0} (i!)^{n_i} n_i!.$$

The group G acts transitively on the set $P(\Gamma)$, and the stabilizer of this action is equal to $\mathrm{Aut}(\Gamma)$. Thus, the number of distinct pairings yielding the graph Γ is

$$\frac{\prod_i (i!)^{n_i} n_i!}{|\mathrm{Aut}(\Gamma)|},$$

and the theorem follows. □

We conclude this preliminary section with a bit of "generatingfunctionology" that motivates the term "partition function" for Z.

Theorem 5 *Let* $Z_0 = \frac{(2\pi)^{d/2}}{\det(B)}$. *Then one has*

$$\log(Z/Z_0) = \sum_{\mathbf{n}=(n_0,n_1,\ldots)} \left(\prod_{i=0}^{\infty} g_i^{n_i} \right) \sum_{\Gamma\in G_c(\mathbf{n})} \frac{\hbar^{b(\Gamma)}}{|Aut(\Gamma)|} F_\Gamma,$$

where the outer summation is over $\mathbf{n}$ *as before, and* $G_c(\mathbf{n})$ *denotes the set of isomorphism classes of connected graphs in* $G(\mathbf{n})$.

The proof of this theorem is a combinatorial exercise and will be omitted. Note that if we begin from this theorem, then the terms in Feynman's Theorem can be viewed as arising from "partitioning" a disconnected graph into its connected components. This explains the terminology for Z in this particular situation.

3 Quantization of Finite-Dimensional Symplectic Vector Spaces

From a physical perspective, geometric quantization arises from an attempt to make sense of the passage from a classical theory to the corresponding quantum theory. The state space in classical mechanics is represented by a symplectic manifold, and the observables (quantities like position and momentum) are given by smooth real-valued functions on that manifold. Quantum mechanics, on the other hand, has a Hilbert space as its state space, and the observables are given by self-adjoint linear operators. Thus, quantization should associate a Hilbert space to each symplectic manifold and a self-adjoint linear operator to each smooth real-valued function, and this process should be functorial with respect to symplectic diffeomorphisms.

By considering certain axioms required of the quantization procedure, we show in Sect. 3.1 that the Hilbert space of states can be viewed as a certain space of functions. Careful study of these axioms leads, in Sect. 3.2, to a representation of a quantized symplectic transformation as an explicit expression in terms of multiplication and differentiation of the coordinates. However, as is explained in Sect. 3.4, it can also be expressed as a certain integral over the underlying vector space. This representation has two advantages. First, the method of Feynman diagrams allows one to re-write it as a combinatorial summation, as is explained in Sect. 3.5. Second, it can be generalized to the case in which the symplectic diffeomorphism is nonlinear. Though the nonlinear case will not be addressed in these notes, we conclude this section with a few comments on nonlinear symplectomorphisms and other possible generalizations of the material developed here.

3.1 The Set-Up

The material of this section is standard in the physics literature, and can be found, for example, in [4] or [10].

3.1.1 Quantization of the State Space

Let V be a real symplectic vector space of dimension $2n$, whose elements are considered to be the classical states. Roughly speaking, elements of the associated Hilbert space of quantum states will be square-integrable functions on V. It is a basic physical principle, however, that quantum states should depend on only half as many variables as the corresponding classical states; there are n position coordinates and n momentum coordinates describing the classical state of a system, whereas a quantum state is determined by either position or momentum alone. Thus, before quantizing V it is necessary to choose a **polarization**, a decomposition into half-dimensional subspaces. Because these two subspaces should be thought of as position and momentum, which mathematically are the zero section and the fiber direction of the cotangent bundle to a manifold, they should be Lagrangian subspaces of V.

The easiest way to specify a polarization in the context of vector spaces is to choose a symplectic basis $e = \{e^\alpha, e_\beta\}_{1\le\alpha,\beta\le n}$. Recall, such a basis satisfies

$$\langle e^\alpha, e^\beta\rangle = \langle e_\alpha, e_\beta\rangle = 0, \quad \langle e^\alpha, e_\beta\rangle = \delta^\alpha_\beta,$$

where $\langle\cdot,\cdot\rangle$ is the symplectic form on V. The polarization may be specified by fixing the subspace $R = \text{Span}\{e_\alpha\}_{1\le\alpha\le n}$ and viewing V as the cotangent bundle T^*R in such a way that $\langle\cdot,\cdot\rangle$ is identified with the canonical symplectic form.[1] An element of V will be written in the basis e as

$$\sum_\alpha p_\alpha e^\alpha + \sum_\beta q^\beta e_\beta.$$

For the remainder of these notes, we will suppress the summation symbol in expressions like this, adopting Einstein's convention that when Greek letters appear both up and down, they are automatically summed over all values of the index. For example, the above summation would be written simply as $p_\alpha e^\alpha + q^\beta e_\beta$.

Let $V \cong \mathbb{R}^{2n}$ be a symplectic vector space with symplectic basis $e = \{e^\alpha, e_\beta\}_{1\le\alpha,\beta\le n}$. Then the **quantization** of (V, e) is the Hilbert space $\mathscr{H}_e$ of square-integrable functions on R which take values in $\mathbb{C}[[\hbar, \hbar^{-1}]]$. Here, $\hbar$ is considered as a formal parameter, although physically, it denotes Plank's constant.

It is worth noting that, while it is necessary to impose square-integrability in order to obtain a Hilbert space, in practice one often needs to consider formal functions on R that are not square-integrable. The space of such formal functions from R to $\mathbb{C}[[\hbar, \hbar^{-1}]]$ is called the **Fock space**.

3.1.2 Quantization of the Observables

Observables in the classical setting are smooth functions $f \in C^\infty(V)$, and the result of a measurement is the value taken by f on a point of V. In the quantum framework, observables are operators U on $\mathscr{H}_e$, and the result of a measurement is an eigenvalue of U. In order to ensure that these eigenvalues are real, we require that the operators be self-adjoint.

There are a few other properties one would like the quantization $Q(f)$ of observables f to satisfy. We give one possible such list below, following Section 3

[1]Note that in this identification, we have chosen for the fiber coordinates to be the *first n* coordinates. It is important to keep track of whether upper indices or lower indices appear first in the ordering of the basis to avoid sign errors.

of [4]. Here, a set of observables is called **complete** if any function that "Poisson-commutes" (that is, has vanishing Poisson bracket) with every element of the set is a constant function. Likewise, a set of operators is called **complete** if any operator that commutes with every one of them is the identity.

The quantization procedure should satisfy:

1. **Linearity**: $Q(\lambda f + g) = \lambda Q(f) + Q(g)$ for all $f, g \in C^\infty(V)$ and all $\lambda \in \mathbb{R}$.
2. **Preservation of constants**: $Q(\mathbf{1}) = \text{id}$.
3. **Commutation**: $[Q(f), Q(g)] = \hbar Q(\{f, g\})$, where $\{\ ,\ \}$ denotes the Poisson bracket.[2]
4. **Irreducibility**: If $\{f_1, \ldots, f_k\}$ is a complete set of observables, then $\{Q(f_1), \ldots, Q(f_k)\}$ is a complete set of operators.

However, it is in general not possible to satisfy all four of these properties simultaneously. In practice, this forces one to restrict to quantizing only a certain complete subset of the observables, or to relax the properties required. We will address this in our particular case of interest shortly.

One complete set of observables on the state space V is given by the coordinate functions $\{p_\alpha, q^\beta\}_{\alpha,\beta=1,\ldots,n}$, and one can determine a quantization of these observables by unpacking conditions (1)–(4) above. Indeed, when f and g are coordinate functions, condition (3) reduces to the **canonical commutation relations (CCR)**, where we write $\hat{x}$ for $Q(x)$:

$$[\hat{p}_\alpha, \hat{p}_\beta] = [\hat{q}^\alpha, \hat{q}^\beta] = 0, \quad [\hat{p}_\alpha, \hat{q}^\beta] = \hbar\delta_\alpha^\beta.$$

The algebra generated by elements $\hat{q}^\alpha$ and $\hat{p}_\beta$ subject to these commutation relations is known as the Heisenberg algebra. Thus, the above can be understood as requiring that the quantization of the coordinate functions defines a representation of the Heisenberg algebra. By Schur's Lemma, condition (4) is equivalent to the requirement that this representation be irreducible. The following definition provides such a representation.

The **quantization** of the coordinate functions[3] is given by

$$\hat{q}^\alpha \Psi = q^\alpha \Psi,$$

$$\hat{p}_\alpha \Psi = \hbar \frac{\partial \Psi}{\partial q^\alpha}$$

for $\Psi \in \mathcal{H}_e$.

---

[2]This convention differs by a factor of i from what is taken in [4], but we choose it to match with what appears in the Gromov-Witten theory literature.

[3]To be precise, these operators do not act on the entire quantum state space $\mathcal{H}_e$, because elements of $\mathcal{H}_e$ may not be differentiable. However, this will not be an issue in our applications, because quantized operators will always act on power series.

In fact, the complete set of observables we will need to quantize for our intended applications consists of the quadratic functions in the Darboux coordinates. To quantize these, order the variables of each quadratic monomial with q-coordinates on the left, and quantize each variable as above. That is:

$$\begin{aligned} Q(q^\alpha q^\beta) &= q^\alpha q^\beta \\ Q(q^\alpha p_\beta) &= \hbar q^\alpha \frac{\partial}{\partial q^\beta} \\ Q(p_\alpha p_\beta) &= \hbar^2 \frac{\partial^2}{\partial q^\alpha \partial q^\beta}. \end{aligned}$$

There is one important problem with this definition: the commutation condition (3) holds only up to an additive constant when applied to quadratic functions. More precisely,

$$[Q(f)Q(g)] = \hbar Q(\{f, g\}) + \hbar^2 C(f, g),$$

where C is the cocycle given by

$$C(p_\alpha p_\beta, q^\alpha q^\beta) = \begin{cases} 1 & \alpha \neq \beta \\ 2 & \alpha = \beta \end{cases} \tag{8}$$

and $C(f, g) = 0$ for any other pair of quadratic monomials f and g. This ambiguity is sometimes expressed by saying that the quantization procedure gives only a projective representation of the Lie algebra of quadratic functions in the Darboux coordinates.

3.1.3 Quantization of Symplectomorphisms

All that remains is to address the issue of functoriality. That is, a symplectic diffeomorphism $T : V \to V$ should give rise to an operator $U_T : \mathscr{H}_e \to \mathscr{H}_{\tilde{e}}$. In fact, we will do something slightly different: we will associate to T an operator $U_T : \mathscr{H}_e \to \mathscr{H}_e$. In certain cases, there is a natural identification between $\mathscr{H}_e$ and $\mathscr{H}_{\tilde{e}}$, so our procedure does the required job. More generally, the need for such an identification introduces some ambiguity into the functoriality of quantization.

Furthermore, we will consider only the case in which T is linear and is the exponential of an infinitesimal symplectic transformation, which is simply a linear transformation whose exponential is symplectic. For applications to Gromov-Witten theory later, such transformation are the only ones we will need to quantize.

The computation of U_T is the content of the next three sections.

3.2 Quantization of Symplectomorphisms via the CCR

The following section closely follows the presentation given in [27].

Let $T : V \to V$ be a linear symplectic isomorphism taking the basis e to a new basis $\tilde{e} = \{\tilde{e}^\alpha, \tilde{e}_\beta\}$ via the transformation

$$\tilde{e}^\alpha = A^\alpha_\beta e^\beta + C^{\beta\alpha} e_\beta,$$

$$\tilde{e}_\alpha = B_{\beta\alpha} e^\beta + D^\beta_\alpha e_\beta.$$

Let $\tilde{p}_\alpha$ and $\tilde{q}^\beta$ be the corresponding coordinate functions for the basis vectors $\tilde{e}^\alpha$ and $\tilde{e}_\beta$. Then $\widehat{\tilde{p}_\beta}$ and $\widehat{\tilde{q}^\alpha}$ are defined in the same way as above. The relation between the two sets of coordinate functions is:

$$p_\alpha = A^\beta_\alpha \tilde{p}_\beta + B_{\alpha\beta} \tilde{q}^\beta \qquad (9)$$
$$q^\alpha = C^{\alpha\beta} \tilde{p}_\beta + D^\alpha_\beta \tilde{q}^\beta,$$

and the same equations give relations between their respective quantizations. We will occasionally make use of the matrix notation

$$p = A\tilde{p} + B\tilde{q}$$
$$q = C\tilde{p} + D\tilde{q}$$

to abbreviate the above.

To define the operator $\mathscr{H}_e \to \mathscr{H}_e$ associated to the transformation T, observe that by inverting the relationship (9) and quantizing, one can view both $\hat{p}_\alpha, \hat{q}^\beta$ and $\widehat{\tilde{p}}_\alpha, \widehat{\tilde{q}^\beta}$ as representations of $\mathscr{H}_e$. The operator U_T will be defined by the requirement that

$$\hat{\tilde{q}}^\alpha U_T = U_T \hat{q}^\alpha \qquad (10)$$

$$\hat{\tilde{p}}_\alpha U_T = U_T \hat{p}_\alpha. \qquad (11)$$

As the computation will show, these equations uniquely specify U_T up to a multiplicative constant.

To obtain an explicit formula for U_T, we restrict as mentioned above to the case in which

$$T = \begin{pmatrix} A & B \\ C & D \end{pmatrix} = \exp \begin{pmatrix} a & b \\ c & d \end{pmatrix}.$$

In other words, if

$$T_t = \begin{pmatrix} A(t) & B(t) \\ C(t) & D(t) \end{pmatrix} = \exp\begin{pmatrix} ta & tb \\ tc & td \end{pmatrix},$$

then $\left(\begin{smallmatrix} a & b \\ c & d \end{smallmatrix}\right)$ is the derivative at $t = 0$ of the family of transformations T_t. The infinitesimal symplectic relations

$$b^T = b, \qquad c^T = c, \qquad a^T = -d \tag{12}$$

follow from the relations defining T by differentiation.

This family of transformations yields a family of bases $\{e_t^\alpha, e_\alpha^t\}$ defined by

$$e_t^\alpha = A(t)^\alpha_\beta e^\beta + C(t)^{\beta\alpha} e_\beta$$
$$e_\alpha^t = B(t)_{\beta\alpha} e^\beta + D(t)^\beta_\alpha e_\beta,$$

with corresponding coordinate functions $\{p_\alpha^t, q_t^\alpha\}$. Note that we obtain the original transformation T, as well as the original basis $\tilde{e}$ and coordinate functions $\tilde{p}_\alpha, \tilde{q}^\beta$, by setting $t = 1$.

Using the fact that

$$T_t^{-1} = \exp\begin{pmatrix} -at & -bt \\ -ct & -dt \end{pmatrix} = \begin{pmatrix} I - ta & -tb \\ -tc & I - td \end{pmatrix} + O(t^2),$$

we obtain the relations

$$e^\alpha = e_t^\alpha - ta^\alpha_\beta e_t^\beta - tc^{\beta\alpha} e_\beta^t + O(t^2),$$
$$e_\alpha = e_\alpha^t - tb_{\beta\alpha} e_t^\beta - td^\beta_\alpha e_\beta^t + O(t^2).$$

This implies

$$p_\alpha^t = p_\alpha - ta^\beta_\alpha p_\beta - tb_{\alpha\beta} q^\beta + O(t^2)$$
$$q_t^\alpha = q^\alpha - tc^{\alpha\beta} p_\beta - td^\alpha_\beta q^\beta + O(t^2)$$

and consequently

$$\hat{p_\alpha^t} = \hat{p_\alpha} - ta^\beta_\alpha \hat{p_\beta} - tb_{\alpha\beta} \hat{q^\beta} + O(t^2) \tag{13}$$
$$\hat{q_t^\alpha} = \hat{q^\alpha} - tc^{\alpha\beta} \hat{p_\beta} - td^\alpha_\beta \hat{q^\beta} + O(t^2).$$

Let $U_t = U_{T_t}$, so that

$$\hat{p^t_\alpha} U_t = U_t \hat{p_\alpha} \tag{14}$$
$$\hat{q^\alpha_t} U_t = U_t \hat{q^\alpha}.$$

Denote by u the **infinitesimal variation of** U_T,

$$u = \left.\frac{d}{dt}\right|_{t=0} U_t.$$

By plugging (13) into the above equations and taking the derivative at $t = 0$, we derive commutation relations satisfied by u:

$$[\hat{p_\alpha}, u] = a_\alpha^\beta \hat{p_\beta} + b_{\alpha\beta}\hat{q^\beta}$$
$$[\hat{q^\alpha}, u] = c^{\alpha\beta} \hat{p_\beta} + d_\beta^\alpha \hat{q^\beta}.$$

These equations allow us to determine the infinitesimal variation of U up to a constant:

$$u = -\frac{1}{2\hbar} c^{\alpha\beta} \hat{p_\alpha}\hat{p_\beta} + \frac{1}{\hbar} a_\alpha^\beta \hat{q^\alpha}\hat{p_\beta} + \frac{1}{2\hbar} b_{\alpha\beta}\hat{q^\alpha}\hat{q^\beta} + C.$$

After identifying $\hat{p_\alpha}$ and $\hat{q^\alpha}$ with the operators $\hbar\frac{\partial}{\partial q^\alpha}$ and q^α, respectively, we obtain

$$u = -\frac{\hbar}{2} c^{\alpha\beta} \frac{\partial}{\partial q^\alpha}\frac{\partial}{\partial q^\beta} + a_\alpha^\beta q^\alpha \frac{\partial}{\partial q^\beta} + \frac{1}{2\hbar} b_{\alpha\beta} q^\alpha q^\beta + C. \tag{15}$$

Expanding (14) with respect to t yields formulas for $\left[\hat{p_\alpha}, \left((\frac{d}{dt})^k U_t\right)\big|_{t=0}\right]$ and $\left[\hat{q^\alpha}, \left((\frac{d}{dt})^k U_t\right)\big|_{t=0}\right]$. Using these, one can check that if T_t takes the form $T_t = \exp\left(t\left(\begin{smallmatrix} a & b \\ c & d\end{smallmatrix}\right)\right)$, then U_t will take the form $U_t = \exp(tu)$, so in particular, $U_T = \exp(u)$. Thus, Eq. (15) gives a general formula for U_T.

The formula simplifies significantly when T takes certain special forms. Let us look explicitly at some particularly simple cases.

Example 6 Consider first the case where $b = c = 0$. In this case, we obtain

$$(U_T\psi)(q) = \exp\left(a_\alpha^\beta q^\alpha \frac{\partial}{\partial q^\beta}\right)\psi(q) = \psi(A^T q).$$

Let us verify this formula in a very easy case. Suppose that a has only one nonzero entry, and that this entry is off the diagonal, so that $a_\alpha^\beta = \delta_\alpha^i \delta_j^\beta$ for some fixed $i \neq j$. Assume furthermore that ψ is a monomial:

$$\psi(q) = \prod_\alpha (q^\alpha)^{c_\alpha}.$$

Then

$$
\begin{aligned}
(U_T\psi)(q) &= \exp\left(a_i^j q^i \frac{\partial}{\partial q^j}\right)\prod (q^\alpha)^{c_\alpha}\\
&= \sum_{k=0}^{c_j} \frac{1}{k!}\left(a_i^j q^i \frac{\partial}{\partial q^j}\right)^k \prod (q^\alpha)^{c_\alpha}\\
&= \sum_{k=0}^{c_j} \frac{c_j!}{(k!)(c_j-k)!}\left(a_i^j\right)^k (q^i)^k (q^j)^{c_j-k} \prod_{\alpha\neq j} (q^\alpha)^{c_\alpha}\\
&= \prod_{\alpha\neq j} (q^\alpha)^{c_\alpha}\left(q^j + a_i^j q^i\right)^{c_j}\\
&= \psi(A^T q),
\end{aligned}
$$

where we use that $A = \exp(a) = I + a$ in this case.

Example 7 In the case where $a = c = 0$, the formula above directly implies[4]

$$(U_T\psi)(q) = \exp\left(\frac{1}{2\hbar} b_{\alpha\beta} q^\alpha q^\beta\right)\psi(q).$$

However, in this case, it is easy to check that $B = b$, and we obtain

$$(U_T\psi)(q) = \exp\left(\frac{1}{2\hbar} B_{\alpha\beta} q^\alpha q^\beta\right)\psi(q).$$

Example 8 Finally, consider the case where $a = b = 0$. Then

$$(U_T\psi)(q) = \exp\left(-\frac{\hbar}{2} C^{\alpha\beta}\frac{\partial}{\partial q^\alpha}\frac{\partial}{\partial q^\beta}\right)\psi(q).$$

Remark 9 It is worth noting that this expression can be evaluated using Feynman diagram techniques, in which each diagram corresponds to a term in the Taylor series expansion of the exponential; we will discuss this further in Sect. 3.5.4.

[4]The analogue of this case in the infinite-dimensional situation is referred to in the Gromov-Witten theory literature as "lower-triangular", although the matrix representing T is in fact upper-triangular in our chosen ordering of the basis. To minimize confusion, we will avoid the terminology "upper-triangular" and "lower-triangular" in these notes.

In fact, the above three examples let us compute U_T for any symplectic matrix $T = \begin{pmatrix} A & B \\ C & D \end{pmatrix} = \exp\begin{pmatrix} a & b \\ c & d \end{pmatrix}$ for which the lower-right submatrix D is invertible.[5] To do so, decompose T as

$$T = \begin{pmatrix} A & B \\ C & D \end{pmatrix} = \begin{pmatrix} I & BD^{-1} \\ 0 & I \end{pmatrix} \cdot \begin{pmatrix} I & 0 \\ DC^T & I \end{pmatrix} \cdot \begin{pmatrix} D^{-T} & 0 \\ 0 & D \end{pmatrix}.$$

Each of the matrices on the right falls into one of the cases we calculated above. Furthermore, the quantization procedure satisfies $U_{T_1 \circ T_2} = U_{T_1} \circ U_{T_2}$ (up to a constant), since both sides satisfy (10) and (11). Thus, the formula for the quantization U_T of any such T is as follows:

$$(U_T\psi)(q) = \exp\left(\frac{1}{2\hbar}(BD^{-1})_{\alpha\beta}q^\alpha q^\beta\right)\exp\left(-\frac{\hbar}{2}(DC^T)^{\alpha\beta}\frac{\partial}{\partial q^\alpha}\frac{\partial}{\partial q^\beta}\right)\psi((D^{-T})^\alpha_\beta q),$$

or, in matrix notation,

$$(U_T\psi)(q) = \exp\left(\frac{1}{2\hbar}(BD^{-1}q)\cdot q\right)\exp\left(-\frac{\hbar}{2}\left(DC^T\frac{\partial}{\partial q}\right)\cdot\frac{\partial}{\partial q}\right)\psi(D^{-1}q).$$

One should be somewhat careful with this expression, since the two exponentials have, respectively, infinitely many negative powers of $\hbar$ and infinitely many positive powers of $\hbar$, so a priori their composition may have some powers of $\hbar$ whose coefficients are divergent series. Avoiding this issue requires one to apply each quantized operator to $\psi(D^{-1}q)$ in turn, verifying at each stage that the coefficient of every power of $\hbar$ converges. A similar issue will arise when dealing with powers of the variable z in the infinite-dimensional setting; see Sect. 5.3.

3.3 *Quantization via Quadratic Hamiltonians*

Before moving on to other expressions for the quantization U_T, let us briefly observe that the formulas obtained in the previous section can be described in a much simpler fashion by referring to the terminology of Hamiltonian mechanics. We have preferred the longer derivation via the **CCR** because it more clearly captures the "obvious" functoriality one would desire from the quantization procedure, but the Hamiltonian perspective is the one that is typically taken in discussions of quantization in Gromov-Witten theory (see, for example, [6] or [16]).

---

[5]Throughout this text, we will assume for convenience that D is invertible. However, if this is not the case, one can still obtain similar formulas by decomposing the matrix differently.

Let $T = \exp(F)$ be a symplectomorphism as above, where

$$F = \begin{pmatrix} a & b \\ c & d \end{pmatrix}$$

is an infinitesimal symplectic transformation. Because the tangent space to a symplectic vector space at any point is canonically identified with the vector space itself, we can view $F : V \to V$ as a vector field on V. If ω is the 2-form giving the symplectic structure, the contraction $\iota_F\omega$ is a 1-form on V. Since V is topologically contractible, we can write $\iota_F\omega = dh_F$ for some function $h_F : V \to \mathbb{R}$. This function is referred to as the **Hamiltonian** of F. Concretely, it is described by the formula

$$h_F(v) = \frac{1}{2}\langle Fv, v\rangle$$

for $v \in V$, where $\langle\ ,\ \rangle$ is the symplectic pairing.

Being a classical observable, the quantization of the function $h_F : V \to \mathbb{R}$ has already been defined. Define the quantization of F by

$$\hat{F} = \frac{1}{\hbar}\hat{h}_F.$$

The quantization of the symplectomorphism T is then defined as

$$U_T = \exp(\hat{F}).$$

It is an easy exercise to check that $\hat{F}$ agrees with the general formula given by (15), so the two definitions of U_T coincide.

One advantage of the Hamiltonian perspective is that it provides a straightforward way to understand the noncommutativity of the quantization procedure for infinitesimal symplectic transformations. Recall, the quantization of quadratic observables obeys the commutation relation

$$[Q(f), Q(g)] = \hbar Q(\{f, g\}) + \hbar^2 C(f, g)$$

for the cocycle C defined in (8). It is easy to check (for example, by working in Darboux coordinates) that the Hamiltonian h_A associated to an infinitesimal symplectic transformation satisfies

$$\{h_A, h_B\} = h_{[A,B]}.$$

Thus, the commutation relation for infinitesimal symplectic transformations is

$$[\hat{A}, \hat{B}] = \widehat{[A, B]} + C(h_A, h_B).$$

For an explicit computation of this cocycle in an (infinite-dimensional) case of particular interest, see Example 1.3.4.1 of [6].

3.4 Integral Formulas

The contents of this section are based on lectures given by Xiang Tang at the RTG Workshop on Quantization at the University of Michigan in December 2011. In a more general setting, the material is discussed in [3].

Our goal is to obtain an alternative expression for U_T of the form

$$(U_T\psi)(q) = \lambda \int_{\mathbb{R}^n}\int_{\mathbb{R}^n} e^{\frac{1}{\hbar}(\phi(q,p')-q'\cdot p')}\psi(q')dq'dp' \tag{16}$$

for a function $\phi : \mathbb{R}^{2n} \to \mathbb{R}$ and a constant $\lambda \in \mathbb{R}[[\hbar, \hbar^{-1}]]$ to be determined. Such operators, since they generalize the Fourier transform of ψ, are known as **Fourier integral operators**.

The advantage of this alternate expression for U_T is twofold. First, they allow quantized operators to be expressed as sums over Feynman diagrams, and this combinatorial expansion will be useful later, especially in Sect. 5.5. Second, the notion of a Fourier integral operator generalizes to the case when the symplectic diffeomorphism is not necessarily linear, as well as to the case of a Lagrangian submanifold of the cotangent bundle that is not the graph of any symplectomorphism; we will comment briefly on these more general settings in Sect. 3.6 below.

To define ϕ, first let Γ_T be the graph of T:

$$\Gamma_T = \left\{(p,q,\tilde{p},\tilde{q}) \;\middle|\; \begin{array}{l} p_\alpha = A_\alpha^\beta \tilde{p}_\beta + B_{\alpha\beta}\tilde{q}^\beta \\ q^\alpha = C^{\alpha\beta}\tilde{p}_\beta + D_\beta^\alpha \tilde{q}^\beta \end{array}\right\} \subset \overline{\mathbb{R}^{2n}} \times \mathbb{R}^{2n}.$$

Here, $\overline{\mathbb{R}^{2n}}$ denotes the symplectic vector space obtained by equipping $\mathbb{R}^{2n}$ with the opposite of the standard symplectic form, so that the symplectic form on the product is given by

$$\langle (p,q,\tilde{p},\tilde{q}),(P,Q,\tilde{P},\tilde{Q})\rangle = \sum_i \left(-p_iQ^i + P_iq^i + \tilde{p}_i\tilde{Q}^i - \tilde{P}_i\tilde{q}^i\right).$$

Under this choice of symplectic form, Γ_T is a Lagrangian submanifold of the product.

There is an isomorphism of symplectic vector spaces

$$\overline{\mathbb{R}^{2n}} \times \mathbb{R}^{2n} \xrightarrow{\sim} T^*(\mathbb{R}^{2n})$$
$$(p,q,\tilde{p},\tilde{q}) \mapsto (q,\tilde{p},p,\tilde{q}),$$

where $T^*(\mathbb{R}^{2n})$ is equipped with the canonical symplectic form

$$\langle (q, p, \pi, \xi), (Q, P, \Pi, \Xi) \rangle = \sum_i \left(q^i \Pi_i + p_i \Xi^i - Q^i \pi_i - P_i \xi^i \right).$$

Thus, one can view Γ_T as a Lagrangian submanifold of the cotangent bundle. Define $\phi = \phi(q, p')$ as the generating function for this submanifold. Explicitly, this says that

$$\left\{ (q, \tilde{p}, p, \tilde{q})) \left| \begin{array}{l} p_\alpha = A_\alpha^\beta \tilde{p}_\beta + B_{\alpha\beta} \tilde{q}^\beta \\ q^\alpha = C^{\alpha\beta} \tilde{p}_\beta + D_\beta^\alpha \tilde{q}^\beta \end{array} \right. \right\} = \left\{ \left(q, p', \frac{\partial \phi}{\partial q}, \frac{\partial \phi}{\partial p'} \right) \right\}.$$

Let us restrict, similarly to Sect. 3.2, to the case in which D is invertible. The relations defining Γ_T can be rearranged to give

$$\tilde{q} = D^{-1} q - D^{-1} C \tilde{p}$$
$$p = B D^{-1} q + D^{-T} \tilde{p}.$$

Therefore, $\phi(q, p')$ is defined by the system of partial differential equations

$$\frac{\partial \phi}{\partial q} = B D^{-1} q + D^{-T} p'$$
$$\frac{\partial \phi}{\partial p'} = D^{-1} q - D^{-1} C p'.$$

These are easily solved; up to an additive constant, one obtains

$$\phi(q, p') = \frac{1}{2} (B D^{-1} q) \cdot q + (D^{-1} q) \cdot p' - \frac{1}{2} (D^{-1} C p') \cdot p'.$$

The constant λ in the definition of U_T is simply a normalization factor, and is given by

$$\lambda = \frac{1}{\hbar^n},$$

as this will be necessary to make the integral formulas match those computed in the previous section. We thus obtain the following definition for U_T:

$$(U_T \psi)(q) = \frac{1}{\hbar^n} \int_{\mathbb{R}^n} \int_{\mathbb{R}^n} e^{\frac{1}{\hbar} \left(\frac{1}{2} (B D^{-1} q) \cdot q + (D^{-1} q) \cdot p' - \frac{1}{2} (D^{-1} C p') \cdot p' - q' \cdot p' \right)} \psi(q') dq' dp'. \tag{17}$$

We should verify that this formula agrees with the one obtained in Sect. 3.2. This boils down to properties of the Fourier transform, which we will define as

$$\hat{\psi}(y) = \int_{\mathbb{R}^n} e^{-ix\cdot y}\psi(x)dx.$$

Under this definition,

$$\begin{aligned}(U_T\psi)(q) &= \frac{e^{\frac{1}{2\hbar}(BD^{-1}q)\cdot q}}{\hbar^n}\int_{\mathbb{R}^n}\int_{\mathbb{R}^n} e^{\frac{1}{\hbar}(D^{-1}q)\cdot p'}e^{-\frac{1}{2\hbar}(D^{-1}Cp')\cdot p'}e^{-\frac{1}{\hbar}q'\cdot p'}\psi(q')dq'dp'\\ &= i^n e^{\frac{1}{2\hbar}(BD^{-1}q)\cdot q}\int_{i\mathbb{R}^n} e^{i(D^{-1}q)\cdot p''}e^{\frac{\hbar}{2}(D^{-1}Cp'')\cdot p''}\hat{\psi}(p'')dp''\\ &= e^{\frac{1}{2\hbar}(BD^{-1}q)\cdot q}\int_{\mathbb{R}^n} e^{iQ\cdot p'''}e^{-\frac{\hbar}{2}(D^{-1}Cp''')\cdot p'''}\widehat{\psi\circ i}(p''')dp'''\\ &= e^{\frac{1}{2\hbar}(BD^{-1}q)\cdot q}\int_{\mathbb{R}^n} e^{iQ\cdot p'''}\left(e^{-\frac{\hbar}{2}(D^{-1}C\frac{\partial}{\partial q'})\cdot\frac{\partial}{\partial q'}}(\psi)\circ i\right)^{\wedge}(p''')dp'''\\ &= e^{\frac{1}{2\hbar}(BD^{-1}q)\cdot q}e^{-\frac{\hbar}{2}\left(DC^T\frac{\partial}{\partial q}\right)\cdot\frac{\partial}{\partial q}}\psi(D^{-1}q),\end{aligned}$$

where we use the changes of variables $\frac{1}{\hbar}p' = ip''$, $ip'' = p'''$, and $D^{-1}q = iQ$. The integral formula therefore matches the one defined via the **CCR**.

3.5 *Expressing Integrals via Feynman Diagrams*

The formula given for $U_T\psi$ in (17) bears a striking resemblance to the type of integral computed by Feynman's theorem, and in this section, we will make the connection precise.

3.5.1 Genus-Modified Feynman's Theorem

In order to apply Feynman's theorem, the entire integrand must be an exponential, so we will assume that

$$\psi(q) = e^{\frac{1}{\hbar}f(q)}.$$

Furthermore, let us assume that $f(q)$ is $\hbar^{-1}$ times a power series in $\hbar$, so ψ is of the form

$$\psi(q) = e^{\sum_{g\geq 0}\hbar^{g-1}\mathcal{F}_g(q)}.$$

Then

$$(U_T\psi)(q) = e^{\frac{1}{2\hbar}(BD^{-1}q)\cdot q}\, \hbar^{-n} \int_{\mathbb{R}^{2n}} e^{-\frac{1}{\hbar}S(p',q')} dp'dq'$$

with

$$S(p',q') = \Big(-(D^{-1}q)\cdot p' + \frac{1}{2}(D^{-1}Cp')\cdot p' + q'\cdot p' \Big) - \sum_{g\geq 0} \hbar^g \mathcal{F}_g(q').$$

Note that if we let $y = (p', q')$, then the bilinear leading term of $S(p', q')$ (in parentheses above) is equal to

$$\frac{1}{2}(D^{-1}Cp')\cdot p' + \frac{1}{2}(q' - D^{-1}q)\cdot p' + \frac{1}{2}p'\cdot(q' - D^{-1}q) = \frac{\mathcal{B}(y - D^{-1}q, y - D^{-1}q)}{2},$$

where $\mathcal{B}(y_1, y_2)$ is the bilinear form given by the block matrix

$$\begin{pmatrix} D^{-1}C & I \\ I & 0 \end{pmatrix}$$

and $q = (0, q) \in \mathbb{R}^{2n}$.

Changing variables, we have

$$(U_T\psi)(q) = e^{\frac{1}{2\hbar}(BD^{-1}q)\cdot q}\, \hbar^{-n} \int_{\mathbb{R}^{2n}} e^{-\frac{1}{\hbar}\left(\frac{-\mathcal{B}(y,y)}{2} - \sum_{g\geq 0}\hbar^g \mathcal{F}_g(q' + D^{-1}q)\right)} dy.$$

Each of the terms $-\mathcal{F}_g(q' + D^{-1}q)$ can be decomposed into pieces that are homogeneous in q':

$$-\mathcal{F}_g(q' + D^{-1}q) = \sum_{m\geq 0} \frac{1}{m!}\left(-\partial^m\, \mathcal{F}_g\big|_{q'=D^{-1}q} \cdot (q')^m \right),$$

where $-\partial^m \mathcal{F}_g|_{q'=-D^{-1}q} \cdot (q')^m$ is short-hand for the m-tensor

$$B_{g,m} = -\sum_{|\mathbf{m}|=m} \frac{m!}{m_1!\cdots m_n!}\, \frac{\partial^m \mathcal{F}_g}{(\partial q_1')^{m_1}\cdots(\partial q_n')^{m_n}}\Bigg|_{q'=D^{-1}q} (q_1')^{m_1}\cdots(q_n')^{m_n},$$

in which the sum is over all n-tuples $\mathbf{m} = (m_1, \ldots, m_n) \in \mathbb{Z}^n_{\geq 0}$ such that $|\mathbf{m}| = m_1 + \cdots + m_n = m$. We consider this as an m-tensor in q' whose coefficients involve a formal parameter q.

Thus, we have expressed the quantized operator as

$$(U_T\psi)(q) = e^{\frac{1}{2\hbar}(BD^{-1}q)\cdot q}\, \hbar^{-n} \int_{\mathbb{R}^{2n}} e^{-\frac{1}{\hbar}\left(\frac{-\mathcal{B}(y,y)}{2} + \sum_{g,m\geq 0} \frac{\hbar^g}{m!} B_{g,m}(q',\ldots,q')\right)} dy.$$

This is essentially the setting in which Feynman's theorem applies, but we must modify Feynman's theorem to allow for the presence of powers of $\hbar$ in the exponent. This is straightforward, but nevertheless interesting, as it introduces a striking interpretation of g as recording the "genus" of vertices in a graph.

Theorem 10 *Let V be a vector space of dimension d, and let*

$$S(x) = \frac{1}{2}B(x,x) + \sum_{g,m\geq 0} \frac{\hbar^g}{m!} B_{g,m}(x,\ldots,x),$$

in which each $B_{g,m}$ is an m-multilinear form and $B_{0,0} = B_{0,1} = B_{0,2} = 0$. Consider the integral

$$Z = \hbar^{-\frac{d}{2}} \int_V e^{-S(x)/\hbar} dx.$$

Then

$$Z = \frac{(2\pi)^{d/2}}{\sqrt{\det(B)}} \sum_{\mathbf{n}=(n_0,n_1,\ldots)} \sum_{\Gamma\in G'(\mathbf{n})} \frac{\hbar^{-\chi_\Gamma}}{|Aut(\Gamma)|} F_\Gamma,$$

where F_Γ is the genus-modified Feynman amplitude, defined below.

Here, $G'(\mathbf{n})$ denotes the set of isomorphism classes of graphs with n_i vertices of valence i for each $i \geq 0$, in which each vertex v is labeled with a genus $g(v) \geq 0$. Given $\Gamma \in G'(\mathbf{n})$, the **genus-modified Feynman amplitude** is defined by the following procedure:

1. Put the m-tensor $-B_{g,m}$ at each m-valent vertex of genus g in Γ.
2. For each edge e of Γ, take the contraction of tensors attached to the vertices of e using the bilinear form B^{-1}. This will produce a number F_{Γ_i} for each connected component Γ_i of Γ.
3. If $\Gamma = \bigsqcup_i \Gamma_i$ is the decomposition of Γ into connected components, define $F_\Gamma = \prod_i F_{\Gamma_i}$.

Furthermore, the **Euler characteristic** of Γ is defined as

$$\chi_\Gamma = -\sum_{v\in V(\Gamma)} g(v) + |V(\Gamma)| - |E(\Gamma)|.$$

Having established all the requisite notation, the proof of the theorem is actually easy.

Proof of Theorem 10 Reiterate the proof of Feynman's theorem to obtain

$$Z = \sum_{\mathbf{n}=(n_{g,m})_{g,m\geq 0}} \left(\prod_{g,m\geq 0} \frac{\hbar^{g n_{g,m} + (\frac{m}{2}-1) n_{g,m}}}{(m!)^{n_{g,m}} n_{g,m}!} \right) \int_V e^{-B(y,y)/2} \prod_{g,m\geq 0} (-B_{g,m})^{n_{g,m}} dy.$$

As before, Wick's theorem shows that the integral contributes the desired summation over graphs, modulo factors coming from over-counting. A similar orbit-stabilizer argument shows that these factors precisely cancel the factorials in the denominator. The power of $\hbar$ is

$$\sum_{g\geq 0} g \left(\sum_{m\geq 0} n_{g,m} \right) + \sum_{m\geq 0} \left(\frac{m}{2} - 1 \right) \left(\sum_{g\geq 0} n_{g,m} \right),$$

which is the sum of the genera of the vertices plus the number of edges minus the number of vertices, or in other words, $-\chi_\Gamma$, as required. □

It should be noted that for the proof of the theorem, there is no particular reason to think of g as recording the genus of a vertex—it is simply a label associated to the vertex that records which of the m-tensors $B_{g,m}$ one attaches. The convenience of the interpretation of g as genus comes only from the fact that it simplifies the power of $\hbar$ neatly.

3.5.2 Feynman Diagram Formula for U_T

We are now ready to give an expression for $U_T\psi$ in terms of Feynman diagrams. In order to apply Theorem 10, we must make one more assumption: that $\mathcal{F}_0$ has no terms of homogeneous degree less than 3 in q'. Assuming this, we obtain the following expression by directly applying the theorem:

$$(U_T\psi)(q) = \frac{(2\pi)^n e^{\frac{1}{2\hbar}(BD^{-1}q)\cdot q}}{\sqrt{\det(D^{-1}C)}} \sum_{\mathbf{n}=(n_0,n_1,\ldots)} \sum_{\Gamma\in G'(\mathbf{n})} \frac{\hbar^{-\chi_\Gamma}}{|\mathrm{Aut}(\Gamma)|} F_\Gamma(q), \tag{18}$$

where $F_\Gamma(q)$ is the genus-modified Feynman amplitude given by placing the m-tensor

$$\sum_{|\mathbf{m}|=m} \frac{m!}{m_1!\cdots m_n!} \left. \frac{\partial^m \mathcal{F}_g}{(\partial q_1')^{m_1}\cdots(\partial q_n')^{m_n}} \right|_{q'=D^{-1}q} (q_1')^{m_1}\cdots(q_n')^{m_n}$$

at each m-valent vertex of genus g in Γ and taking the contraction of tensors using the bilinear form

$$\begin{pmatrix} D^{-1}C & I \\ I & 0 \end{pmatrix}^{-1} = -\begin{pmatrix} 0 & I \\ I & D^{-1}C \end{pmatrix}.$$

In fact, since this bilinear form is only ever applied to vectors of the form $(0, q')$, we are really only taking contraction of tensors using the bilinear form $-D^{-1}C$ on $\mathbb{R}^n$.

3.5.3 Connected Graphs

Recall from Theorem 5 that the logarithm of a Feynman diagram sum yields the sum over only connected graphs. That is:

$$(U_T\psi)(q) = \frac{(2\pi)^n e^{\frac{1}{2\hbar}(BD^{-1}q)\cdot q}}{\sqrt{\det(D^{-1}C)}} \exp\left(\sum_{\Gamma \in G'_c} \frac{\hbar^{-\chi_\Gamma}}{|\mathrm{Aut}(\Gamma)|} F_\Gamma(q) \right),$$

where G'_c denotes the set of isomorphism classes of *connected* genus-labeled graphs. Thus, writing

$$\overline{\mathcal{F}}_g(q) = \sum_{\Gamma \in G'_c(g)} \frac{1}{|\mathrm{Aut}(\Gamma)|} F_\Gamma(q)$$

with $G'_c(g)$ collecting connected graphs of genus g, we have:

$$(U_T\psi)(q) = \frac{(2\pi)^n e^{\frac{1}{2\hbar}(BD^{-1}q)\cdot q}}{\sqrt{\det(D^{-1}C)}} \exp\left(\sum_{g \geq 0} \hbar^{g-1} \overline{\mathcal{F}}_g(q) \right).$$

For those familiar with Gromov-Witten theory, this expression should be salient—we will return to it in Sect. 5 of the book.

3.5.4 Another Diagram Expansion

At this point, we can return to a remark made previously (Remark 9), regarding the computation of $U_T\psi$ in the case where a quadratic differential operator appears in the formula. In that case, we may express the quantization formula obtained via the **CCR** as a graph sum in a rather different way. We explain how to do this below, and show that we ultimately obtain the same graph sum as that which arises from applying Feynman's Theorem to the integral operator.

Let

$$T = \begin{pmatrix} I & 0 \\ C & I \end{pmatrix}.$$

We showed in Sect. 3.2 that

$$(U_T\psi)(q) = \exp\left(-\frac{\hbar}{2}C^{\alpha\beta}\frac{\partial}{\partial q^\alpha}\frac{\partial}{\partial q^\beta}\right)\psi(q).$$

Suppose that $\psi(q) = e^{\sum_{g\geq 0}\hbar^{g-1}\mathcal{F}_g(q)}$ as above and expand both exponentials in Taylor series. Then $(U_T\psi)(q)$ can be expressed as:

$$\sum_{\{i_{\alpha,\beta}\},\{\ell_g\}} \frac{\hbar^{\sum i_{\alpha\beta}+\sum \ell_g(g-1)}}{\prod i_{\alpha\beta}!\prod \ell_g!}\prod_{\alpha,\beta}\left(-\frac{C^{\alpha\beta}}{2}\right)^{i_{\alpha\beta}}\left(\frac{\partial}{\partial q^\alpha}\frac{\partial}{\partial q^\beta}\right)^{i_{\alpha\beta}}\prod_g(\mathcal{F}_g(q))^{\ell_g}. \tag{19}$$

Whenever a product of quadratic differential operators acts on a product of functions, the result can be written as a sum over diagrams. As an easy example, suppose one wishes to compute

$$\frac{\partial^2}{\partial x\partial y}(fgh)$$

for functions f, g, h in variables x and y. The product rule gives nine terms, each of which can be viewed as a way of attaching an edge labeled

to a collection of vertices labeled f, g, and h. (We allow both ends of an edge to be attached to the same vertex.)

Applying this general principle to the expression (19), one can write each of the products $\prod(-\frac{C^{\alpha\beta}}{2})^{i_{\alpha\beta}}\prod\left(\frac{\partial}{\partial q^\alpha}\frac{\partial}{\partial q^\beta}\right)^{i_{\alpha\beta}}\mathcal{F}_g(q)$ as a sum over graphs obtained by taking ℓ_g vertices of genus g for each g, with vertices of genus g labeled $\mathcal{F}_g$, and attaching $i_{\alpha\beta}$ edges labeled

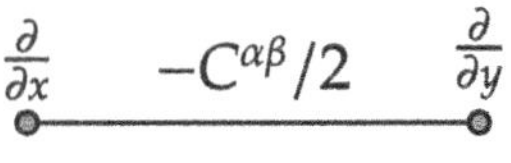

in all possible ways. Each possibility gives a graph $\hat{\Gamma}$. It is a combinatorial exercise to check that the contributions from all choices of $\hat{\Gamma}$ combine to give a factor of $\frac{\prod i_{\alpha\beta}!\cdot\prod \ell_g!}{\mathrm{Aut}(\hat{\Gamma})|}$.

Thus, we have expressed $(U_T\psi)(q)$ as

$$\sum_{\{i_{\alpha\beta}\},\{\ell_g\},\hat{\Gamma}} \frac{\hbar^{-\chi_{\hat{\Gamma}}}}{|\mathrm{Aut}(\hat{\Gamma})|} G_{\mathbf{i},\boldsymbol{\ell},\hat{\Gamma}}(q),$$

where $G_{\mathbf{i},\boldsymbol{\ell},\hat{\Gamma}}(q)$ is obtained by way of the above procedure. This is essentially the Feynman amplitude computed previously, but there is one difference: $G_{\mathbf{i},\boldsymbol{\ell},\Gamma}(q)$ is computed via edges whose two ends are *labeled*, whereas Feynman diagrams are unlabeled. By summing up all possible labelings of the same unlabeled graph Γ, we can rewrite this as

$$(U_T\psi)(q) = \sum_{\Gamma} \frac{\hbar^{-\chi_\Gamma}}{|\mathrm{Aut}(\Gamma)|} G_\Gamma,$$

where G_Γ is the Feynman amplitude computed by placing the m-tensor

$$\sum_{|\mathbf{m}|=m} \frac{m!}{m_1!\cdots m_n!} \frac{\partial^m \mathcal{F}_g}{(\partial q_1')^{m_1}\cdots(\partial q_n')^{m_n}}\bigg|_{q'=q} (q_1')^{m_1}\cdots(q_n')^{m_n}$$

at each m-valent vertex of genus g in Γ and taking the contraction of tensors using the bilinear form $-\frac{1}{2}(C + C^T) = -C$. Up to a multiplicative constant, which we can ignore, this matches the Feynman diagram expansion obtained previously.

3.6 Generalizations

As remarked in Sect. 3.4, one advantage of the integral formula representation of a quantized operator is that it generalizes in at least two ways beyond the cases considered here.

First, if T is a symplectic diffeomorphism that is not necessarily linear, one can still define ϕ as the generating function of the graph of T, and under this definition, the integral in (16) still makes sense. Thus, in principle, integral formulas allow one to define the quantization of an arbitrary symplectic diffeomorphism. As it turns out, the formula in (16) is no longer quite right in this more general setting; the constant λ should be allowed to be a function $b = b(q, p', \hbar)$ determined by T and its derivatives. Nevertheless, an integral formula can still be obtained. This is very important from a physical perspective, since the state space of classical mechanics is typically a nontrivial symplectic manifold. While one can reduce to the case of $\mathbb{R}^{2n}$ by working locally, the quantization procedure should be functorial with respect to arbitrary symplectic diffeomorphisms, which can certainly be nonlinear even in local coordinates.

A second possible direction for generalization is that the function ϕ need not be the generating function of the graph of a symplectomorphism at all. Any Lagrangian submanifold $L \subset T^*(\mathbb{R}^{2n})$ has a generating function, and taking ϕ to be the generating function of this submanifold, (16) gives a formula for the quantization of L.

4 Interlude: Basics of Gromov-Witten Theory

In order to apply the formulas for $U_T\psi$ to obtain results in Gromov-Witten theory, it is necessary to quantize infinite-dimensional symplectic vector spaces. Thus, we devote Sect. 5 to the infinite-dimensional situation, discussing how to adapt the finite-dimensional formulas and how to avoid issues of convergence. Before doing this, however, we pause to give a brief overview of the basics of Gromov-Witten theory. Although this material will not be strictly necessary until Sect. 6, we include it now to motivate our interest in the infinite-dimensional case and the specific assumptions made in the next section.

4.1 Definitions

The material of this section can be found in any standard reference on Gromov-Witten theory, for example [7] or [19].

Let X be a projective variety. Roughly speaking, the Gromov-Witten invariants of X encode the number of curves passing through a prescribed collection of subvarieties. In order to define these invariants rigorously, we will first need to define the moduli space $\overline{\mathcal{M}}_{g,n}(X, d)$ of stable maps.

Definition 11 A genus-g, n-pointed **pre-stable curve** is an algebraic curve C with $h^1(C, \mathcal{O}_C) = g$ and at worst nodal singularities, equipped with a choice of n distinct ordered marked points $x_1, \ldots, x_n \in C$.

Fix non-negative integers g and n, and a cycle $d \in H_2(X; \mathbb{Z})$.

Definition 12 A **pre-stable map** of genus g and degree d with n marked points is an algebraic map $f : C \to X$ whose domain is a genus-g, n-pointed pre-stable curve, and for which $f_*[C] = d$. Such a map is **stable** if it has only finitely many automorphisms as a pointed map; concretely, this means that every irreducible component collapsed to a point by f has at least three special points (marked points or nodes) if its genus is zero and at least one special point if its genus is one.

As eluded to in this definition, there is a suitable notion of isomorphism of stable maps. Namely, stable maps $f : C \to X$ and $f' : C' \to X$ are **isomorphic** if there is an isomorphism of curves $s : C \to C'$ which preserves the markings and satisfies $f' \circ s = f$.

There is a moduli space $\overline{\mathcal{M}}_{g,n}(X, d)$ whose points are in bijection with isomorphism classes of stable maps of genus g and degree d with n marked points. To be more precise, $\overline{\mathcal{M}}_{g,n}(X, d)$ is only a coarse moduli scheme, but it can be given the structure of a Deligne-Mumford stack, and with this extra structure it is a fine moduli stack. In general, this moduli space is singular and may possess components of different dimensions. However, there is always an "expected" or "virtual" dimension, denoted vdim, and a class $[\overline{\mathcal{M}}_{g,n}(X, d)]^{\text{vir}} \in H_{2\text{vdim}}(\overline{\mathcal{M}}_{g,n}(X, d))$, called the **virtual fundamental class**, which plays the role of the fundamental class for the purpose of intersection theory. The following theorem collects some of the important (and highly non-trivial) properties of this moduli space.

Theorem 13 *There exists a compact moduli space $\overline{\mathcal{M}}_{g,n}(X, d)$, of virtual dimension equal to*

$$vdim = (\dim(X) - 3)(1 - g) + \int_d c_1(T_X) + n.$$

It admits a virtual fundamental class $[\overline{\mathcal{M}}_{g,n}(X, d)]^{vir} \in H_{2vdim}(\overline{\mathcal{M}}_{g,n}(X, d))$.

Virtual dimension can be given a precise meaning in terms of deformation theory, which we will omit. In certain easy cases, though, the moduli space is smooth and pure-dimensional, and in these cases the virtual dimension is simply the ordinary dimension, and the virtual fundamental class is the ordinary fundamental class. For example, this occurs when $g = 0$ and X is convex (such as the case $X = \mathbb{P}^r$) or when $g = 0$ and $d = 0$.

Gromov-Witten invariants will be defined as integrals over the moduli space of stable maps. The classes we will integrate will come from two places. First, there are **evaluation maps**

$$\text{ev}_i : \overline{\mathcal{M}}_{g,n}(X, d) \to X$$

for $i = 1, \ldots, n$, defined by sending $(C; x_1, \ldots, x_n; f)$ to $f(x_i)$. In fact, these set-theoretic maps are morphisms of schemes (or stacks). Second, there are ψ **classes**. To define these, let $\mathcal{L}_i$ be the line bundle on $\overline{\mathcal{M}}_{g,n}(X, d)$ whose fiber over a point $(C; x_1, \ldots, x_n; f)$ is the cotangent line[6] to the curve C at x_i. Then

$$\psi_i = c_1(\mathcal{L}_i)$$

for $i = 1, \ldots, n$.

[6]Of course, this is only a heuristic definition, as one cannot specify a line bundle by prescribing its fibers. To be more precise, one must consider the universal curve $\pi : \mathcal{C} \to \overline{\mathcal{M}}_{g,n}(X, d)$. This carries a relative cotangent line bundle ω_π. Furthermore, there are sections $s_i : \overline{\mathcal{M}}_{g,n}(X, d) \to \mathcal{C}$ sending $(C; x_1, \ldots, x_n; f)$ to $x_i \in C \subset \mathcal{C}$. We define $\mathcal{L}_i = s_i^* \omega_\pi$.

Definition 14 Fix cohomology classes $\gamma_1, \ldots, \gamma_n \in H^*(X)$ and integers $j_1, \ldots, j_n \in \mathbb{Z}_{\geq 0}$. The corresponding **Gromov-Witten invariant** (or **correlator**) is

$$\langle \gamma_1 \psi_1^{j_1}, \cdots, \gamma_n \psi_n^{j_n} \rangle_{g,n,d}^X = \int_{[\overline{\mathcal{M}}_{g,n}(X,d)]^{vir}} ev_1^*(\gamma_1)\psi_1^{j_1} \cdots ev_n^*(\gamma_n)\psi_n^{j_n}.$$

Often in what follows, the indices on ψ class insertions will be dropped. We will also sometimes make use of the physics notation

$$\langle \tau_{j_1}(\gamma_1), \cdots, \tau_{j_n}(\gamma_n) \rangle_{g,n,d}^X$$

for the above invariant, in which τ is a formal symbol recording the powers of ψ.

While the enumerative significance of this integral is not immediately obvious, there is an interpretation in terms of curve-counting in simple cases. Indeed, suppose that $\overline{\mathcal{M}}_{g,n}(X,d)$ is smooth and its virtual fundamental class is equal to its ordinary fundamental class. Suppose, further, that the classes γ_i are Poincaré dual to transverse subvarieties $Y_i \subset X$, and that $j_i = 0$ for all i. Then the Gromov-Witten invariant above is equal to the number of genus-g, n-pointed curves in X whose first marked point lies on Y_1, whose second marked point lies on Y_2, et cetera. Thus, the invariant indeed represents (in some sense) a count of the number of curves passing through prescribed subvarieties.

In order to encode these invariants in a notationally parsimonious way, we write

$$\mathbf{a}^i(z) = a_0^i + a_1^i z + a_2^i z^2 + \cdots$$

for $a_j^i \in H^*(X)$. Then, given $\mathbf{a}^1, \ldots, \mathbf{a}^n \in H^*(X)[[z]]$, define

$$\langle \mathbf{a}^1(\psi), \ldots, \mathbf{a}^n(\psi) \rangle_{g,n,d}^X = \int_{[\overline{\mathcal{M}}_{g,n}(X,d)]^{\text{vir}}} \left(\sum_{j=0}^{\infty} ev_1^*(a_j^1)\psi_1^j \right) \cdots \left(\sum_{j=0}^{\infty} ev_n^*(a_j^n)\psi_n^j \right).$$

Arbitrary Gromov-Witten invariants of X are determined by those in which every insertion is the same. Thus, making use of the above notation, we can describe all of the genus-g invariants of X via the generating function

$$\mathcal{F}_X^g(\mathbf{t}(z)) = \sum_{n,d \geq 0} \frac{Q^d}{n!} \langle \mathbf{t}(\psi), \ldots, \mathbf{t}(\psi) \rangle_{g,n,d}^X,$$

which is a formal function of the variable $\mathbf{t}(z) \in H^*(X)[[z]]$ taking values in the **Novikov ring** $\mathbb{C}[[Q]]$. Introducing another parameter $\hbar$ to record the genus, we can combine all of the genus-g generating functions into one:

$$\mathcal{D}_X = \exp\left(\sum_g \hbar^{g-1} \mathcal{F}_X^g \right).$$

This is referred to as the **total descendent potential** of X.

It is convenient to sum invariants with certain fixed insertions over all ways of adding additional insertions, as well as all choices of degree. Thus, we define:

$$\langle\langle \mathbf{a}^1(\psi), \ldots, \mathbf{a}^n(\psi)\rangle\rangle^X_{g,n}(s) = \sum_{m,d} \frac{Q^d}{m!}\langle \mathbf{a}^1(\psi), \ldots, \mathbf{a}^n(\psi), s, \ldots, s\rangle^X_{g,n+m,d}$$

for specified $s \in H^*(X)$.

4.2 Basic Equations

In practice, Gromov-Witten invariants are usually very difficult to compute by hand. Instead, calculations are typically carried out combinatorially by beginning from a few easy cases and applying a number of relations. We state those relations in this section. The statements of the relations can be found, for example, in [19], while their expressions as differential equations are given in [23].

4.2.1 String Equation

The string equation addresses invariants in which one of the insertions is $1 \in H^0(X)$ with no ψ classes. It states:

$$\langle \tau_{a_1}(\gamma_1), \cdots, \tau_{a_n}(\psi_n)\ 1\rangle^X_{g,n+1,d} = \sum_{i=1}^{n} \langle \tau_{a_1}(\gamma_1), \cdots, \tau_{a_{i-1}}(\gamma_{i-1}), \tau_{a_i-1}(\gamma_i), \tau_{a_{i+1}}(\gamma_{i+1}), \cdots, \tau_{a_n}(\gamma_n)\rangle^X_{g,n,d} \tag{20}$$

whenever $\overline{\mathcal{M}}_{g,n}(X, d)$ is nonempty.

The proof of this equation relies on a result about pullbacks of ψ classes under the forgetful morphism $\overline{\mathcal{M}}_{g,n+1}(X, d) \to \overline{\mathcal{M}}_{g,n}(X, d)$ that drops the last marked point. Because this morphism involves contracting irreducible components of C that become unstable after the forgetting operation, ψ classes on the target do not pull back to ψ classes on the source. However, there is an explicit comparison result, and this is the key ingredient in the proof of (20). See [19] for the details.

It is a basic combinatorial fact that differentiation of the generating function with respect to the variable t^i_j corresponds to adding an additional insertion of $\tau_j(\phi_i)$. Starting from this, one may express the string equation in terms of a differential equation satisfied by the Gromov-Witten generating function. To see this, fix a basis $\{\phi_1, \ldots, \phi_k\}$ for $H^*(X)$, and write

$$t_j = t^i_j\phi_i.$$

Then

$$\begin{aligned}\frac{\partial}{\partial t_0^1}\sum_g \hbar^{g-1}\mathcal{F}_X^g &= \sum_{g,n,d}\frac{Q^d\hbar^{g-1}}{n!}\langle \mathbf{t}(\psi),\ldots,\mathbf{t}(\psi),1\rangle_{g,n+1,d}^X\\ &= \sum_{g,n,d}\frac{Q^d\hbar^{g-1}}{(n-1)!}\left\langle \mathbf{t}(\psi),\ldots,\mathbf{t}(\psi),\sum_{i,j}t_{j+1}^i\tau_j(\phi_i)\right\rangle_{g,n,d}^X\\ &\quad+\frac{1}{2\hbar}\langle \mathbf{t}(\psi),\mathbf{t}(\psi),1\rangle_{0,3,0}^X+\langle 1\rangle_{1,1,0}\\ &=\sum_{i,j}t_{j+1}^i\frac{\partial}{\partial t_j^i}\sum_g\hbar^{g-1}\mathcal{F}_X^g+\frac{1}{2\hbar}\langle \mathbf{t}(\psi),\mathbf{t}(\psi),1\rangle_{0,3,0}^X+\langle 1\rangle_{1,1,0}.\end{aligned}$$

The "exceptional" terms at the end arise from reindexing the summation, because the moduli spaces $\overline{\mathcal{M}}_{g,n}(X,d)$ do not exist for $(g,n,d)=(0,2,0)$ or $(1,0,0)$. The first exceptional term is equal to

$$\frac{1}{2\hbar}\langle t_0,t_0\rangle^X,$$

where $\langle\ ,\ \rangle^X$ denotes the Poincaré pairing on X, because the ψ classes are trivial on $\overline{\mathcal{M}}_{0,3}(X,0)$. The second exceptional term vanishes for dimension reasons.

Taking the coefficient of $\hbar^{-1}$, we find that $\mathcal{F}_X^0$ satisfies the following differential equation:

$$\frac{\partial\mathcal{F}_X^0}{\partial t_0^1}=\frac{1}{2}\langle t_0,t_0\rangle^X+\sum_{i,j}t_{j+1}^i\frac{\partial\mathcal{F}_X^0}{\partial t_j^i}. \tag{21}$$

Before we continue, let us remark that the total-genus string equation can be presented in the following alternative form:

$$\sum_{g,n,d}\frac{Q^d\hbar^{g-1}}{(n-1)!}\langle 1,\mathbf{t}(\psi),\ldots,\mathbf{t}(\psi)\rangle_{g,n,d}^X=\sum_{g,n,d}\frac{Q^d\hbar^{g-1}}{(n-1)!}\left\langle\left[\frac{\mathbf{t}(\psi)}{\psi}\right]_+,\mathbf{t}(\psi),\ldots,\mathbf{t}(\psi)\right\rangle_{g,n,d}^X+\frac{1}{2\hbar}\langle t_0,t_0\rangle^X. \tag{22}$$

This expression will be useful later.

4.2.2 Dilaton Equation

The dilaton equation addresses the situation in which there is an insertion of $1 \in H^*(X)$ with a first power of ψ attached to it:

$$\langle \tau_{a_1}(\gamma_1), \cdots, \tau_{a_n}\tau_1(T_0)\rangle^X_{g,n,d} = (2g-2+n)\langle \tau_{a_1}(\gamma_1), \ldots, \tau_{a_n}(\gamma_n)\rangle^X_{g,n,d}.$$

Again, it can be expressed as a differential equation on the generating function. In genus zero, the equation is:

$$\frac{\partial \mathcal{F}^0_X}{\partial t^1_1} = \sum_{\substack{1\leq i\leq k\\ j\geq 0}} t^i_j \frac{\partial \mathcal{F}^0_X}{\partial t^i_j} - 2\mathcal{F}^0_X. \tag{23}$$

The proof is similar to the above, so we omit it.

A simple but extremely important device known as the **dilaton shift** allows us to express this equation in a simpler form. Define a new parameter $\mathbf{q}(z) = q_0 + q_1 z + \cdots \in H^*(X)[[z]]$ by

$$\mathbf{q}(z) = \mathbf{t}(z) - z, \tag{24}$$

so that $q_i = t_i$ for $i \neq 1$ and $q_1 = t_1 - 1$. If we perform this change of variables, then the dilaton equation says precisely that

$$\sum_{\substack{1\leq i\leq k\\ j\geq 0}} q^i_j \frac{\partial \mathcal{F}^0_X}{\partial q^i_j} = 2\mathcal{F}^0_X,$$

or in other words that $\mathcal{F}^0_X(\mathbf{q}(z))$ is *homogeneous* of degree two.

4.2.3 Topological Recursion Relations

A more general equation relating Gromov-Witten invariants to ones with lower powers of ψ is given by the topological recursion relations. In genus zero, the relation is:

$$\langle\langle \tau_{a_1+1}(\gamma_1), \tau_{a_2}(\gamma_2), \tau_{a_3}(\gamma_3)\rangle\rangle^X_{0,3}(\tau)$$
$$= \sum_a \langle\langle \tau_{a_1}(\gamma_1), \phi_a\rangle\rangle^X_{0,2} \langle\langle \phi^a, \tau_{a_2}(\gamma_2), \tau_{a_3}(\gamma_3)\rangle\rangle^X_{0,3},$$

where, as above, $\{\phi_a\}$ is a basis for $H^*(X)$, and $\{\phi^a\}$ denotes the dual basis under the Poincaré pairing. There are also topological recursion relations in higher genus (see [11, 14, 15]), but we omit them here as they are more complicated and not necessary for our purposes.

In terms of a differential equation, the genus-zero topological recursion relations are given by

$$\frac{\partial^3 \mathcal{F}_X^0}{\partial t_{j_1}^{i_1} \partial t_{j_2}^{i_2} \partial t_{j_3}^{i_3}} = \sum_{\mu,\nu} \frac{\partial^2 \mathcal{F}_X^0}{\partial t_{j_1}^{i_1} \partial t_0^{\mu}} g^{\mu\nu} \frac{\partial \mathcal{F}_X^0}{\partial t_{j_2}^{i_2} \partial t_{j_3}^{i_3} \partial t_0^{\nu}}. \tag{25}$$

Here, we use $g_{\mu\nu}$ to denote the matrix for the Poincaré pairing on $H^*(X)$ in the basis $\{\phi_\alpha\}$ and $g^{\mu\nu}$ to denote the inverse matrix.

4.2.4 Divisor Equation

The divisor equation describes invariants in which one insertion lies in $H^2(X)$ (with no ψ classes) in terms of invariants with fewer insertions. For $\rho \in H^2(X)$, it states:

$$\langle \tau_{a_1}(\gamma_1), \cdots, \tau_{a_n}(\gamma_n)\ \rho\rangle_{g,n+1,d}^X = \langle \rho, d\rangle \langle \tau_{a_1}(\gamma_1), \cdots, \tau_{a_n}(\gamma_n)\rangle_{g,n,d}^X$$

$$+ \sum_{i=1}^{n} \langle \tau_{a_1}(\gamma_1), \cdots, \tau_{a_{i-1}}(\gamma_{i-1}), \tau_{a_i - 1}(\gamma_i \rho), \tau_{a_{i+1}}(\gamma_{i+1}), \cdots, \tau_{a_n}(\gamma_n)\rangle_{g,n,d}^X,$$

or equivalently,

$$\langle \mathbf{a}_1(\psi), \ldots, \mathbf{a}_{n-1}(\psi), \rho\rangle_{g,n,d}^X = \langle \rho, d\rangle \langle \mathbf{a}_1(\psi), \ldots, \mathbf{a}_{n-1}(\psi)\rangle_{g,n-1,d}^X \tag{26}$$

$$+ \sum_{i=1}^{n-1} \left\langle \mathbf{a}_1(\psi), \ldots, \left[\frac{\rho \mathbf{a}_i(\psi)}{\psi}\right]_+, \ldots, \mathbf{a}_{n-1}(\psi)\right\rangle_{g,n-1,d}^X.$$

This equation, too, can be expressed as a differential equation on the generating function. The resulting equation is not needed for the time being, but we will return to it in Sect. 6.

4.3 Axiomatization

Axiomatic Gromov-Witten theory attempts to formalize the structures which arise in a genus-zero Gromov-Witten theory. One advantage of such a program is that any properties proved in the framework of axiomatic Gromov-Witten theory will necessarily hold for any of the variants of Gromov-Witten theory that share the same basic properties, such as the orbifold theory or FJRW theory. See [23], for a more detailed exposition of the subject of axiomatization.

Let H be an arbitrary $\mathbb{Q}$-vector space equipped with a distinguished element 1 and a nondegenerate inner product $(\, , \,)$. Let

$$\mathbb{H} = H((z^{-1})),$$

which is a symplectic vector space with symplectic form Ω defined by

$$\Omega(f, g) = \mathrm{Res}_{z=0}\Big((f(-z), g(z)) \Big).$$

An arbitrary element of $\mathbb{H}$ can be expressed as

$$\sum_{k \geq 0} p_{k,\alpha} \phi^{\alpha} (-z)^{-1-k} + \sum_{\ell \geq 0} q_{\ell}^{\beta} \phi_{\beta} z^{\ell}, \tag{27}$$

in which $\{\phi_1, \ldots, \phi_d\}$ is a basis for H with $\phi_1 = 1$. Define a subspace $\mathbb{H}_+ = H[[z]]$ of $\mathbb{H}$, which has coordinates q_j^i. Elements of $\mathbb{H}_+$ are identified with $\mathbf{t}(z) \in H[[z]]$ via the dilaton shift

$$t_j^i = q_j^i + \delta^{i1} \delta_{j1}.$$

Definition 15 A **genus-zero axiomatic theory** is a pair $(\mathbb{H}, G_0)$, where $\mathbb{H}$ is as above and $G_0 = G_0(\mathbf{t})$ is a formal function of $\mathbf{t}(z) \in H[[z]]$ satisfying the differential equations (21), (23), and (25).

In the case where $H = H^*(X; \Lambda)$ equipped with the Poincaré pairing and $G_0 = \mathcal{F}_X^0$, one finds that the genus-zero Gromov-Witten theory of X is an axiomatic theory.

Note that we require an axiomatic theory to satisfy neither the divisor equation (for example, this fails in orbifold Gromov-Witten theory), nor the WDVV equations, both of which are extremely useful in ordinary Gromov-Witten theory. While these properties are computationally desirable for a theory, they are not necessary for the basic axiomatic framework.

5 Quantization in the Infinite-Dimensional Case

Axiomatization reduces the relevant structures of Gromov-Witten theory to a special type of function on an infinite-dimensional symplectic vector space

$$\mathbb{H} = H((z^{-1})). \tag{28}$$

As we will see in Sect. 6, the actions of quantized operators on the quantization of $\mathbb{H}$ have striking geometric interpretations in the case where $H = H^*(X; \Lambda)$ for a projective variety X. Because our ultimate goal is the application of quantization to the symplectic vector space (28), we will assume throughout this section that the infinite-dimensional symplectic vector space under consideration has that form.

5.1 *The Symplectic Vector Space*

Let H be a vector space of finite dimension equipped with a nondegenerate inner product $(\, , \,)$ and let $\mathbb{H}$ be given by (28). As explained in Sect. 4.3, $\mathbb{H}$ is a symplectic vector space under the symplectic form Ω defined by

$$\Omega(f, g) = \mathrm{Res}_{z=0}\Big((f(-z), g(z)) \Big).$$

The subspaces

$$\mathbb{H}_+ = H[z]$$

and

$$\mathbb{H}_- = z^{-1}H[[z^{-1}]]$$

are Lagrangian. A choice of basis $\{\phi_1, \ldots, \phi_d\}$ for H yields a symplectic basis for $\mathbb{H}$, in which the expression for an arbitrary element in Darboux coordinates is

$$\sum_{k\geq 0} p_{k,\alpha}\phi^\alpha(-z)^{-1-k} + \sum_{\ell\geq 0} q_\ell^\beta \phi_\beta z^\ell.$$

Here, as before, $\{\phi^\alpha\}$ denotes the dual basis to $\{\phi_\alpha\}$ under the pairing $(\, , \,)$. We can identify $\mathbb{H}$ as a symplectic vector space with the cotangent bundle $T^*\mathbb{H}_+$.

Suppose that $T : \mathbb{H} \to \mathbb{H}$ is an endomorphism of the form

$$T = \sum_m B_m z^m, \tag{29}$$

where $B_m : H \to H$ are linear transformations. Let T^* denote the endomorphism given by taking the adjoint B_m^* of each transformation B_m with respect to the pairing. Then the symplectic adjoint of T is

$$T^\dagger(z) = T^*(-z) = \sum_m B_m^*(-z)^m.$$

As usual, a symplectomorphism is an endomorphism T of $\mathbb{H}$ that satisfies $\Omega(Tf, Tg) = \Omega(f, g)$ for any $f, g \in \mathbb{H}$. When T has the form (29), one can check that this is equivalent to the condition

$$T^*(-z)T(z) = \mathrm{Id}.$$

We will specifically be considering symplectomorphisms of the form

$$T = \exp(A)$$

in which A also has the form (29). In this case, we will require that A is an infinitesimal symplectic transformation, or in other words that $\Omega(Af, g) + \Omega(f, Ag) = 0$ for any $f, g \in \mathbb{H}$. Using the expression for A as a power series as in (29), one finds that this condition is equivalent to $A^*(-z) + A(z) = 0$, which in turn implies that

$$A_m^* = (-1)^{m+1} A_m. \tag{30}$$

There is another important restriction we must make: the transformation A (and hence T) will be assumed to contain *either* only nonnegative powers of z *or* only nonpositive powers of z. In the Gromov-Witten theory literature, a transformation

$$R = \sum_{m \geq 0} R_m z^m$$

with only nonnegative powers of z is typically referred to as upper-triangular, while a transformation

$$S = \sum_{m \leq 0} S_m z^m$$

with only nonpositive powers of z is referred to as lower-triangular. We will avoid this terminology for the most part, because (as remarked previously) it disagrees with the ordering of basis elements used in Sect. 3; however, we will occasionally refer to these two situations collectively as "upper-triangular and lower-triangular", simply meaning that positive and negative powers of z do not both appear.

The reason for restricting to upper- and lower-triangular transformations is that if A had both positive and negative powers of z, then exponentiating A would yield a series in which a single power of z could have a nonzero contribution from infinitely many terms, and the result would not obviously be convergent. There are still convergence issues to be addressed when one composes the two types of operators, but we defer discussion of this to Sect. 5.3.

5.2 *Quantization via Quadratic Hamiltonians*

Although in the finite-dimensional case we needed to start by choosing a symplectic basis, by expressing our symplectic vector space as $\mathbb{H} = H((z^{-1}))$ we have already implicity chosen a polarization $\mathbb{H} = \mathbb{H}_+ \oplus \mathbb{H}_-$. Thus, the quantization of the symplectic vector space $\mathbb{H}$ should be thought of as the Hilbert space $\mathscr{H}$ of square-integrable functions on $\mathbb{H}_+$ with values in $\mathbb{C}[[\hbar, \hbar^{-1}]]$. As in the finite-dimensional case, we will sometimes in practice allow $\mathscr{H}$ to contain formal functions that are not square-integrable, and the space of all such formal functions will be referred to as the Fock space.

5.2.1 The Quantization Procedure

Observables, which classically are functions on $\mathbb{H}$, are quantized the same way as in the finite-dimensional case, by setting

$$\hat{q}_k^\alpha = q_k^\alpha$$

$$\hat{p}_{k,\alpha} = \hbar \frac{\partial}{\partial q_k^\alpha},$$

and quantizing an arbitrary analytic function by expanding it in a Taylor series and ordering the variables within each monomial in the form $q_{k_1}^{\alpha_1} \cdots q_{k_n}^{\alpha_n} p_{\ell_1,\beta_1} \cdots p_{\ell_m,\beta_m}$.

In order to quantize a symplectomorphism $T = \exp(A)$ of the form discussed in the previous section, we will mimic the procedure discussed in Sect. 3.3. Namely, define a function h_A on $\mathbb{H}$ by

$$h_A(f) = \frac{1}{2}\Omega(Af, f).$$

Since h_A is a classical observable, it can be quantized by the above formula. We define the quantization of A via

$$\hat{A} = \frac{1}{\hbar}\widehat{h_A},$$

and U_T is defined by

$$U_T = \exp(\hat{A}).$$

The next section is devoted to making this formula more explicit.

5.2.2 Basic Examples

Before turning to the case of a general upper- or lower-triangular symplectomorphism, we will begin with two simple but crucial examples. These computations follow Example 1.3.3.1 of [6].

Example 16 Suppose that the infinitesimal symplectic transformation A is of the form

$$A = A_m z^m,$$

where $A_m : H \to H$ is a linear transformation and $m > 0$.

To compute $\hat{A}$, one must first compute the Hamiltonian h_A. Let

$$f(z) = \sum_{k\geq 0} p_{k,\alpha}\phi^\alpha(-z)^{-1-k} + \sum_{\ell\geq 0} q_\ell^\beta \phi_\beta z^\ell \in \mathbb{H}.$$

Then

$$\begin{aligned} h_A(f) &= \frac{1}{2}\Omega(Af, f) \\ &= \frac{1}{2}\operatorname{Res}_{z=0}\Bigg((-z)^m \sum_{k_1\geq 0} p_{k_1,\alpha}(A_m\phi^\alpha)\, z^{-1-k_1} + (-z)^m \sum_{\ell_1\geq 0} q_{\ell_1}^\beta (A_m\phi_\beta)(-z)^{\ell_1}, \\ &\qquad \sum_{k_2\geq 0} p_{k_2,\alpha}\phi^\alpha(-z)^{-1-k_2} + \sum_{\ell_2\geq 0} q_{\ell_2}^\beta \phi_\beta z^{\ell_2}\Bigg) \end{aligned}$$

Since only the z^{-1} terms contribute to the residue, the right-hand side is equal to

$$\begin{aligned} &\frac{1}{2}\sum_{k\geq 0}^{m-1}(-1)^k p_{k,\alpha} p_{m-k-1,\beta}(A_m\phi^\alpha, \phi^\beta) + \frac{1}{2}\sum_{k\geq 0}(-1)^m p_{m+k,\alpha} q_k^\beta (A_m\phi^\alpha, \phi_\beta) \\ &\qquad - \frac{1}{2}\sum_{k\geq 0} q_k^\beta p_{m+k,\alpha}(A_m\phi_\beta, \phi^\alpha). \end{aligned}$$

By (30), we have

$$(\phi^\alpha, A_m\phi_\beta) = (-1)^{m+1}(A_m\phi^\alpha, \phi_\beta).$$

Thus, denoting $(A_m)^\alpha_\beta = (A_m\phi^\alpha, \phi_\beta)$ and $(A_m)^{\alpha\beta} = (A_m\phi^\alpha, \phi^\beta)$, we can write

$$h_A(f) = \frac{1}{2}\sum_{k=0}^{m-1}(-1)^k p_{k,\alpha} p_{m-k-1,\beta}(A_m)^{\alpha\beta} + (-1)^m \sum_{k\geq 0} p_{m+k,\alpha} q_k^\beta (A_m)^\alpha_\beta.$$

This implies that

$$\hat{A} = \frac{\hbar}{2}\sum_{k=0}^{m-1}(-1)^k (A_m)^{\alpha\beta}\frac{\partial}{\partial q_k^\alpha}\frac{\partial}{\partial q_{m-k-1}^\beta} + (-1)^m \sum_{k\geq 0}(A_m)^\alpha_\beta q_k^\beta \frac{\partial}{\partial q_{m+k}^\alpha}.$$

Example 17 Similarly to the previous example, let

$$A = A_m z^m,$$

this time with $m < 0$. Let $(A_m)_{\alpha\beta} = (A_m \phi_\alpha, \phi_\beta)$. An analogous computation shows that

$$\hat{A} = \frac{1}{2\hbar} \sum_{k=0}^{-m-1} (-1)^{k+1} (A_m)_{\alpha\beta} q^\alpha_{-m-k-1} q^\beta_k + (-1)^m \sum_{k \geq -m} (A_m)^\alpha_\beta q^\beta_k \frac{\partial}{\partial q^\alpha_{k+m}}.$$

Example 18 More generally, let

$$A = \sum_{m<0} A_m z^m$$

be an infinitesimal symplectic endomorphism. Then the quadratic Hamiltonian associated to A is

$$h_A(f) = \frac{1}{2} \sum_{k,m} (-1)^{m+1} (A_{-k-m-1})_{\alpha\beta} q^\alpha_k q^\beta_m + \sum_{k,m} (-1)^m (A_m)^\alpha_\beta p_{k+m,\alpha} q^\beta_k.$$

Thus, after quantization, we obtain

$$\hat{A} = \frac{1}{2\hbar} \sum_{k,m} (-1)^{m+1} (A_{-k-m-1})_{\alpha\beta} q^\alpha_k q^\beta_m + \sum_{k,m} (-1)^m (A_m)^\alpha_\beta q^\beta_k \frac{\partial}{\partial q^\alpha_{k+m}}. \tag{31}$$

It is worth noting that some of the fairly complicated expressions appearing in these formulas can be written more succinctly, as discussed at the end of Example 1.3.3.1 of [6]. Indeed, the expression

$$\partial_A := \sum_k (A_m)^\alpha_\beta q^\beta_k \frac{\partial}{\partial q^\alpha_{k+m}} \tag{32}$$

that appears in $\hat{A}$ for both $m > 0$ and $m < 0$ acts on $\mathbf{q}(z) \in \mathbb{H}_+$ by

$$\left(\sum_k (A_m)^\alpha_\beta q^\beta_k \frac{\partial}{\partial q^\alpha_{k+m}} \right) \left(\sum_\ell q^\gamma_\ell \phi_\gamma z^\ell \right) = \left[\sum_k (A_m)^\alpha_\beta q^\beta_k \phi_\alpha z^{k+m} \right]_+,$$

where $[\cdot]_+$ denotes the power series truncation with only nonnegative powers of z. In other words, if $\mathbf{q} = \sum_\ell q^i_\ell \phi_i z^\ell$, then

$$\partial_A \mathbf{q} = [A\mathbf{q}]_+. \tag{33}$$

By the same token, consider the expression

$$\sum_{k\geq 0}(-1)^k (A_m)^{\alpha\beta}\frac{\partial}{\partial q_k^\alpha}\frac{\partial}{\partial q_{m-k-1}^\beta}$$

appearing in $\hat{A}$ for $m > 0$. The quadratic differential operator $\frac{\partial}{\partial q_k^\alpha}\frac{\partial}{\partial q_{m-k-1}^\beta}$ can be thought of as a bivector field on $\mathbb{H}_+$, and since $\mathbb{H}_+$ is a vector space, a bivector field can be identified with a tensor product of two maps $\mathbb{H}_+ \to \mathbb{H}_+$. Specifically, $\frac{\partial}{\partial q_k^\alpha}\frac{\partial}{\partial q_{m-k-1}^\beta}$ is the bivector field corresponding to the constant map

$$\phi_\alpha z_+^k \otimes \phi_\beta z_-^{m-k-1} \in H[z_+] \otimes H[z_-] \cong \mathbb{H}_+ \otimes \mathbb{H}_+.$$

Using the identity

$$\sum_{k=0}^{m-1}(-1)^k z_+^k z_-^{m-1-k} = \frac{z_+^m + (-1)^{m-1} z_-^m}{z_+ + z_-},$$

then, it follows that

$$\sum_{k\geq 0}(-1)^k (A_m)^{\alpha\beta}\frac{\partial}{\partial q_k^\alpha}\frac{\partial}{\partial q_{m-k-1}^\beta} = \left[\frac{A(z_+) + A^*(z_-)}{z_+ + z_-}\right]_+, \tag{34}$$

where we use the pairing to identify $(A_m)^{\alpha\beta}\phi_\alpha \otimes \phi_\beta \in \mathbb{H}_+ \otimes \mathbb{H}_+$ with $A_m \in \text{End}(H)$. Here, the power series truncation is included to ensure that (34) is trivially valid when m is negative.

The modified expressions (33) and (34) can be useful in recognizing the appearance of a quantized operator in computations. For example, (33) will come up in Proposition 19 below, and both (33) and (34) arise in the context of Gromov-Witten theory in Theorem 1.6.4 of [6].

5.2.3 Formulas for the General Cases

Extending the above two examples carefully, one obtains formulas for the quantization of any upper-triangular or lower-triangular symplectomorphism. The following results are quoted from [16]; see also [24] for an exposition.

Proposition 19 *Let S be a symplectomorphism of $\mathbb{H}$ of the form $S = \exp(A)$, with*

$$S(z) = I + S_1/z + S_2/z^2 + \cdots.$$

Define a quadratic form W_S on $\mathbb{H}_+$ by the equation

$$W_S(\mathbf{q}) = \sum_{k,\ell\geq 0} (W_{k\ell}\mathbf{q}_k, \mathbf{q}_\ell),$$

where $\mathbf{q}_k = q_k^\alpha \phi_\alpha$ and $W_{k\ell}$ is defined by

$$\sum_{k,\ell\geq 0} \frac{W_{k\ell}}{z^k w^\ell} = \frac{S^*(w)S(z) - I}{w^{-1} + z^{-1}}.$$

Then the quantization of S^{-1} acts on the Fock space by

$$(U_{S^{-1}}\Psi)(\mathbf{q}) = \exp\left(\frac{W_S(\mathbf{q})}{2\hbar}\right)\Psi([S\mathbf{q}]_+)$$

for any function Ψ of $\mathbf{q} \in \mathbb{H}_+$. Here, as above, $[S\mathbf{q}]_+$ denotes the truncation of $S(z)\mathbf{q}$ to a power series in z.

Proof Let $A = \sum_{m<0} A_m z^m$. Introduce a real parameter t and denote

$$G(t, \mathbf{q}) = e^{-t\hat{A}}\Psi(\mathbf{q}).$$

Define a t-dependent analogue of W_S via

$$W_t(\mathbf{q}) := \sum_{k,\ell\geq 0} (W_{k\ell}(t)\mathbf{q}_k, \mathbf{q}_\ell), \tag{35}$$

where

$$\sum_{k,\ell\geq 0} \frac{W_{k\ell}(t)}{z^k w^\ell} = \frac{e^{t\cdot A^*(w)}e^{t\cdot A(z)} - 1}{z^{-1} + w^{-1}}.$$

Note that $W_{k\ell}(t) = W_{\ell k}^*(t)$.

We will prove that

$$G(t, \mathbf{q}) = \exp\left(\frac{W_t(\mathbf{q})}{2\hbar}\right)\psi\left(\left[e^{tA}\mathbf{q}\right]_+\right) \tag{36}$$

for all t. The claim follows by setting $t = 1$.

To prove (36), let

$$g(t, \mathbf{q}) = \log(G(t, \mathbf{q}))$$

and write $\Psi = \exp(f)$. Then, taking logarithms, it suffices to show

$$g(t, \mathbf{q}) = \frac{W_t(\mathbf{q})}{2\hbar} + f\left(\left[e^{tA}\mathbf{q}\right]_+\right). \tag{37}$$

Notice that

$$\begin{aligned}\frac{d}{dt}G(t, \mathbf{q}) &= -\hat{A}\, G(t, \mathbf{q})\\ &= \frac{1}{2\hbar}\sum_{k,\ell}(-1)^{\ell}(A_{-k-\ell-1})_{\alpha\beta}q_k^{\alpha}q_{\ell}^{\beta}G(t, \mathbf{q})\\ &\quad+\sum_{k,\ell}(-1)^{\ell-1}(A_{\ell})^{\alpha}_{\beta}q_k^{\beta}\frac{\partial}{\partial q^{\alpha}_{k+\ell}}G(t, \mathbf{q}),\end{aligned}$$

using Example (18). This implies that $g(t, \mathbf{q})$ satisfies the differential equation

$$\frac{d}{dt}g(t, \mathbf{q}) = \frac{1}{2\hbar}\sum_{k,\ell}(-1)^{\ell}A_{-k-\ell-1,\alpha\beta}q_k^{\alpha}q_{\ell}^{\beta} + \sum_{k,\ell}(-1)^{\ell-1}(A_{\ell})^{\alpha}_{\beta}q^{\beta}_{k+\ell}\frac{\partial g}{\partial q_k^{\alpha}} \tag{38}$$

We will prove that the right-hand side of (37) satisfies the same differential equation.

The definition of $W_{k\ell}(t)$ implies that

$$\frac{d}{dt}W_{k\ell}(t) = \sum_{\ell'=0}^{\ell}A^{*}_{\ell'-\ell}W_{k\ell'}(t) + \sum_{k'=0}^{k}W_{k'\ell}(t)A_{k'-k} + (-1)^{k}A_{-k-\ell-1}.$$

Therefore,

$$\begin{aligned}\frac{1}{2\hbar}\frac{d}{dt}W_t(\mathbf{q}) &= \frac{1}{2\hbar}\sum_{k,\ell}\Bigg(\sum_{\ell'}(A^{*}_{\ell'-\ell}W_{k\ell'}(t)\mathbf{q}_k, \mathbf{q}_{\ell}) + \sum_{k'}(W_{k'\ell}(t)A_{k'-k}\mathbf{q}_k, \mathbf{q}_{\ell})\\ &\qquad + (-1)^{k}(A_{-k-\ell-1}\mathbf{q}_k, \mathbf{q}_{\ell})\Bigg)\\ &= \frac{1}{2\hbar}\sum_{k,\ell}\left(2\sum_{\ell'}(W_{k\ell'}(t)\mathbf{q}_k, A_{\ell'-\ell}\mathbf{q}_{\ell}) + (-1)^{k}(A_{-k-\ell-1}\mathbf{q}_{\ell}, \mathbf{q}_k)\right)\\ &= \frac{1}{2\hbar}\sum_{k,\ell}\Bigg(2\sum_{\ell'}(-1)^{\ell'-\ell-1}(A_{\ell'-\ell})^{\alpha}_{\beta}q_{\ell}^{\beta}(W_{k\ell'}(t)\mathbf{q}_k, \phi_{\alpha})\\ &\qquad (-1)^{k}(A_{-k-\ell-1}\mathbf{q}_{\ell}, \mathbf{q}_k)\Bigg)\end{aligned}$$

$$= \frac{1}{2\hbar}\sum_{k,\ell}(-1)^{\ell}(A_{-k-\ell-1})_{\alpha\beta}q_k^{\alpha}q_{\ell}^{\beta}$$
$$+\frac{1}{\hbar}\sum_{k,\ell,\ell'}(-1)^{\ell'-\ell-1}(A_{\ell'-\ell})^{\alpha}_{\beta}q_{\ell}^{\beta}\left(W_{k\ell'}(t)\mathbf{q}_k, \frac{\partial}{\partial q_{\ell'}^{\alpha}}\mathbf{q}_{\ell'}\right)$$
$$= \frac{1}{2\hbar}\sum_{k,\ell}(-1)^{\ell}(A_{-k-\ell-1})_{\alpha\beta}q_k^{\alpha}q_{\ell}^{\beta} + \sum_{k,\ell}(-1)^{\ell-1}(A_{\ell})^{\alpha}_{\beta}q_k^{\beta}\frac{\partial}{\partial^{\alpha}_{k+\ell}}\left(\frac{W_t(\mathbf{q})}{2\hbar}\right).$$

Furthermore, using Eq. (33), it can be shown that

$$\frac{df\left(\left[e^{tA}\mathbf{q}\right]_+\right)}{dt} = \sum_{k,\ell}(-1)^{\ell-1}(A_{\ell})^{\alpha}_{\beta}q_k^{\beta}\frac{\partial}{\partial q^{\alpha}_{k+\ell}}f\left(\left[e^{tA}\mathbf{q}\right]_+\right).$$

Thus, both sides of (37) satisfy the same differential equation. Since they agree when $t = 0$, and each monomial in $\mathbf{q}$ and $\hbar$ depends polynomially on t, it follows that the two sides of (37) are equal. □

The other case has an analogous proposition, but we omit the proof.

Proposition 20 *Let R be a symplectomorphism of $\mathbb{H}$ of the form $R = \exp(B)$, with*

$$R(z) = I + R_1 z + R_2 z^2 + \cdots .$$

Define a quadratic form V_R on $\mathbb{H}_-$ by the equation

$$V_R(p_0(-z)^{-1} + p_1(-z)^{-2} + p_2(-z)^{-3} + \cdots) = \sum_{k,\ell\geq 0}(p_k, V_{k\ell}p_{\ell}),$$

where $V_{k\ell}$ is defined by

$$\sum_{k,\ell\geq 0}(-1)^{k+\ell}V_{k\ell}w^k z^{\ell} = \frac{R^*(w)R(z) - I}{z+w}.$$

Then the quantization of R acts on the Fock space by

$$(U_R\Psi)(\mathbf{q}) = \left[\exp\left(\frac{\hbar V_R(\partial_{\mathbf{q}})}{2}\right)\Psi\right](R^{-1}\mathbf{q}),$$

where $V_R(\partial_{\mathbf{q}})$ is the differential operator obtained from $V_R(\mathbf{p})$ by replacing p_k by $\frac{\partial}{\partial q_k}$.

5.3 Convergence

In Sect. 3, we expressed the quantization of an arbitrary symplectic transformation by decomposing it into a product of upper-triangular and lower-triangular transformations, each of whose quantizations was known. In the infinite-dimensional setting, however, such a decomposition is problematic, because composing a series containing infinitely many nonnegative powers of z with one containing infinitely many nonpositive powers will typically yield a divergent series. This is why we have only defined quantization for upper-triangular or lower-triangular operators, not products thereof.

It is possible to avoid unwanted infinities if one is vigilant about each application of a quantized operator to an element of $\mathcal{H}$. For example, while the symplectomorphism $S \circ R$ may not be defined, it is possible that $(\hat{S} \circ \hat{R})\Psi$ makes sense for a given $\Psi \in \mathcal{H}$ if $\hat{S}(\hat{R}\Psi)$ has a convergent contribution to each power of z. This verification can be quite complicated; see Chapter 9, Section 3 of [24] for an example.

5.4 Feynman Diagrams and Integral Formulas Revisited

In this section, we will attempt to generalize the integral formulas and their resulting Feynman diagram expansions computed in Sect. 3 to the infinite-dimensional case. This is only interesting for symplectomorphisms with nonnegative powers of z, since for transformations with nonpositive powers, the Feynman amplitude of any graph with at least one edge vanishes.

To start, we must compute the analogues of the matrices A, C, and D that describe the transformation in Darboux coordinates. Suppose that

$$R = \sum_{m \geq 0} R_m z^m$$

is a symplectomorphism. If

$$e^{k,\alpha} = \phi^{\alpha}(-z)^{-1-k}, \quad e_{\alpha}^{\ell} = \phi_{\alpha} z^{\ell},$$

then it is easily check that

$$\widetilde{e}^{k,\alpha} = R \cdot e^{k,\alpha} = \sum_{k' \geq 0} (-1)^{k-k'} (R_{k-k'})^{\alpha}_{\gamma} e^{k',\gamma} + \sum_{\ell' \geq 0} (-1)^{-1-k} (R_{\ell'+k+1})^{\alpha\gamma} e_{\gamma}^{\ell'},$$

$$\widetilde{e}_{\beta}^{\ell} = R(e_{\beta}^{\ell}) = \sum_{\ell' \geq 0} (R^{*}_{\ell'-\ell})^{\gamma}_{\beta} e_{\gamma}^{\ell'}.$$

Let $\widetilde{p}_{k,\alpha}$ and $\widetilde{q}_\ell^\beta$ be defined by

$$\sum_{k\geq 0} p_{k,\alpha}e^{k,\alpha} + \sum_{\ell\geq 0} q_\ell^\beta e_\beta^\ell = \sum_{k\geq 0} \widetilde{p}_{k,\alpha}\widetilde{e}^{k,\alpha} + \sum_{\ell\geq 0} \widetilde{q}_\ell^\beta \widetilde{e}_\beta^\ell.$$

Then the relations among these coordinates are:

$$p_{k,\alpha} = \sum_{k'\geq 0}(-1)^{k'-k}(R_{k'-k})_\alpha^\gamma \widetilde{p}_{k',\gamma},$$

$$q_\ell^\beta = \sum_{k'\geq 0}(-1)^{-1-k'}(R_{\ell+k'+1})^{\gamma\beta}\widetilde{p}_{k',\gamma} + \sum_{\ell'\geq 0}(R^*_{\ell-\ell'})_\gamma^\beta \widetilde{q}_{\ell'}^\gamma.$$

That is, if we define matrices $\mathscr{A}$, $\mathscr{C}$, and $\mathscr{D}$ by

$$(\mathscr{A}^*)_{(k,\alpha)}^{(k',\gamma)} = (-1)^{k'-k}(R_{k'-k})_\alpha^\gamma,$$

$$\mathscr{C}_{(\ell,\beta),(k',\gamma)} = (-1)^{-1-k'}(R_{\ell+k'+1})^{\gamma\beta},$$

$$\mathscr{D}_{(\ell',\gamma)}^{(\ell,\beta)} = (R^*_{\ell-\ell'})_\gamma^\beta,$$

then the coordinates are related by

$$\mathbf{p} = \mathscr{A}\widetilde{\mathbf{p}}$$
$$\mathbf{q} = \mathscr{C}\widetilde{\mathbf{p}} + \mathscr{D}\widetilde{\mathbf{q}}.$$

These matrices have rows and columns indexed by

$$(k,\alpha) \in \mathbb{Z}^{\geq 0} \times \{1,\ldots,d\},$$

but the entries vanish when k is sufficiently large.

As in Sect. 3.4, the integral formula will be expressed in terms of a function $\phi : \mathbb{H} \to \mathbb{R}$ defined by

$$\phi(q,p') = (\mathscr{D}^{-1}q)\cdot p' - \frac{1}{2}(\mathscr{D}^{-1}\mathscr{C}p')\cdot p'.$$

It is not necessary to invert $\mathscr{D}$, as one has $\mathscr{D}^{-1} = \mathscr{A}^T$ in this case. Thus, the above gives an explicit formula for ϕ.

Equipped with this, we would like to define

$$(U_R\psi)(q) = \lambda \int e^{\frac{1}{\hbar}(\phi(q,p')-q'\cdot p')}\psi(q')dq'dp', \tag{39}$$

where λ is an appropriate normalization constant. The problem with this, though, is that the domain of the variables $q = q_\ell^\beta$ and $p' = p'_{k,\alpha}$ over which we integrate is an infinite-dimensional vector space. We have not specified a measure on this space, so it is not clear that (39) makes sense.

Our strategy for making sense of (39) will be to *define* it by its Feynman diagram expansion, as given by (18). Modulo factors of 2π, which are irrelevant because U_R is defined only up to a real multiplicative constant, the answer is:

$$(U_R\psi)(q) = \frac{1}{\sqrt{\det(\mathscr{D}^{-1}\mathscr{C})}} \sum_{\mathbf{n}=(n_0,n_1,\ldots)} \sum_{\Gamma \in G'(\mathbf{n})} \frac{\hbar^{-\chi_\Gamma}}{|\mathrm{Aut}(\Gamma)|} F_\Gamma(q),$$

where, as before, $G'(\mathbf{n})$ is the set of isomorphism classes of genus-labeled Feynman diagrams, and $F_\Gamma(q)$ is the genus-modified Feynman amplitude given by placing the m-tensor

$$\sum_{\substack{\mathbf{m} \\ \sum m_{\ell,\beta}=m}} \frac{m!}{\prod m_{\ell,\beta}!} \frac{\partial^m \mathcal{F}_g(s)}{\prod (\partial s_\beta^\ell)^{m_{\ell,\beta}}} \Bigg|_{s=\mathscr{D}^{-1}q} \cdot \prod (s_\ell^\beta)^{m_{\ell,\beta}}$$

at each m-valent vertex of genus g and taking the contraction of tensors using the bilinear form $-\mathscr{D}^{-1}\mathscr{C}$. Here, as before, $\mathcal{F}_g$ is defined by the expansion

$$\psi(q) = e^{\sum_{g \geq 0} \hbar^{g-1} \mathcal{F}_g(q)}.$$

Note that $\det(\mathscr{D}^{-1}\mathscr{C})$ is well-defined because, although these matrices have indices ranging over an infinite indexing set, they are zero outside of a finite range.

5.5 Semi-classical Limit

The most important feature of quantization for our purposes that it relates higher genus information to genus-zero information. We will return to this principle in the next section in the specific context of Gromov-Witten theory. Before we do so, however, let us give a precise statement of this idea in the abstract setting of symplectic vector spaces.

Let $\mathbb{H}$ be a symplectic vector space, finite- or infinite-dimensional. Fix a polarization $\mathbb{H} = \mathbb{H}_+ \oplus \mathbb{H}_-$, and suppose that for each $g \geq 0$,

$$\mathcal{F}_X^g : \mathbb{H}_+ \to \mathbb{R}, \quad \mathcal{F}_Y^g : \mathbb{H}_+ \to \mathbb{R}$$

are functions on $\mathbb{H}_+$. Package each of these two collections into the total descendent potentials:

$$\mathcal{D}_X = \exp\left(\sum_{g\geq 0} \hbar^{g-1}\mathcal{F}_X^g\right), \quad \mathcal{D}_Y = \exp\left(\sum_{g\geq 0} \hbar^{g-1}\mathcal{F}_Y^g\right).$$

Let $\mathcal{L}_X$ and $\mathcal{L}_Y$ be the Lagrangian subspaces of $\mathbb{H}$ that, under the identification of $\mathbb{H}$ with $T^*\mathbb{H}_+$, coincide with the graphs of $d\mathcal{F}_X^0$ and $d\mathcal{F}_Y^0$, respectively. That is,

$$\mathcal{L}_X = \{(p, q) \mid p = d_q\mathcal{F}_X^0\} \subset \mathbb{H},$$

and similarly for $\mathcal{L}_Y$.

Theorem 21 *Let T be a symplectic transformation such that*

$$U_T\mathcal{D}_X = \mathcal{D}_Y.$$

Then

$$T(\mathcal{L}_X) = \mathcal{L}_Y.$$

*(The passage from D^X to $\mathcal{L}_X$ is sometimes referred to as a **semi-classical limit**.)*

Proof We will prove this statement in the finite-dimensional setting, but all of our arguments should carry over with only notational modifications to the infinite-dimensional case. To further simplify, it suffices to prove the claim when T is of one of the three basic types considered in Sect. 3.2.

Case 1: $T = \begin{pmatrix} I & B \\ 0 & I \end{pmatrix}$ Using the explicit formula for U_T obtained in Sect. 3, the assumption $U_T\mathcal{D}_X = \mathcal{D}_Y$ in this case can be written as

$$\exp\left(\sum_{g\geq 0} \hbar^{g-1}\mathcal{F}_Y^g\right) = \exp\left(\frac{1}{2\hbar}B_{\alpha\beta}q^\alpha q^\beta\right)\exp\left(\sum_{g\geq 0} \hbar^{g-1}\mathcal{F}_X^g\right).$$

Taking logarithms of both sides, picking out the coefficient of $\hbar^{-1}$, and taking derivatives with respect to q, we find

$$d_q\mathcal{F}_Y^0 = Bq + d_q\mathcal{F}_X^0. \tag{40}$$

Now, choose a point $(\overline{p}, \overline{q}) \in \mathcal{L}_X$, so that $d_{\overline{q}}\mathcal{F}_X^0 = \overline{p}$. Explicitly, the point in question is $\overline{p}_\alpha e^\alpha + \overline{q}^\alpha e_\alpha$, so its image under T is

$$\overline{p}_\alpha \widetilde{e^\alpha} + \overline{q}^\alpha \widetilde{e_\alpha} = (\overline{p} + B\overline{q})_\alpha e^\alpha + \overline{q}^\alpha e_\alpha,$$

using the expressions for $\widetilde{e^{\alpha}}$ and $\widetilde{e_{\alpha}}$ in terms of the e basis obtained in Sect. 3. Thus, the statement that $T(\overline{p}, \overline{q}) \in \mathcal{L}_Y$ is equivalent to

$$d_{\overline{q}}\mathcal{F}_Y^0 = \overline{p} + B\overline{q}.$$

Since $\overline{p} = d_{\overline{q}}\mathcal{F}_X^0$ by assumption, this is precisely Eq. (40). This proves that $T(\mathcal{L}_X) \subset \mathcal{L}_Y$, and the reverse inclusion follows from the analogous claim applied to T^{-1}.

Case 2: $T = \begin{pmatrix} A & 0 \\ 0 & A^{-T} \end{pmatrix}$ This case is very similar to the previous one, so we omit the proof.

Case 3: $T = \begin{pmatrix} I & 0 \\ C & I \end{pmatrix}$ Consider the Feynman diagram expression for $U_T\mathcal{D}_X$ obtained in Sect. 3.5. Up to a constant factor, the assumption $U_T\mathcal{D}_X = \mathcal{D}_Y$ in this case becomes

$$\exp\left(\sum_{g \geq 0} \hbar^{g-1}\mathcal{F}_Y^g(q)\right) = \sum_{\Gamma} \frac{\hbar^{-\chi_\Gamma}}{|\mathrm{Aut}(\Gamma)|} F_\Gamma^X(q), \tag{41}$$

where $F_\Gamma^X(q)$ is the Feynman amplitude given by placing the m-tensor

$$\sum_{|\mathbf{m}|=m!} \frac{1}{m_1! \cdots m_n!} \frac{\partial^m \mathcal{F}_g^X}{(\partial q_1')^{m_1} \cdots (\partial q_n')^{m_n}}\Bigg|_{q'=q} (q_1')^{m_1} \cdots (q_n')^{m_n}$$

at each m-valent vertex of genus g in Γ and taking the contraction of tensors using the bilinear form $-C$.

Recall that if one takes the logarithm of a sum over Feynman diagrams, the result is a sum over connected graphs only, and in this case, $-\chi_\Gamma = g - 1$. Thus, if one takes the logarithm of both sides of (41) and picks out the coefficient of $\hbar^{-1}$, the result is

$$\mathcal{F}_Y^0(q) = \sum_{\Gamma \text{ connected, genus } 0} \frac{F_\Gamma(q)}{|\mathrm{Aut}(\Gamma)|}.$$

Let us give a more explicit formulation of the definition of $F_\Gamma(q)$. For convenience, we adopt the notation of Gromov-Witten theory and write

$$\mathcal{F}_0^X(q) = \langle\,\rangle + \langle e_\alpha \rangle^X q^\alpha + \frac{1}{2}\langle e_\alpha, e_\beta \rangle^X q^\alpha q^\beta + \cdots = \langle\langle\,\rangle\rangle^X(q),$$

where $\langle\langle \phi_1, \ldots, \phi_n \rangle\rangle^X(q) := \sum_{k \geq 0} \frac{1}{k!} \langle \phi_1, \ldots, \phi_n, q, \ldots, q \rangle^X$ (k copies of q) and the brackets are defined by the above expansion. Then derivatives of $\mathcal{F}_X^0$ are given by adding insertions to the double bracket. It follows that

$$F_\Gamma^X(q) = \sum_{\{i_h\}} \prod_{v \in V(\Gamma)} \left\langle\left\langle \prod_{h \in H(v)} e_{i_h} \right\rangle\right\rangle^X (q) \prod_{e=(a,b) \in E(\Gamma)} (-C^{i_a i_b}).$$

Here, $V(\Gamma)$ and $E(\Gamma)$ denote the vertex sets and edge sets of Γ, respectively, while $H(v)$ denotes the set of half-edges associated to a vertex v. The summation is over all ways to assign an index $i_h \in \{1, \ldots, d\}$ to each half-edge h, where d is equal to the dimension of $\mathbb{H}_+$. For an edge e, we write $e = (a, b)$ if a and b are the two half-edges comprising e.

Thus, we have re-expressed the relationship between $\mathcal{F}_0^Y$ and $\mathcal{F}_0^X$ as

$$\mathcal{F}_0^Y(q) = \sum_{\Gamma \text{ connected, genus } 0} \frac{1}{|\mathrm{Aut}(\Gamma)|} \sum_{\{i_h\}} \prod_{v \in V(\Gamma)} \left\langle\left\langle \prod_{h \in H(v)} e_{i_h} \right\rangle\right\rangle^X (q) \prod_{e=(a,b) \in E(\Gamma)} (-C^{i_a i_b}).$$

Now, to prove the claim, choose a point $(\overline{p}, \overline{q}) \in \mathcal{L}_Y$. We will prove that $T^{-1}(\overline{p}, \overline{q}) \in \mathcal{L}_X$. Applying the same reasoning used in Case 1 to the inverse matrix $T^{-1} = \begin{pmatrix} I & 0 \\ -C & I \end{pmatrix}$, we find that

$$T^{-1}(\overline{p}, \overline{q}) = \overline{p}_\alpha e^\alpha + (-C\overline{p} + \overline{q})^\alpha e_\alpha.$$

Therefore, the claim is equivalent to

$$d_{\overline{q}} \mathcal{F}_0^Y = d_{-C d_{\overline{q}} \mathcal{F}_0^Y + \overline{q}} \mathcal{F}_0^X.$$

From the above, one finds that the ith component of the vector $d_{\overline{q}} \mathcal{F}_Y^0$ is equal to

$$\sum_\Gamma \frac{1}{|\mathrm{Aut}(\Gamma)|} \sum_{\{i_h\}, w \in V(\Gamma)} \left\langle\left\langle e_i \prod_{h \in H(w)} e_{i_h} \right\rangle\right\rangle^X (\overline{q}) \prod_{v \neq w \in V(\Gamma)} \left\langle\left\langle \prod_{h \in H(v)} e_{i_h} \right\rangle\right\rangle^X (\overline{q}) \prod_{e=(a,b) \in E(\Gamma)} (-C^{i_a i_b}).$$

On the other hand, the same equation shows that the ith component of the vector $d_{-C d_{\overline{q}} \mathcal{F}_0^Y + \overline{q}} \mathcal{F}_X^0$ is equal to

$$\begin{aligned}
&\langle\langle e_i \rangle\rangle^X (-C d_{\overline{q}} \mathcal{F}_0^Y + \overline{q}) \\
&= \langle\langle e_i \rangle\rangle^X \left(\sum_\Gamma \frac{1}{|\mathrm{Aut}(\Gamma)|} \sum_{\{i_h\}, w \in V(\Gamma), j, k} \left\langle\left\langle e_j \prod_{h \in H(w)} e_{i_h} \right\rangle\right\rangle^X (\overline{q}) \prod_{v \neq w \in V(\Gamma)} \left\langle\left\langle \prod_{h \in H(v)} e_{i_h} \right\rangle\right\rangle^X (\overline{q}) \right. \\
&\qquad \left. \prod_{e=(a,b) \in E(\Gamma)} (-C^{i_a i_b})(-C)^{kj} e_k + q^\ell e_\ell \right)
\end{aligned}$$

$$= \sum_{\Gamma_1,\ldots,\Gamma_n} \frac{1}{n!|\mathrm{Aut}(\Gamma_1)|\cdots|\mathrm{Aut}(\Gamma_n)|} \sum_{\substack{\{i_h\} \\ w_1\in V(\Gamma_1),\ldots,w_n\in V(\Gamma_n) \\ j_1,\ldots,j_n \\ k_1,\ldots,k_n \\ \ell_1,\ldots,\ell_m}} \prod_{c=1}^{n} \left\langle\left\langle e_{j_c} \prod_{h\in H(w_c)} e_{i_h} \right\rangle\right\rangle^X (\overline{q})$$

$$\prod_{v\neq w_1,\ldots,w_n} \left\langle\left\langle \prod_{h\in H(v)} e_{i_h} \right\rangle\right\rangle^X (\overline{q}) \prod_{e=(a,b)} (-C^{i_a i_b})(-C)^{k_c j_c} \frac{\langle e_i, e_{k_1}, \ldots, e_{k_n}, q^{\ell_1} e_{\ell_1}, \ldots, q^{\ell_m} e_{\ell_m} \rangle}{m!}$$

$$= \sum_{\Gamma_1,\ldots,\Gamma_n} \frac{1}{n!|\mathrm{Aut}(\Gamma_1)|\cdots|\mathrm{Aut}(\Gamma_n)|} \sum_{\substack{\{i_h\} \\ w_1\in V(\Gamma_1),\ldots,w_n\in V(\Gamma_n) \\ j_1,\ldots,j_n \\ k_1,\ldots,k_n}} \prod_{c=1}^{n} \left\langle\left\langle e_{j_c} \prod_{h\in H(w_c)} e_{i_h} \right\rangle\right\rangle^X (\overline{q})$$

$$\prod_{v\neq w_1,\ldots,w_n} \left\langle\left\langle \prod_{h\in H(v)} e_{i_h} \right\rangle\right\rangle^X (\overline{q}) \prod_{e=(a,b)} (-C^{i_a i_b})(-C)^{k_c j_c} \langle\langle e_i, e_{k_1}, \ldots, e_{k_n} \rangle\rangle(\overline{q})$$

Upon inspection, this is equal to the sum of all ways of starting with a distinguished vertex (where e_i is located) and adding n spokes labeled $k_1, \ldots, k_n$, then attaching n graphs $\Gamma_1, \ldots, \Gamma_n$ to this vertex via half-edges labeled $j_1, \ldots, j_n$. This procedures yields all possible graphs with a distinguished vertex labeled by e_i—the same summation that appears in the expression for $d_{\overline{q}}\mathcal{F}_0^Y$. Each total graph appears in multiple ways, corresponding to different ways of partitioning it into subgraphs labeled $\Gamma_1, \ldots, \Gamma_n$. However, it is a combinatorial exercise to verify that, with this over-counting, the automorphism factor in front of each graph Γ is precisely $\frac{1}{|\mathrm{Aut}(\Gamma)|}$.

Thus, we find that $d_{-Cd_{\overline{q}}\mathcal{F}_0 Y+\overline{q}}\mathcal{F}_0^X = d_{\overline{q}}\mathcal{F}_0^Y$, as required. □

6 Applications of Quantization to Gromov-Witten Theory

In this final section, we will return to the situation in which $\mathbb{H} = H^*(X; \Lambda)((z^{-1}))$ for X a projective variety. In Sect. 6.1, we show that many of the basic equations of Gromov-Witten theory can be expressed quite succinctly as equations satisfied by the action of a quantized operator on the total descendent potential. More strikingly, according to Givental's conjecture, there is a converse in certain special cases to the semi-classical limit statement explained in Sect. 5.5; we discuss this in Sect. 6.2 below. In Sect. 6.3, we briefly outline the machinery of twisted Gromov-Witten theory developed by Coates and Givental. This is a key example of the way quantization can package complicated combinatorics into a manageable formula.

Ultimately, these notes only scratch the surface of the applicability of the quantization machinery to Gromov-Witten theory. There are many other interesting directions in this vein, so we conclude the book with a brief overview of some of the other places in which quantization arises. The interested reader can find much more in the literature.

6.1 *Basic Equations via Quantization*

Here we give a simple application of quantization as a way to rephrase some of the axioms of Gromov-Witten theory. This section closely follows Examples 1.3.3.2 and 1.3.3.3 of [6].

6.1.1 String Equation

Recall from (22) that the string equation can be expressed as follows:

$$\sum_{g,n,d} \frac{Q^d \hbar^{g-1}}{(n-1)!} \langle 1, \mathbf{t}(\psi), \ldots, \mathbf{t}(\psi) \rangle^X_{g,n,d} = \sum_{g,n,d} \frac{Q^d \hbar^{g-1}}{(n-1)!} \left\langle \left[\frac{\mathbf{t}(\psi)}{\psi} \right]_+, \mathbf{t}(\psi), \ldots, \mathbf{t}(\psi) \right\rangle^X_{g,n,d} + \frac{1}{2\hbar} \langle t_0, t_0 \rangle.$$

Applying the dilaton shift (24), this is equivalent to

$$-\frac{1}{2\hbar}(q_0, q_0) - \sum_{g,n,d} \frac{Q^d \hbar^{g-1}}{(n-1)!} \left\langle \left[\frac{\mathbf{q}(\psi)}{\psi} \right]_+, \mathbf{t}(\psi), \ldots, \mathbf{t}(\psi) \right\rangle^X_{g,n,d} = 0.$$

From Example 17 with $m = -1$, one finds

$$\begin{aligned} \widehat{\frac{1}{z}} &= -\frac{1}{2\hbar}\langle q_0, q_0 \rangle - \sum_k q^\alpha_{k+1} \frac{\partial}{\partial q^\alpha_k} \\ &= -\frac{1}{2\hbar}\langle q_0, q_0 \rangle - \partial_{1/z}. \end{aligned}$$

Thus, applying (33), it follows that the string equation is equivalent to

$$\widehat{\frac{1}{z}} \mathcal{D}_X = 0.$$

6.1.2 Divisor Equation

In a similar fashion, the divisor equation can be expressed in terms of a quantized operator. Summing over g, n, d and separating the exceptional terms, Eq. (26) can be stated as

$$\begin{aligned}\sum_{g,n,d} \frac{Q^d \hbar^{g-1}}{(n-1)!} \langle \mathbf{t}(\psi), \ldots, \mathbf{t}(\psi), \rho\rangle^X_{g,n,d} &= \sum_{g,n,d} \frac{Q^d \hbar^{g-1}}{n!} (\rho, d) \langle \mathbf{t}(\psi), \ldots, \mathbf{t}(\psi)\rangle_{g,n,d} \\ &+ \sum_{g,n,d} \frac{Q^d \hbar^{g-1}}{(n-1)!} \left\langle \left[\frac{\rho \mathbf{t}(\psi)}{\psi}\right]_+, \mathbf{t}(\psi), \ldots, \mathbf{t}(\psi)\right\rangle^X_{g,n,d} \\ &+ \frac{1}{2\hbar} \langle \mathbf{t}(\psi), \mathbf{t}(\psi), \rho\rangle_{0,3,0} + \langle \rho \rangle_{1,1,0}. \end{aligned} \tag{42}$$

The left-hand side and the second summation on the right-hand side combine to give

$$-\sum_{g,n,d} \frac{Q^d \hbar^{g-1}}{(n-1)!} \left\langle \left[\frac{\rho(\mathbf{t}(\psi) - \psi)}{\psi}\right]_+, \mathbf{t}(\psi), \ldots, \mathbf{t}(\psi)\right\rangle^X_{g,n,d}.$$

After the dilaton shift, this is equal to $\partial_{\rho/z}(\sum \hbar^{g-1} \mathcal{F}^X_g)$.

As for the first summation, let $\tau_1, \ldots, \tau_r$ be a choice of basis for $H_2(X; \mathbb{Z})$, which yields a set of generators $Q_1, \ldots, Q_r$ for the Novikov ring. Write

$$\rho = \sum_{i=1}^r \rho_i \tau^i$$

in the dual basis $\{\tau^i\}$ for $\{\tau_i\}$. Then the first summation on the right-hand side of (42) is equal to

$$\sum_{i=1}^r \rho_i Q_i \frac{\partial}{\partial Q_i} \left(\sum_g \hbar^{g-1} \mathcal{F}^X_g\right).$$

The first exceptional term is computed as before:

$$\langle \mathbf{t}(\psi), \mathbf{t}(\psi), \rho\rangle_{0,3,0} = (q_0 \rho, q_0).$$

The second exceptional term is more complicated in this case. We require the fact that

$$\overline{\mathcal{M}}_{1,1}(X, 0) \cong X \times \overline{\mathcal{M}}_{1,1},$$

and that under this identification

$$[\overline{\mathcal{M}}_{1,1}(X,d)]^{\mathrm{vir}} = e(T_X \times \mathcal{L}_1^{-1}) = e(T_X) - \psi_1 c_{D-1}(T_X),$$

where $\mathcal{L}_1$ is the cotangent line bundle (whose first Chern class is ψ_1) and $D = \dim(X)$. Thus,

$$\langle \rho \rangle_{1,1,0}^X = \int_X e(T_X)\cdot\rho - \int_{\overline{\mathcal{M}}_{1,1}} \psi_1 \int_X c_{D-1}(T_X)\cdot\rho = -\frac{1}{24}\int_X c_{D-1}(T_X)\cdot\rho.$$

Putting all of these pieces together, we can express the divisor equation as

$$-\frac{1}{2\hbar}(q_0\rho, q_0) - \partial_{\rho/z}\left(\sum_g \hbar^{g-1}\mathcal{F}_g^X\right)$$
$$= \left(\sum_i \rho_i Q_i \frac{\partial}{\partial Q_i} - \frac{1}{24}\int_X c_{D-1}(T_X)\cdot\rho\right)\left(\sum_g \hbar^{g-1}\mathcal{F}_g^X\right),$$

or in other words, as

$$\widehat{\left(\frac{\rho}{z}\right)}\cdot\mathcal{D}_X = \left(\sum_i \rho_i Q_i \frac{\partial}{\partial Q_i} - \frac{1}{24}\int_X c_{D-1}(T_X)\cdot\rho\right)\mathcal{D}_X.$$

It should be noted that the left-hand side of this equality makes sense because multiplication by ρ is a self-adjoint linear transformation on $H^*(X)$ under the Poincaré pairing, and hence multiplication by ρ/z is an infinitesimal symplectic transformation.

6.2 Givental's Conjecture

The material in this section can be found in [17] and [23].

Recall from Sect. 4.3 that an axiomatic genus zero theory is a symplectic vector space $\mathbb{H} = H((z^{-1}))$ together with a formal function $G_0(\mathbf{t})$ satisfying the differential equations corresponding to the string equation, dilaton equation, and topological recursion relations in genus zero.

The **symplectic** (or **twisted**) **loop group** is defined as the set $\{M(z)\}$ of End(H)-valued formal Laurent series in z^{-1} satisfying the symplectic condition $M^*(-z)M(z) = I$. There is an action of this group on the collection of axiomatic genus zero theories. To describe the action, it is helpful first to reformulate the definition of an axiomatic theory in a more geometric, though perhaps less transparent, way.

Associated to an axiomatic genus zero theory is a Lagrangian subspace

$$\mathcal{L} = \{(\mathbf{p}, \mathbf{q}) \mid \mathbf{p} = d_{\mathbf{q}} G_0\} \subset \mathbb{H},$$

where $(\mathbf{p}, \mathbf{q})$ are the Darboux coordinates on $\mathbb{H}$ defined by (27) and we are identifying $\mathbb{H} \cong T^*\mathbb{H}_+$ by way of this polarization. Note, here, that $G_0(\mathbf{t})$ is identified with a function of $\mathbf{q} \in \mathbb{H}_+$ via the dilaton shift, and that this is the same Lagrangian subspace discussed in Sect. 5.5.

According to Theorem 1 of [18], a function $G_0(\mathbf{t})$ satisfies the requisite differential equations if and only if the corresponding Lagrangian subspace $\mathcal{L}$ is a Lagrangian *cone* with the vertex at the origin satisfying

$$\mathcal{L} \cap T_{\mathbf{f}}\mathcal{L} = zT_{\mathbf{f}}\mathcal{L}$$

for each $\mathbf{f} \in \mathcal{L}$.

The symplectic loop group can be shown to preserve these properties. Thus, an element T of the symplectic loop group acts on the collection of axiomatic theories by sending the theory with Lagrangian cone $\mathcal{L}$ to the theory with Lagrangian cone $T(\mathcal{L})$.

There is one other equivalent formulation of the definition of axiomatic genus-zero theories, in terms of abstract Frobenius manifolds. Roughly speaking, a Frobenius manifold is a manifold equipped with a product on each tangent space that gives the tangent spaces the algebraic structure of Frobenius algebras. A Frobenius manifold is called **semisimple** if, on a dense open subset of the manifold, these algebras are semisimple. This yields a notion of semisimplicity for axiomatic genus zero theories. Given this, we can formulate the statement of the symplectic loop group action more precisely:

Theorem 22 ([18]) *The symplectic loop group acts on the collection of axiomatic genus-zero theories. Furthermore, the action is transitive on the semisimple theories of a fixed rank N.*

Here, the **rank** of a theory is the rank of H.

The genus-zero Gromov-Witten theory of a collection of N points gives a semisimple axiomatic theory of rank N, which we denote by H_N. The theorem implies that any semisimple axiomatic genus-zero theory $T = (\mathbf{H}, G_0)$ can be obtained from H_N by the action of an element of the twisted loop group. Via the process of Birkhoff factorization, such a transformation can be expressed as $S \circ R$ in which S has only nonpositive powers of z and R has only nonnegative powers.

Definition 23 The **axiomatic τ-function** of an axiomatic theory T is defined by

$$\tau_G^T = \hat{S}(\hat{R}\mathcal{D}_N),$$

where $S \circ R$ is the element of the symplectic loop group taking the theory H_N of N points to T, and $\mathcal{D}_N$ is the total descendent potential for the Gromov-Witten theory of N points.

If T is in fact the genus-zero Gromov-Witten theory of a space X, then we have two competing definitions of the higher-genus potential: $\mathcal{D}_X$ and τ_G^T. Givental's conjecture is the statement that these two agree:

Conjecture 24 (Givental's Conjecture [17]) If T is the semisimple axiomatic theory corresponding to the genus-zero Gromov-Witten theory of a projective variety X, then $\tau_G^T = \mathcal{D}_X$.

In other words, the conjecture posits that in the semisimple case, if an element of the symplectic loop group matches two genus zero theories, then its quantization matches their total descendent potentials. Because the action of the symplectic loop group is transitive on semisimple theories, this amounts to a classification of all higher-genus theories for which the genus-zero theory is semisimple.

Givental proved his conjecture in case X admits a torus action and the total descendent potentials are taken to be the equivariant Gromov-Witten potentials. In 2005 Teleman announced a proof of the conjecture in general:

Theorem 25 ([28]) *Givental's conjecture holds for any semisimple axiomatic theory.*

One important application of Givental's conjecture is the proof of the Virasoro conjecture in the semisimple case. The conjecture states:

Conjecture 26 For any projective manifold X, there exist "Virasoro operators" $\{\widehat{L_m^X}\}_{m\geq -1}$ satisfying the relations

$$[\widehat{L_m^X}, \widehat{L_n^X}] = (m-n)\widehat{L_{m+n}^X}, \tag{43}$$

such that

$$\widehat{L_m^X}\mathcal{D}_X = 0$$

for all $m \geq -1$.

In the case where X is a collection of N points, the conjecture holds by setting $\widehat{L_m^X}$ equal to the quantization of

$$L_m := -z^{-1/2}D^{m+1}z^{-1/2},$$

where

$$D := z\left(\frac{d}{dz}\right)z = z^2\frac{d}{dz} + z.$$

The resulting operators $\{\widehat{L_m}\}_{m\geq -1}$ are the same as N copies of those used in Witten's conjecture [29], and the relations (43) indeed hold for these operators.

Thus, by Witten's conjecture, the Virasoro conjecture holds for any semisimple Gromov-Witten theory by setting

$$\widehat{L_m^X} = \widehat{S}(\widehat{R}\widehat{L_m}\widehat{R}^{-1})\widehat{S}^{-1}$$

for the transformation $S \circ R$ taking the theory of N points to the Gromov-Witten theory of X.[7]

6.3 Twisted Theory

The following is due to Coates and Givental; we refer the reader to the exposition presented in [6].

Let X be a projective variety equipped with a holomorphic vector bundle E. Then E induces a K-class on $\overline{\mathcal{M}}_{g,n}(X, d)$,

$$E_{g,n,d} = \pi_* f^* E \in K^0(\overline{\mathcal{M}}_{g,n}(X, d)),$$

where

$$\begin{array}{ccc} \mathscr{C} & \xrightarrow{\ f\ } & X \\ {\scriptstyle \pi}\big\downarrow & & \\ \overline{\mathcal{M}}_{g,n}(X, d), & & \end{array}$$

is the universal family over the moduli space. Consider an invertible multiplicative characteristic class

$$c : K^0(\overline{\mathcal{M}}_{g,n}(X, d)) \to H^*(\overline{\mathcal{M}}_{g,n}(X, d)).$$

Any such class can be written in terms of Chern characters

$$c(\cdot) = \exp\left(\sum_{k\geq 0} s_k \mathrm{ch}_k(\cdot)\right),$$

for some parameters s_k.

[7] In fact one must check that $\widehat{L_m^X}$ defined this way agrees with the Virasoro operators of the conjecture, but this can be done.

A **twisted Gromov-Witten invariant** is defined as

$$\langle \tau_1(\gamma_1)\cdots\tau_n(\gamma_n);\, c(E_{g,n,d})\rangle^X_{g,n,d} = \int_{[\overline{\mathcal{M}}_{g,n}(X,d)]^{\mathrm{vir}}} ev_1^*(\gamma_1)\psi_1^{a_1}\cdots ev_n^*(\gamma_n)\psi_n^{a_n} c(E_{g,n,d}).$$

These fit into a twisted genus-g potential $\mathcal{F}^g_{c,E}$ and a twisted total descendent potential $\mathcal{D}_{c,E}$ in just the way that the usual Gromov-Witten invariants do.

There is also a Lagrangian cone $\mathcal{L}_{c,E}$ associated to the twisted theory, but a bit of work is necessary in order to define it. The reason for this is that the Poincaré pairing on $H(X;\Lambda)$ should be given by three-point correlators. As a result, when we replace Gromov-Witten invariants by their twisted versions we must modify the Poincaré pairing, and hence the symplectic structure on $\mathbb{H}$, accordingly. Denote this modified symplectic vector space by $\mathbb{H}_{c,E}$. There is a symplectic isomorphism

$$\mathbb{H}_{c,E} \to \mathbb{H}$$

$$x \mapsto \sqrt{c(E)}x.$$

We define the Lagrangian cone $\mathcal{L}_{c,E}$ of the twisted theory by

$$\mathcal{L}_{c,E} = \sqrt{c(E)}\cdot\{(\mathbf{p},\mathbf{q}) \mid \mathbf{p} = d_{\mathbf{q}}\mathcal{F}^0_{c,E}\} \subset \mathbb{H},$$

where we use the usual dilaton shift to identify $\mathcal{F}^0_{c,E}(\mathbf{t})$ with a function of $\mathbf{q} \in (\mathbb{H}_{c,E})_+$.

The quantum Riemann-Roch theorem of Coates-Givental gives an expression for $\mathcal{D}_{c,E}$ in terms of a quantized operator acting on the untwisted Gromov-Witten descendent potential $\mathcal{D}_X$ of X:

Theorem 27 ([6]) *The twisted descendent potential is related to the untwisted descendent potential by*

$$\exp\left(-\frac{1}{24}\sum_{\ell>0} s_{\ell-1}\int_X ch_\ell(E)c_{D-1}(T_X)\right)\exp\left(\frac{1}{2}\int_X e(X)\wedge\left(\sum_{j\geq 0} s_j ch_j(E)\right)\right)\mathcal{D}_{c,E}$$

$$=\exp\left(\sum_{\substack{m>0\\ \ell\geq 0}} s_{2m-1+\ell}\frac{B_{2m}}{(2m)!}(ch_\ell(E)z^{2m-1})^\wedge\right)\exp\left(\sum_{\ell>0} s_{\ell-1}(ch_\ell(E)/z)^\wedge\right)\mathcal{D}_X.$$

Here B_{2m} are the Bernoulli numbers.

The basic idea of this theorem is to write

$$\begin{aligned} c(E_{g,n,d}) &= \exp\left(\sum_{k\geq 0} s_k \mathrm{ch}_k(R\pi_* f^* E)\right) \\ &= \exp\left(\sum_{k\geq 0} s_k \left(\pi_*(\mathrm{ch}(f^*E)\mathrm{Td}^\vee(T_\pi))\right)_k\right), \end{aligned}$$

using the Grothendieck-Riemann-Roch formula. A geometric theorem expresses $\mathrm{Td}^\vee(T_\pi)$ in terms of ψ classes on various strata of the moduli space, and the rest of the proof of Theorem 27 is a difficult combinatorial exercise in keeping track of these contributions.

Taking a semi-classical limit and applying the result discussed in Sect. 5.5 of the previous section, we obtain:

Corollary 28 *The Lagrangian cone $\mathcal{L}_{c,E}$ satisfies*

$$\mathcal{L}_{c,E} = \exp\left(\sum_{m\geq 0}\sum_{0\leq \ell\leq D} s_{2m-1+\ell}\frac{B_{2m}}{(2m)!}ch_\ell(E)z^{2m-1}\right)\mathcal{L}_X.$$

This theorem and its corollary are extremely useful even when one is only concerned with the genus zero statement. For example, it is used in the proof of the Landau-Ginzburg/Calabi-Yau correspondence in [5]. In that context, the invariants under consideration, known as FJRW invariants, are given by twisted Gromov-Witten invariants *only* in genus zero. Thus, Theorem 27 actually says nothing about higher-genus FJRW invariants. Nevertheless, an attempt to directly apply Grothendieck-Riemann-Roch in genus zero to obtain a relationship between FJRW invariants and untwisted invariants is combinatorially unmanageable; thus, the higher-genus statement, while not directly applicable, can be viewed as a clever device for keeping track of the combinatorics of the Grothendieck-Riemann-Roch computation.

6.4 Concluding Remarks

There are a number of other places in which quantization proves useful for Gromov-Witten theory. For example, it was shown in [13, 20, 22] that relations in the so-called tautological ring, an important subring of $H^*(\overline{\mathcal{M}}_{g,n}(X,\beta))$, are invariant under the action of the symplectic loop group. This was used to give one proof of Givental's Conjecture in genus $g \leq 2$ (see [9, 30]), and can also be used to derive tautological relations (see [1, 2, 21]). We refer the interested reader to [23] for a summary of these and other applications of quantization with more complete references.

Acknowledgements The authors are indebted to Huai-Liang Chang, Wei-Ping Li, and Yongbin Ruan for organizing the workshop at which these notes were presented. Special thanks are due to Yongbin Ruan for his constant guidance and support, as well as to Yefeng Shen for working through the material with us and assisting in the creation of these notes. Y.P. Lee and Xiang Tang both gave extremely helpful talks at the RTG Workshop on Quantization organized by the authors in December 2011 at the University of Michigan. Useful comments on earlier drafts were provided by Pedro Acosta and Weiqiang He. A number of existing texts were used in the authors' learning process; we have attempted to include references to the literature throughout these notes, but we apologize in advance for any omissions. This work was supported in part by FRG grant DMS 1159265 RTG and NSF grant 1045119 RTG.

References

1. Arcara, D., Lee, Y.-P.: Tautological equations in $\overline{M}_{3,1}$ via invariance conjectures (2005). arXiv preprint: math/0503184
2. Arcara, D., Lee, Y.-P.: Tautological equations in genus 2 via invariance constraints. Bull. Inst. Math. Acad. Sinica **2**(1), 1 (2007)
3. Bates, S., Weinstein, A.: Lectures Notes on the Geometry of Quantization. American Mathematical Society, Providence (1997)
4. Blau, M.: Symplectic geometry and geometric quantization. Found at http://www.ictp.trieste.it/~mblau/lecturesGQ.ps.gz
5. Chiodo, A., Ruan, Y.: Landau–Ginzburg/Calabi–Yau correspondence for quintic three-folds via symplectic transformations. Invent. Math. **182**(1), 117–165 (2010)
6. Coates, T.: Riemann-Roch theorems in Gromov-Witten theory. Ph.D. thesis, University of California at Berkeley (1997). Also available at http://math.harvard.edu/~tomc/thesis.pdf
7. Cox, D.A., Katz, S.: Mirror Symmetry and Algebraic Geometry. American Mathematical Society, Providence (1999)
8. Da Silva, A.C.: Lectures on Symplectic Geometry. Springer, Berlin (2001)
9. Dubrovin, B., Zhang, Y.: Frobenius manifolds and Virasoro constraints. Sel. Math. **5**(4), 423–466 (1999)
10. Echeverría-Enríquez, A., Lecanda, M.C.M., Roy, N.R., Victoria-Monge, C.: Mathematical foundations of geometric quantization. Extracta Math. **13**(2), 135–238 (1998)
11. Eguchi, T., Xiong, C.: Quantum cohomology at higher genus. Adv. Theor. Math. Phys. **2**, 219–229 (1998), hep-th/9801010
12. Etingof, P.: Mathematical ideas and notions of quantum field theory (2002). Found at http://www-math.mit.edu/~etingof/lect.ps
13. Faber, V, Shadrin, S., Zvonkine, D.: Tautological relations and the r-spin Witten conjecture (2006). arXiv preprint: math/0612510
14. Getzler, E.: Intersection theory on $\overline{M}_{1,4}$ and elliptic Gromov-Witten invariants. J. Am. Math. Soc. **10**(4), 973–998 (1997)
15. Getzler, E.: Topological recursion relations in genus 2. In: Integrable Systems and Algebraic Geometry, pp. 73–106. World Scientific Publishing, Singapore (1998)
16. Givental, A.: Gromov-Witten invariants and quantization of quadratic hamiltonians. Mosc. Math. J **1**(4), 551–568 (2001)
17. Givental, A.: Semisimple frobenius structures at higher genus. Int. Math. Res. Not. **2001**(23), 1265–1286 (2001)

18. Givental, A.: Symplectic geometry of Frobenius structures. In: Frobenius Manifolds, pp. 91–112. Springer, Cham (2004)
19. Hori, K., Katz, S., Pandharipande, R., Thomas, R., Vakil, R.: Mirror Symmetry. Clay Mathematics Institute, Peterborough (2003)
20. Lee, Y.-P.: Witten's conjecture, Virasoro conjecture, and invariance of tautological equations (2003). arXiv preprint: math.AG/0311100
21. Lee, Y.-P.: Invariance of tautological equations I: conjectures and applications. J. Eur. Math. Soc. **10**(2), 399–413 (2008)
22. Lee, Y.-P.: Invariance of tautological equations II: Gromov-Witten theory. J. Am. Math. Soc. **22**(2), 331–352 (2009)
23. Lee, Y.-P.: Notes on axiomatic Gromov–Witten theory and applications. In: Algebraic Geometry—Seattle 2005, pp. 309–323. Proceedings of Symposia in Pure Mathematics, vol. 80, Part 1. American Mathematical Society, Providence (2009)
24. Lee, Y.-P., Pandharipande, R.: Frobenius manifolds, Gromov-Witten theory, and Virasoro constraints (2004). Found at http://www.math.ethz.ch/~rahul/Part1.ps
25. McDuff, D., Salamon, D.: Introduction to Symplectic Topology. Oxford University Press, Oxford (1998)
26. Ritter, W.G.: Geometric quantization (2002). arXiv preprint: math-ph/0208008
27. Schwarz, A., Tang, X.: Quantization and holomorphic anomaly. J. High Energy Phys. **2007**(03), 062 (2007)
28. Teleman, C.: The structure of 2d semi-simple field theories. Invent. Math. **188**(3), 525–588 (2012)
29. Witten, E.: Two-dimensional gravity and intersection theory on moduli space. Surv. Differ. Geom. **1**(243), 74 (1991)
30. Xiaobo, L., Gang, T.: Quantum product on the big phase space and the Virasoro conjecture. Adv. Math. **169**(2), 313–375 (2002)

Some Classical/Quantum Aspects of Calabi-Yau Moduli

Si Li

Contents

Abstract We review some classical and quantum geometry of Calabi-Yau moduli related to B-model aspects of closed string mirror symmetry. This note comes out of the author's lectures in the workshop "B-model aspects of Gromov-Witten theory" held at University of Michigan in 2013.

1 Introduction

Mirror symmetry is a physics-motivated duality between symplectic geometry (or the *A-model*) and complex geometry (or the *B-model*). In contrast to the A-model, Calabi-Yau condition is necessary for a well-defined B-model. In this article we discuss several aspects of local geometry on the moduli space in the B-model related to closed string mirror symmetry, focusing on compact Calabi-Yau models and Landau-Ginzburg models.

This article consists of two main parts: classical geometry (or the genus zero theory) and quantum geometry (or the higher genus theory). The geometry of genus zero theory can be summarized as defining the Frobenius manifold structure [14] on the local moduli space of Calabi-Yau geometry. It originated (called the *flat structure*) around early 1980s from K. Saito's theory of primitive forms [32, 33] in his study of period integrals over vanishing cycles associated to an isolated

S. Li (✉)
Yau Mathematical Sciences Center, Tsinghua University, Beijing, China
e-mail: sili@mail.tsinghua.edu.cn

E. Clader, Y. Ruan (eds.), *B-Model Gromov-Witten Theory*, Trends in Mathematics,
https://doi.org/10.1007/978-3-319-94220-9_4

singularity. This has now become the geometric content of Landau-Ginzburg B-model encoding the genus zero correlation functions. K. Saito's construction was extended by Barannikov and Kontsevich [5] to compact Calabi-Yau models via tools of deformation theory, and packaged into the framework of variation of semi-infinite Hodge structures [2, 3]. The first part will be mainly reviewing this classical story. The quantum B-model on Calabi-Yau manifolds has a candidate in physics via the quantization of a gauge theory [6] (*Kodaira-Spencer gauge theory*) whose classical limit describes the deformation of complex structures. Geometrically, such quantization can be obtained as the infinite dimensional Weyl quantization with the help of renormalization techniques in quantum field theory [11]. This is a realization of the topological B-twisted closed string field theory in the sense of Zwiebach [38]. The second part will be focused on explaining this subject.

2 Classical Geometry

2.1 Deformation Theory on Calabi-Yau and Local Moduli

We start with the deformation theory on Calabi-Yau manifolds via polyvector fields following [5].

2.1.1 Polyvector Fields

Let X be a compact Calabi-Yau manifold of dimension d. Ω_X will be a fixed holomorphic volume form which is unique up to a constant. We consider

$$\mathrm{PV}(X) = \bigoplus_{0\leq i,j\leq d} \mathrm{PV}^{i,j}(X), \quad \mathrm{PV}^{i,j}(X) = \mathcal{A}^{0,j}(X, \wedge^i T_X)$$

the space of polyvector fields on X. Here T_X is the holomorphic tangent bundle, and $\mathcal{A}^{0,j}(X, \wedge^i T_X)$ is the space of smooth $(0, j)$-forms valued in $\wedge^i T_X$. $\mathrm{PV}(X)$ is a differential bi-graded commutative algebra: the differential is

$$\bar{\partial} : \mathrm{PV}^{i,j}(X) \to \mathrm{PV}^{i,j+1}(X),$$

and the algebra structure arises from wedge product. Our degree convention is that elements of $\mathrm{PV}^{i,j}(X)$ are of degree $j - i$. The graded-commutativity says

$$\alpha\beta = (-1)^{|\alpha||\beta|}\beta\alpha$$

where $|\alpha|$, $|\beta|$ denote the degree of α, β respectively. Ω_X induces an identification between the space of polyvector fields and differential forms

$$\mathrm{PV}^{i,j}(X) \overset{\lrcorner\Omega_X}{\cong} \mathcal{A}^{d-i,j}(X)$$
$$\alpha \to \alpha\lrcorner\Omega_X$$

where $\lrcorner$ is the contraction, and $\mathcal{A}^{i,j}(X)$ denotes smooth differential forms of type (i, j). The holomorphic de Rham differential ∂ on forms defines an operator on $\mathrm{PV}(X)$ via the above isomorphism, which we still denote by

$$\partial : \mathrm{PV}^{i,j}(X) \to \mathrm{PV}^{i-1,j}(X)$$

i.e.

$$(\partial\alpha)\lrcorner\Omega_X \equiv \partial(\alpha\lrcorner\Omega_X), \quad \alpha \in \mathrm{PV}(X).$$

The definition of ∂ doesn't depend on the choice of Ω_X on compact Calabi-Yau manifolds. It induces a bracket on polyvector fields (Bogomolov-Tian-Todorov lemma)

$$\{\alpha, \beta\} := \partial(\alpha\beta) - (\partial\alpha)\beta - (-1)^{|\alpha|}\alpha(\partial\beta)$$

which coincides with the Schouten-Nijenhuis bracket (up to a sign). The fundamental algebraic structures of polyvector fields on Calabi-Yau geometry can be summarized by saying that the tuple $\{(\mathrm{PV}(X), \bar{\partial}), \wedge, \partial, \{-,-\}\}$ defines a differential Gerstenhaber-Batalin-Vilkovisky (GBV) algebra.

We can integrate polyvector fields by the *trace map* $\mathrm{Tr} : \mathrm{PV}(X) \to \mathbb{C}$

$$\mathrm{Tr}(\alpha) := \int_X (\alpha\lrcorner\Omega_X) \wedge \Omega_X. \tag{2.1}$$

This is only non-vanishing on $\mathrm{PV}^{d,d}(X)$. Let $\langle -, - \rangle$ be the induced pairing $\mathrm{PV}(X) \otimes \mathrm{PV}(X) \to \mathbb{C}$

$$\alpha \otimes \beta \to \langle \alpha, \beta \rangle \equiv \mathrm{Tr}(\alpha\beta).$$

It is easy to see that $\bar{\partial}$ is (graded) skew self-adjoint for this pairing and ∂ is (graded) self-adjoint.

2.1.2 Deformation of Complex Structures

We are interested in the moduli space of complex structures on compact Calabi-Yau manifolds. The main local result is the smoothness of the moduli space

(Bogomolov-Tian-Todorov Theorem), which is also a direct consequence of the differential GBV structure.

Let us fix a choice of Kähler metric on X. Locally, the deformation space of complex structure of X can be described by the space

$$\mathcal{M}^{cx} := \left\{ \mu \in \mathrm{PV}^{1,1}(X), \|\mu\| < \epsilon \,\middle|\, \bar{\partial}\mu + \frac{1}{2}\{\mu, \mu\} = 0, \bar{\partial}^*\mu = 0 \right\},$$

where ϵ is a sufficiently small number. Let $\mu_1 \in H^1(X, T_X)$ be a harmonic element with respect to the Kähler metric. μ_1 represents a tangent vector of the moduli space at the point X, i.e. a first order deformation. It can be extended to a genuine deformation

$$\mu_t = \sum_{k=1}^{\infty} t^k \mu_k \in \mathrm{PV}^{1,1}(X), \quad |t| << 1$$

by solving recursively (in order of powers of t)

$$\bar{\partial}\mu_t = -\frac{1}{2}\{\mu_t, \mu_t\}, \quad \bar{\partial}^*\mu_t = 0,$$

or equivalently by solving

$$\bar{\partial}\mu_i = -\frac{1}{2}\sum_{k=1}^{i-1}\{\mu_i, \mu_{k-i}\}, \quad i > 1, \quad \bar{\partial}^*\mu_i = 0.$$

For a general complex manifold, the harmonic part of the RHS may not be vanishing, representing the obstructions for solving the above equation. However, this does not happen for Calabi-Yau manifolds. Indeed, we can solve μ_t with the additional property that $\partial\mu = 0$. Suppose we have solved μ_k for $k < i$, with $\bar{\partial}^*\mu_k = \partial\mu_k = 0$. Bogomolov-Tian-Todorov lemma implies that

$$\{\mu_k, \mu_{i-k}\} = \partial(\mu_k \wedge \mu_{i-k}),$$

which has no harmonic component. It follows that μ_i can be solved by

$$\mu_i = -\frac{1}{2}\bar{\partial}^* G\partial(\sum_{k=1}^{i-1}\mu_i \wedge \mu_{k-i})$$

which satisfies $\bar{\partial}^*\mu_i = \partial\mu_i = 0$. Here $G = \frac{1}{\Delta}$ is the Green's operator for the Laplacian $\Delta = \bar{\partial}\bar{\partial}^* + \bar{\partial}^*\bar{\partial}$ acting on $\mathrm{PV}(X)$. It can be further shown that the power series μ_t is convergent given t sufficiently small. This implies that the local deformation of complex structures on Calabi-Yau manifolds is unobstructed.

2.1.3 Extended Deformation Space and the Formality Theorem

We can consider the extended deformation space $\mathcal{M}$ [5] by solving

$$\bar{\partial}\mu + \frac{1}{2}\{\mu, \mu\} = 0$$

modulo gauge equivalence. Here μ is allowed to be polyvectors of all types. By the same argument as above, the deformation problem is unobstructed.

Remark 1 In this paper, we treat $\mathcal{M}$ as a formal graded manifold [5].

In [5], Barannikov and Kontsevich have introduced a remarkable way to organize the above argument via the *Formality Theorem*. The deformation problem is controlled by the differential graded Lie algebra (DGLA)

$$(\mathrm{PV}(X), \bar{\partial}, \{,\}).$$

There are two closely related DGLA's. The first one is

$$(\ker \partial, \bar{\partial}, \{,\}),$$

where $\ker \partial \subset \mathrm{PV}(X)$ is the subspace of polyvector fields annihilated by ∂. Bogomolov-Tian-Todorov lemma implies that $\{,\}$ is a well-defined Lie bracket on $\ker \partial$. In fact,

$$\{,\} : \ker \partial \times \ker \partial \to \mathrm{im}\, \partial \subset \ker \partial.$$

The second DGLA is

$$(\mathbb{H}, 0, 0)$$

where $\mathbb{H} \subset \mathrm{PV}(X)$ is the subspace of harmonic elements. We associate the trivial differential and Lie bracket. There is a well-defined diagram of morphisms of DGLA's

$$(\mathrm{PV}(X), \bar{\partial}, \{,\}) \xleftarrow{\ j\ } (\ker \partial, \bar{\partial}, \{,\}) \xrightarrow{\ \pi\ } (\mathbb{H}, 0, 0)$$

where j is the natural embedding, and π is the harmonic projection. By Hodge theory, both j and π induce isomorphisms on the cohomology of the differential complex, hence quasi-isomorphisms of DGLA's. Since quasi-isomorphisms can be inverted via L_∞ morphisms, we obtain the following Formality Theorem

Theorem 2 ([5]) *The DGLA* $(\mathrm{PV}(X), \bar{\partial}, \{,\})$ *is* L_∞ *quasi-isomorphic to the DGLA* $(\mathbb{H}, 0, 0)$ *of its cohomology.*

Quasi-isomorphic DGLA's have equivalent moduli functors. It follows that the extended deformation space is smooth, being locally parametrized by $\mathbb{H}$.

2.2 Generalized Period Map and Frobenius Manifold Structure

There is a line bundle $\mathcal{L}$ over $\mathcal{M}_X^{cx}$ whose fiber parametrizes the holomorphic volume forms. It gives rise to the period map (locally)

$$\mathcal{M}^{cx} \to \mathbb{P}(H^n(X, \mathbb{C}))$$

by sending $[X_t] \in \mathcal{M}^{cx}$ to the line in $H^n(X, \mathbb{C})$ representing the fiber of $\mathcal{L}$.

Period map here can be viewed as varying the holomorphic volume form along with the deformation of the complex structure. The choice of the deformation of the pair (X, Ω_X) can be described by a pair $(\mu, \rho) \in \mathrm{PV}^{1,1}(X) \oplus \mathrm{PV}^{0,0}(X)$ as follows. μ defines a deformation of complex structure solving

$$\bar{\partial}\mu + \frac{1}{2}\{\mu, \mu\} = 0.$$

It is easy to see that $e^{\mu} \lrcorner \Omega_X$ is of type $(n, 0)$ in the new complex structure μ. It differs from the new holomorphic volume form by a factor e^{ρ}, which solves the equation

$$d(e^{\rho} e^{\mu} \lrcorner \Omega_X) = 0.$$

This can be also read by

$$\bar{\partial}\mu + \frac{1}{2}\{\mu, \mu\} = 0, \quad \bar{\partial}\rho + \partial\mu + \{\mu, \rho\} = 0,$$

or simply

$$Q(\mu + z\rho) + \frac{1}{2}\{\mu + z\rho, \mu + z\rho\} = 0,$$

where $Q = \bar{\partial} + z\partial$ and z is a formal parameter.

Barannikov [2, 3] extended the period map to the "generalized period" on the extended moduli space $\mathcal{M}$. It can be viewed as the compact Calabi-Yau analogue of Saito's primitive period map [33] for isolated singularities. We briefly review his construction here. Consider the new DGLA

$$(\mathrm{PV}(X)[[z]], Q, \{,\}).$$

Remark 3 The formal variable z is the same as $\hbar$ in [2, 3].

Notation 4 Given a vector space A, $A[[z]]$ ($A((z))$ respectively) will denote the formal power series (Laurent series respectively) in z valued in A. $A[[\mathbf{u}]]$ will denote the formal power series in $\mathbf{u} = \{u^\alpha\}$ valued in A. If both sets of variables are involved, the topology is understood as follows: $A((z))[[\mathbf{u}]] \equiv B[[\mathbf{u}]]$ for $B = A((z))$, while $A[[\mathbf{u}]]((z)) \equiv C((z))$ for $C = A[[\mathbf{u}]]$, etc.

There exists universal solutions [2] (modulo gauge equivalence)

$$\mu(u, z) = \sum_\alpha \mu_\alpha(z) u^\alpha + \frac{1}{2} \sum_{\alpha,\beta} \mu_{\alpha\beta}(z) u^\alpha u^\beta + \cdots \in \mathrm{PV}(X)[[z]][[\mathbf{u}]]$$

to the associated Maurer-Cantan equation

$$Q\mu(u, z) + \frac{1}{2}\{\mu(u, z), \mu(u, z)\} = 0,$$

where u^α are the deformation parameters as coordinates on $\mathcal{M}$, and $\mu_\alpha(z)$ forms a $\mathbb{C}[[z]]$-basis of $H^*(\mathrm{PV}(X)[[z]], Q)$. It is direct to check that the Maurer-Cartan equation is formally equivalent to

$$Qe^{\mu(u,z)/z} = 0.$$

Note that in our notation, $e^{\mu(u,z)/z} \in \mathrm{PV}(X)((z))[[\mathbf{u}]]$.

Notation 5 Given $\mu \in \mathrm{PV}(X)[[z]]$ with $Q\mu = 0$, we will use $[\mu]$ to represent its cohomology class in $H^*(\mathrm{PV}(X)[[z]], Q)$. Similar notations apply to other cohomologies.

Let us define an isomorphism

$$\Gamma_\Omega : \mathrm{PV}(X)((z)) \to \mathcal{A}(X)((z)), \quad z^k\alpha \to z^{k+i-1}\alpha \lrcorner \Omega_X, \quad \alpha \in \mathrm{PV}^{i,j}(X). \tag{2.2}$$

It transfers Q to the de Rham differential

$$\Gamma_\Omega \circ Q = d \circ \Gamma_\Omega.$$

As a result, the universal solutions $\mu(u, z)$ defines a cohomology class

$$\Gamma_\Omega(\left[ze^{\mu(u,z)/z}\right]) \in H^*(X, \mathbb{C})((z))[[\mathbf{u}]].$$

Definition 6 For simplicity, let us denote from now on by

$$S(X) := \mathrm{PV}(X)((z)), \quad S_+(X) := \mathrm{PV}(X)[[z]], \quad S_-(X) := z^{-1}\,\mathrm{PV}(X)[z^{-1}].$$

Lemma 7 *Under the isomorphism Γ_Ω, we have*

$$\Gamma_\Omega(S_+(X)) = \prod_{p\in\mathbb{Z}} z^{d-p+1} F^p \mathcal{A}(X),$$

where $F^p\mathcal{A}(X) = \mathcal{A}^{\geq p,*}(X)$. At the cohomology level we have an isomorphism

$$\Gamma_\Omega : H^*(S_+(X), Q) \stackrel{\cong}{\to} \prod_{p\in\mathbb{Z}} z^{d-p+1} F^p H^*(X, \mathbb{C}).$$

Similarly

$$\Gamma_\Omega : H^*(S(X), Q) \stackrel{\cong}{\to} H^*(X, \mathbb{C})((z)).$$

Definition 8 We define a symplectic pairing on $S(X)$ by

$$\omega(f(z)\alpha, g(z)\beta) := \mathrm{Res}_{z=0}\,(f(z)g(-z)dz)\,\mathrm{Tr}(\alpha\beta).$$

The differential Q is (graded) skew-symmetric with respect to the symplectic pairing ω. Therefore ω descends to define a symplectic pairing on the cohomology $H^*(S(X), Q)$, where $H^*(S_+(X), Q)$ becomes an isotropic subspace.

Definition 9 An opposite filtration of $H^*(S(X), Q)$ is a linear isotropic subspace $\mathcal{L} \subset H^*(S(X), Q)$ such that

(1) $H^*(S(X), Q) = H^*(S_+(X), Q) \oplus \mathcal{L}$,
(2) $\mathcal{L}$ is preserved by the operator $z^{-1} : H^*(S(X), Q) \to H^*(S(X), Q)$.

The subspaces $z^k H^*(S_+(X), Q) \subset H^*(S_+(X), Q)$, $k \geq 0$, defines a decreasing filtration, whose associated graded space is

$$\mathrm{Gr}\, H^*(S_+(X), Q) \cong H^*(X, \wedge^* T_X)[[z]].$$

It is easy to see that under Γ_Ω, this filtration can be identified with the Hodge filtration, and $\mathcal{L}$ is equivalent to an opposite splitting filtration. Given an opposite filtration $\mathcal{L}$, it defines us a splitting projection

$$\pi_+^{\mathcal{L}} : H^*(S(X), Q) \to H^*(S_+(X), Q),$$

and an isomorphism of vector spaces

$$H^*(S_+(X), Q)/zH^*(S_+(X), Q) \cong H^*(S_+(X), Q) \cap z\mathcal{L},$$

which further induces an isomorphism of $\mathbb{C}[[z]]$-modules

$$\mathrm{Gr}\, H^*(S_+(X), Q) \cong H^*(S_+(X), Q).$$

Definition 10 $\mathcal{L}$ leads to a choice of $\mathbb{C}[[z]]$-basis of $H^*(S_+(X), Q)$ by $H^*(S_+(X), Q) \cap z\mathcal{L}$. We will let $\{\mu_\alpha^{\mathcal{L}}\}_\alpha$ denote such a basis that

$$H^*(S_+(X), Q) \cap z\mathcal{L} = \mathrm{Span}_{\mathbb{C}}\{[\mu_\alpha^{\mathcal{L}}]\}.$$

Proposition 11 *Given an opposite filtration $\mathcal{L}$, there exists a universal solution of the form*

$$\mu^{\mathcal{L}}(\tau, z) = \sum_\alpha \mu_\alpha^{\mathcal{L}} \tau_z^\alpha + \frac{1}{2} \sum_{\alpha,\beta} \mu_{\alpha\beta}(z) \tau_z^\alpha \tau_z^\beta + \cdots \in \mathrm{PV}(X)[[z]][[\boldsymbol{\tau}]],$$

where $\boldsymbol{\tau} = \{\tau^\alpha\}$ are coordinates on $\mathcal{M}$, $\tau_z^\alpha = \tau^\alpha + O(\tau^2) \in \mathbb{C}[[z]][[\boldsymbol{\tau}]]$ such that

$$\pi_+^{\mathcal{L}}\left(\left[ze^{\mu^{\mathcal{L}}(\tau,z)/z} - z\right]\right) = \sum_\alpha \mu_\alpha^{\mathcal{L}} \tau^\alpha.$$

Proof Up to a (z-dependent) linear change of coordinates on u, we can assume that the universal solution $\mu(u, z)$ is of the form

$$\mu(u, z) = \sum_\alpha \mu_\alpha^{\mathcal{L}} u^\alpha + O(u^2).$$

Consider the projection

$$\pi_+^{\mathcal{L}}\left(\left[ze^{\mu(u,z)/z} - z\right]\right) \in H^*(S_+(X), Q)[[\mathbf{u}]].$$

Since $\mu_\alpha^{\mathcal{L}}$ forms a $\mathbb{C}[[z]]$-basis of $H^*(S_+(X), Q)$, we can write

$$\pi_+^{\mathcal{L}}\left(\left[ze^{\mu(u,z)/z} - z\right]\right) = \sum_\alpha \mu_\alpha^{\mathcal{L}} \tau^\alpha(u, z)$$

where

$$\tau^\alpha(u, z) = u^\alpha + O(u^2) \in \mathbb{C}[[z, \mathbf{u}]].$$

In particular, u^α can be solved in terms of τ^α, z by

$$u^\alpha(\tau, z) = \tau^\alpha + O(\tau^2) \in \mathbb{C}[[z, \boldsymbol{\tau}]].$$

Then $\mu^{\mathcal{L}}(\tau, z) = \mu(u(\tau, z), z)$. □

In particular, we find the relation

$$\left[ze^{\mu^{\mathcal{L}}(\tau,z)/z}\right] \in z + \sum_{\alpha} \left[\mu_\alpha^{\mathcal{L}}\right] \tau^\alpha + \mathcal{L}[[\boldsymbol{\tau}]], \tag{2.3}$$

where $\mathcal{L} = z^{-1}\mathrm{Span}_{\mathbb{C}}\{[\mu_\alpha^{\mathcal{L}}]\}[z^{-1}]$.

Definition 12 ([2]) Given an opposite filtration $\mathcal{L}$, we define the generalized period map

$$\Pi^{\mathcal{L}} : \mathcal{M} \to H^*(X, \mathbb{C})$$

as the map of formal (graded) manifolds from $(\mathcal{M}, 0)$ to $(H^*(X, \mathbb{C}), \Omega_X)$ by

$$\tau^\alpha \to \Gamma_\Omega(\left[ze^{\mu^{\mathcal{L}}(\tau,z)/z}\right])|_{z=1}.$$

It is easy to see that $\Pi^{\mathcal{L}}$ is an isomorphism of formal graded manifolds.

2.2.1 Frobenius Manifold Structure

Now we explain Barannikov's formulation [2, 3] of Frobenius manifold structure on $\mathcal{M}$ associated to an opposite filtration.

Definition 13 Let $\mathcal{H} \equiv H^*(S(X), Q)$, and let $\mathcal{H}_{\mathcal{M}}^{(0)} \subset \mathcal{H}[[\boldsymbol{\tau}]]$ be the free $\mathbb{C}[[z]][[\boldsymbol{\tau}]]$-module generated by $[z\partial_{\tau^\alpha} e^{\mu(\tau,z)/z}]$. The symplectic pairing ω extends $\mathbb{C}[[\boldsymbol{\tau}]]$-linearly to

$$\omega : \mathcal{H}[[\boldsymbol{\tau}]] \otimes_{\mathbb{C}[[\boldsymbol{\tau}]]} \mathcal{H}[[\boldsymbol{\tau}]] \to \mathbb{C}[[\boldsymbol{\tau}]]$$

which we denote by the same symbol. $\boldsymbol{\tau}$ is the coordinate defined in Proposition 11.

Lemma 14 $\mathcal{H}[[\boldsymbol{\tau}]] = \mathcal{H}_{\mathcal{M}}^{(0)} \oplus \mathcal{L}[[\boldsymbol{\tau}]]$. *Moreover, this is an isotropic decomposition, i.e.* $\omega(\mathcal{H}_{\mathcal{M}}^{(0)}, \mathcal{H}_{\mathcal{M}}^{(0)}) = 0$.

Proof The decomposition $\mathcal{H}[[\boldsymbol{\tau}]] = \mathcal{H}_{\mathcal{M}}^{(0)} \oplus \mathcal{L}[[\boldsymbol{\tau}]]$ follows from (2.3) since $\left[z\partial_{\tau^\alpha} e^{\mu(\tau,z)/z}\right] \in \mu_\alpha^{\mathcal{L}} + \mathcal{L}[[\boldsymbol{\tau}]]$. To see $\mathcal{H}_{\mathcal{M}}^{(0)}$ is an isotropic subspace,

$$\begin{aligned}
&\omega(a(z)z\partial_{\tau^\alpha} e^{\mu(\tau,z)/z}, b(z)z\partial_{\tau^\beta} e^{\mu(\tau,z)/z}) \\
&= \mathrm{Res}_{z=0} \mathrm{Tr}(a(z)b(-z)\partial_{\tau^\alpha}\mu(\tau, z)\partial_{\tau^\beta}\mu(\tau, -z)e^{(\mu(\tau,z)-\mu(\tau,-z))/z})dz.
\end{aligned}$$

If $a(z), b(z)$ contains only non-negative powers of z, the expression inside Tr has only non-negative powers of z whose residue vanishes. □

Lemma 15 (Transversality) $\partial_{\tau^\alpha} : \mathcal{H}^{(0)}_{\mathcal{M}} \to z^{-1}\mathcal{H}^{(0)}_{\mathcal{M}}$.

Proof By Lemma 14, we only need to show that $\omega(z\partial_{\tau^\alpha}\mathcal{H}^{(0)}_{\mathcal{M}}, \mathcal{H}^{(0)}_{\mathcal{M}}) = 0$, which follows from a similar calculation as in Lemma 14. □

Corollary 16 *There exists* $A^{\gamma}_{\alpha\beta}(\boldsymbol{\tau}) \in \mathbb{C}[[\boldsymbol{\tau}]]$ *such that*

$$(\partial_{\tau^\alpha}\partial_{\tau^\beta} - z^{-1}A^{\gamma}_{\alpha\beta}(\boldsymbol{\tau})\partial_{\tau^\gamma})\left[e^{\mu(\tau,z)/z}\right] = 0.$$

Proof By Eq. (2.3) and Lemma 14, $[z\partial_{\tau^\alpha}e^{\mu(\tau,z)/z}]$ forms a $\mathbb{C}[[\boldsymbol{\tau}]]$-basis of $\mathcal{H}^{(0)}_{\mathcal{M}} \cap z\mathcal{L}[[\boldsymbol{\tau}]]$. By Eq. (2.3) and Lemma 15,

$$z\partial_{\tau^\beta}[z\partial_{\tau^\alpha}e^{\mu(\tau,z)/z}] \in \mathcal{H}^{(0)}_{\mathcal{M}} \cap z\mathcal{L}[[\boldsymbol{\tau}]],$$

hence a $\mathbb{C}[[\boldsymbol{\tau}]]$-linear combination of $\{[z\partial_{\tau^\alpha}e^{\mu(\tau,z)/z}]\}_\alpha$. □

The following corollary is a direct consequence.

Corollary 17 *The generalized period satisfies*

$$(\partial_{\tau^\alpha}\partial_{\tau^\beta} - A^{\gamma}_{\alpha\beta}(\boldsymbol{\tau})\partial_{\tau^\gamma})\Pi^{\mathcal{L}} = 0.$$

Let us define a metric by

$$g_{\alpha\beta} := \omega(\partial_{\tau^\alpha}e^{\mu(\tau,z)/z}, z\partial_{\tau^\beta}e^{\mu(\tau,z)/z}).$$

Lemma 18 $g_{\alpha\beta}$ *is a non-degenerate constant matrix.*

Proof It follows from Eq. (2.3) that $g_{\alpha\beta} = \mathrm{Tr}(\mu^{\mathcal{L}}_\alpha \wedge \mu^{\mathcal{L}}_\beta)$. □

Corollary 19 *Let* $A_{\alpha\beta\gamma}(\boldsymbol{\tau}) := \sum_\delta A^{\delta}_{\alpha\beta}(\boldsymbol{\tau})g_{\delta\gamma}$. *Then* $A_{\alpha\beta\gamma}(\boldsymbol{\tau})$ *is (graded) symmetric in* α, β, γ.

Proof This follows from $\partial_\gamma g_{\alpha\beta} = 0$. □

Lemma 20 $A^{\gamma}_{\alpha\beta}(\boldsymbol{\tau}) \in \mathbb{C}[[\boldsymbol{\tau}]]$ *satisfies the WDVV equation.*

Proof Define the Dubrovin connection

$$\nabla_{\tau^\alpha} = \partial_{\tau^\alpha} - z^{-1}A_\alpha,$$

where A_α is the $\mathbb{C}[[z]][[\tau]]$-linear transformation on $\mathcal{H}^{(0)}_{\mathcal{M}}$ defined on the basis by

$$A_\alpha : \left[z\partial_{\tau^\beta}e^{\mu(\tau,z)/z}\right] \to \sum_\gamma A^{\gamma}_{\alpha\beta}\left[z\partial_{\tau^\gamma}e^{\mu(\tau,z)/z}\right].$$

Then $\nabla\left[z\partial_{\tau^\alpha}e^{\mu(\tau,z)/z}\right] = 0$ on the basis. The WDVV equation is equivalent to $\nabla^2 = 0$, which follows from the curvature condition. □

The properties above can be summarized as follows. The triple $(\partial_{\tau^\alpha}, A_{\alpha\beta}^{\gamma}, g_{\alpha\beta})$ defines a (formal) Frobenius manifold structure on $\mathcal{M}$, with τ^α being the flat coordinates. In particular, there exists a function $\mathcal{F}_0^{\mathcal{L}}(\boldsymbol{\tau})$ satisfying

$$A_{\alpha\beta\gamma}(\boldsymbol{\tau}) = \partial_{\tau^\alpha}\partial_{\tau^\beta}\partial_{\tau^\gamma}\mathcal{F}_0^{\mathcal{L}}(\boldsymbol{\tau}).$$

There also exists the Euler vector field and identity vector field.

$\mathcal{F}_0^{\mathcal{L}}(\boldsymbol{\tau})$ is called the prepotential, encoding the genus zero correlation functions in the Calabi-Yau B-model. It depends on the choice of the opposite filtration $\mathcal{L}$. When X is around the large complex limit, the degeneration leads to an opposite Monodromy weight filtration, and $\mathcal{F}_0^{\mathcal{L}}(\boldsymbol{\tau})$ is identified with the genus zero Gromov-Witten invariants of the mirror Calabi-Yau for a large class of examples [2, 16, 29].

2.3 *Landau-Ginzburg Model*

Now we move to the Landau-Ginzburg B-model. We will focus on an isolated singularity defined by a weighted homogeneous polynomial

$$f : X = \mathbb{C}^n \to \mathbb{C}, \quad f(\lambda^{q_1}x_1, \cdots, \lambda^{q_n}x_n) = \lambda f(x_1, \cdots, x_n).$$

q_i are called the weights of x_i, and the central charge of f is defined by

$$\hat{c}_f = \sum_i (1 - 2q_i).$$

Associated to f, K. Saito has introduced the concept of a primitive form [33], which induces a Frobenius manifold structure (originally called a flat structure) on the local universal deformation space of f. The construction of primitive forms for arbitrary isolated singularities is later fully established by M. Saito [34]. See also [4, 12, 13, 37] for generalizations to certain class of Laurent polynomials. This gives rise to the genus zero correlation functions in the Landau-Ginzburg B-model. The generalized period map for compact Calabi-Yau manifolds can be viewed as the analogue of primitive period map. See [35] also for a summary of primitive form in the context of mirror symmetry.

In this rest of this section, we will give a brief review of primitive forms. Our presentation will base on the work [28], which exhibits a unified geometry of Landau-Ginzburg and Calabi-Yau models. We will also describe the perburbative formula of primitive forms [28] which is fully developed in [20, 27] to prove the mirror symmetry conjecture between Landau-Ginzburg models.

2.3.1 Universal Unfolding

The DGLA controlling the deformation theory has a natural twisting in the Landau-Ginzburg case

$$\left(\mathrm{PV}(X), \bar{\partial}_f, \{,\}\right), \quad \bar{\partial}_f = \bar{\partial} + df\lrcorner,$$

where $df\lrcorner$ is the contraction with the holomorphic 1-form df. We will be also working with a subcomplex

$$\mathrm{PV}_c(X) \subset \mathrm{PV}(X)$$

of polyvector fields with compact support. Since $X = \mathbb{C}^n$ is Stein, we have

Lemma 21 *The embedding* $(\mathrm{PV}_c(X), \bar{\partial}_f) \hookrightarrow (\mathrm{PV}(X), \bar{\partial}_f)$ *is quasi-isomorphic. The cohomology is given by*

$$H^*(\mathrm{PV}(X), \bar{\partial}_f) \cong \mathrm{Jac}_0(f),$$

where $\mathrm{Jac}_{\mathbf{0}}(f) = \mathbb{C}\{x^i\}/\{\partial_i f\}$ *is the Milnor ring of the isolated singularity.*

It follows that in the Landau-Ginzburg case, the universal solutions of the associated Maurer-Cartan equation is greatly simplified, and can be represented as a deformation of $f(\mathbf{x})$ via the universal unfolding:

$$F : \mathbb{C}^n \times \mathbb{C}^\mu \to \mathbb{C}, \quad F(\mathbf{x}, \mathbf{s}) := f(\mathbf{x}) + \sum_{\alpha=1}^{\mu} s_\alpha \phi_\alpha(\mathbf{x}), \quad \mathbf{s} = (s_1, \cdots, s_\mu).$$

where $\mu = \dim_{\mathbb{C}} \mathrm{Jac}_{\mathbf{0}}(f)$, and $\{\phi_\alpha(\mathbf{x})\}$ is a basis of $\mathrm{Jac}_{\mathbf{0}}(f)$.

In the case f being weighted homogenous, we can further assume that ϕ_α are all weighted homogeneous with increasing degrees

$$0 = \deg(\phi_1) \leq \deg(\phi_2) \leq \cdots \leq \deg(\phi_\mu) = \hat{c}_f, \quad \text{where } \deg(x_i) = q_i.$$

We will extend our weight degree assignment to the deformation parameter

$$\deg(s_\alpha) := 1 - \deg(\phi_\alpha)$$

such that F becomes weighted homogeneous of total degree 1. Let us denote by

$$\mathcal{M} := (\mathbb{C}^\mu, \mathbf{0})$$

the germ around $\mathbf{0} \in \mathbb{C}^\mu$, parametrizing the local deformation space. $\{s_\alpha\}$ is viewed as a coordinate system on $\mathcal{M}$.

Let $\Omega := dx^1 \wedge \cdots \wedge dx^n$ be our fixed holomorphic volume form. Let $\Omega^k_{X,\mathbf{0}}$ be the germ of holomorphic k-forms at $\mathbf{0}$.

Definition 22 $\Omega_f := \Omega^n_{X,\mathbf{0}}/df \wedge \Omega^{n-1}_{X,\mathbf{0}}$.

With our choice of Ω, we can identify

$$\mathrm{Jac}_{\mathbf{0}}(f) \to \Omega_f, \quad [\phi] \to [\phi\Omega].$$

There exists a classical residue pairing defined on Ω_f:

$$\eta_f : \Omega_f \otimes \Omega_f \to \mathbb{C}.$$

This has an alternate geometric description as follows. Recall the trace map

$$\mathrm{Tr} : \mathrm{PV}_c(X) \to \mathbb{C}, \quad \mu \to \int_X \mu \lrcorner \Omega \wedge \Omega$$

is well-defined on $\mathrm{PV}_c(X)$. It is easy to see that it descends to cohomologies

$$\mathrm{Tr} : H^*(\mathrm{PV}_c(X), \bar{\partial}_f) \to \mathbb{C}.$$

Proposition 23 ([28]) *Let* $\iota : H^*(\mathrm{PV}_c(X), \bar{\partial}_f) \to H^*(\mathrm{PV}(X), \bar{\partial}_f)$ *denote the isomorphism as in Lemma 21. Then the residue pairing is related to the trace map by*

$$\eta_f([\phi_1\Omega], [\phi_2\Omega]) = \mathrm{Tr}(\iota^{-1}([\phi_1]) \wedge \iota^{-1}([\phi_2])), \quad \forall [\phi_i] \in \mathrm{Jac}_{\mathbf{0}}(f).$$

2.3.2 Brieskorn Lattice and Higher Residues

Analogous to the Calabi-Yau case, we consider the following extended DGLA

$$(\mathrm{PV}(X)[[z]], Q_f, \{,\}), \quad Q_f := \bar{\partial}_f + z\partial_\Omega,$$

where ∂_Ω is defined with respect to the volume form Ω.

Definition 24 ([32]) Define $\mathcal{H}^{(0)}_f := \Omega^n_{X,\mathbf{0}}[[z]]/(df + zd)\Omega^{n-1}_{X,\mathbf{0}}$ the (formally completed) *Brieskorn lattice* associated to f.

Lemma 25 ([28]) *The embedding* $(\mathrm{PV}_c(X)[[z]], Q_f) \hookrightarrow (\mathrm{PV}(X)[[z]], Q_f)$ *is a quasi-isomorphism. It induces isomorphisms*

$$H^*(\mathrm{PV}(X)[[z]], Q_f) \cong H^0(\mathrm{PV}(X)[[z]], Q_f) \overset{z\Gamma_\Omega}{\cong} \mathcal{H}^{(0)}_f,$$

where Γ_Ω *is defined the same as in* (2.2).

There is a similar semi-infinite Hodge filtration on $\mathcal{H}_f^{(0)}$ given by $\mathcal{H}_f^{(-k)} := z^k \mathcal{H}_f^{(0)}$, with graded pieces

$$\mathcal{H}_f^{(-k)}/\mathcal{H}_f^{(-k-1)} \cong \Omega_f.$$

In particular, $\mathcal{H}_f^{(0)}$ is a free $\mathbb{C}[[z]]$-module of rank μ. We will also denote the extension to Laurent series by

$$\mathcal{H}_f := \mathcal{H}_f^{(0)} \otimes_{\mathbb{C}[[z]]} \mathbb{C}((z)).$$

There is a natural $\mathbb{Q}$-grading on $\mathcal{H}_f^{(0)}$ defined by assigning the weight degrees

$$\deg(x_i) = q_i, \quad \deg(dx_i) = q_i, \quad \deg(z) = 1.$$

For a homogeneous element of the form $\varphi = z^k g(x_i) dx_1 \wedge \cdots \wedge dx_n$, we define

$$\deg(\varphi) = \deg(g) + k + \sum_i q_i.$$

In [32], K. Saito constructed a *higher residue pairing*

$$K_f : \mathcal{H}_f^{(0)} \otimes \mathcal{H}_f^{(0)} \to z^n \mathbb{C}[[z]]$$

which satisfies the following properties

1. K_f is equivariant with respect to the $\mathbb{Q}$-grading, i.e.,

$$\deg(K_f(\alpha, \beta)) = \deg(\alpha) + \deg(\beta)$$

   for homogeneous elements $\alpha, \beta \in \mathcal{H}_f^{(0)}$.
2. $K_f(\alpha, \beta) = (-1)^n \overline{K_f(\beta, \alpha)}$, where the $-$ operator takes $z \to -z$.
3. $K_f(v(z)\alpha, \beta) = K_f(\alpha, v(-z)\beta) = v(z) K_f(\alpha, \beta)$ for $v(z) \in \mathbb{C}[[z]]$.
4. The leading z-order of K_f defines a pairing

$$\mathcal{H}_f^{(0)}/z\mathcal{H}_f^{(0)} \otimes \mathcal{H}_f^{(0)}/z\mathcal{H}_f^{(0)} \to \mathbb{C}, \quad \alpha \otimes \beta \mapsto \lim_{z \to 0} z^{-n} K_f(\alpha, \beta)$$

   which coincides with the usual residue pairing

$$\eta_f : \Omega_f \otimes \Omega_f \to \mathbb{C}.$$

The last property implies that K_f defines a semi-infinite extension of the residue pairing, which explains the name "higher residue". An alternate way to understand the higher residue pairing is through the trace map in the spirit of Proposition 23. Let us define a pairing

$$\tilde{K}_f : \mathrm{PV}_c(X)[[z]]\times\mathrm{PV}_c(X)[[z]] \to z^n\mathbb{C}[[z]], \quad \tilde{K}_f(f(z)\alpha, g(z)\beta) = z^n f(z)g(-z)\,\mathrm{Tr}(\alpha\beta).$$

It is easy to see that $\tilde{K}_f$ descends to $H^*(\mathrm{PV}_c(X)[[z]], Q_f)$ which is canonically isomorphic to $H^*(\mathrm{PV}(X)[[z]], Q_f)$.

Proposition 26 ([28]) *$\tilde{K}_f$ coincides with K_f under the isomorphism $H^*(\mathrm{PV}(X)[[z]], Q_f) \cong \mathcal{H}_f^{(0)}$ as in Lemma 25.*

The Brieskorn lattice and the higher residue pairing can be extended to the family case on the germ $\mathcal{M}$ associated to the unfolding F. We have

$$\mathcal{H}_F^{(0)} := \Omega^n_{X\times\mathcal{M}/\mathcal{M},\mathbf{0}}[[z]]/(dF + zd)\Omega^{n-1}_{X\times\mathcal{M}/\mathcal{M},\mathbf{0}}$$

where $\Omega^*_{X\times\mathcal{M}/\mathcal{M},\mathbf{0}}$ is the germ of the sheaf of relative holomorphic differential forms at $\mathbf{0}$. It can be viewed as a free sheaf of rank μ on $\mathcal{M}\times\hat{\Delta}$, where $\hat{\Delta}$ is the formal disk with parameter z. $\mathcal{H}_F^{(0)}$ is equipped with a flat Gauss-Manin connection on $\mathcal{M}\times\hat{\Delta}$, denoted by ∇^{GM}. The higher residue pairing extends to

$$K_F : \mathcal{H}_F^{(0)} \otimes_{\mathcal{O}_\mathcal{M}} \mathcal{H}_F^{(0)} \to z^n\mathcal{O}_\mathcal{M}[[z]]$$

satisfying the following properties

1. $K_F(s_1, s_2) = (-1)^n\overline{K_F(s_2, s_1)}$, where $-$ is the operator $z \to -z$.
2. $K_F(g(z)s_1, s_2) = K_F(s_1, g(-z)s_2) = g(z)K_F(s_1, s_2)$ for any $g \in \mathcal{O}_\mathcal{M}[[z]]$.
3. $\partial_V K_F(s_1, s_2) = K_F(\nabla_V^{GM}s_1, s_2) + K_F(s_1, \nabla_V^{GM}s_2)$ for any $V \in T_\mathcal{M}$.
4. $z\partial_z K_F(s_1, s_2) = K_F(\nabla_{z\partial_z}^{GM}s_1, s_2) + K_F(s_1, \nabla_{z\partial_z}^{GM}s_2)$.
5. The induced pairing

$$\mathcal{H}_F^{(0)}/z\mathcal{H}_F^{(0)} \otimes_{\mathcal{O}_\mathcal{M}} \mathcal{H}_F^{(0)}/z\mathcal{H}_F^{(0)} \to \mathcal{O}_\mathcal{M}$$

coincides with the classical residue pairing.

2.3.3 Primitive Forms

Definition 27 A section $\zeta \in \mathcal{H}_F^{(0)}$ is called a *primitive form* if it satisfies the following conditions:

(1) (Primitivity) The section ζ induces an $\mathcal{O}_\mathcal{M}$-module isomorphism

$$z\nabla^{GM}\zeta : T_\mathcal{M} \to \mathcal{H}_F^{(0)}/z\mathcal{H}_F^{(0)}; \quad V \mapsto z\nabla_V^{GM}\zeta.$$

(2) (Orthogonality) For any local sections V_1, V_2 of $T_{\mathcal{M}}$,

$$K_F\left(\nabla^{GM}_{V_1}\zeta, \nabla^{GM}_{V_2}\zeta\right) \in z^{n-2}\mathcal{O}_{\mathcal{M}}.$$

(3) (Holonomicity) For any local sections V_1, V_2, V_3 of $T_{\mathcal{M}}$,

$$K_F\left(\nabla^{GM}_{V_1}\nabla^{GM}_{V_2}\zeta, \nabla^{GM}_{V_3}\zeta\right) \in z^{n-3}\mathcal{O}_{\mathcal{M}} \oplus z^{n-2}\mathcal{O}_{\mathcal{M}};$$
$$K_F\left(\nabla^{GM}_{z\partial_z}\nabla^{GM}_{V_1}\zeta, \nabla^{GM}_{V_2}\zeta\right) \in z^{n-3}\mathcal{O}_{\mathcal{M}} \oplus z^{n-2}\mathcal{O}_{\mathcal{M}}.$$

(4) (Homogeneity) There is a constant $r \in \mathbb{C}$ such that

$$\left(\nabla^{\Omega}_{z\partial_z} + \nabla^{\Omega}_{E}\right)\zeta = r\zeta.$$

where E is the Euler vector field. In the case of weighted homogeneous singularity, we have $r = \sum_i q^i$.

The space of primitive forms has a geometric description. Let us extend the higher residue pairing to

$$K_f : \mathcal{H}_f \otimes \mathcal{H}_f \to \mathbb{C}((z)).$$

This defines a symplectic pairing ω_f on $\mathcal{H}_f$ by

$$\omega_f(\alpha, \beta) := \mathrm{Res}_{z=0}\, z^{-n} K_f(\alpha, \beta) dz,$$

with $\mathcal{H}_f^{(0)}$ being an isotropic subspace. Following [33],

Definition 28 A good section σ is a splitting of the quotient $\mathcal{H}_f^{(0)} \to \Omega_f$:

$$\sigma : \Omega_f \to \mathcal{H}_f^{(0)},$$

such that: (1) σ preserves the $\mathbb{Q}$-grading; (2) $K_f(\mathrm{Im}(\sigma), \mathrm{Im}(\sigma)) \subset z^n\mathbb{C}$. A basis of the image $\mathrm{Im}(\sigma)$ of a good section σ will be called a good basis of $\mathcal{H}_f^{(0)}$.

Definition 29 A good opposite filtration $\mathcal{L}$ is defined by a splitting

$$\mathcal{H}_f = \mathcal{H}_f^{(0)} \oplus \mathcal{L}$$

such that: (1) $\mathcal{L}$ preserves the $\mathbb{Q}$-grading; (2) $\mathcal{L}$ is an isotropic subspace; (3) $z^{-1} : \mathcal{L} \to \mathcal{L}$.

Remark 30 Here for f being weighted homogeneous, (1) is equivalent to the conventional condition that $\nabla^{GM}_{z\partial_z}$ preserves $\mathcal{L}$ (see e.g. [28] for an exposition).

The above two definitions are equivalent. In fact, a good opposite filtration $\mathcal{L}$ defines the splitting $\sigma : \Omega_f \xrightarrow{\cong} \mathcal{H}_f^{(0)} \cap z\mathcal{L}$. Conversely, a good section σ gives rise to the good opposite filtration $\mathcal{L} = z^{-1}\mathrm{Im}(\sigma)[z^{-1}]$. As shown in [33, 34], the primitive forms associated to weighted homogeneous singularities are in one-to-one correspondence with good sections (up to a nonzero scalar). We remark that for general isolated singularities, we need the notion of *very good sections* [34, 36] in order to incorporate with the monodromy.

Theorem 31 ([33]) *The space of primitive forms of f up to rescaling by a constant is isomorphic to the space of good sections.*

Remark 32 The generalization of this identification to arbitrary isolated singularities is established by M. Saito [34, 36].

2.4 Perturbative Theory of Primitive Forms

In this subsection, we describe the algebraic algorithm [27, 28] to compute the primitive form, flat coordinates and the prepotential with respect to a good basis.

We start with a good basis $\{[\phi_\alpha \Omega]\}_{\alpha=1}^{\mu}$ of $\mathcal{H}_f^{(0)}$, where $\{\phi_\alpha\}_{\alpha=1}^{\mu}$ are weighted homogeneous polynomials in $\mathbb{C}[\mathbf{x}]$ that represent a basis of $\mathrm{Jac}_{\mathbf{0}}(f)$ and $\phi_1 = 1$.

2.4.1 The Exponential Map

Let F be a local universal unfolding of $f(\mathbf{x})$

$$F(\mathbf{x}, \mathbf{s}) := f(\mathbf{x}) + \sum_{\alpha=1}^{\mu} s_\alpha \phi_\alpha(\mathbf{x}), \quad \mathbf{s} = (s_1, \cdots, s_\mu).$$

Let $B := \mathrm{Span}_{\mathbb{C}}\{[\phi_\alpha \Omega]\} \subset \mathcal{H}_f^{(0)}$ be spanned by the chosen good basis. Then

$$\mathcal{H}_f^{(0)} = B[[z]], \quad \mathcal{H}_f = B((z)).$$

Let $B_F := \mathrm{Span}_{\mathbb{C}}\{\phi_\alpha \Omega\}$ be the vector space spanned by the forms $\phi_\alpha \Omega$. We use a different notation to distinguish it with B, since B_F should be viewed as a subspace of the Brieskorn lattice for the unfolding F. See [27, 28] for more details. Consider the following exponential operator [27, 28]

$$e^{(F-f)/z} : B_F \to B((z))[[\mathbf{s}]]$$

defined as a $\mathbb{C}$-linear map on the basis of B_F as follows. Let

$$\mathbb{C}[\mathbf{s}]_k := \mathrm{Sym}^k(\mathrm{Span}_{\mathbb{C}}\{s_1, \cdots, s_\mu\})$$

denote the space of k-homogeneous polynomial in $\mathbf{s}$ (not to be confused with the weighted homogeneous polynomials). As elements in $\mathcal{H}_f \otimes \mathbb{C}[\mathbf{s}]_k$, we can decompose

$$[z^{-k}(F-f)^k \phi_\alpha \Omega] = \sum_{m \geq -k} \sum_\beta h^{(k)}_{\alpha\beta,m} z^m [\phi_\beta \Omega],$$

where $h^{(k)}_{\alpha\beta,m} \in \mathbb{C}[\mathbf{s}]_k$. Then we define

$$e^{(F-f)/z}(\phi_\alpha \Omega) := \sum_{k=0}^{\infty} \sum_\beta \sum_{m \geq -k} h^{(k)}_{\alpha\beta,m} \frac{z^m}{k!} [\phi_\beta \Omega] \in B((z))[[\mathbf{s}]].$$

The exponential map extends to a $\mathbb{C}((z))[[\mathbf{s}]]$-linear isomorphism

$$e^{(F-f)/z} : B_F((z))[[\mathbf{s}]] \to B((z))[[\mathbf{s}]],$$

which plays the role of parallel transport with respect to the Gauss-Manin connection. Let

$$K_f : B((z))[[\mathbf{s}]] \times B((z))[[\mathbf{s}]] \to \mathbb{C}((z))[[\mathbf{s}]]$$

also denote the $\mathbb{C}[[\mathbf{s}]]$-linear extension of the higher residue pairing to $\mathcal{H}_f[[\mathbf{s}]]$.

Lemma 33 ([27, 28]) *For any $\varphi_1, \varphi_2 \in B_F$, we have*

$$K_f(e^{(F-f)/z}\varphi_1, e^{(F-f)/z}\varphi_2) \in z^n \mathbb{C}[[z, \mathbf{s}]]$$

In particular, $e^{(F-f)/z}$ maps $B_F[[z]]$ to an isotropic subspace of $\mathcal{H}_f[[\mathbf{s}]]$.

Theorem 34 ([27, 28]) *Given a good basis $\{[\phi_\alpha d^n \mathbf{x}]\}_{\alpha=1}^{\mu} \subset \mathcal{H}_f^{(0)}$, there exists a unique pair $(\zeta, \mathcal{J})$ satisfying the following: (1) $\zeta \in B_F[[z]][[\mathbf{s}]]$, (2) $\mathcal{J} \in [\Omega] + z^{-1}B[z^{-1}][[\mathbf{s}]] \subset \mathcal{H}_f[[s]]$, and*

$$e^{(F-f)/z}\zeta = \mathcal{J}. \qquad (\star)$$

Moreover, both ζ and $\mathcal{J}$ are homogeneous of weight $\sum_i q_i$.

This is the analogue of (2.3) for Calabi-Yau. $\zeta(\mathbf{s})$ can be solved recursively with respect to the order in $\mathbf{s}$. We refer to [27] for details, and to [28] for a compact formula of this algorithm. The decomposition is a formal solution of the Riemann-Hilbert-Birkhoff problem for primitive forms [33]. The volume form

$\zeta = \sum_{k=0}^{\infty} \sum_{\alpha} \zeta_{(k)}^{\alpha} [\phi_\alpha \Omega]$ gives the power series expansion of a representative of the primitive form associated to the good basis $\{[\phi_\alpha d^n \mathbf{x}]\}_{\alpha=1}^{\mu}$.

2.4.2 Flat Coordinates and Potential Function

Let $(\zeta, \mathcal{J})$ be the unique solution of $(\star)$. ζ represents the power series expansion of a primitive form. However for the purpose of mirror symmetry, it is more convenient to work with $\mathcal{J}$, which plays the role of Givental's J-function (see [17] for an introduction). This allows us to read off the flat coordinates and the potential function of the associated Frobenius manifold structure.

With the embedding $z^{-1}\mathbb{C}[z^{-1}][[\mathbf{s}]] \hookrightarrow z^{-1}\mathbb{C}[[z^{-1}]][[\mathbf{s}]]$, we decompose

$$\mathcal{J} = [d^n x] + \sum_{m=-1}^{-\infty} z^m \mathcal{J}_m, \quad \text{where } \mathcal{J}_m = \sum_{\alpha} \mathcal{J}_m^{\alpha} [\phi_\alpha \Omega], \ \mathcal{J}_m^{\alpha} \in \mathbb{C}[[\mathbf{s}]].$$

We denote the z^{-1}-term by

$$t_\alpha(\mathbf{s}) := \mathcal{J}_{-1}^{\alpha}(\mathbf{s}).$$

It is easy to see that $t_\alpha = s_\alpha + O(\mathbf{s}^2)$ and is homogeneous of the same weight as s_α. Therefore t_α defines a set of new homogeneous local coordinates on the (formal) deformation space of f.

Proposition 35 *The function $\mathcal{J} = \mathcal{J}(\mathbf{s}(\mathbf{t}))$ in coordinates t_α satisfies*

$$\partial_{t_\alpha} \partial_{t_\beta} \mathcal{J} = z^{-1} \sum_{\gamma} A_{\alpha\beta}^{\gamma}(\mathbf{t}) \partial_{t_\gamma} \mathcal{J}$$

for some homogeneous $A_{\alpha\beta}^{\gamma}(\mathbf{t}) \in \mathbb{C}[[\mathbf{t}]]$ of weighted degree $\deg \phi_\alpha + \deg \phi_\beta - \deg \phi_\gamma$. Moreover, for any $\alpha, \beta, \gamma, \delta$,

$$\partial_{t_\alpha} A_{\beta\gamma}^{\delta} = \partial_{t_\beta} A_{\alpha\gamma}^{\delta}, \quad \sum_{\sigma} A_{\alpha\sigma}^{\delta} A_{\beta\gamma}^{\sigma} = \sum_{\sigma} A_{\beta\sigma}^{\delta} A_{\alpha\gamma}^{\sigma}$$

Lemma 36 *In terms of the coordinates t_α, we have*

$$K_f(z\partial_{t_\alpha} \mathcal{J}, z\partial_{t_\beta} \mathcal{J}) = z^n g_{\alpha\beta}.$$

Here $g_{\alpha\beta}$ is the constant equal to the residue pairing $\eta_f(\phi_\alpha \Omega, \phi_\beta \Omega)$.

Similarly, the triple $(\partial_{t_\alpha}, A^{\gamma}_{\alpha\beta}, g_{\alpha\beta})$ defines a (formal) Frobenius manifold structure on a neighborhood S of the origin with $\{t_\alpha\}$ being the flat coordinates, together with the potential function $\mathcal{F}_0(\mathbf{t})$ satisfying

$$A_{\alpha\beta\gamma}(\mathbf{t}) = \partial_{t_\alpha}\partial_{t_\beta}\partial_{t_\gamma}\mathcal{F}_0(\mathbf{t}).$$

It is not hard to see that $\mathcal{F}_0(\mathbf{t})$ is homogeneous of degree $3 - \hat{c}_f$. The potential function $\mathcal{F}_0(\mathbf{t})$ can also be computed perturbatively.

Remark 37 ζ is in fact an analytic primitive form [28]. Therefore, both t_α and $\mathcal{F}_0(\mathbf{t})$ are analytic functions of $\mathbf{s}$ at the germ $\mathbf{s} = 0$.

Remark 38 The closed formula of primitive forms for weighted homogenous singularities only exists for ADE ($\hat{c}_f < 1$) and simple elliptic singularities $\hat{c}_f = 1$ [33]. They can be easily obtained via the perburbative method [28]. For $\hat{c}_f > 1$, expressions of primitive forms are unknown and this has become long one of the major obstacles toward understanding mirror symmetry between Landau-Ginzburg models. It turns out that the perturbative formula, together with the WDVV equation, is enough to compute the full data of the Landau-Ginzburg B-model. The first non-trivial examples are Arnold's unimodular exceptional singularities, whose mirror symmetry with FJRW-theory [15] (Landau-Ginzburg A-model) is established [27] via the perburbative method. Such mirror symmetry between singularity theories is fully established in [20] for almost all weighted homogeneous polynomials when Landau-Ginzburg mirrors exist.

3 Quantum Geometry

In this section, we will quantize the symplectic structure that appears in the previous section for the generalized period maps, or the primitive forms. We analyze Givental's symplectic loop space formalism in the context of B-model geometry, and explain the Fock space construction via the renormalization techniques of gauge theory. It leads to the quantum BCOV theory developed in [11, 23]. This is parallel to another categorical approach [8, 9, 22] to the quantum B-model associated to a Calabi-Yau categories of D-branes. Our quantum field theory approach has the advantage of manifest physics intuitions and is related to methods of background symmetries and integrable hierarchies.

3.1 A Toy Model of Weyl Quantization

3.1.1 Weyl Algebra and Fock Space

Let us recall the construction of the Fock module for a finite dimensional dg symplectic vector space (V, ω, d), where ω is the symplectic pairing on V, and d is the differential which is skew self-adjoint with respect to ω. Let

$$\mathcal{W}(V) := \prod_{n \geq 0} (V^*)^{\otimes n}[[\hbar]]/ \sim$$

be the (formal) Weyl algebra of V, which is the pro-free dg algebra generated by the linear dual V^* and a formal parameter $\hbar$, subject to the relation

$$[a, b] \sim \hbar\omega^{-1}(a, b), \quad \forall a, b \in V^*.$$

Here $\omega^{-1} \in \wedge^2 V$ is the inverse of ω, and $[a, b] := a \otimes b \mp b \otimes a$ is the graded commutator in the tensor algebra generated by V^*. Let V_+ be a Lagrangian subcomplex of V, and $Ann(V_+) \subset V^*$ be the annihilator of V_+. Then the Fock module $\mathcal{F}ock(V_+)$ is defined to be the quotient

$$\mathcal{F}ock(V_+) := \mathcal{W}(V)/\mathcal{W}(V)Ann(V_+).$$

Since V_+ is preserved by the differential, $\mathcal{F}ock(V_+)$ naturally inherits a dg structure from d. We will denote it by $\hat{d}$.

Let us choose a complementary linear Lagrangian subspace $V_- \subset V$ such that

$$V = V_+ \oplus V_-.$$

V_- may not be preserved by the differential. It allows us to formally identify

$$V \cong T^*(V_+)$$

Let

$$\mathcal{O}(V_+) = \prod_{n \geq 0} \mathrm{Sym}^n(V_+^*)$$

be the space of formal functions on the graded vector space V_+. V_- defines a splitting of the map $V^* \to V_+^*$, hence a morphism

$$\begin{array}{ccc} \mathcal{O}(V_+)[[\hbar]] & \longrightarrow & \mathcal{W}(V) \\ & \searrow^{\cong} & \downarrow \\ & & \mathcal{F}ock(V_+) \end{array}$$

which identifies the Fock module with the algebra $\mathcal{O}(V_+)[[\hbar]]$. The differential $\hat{d}$ can be described as follows. Let $\pi_+ : V \to V_+$ be the projection corresponding to the splitting $V = V_+ \oplus V_-$. Consider $(d \otimes 1)\omega^{-1}$, which is an element of $V \otimes V$. Let P be the projection

$$P = \pi_+ \otimes \pi_+ \left((d \otimes 1)\omega^{-1}\right) \in V_+ \otimes V_+$$

and it is easy to see that $P \in \mathrm{Sym}^2(V_+)$. Let $\partial_P : \mathcal{O}(V_+) \to \mathcal{O}(V_+)$ be the operator of contracting with the symmetric 2-tensor P

$$\partial_P : \mathrm{Sym}^n(V_+^*) \to \mathrm{Sym}^{n-2}(V_+^*).$$

Lemma 39 *Under the isomorphism $\mathcal{F}ock(V_+) \cong \mathcal{O}(V_+)[[\hbar]]$, $\hat{d}$ takes the form*

$$\hat{d} = d + \hbar\partial_P$$

where d here is the induced differential on $\mathcal{O}(V_+)$ from d on V_+.

∂_P will be called a BV operator. It induces a bracket on $\mathcal{O}(V_+)$ by

$$\{\Phi_1, \Phi_2\}_P := \partial_P(\Phi_1\Phi_2) - (\partial_P\Phi_1)\Phi_2 - (-1)^{|\Phi_1|}\Phi_1\partial_P\Phi_2, \quad \Phi_i \in \mathcal{O}(V_+).$$

Here $|\Phi|$ is the cohomology degree of Φ. We will also need a slightly larger Fock space given by

$$\mathcal{F}ock^+(V_+) := \prod_{k=0}^{\infty}\Big(\bigoplus_{\substack{m\geq 0, n\in\mathbb{Z}\\ m+2n=k}} \mathrm{Sym}^m(V_+^*)\hbar^n\Big),$$

i.e. we allow negative powers of $\hbar$ in an appropriate topology.

3.1.2 Lagrangian and Quantization

In the classical geometry, we are interested in a Lagrangian submanifold $\mathcal{L}$ of V. Under the isomorphism

$$V \cong T^*(V_+),$$

$\mathcal{L}$ can be represented (locally) as a graph $\mathcal{L} = \mathrm{Graph}(dF_0)$. We impose a symmetry condition that d is tangent to $\mathcal{L}$, where we treat d as defining a square-zero vector field on V. This can be viewed as an infinitesimal gauge symmetry.

Lemma 40 *d being tangent to $\mathcal{L}$ is equivalent to the following equation for F_0*

$$dF_0 + \frac{1}{2}\{F_0, F_0\}_P = 0.$$

This is called the *classical master equation*. It says that $d + \{F_0, -\}_P$ defines a new square-zero vector field on V_+. Geometrically, let

$$\pi_+|_{\mathcal{L}} : \mathcal{L} \to V_+.$$

Then $d + \{F_0, -\}_P = (\pi_+|_{\mathcal{L}})_*(d)$ is the push-forward of the vector field d on $\mathcal{L}$.

In the quantum theory, we are interested in a vector $|F\rangle \in \mathcal{F}ock^+(V_+)$ satisfying the "gauge invariance condition": $d|F\rangle = 0$. To relate $|F\rangle$ to $\mathcal{L}$ in the $\hbar \to 0$ classical limit, we consider $|F\rangle$ of the form represented by $e^{F/\hbar}$

$$|F\rangle \leftrightarrow e^{F/\hbar}, \quad F = \sum_{g \geq 0} \hbar^g F_g \in \mathcal{O}(V_+)[[\hbar]].$$

By Lemma 39, the gauge invariance becomes $(d + \hbar\partial_P)e^{F/\hbar} = 0$, or equivalently

$$(d + \hbar\partial_P)F + \frac{1}{2}\{F, F\}_P = 0. \tag{3.1}$$

This is called the *quantum master equation*.

In summary of our toy model, the quantization scheme quantizes the Lagrangian $\mathcal{L}$ to a state $|F\rangle$ in the Fock space. Equivalently, it quantizes F_0 which satisfies the classical master equation to $F = F_0 + \hbar F_1 + \cdots$ which satisfies the quantum master equation.

3.2 Symplectic Geometry and BCOV Theory

Following Givental's symplectic formulation [18, 19] of Gromov-Witten theory in the A-model and the parallel Barannikov's work [1, 2] in the B-model, our dg symplectic vector space is (note that our degree assignment in this article differs from that in [11])

$$\mathcal{S}(X) = \mathrm{PV}(X)((z)),$$

with differential $Q = \bar{\partial} + z\partial$ and symplectic pairing ω by Definition 8.

In [6], Bershadsky et al. introduced a gauge theory for polyvector fields on Calabi-Yau three-folds. This is further extended to arbitrary Calabi-Yau manifolds in [11]. The space of fields of the BCOV theory is

$$S_+(X) \equiv \mathrm{PV}(X)[[z]]$$

which is a linear isotropic subspace of $S(X)$. The classical action functional of the BCOV theory can be constructed from the following Lagrangian (the embedding is in the sense of formal scheme via functor of points on Artinian rings [11])

$$\mathcal{L}_X = \left\{ z(e^{\mu/z} - 1) | \mu \in S_+(X) \right\} \subset S(X).$$

This can be viewed as the lifting of that in Proposition 11 to the cochain level. The geometry of $\mathcal{L}_X$ can be described by the following

Lemma 41 ([11]) *$\mathcal{L}_X$ is a formal Lagrangian submanifold of $\mathcal{S}(X)$, preserved by the differential $Q = \bar{\partial} + z\partial$. Moreover, $\mathcal{L}_X + z$ is a Lagrangian cone preserved by the infinitesimal symplectomorphism of $S(X)$ given by multiplying by z^{-1}.*

Remark 42 $\mathcal{L}_X + z$ is called the dilaton shift of $\mathcal{L}_X$ [18].

Consider the splitting

$$S(X) = S_+(X) \oplus S_-(X) \tag{3.2}$$

where recall $S_-(X) = z^{-1}\,\mathrm{PV}(X)[z^{-1}]$. It allows us to formally identify

$$S(X) \cong T^*(S_+(X)).$$

The generating functional $\mathbf{F}_{\mathcal{L}_X}$ is a formal function on $S_+(X)$ such that

$$\mathcal{L}_X = \mathrm{Graph}(d\mathbf{F}_{\mathcal{L}_X}).$$

The explicit formula is worked out in [11]

Proposition 43 ([11]) $\mathbf{F}_{\mathcal{L}_X}(\mu) = \mathrm{Tr}\,\langle e^{\mu}\rangle_0$, *where*

$$\langle - \rangle_0 : \mathrm{Sym}(\mathrm{PV}(X)[[z]]) \to \mathrm{PV}(X)$$

is given by intersection of ψ-classes over the moduli space of marked rational curves

$$\left\langle \alpha_1 z^{k_1}, \cdots, \alpha_n z^{k_n} \right\rangle_0 := \alpha_1 \cdots \alpha_n \int_{\overline{M}_{0,n}} \psi_1^{k_1} \cdots \psi_n^{k_n} = \binom{n-3}{k_1, \cdots, k_n} \alpha_1 \cdots \alpha_n$$

Definition 44 ([11]) The classical BCOV interaction is defined to be the formal local functional on $S_+(X)$ given by $\mathbf{F}_{\mathcal{L}_X}$.

Remark 45 Our definition of BCOV interaction extends the original Kodaira-Spencer interaction in [6] by turning on the "gravitational descendants" z. It is also equivalent to that used by Losev-Shadrin-Shneiberg [30] in the discussion of finite dimensional toy models of Hodge field theory.

We can transfer the geometry of the Lagrangian $\mathcal{L}_X$ into properties of $\mathbf{F}_{\mathcal{L}_X}$.

Proposition 46 ([11]) $\mathbf{F}_{\mathcal{L}_X}$ *satisfies the* classical master equation

$$Q\mathbf{F}_{\mathcal{L}_X} + \frac{1}{2}\left\{\mathbf{F}_{\mathcal{L}_X}, \mathbf{F}_{\mathcal{L}_X}\right\} = 0$$

where Q is the induced derivation on the functionals of $S_+(X)$, and $\{-,-\}$ is the Poisson bracket on local functionals induced from the distribution representing the operator ∂ (see Remark 52).

This is equivalent to that $\mathcal{L}_X$ is preserved by Q (See Lemma 40 for an explanation in the toy model). The classical master equation implies that $Q + \left\{\mathbf{F}_{\mathcal{L}_X}, -\right\}$ is a square-zero operator acting on local functionals. In physics terminology, it generates the gauge symmetry, and defines the gauge theory in the Batalin-Vilkovisky formalism.

3.3 Givental's Formalism via Renormalization

The dg symplectic vector space related to the BCOV theory is $(S(X), \omega, Q)$. If we run the machine to construct the Fock space as in the previous section, we immediately run into trouble: $\mathrm{PV}(X)$ is infinite dimensional! This is a well-known phenomenon in quantum field theory, which is related to the difficulty of ultra-violet divergence. The standard way of solving this is to use the renormalization technique. We will follow the approach developed in [10].

3.3.1 Functionals on the Fields

Let $S_+(X)^{\otimes n}$ be the completed projective tensor product of n copies of $S_+(X)$. It can be viewed as the space of smooth polyvector fields on X^n with a formal variable z for each factor. Let

$$\mathcal{O}^{(n)}(S_+(X)) = \mathrm{Hom}\left(S_+(X)^{\otimes n}, \mathbb{C}\right)_{S_n}$$

denote the space of continuous linear maps (distributions), and the subscript S_n denotes taking S_n coinvariants. $\mathcal{O}^{(n)}(S_+(X))$ will be the space of homogeneous degree n functionals on the space of fields $S_+(X)$, playing the role of $\mathrm{Sym}^n(V^*)$ in our toy model. We will also let

$$\mathcal{O}^{(n)}_{loc}(S_+(X)) \subset \mathcal{O}^{(n)}(S_+(X))$$

be the subspace of local functionals, i.e. those of the form given by the integration of a Lagrangian density

$$\int_X \mathcal{L}(\mu), \quad \mu \in S_+(X).$$

Definition 47 The algebra of functionals $\mathcal{O}(S_+(X))$ on $S_+(X)$ is defined to be

$$\mathcal{O}(S_+(X)) = \prod_{n\geq 0} \mathcal{O}^{(n)}(S_+(X))$$

and the space of local functionals is defined to be the subspace

$$\mathcal{O}_{loc}(S_+(X)) = \prod_{n\geq 0} \mathcal{O}^{(n)}_{loc}(S_+(X))$$

3.3.2 Effective Fock Space

Let g be a Kähler metric on X. Let

$$K^g_L \in \mathrm{PV}(X) \otimes \mathrm{PV}(X), \quad L > 0$$

be the heat kernel for the operator $e^{-L[\bar\partial,\bar\partial^*]}$, where $\bar\partial^*$ is the adjoint of $\bar\partial$ with respect to the metric g and $\left[\bar\partial, \bar\partial^*\right] = \bar\partial\bar\partial^* + \bar\partial^*\bar\partial$ is the Laplacian acting on $\mathrm{PV}(X)$. It is a smooth polyvector field on $X \times X$ defined by the equation

$$\left(e^{-L[\bar\partial,\bar\partial^*]}\alpha\right)(x) = \int_X \left(K^g_L(x,y)\alpha(y) \vdash \Omega_X(y)\right) \wedge \Omega_X(y)$$

where we have chosen coordinates (x, y) on $X \times X$, and we integrate over the second copy of X using the trace map.

Definition 48 The effective inverse $\omega^{-1}_{g,L}$ for the symplectic form ω is defined to be the kernel

$$\omega^{-1}_{g,L} = \sum_{k\in\mathbb{Z}} K^g_L(-z)^k \otimes z^{-k-1} \in S(X) \otimes S(X), \quad L > 0.$$

Note that $\lim_{L\to 0} K_L^g$ is the delta-function distribution, which is no longer a smooth polyvector field, hence not an element of $S(X) \otimes S(X)$. $\omega_{g,L}^{-1}$ can be viewed as the regularization of ω^{-1} in the infinite dimensional setting.

Let $S(X)^*$ be the continuous linear dual of $S(X)$ (distributions on $S(X)$ with extra care on the z-adic topology. See [11] for more details).

Definition 49 The effective Weyl algebra $\mathcal{W}(S(X), g, L)$ is the quotient of the completed tensor algebra

$$\left(\prod_{n\geq 0} \left(S(X)^*\right)^{\otimes n}\right) \otimes \mathbb{C}[[\hbar]]$$

by the topological closure of the two-sided ideal generated by

$$[a, b] - \hbar \left\langle \omega_{g,L}^{-1}, a \otimes b \right\rangle, \quad L > 0$$

for $a, b \in S(X)^*$. Here $\langle , \rangle$ is the natural pairing between $S(X)$ and its dual.

Similarly, the Fock space can also be defined using the regularized kernel $\omega_{g,L}^{-1}$.

Definition 50 The effective Fock space $\mathcal{F}ock\,(S_+(X), g, L)$ is the quotient of $\mathcal{W}(\mathcal{S}(X))$ by the left ideal generated topologically by the subspace

$$Ann(S_+(X), g, L) \subset S(X)^*.$$

Similar to the finite dimensional case, the splitting $S(X) = S_+(X) \oplus S_-(X)$ gives the identification

$$\mathcal{F}ock\,(S_+(X), g, L) \cong \mathcal{O}(S_+(X))[[\hbar]].$$

We refer to [11] for detailed discussions.

3.3.3 Effective BV Formalism

We would like to understand the quantized operator $\hat{Q}_L$ for Q acting on the Fock space represented by the above identification. This is completely similar to the toy model. Let

$$(\partial \otimes 1) K_L^g \in \mathrm{Sym}^2(\mathrm{PV}(X))$$

be the kernel for the operator $\partial e^{-L[\bar{\partial}, \bar{\partial}^*]}$. It can be viewed as the projection of $(Q \otimes 1)\omega_{L,g}^{-1} \in \mathrm{Sym}^2(S(X))$ to $\mathrm{Sym}^2(S_+(X))$.

Definition 51 We define the effective BV operator

$$\Delta_L : \mathcal{O}(S_+(X)) \to \mathcal{O}(S_+(X))$$

as the operator of contracting with the smooth kernel $(\partial \otimes 1)K_L^g$.

Since $\Delta_L : \mathcal{O}^{(n)}(S_+(X)) \to \mathcal{O}^{(n-2)}(S_+(X))$, it could be viewed as an order two differential operator on the infinite dimensional vector space $S_+(X)$. Note that Δ_L has odd cohomology degree, and $(\Delta_L)^2 = 0$. It defines a Batalin-Vilkovisky structure on $\mathcal{O}(S_+(X))$, with the Batalin-Vilkovisky bracket defined by

$$\{S_1, S_2\}_L = \Delta_L (S_1 S_2) - (\Delta_L S_1) S_2 - (-1)^{|S_1|} S_1 (\Delta_L S_2), \quad L > 0.$$

Remark 52 If $S_1, S_2 \in \mathcal{O}_{loc}(\mathcal{E}(X))$, then $\lim_{L \to 0} \{S_1, S_2\}_L$ is well-defined, which is precisely the Poisson bracket in Proposition 46.

Proposition 53 ([11]) *Under the isomorphism* $\mathcal{F}ock\,(\mathcal{S}_+(X), g, L) \cong \mathcal{O}(S_+(X))[[\hbar]]$, *the induced differential* $\hat{Q}_L$ *is* $\hat{Q}_L = Q + \hbar\Delta_L$.

The proof is similar to Lemma 39.

3.3.4 Renormalization Group Flow and Homotopy Equivalence

We need to specify a choice of the metric g and a positive number $L > 0$ to construct the Fock space $\mathcal{F}ock\,(S_+(X), g, L)$. However, we are in a bit better situation. The general machinery of renormalization theory in [10] allows us to show that the effective Fock spaces are independent of the choice of g and L up to homotopy. This is discussed in detail in [11]. We will discuss here the homotopy between different choices of the scale L, which is related to the renormalization group flow in quantum field theory.

Definition 54 The effective propagator is defined to be the smooth kernel

$$P_\epsilon^L = \int_\epsilon^L du(\bar{\partial}^*\partial \otimes 1)K_u^g \in \mathrm{Sym}^2(\mathrm{PV}(X)), \quad L > \epsilon > 0 \tag{3.3}$$

representing the operator $\bar{\partial}^*\partial e^{-L[\bar{\partial},\bar{\partial}^*]}$.

Lemma 55 *As an operator on* $\mathcal{O}(S_+(X))[[\hbar]]$,

$$\hat{Q}_L = e^{\hbar\partial_{P_\epsilon^L}} \hat{Q}_\epsilon e^{-\hbar\partial_{P_\epsilon^L}}$$

where $\partial_{P_\epsilon^L} : \mathcal{O}(\mathcal{E}(X)) \to \mathcal{O}(\mathcal{E}(X))$ *is the contraction by the smooth kernel* P_ϵ^L.

It follows from this lemma that $e^{\hbar \partial_{P_\epsilon^L}}$ defines the chain homotopy

$$e^{\hbar \partial_{P_\epsilon^L}} : (\mathcal{O}(S_+(X))[[\hbar]], Q + \hbar \Delta_\epsilon) \to (\mathcal{O}(S_+(X))[[\hbar]], Q + \hbar \Delta_L)$$

between Fock spaces defined at scales ϵ and L. It defines a flow on the space of functionals on the fields, which is called the renormalization group flow in [10] following the physics terminology.

Proposition 56 ([11]) *The cohomology $H^*(\mathcal{F}ock(S_+(X), g, L), \hat{Q}_L)$ is independent of g and L. There are canonical isomorphisms*

$$H^*(\mathcal{F}ock(S_+(X), g, L), \hat{Q}_L) \cong \mathcal{F}ock\left(H^*(S_+(X), Q)\right)$$

where $\mathcal{F}ock\,(H^*(S_+(X), Q))$ is the Fock space for the Lagrangian subspace $H^*(S_+(X), Q)$ of the symplectic space $(H^*(S(X), Q), \omega)$

Remark 57 $\mathcal{F}ock\,(H^*(S_+(X), Q))$ is the mirror of the Fock space of de Rham cohomology classes for Gromov-Witten theory discussed in [7].

3.4 Quantum BCOV Theory

3.4.1 Perturbative Quantization

Definition 58 ([11]) A perturbative quantization of BCOV theory on X is given by a family of functionals

$$\mathbf{F}[L] = \sum_{g \geq 0} \hbar^g \mathbf{F}_g[L] \in \mathcal{O}(S_+(X))[[\hbar]]$$

for each $L \in \mathbb{R}_{>0}$, satisfying the following properties

(1) The renormalization group flow equation

$$\mathbf{F}[L] = W\left(P_\epsilon^L, \mathbf{F}[\epsilon]\right)$$

for all $L > \epsilon > 0$. Here $W\left(P_\epsilon^L, \mathbf{F}[\epsilon]\right)$ is the connected Feynman graph integrals (connected graphs) with propagator P_ϵ^L (3.3) and vertices $\mathbf{F}[\epsilon]$. This is equivalent to

$$e^{\mathbf{F}[L]/\hbar} = e^{\hbar \frac{\partial}{\partial P_\epsilon^L}} e^{\mathbf{F}[\epsilon]/\hbar}$$

(2) The quantum master equation holds

$$Q\mathbf{F}[L] + \hbar\Delta_L \mathbf{F}[L] + \frac{1}{2}\{\mathbf{F}[L], \mathbf{F}[L]\}_L = 0, \quad \forall L > 0.$$

This is equivalent to

$$(Q + \hbar\Delta_L)\, e^{\mathbf{F}[L]/\hbar} = 0$$

(3) The locality axiom, as in [10]. This says that $\mathbf{F}[L]$ has a small L asymptotic expansion in terms of local functionals.

(4) The classical limit condition

$$\lim_{L\to 0}\lim_{\hbar\to 0} \mathbf{F}[L] \equiv \lim_{L\to 0} \mathbf{F}_0[L] = \mathbf{F}_{\mathcal{L}_X}.$$

(5) Degree axiom and Hodge weight axiom (see [11]).

3.4.2 Higher Genus B-Model

Given a quantization $\{\mathbf{F}[L]\}_{L>0}$ of the BCOV theory, we obtain a state $\left[e^{\mathbf{F}[L]/\hbar}\right]$ in the Fock space $\mathcal{F}ock^+ (H^*(S_+(X)))$ by Proposition 56. We will denote it by $Z_{\mathbf{F}}$. Let us choose an opposite filtration $\mathcal{L}$ (Definition 9), which induces isomorphisms

$$H^*(S(X), Q) \cong H^*(X, \wedge^* T_X)((z)), \quad H^*(S_+(X), Q) \cong H^*(X, \wedge^* T_X)[[z]].$$

In particular, it induces a natural identification

$$\Phi_{\mathcal{L}} : \mathcal{F}ock(H^*(S(X))) \xrightarrow{\cong} \mathcal{O}(H^*(X, \wedge^* T_X)[[z]])[[\hbar]].$$

Definition 59 Let $\mathbf{F}$ be a quantization of the BCOV theory on X, and $\mathcal{L}$ be an opposite filtration of $H^*(S_+(X), Q)$. Let $\alpha_1, \cdots, \alpha_n \in H^*(X, \wedge^* T_X)$. The *correlation functions* associated to $\mathbf{F}, \mathcal{L}$ is defined to be

$$\mathbf{F}_X^{B,\mathcal{L}}\left(z^{k_1}\alpha_1, \cdots, z^{k_n}\alpha_n\right) := \left(\frac{\partial}{\partial z^{k_1}\alpha_1} \cdots \frac{\partial}{\partial z^{k_n}\alpha_n}\right) \hbar \log \Phi_{\mathcal{L}}\left(Z_{\mathbf{F}}\right)(0) \in \mathbb{C}[[\hbar]].$$

Here the superscript "B" refers to the B-model. We can further decompose $\mathbf{F}_X^{B,\mathcal{L}} = \sum_{g\geq 0} \hbar^g \mathbf{F}_{g,X}^{B,\mathcal{L}}$. Then $\mathbf{F}_{g,X}^{B,\mathcal{L}}$ will be the candidate for the higher genus B-model invariants on X. It is conjectured in [11] that there exists a canonical quantization $\mathbf{F}$ (up to homotopy) of the BCOV theory on X which is mirror to the Gromov-Witten theory on the mirror Calabi-Yau manifold. This proves to be the case for X being an elliptic curve [25, 26].

3.4.3 The Opposite Filtrations

There are two natural opposite filtrations of $H^*(\mathcal{S}(X), Q)$.

The first one is given by the complex conjugate splitting of the Hodge filtration, which we denote by $\mathcal{L}_{\bar{X}}$. In this case the correlation function $\mathbf{F}_X^{B,\mathcal{L}_{\bar{X}}}$ can be realized explicitly as follows. Consider the limit

$$\mathbf{F}[\infty] = \lim_{L\to\infty} \mathbf{F}[L]$$

which is well-defined since X is compact, hence P_L^∞ is smooth. The quantum master equation at $L = \infty$ says that

$$Q\mathbf{F}[\infty] = 0$$

as $\lim_{L\to\infty} \Delta_L = 0$. It follows that $\mathbf{F}[\infty]$ descends to a functional on $H^*(\mathcal{S}_+(X), Q)$

$$\mathbf{F}[\infty] \in H^*(\mathcal{O}(\mathcal{S}_+(X))[[\hbar]], Q) \cong \mathcal{O}(H^*(\mathcal{S}_+(X), Q))[[\hbar]].$$

The choice of the Kähler metric induces isomorphisms

$$H^*(\mathcal{S}(X), Q) \cong H^*(X, \wedge^* T_X)((z)), \quad H^*(\mathcal{S}_+(X), Q) \cong H^*(X, \wedge^* T_X)[[z]]$$

via Hodge theory, hence defining an opposite filtration which is precisely $\mathcal{L}_{\bar{X}}$. Then

$$\mathbf{F}_X^{B,\mathcal{L}_{\bar{X}}} = \mathbf{F}[\infty].$$

The second choice is relevant for mirror symmetry, which is defined near a large complex limit in the moduli space of complex structures on X. Near any such large complex limit point, there is an associated monodromy weight filtration $\mathcal{W}$ which splits the Hodge filtration. Then the correlation function

$$\mathbf{F}_{g,n,X}^{B,\mathcal{W}} : \mathrm{Sym}^n \left(H^*(X, \wedge^* T_X)[[z]]\right) \to \mathbb{C}$$

will be the mirror of the descendant Gromov-Witten invariants

$$\langle - \rangle_{g,n,X^\vee}^{GW} : \mathrm{Sym}^n \left(H^*(X^\vee, \mathbb{C})[[z]]\right) \to \mathbb{C}$$

on the mirror Calabi-Yau $X^\vee$ under the mirror map.

Note that $F_X^{B,\mathcal{L}_{\bar{X}}}$ doesn't vary holomorphically due to the complex conjugate splitting $\mathcal{L}_{\bar{X}}$. This is the famous holomorphic anomaly discovered in [6]. Given a large complex limit point, the natural way to retain holomorphicity is to consider

$\mathbf{F}_{g\ X}^{B,\mathcal{W}}$, which is usually denoted in physics literature by

$$\mathbf{F}_{g,X}^{B,\mathcal{W}} \equiv \lim_{\bar{\tau}\to\infty} \mathbf{F}_{X}^{B,\mathcal{L}_{\bar{X}}}$$

as the "$\bar{\tau} \to \infty$-limit" [6] near the large complex limit.

3.4.4 Higher Genus Mirror Symmetry

The mirror symmetry for elliptic curves is easy to describe. Let E represent an elliptic curve. In the A-model, we have the moduli of (complexified) Kähler class $[\omega] \in H^2(E, \mathbb{C})$ parametrized by the symplectic volume

$$q = \mathrm{Tr}\, \omega$$

where the trace map in the A-model is given by the integration $\mathrm{Tr} = \int_E$. The mirror in the B-model is the elliptic curve $\mathrm{E}_\tau = \mathbb{C}/\mathbb{Z} \oplus \mathbb{Z}\tau$, with complex structure τ related to q by the mirror map

$$q = e^{2\pi i \tau}.$$

Let

$$\Phi_\tau : \bigoplus_{i,j} H^i(E, \wedge^j T_E^*)[-i-j] \to \bigoplus_{i,j} H^i(\mathrm{E}_\tau, \wedge^j T_{\mathrm{E}_\tau})[-i-j]$$

be the unique isomorphism of commutative bigraded algebras which is compatible with the trace on both sides. This is z-linearly extended to an isomorphism

$$\Phi_\tau : H^*(E, \mathbb{C})[[z]] \to H^*(E_\tau, \wedge^* T_{E_\tau})[[z]].$$

The canonical quantization of BCOV theory was analyzed in [25], and the explicit solution was presented in [26] via vertex algebra techniques. This leads to the establishment of higher genus mirror symmetry on elliptic curves.

Theorem 60 ([25]) *For all $\alpha_1, \cdots, \alpha_n \in H^*(E, \mathbb{C})[[z]]$, the A-model descendant Gromov-Witten invariants on E can be identified with the B-model BCOV correlation functions*

$$\sum_d q^d \langle \alpha_1, \cdots, \alpha_n \rangle_{g,n,d}^{GW(E)} = \lim_{\bar{\tau}\to\infty} \mathbf{F}_{E_\tau}^{B,\mathcal{L}_{\bar{E}_\tau}} (\Phi_\tau(\alpha_1), \cdots, \Phi_\tau(\alpha_n))$$

where the large complex limit is taken to be $\mathrm{Im}\tau \to \infty$ *on the upper half plane* $\mathbb{H}$.

It is proved in [24, 25] that the correlation functions for $\mathbf{F}_{\mathrm{E}_\tau}^{B,\mathcal{L}_{\mathrm{E}_\tau}^{\bar{\tau}}}$, before taking the $\bar{\tau} \to \infty$ limit, are almost holomorphic modular forms exhibiting mild anti-holomorphic dependence on $\bar{\tau}$. On the other hand, the correlation functions of Gromov-Witten theory are given by quasi-modular forms [31]. In this example, the $\bar{\tau} \to \infty$ limit is the well-known identification between almost holomorphic modular forms and quasi-modular forms [21].

Acknowledgements The authors would like to thank the organizers and participants of the workshop on B-model aspects of Gromov-Witten theory, and the hospitality of the mathematics department at University of Michigan.

References

1. Barannikov, S.: Extended moduli spaces and mirror symmetry in dimensions n > 3, ProQuest LLC, Ann Arbor, MI. Thesis (Ph.D.), University of California, Berkeley (1999)
2. Barannikov, S.: Quantum periods, I. Semi-infinite variations of Hodge structures. Int. Math. Res. Not. **23**, 1243–1264 (2001)
3. Barannikov, S.: Non-commutative periods and mirror symmetry in higher dimensions. Commun. Math. Phys. **228**(2), 281–325 (2002)
4. Barannikov, S.: Semi-infinite Hodge structures and mirror symmetry for projective spaces (2001). arXiv:math.AG/0010157
5. Barannikov, S., Kontsevich, M.: Frobenius manifolds and formality of Lie algebras of polyvectorfields. Int. Math. Res. Not. **4**, 201–215 (1998)
6. Bershadsky, M., Cecotti, S., Ooguri, H., Vafa, C.: Kodaira-Spencer theory of gravity and exact results for quantum string amplitudes. Commun. Math. Phys. **165**(2), 311–427 (1994). MR1301851(95f:32029)
7. Coates, T., Givental, A.B.: Quantum Riemann-Roch, Lefschetz and Serre. Ann. Math. (2) **165**(1), 15–53 (2007)
8. Costello, K.: Topological conformal field theories and Calabi-Yau categories. Adv. Math. **210**(1), 165–214 (2007)
9. Costello, K.: The partition function of a topological field theory. J. Topol. **2**(4), 779–822 (2009)
10. Costello, K.: Renormalization and Effective Field Theory. Mathematical Surveys and Monographs, vol. 170. American Mathematical Society, Providence (2011). MR2778558
11. Costello, K., Li, S.: Quantum BCOV theory on Calabi-Yau manifolds and the higher genus Bmodel (2011). arXiv:1201.4501 [math.QA]
12. Douai, A., Sabbah, C.: Gauss-Manin systems, Brieskorn lattices and Frobenius structures, I. In: Proceedings of the International Conference in Honor of Frédéric Pham (Nice, 2002), pp. 1055–1116 (2003)
13. Douai, A., Sabbah, C.: Gauss-Manin systems, Brieskorn lattices and Frobenius structures, II. In: Frobenius Manifolds. Aspects of Mathematics, vol. E36, pp. 1–18. Vieweg, Wiesbaden (2004)
14. Dubrovin, B.: Geometry of 2D topological field theories. In: Integrable Systems and Quantum Groups (Montecatini Terme, 1993). Lecture Notes in Mathematics, vol. 1620, pp. 120–348. Springer, Berlin (1996)
15. Fan, H., Jarvis, T., Ruan, Y.: The Witten equation, mirror symmetry, and quantum singularity theory. Ann. Math. (2) **178**(1), 1–106 (2013)
16. Givental, A.B.: A mirror theorem for toric complete intersections. In: Topological Field Theory, Primitive Forms and Related Topics (Kyoto, 1996). Progress in Mathematics, vol. 160, pp. 141–175. Birkhäuser, Boston (1998)

17. Givental, A.B.: A tutorial on quantum cohomology. In: Symplectic Geometry and Topology (Park City, UT, 1997). IAS/Park City Mathematics Series, vol. 7, pp. 231–264. American Mathematical Society, Providence (1999)
18. Givental, A.B.: Gromov-Witten invariants and quantization of quadratic Hamiltonians. Mosc. Math. J. **1**(4), 551–568, 645 (2001). Dedicated to the memory of I. G. Petrovskii on the occasion of his100th anniversary
19. Givental, A.B.: Symplectic geometry of Frobenius structures. In: Frobenius Manifolds. Aspects of Mathematics, vol. E36, pp. 91–112. Vieweg, Wiesbaden (2004)
20. He, W., Li, S., Shen, Y., Webb, R.: Landau-Ginzburg mirror symmetry conjecture (2015). arXiv:math.AG/1503.01757
21. Kaneko, M., Zagier, D.: A generalized Jacobi theta function and quasimodular forms. In: The Modulispace of Curves (Texel Island, 1994). Progress in Mathematics, vol. 129, pp. 165–172. Birkhäuser, Boston (1995)
22. Kontsevich, M., Soibelman, Y.: Notes on A_∞-algebras, A_∞-categories and non-commutative geometry. In: Homological Mirror Symmetry. Lecture Notes in Physics, vol. 757, pp. 153–219. Springer, Berlin (2009). MR2596638
23. Li, S.: Calabi-Yau geometry and higher genus mirror symmetry. Thesis, Harvard (2011)
24. Li, S.: Feynman graph integrals and almost modular forms. Commun. Number Theory and Phys. **6**, 129–157 (2012)
25. Li, S.: BCOV theory on the elliptic curve and higher genus mirror symmetry (2011). arXiv:1112.4063 [math.QA]
26. Li, S.: Vertex algebras and quantum master equation (2016). arXiv:1612.01292 [math.QA]
27. Li, C., Li, S., Saito, K., Shen, Y.: Mirror symmetry for exceptional unimodular singularities. J. Eur. Math. Soc. **19**(4), 1189–1229 (2017)
28. Li, C., Li, S., Saito, K.: Primitive forms via polyvector fields (2013). arXiv: math.AG/1311.1659
29. Lian, B.H., Liu, K.F., Yau, S.T.: Mirror principle I. Asian J. Math. **1**(4), 729–763 (1997)
30. Losev, A., Shadrin, S., Shneiberg, I.: Tautological relations in Hodge field theory. Nucl. Phys. B **786**(3), 267–296 (2007)
31. Okounkov, A., Pandharipande, R.: Virasoro constraints for target curves. Invent. Math. **163**(1), 47–108 (2006)
32. Saito, K.: The higher residue pairings $K_F^{(k)}$ for a family of hypersurface singular points. In: Singularities, Part 2. Proceedings of Symposia in Pure Mathematics (Arcata, Calif., 1981), pp. 441–463 (1983)
33. Saito, K.: Period mapping associated to a primitive form. Publ. Res. Inst. Math. Sci. **19**(3), 1231–1264 (1983)
34. Saito, M.: On the structure of Brieskorn lattice. Ann. Inst. Fourier (Grenoble) **39**(1), 27–72 (1989)
35. Saito, K.: From primitive form to mirror symmetry (2014). arXiv: math.AG/1408.4208
36. Saito, M.: On the structure of Brieskorn lattices, II (2013). arXiv: math.AG/1312.6629
37. Takahashi, A.: Primitive forms, topological LG models coupled to gravity and mirror symmetry (1998). arXiv: math.AG/9802059
38. Zwiebach, B.: Closed string field theory: quantum action and the Batalin-Vilkovisky master equation. Nucl. Phys. B **390**(1), 33–152 (1993)

Eynard-Orantin B-Model and Its Application in Mirror Symmetry

Bohan Fang

Contents

Abstract We describe the Eynard-Orantin recursive algorithm on a spectral curve, and give a biased survey on its roles as B-models which predict various higher genus A-model invariants via mirror symmetry.

1 Introduction

The Eynard-Orantin topological recursion is a recursive algorithm from the matrix model theory [24]. Mathematically speaking, it starts with an affine plane curve Σ with a choice of a fundamental normalized differential of the second kind, and then the algorithm recursively produces a series of symmetric meromorphic forms $\omega_{g,n}$ on the product of n copies of Σ. These $\omega_{g,n}$ are called B-model higher genus invariants. They are genus g correlators with n boundary components. We will survey two aspects of this recursive algorithm—its relation to a quantization of a semisimple Frobenius manifold, and its role in mirror symmetry.

B. Fang (✉)
Beijing International Center for Mathematical Research, Peking University, Beijing, China

E. Clader, Y. Ruan (eds.), *B-Model Gromov-Witten Theory*, Trends in Mathematics,
https://doi.org/10.1007/978-3-319-94220-9_5

1.1 Relation to Givental's Quantization and Abstract Frobenius Structure

The Eynard-Orantin recursion is a recursive algorithm in computing higher genus invariants of a Frobenius manifold, as shown in [19]. They show that we can define the recursion formally around each ramification point using the data from a calibrated Frobenius manifold, the recursion is equivalent to Givental's quantization [39, 40]. Another important theorem of [19] is that they express Eynard-Orantin higher genus invariants as graph sums. This allows us to compare with the graph sum formula of Gromov-Witten invariants, which is essential in the proof of mirror symmetry involving Eynard-Orantin recursion.

1.2 Eynard-Orantin Recursion as B-Model in Mirror Symmetry

Mirror symmetry is the equivalence between the A-model (about the Kähler structure of a manifold) and the B-model of its mirror (about the complex structure). When the spectral curve is the B-model mirror to some A-model, like topological strings on a toric Calabi-Yau threefold, the Eynard-Orantin B-model invariants predict the A-model strings correctly. This is the Bouchard-Klemm-Mariño-Pasquetti (BKMP, [11, 12, 55]) remodeling conjecture, proved recently in [25, 31, 32]. There are also various mirror symmetry statements along this line, as long as one can have a mirror curve as the B-model, e.g. the case of $\mathbb{P}^1$ (Norbury-Scott conjecture [19, 30, 59]), Bouchard-Mariño conjecture in various settings [9, 10, 13].

These Eynard-Orantin higher genus invariants $\omega_{g,n}$ enjoy many nice properties. In [24], the authors discuss the variation of $\omega_{g,n}$ with respect to the moduli of the spectral curves. Also, the fundamental normalized differential of the second kind depends on the choice of a Torelli marking. It changes under a modular transformation. The modularity property of Gromov-Witten invariants for toric Calabi-Yau threefolds follows from the BKMP conjecture and the modularity property of the Eynard-Orantin B-model invariants $\omega_{g,n}$.

1.3 Structure of This Paper

We first review the definition of Eynard-Orantin topological recursion in Sect. 2. In Sect. 3 we will state the equivalence between the Eynard-Orantin topological recursion on a formal spectral curve and Givental's quantization on a Frobenius manifold. In Sect. 4 we will review the applications of Eynard-Orantin recursions to all genera mirror symmetry. The last section is about the modularity property of Gromov-Witten invariants from the modularity of Eynard-Orantin invariants, via the mirror symmetry statement introduced in Sect. 4.

This survey is far from covering the vast scope of the Eynard-Orantin topological recursion, which is a very active field of research of late. We do not systematically cover the fundamental properties of the Eynard-Orantin recursion, like the variations of $\omega_{g,n}$ with respect to the moduli of spectral curves [24]. There are many other fantastic applications of the recursion, not necessarily along the line of mirror symmetry for toric Calabi-Yau threefolds, like Weil-Pertersson volume [22, 23, 65]. The recent progress on "quantum curves", and the application of Eynard-Orantin recursion to non-semisimple situations by taking non-semisimple limits, are also beyond the reach of this survey.

2 Spectral Curve and Eynard-Orantin Recursion

2.1 Spectral Curves

Let Σ be a smooth affine algebraic curve in $(\mathbb{C}^*)^2$. The coordinate Y maps Σ into the second component of $(\mathbb{C}^*)^2$. It is a holomorphic function on Σ. Let $Y = e^{-y}$. We denote the covering map π_Y

$$\pi_Y : \mathbb{C}^* \times \mathbb{C} \to (\mathbb{C}^*)^2,$$
$$(a, y) \mapsto (a, e^{-y}).$$

Let $\widetilde{\Sigma}$ be the lift of Σ under this map, and let $\overline{\Sigma}$ be a choice of smooth compactification of Σ, which is a compact Riemann surface.[1]

Recall that a Torelli marking on $\overline{\Sigma}$ is a choice of cycles $A_1, \ldots, A_{\mathfrak{g}}, B_1, \ldots, B_{\mathfrak{g}}$ in $H_1(\overline{\Sigma}; \mathbb{C})$, such that $A_i \cap B_j = \delta_{i,j}$ and $A_i \cap A_j = B_i \cap B_j = 0$, where $\mathfrak{g}$ is the genus of $\overline{\Sigma}$.[2] Given such a marking, following the notions [34], we define the fundamental normalized differential of the second kind (a.k.a. Bergman kernel in Eynard-Orantin [24]).

Definition 1 The *fundamental normalized differential of the second kind* (abbreviated as *fundamental differential* in this paper) associated to a Torelli marking on $\overline{\Sigma}$ is the symmetric meromorphic form on $(\overline{\Sigma})^2$ satisfying the following conditions.

- The only pole is the double pole along the diagonal, i.e. given any local coordinate ζ near a point $p \in \overline{\Sigma}$, the differential B has the following form near $(p, p) \in (\overline{\Sigma})^2$

$$B = \frac{d\zeta_1 d\zeta_2}{(\zeta_1 - \zeta_2)^2} + \text{holomorphic part}.$$

[1] We reserve the variable $X = e^{-x}$ for other purposes. In many but not all examples, it will be the first coordinate.

[2] We allow such cycles to be non-geometric, i.e. elements in $H_1(\overline{\Sigma}; \mathbb{C})$.

- It is normalized by the choice of A-cycles

$$\int_{q \in A_i} B(p, q) = 0, \text{ for } i = 1, \ldots, \mathfrak{g}.$$

Remark 2 The pairing of cycles in $H_1(\overline{\Sigma}; \mathbb{C})$ turns it into a symplectic vector space. The subspace spanned by A-cycles is a Lagrangian subspace. The fundamental differential B only depends on the choice of this Lagrangian subspace.

Definition 3 A spectral curve $\boldsymbol{\Sigma} = (\Sigma, x, B)$ consists of the following data:

- a smooth affine algebraic curve Σ in $(\mathbb{C}^*)^2$ together with a Torelli marking on $\overline{\Sigma}$;
- a holomorphic Morse function (superpotential) x from the universal cover of Σ to $\mathbb{C}^*$, such that dx descends to a meromorphic form on $\overline{\Sigma}$ with poles in $\overline{\Sigma} \setminus \Sigma$;
- a fundamental normalized differential of the second kind B on $\overline{\Sigma}$ with respect to such choice of A-cycles.

Remark 4 In the applications of the Eynard-Orantin recursion, very often $X = e^{-x}$ is the first coordinate of the affine curve Σ.

Fix a spectral curve $\boldsymbol{\Sigma}$. Since x is Morse, the critical points where $dx = 0$ form a finite set $\{p_\alpha : \alpha \in I_{\boldsymbol{\Sigma}}\}$. Define the Liouville form $\Phi = ydx = -\log Y dx$. It is a well-defined holomorphic form on $\widetilde{\Sigma}$, and is meromorphic on the smooth completion of $\widetilde{\Sigma}$.

At each critical point p_α, we define the local coordinate ζ_α by

$$x = \zeta_\alpha^2 + x_{0,\alpha},$$

where $x_{0,\alpha}$ is the critical value of x at p_α. For any p near p_α, let $\bar{p}$ be the point on Σ such that $\zeta_\alpha(\bar{p}) = -\zeta_\alpha(p)$.

The central topic of this survey, Eynard-Orantin's topological recursion, is essentially defined around each critical point of x on the spectral curve. Following [19], we define formal spectral curves below.[3]

Definition 5 A formal spectral curve $\mathfrak{C}$ is a disjoint union of $\{C_\alpha\}_{\alpha \in I_{\mathfrak{C}}}$ where each $C_\alpha = \mathrm{Spec}\mathbb{C}[[\zeta_\alpha]]$, together with following information.

- A function $y_\alpha = \sum_{i=0}^{\infty} h_i^\alpha (\zeta_\alpha)^i$ on C_α where $h_1^\alpha \neq 0$.
- A holomorphic Morse function $x_\alpha = x_{0,\alpha} + \zeta_\alpha^2$ on C_α.

[3]In [19], Eynard-Orantin recursions on such formal spectral curves are called local topological recursions.

- The "fundamental normalized differential of the second kind" $B^{\alpha,\beta} \in \Gamma(T^*(C_\alpha \times C_\beta \setminus C_{\alpha,\beta}))$

$$B^{\alpha,\beta}(\zeta_\alpha, \zeta_\beta) = \delta_{\alpha,\beta}\frac{d\zeta_\alpha \otimes d\zeta_\beta}{(\zeta_\alpha - \zeta_\beta)^2} + \sum_{i,j\geq 0} B^{\alpha,\beta}_{i,j}(\zeta_\alpha)^i(\zeta_\beta)^j d\zeta_\alpha \otimes d\zeta_\beta,$$

where $C_{\alpha,\beta} \cong \mathrm{Spec}\mathbb{C}[[\zeta]]$ is the diagonal. We require $B^{\alpha,\beta}_{i,j} = B^{\beta,\alpha}_{i,j}$.

Any spectral curve induces a formal spectral curve. We will consider the recursions for both actual and formal spectral curves in the next subsection.

2.2 *Eynard-Orantin's Topological Recursion*

Definition 6 The Eynard-Orantin recursive algorithm defines a sequence of symmetric meromorphic forms $\omega_{g,n}$ on $(\overline{\Sigma})^n$ for $g \in \mathbb{Z}_{\geq 0}, n \in \mathbb{Z}_{>0}$ as follows.

- Initial conditions:

$$\omega_{0,1} = 0, \quad \omega_{0,2} = B.$$

- Recursive algorithm:

$$\omega_{g,n}(p_1, \ldots, p_n)= \sum_{p'\in I_\Sigma} \mathrm{Res}_{p=p'}\frac{\int_{\xi=p}^{\bar{p}} B(p_n, \xi)}{2(\Phi(p) - \Phi(\bar{p}))}\big(\omega_{g-1,n+1}(p, \bar{p}, p_1, \ldots, p_{n-1})$$
$$+ \sum_{g_1+g_2=g} \sum_{I\sqcup J=\{1,\ldots,n-1\}} \omega_{g_1,|I|+1}(p, p_I)\omega_{g_2,|J|+1}(\bar{p}, p_J)\big).$$

Proposition 7 *When $2g - 2 + n > 0$, the poles of $\omega_{g,n}$ are at $dx_i = 0$ (critical points), where dx_i is the differential of the superpotential on i-th copy of $(\Sigma)^n$.*

Proof The proof is in Appendix A of [24]. We repeat it here. Assume the statement is correct for all (g, n) such that $g < g_0$ or $g = g_0, n < n_0$ where $2g_0 - 2 + n_0 > 0$ Then by the recursion, if $p_1, \ldots, p_n$ are not any critical point, all ω_{g_0-1,n_0+1}, $\omega_{g_1,|I|+1}$ and $\omega_{g_2,|J|+1}$ on the RHS are not at their poles, and the residues are well-defined. Therefore ω_{g_0,n_0} is finite at $(p_1, \ldots, p_n)$. Notice that we can always make the contour around p' small enough to avoid $p_1, \ldots, p_{n-1}$ such that we would not encounter the diagonal pole of $\omega_{0,2}$.

Definition 6 also applies to any formal spectral curve $\mathfrak{C}=(\{C_\alpha\}, \{x_\alpha\}, \{y_\alpha\}, B^{\alpha,\beta})$, and produces a sequence of meromorphic symmetric differential n-forms $\omega_{g,n}$ on $(\sqcup_{\alpha\in I_{\mathfrak{C}}} C_\alpha)^n$.

2.3 Differential Forms on Spectral Curves

For any spectral curve $\mathbf{\Sigma}$, we define *the preferred basis of differentials of the second kind* as below

$$\theta_d^\alpha(p) := (2d-1)!!2^{-d}\mathrm{Res}_{p'\to p_\alpha} B(p,p')\zeta_\alpha(p')^{-2d-1}.$$

The form θ_d^α satisfies the following properties.

- θ_d^α is a meromorphic 1-form on $\overline{\Sigma}$ with a single pole of order $2d+2$ at p_α.
- In local coordinate

$$\theta_d^\alpha = (-\frac{(2d+1)!!}{2^d \zeta_\alpha^{2d+2}} + \text{holomorphic part})d\zeta_\alpha.$$

- $\int_{A_i}\theta_d^\alpha = 0$ for $i = 1, \dots \mathfrak{g}$.

For $k > 0$, we define

$$\hat{\xi}_{\alpha,k} = (-1)^k(\frac{d}{dx})^{k-1}(\frac{\theta_0^\alpha}{dx}),$$

which is a meromorphic function on $\overline{\Sigma}_q$. As a convention, we write $d\hat{\xi}_{\alpha,0} = \theta_0^\alpha$, although $\hat{\xi}_{\alpha,0}$ is not a well defined global meromorphic function on $\overline{\Sigma}_q$.

Similarly, for any formal spectral curve $\mathfrak{C} = \{C_\beta\}_{\beta\in I_{\mathfrak{C}}}$, we define these meromorphic forms $\theta_{d,\beta}^\alpha(\zeta_\beta)$ on C_β

$$\theta_{d,\beta}^\alpha(\zeta_\beta) := (2d-1)!!2^{-d}\mathrm{Res}_{\zeta_\alpha\to 0} B^{\alpha,\beta}(\zeta_\alpha,\zeta_\beta)(\zeta_\alpha)^{-2d-1}.$$

We have

$$\theta_{d,\beta}^\alpha = (-\frac{\delta_{\alpha,\beta}(2d+1)!!}{2^d \zeta_\beta^{2d+2}} + \text{holomorphic part})d\zeta_\beta.$$

We also define

$$\hat{\xi}_{\alpha,\beta,k} = (-1)^k(\frac{d}{dx})^{k-1}(\frac{\theta_{0,\beta}^\alpha}{dx}), \quad d\hat{\xi}_{\alpha,\beta,0} = \theta_{0,\beta}^\alpha.$$

3 Identification of Eynard-Orantin's Recursion with Givental's Quantization

3.1 Frobenius Manifold

In this section, we explain the equivalence of Givental's quantization of a semisimple Frobenius manifold and the corresponding Eynard-Orantin recursion.

Definition 8 A Frobenius algebra $(V, \star)$ is a finite-dimensional associative algebra V over a field k with unit $\mathbf{1}$ equipped with a non-degenerate bilinear pairing $(,): V \times V \to k$ such that $(a \star b, c) = (a, b \star c)$.

We fix the dimension of the Frobenius algebra (or later, manifold) in discussion to be N. A Frobenius algebra is *semisimple* if it has a basis $\{\phi_\alpha\}_{\alpha=1,\ldots,N}$ such that $\phi_\alpha \star \phi_\beta = \delta_{\alpha,\beta}\phi_\alpha$. Such basis is unique up to a permutation.

Definition 9 A Frobenius manifold V is a k-manifold with a flat pseudo-Riemannian metric $(,)$ with the following properties.

- Locally there is a function F whose third covariant derivative F_{abc} at q defines a product $\star_q$ on the tangent by $F_{abc}|_q = (\partial_a \star_q \partial_b, \partial_c)$, such that each tangent space at a point q is a Frobenius algebra with the product $\star_q$ and the pairing from the Riemannian metric.
- The vector field the of unit $\mathbf{1}$ is covariantly constant and preserves the multiplication.

A Frobenius manifold V is *generically semisimple* if for generic $q \in V$, T_qV is semisimple. We sometimes write $\star$ instead of $\star_q$ when the context is clear.

Let $\tau^a, a = 1, \ldots, N$ be flat coordinates on a Frobenius manifold, and let $H_a = \frac{\partial}{\partial \tau^a}$ be the corresponding frames in the tangent bundle. The *quantum connection* ∇ is given as follows

$$\nabla_a = z\partial_a - H_a\star$$

The *quantum differential equation (QDE)* is

$$\nabla\eta = 0$$

The QDE is a system of first-order differential equations, and a choice of fundamental solutions $S_\tau = (\eta_1(\tau), \ldots, \eta_N(\tau))$ is called an *S-calibration*.

Definition 10 Around a semisimple point $p \in V$ (we assume $\tau(p) = \tau_0$), we define the following notions.

- Canonical basis $\phi_\alpha(\tau)$ such that $\phi_\alpha(\tau) \star \phi_\beta(\tau) = \delta_{\alpha,\beta}$. We have

$$(\phi_\alpha(\tau), \phi_\beta(\tau)) = \frac{\delta_{\alpha,\beta}}{\Delta^\alpha(\tau)}.$$

- Canonical coordinates $u^\alpha(\tau)$ such that $\frac{\partial}{\partial u^\alpha(\tau)} = \phi_\alpha(\tau)$. They are fixed up to constants.
- Flat basis ϕ_α which is the parallel transport (according to the Levi-Civita connection of the Riemannian metric on V) of $\phi_\alpha(\tau_0)$ at $\tau = 0$. We also denote $\Delta^\alpha = \Delta^\alpha(\tau_0)$.
- Normalized basis $\hat{\phi}_\alpha(\tau) = \phi_\alpha(\tau)\sqrt{\Delta^\alpha(\tau)}$; $\hat{\phi}_\alpha = \phi_\alpha\sqrt{\Delta^\alpha}$.
- The dual basis $\{\phi^\alpha\}$ to $\{\phi_\alpha\}$, and the dual basis $\{\phi^\alpha(\tau)\}$ to $\{\phi_\alpha(\tau)\}$. The normalized basis are self-dual.

Theorem 11 *Around a semisimple point $p \in V$, there exists an S-calibration $S_\tau = (\eta_1(\tau), \ldots, \eta_N(\tau))$. Each $\eta_\alpha(\tau) = \sum_{a=1}^N (S_\tau)_a^{\hat{\alpha}} H^a$ where H^a is the dual basis to H_a. One can decompose S_τ as following*

$$(S_\tau)_a^{\hat{\alpha}} = (\Psi_\tau)_a^\beta R_\tau(z)_\beta^\alpha e^{\frac{u^\alpha(\tau)}{z}}.$$

Here Ψ_τ is the transition matrix from $\hat{\phi}_\alpha(\tau)$ to H_a

$$H_a = \sum_{a=1}^N (\Psi_\tau)_a^\alpha \hat{\phi}_\alpha(\tau),$$

and

$$(R_\tau)_\beta^\alpha(z) = \delta_\beta^\alpha + O(z)$$

is a formal power series in z, and it is unitary

$$(R_\tau)_\alpha^\gamma(z)(R_\tau)_\beta^\gamma(-z) = \delta_{\alpha\beta}.$$

Furthermore, R_τ it is uniquely determined by up to a right multiplication of $\exp(\sum_{i=1}^\infty a_{2i-1}z^{2i-1})$ where a_{2i-1} are constant diagonal matrices.

Let S_τ be an S-calibration. Define an operator $\mathcal{S}_\tau : T_\tau V \to T_\tau V$ by

$$(S_\tau)_a^{\hat{\alpha}} = (H_a, \mathcal{S}_\tau(\hat{\phi}^\alpha)).$$

Define the matrix

$$(S_\tau)_{\overline{\alpha}}^{\hat{\beta}} = (\hat{\phi}_\beta(\tau), \mathcal{S}_\tau(\phi_\alpha)).$$

3.2 Quantizations of a Generically Semi-Simple Frobenius Manifold

We will introduce Givental's quantization for semi-simple Frobenius manifolds. When the Frobenius manifold comes from genus 0 Gromov-Witten theory of a toric manifold, this quantization matches higher genus Gromov-Witten invariants. First we introduce the following notations

$$\mathbf{u}(z) = (\mathbf{u}_1(z), \mathbf{u}_2(z), \dots), \quad \mathbf{u}_j(z) = \sum_{a=1}^{N} \mathbf{u}_j^a(z) H_a,$$

$$\mathbf{u}_j^a(z) = (u_j^a)_0 + (u_j^a)_1 z + (u_j^a)_2 z^2 + \dots,$$

$$\mathbf{t}(z) = \sum_{a=1}^{N} \mathbf{t}^a(z) H_a, \quad \mathbf{t}^a(z) = (t^a)_0 + (t^a)_1 z + (t_2^a) z^2 + \dots.$$

Example 12 Let X be a smooth toric manifold over $\mathbb{C}$, and $\mathbb{T} \subset X$ be the dense open torus in X. The equivariant quantum cohomology $QH^*_{\mathbb{T}}(X; \mathcal{Q})$ is a Frobenius algebra over the fractional field $\mathcal{Q}$ of $H^*_{\mathbb{T}}(\mathrm{pt}; \mathbb{C})$, and it is semisimple around the origin. When X is compact, the non-equivariant quantum cohomology $QH^*(X; \mathbb{C})$ is semisimple generically. It is not necessarily semisimple when the Kähler parameter is zero, i.e. the ordinary non-equivariant cohomology algebra is not necessarily semisimple. We recall the definition of equivariant Gromov-Witten invariants for X below. We do not assume X is compact in the equivariant setting.

Let $\overline{\mathcal{M}}_{g,n}(X; \beta)$ be the moduli of the stable maps from a genus g, n-marked curve to X in class $\beta \in H_2(X; \mathbb{Z})$. Recall that ψ-class $\psi_i = c_1(\mathbb{L}_i)$ where $\mathbb{L}_i$ is formed by cotangent lines at i-th marked point on $\overline{\mathcal{M}}_{g,n}(X; \beta)$. Let $\bar{\psi}_i = \pi^* \psi_{\mathrm{pt},i}$, where $\psi_{\mathrm{pt},i}$ is the i-th ψ-class on the moduli space of curves $\overline{\mathcal{M}}_{g,n}$, and $\pi : \overline{\mathcal{M}}_{g,n}(X; \beta) \to \overline{\mathcal{M}}_{g,n}$ is the forgetful map.

The $\mathbb{T}$-equivariant genus g degree d Gromov-Witten invariants of X are defined by

$$\langle \gamma_1 \hat{\psi}_1^{a_1}, \cdots, \gamma_n \hat{\psi}_n^{a_n} \rangle_{g,n,\beta}^{X,\mathbb{T}} = \int_{[\overline{\mathcal{M}}_{g,n}(X,d)^{\mathbb{T}}]^{w,\mathbb{T}}} \frac{\iota^*\left(\prod_{j=1}^n \mathrm{ev}_j^*(\gamma_j)(\hat{\psi}_j^{\mathbb{T}})^{a_j}\right)}{e_{\mathbb{T}}(N^{\mathrm{vir}})} \in \mathcal{Q}.$$

where the weighted virtual fundamental class $[\overline{\mathcal{M}}_{g,n}(X,d)^{\mathbb{T}}]^{w,\mathbb{T}}$ [1, 2] (resp. the virtual normal bundle N^{vir} of $\overline{\mathcal{M}}_{g,n}(X,d)^{\mathbb{T}}$ in $\overline{\mathcal{M}}_{g,n}(X,d)$) is defined by the fixed (resp. moving) part of the restriction to $\overline{\mathcal{M}}_{g,n}(X,d)^{\mathbb{T}}$ of the $\mathbb{T}$-equivariant perfect obstruction theory on $\overline{\mathcal{M}}_{g,n}(X,d)$ [41], and $\iota^* : H^*_{\mathbb{T}}(\overline{\mathcal{M}}_{g,n}(X,d); \mathbb{Q}) \to H^*_{\mathbb{T}}(\overline{\mathcal{M}}_{g,n}(X,d)^{\mathbb{T}}; \mathbb{Q})$ is induced by the inclusion map $\iota : \overline{\mathcal{M}}_{g,n}(X,d)^{\mathbb{T}} \hookrightarrow \overline{\mathcal{M}}_{g,n}(X,d)$. Here $\hat{\psi} = \psi$ or $\bar{\psi}$, and these invariants are called ancestors or descendants, respectively.

Let $\tau \in H^*_{\mathbb{T}}(X; \mathbb{Q})$. Define the ancestor/descendant ($\hat{\psi} = \bar{\psi}$ or ψ) potential with primary insertions (we suppress the torus symbol $\mathbb{T}$ from here in the Gromov-Witten notations for closed Gromov-Witten invariants)

$$\langle\langle \mathbf{u}_1(\hat{\psi}_1), \ldots, \mathbf{u}_n(\hat{\psi}_n)\rangle\rangle^X_{g,n} = \sum_{m=0}^{\infty}\sum_{\beta \geq 0} \frac{\langle \mathbf{u}_1(\hat{\psi}_1), \cdots, \mathbf{u}_n(\hat{\psi}_n), \tau^m\rangle^{X,\mathbb{T}}_{g,n+m,\beta}}{m!}.$$

We always assume this sum converges for a suitable domain of τ.[4]

The quantum cohomology is defined by

$$(a *_\tau b, c) = \langle\langle a, b, c\rangle\rangle^X_{0,3}, \quad a, b, c \in H^*_{\mathbb{T}}(X; \mathbb{Q}).$$

Let $\mathbf{t} = \mathbf{u}_1 = \mathbf{u}_2 = \mathbf{u}_3 = \ldots$. We define

$$F^X_{g,n}(\mathbf{t}) = \langle\langle \mathbf{t}, \ldots, \mathbf{t}\rangle\rangle^X_{g,n}, \quad F^X_g = \langle\langle\rangle\rangle^X_{g,0}.$$

Here $F^X_g = F^X_g(\tau)$ is a function of τ.

Fix a generically semisimple Frobenius manifold V with $\dim V = N$. Given two S-calibration S_τ and $\widetilde{S}_\tau$ where $\widetilde{S}_\tau$ allows such a decomposition

$$(\widetilde{S}_\tau)^\alpha_a = (\Psi_\tau)^\beta_a (R_\tau)^\alpha_\beta e^{\frac{u^\alpha}{z}},$$

we will describe the graph sum formula for higher genus descendant and ancestor potentials with these choices of S-calibrations. Let Γ be a connected graph. We introduce the following notations.

1. $V(\Gamma)$ is the set of vertices in Γ.
2. $E(\Gamma)$ is the set of edges in Γ.
3. $H(\Gamma)$ is the set of half edges in Γ.
4. $L^o(\Gamma)$ is the set of ordinary leaves in Γ. The ordinary leaves are ordered: $L^o(\Gamma) = \{l_1, \ldots, l_n\}$ where n is the number of ordinary leaves.
5. $L^1(\Gamma)$ is the set of dilaton leaves in Γ. The dilaton leaves are unordered.

We also introduce the following labels:

1. (genus) $g : V(\Gamma) \to \mathbb{Z}_{\geq 0}$.
2. (marking) $\alpha : V(\Gamma) \to \{1, \ldots, N\}$. This induces $\alpha : L(\Gamma) = L^o(\Gamma) \cup L^1(\Gamma) \to \{1, \ldots, N\}$, as follows: if $l \in L(\Gamma)$ is a leaf attached to a vertex $v \in V(\Gamma)$, define $\alpha(l) = \alpha(v)$.
3. (height) $k : H(\Gamma) \to \mathbb{Z}_{\geq 0}$.

[4]It should converge near "large radius limit" τ_0. We decompose $\tau = \tau' + \tau''$, $\tau' \in H^2_{\mathbb{T}}(X; \mathbb{Q})$ and $\tau'' \in H^{\neq 2}_{\mathbb{T}}(X; \mathbb{Q})$. Here $\tau'_0 = -\infty$ and $\tau''_0 = 0$. This fact allows us to avoid Novikov variables. It is a highly non-trivial statement (see [45]). A common practice is to utilize Novikov variables first, and the convergence follows from the B-model *after* establishing a mirror symmetry statement.

Given an edge e, let $h_1(e), h_2(e)$ be the two half edges associated to e. The order of the two half edges does not affect the graph sum formula in this paper. Given a vertex $v \in V(\Gamma)$, let $H(v)$ denote the set of half edges emanating from v. The valency of the vertex v is equal to the cardinality of the set $H(v)$: $\mathrm{val}(v) = |H(v)|$. A labeled graph $\vec{\Gamma} = (\Gamma, g, \alpha, k)$ is *stable* if

$$2g(v) - 2 + \mathrm{val}(v) > 0$$

for all $v \in V(\Gamma)$.

Let $\mathbf{\Gamma}(V)$ denote the set of all stable labeled graphs $\vec{\Gamma} = (\Gamma, g, \alpha, k)$, associated to the Frobenius manifold V. The genus of a stable labeled graph $\vec{\Gamma}$ is defined to be

$$g(\vec{\Gamma}) := \sum_{v \in V(\Gamma)} g(v) + |E(\Gamma)| - |V(\Gamma)| + 1 = \sum_{v \in V(\Gamma)} (g(v) - 1) + (\sum_{e \in E(\Gamma)} 1) + 1.$$

Define

$$\mathbf{\Gamma}_{g,n}(V) = \{\vec{\Gamma} = (\Gamma, g, \alpha, k) \in \mathbf{\Gamma}(V) : g(\vec{\Gamma}) = g, |L^o(\Gamma)| = n\}.$$

We assign weights to leaves, edges, and vertices of a labeled graph $\vec{\Gamma} \in \mathbf{\Gamma}(V)$ as follows.

1. *Ordinary leaves.* To each ordinary leaf $l_j \in L^o(\Gamma)$ with $\alpha(l_j) = \alpha \in \{1, \dots, N\}$ and $k(l) = k \in \mathbb{Z}_{\geq 0}$, we define two kinds of weight:

$$(\mathcal{L}^{\mathbf{u}}_d)^\alpha_k(l_j) = [z^k]\Big(\sum_{\alpha,a=1}^{N} \Big(\mathbf{u}^a_j(z) S^{\hat{\beta}}{}_a(z)\Big)_+ R(-z)_\beta^{\ \alpha}\Big),$$

$$(\mathcal{L}^{\mathbf{u}}_a)^\alpha_k(l_j) = [z^k]\Big(\sum_{\alpha,a=1}^{N} \Big(\mathbf{u}^a_j(z) \Psi_a^{\ \beta}\Big) R(-z)_\beta^{\ \alpha}\Big).$$

The notion $()_+$ discards negative powers of z, i.e. $(\sum_{n \in \mathbb{Z}} a_n z^n)_+ = \sum_{n \geq 0} a_n z^n$.

2. *Dilaton leaves.* To each dilaton leaf $l \in L^1(\Gamma)$ with $\alpha(l) = \alpha \in I_\Sigma$ and $2 \leq k(l) = k \in \mathbb{Z}_{\geq 0}$, we assign

$$(\mathcal{L}^1)^\alpha_k(l) = [z^{k-1}]\Big(-\sum_{\beta=1}^{N} \frac{1}{\sqrt{\Delta^\beta(\tau)}} R_\beta^{\ \alpha}(-z)\Big).$$

3. *Edges.* To an edge connecting two vertices marked by $\alpha, \beta \in \{1, \dots, N\}$ and with heights k and l at its two half-edges, we assign

$$\mathcal{E}^{\alpha,\beta}_{k,l}(e) = [z^k w^l]\Big(\frac{1}{z+w}(\delta_{\alpha,\beta} - \sum_{\gamma=1}^{N} R_\gamma^{\ \alpha}(-z) R_\gamma^{\ \beta}(-w)\Big).$$

4. *Vertices.* To a vertex v with genus $g(v) = g \in \mathbb{Z}_{\geq 0}$ and with marking $\alpha(v) = \alpha$, with n ordinary leaves and half-edges attached to it with heights $k_1, \ldots, k_n \in \mathbb{Z}_{\geq 0}$ and m more dilaton leaves with heights $k_{n+1}, \ldots, k_{n+m} \in \mathbb{Z}_{\geq 0}$, we assign

$$(\sqrt{\Delta^{\alpha(v)}(\tau)})^{2g(v)-2+\mathrm{val}(v)}\langle \prod_{h\in H(v)} \tau_{k(h)}\rangle_{g(v)} = \left(\sqrt{\Delta^{\alpha}(\tau)}\right)^{2g(v)-2+\mathrm{val}(v)}$$

$$\int_{\overline{\mathcal{M}}_{g,n+m}} \psi_1^{k_1} \cdots \psi_{n+m}^{k_{n+m}}.$$

Define the weight of a labeled graph $\vec{\Gamma} \in \mathbf{\Gamma}(V)$ to be (the letter F means "Frobenius")

$$w^{\mathbf{u}}_{F,\bullet}(\vec{\Gamma}) = \prod_{v\in V(\Gamma)} (\sqrt{\Delta^{\alpha(v)}(\tau)})^{2g(v)-2+\mathrm{val}(v)}\langle \prod_{h\in H(v)} \tau_{k(h)}\rangle_{g(v)} \prod_{e\in E(\Gamma)} \mathcal{E}^{\alpha(v_1(e)),\alpha(v_2(e))}_{k(h_1(e)),k(h_2(e))}(e) \tag{1}$$

$$\cdot \prod_{j=1}^{n} (\mathcal{L}^{\mathbf{u}}_{\bullet})^{\alpha(l_n)}_{k(l_j)}(l_j) \prod_{l\in L^1(\Gamma)} (\mathcal{L}^1)^{\alpha(l)}_{k(l)}(l),$$

where $\bullet = a$ or d.

Definition 13 Suppose that $2g - 2 + n > 0$. Define the ancestor potential

$$\langle\langle \mathbf{u}_1(\bar{\psi}_1), \ldots, \mathbf{u}_n(\bar{\psi}_n)\rangle\rangle^V_{g,n} = \sum_{\vec{\Gamma}\in\mathbf{\Gamma}_{g,n}(V)} \frac{w^{\mathbf{u}}_{F,a}(\vec{\Gamma})}{|\mathrm{Aut}(\vec{\Gamma})|}.$$

and the descendant potential

$$\langle\langle \mathbf{u}_1(\psi_1), \ldots, \mathbf{u}_n(\psi_n)\rangle\rangle^V_{g,n} = \sum_{\vec{\Gamma}\in\mathbf{\Gamma}_{g,n}(V)} \frac{w^{\mathbf{u}}_{F,d}(\vec{\Gamma})}{|\mathrm{Aut}(\vec{\Gamma})|}.$$

Remark 14 The ψ-classes and $\bar{\psi}$-classes here are just notations. If X is a toric manifold, and $V = QH^*_{\mathbb{T}}(X; \mathcal{Q})$, one may choose

$$(S_\tau)^{\hat{\alpha}}_a = (H_a, \hat{\phi}_\alpha) + \langle\langle H_a, \frac{\hat{\phi}_\alpha}{z-\psi}\rangle\rangle^X_{0,2},$$

and

$$(\widetilde{S}_\tau)^{\alpha}_a = (S_\tau)^{\alpha}_a \cdot \exp(-\sum_{n=1}^{\infty} \frac{B_{2n}}{2n(2n-1)}(\frac{z}{\chi^\alpha})^{2n-1}).$$

Givental [39, 40] shows the following (when $2g-2+n>0$)

$$\langle\langle \mathbf{u}_1(\bar{\psi}_1),\ldots,\mathbf{u}_n(\bar{\psi}_n)\rangle\rangle^V_{g,n} = \langle\langle \mathbf{u}_1(\bar{\psi}_1),\ldots,\mathbf{u}_n(\bar{\psi}_n)\rangle\rangle^X_{g,n},$$
$$\langle\langle \mathbf{u}_1(\psi_1),\ldots,\mathbf{u}_n(\psi_n)\rangle\rangle^V_{g,n} = \langle\langle \mathbf{u}_1(\psi_1),\ldots,\mathbf{u}_n(\psi_n)\rangle\rangle^X_{g,n}.$$

Let $\mathbf{u}_1=\mathbf{u}_2=\cdots=\mathbf{t}$. One can define the total ancestor potential

$$\mathcal{A}_\tau(\mathbf{t}) = \exp\Big(\sum_{g,n=0}^{\infty}\frac{\hbar^{g-1}}{n!}\langle\langle \mathbf{t}(\bar{\psi}_1),\ldots,\mathbf{t}(\bar{\psi}_n)\rangle\rangle^V_{g,n}\Big).$$

The graph sum formula for the ancestor potentials is another form of the following Givental's quantization process [39] (without $(g,n)=(1,0)$ information,which is captured by $C(\tau)$ and not defined here)

$$e^{F_1(\tau)}\mathcal{A}_\tau(\mathbf{t}) = e^{C(\tau)}\widehat{\Psi}\widehat{R}(z)e^{\frac{\widehat{U}}{z}}\prod_{a=1}^{N}\mathcal{T}(\mathbf{t}^a).$$

For $a=1,\ldots,N$, $\mathcal{T}(\mathbf{t}^a)$ is the Kontsevich tau-function

$$\mathcal{T}(\mathbf{t}^a) = \exp\Big(\sum_{g,n=0}^{\infty}\frac{\hbar^{g-1}}{n!}\langle \mathbf{t}^a(\psi_{\mathrm{pt},1}),\ldots,\mathbf{t}^a(\psi_{\mathrm{pt},n})\rangle^{\mathrm{pt}}_{g,n}\Big),$$

Let

$$\mathcal{D}(\mathbf{t}) = \exp\Big(\sum_{g,n=0}^{\infty}\frac{\hbar^{g-1}}{n!}\langle \mathbf{t}(\psi_1),\ldots,\mathbf{t}(\psi_n)\rangle^X_{g,n}\Big).$$

It does not depend on τ. Givental's quantization formula says

$$\mathcal{D}(\mathbf{t}) = e^{C(\tau)}\widehat{S}_\tau^{-1}(z)\mathcal{A}_\tau(\mathbf{t}).$$

It is a consequence of the graph sum formula for the descendant potential.

3.3 A Graph Sum Formula for Eynard-Orantin Recursions

Dunin-Barkowski–Orantin–Shadrin–Spitz express the Eynard-Orantin higher genus differential forms in terms of a graph sum in [19], and then compare with Givental's quantized descendant potentials.

We expand the fundamental normalized differential around (p_α, p_β) where p_α, p_β are critical points of x.

$$B(\zeta_\alpha, \zeta_\beta) = (\frac{\delta_{\alpha,\beta}}{(\zeta_\alpha - \zeta_\beta)^2} + \sum_{k,l\in\mathbb{Z}_{\geq 0}} B^{\alpha,\beta}_{k,l}(\zeta_\alpha)^k(\zeta_\beta)^l)d\zeta_\alpha d\zeta_\beta.$$

In case of a formal spectral curve, the fundamental normalized differential is defined by these coefficients $B^{\alpha,\beta}_{i,j}$. Define the propagators

$$\check{B}^{\alpha,\beta}_{k,l} = \frac{(2k-1)!!(2l-1)!!}{2^{k+l+1}} B^{\alpha,\beta}_{2k,2l},$$

and

$$\check{h}^\alpha_k = 2(2k-1)!!h^\alpha_{2k-1}.$$

Here we quote a lemma [21, Equation (D.4)] on the relation between $\hat{\xi}_{\alpha,k}$ and θ^α_k.

Lemma 15

$$\theta^\alpha_k = d\hat{\xi}_{\alpha,k} - \sum_{i=0}^{k-1}\sum_\beta \hat{B}^{\alpha,\beta}_{k-1-i,0} d\hat{\xi}_{\beta,i}.$$

Here β sums over $I_{\boldsymbol{\Sigma}}$ or $I_{\mathfrak{C}}$ for any spectral curve $\boldsymbol{\Sigma}$ or formal spectral curve $\mathfrak{C}$.

Similarly to $\boldsymbol{\Gamma}(V)$ we define the set of decorated stable graph $\boldsymbol{\Gamma}(\boldsymbol{\Sigma})$ (or $\boldsymbol{\Gamma}(\mathfrak{C})$ if we are working with a formal spectral curve)—the only difference is the marking as below.

(2)' (marking) $\alpha : V(\Gamma) \to I_{\boldsymbol{\Sigma}}$ (or $I_{\mathfrak{C}}$). We also define the marking of leaf $\alpha(l)$ to be the marking of the vertex it attaches to.

Given a labeled graph $\vec{\Gamma} \in \boldsymbol{\Gamma}_{g,n}(\boldsymbol{\Sigma})$ with $L^o(\Gamma) = \{l_1, \ldots, l_n\}$, we define its weight to be (the letter S means "spectral curves")

$$w^{\mathbf{p}}_S(\vec{\Gamma}) = (-1)^{g(\vec{\Gamma})-1} \prod_{v\in V(\Gamma)} \left(\frac{h_1^{\alpha(v)}}{\sqrt{-2}}\right)^{2-2g-\mathrm{val}(v)} \langle \prod_{h\in H(v)} \tau_{k(h)}\rangle_{g(v)} \prod_{e\in E(\Gamma)} \check{B}^{\alpha(v_1(e)),\alpha(v_2(e))}_{k(e),l(e)}$$
$$\cdot \prod_{j=1}^n (\check{\mathcal{L}}^{\mathbf{p}})^{\alpha(l_j)}_{k(l_j)}(l_j) \prod_{l\in L^1(\Gamma)} (-\frac{1}{\sqrt{-2}})\check{h}^{\alpha(l)}_{k(l)}. \tag{2}$$

Here $\mathbf{p} = (p_1, \ldots, p_n) \in (\overline{\Sigma})^n$ in case of an actual spectral curve, and the ordinary leaf is

$$(\check{\mathcal{L}}^{\mathbf{p}})^\alpha_k(l_j) = -\frac{1}{\sqrt{-2}}\theta^\alpha_k(p_j).$$

When we are working with a formal curve, $\mathbf{p} = (\zeta_{\beta_1}, \ldots, \zeta_{\beta_n}) \in C_{\beta_1} \times \cdots \times C_{\beta_n}$. The ordinary leaf is

$$(\check{\mathcal{L}}^{\mathbf{p}})^{\alpha}_{k}(l_j) = -\frac{1}{\sqrt{-2}}\theta^{\alpha}_{k,\beta_j}(\zeta_{\beta_j}).$$

The graph sum formula of $\omega_{g,n}$ is the following.

Theorem 16 (Dunin-Barkowski–Orantin–Shadrin–Spitz [19])

$$\omega_{g,n}(\mathbf{p}) = \sum_{\vec{\Gamma}\in\mathbf{\Gamma}(\mathbf{\Sigma})} \frac{w^{\mathbf{p}}_{S}(\vec{\Gamma})}{\mathrm{Aut}(\vec{\Gamma})}.$$

Starting from a generically semisimple Frobenius manifold V with an S and $\widetilde{S}$-calibration, around a semisimple point $p \in V$, we define a family of formal spectral curves $\mathfrak{C}_V(\tau) = \{C_{V,\beta} = \mathrm{Spec}\mathbb{C}[[z_\beta]])\}_{\beta=1}^{N}$, together with the following information

$$h^{\alpha}_{2k-1}(\tau) = [z^{k-1}]\left(\sum_{\beta=1}^{N} \frac{\sqrt{-2}}{(2k-1)!!2^{k-1}\sqrt{\Delta^{\alpha}(\tau)}}(R_\tau)_{\alpha'}^{\ \alpha}(-z)\right), \ k \geq 0.$$

$$B^{\alpha,\beta}_{2k,2l}(\tau) = \frac{2^{k+l+1}}{(2k-1)!!(2l-1)!!}[z^k w^l]\left(\frac{1}{z+w}(\delta_{\alpha,\beta} - \sum_{\gamma=1}^{N}(R_\tau)_{\gamma}^{\ \alpha}(-z)(R_\tau)_{\gamma}^{\ \beta}(-w)\right), \ k, l \geq 0.$$

Notice that they only depend on R_τ, which comes from factorizing $\widetilde{S}$. Even coefficients of h^{α}_{k} and odd coefficients of $B^{\alpha,\beta}_{k,l}$ could be arbitrarily chosen.

Define

$$\overline{\boldsymbol{u}}^{\alpha}_{j}(z) = \sum_{z=0}^{\infty}(\overline{u}^{\alpha}_{j})_k z^k = \sum_{b=1}^{N}\mathbf{u}^{a}_{j}(z)\Psi_{a}^{\ \alpha},$$

$$\widetilde{\mathbf{u}}^{\alpha}_{j}(z) = \sum_{z=0}^{\infty}(\widetilde{u}^{\alpha}_{j})_k z^k = \sum_{b=1}^{N}\left(S^{\hat{\alpha}}_{\ \overline{b}}(z)\mathbf{u}^{b}_{j}(z)\right)_{+},$$

where

$$S^{\hat{\alpha}}_{\ \overline{b}}(z) = (\hat{\phi}^{\alpha}(\tau), \mathcal{S}(H_b)).$$

Theorem 17 ([19]) *When* $2g-2+n>0$,

$$\omega^{\mathfrak{C}_V(\tau)}_{g,n}(z^1_{\beta_1}, \ldots, z^n_{\beta_n}) = (-1)^{g-1+n}\langle\langle \mathbf{u}_1(\bar{\psi}), \ldots, \mathbf{u}_n(\bar{\psi})\rangle\rangle^{V}_{g,n}|_{(\overline{u}^{\alpha}_{j})_k = d\hat{\xi}^{\alpha}_{k,\beta_j}(z^j_{\beta_j})},$$

$$\omega^{\mathfrak{C}_V(\tau)}_{g,n}(z^1_{\beta_1}, \ldots, z^n_{\beta_n}) = (-1)^{g-1+n}\langle\langle \mathbf{u}_1(\psi), \ldots, \mathbf{u}_n(\psi)\rangle\rangle^{V}_{g,n}|_{(\widetilde{u}^{\alpha}_{j})_k = d\hat{\xi}^{\alpha}_{k,\beta_j}(z^j_{\beta_j})}.$$

3.4 Oscillatory Integrals on the Spectral Curves

Let Σ be a spectral curve, and γ_α be the Lefschetz thimble in Σ with respect to x such that p_α is the only critical point in γ_α and

$$x(\gamma_\alpha) = [x_{0,\alpha}, +\infty).$$

We define $\check{R}(z)$ as the power series in the following asymptotic expansion.

$$\int_{\gamma_\alpha} e^{-\frac{x}{z}} \theta_0^\beta \sim 2\sqrt{\frac{\pi}{z}} e^{-\frac{x_{0,\alpha}}{z}} \check{R}_\beta{}^\alpha(z).$$

Notice that this definition is also well-defined for formal spectral curves. We have the following property for $\check{B}_{k,l}^{\alpha,\beta}$ [22, Equation (B.9)]:

$$\check{B}_{k,l}^{\alpha,\beta} = [z^k w^l] \left(\frac{1}{z+w} (\delta_{\alpha,\beta} - \sum_{\gamma \in I_\Sigma} \check{R}_\gamma{}^\alpha(z) \check{R}_\gamma{}^\beta(w)) \right).$$

We consider the space of differential forms spanned by θ_0^α, denoted by $\check{V}_\tau$. It is isomorphic to $T_\tau V$ by $\theta_0^\alpha \mapsto \frac{\hat{\phi}^\alpha(\tau)}{\sqrt{-2}}$. Denote this isomorphism by $\mathfrak{r}$. By [25, Appendix D], the differential form

$$d(\frac{dy}{dx}) = \frac{1}{2} \sum_{\beta=1}^{N} h_1^\beta(\tau) \theta_0^\beta.$$

We have the following correspondence table between the Frobenius manifold V and the family of formal spectral curves $\mathfrak{C}_V(\tau)$.

Frobenius manifold V	Correspondence	Family of formal spectral curves $\mathfrak{C}_V(\tau)$
Dimension N	$=$	# of formal disks
R-matrix $R_\alpha{}^\beta(z)$	$=$	$\check{R}_\alpha{}^\beta(-z)$
Propagator $\mathcal{E}_{i,j}^{\alpha,\beta}$	$=$	Propagator $\check{B}_{i,j}^{\alpha,\beta}$
Canonical coordinate u^α	$=$	Critical value $x_{0,\alpha}$
S-matrix $\widetilde{S}_{\underline{\hat{\beta}}}^{\hat{\alpha}} = (\hat{\phi}_\beta(\tau), S_\tau(\hat{\phi}^\alpha))$	$\sim$	Oscillatory integral $\frac{1}{2\sqrt{\pi z}} \int_{\gamma_\alpha} e^{-\frac{x}{z}} \theta_0^\beta$
Meromorphic form $\frac{\theta_0^\alpha}{\sqrt{-2}}$	$\overset{\mathfrak{r}}{\mapsto}$	Canonical basis $\hat{\phi}^\alpha(\tau)$
$d(\frac{dy}{dx})$	$\overset{\mathfrak{r}}{\mapsto}$	Identity $\mathbf{1}$
$\sqrt{\Delta^\alpha(\tau)}$	$=$	$\frac{-\sqrt{-2}}{h_1^\alpha(\tau)}$

4 Applications of Eynard-Orantin Recursion: Mirror Symmetry

Mirror symmetry relates the A-model theory on a target space to the B-model theory on its mirror space. Gromov-Witten invariants are a typical type of A-model invariants. In order to apply the recursion algorithm and to use Eynard-Orantin higher genus invariants $\omega_{g,n}$ to predict Gromov-Witten invariants, we need a mirror B-model in the form of a spectral curve. When the target space is a 1-dimensional toric variety, like $\mathbb{P}^1$, the mirror Landau-Ginzburg model is a superpotential on $\mathbb{C}^*$. After suitable compactification, one may directly regard this as a spectral curve. Another (much bigger) class of examples is toric Calabi-Yau 3-orbifolds. Their mirrors, although 3-dimensional, could be reduced to mirror curves by dimensional reduction. Lying at the intersection of these two classes is the Lambert curve, which could be regarded as $\mathbb{P}^1$ in the large radius limit, or as $\mathbb{C}^3$ with limiting equivariant data (large framing limit). The relations among these examples is summarized in the following diagram.

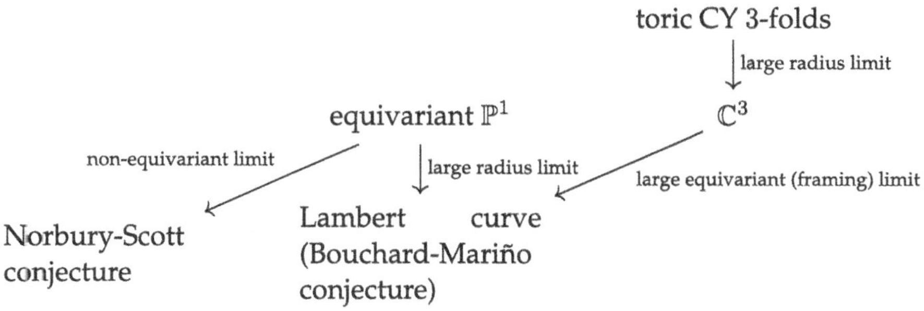

4.1 *Airy Curve*

Let's look at the easiest case, which is roughly "mirror symmetry of a point". The Airy curve, in our notation, is a formal curve $\mathfrak{C} = (C, y, x, B)$ where

$$C = \mathrm{Spec}[[\zeta]],$$

$$y = \zeta, \ x = \zeta^2,$$

$$B = \frac{d\zeta_1 d\zeta_2}{(\zeta_1 - \zeta_2)^2}.$$

Remark 18 We may regard this curve as the parabola $x = y^2$ in $\mathbb{C}^2$, which is the Airy curve in the usual sense. The fundamental normalized differential B is the unique one on its compactification $\mathbb{P}^1$.

Once we run the Eynard-Orantin recursion for the spectral curve $\mathfrak{C}$, we have the following theorem.

Theorem 19

$$\omega_{g,n}(\zeta_1,\ldots,\zeta_n) = (-2)^{2-2g-n} \sum_{d_1+\cdots+d_n=d_{g,n}} \prod_{i=1}^{n} \frac{(2d_i+1)!!d\zeta_i}{\zeta_i^{2d_i+2}} \langle \prod_{i=1}^{n} \psi_i^{d_i} \rangle.$$

This theorem is a direct consequence of the graph sum formula for Eynard-Orantin recursions (Theorem 16). There is only one critical point of x, labeled by 1. It is straightforward to check that all propagators $\check{B}^{1,1}_{i,j} = 0$ for all i, j. The differential forms

$$\theta^1_k(\zeta) = -\frac{(2k+1)!!d\zeta}{2^k\zeta^{2k+2}},$$

and $h^1_1 = 1$. There are no dilaton leafs since all of them are zero.

4.2 *Lambert Curve*

Lambert curve is given by $\mathbf{\Sigma} = (\Sigma, x, B)$ where

$$\Sigma = \{0\} \times \mathbb{C}^* \in (\mathbb{C}^*)^2,\ \overline{\Sigma} \cong \mathbb{P}^1$$

$$x = e^{-y} + y,$$

$$B = \frac{dY_1 dY_2}{(Y_1 - Y_2)^2}.$$

Here $X = e^{-x}$, $Y = e^{-y}$. The only branch point is at $Y = 1$. At $Y = 0$, the value of X is well defined. Using the Eynard-Orantin recursion, we construct $\omega_{g,n}$ as symmetric meromorphic n-form on $(\mathbb{P}^1)^n$. Notice that $\omega_{g,n}$ is smooth at $Y = 0$, and can be expanded in series by X.

Lambert curve predicts *Hurwitz numbers* on the A-side. Consider ramified covers of $\mathbb{P}^1$ by a genus g curve with a specified ramification profile at a special point on $\mathbb{P}^1$. All other branch points in $\mathbb{P}^1$ are simple and fixed. The ramification profile is given by a partition μ of length $n := \ell(\mu)$. The number of such covers is denoted by $H_{g,\mu}$.

We collect all Hurwitz numbers at fixed genus g for all ramification profiles μ of the same length $n = \ell(\mu)$ into a generating function

$$H_g(X_1,\ldots,X_n) = \sum_{\ell(\mu)=n} \frac{m_\mu(X)|\mathrm{Aut}(\mu)|\prod_{i=1}^n \mu_i H_{g,\mu}}{(2g-2+n+|\mu|)!},$$

where $m_\mu(X)$ is a monomial symmetric function in $X_1, \ldots, X_n$ defined by

$$m_\mu(X) = \frac{1}{|\mathrm{Aut}(\mu)|} \sum_{\sigma \in S_n} \prod_{i=1}^{n} (X_{\sigma(i)})^{\mu_i - 1}$$

Here S_n is the permutation group.

The Bouchard-Mariño conjecture says the following [10].

Theorem 20 (Bouchard-Mariño Conjecture) *When* $2g - 2 + n > 0$,

$$H_g(X_1, \ldots, X_n) dX_1 \ldots dX_n = \omega_{g,n}.$$

The right side should be understood as an power series expansion at $X = 0$.

We omit the unstable cases $(g, n) = (0, 1), (0, 2)$ for simplicity here. This theorem is proved in [9, 27, 58]. Here we introduce the ELSV formula [20, 42] for later use. This relates Hurwitz numbers to Hodge integrals, which are more relevant to A-model GW theory in mirror symmetry.

Theorem 21 (ELSV Formula)

$$H_{g,\mu} = \prod_{i=1}^{n} \frac{\mu_i^{\mu_i}}{\mu_i!} \int_{\overline{\mathcal{M}}_{g,n}} \frac{\Lambda_g^\bullet(1)}{(1 - \mu_1 \psi_1) \ldots (1 - \mu_n \psi_n)}$$

Here $\Lambda_g^\bullet(u) = u^g - \lambda_1 u^{g-1} + \cdots + (-1)^g \lambda_g$, and $\lambda_i = c_i(E)$ where E is the Hodge bundle. We will see that from this formula, the Bouchard-Mariño conjecture is a consequence of all genera equivariant mirror symmetry for $\mathbb{P}^1$ (Sect. 4.3), and also a consequence of the BKMP conjecture for $\mathbb{C}^3$ (Sect. 4.4).

4.3 Projective Line

Let $X = \mathbb{P}^1$. Its mirror is a Landau-Ginzburg model

$$W(Y) = t_0 + Y + \frac{e^{t_1}}{Y}.$$

To capture the equivariant data of $\mathbb{P}^1$, we use a modified *equivariant* superpotential

$$\widetilde{W} = W + \mathsf{w}_1 \log Y + \mathsf{w}_2 \log \frac{e^{t_1}}{Y}.$$

The 2-torus $\mathbb{T}$ acts by turning homogeneous coordinates $(s_1, s_2) \cdot (z_1 : z_2) = (s_1 z_1 : s_2 z_2)$. The characters w_i are basis in the character lattice $\mathsf{w}_i : (s_1, s_2) \mapsto s_i \in$

$\mathbb{C}^*$. Let $x = \widetilde{W}$, and $\overline{\Sigma} = \mathbb{P}^1$ where $(1 : Y)$ is its coordinate. There is only one fundamental differential

$$B = \frac{dY_1 dY_2}{(Y_1 - Y_2)^2}.$$

since there is no choice of A-cycles. The spectral curve is $\mathbf{\Sigma} = (\Sigma, x, B)$. The all genera mirror symmetry for $\mathbb{P}^1$ is the following theorem [30].

Theorem 22 (Fang-Liu-Zong) *Let* $\mathbf{t} = t_0\mathbf{1} + t_1 H$ *where* $\mathbf{1}$ *is the unit in* $H^0_{\mathbb{T}}(\mathbb{P}^1; \mathbb{C})$ *and* H *is the equivariant lift of the hyperplane class whose restriction at two* $\mathbb{T}$*-fixed points gives* w_1 *and* w_2 *respectively.*

$$\omega_{g,n}|_{\frac{1}{\sqrt{-2}} d\xi_{\alpha,k}(Y_j) = (\tilde{u})^\alpha_k} = (-1)^{g-1+n} F^{\mathbb{P}^1,\mathbb{T}}_{g,n}(\mathbf{u}_1, \ldots, \mathbf{u}_n, \mathbf{t}).$$

Since the proof utilizes the same idea as in the proof of the BKMP conjecture which will be discussed in more details (see Sect. 4.5), we only briefly remark a few words here. Notice the similarity between this theorem and Theorem 17—the right side is the actual Gromov-Witten potential, while the one in Theorem 17 comes from the quantization for the Frobenius manifold. They agree as shown in [39, 40].

We mention that taking the non-equivariant limit $\mathsf{w} \to 0$ and when there is no primary insertions, this theorem leads to the Norbury-Scott conjecture [19, 59].

Theorem 23 (Norbury-Scott) *Near* $Y = 0$, *in the non-equivariant limit* ($\mathsf{w}_1 = \mathsf{w}_2 = 0$, $t_0 = 0$), $x^{-1} = (Y + \frac{e^{t_1}}{Y})^{-1}$ *is a coordinate such that one can expand* $\omega_{g,n}$ *in power series*

$$\omega_{g,n} = (-1)^{g-1+n} \sum_{a_1,\ldots,a_n \in \mathbb{Z}_{\geq 0}} \langle\langle \tau_{a_1}(H) \ldots \tau_{a_n}(H) \rangle\rangle^{\mathbb{P}^1}_{g,n} \prod_{j=1}^{n} \frac{(a_j+1)!}{x^{a_j+2}} dx_j.$$

Remark 24 The divisor equation says ($q = e^{t_1}$)

$$\langle\langle \tau_{a_1}(H) \ldots \tau_{a_n}(H) \rangle\rangle^{\mathbb{P}^1}_{g,n} = q^{\frac{1}{2}\sum_{i=1}^{n} a_i + 1 - g} \langle \tau_{a_1}(H) \ldots \tau_{a_n}(H) \rangle^{\mathbb{P}^1}_{g,n}.$$

The Norbury-Scott conjecture corresponds to setting $q = 1$, i.e. $t^1 = 0$.

Taking the large radius limit of the equivariant mirror theorem (Theorem 22) for $\mathbb{P}^1$, one could recover the Bouchard-Mariño conjecture. The superpotential becomes $x = Y + \mathsf{w}_1 \log Y$ by setting $q = 0$. Setting $\mathsf{w}_1 = -1$ turns this into a Lambert curve. The localization calculation of $F^{\mathbb{P}^1,\mathbb{T}}$ in the limit produces the Hodge integrals, and ELSV formula turns it into the desired generating function involving Hurwitz numbers, as shown in [30, Section 5].

4.4 *Mirror of* $\mathbb{C}^3$ *(Topological Vertex)*

Let's switch gears and proceed to toric Calabi-Yau threefolds. The mirror of a toric 3-(orbi)fold, by the construction of Givental [38], is a Landau-Ginzburg model $W : (\mathbb{C}^*)^3 \to \mathbb{C}$. A Calabi-Yau should have a Calabi-Yau mirror. A special feature of a toric Calabi-Yau variety is that its mirror's superpotential $W = H(X, Y)Z$, $(X, Y, Z) \in (\mathbb{C}^*)^3$. As pointed out in [44], the Calabi-Yau mirror is $\{H(X, Y) = uv,\ u, v \in \mathbb{C},\ X, Y \in \mathbb{C}^*\} \subset (\mathbb{C}^*)^2 \times \mathbb{C}^2$. Furthermore, this Calabi-Yau mirror can be reduced to a mirror curve $\{H(X, Y) = 0\} \subset (\mathbb{C}^*)^2$. All these different mirrors should be equivalent, carrying the same B-model information.

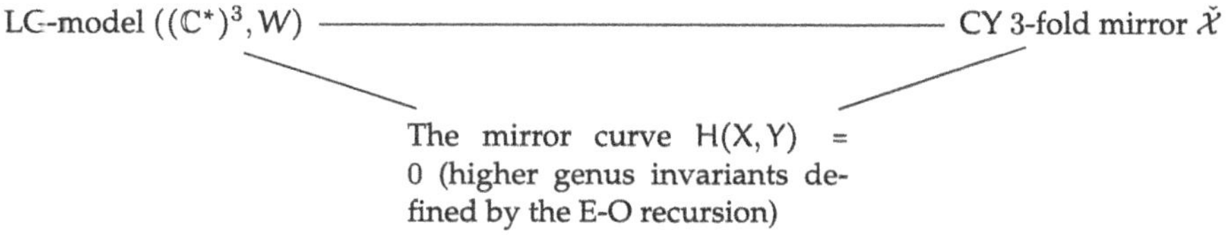

The simplest toric Calabi-Yau threefold is $\mathbb{C}^3$, equipped with the Calabi-Yau form $dZ_1 \wedge dZ_2 \wedge dZ_3$ where $(Z_1, Z_2, Z_3) \in \mathbb{C}^3$ are the coordinates. Its mirror is a 3-dimensional Landau-Ginzburg model $(\mathbb{C}^*)^3$, with a superpotential $W = XZ + YZ + Z = H(X, Y)Z$, where $X, Y, Z \in \mathbb{C}^*$ [35]. The mirror curve is $\{H(X, Y) = 0\}$ as an affine plane curve in $(\mathbb{C}^*)^2$.

We want to consider *open* Gromov-Witten invariants of $\mathbb{C}^3$, which count holomorphic maps from bordered Riemann surfaces to $\mathbb{C}^3$ mapping boundaries to a Lagrangian submanifold L (an A-brane). The construction of such invariants is very complicated. Here we require that L is a so-called *Aganagic-Vafa* brane. This gives a very important class of open Gromov-Witten invariants. They play central roles in many interesting topics involving mirror symmetry and the theory of topological vertex [3–5].

In this particular example $\mathbb{C}^3$, an Aganagic-Vafa brane L is given by

$$L = \{(Z_1, Z_2, Z_3) \in \mathbb{C}^3 : |Z_1|^2 - |Z_2|^2 = c, |Z_2|^2 - |Z_3|^2 = 0, \mathrm{Arg}(Z_1 Z_2 Z_3) = \mathrm{const}\} \cong S^1 \times \mathbb{R}^2.$$

It is a Harvey-Lawson special Lagrangian [43], and c is its "open Kähler parameter". Let $\mu = \{\mu_1, \ldots, \mu_n\}$ be a partition of length $\ell(\mu) = n$. Naïvely, we denote the number $N_{g,n,\mu}^{\mathbb{C},L}$ by the counting of the holomorphic maps described below.

$$\left\{\begin{matrix} (C, \partial C), \text{ where the genus of} \\ C \text{ is } g, \text{ and } \partial C \text{ has } n \text{ com-} \\ \text{ponents} \end{matrix}\right\} \xrightarrow{f} (\mathbb{C}^3, L),$$

The winding number of each boundary component is given by μ_i, $i = 1, \ldots, n$. The definition and computation of such maps in symplectic and algebraic settings can be found in [46, 48, 52–54].

A common phenomenon in open string counting is that the moduli space of such maps has codimension 1 boundaries (walls). The counting changes across the wall, thus depends on a particular choice of chamber in the moduli space. In case that L is an Aganagic-Vafa brane in a toric Calabi-Yau threefold, the result depends on a *framing* parameter, an integer $f \in \mathbb{Z}$. We denote this number by $N_{g,n,\mu}^{\mathbb{C}^3,L,f}$.

A simple way to understand this $N_{g,n,\mu}^{\mathbb{C}^3,L,f}$ is to write down the localization formula—assuming one can actually do the localization (see e.g. [47]). It turns out that the answer we get depends on the torus we choose to localize, unlike the case of closed Gromov-Witten invariants. Denote the Calabi-Yau torus by $\mathbb{T}' = \{(Z_1, Z_2, Z_3) \in (\mathbb{C}^*)^3, Z_1Z_2Z_3 = 1\}$, which preserves the Calabi-Yau form. Let w_1 and w_2 be the following character in $\mathrm{Hom}(\mathbb{T}', \mathbb{C}^*)$

$$\mathsf{w}_1(Z_1, Z_2, \frac{1}{Z_1Z_2}) = Z_1, \quad \mathsf{w}_2(Z_1, Z_2, \frac{1}{Z_1Z_2}) = Z_2.$$

If we choose $\mathbb{T}'_f = \mathrm{Ker}(\mathsf{w}_2 - f\mathsf{w}_1) \subset \mathbb{T}'$, we will get $N_{g,n,\mu}^{\mathbb{C}^3,L,f}$ by the localization formula.[5] We can assemble these $N_{g,n,\mu}^{\mathbb{C}^3,L,f}$ into a generating function

$$F_{g,n}^{\mathbb{C}^3,L,f} = \sum_{\mu_1,\ldots,\mu_n=1}^{\infty} N_{g,n,\mu}^{\mathbb{C}^3,L,f} X_1^{\mu_1} \ldots X_n^{\mu_n}.$$

The mirror B-model starts from the reparametrized mirror curve

$$H_f(X, Y) = X^{-f}Y + Y + 1.$$

This defines an affine plane curve $\Sigma \subset (\mathbb{C}^*)^2$ whose compactification $\overline{\Sigma} \cong \mathbb{P}^1$. Define the superpotential and the fundamental differential

$$W = x, \quad B(Y_1, Y_2) = \frac{dY_1dY_2}{(Y_1 - Y_2)^2}.$$

Here $X = e^{-x}$ and $Y = e^{-y}$. The moment map and the mirror curve is illustrated in Fig. 1.

After running the Eynard-Orantin recursion, we get a sequence of $\omega_{g,n}$. The famous BKMP remodeling conjecture [11, 55] asserts the following.

[5] If one insists on algebraic geometry, we can use relative Gromov-Witten invariants as the definition. This involves partially compactifying $\mathbb{C}^3$ into the total space of $\mathcal{O}_{\mathbb{P}^1}(-1-f) \oplus \mathcal{O}_{\mathbb{P}^1}(f)$, and define $N_{g,n,\mu}^{\mathbb{C}^3,L,f}$ as the relative Gromov-Witten invariants on this space relative to the fiber divisor at the infinity in $\mathbb{P}^1$. The tangency condition at the divisor is given by μ. (See [48, 53, 54].)

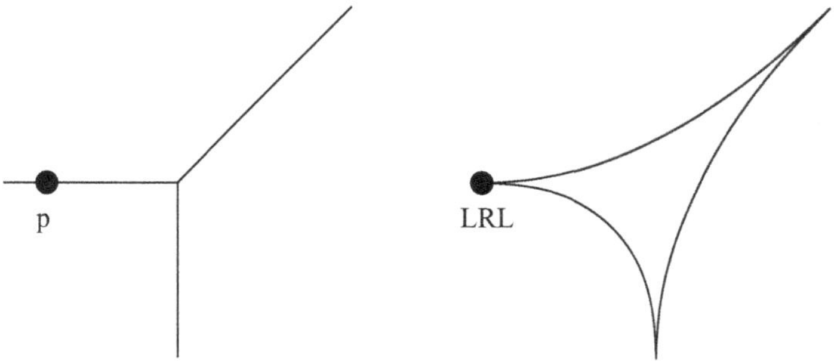

Fig. 1 Under the moment map of the real Calabi-Yau torus $\mathbb{T}'_{\mathbb{R}}$, the toric graph is the image of $\mathbb{T}'$-invariant 1-dimensional subvariety (left). The image of the Aganagic-Vafa brane is a point on the toric graph. The mirror curve (right) is $\mathbb{P}^1$ with three punctures, and its compactification is $\mathbb{P}^1$. The large radius point is where $X = 0$

Theorem 25 ([14, 67, 68]) *For* $g \geq 0$, $n \geq 1$, $2g - 2 + n > 0$,

$$F_{g,n}^{\mathbb{C}^3,L,f} = \int^{X_1} \cdots \int^{X_n} \omega_{g,n}.$$

Remark 26 The cases $(g, n) = (0, 1), (0, 2)$ have special forms which are omitted here for simplicity. There are also predictions on the free energies F_g which are the generating functions of closed GW invariants based on $\omega_{g,n}$. We will discuss F_g in general in the next subsection.

We will postpone the discussion of the proof to the next subsection. The following theorem relates Hurwitz numbers to $F_{g,n}^{\mathbb{C}^3,L,f}$ in the large framing limit.

Theorem 27 ([13])

$$\lim_{f\to\infty} (-1)^n f^{2-2g+n} F_{g,n}^{\mathbb{C}^3,L,f}(\frac{X_1}{f}, \ldots, \frac{X_n}{f}) = H_g(X_1, \ldots, X_n).$$

We have a localization formula for $F_{g,n}^{\mathbb{C}^3,L,f}$. One can write it as a triple Hodge integral with some disk factors (elementary functions). The proof of this theorem is a direct calculation, in which one takes the limit $f \to \infty$ in the triple Hodge integral (cf. ELSV formula 21)

$$\Lambda_g^\bullet(1)\Lambda_g^\bullet(f)\Lambda_g^\bullet(-1-f) = \Lambda_g^\bullet(1)(-1)^{g-1} f^{2g-2}(1 + O(\frac{1}{f})).$$

On the other hand, the mirror curve

$$XY^{-f} + Y + 1 = 0$$

reduces to the Lambert curve $X' = Y'e^{-Y'}$ under the change of variable

$$X = -(-1)^f \frac{X'}{f}, \quad Y = -1 + \frac{Y'}{f}.$$

and taking limit $f \to \infty$. Theorem 20 is a consequence of Theorem 25 after one takes limit $f \to \infty$, and rewrites the open GW potential in the Hurwitz potential by Theorem 27.

4.5 Mirror of a Semi-Projective Toric Calabi-Yau Threefold

A toric Calabi-Yau threefold X is a Calabi-Yau 3-dimensional manifold (or more generally, an orbifold) with a Zariski open and dense algebraic torus $\mathbb{T} \cong (\mathbb{C}^*)^3$. The action of $\mathbb{T}$ on itself extends to X. For simplicity, we require X is a smooth manifold, and will remark briefly on orbifolds in Sect. 4.6. We also require that X is semi-projective, i.e. it is projective over its affinization. The last condition is equivalent to that the union of all cones defining X is convex in $\mathbb{R}^3$. Let $\mathbb{T}'$ be the 2-dimensional subtorus preserving the Calabi-Yau form.

Let $N = \mathrm{Hom}(\mathbb{C}^*, \mathbb{T})$ and $M = \mathrm{Hom}(\mathbb{T}, \mathbb{C}^*) = N^\vee$. The Calabi-Yau torus $\mathbb{T}' = \mathrm{Ker}(\mathsf{w}_3)$ for some $\mathsf{w}_3 \in M$. Being a Calabi-Yau threefold, the fan data to define X is the cone with vertex at the origin over a triangulated integral convex polytope Δ_X on $\{\mathsf{w}_3(y) = 1 | y \in N_\mathbb{R}\}$. If this triangulation cannot be further refined, i.e. each triangle has area $\frac{1}{2}$, the resulting X is a smooth manifold (see Fig. 2).

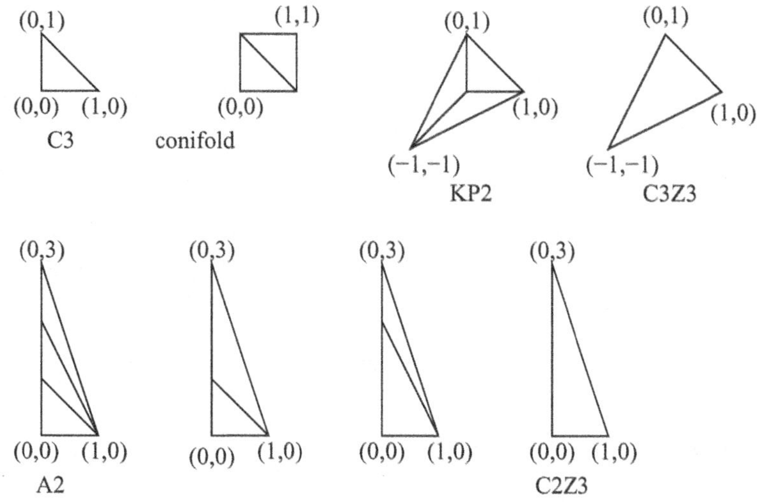

Fig. 2 Defining polytopes of some toric CY 3-(orbi)folds

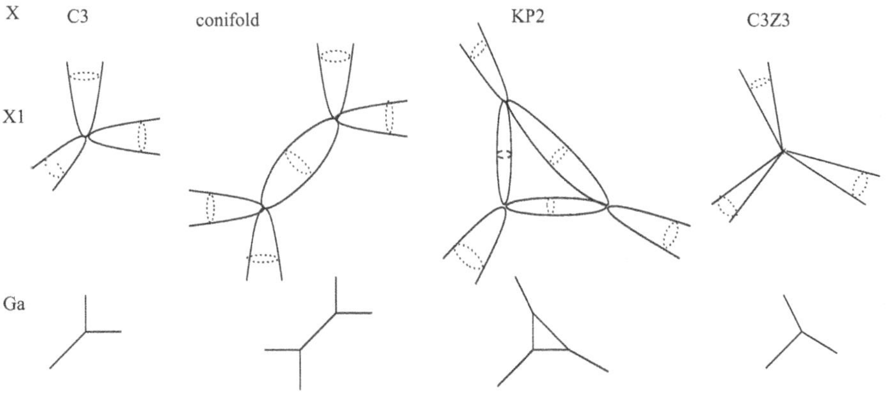

Fig. 3 1-Dimensional $\mathbb{T}'$-invariant subvarieties and toric graphs. We use $\mathcal{X}$ since some are orbifolds

The action of the real torus $\mathbb{T}'_{\mathbb{R}} \subset \mathbb{T}'$ is Hamiltonian, and we can consider the image of all 1-dimensional $\mathbb{T}'$-invariant subvarieties in X under the moment map μ'. Such image is called the toric graph of X (Fig. 3).

An Aganagic-Vafa brane is a Lagrangian 3-dimensional submanifold in X, given by the following condition

$$L \subset \mu'^{-1}(p), \quad \mathrm{Arg}(Z_1 \dots Z_{p+3}) = \text{const on } L.$$

where $(Z_1, \dots, Z_{p+3}) \in \mathbb{C}^{p+3}$ are homogeneous coordinates, and p is a non-vertex point on the toric graph. When p is on a finite segment, L is called an *inner braner*; when p is on a ray, L is called an *outer brane*. We restrict to the case of an outer brane L for simplicity. By suitable arrangement (by some $SL(2; \mathbb{Z})$-action and translation of the defining polytope), we always assume that vertex containing p consists of half-edges in the direction $(-1, 0)$, $(0, -1)$ and $(1, 1)$, and p is on the half-edge in the direction $(-1, 0)$. This half-edge is a ray since L is outer.

Similar to the case of $\mathbb{C}^3$, we consider the open Gromov-Witten invariant $N^{X,L,f}_{g,n,\beta,\mu}$ which counts the maps of the bordered Riemann surface in the topological type (g, n) to the target $(X, (L, f))$ in the curve class β. They form a generating function

$$F^{X,L,f}_{g,n} = \sum_{\mu_1,\dots,\mu_n=1}^{\infty} N^{X,L,f}_{g,n,\beta,\mu} \hat{X}_1^{\mu_1} \dots \hat{X}_n^{\mu_n} Q^{\beta}.$$

Here we use $\hat{X}$ as open variables since they might differ from X in B-model by an open mirror map. The Kähler parameter $Q^{\beta} = \prod_{a=1}^{p} Q_a{}^{\langle p_a, \beta \rangle}$, where p_a form an integral basis in the Kähler cone, and we let $Q_a = e^{-\tau_a}$. The B-model is a mirror curve Σ_q

$$H(X, Y) = XY^{-f} + Y + 1 + \sum_{a=1}^{N-3} q_a X^{m_a} Y^{n_a - f m_a} = 0.$$

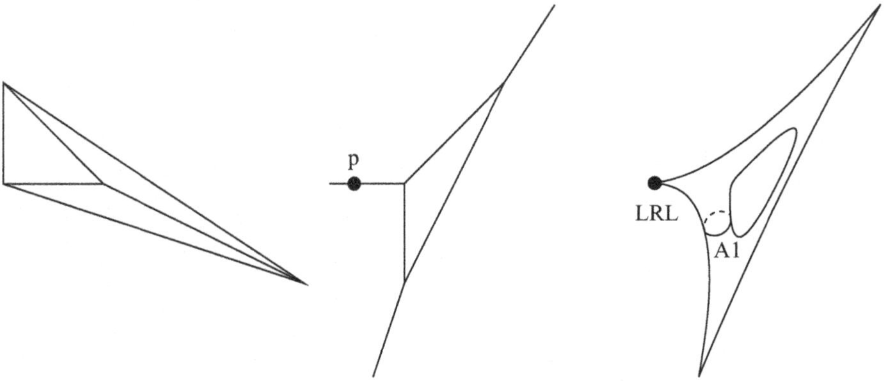

Fig. 4 Defining polytope, toric graph and mirror curve of $\mathcal{O}_{\mathbb{P}^2}(-3)$. Notice that we've arranged the defining polytope and the toric graph in the desired form such that the half edges of the vertex adjacent to p are in the desired direction. The point LRL on the mirror curve is the B-model large radius point, and the period integral around cycle A_1 gives the mirror map

In the equation, $q_1, \ldots, q_{N-3}$ are complex parameters, and $q \to 0$ at the large radius point. The number of complex parameter is $N-3$, where N is the number of integer points inside the defining polytope of X. Under mirror symmetry, N is also the dimension of the equivariant (quantum) cohomology. The integer points inside the defining polytope are denoted by (m_a, n_a). The tropicalization of this curve reduces back to the toric graph (see Fig. 4 as an example). It is also an SYZ dual to the union of 1-dimensional $\mathbb{T}'$-invariant subvarieties. Depending on the choice of the Aganagic-Vafa brane, there is a large radius limit point (the LRL-point in Fig. 4) on the mirror curve where $X = 0$. We specify $e^{-x} = X$, $e^{-y} = Y$ (so at LRL, $x = \infty$ and thus the name large radius). The Landau-Ginzburg superpotential W and its equivariant version $\widetilde{W}$ are

$$W = H(X, Y)Z, \quad \widetilde{W} = W - \log X.$$

The open-closed mirror map is the following

$$\begin{aligned} \tau_a = \tau_a(q) &= \log q_a + h_a(q), \ a = 1, \ldots, N-3 \\ \log \hat{X}_i &= \log X_i + h_0(q), \end{aligned} \tag{3}$$

where $h_a(q) = O(q)$ for $a = 1, \ldots, N-3$. In particular, there are certain choices of geometric cycles $A_a \in H_1(\Sigma_q; \mathbb{Z}), a = 1, \ldots, N-3$, which can be lifted to cycles in $H_1(\widetilde{\Sigma}_q; \mathbb{Z})$, such that

$$\tau_a = \int_{A_a} y dx.$$

The closed part of this map (first line of Eq. (3)) maps q to a Kähler class $\tau \in H^2(X)$. Furthermore there is another cycle $A_0(X)$ which lifts to a path $\widetilde{A}_0(X)$ in the universal cover of Σ_q, such that

$$\log \hat{X} = \int_{\widetilde{A}_0(X)} y dx = \log X + h_0(q).$$

The genus 0 mirror symmetry for closed descendant Gromov-Witten theory is the celebrated toric mirror theorem of Givental and Lian-Liu-Yau [36, 38, 50, 51], in which closed mirror maps are also explicitly given. The open-closed mirror maps are explicitly computed in [49, 57], and the mirror symmetry for disk invariants is conjectured in [3, 4] and proved in [28, 33] under these mirror maps.

The cycles $A_a, a = 1, \ldots, N-3$ induce a Lagrangian subspace of $H_1(\overline{\Sigma}_q; \mathbb{C})$, and thus defines a fundamental differential form B. Define the spectral curve $\boldsymbol{\Sigma}_q = (\Sigma_q, x, B)$. The Eynard-Orantin recursion gives a sequence of higher genus B-model invariants $\omega_{g,n}$. The BKMP remodeling conjecture says

Theorem 28 (Fang-Liu-Zong, [31, 32]) *When $2g-2+n > 0$, $g \geq 0$, $n \geq 1$,*

$$\int^{X_1} \cdots \int^{X_n} \omega_{g,n} = F_{g,n}^{X,L,f}(\hat{X}_1, \ldots, \hat{X}_n).$$

In this theorem, we understand that $\omega_{g,n}$ as power series in X around the large radius limit point. When $2g-2 > 0$, closed free energy is predicted by the following formula

$$F_g^X = \sum_{p_0 \in I_{\boldsymbol{\Sigma}_q}} \mathrm{Res}_{p=p_0} \omega_{g,1}(p) \widetilde{\Phi}(p),$$

where $I_{\boldsymbol{\Sigma}_q}$ is the set of ramification points, and $d\widetilde{\Phi}(p) = \Phi$ is a function locally defined around each ramification point.

4.5.1 Sketch of the BKMP Remodeling Conjecture: Graph Sums

We illustrate the idea of using graph sums to give a proof of this conjecture. As we discussed in Sect. 3, the B-model side could be written as the following graph sums

$$\int^{X_1} \cdots \int^{X_n} \omega_{g,n} = \sum_{\vec{\Gamma} \in \boldsymbol{\Gamma}_{g,n}(\Sigma_q)} \frac{w_{B,O}^X(\vec{\Gamma})}{\mathrm{Aut}(\vec{\Gamma})}. \tag{4}$$

The only difference between $w_{B,O}^X(\vec{\Gamma})$ and $w_S^{\mathbf{P}}(\vec{\Gamma})$ in Eq. (2) is that the ordinary leaf term $(\check{\mathcal{L}}^{\mathbf{P}})_k^\alpha(l_j) = -\frac{1}{\sqrt{2}} \theta_k^\alpha(p_j)$ is replaced by its integral

$$(\check{\mathcal{L}}^O)_k^\alpha(l_j) = -\frac{1}{\sqrt{2}} \int^{X_j} \theta_k^\alpha \tag{5}$$

The first step to deal with the A-model side in Theorem 28 is to reduce it to closed descendant Gromov-Witten invariants by the technique of localization, as done in [28, Proposition 3.4] and [33, Proposition 3.1, 3.2].

By this localization formula, we have the graph sum formula

$$F_{g,n}^{X,L,f} = \sum_{\vec{\Gamma}\in\mathbf{\Gamma}(V(X))} \frac{w^{\hat{X}}_{A,O}(\vec{\Gamma})}{\mathrm{Aut}(\vec{\Gamma})}. \tag{6}$$

The only difference between $w^{\hat{X}}_{A,O}(\vec{\Gamma})$ and $w^{\mathbf{u}}_{F,\bullet}(\vec{\Gamma})$ in Eq. (1) is that the ordinary leaf term $(\mathcal{L}^{\mathbf{u}}_{d})^{\alpha}_{k}(l_j)$ is replaced by

$$(\mathcal{L}^{O})^{\alpha}_{k}(l_j) = [z^k](\sum_{\beta,\gamma=1}^{N} \left(\tilde{\xi}^{\beta}(z, X_j) S^{\hat{\gamma}}_{\hat{\beta}}(z)\right)_+ R(-z)_{\gamma}^{\beta}). \tag{7}$$

Roughly speaking, $\tilde{\xi}^{\beta}(z, X)$ is the generating function counting 1 interior-pointed holomorphic disks mapped to (X, L) with no curve class but all winding numbers. The class ϕ^{β} is inserted in the interior. In order to compare the graph sum formulae (4) and (6), we need to identify $\mathbf{\Gamma}(V(X))$ and $\mathbf{\Gamma}(\mathbf{\Sigma}_q)$ first, and then we will identify the contribution from each graph $w^{\hat{X}}_{A,O}(\vec{\Gamma})$ and $w^{X}_{B,O}(\vec{\Gamma})$. The sets $\mathbf{\Gamma}(V(X))$ and $\mathbf{\Gamma}(\mathbf{\Sigma}_q)$ are just sets of stable decorated graphs, and the part of the decoration that depends on $V(X)$ or $\mathbf{\Sigma}_q$ is the labeling of a vertex by a canonical basis of the Frobenius algebra $V(X)$ for $\mathbf{\Gamma}(V(X))$, or a ramification point of x in the case of $\mathbf{\Gamma}(\mathbf{\Sigma}_q)$. The mirror theorem of semi-positive toric manifolds [36, 38, 50, 51], or later of semi-positive toric orbifolds [17], says the following.

Theorem 29

$$\mathrm{Jac}(\widetilde{W}) \cong QH^{*}_{\mathbb{T}'_f}(X)$$

in the small phase space $\tau \in H^2(X)$ and under the closed mirror map (3).

The Jacobian ring is

$$\mathrm{Jac}(\widetilde{W}) = \frac{\mathbb{C}[X^{\pm}, Y^{\pm}, Z^{\pm}]}{\langle \frac{\partial \widetilde{W}}{\partial X}, \frac{\partial \widetilde{W}}{\partial Y}, \frac{\partial \widetilde{W}}{\partial Z} \rangle}.$$

It is a Frobenius algebra for given q. Here we refrain from saying that it is a Frobenius *manifold*, i.e. we do not give a metric and discuss the flatness condition. This theorem already identifies the canonical basis of both sides.

The canonical basis of left hand side $\mathrm{Jac}(\widetilde{W})$ consists of functions taking value 1 on one critical point of $\widetilde{W}_{\mathbb{T}'}$ and vanish on other critical points. A critical point (X_0, Y_0, Z_0), by direct calculation in [32], is the solution to the following equation

$$H(X, Y) = 0, \quad \frac{\partial H}{\partial Y} = 0, \quad ZX\frac{\partial H}{\partial X} = 1.$$

We see the critical points of $\widetilde{W}$ has a 1-to-1 correspondence to the ramification points of $x : \Sigma_q \to \mathbb{C}^*$.

Once we identify the set of stable decorated graphs, after looking at the weights (1) and (2) (the ordinary leaf terms should be replaced by (7) and (5), as discussed before), we need to show the following:

- Show that $R_\beta^\alpha(-z) = \check{R}_\beta^\alpha(z)$. Notice that this matches vertices, edges and dilaton leafs. Both R and $\check{R}$ are obtained through the decomposition of S-matrices. Dubrovin and Givental's results [18, 37, 40] on this decomposition ensures the uniqueness of R-matrices up to a constant matrix, which can be fixed at the large radius limit ($q = 0$). When $q = 0$, A-side R is computed by the quantum Riemann-Roch [16, 64] and B-side $\check{R}$ is computed by direct integral which produces triple Gamma functions [31], in which we show that they match.
- Show that open leafs (7) and (5) match. By localization, $(\widetilde{\xi}^\beta(z, X_j)S_{\overline{\beta}}^{\hat{\gamma}}(z))_+$ in Eq. (7) is the generating functions of $\hat{\phi}_\gamma(\tau)$ interiorly inserted disk invariants with all winding numbers and in all possible curve classes. It is shown in [28, 33] that this corresponds to $\xi^\gamma(z, X)$, which is the B-model counting of disk invariants with canonical basis inserted.

4.6 Remarks on Orbifolds

Once we adopt the orbifold Gromov-Witten invariants [1, 2, 15], there is no essential difference to state the BKMP conjecture when the toric Calabi-Yau threefold is a toric orbifold in the sense of [8]. When the defining polytope contains a triangle with areas larger than $\frac{1}{2}$, it defines a toric orbifold (with non-trivial orbifold structure). Some examples in Fig. 2 are orbifolds.

In this paper, all quotients of a smooth variety by a finite group are stacky. Let $\mathcal{X}_1 = \mathbb{C}^3/\mathbb{Z}_3$ and $\mathcal{X}_2 = \mathbb{C}^2/\mathbb{Z}_3 \times \mathbb{C}$. For $\mathcal{X}_1$, the generator of $\mathbb{Z}_3$ acts diagonally; while for $\mathcal{X}_2$ the generator of $\mathbb{Z}_3$ acts on each piece of $\mathbb{C}^2$ with opposite non-trivial weights. The Aganagic-Vafa Lagrangian brane L is stacky for $\mathcal{X}_2$ (Fig. 5).

The mirror curves are

$$\mathcal{X}_1 : \quad X^3Y^{-1-3f} + Y + 1 + qXY^{-f} = 0,$$

$$\mathcal{X}_2 : \quad XY^{-f} + Y^3 + 1 + q_1Y + q_2Y^2 = 0.$$

When the Aganagic-Vafa brane L is not stacky, the orbifold BKMP conjecture is conjectured by [12]. It takes the same form as in Theorem 28. One needs to make some adjustment for gerby legs (stacky Lagrangian), as in [31, 32].

The topological vertex algorithm works efficiently for smooth toric Calabi-Yau threefolds [5, 53, 54, 56] and is extended to hard Lefschetz orbifolds [60–62, 69]. However, this algorithm fails when the orbifold $\mathcal{X}$ is non-hard Lefschetz. The affine orbifold $\mathcal{X}_1 = \mathbb{C}^3/\mathbb{Z}_3$ is non-hard Lefschetz—the only vertex in the toric

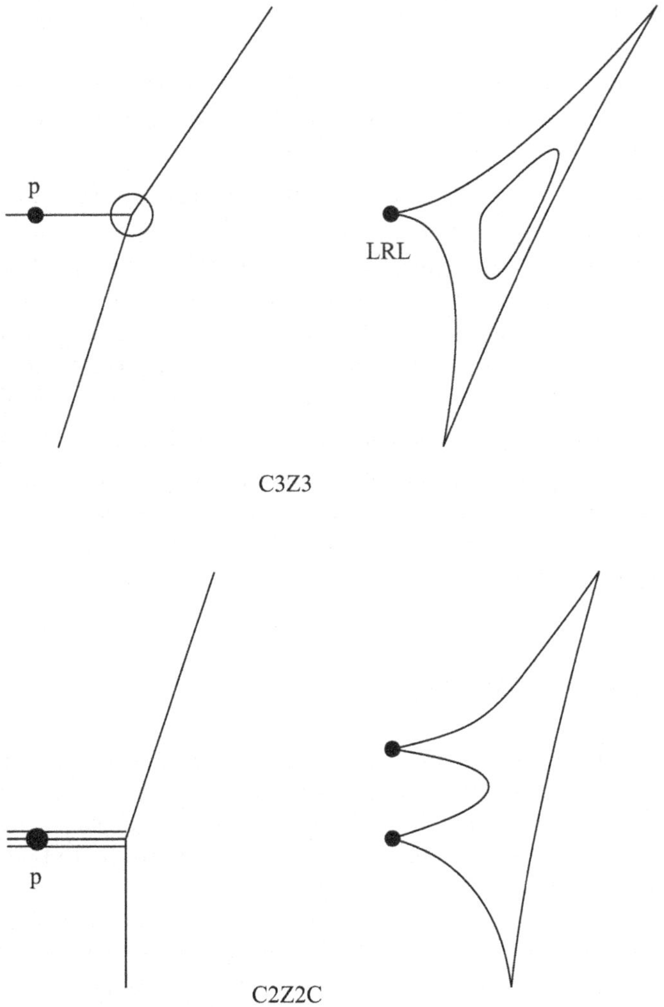

Fig. 5 Toric graph and mirror curves of $\mathbb{C}^3/\mathbb{Z}_3$ and $\mathbb{C}^2/\mathbb{Z}_3 \times \mathbb{C}$

graph corresponds to a higher genus part of the mirror curve. Thus orbifold BKMP conjecture is powerful in the sense that it first provides an effective algorithm.

The case of affine toric Calabi-Yau threefolds ($\mathbb{C}^3/G$ for a Calabi-Yau action by the abelian group G) is proved in [31], and the general toric Calabi-Yau 3-orbifolds is proved in [32]. For affine cases, a particular complication compared to smooth cases is the A-side orbifold Riemann-Roch calculation [29, 64]. For a general toric Calabi-Yau 3-orbifold, one cannot rely on Givental's quantization [39, 40] since his technique is restricted to the smooth situation if not modified extensively. The paper [32] uses Zong's thesis [70], which relies on Teleman's groundbreaking work [63].

There is also an orbifold version of Bouchard-Mariño formula [13]. As shown in [13], it is the large framing limit of the BKMP conjecture for an affine toric Calabi-Yau 3-orbifold. It should also be the large radius limit of an all genera mirror symmetry statement about an orbifold $\mathbb{P}^1$. However this is not addressed in any literature as far as the author knows.

5 Modularity of the Topological Recursion and Its Application

In this section, we will briefly review the modular invariance of $\omega_{g,n}$, and then as an application, illustrate how BKMP remodeling conjecture implies the modularity of Gromov-Witten invariants through an example. The modular invariance is a property emanating from the modular transformation of the fundamental differential $B(p,q)$ on an actual Riemann surface, and thus it is not a feature of formal spectral curves.

5.1 *Modular Invariance of Fundamental Normalized Differentials of the Second Kind*

Let (Σ, x, B) be a spectral curve, and $\overline{\Sigma}$ be its compactification. We fix two sets of Torelli markings

$$(A_k, B_k), (A'_k, B'_k), \ k = 1, \ldots, \mathfrak{g}$$

on $\overline{\Sigma}$. They differ by an $\mathrm{Sp}(2\mathfrak{g}; \mathbb{Z})$ transformation

$$\begin{pmatrix} A \\ B \end{pmatrix} = \begin{pmatrix} a & b \\ c & d \end{pmatrix} \begin{pmatrix} A' \\ B' \end{pmatrix},$$

where a, b, c, d are $\mathfrak{g} \times \mathfrak{g}$ matrices, and $\begin{pmatrix} a & b \\ c & d \end{pmatrix} \in \mathrm{Sp}(2\mathfrak{g}; \mathbb{Z})$. Let $\theta_k, \ k = 1, \ldots, \mathfrak{g}$ be linearly independent holomorphic forms on $\overline{\Sigma}$ given by the Torelli marking (A_k, B_k), i.e.

$$\int_{A_i} \theta_j = \delta_{ij}.$$

The period matrix τ_{ij} is given by

$$\tau_{ij} = \int_{B_j} \theta_i.$$

We know $\mathrm{Im}(\tau) > 0$ (positive definite), and $\tau_{ij} = \tau_{ji}$.

Define the modified cycles

$$A_i(\tau) = A_i - \sum_j \kappa_{ij} B_j(\tau), \quad B_i(\tau) = B_i - \sum_j \tau_{ij} A_j.$$

Here

$$\kappa_{ij}(\tau, \overline{\tau}) = \frac{1}{\overline{\tau} - \tau}$$

is a $\mathfrak{g} \times \mathfrak{g}$ matrix function of τ (not holomorphic). As a convention in this section, we denote the fundamental differential associated to the A-cycles A_i by B_A, and the fundamental differential associated to the modified A-cycles $A_i(\tau)$ by $B_{A(\tau)}$.

By direct calculation, Eynard-Orantin show that in [24]

$$B_{A(\tau)} = B_A + 2\pi\sqrt{-1} \sum_{i,j=1}^{\mathfrak{g}} \theta_i \kappa(\tau, \overline{\tau}) \theta_j.$$

They also show that

$$B_{A'} = B_A + 2\pi\sqrt{-1} \sum_{i,j} \theta_i \hat{\kappa}_{ij}(\tau) \theta_j,$$

where $(\hat{\kappa}_{ij}) = bJ$ and $J = (d - \tau b)^{-1}$. Here τ' is the period matrix fixed by the Torelli marking (A'_k, B'_k).

$$\tau'_{ij} = \int_{B'_j} \theta'_i, \quad \int_{A'_j} \theta'_j = \delta_{ij}.$$

We have

$$\tau' = \frac{\tau a - c}{d - \tau b}, \quad \theta'_i = \sum_{ij} J_j \theta_j.$$

The fact that

$$J^t \kappa(\tau') J + \hat{\kappa}(\tau) = \frac{1}{\overline{\tau} - \tau}$$

implies

$$B_{A'(\tau')} = B_{A(\tau)}.$$

Proposition 30 (Eynard-Orantin) *Given any Torelli marking* (A_k, B_k) *for* $k = 1, \ldots, \mathfrak{g}$, *the modified fundamental differential* $B_{A(\tau)}$ *given by the modified Torelli marking* $(A_k(\tau), B_k(\tau))$ *is independent of the choice of* (A_k, B_k).

This property implies that given a fixed spectral curve $\boldsymbol{\Sigma}$, we have a preferred choice of the fundamental differential $B_{A(\tau)}$ independent of the choice of the A-cycles. We denote this by $\widetilde{B}$. Moreover, under the limit $\mathrm{Im}\tau \to \infty$, $\widetilde{B} \to B_A$.

From the explicit expression of the Eynard-Orantin recursion (Definition 6), for any spectral curve $\boldsymbol{\Sigma}$, we can define its modified B-model invariants $\widetilde{\omega}_{g,n}$ based on this modified fundamental differential $\widetilde{B}$.

5.2 *Modularity of* $\mathcal{O}_{\mathbb{P}^2}(-3)$*: An Example*

The modular invariance of $\widetilde{B}$ and $\widetilde{\omega}$, together with the BKMP remodeling conjecture 28, implies the modularity of the Gromov-Witten invariants of toric Calabi-Yau 3-(orbi)folds. This is a long-expected property of GW invariants. It follows naturally from the modularity of mirror curves from the view point of the remodeling conjecture. We illustrate by an example.

5.2.1 Family of Mirror Curves

Let $X = \mathcal{O}_{\mathbb{P}^2}(-3)$. Its fan is the cone over the defining polytope Δ, as shown in Fig. 4 in Sect. 4.5.

Its secondary stacky fan $\mathfrak{S}$ is a complete fan in $\mathbb{R}$, as shown in Fig. 6. The generators of is 1-cones are

$$b_1 = 1, \quad b_2 = 1, \quad b_3 = 1, \quad b_4 = -3.$$

The toric orbifold $\mathcal{M}_B \cong \mathbb{P}(1, 3)$ defined by $\mathfrak{S}$ is the moduli space of the B-model, or conjecturally, is the stringly Kähler moduli space of the mirror A-model on X. Denote the stacky torus fixed point by $\mathfrak{p}_{\mathrm{orb}}$ and the non-stacky smooth torus fixed point by $\mathfrak{p}_{\mathrm{LRL}}$.

We now define the following extended secondary fan $\widetilde{\mathfrak{S}}$ as a complete fan in $\mathbb{R}^3$ as in Fig. 7. The generators of its 1-cones are

$$\widetilde{b}_1 = (0, 0, 1), \quad \widetilde{b}_2 = (-1, 0, 1), \quad \widetilde{b}_3 = (0, -1, 1), \quad \widetilde{b}_4 = (-1, -1, -3),$$
$$\widetilde{b}_5 = (1, 1, 0), \quad \widetilde{b}_6 = (-2, 1, 0), \quad \widetilde{b}_7 = (1, -2, 0).$$

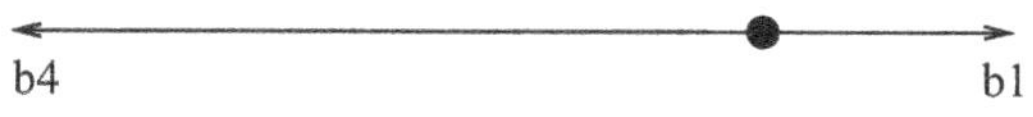

Fig. 6 The secondary fan of $\mathcal{O}_{\mathbb{P}^2}(-3)$

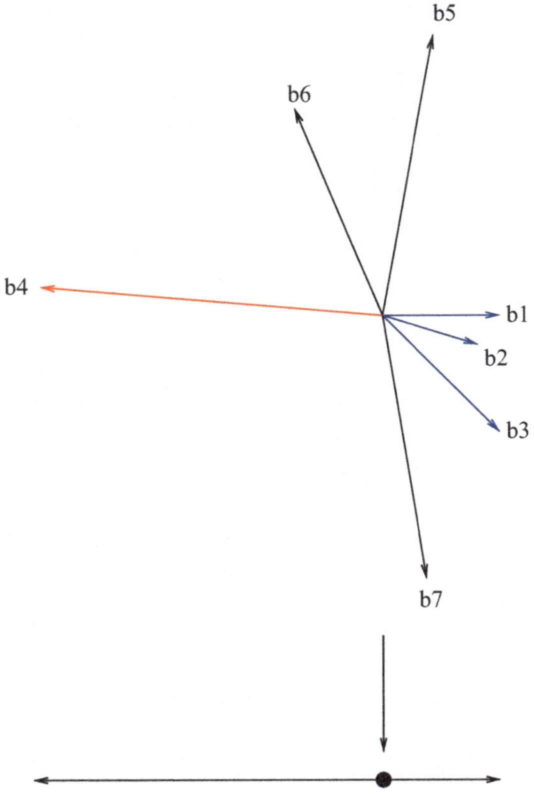

Fig. 7 The extended secondary fan of $\mathcal{O}_{\mathbb{P}^2}(-3)$. The third coordinates of the generators $\widetilde{b}_i$ are the same if they are in the same color. The rays $\widetilde{b}_5, \widetilde{b}_6, \widetilde{b}_7$ form the toric graph of $\mathbb{C}^3/\mathbb{Z}_3$. There is an obvious fan map $\widetilde{\mathfrak{S}} \to \mathfrak{S}$

The top dimensional cones are spanned by $\widetilde{b}_i$ where i ranges from the following index sets

$$\{4,5,6\},\ \{4,6,7\},\ \{4,5,7\},\ \{5,1,2\},\ \{5,1,3\},$$
$$\{6,1,2\},\ \{6,2,3\},\ \{7,2,3\},\ \{7,1,3\},\ \{1,2,3\}.$$

The 2-cones are faces of 3-cones. We denote the toric orbifold associated to the fan $\widetilde{\mathfrak{S}}$ by $\widetilde{\mathcal{M}}_B$ (Fig. 7).

There is an obvious fan map $\widetilde{\mathfrak{S}} \to \mathfrak{S}$ which forgets the first two factors. It induces a toric map $\pi : \widetilde{\mathcal{M}}_B \to \mathcal{M}_B$. The fiber $\pi^{-1}(\mathfrak{p})$ for $\mathfrak{p} \neq \mathfrak{p}_{LRL}$ is a toric orbifold defined by the stacky fan given by $\widetilde{b}_5, \widetilde{b}_6, \widetilde{b}_7$ (on $\mathbb{R}^2$). It is isomorphic to $\mathbb{P}^2/\mathbb{Z}_3$. Over the smooth torus fixed point, the fiber $\pi^{-1}(\mathfrak{p}_{\mathrm{LRL}})$ is three $\mathbb{P}^2$ intersecting along three $\mathbb{P}^1$ with normal crossing singularities (see Fig. 8). If one intersects the fan $\widetilde{\mathfrak{S}}$ by a vertical plane, at different horizontal position, we get the fan of each fiber toric surface. See Fig. 8.

We understand X, Y, q as characters in $\mathrm{Hom}(\mathbb{T}_B, \mathbb{C}^*) = M_B := N_B^\vee$, where $\mathbb{T}_B$ is the open dense 3-torus in $\widetilde{\mathcal{M}}_B$, and $N_B \cong \mathbb{Z}^3$ is the lattice that $\widetilde{b}_i$ belong to. Then

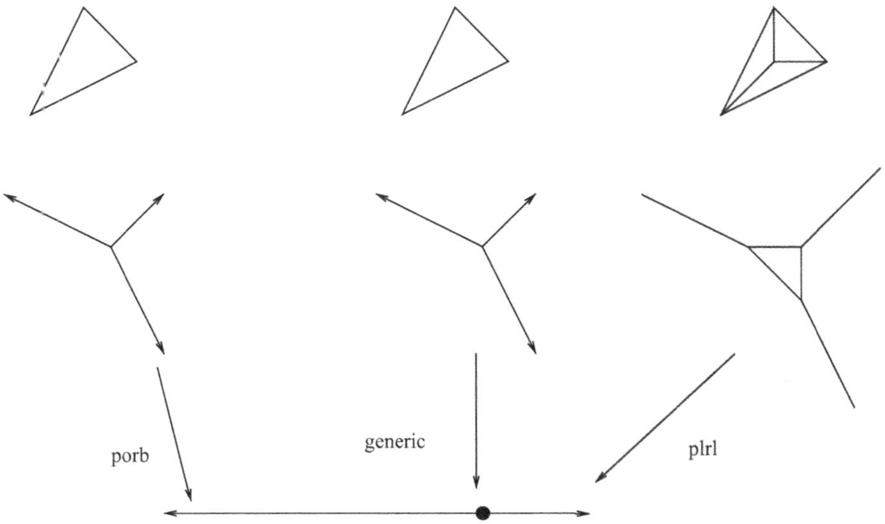

Fig. 8 Over $\mathcal{M}_B$, we have a family of toric surfaces given by π. When $\mathfrak{p} \neq p_{\mathrm{LRL}}$, the fiber $\pi^{-1}(\mathfrak{p}) \cong \mathbb{P}^2/\mathbb{Z}_3$, given by the stacky fan spanned by $\widetilde{b}_5, \widetilde{b}_6, \widetilde{b}_7$. Over p_{LRL}, the toric surface degenerates to a normal crossing of three $\mathbb{P}^2$, as shown by the "fan" and the polytope. The first rows are polytopes and the second rows are fans for fiber toric surfaces at different points in $\mathcal{M}_B$

X, Y, q corresponds to $(1, 0, 0)$, $(0, 1, 0)$ and $(0, 0, 1)$ in M_B respectively. They are sections of a line bundle $\widetilde{\mathcal{L}} = \mathcal{O}_{\widetilde{\mathcal{M}}_B}(\sum_{i=1}^{6} D_i)$. We define a section $H \in \Gamma(\widetilde{\mathcal{L}})$

$$H = X + Y + 1 + qX^{-1}Y^{-1}.$$

We define the compactified global mirror curve $\widetilde{\Sigma} = H^{-1}(0) \subset \widetilde{\mathcal{M}}_B$. It is parametrized over $\mathcal{M}_B$ by $\pi_{\widetilde{\Sigma}} = \pi|_{\widetilde{\Sigma}} : \widetilde{\Sigma} \to \mathcal{M}_B$. For any $\mathfrak{p} \in \mathcal{M}_B$, the fiber $\pi_{\widetilde{\Sigma}}^{-1}(\mathfrak{p})$ is a compact (possibly singular) curve. Let $\mathcal{M}_{B,0}$ be the part of $\mathcal{M}_B$ where the fiber curves are smooth. As shown in Fig. 9, $\mathfrak{p}_{\mathrm{LRL}} \notin \mathcal{M}_{B,0}$ and $\mathfrak{p}_{\mathrm{orb}} \in \mathcal{M}_{B,0}$. There is another point other than $\mathfrak{p}_{\mathrm{LRL}}$ not in $\mathcal{M}_{B,0}$. The fiber has one nodal singularity. This point is called the conifold point $\mathfrak{p}_{\mathrm{con}}$. Thus $\mathcal{M}_{B,0} = \mathcal{M}_B \setminus \{\mathfrak{p}_{\mathrm{LRL}}, \mathfrak{p}_{\mathrm{con}}\}$.

5.2.2 Modularity

The monodromies of the Gauss-Manin connection on the local system $H^1(\Sigma_q; \mathbb{C}) \cong H_1(\Sigma_q; \mathbb{C})$ over $\mathcal{M}_{B,0}$ around $\mathfrak{p}_{\mathrm{LRL}}$ and p_{con} (as computed in [6]) gives the *modular group* Γ of this local system. It is a normal subgroup of the symplectic group $SL(2; \mathbb{Z})$ of index 3.

Over $\mathcal{M}_{B,0}$, we have a smooth family of mirror curves, and the coordinates X, Y are well defined. So X, Y are invariant under the action of the modular group Γ. If

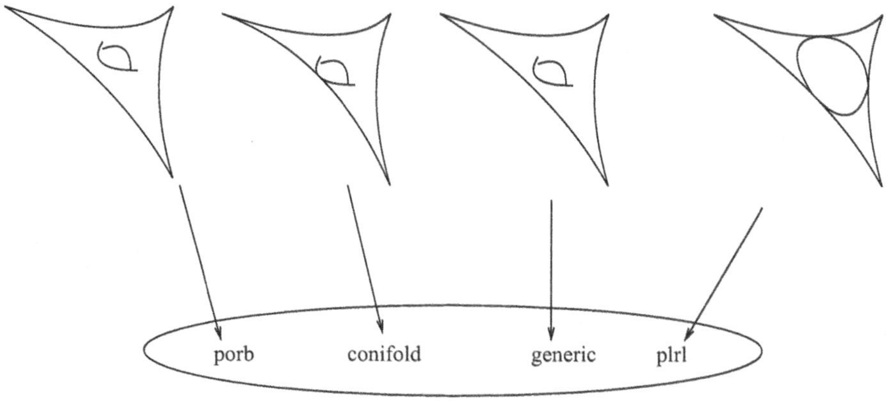

Fig. 9 Over $\mathcal{M}_B$, we have a family of compactified mirror curves $\widetilde{\Sigma}$. At p_{con} and p_{LRL} the mirror curves are singular. As before, the sharp ends in the mirror curve picture are the punctures on the mirror curve. After compactification, they become compact curves in $\pi^{-1}(\mathfrak{p})$. All puncture points are smooth

we use the modified fundamental differential $\widetilde{B}$ to define the higher genus B-model invariants $\widetilde{\omega}_{g,n}$, then they are all well-defined global invariants on $\widetilde{\Sigma}|_{\mathcal{M}_{B,0}}$. In other words, if one uses Torelli-marking-sensitive coordinates τ to express these $\widetilde{\omega}_{g,n}$, they are invariant under the action of the modular group Γ.

Using the mirror map (3) we define the open potential in the holomorphic polarization under A-model flat coordinates.

$$\widetilde{F}^{X,L,f}_{g,n}(\hat{X}_1,\ldots,\hat{X}_n,Q)=\int^{X_1}\cdots\int^{X_n}\widetilde{\omega}_{g,n}.$$

The A-model coordinate $Q = Q(\mathfrak{p})$ is well-defined around the LRL point, and is related to B-model coordiante q around the LRL point under the closed mirror map. The open potential $\widetilde{F}^{X,L,f}_{g,n}$ has non-holomorphic dependence on Q, in contrast to the name "holomorphic polarization". Under the holomorphic limit

$$\lim_{\mathrm{Im}\tau\to\infty}\widetilde{\omega}_{g,n}=\omega_{g,n}.$$

With the BKMP remodeling conjecture (Theorem 28), we have for $2g-2+n>0$ and $n\geq 1$

$$\lim_{\mathrm{Im}\tau\to\infty}\widetilde{F}^{X,L,f}_{g,n}=F^{X,L,f}_{g,n}. \tag{8}$$

If one defines

$$\widetilde{F}^X_g=\frac{1}{2-2g}\sum_{p_0\in I_{\Sigma_q}}\mathrm{Res}_{p=p_0}\widetilde{\omega}_{g,1}(p)\widetilde{\Phi}(p),$$

then for $g \geq 2$

$$\lim_{\mathrm{Im}\tau \to \infty} \widetilde{F}_g^X = F_g^X.$$

The potential $\widetilde{F}_{g,n}^{X,L,f}$ and $\widetilde{F}_g^X$ are globally defined over $\mathcal{M}_B$, although their expansions in Q are only defined around $\mathfrak{p}_{\mathrm{LRL}}$ since Q is a flat coordinate around $\mathfrak{p}_{\mathrm{LRL}}$. Their dependence on $\mathfrak{p} \in \mathcal{M}_B$ is not holomorphic.

Theorem 31 *The Gromov-Witten potential F_g^X can be completed into an analytic function $\widetilde{F}_g^X$, which under the mirror map* (3) *is globally defined on $\mathcal{M}_B$. If $\mathcal{M}_B$ is a modular curve, e.g. when $X = \mathcal{O}_{\mathbb{P}^2}(-3)$, the function $\widetilde{F}_g^X$ is a function of τ and modular invariant.*

Remark 32 The theorem also holds for unstable cases $(g, n) = (0, 0), (0, 1), (0, 2), (1, 0)$ but we need to treat these cases separately. We did not very clearly spell out what this "anti-holomorphic completion" is, as it should be stronger than (8). Indeed, $\widetilde{\omega}_{g,n}$ can be written as a polynomial in $\frac{1}{\mathrm{Im}\tau}$ with holomorphic coefficients [24, 26]. The lowest order of $\mathrm{Im}\tau$ is $2-2g$, and each coefficient in non-holomorphic terms are given by combinations of $\omega_{g',n}$, $g' < g$ in a graph sum formula. The BKMP conjecture allows us to say the same—$\widetilde{F}_g^X$ is a polynomial in $\frac{1}{\mathrm{Im}\tau}$ with the highest power term $(\frac{1}{\mathrm{Im}\tau})^{2g-2}$ and holomorphic term F_g^X. Each coefficient in the non-holomorphic terms are given by $F_{g'}^X$ in a graph sum formula where $g' < g$.

Remark 33 One could use the modularity property to compute higher genus Gromov-Witten invariants for certain toric Calabi-Yau 3-(orbi)folds, thanks to the complete structure theorem of almost holomorphic modular forms. See [6, 7, 66] for numerical calculations and closed formulae for some $\widetilde{F}_g^X$ and F_g^X.

Acknowledgements The author would like to thank Chiu-Chu Melissa Liu and Zhengyu Zong for the exciting and pleasant collaboration—this survey is based on the joint works with them. He also thanks Yongbin Ruan for helpful discussion. BF is partially supported by a start-up grant at Peking University and the Recruitment Program for Global Experts in China.

References

1. Abramovich, D., Graber, T., Vistoli, A.: Algebraic orbifold quantum products. In: Orbifolds in Mathematics and Physics (Madison, WI, 2001). Contemporary Mathematics, vol. 310, pp. 1–24. American Mathematical Society, Providence (2002)
2. Abramovich, D., Graber, T., Vistoli, A.: Gromov-Witten theory of Deligne-Mumford stacks. Amer. J. Math. **130**(5), 1337–1398 (2008)
3. Aganagic, M., Vafa, C.: Mirror symmetry, D-branes and counting holomorphic discs (2000). arXiv:hep-th/0012041
4. Aganagic, M., Klemm, A., Vafa, C.: Disk instantons, mirror symmetry and the duality web. Z. Naturforsch. **A 57**, 1–28 (2002)

5. Aganagic, M., Klemm, A., Mariño, M., Vafa, C.: The topological vertex. Commun. Math. Phys. **254**(2), 425–478 (2005)
6. Aganagic, M., Bouchard, V., Klemm, A.: Topological strings and (almost) modular forms. Commun. Math. Phys. **277**(3), 771–819 (2008)
7. Alim, M., Scheidegger, E., Yau, S.-T., Zhou, J.: Special polynomial rings, quasi modular forms and duality of topological strings. Adv. Theor. Math. Phys. **18**(2), 401–467 (2014)
8. Borisov, L., Chen, L., Smith, G.: The orbifold Chow ring of toric Deligne-Mumford stacks. J. Am. Math. Soc. **18**(1), 193–215 (2005)
9. Borot, G., Eynard, B., Mulase, M., Safnuk, B.: A matrix model for simple Hurwitz numbers, and topological recursion. J. Geom. Phys. **61**(2), 522–540 (2011)
10. Bouchard, V., Mariño, M.: Hurwitz numbers, matrix models and enumerative geometry. In: From Hodge Theory to Integrability and TQFT tt*-Geometry, pp. 263–283. Proceedings of Symposia in Pure Mathematics, vol. 78. American Mathematical Society, Providence (2008)
11. Bouchard, V., Klemm, A., Mariño, M., Pasquetti, S.: Remodeling the B-model. Commun. Math. Phys. **287**(1), 117–178 (2009)
12. Bouchard, V., Klemm, A., Mariño, M., Pasquetti, S.: Topological open strings on orbifolds. Commun. Math. Phys. **296**(3), 589–623 (2010)
13. Bouchard, V., Hernández Serrano, D., Liu, X., Mulase, M.: Mirror symmetry for orbifold Hurwitz numbers. J. Differ. Geom. **98**(3), 375–423 (2014)
14. Chen, L.: Bouchard-Klemm-Marino-Pasquetti conjecture for $\mathbb{C}^3$ (2009). arXiv:0910.3739
15. Chen, W., Ruan, Y.: Orbifold Gromov-Witten theory. In: Orbifolds in Mathematics and Physics. Contemporary Mathematics, vol. 310, pp. 25–85. American Mathematical Society, Providence (2002)
16. Coates, T., Givental, A.: Quantum Riemann-Roch, Lefschetz and Serre. Ann. Math. (2) **165**(1), 15–53 (2007)
17. Coates, T., Corti, A., Iritani, H., Tseng, H.-H.: A mirror theorem for toric stacks. Compos. Math. **151**(10), 1878–1912 (2015)
18. Dubrovin, B.: Geometry of 2D topological field theories. In: Integrable Systems and Quantum Groups (Montecatini Terme, 1993). Lecture Notes in Mathematics, vol. 1620, pp. 120–348. Springer, Berlin (1996)
19. Dunin-Barkowski, P., Orantin, N., Shadrin, S., Spitz, L.: Identification of the Givental formula with the special curves topological recursion procedure. Commun. Math. Phys. **328**(2), 669–700 (2014)
20. Ekedahl, T., Lando, S., Shapiro, M., Vainshtein, A.: Hurwitz numbers and intersections on moduli spaces of curves. Invent. Math. **146**(2), 297–327 (2001)
21. Eynard, B.: Invariants of spectral curves and intersection theory of moduli spaces of complex curves. Commun. Number Theory Phys. **8**(3), 541–588 (2014)
22. Eynard, B.: Intersection number of spectral curves (2011). arXiv:1104.0176
23. Eynard, B., Nicolas, O.: Weil-Petersson volume of moduli spaces, Mirzakhani's recursion and matrix models (2007). arXiv:0705.3600
24. Eynard, B., Orantin, N.: Invariants of algebraic curves and topological expansion. Commun. Number Theory Phys. **1**(2), 347–452 (2007)
25. Eynard, E., Orantin, N.: Computation of open Gromov-Witten invariants for toric Calabi-Yau 3-folds by topological recursion, a proof of the BKMP conjecture. Commun. Math. Phys. **337**(2), 483–567 (2015)
26. Eynard, B., Marño, M., Orantin, N.: Holomorphic anomaly and matrix models. J. High Energy Phys. (6) **2007**, 058, 20 (2007)
27. Eynard, B., Mulase, M., Safnuk, B.: The Laplace transform of the cut-and-join equation and the Bouchard-Mariño conjecture on Hurwitz numbers. Publ. Res. Inst. Math. Sci. **47**(2), 629–670 (2011)
28. Fang, B., Liu, C.-C.M.: Open Gromov-Witten invariants of toric Calabi-Yau 3-folds. Commun. Math. Phys. **323**(1), 285–328 (2013)
29. Fang, B., Liu, C.-C.M., Zong, Z.: Equivariant Gromov-Witten theory of affine smooth toric Deligne-Mumford stacks. Int. Math. Res. Not. IMRN **2016**(7), 2127–2144 (2016)

30. Fang, B., Liu, C.-C.M., Zong, Z.: The Eynard-Orantin recursion and equivariant mirror symmetry for the projective line. Geom. Topol. **21**(4), 2049–2092 (2017)
31. Fang, B., Liu, C.-C.M., Zong, Z.: All genus open-closed mirror symmetry for affine toric Calabi-Yau 3-orbifolds (2013). arXiv:1310.4818
32. Fang, B., Liu, C.-C.M., Zong, Z.: On the remodeling conjecture for toric Calabi-Yau 3-orbifolds (2016). arXiv:1604.07123
33. Fang, B., Liu, C.-C.M., Tseng, H.-H.: Open-closed Gromov-Witten invariants of 3-dimensional Calabi-Yau smooth toric DM stacks (2012). arXiv:1212.6073
34. Fay, J.D.: Theta Functions on Riemann Surfaces. Lecture Notes in Mathematics, vol. 352. Springer, Berlin (1973)
35. Givental, A.B.: Homological geometry and mirror symmetry. In: Proceedings of the International Congress of Mathematicians (Zrich, 1994), vol. 1, 2, pp. 472–480. Birkhuser, Basel (1995)
36. Givental, A.B.: Equivariant Gromov-Witten invariants. Int. Math. Res. Not. **1996**(13), 613–663 (1996)
37. Givental, A.: Elliptic Gromov-Witten invariants and the generalized mirror conjecture. In: Integrable Systems and Algebraic Geometry (Kobe/Kyoto, 1997), pp. 107–155. World Scientific Publishing, River Edge (1998)
38. Givental, A.B.: A mirror theorem for toric complete intersections. In: Topological Field Theory, Primitive Forms and Related Topics (Kyoto, 1996). Progress in Mathematics, vol. 160, pp. 141–175. Birkhäuser, Boston (1998)
39. Givental, A.B.: Gromov-Witten invariants and quantization of quadratic hamiltonians. Mosc. Math. J. **1**(4), 551–568 (2001)
40. Givental, A.B.: Semisimple Frobenius structures at higher genus. Int. Math. Res. Not. **2001**(23), 1265–1286 (2001)
41. Graber, T., Pandharipande, R.: Localization of virtual classes. Invent. Math. **135**(2), 487–518 (1999)
42. Graber, T., Vakil, R.: Hodge integrals and Hurwitz numbers via virtual localization. Compos. Math. **135**(1), 25–36 (2003)
43. Harvey, R., Lawson, H.B. Jr.: Calibrated geometries. Acta Math. **148**, 47–157 (1982)
44. Hori, K., Vafa, C.: Mirror symmetry (2000). arXiv:0002222
45. Iritani, H.: Convergence of quantum cohomology by quantum Lefschetz. J. Reine Angew. Math. **610**, 29–69 (2007)
46. Katz, S., Liu, C.-C.M.: Enumerative geometry of stable maps with Lagrangian boundary conditions and multiple covers of the disc. Adv. Theor. Math. Phys. **5**(1), 1–49 (2001)
47. Klemm, A., Zaslow, E.: Local mirror symmetry at higher genus. In: Winter School on Mirror Symmetry, Vector Bundles and Lagrangian Submanifolds (Cambridge, MA, 1999). AMS/IP Studies in Advanced Mathematics, vol. 23, pp. 183–207. American Mathematical Society, Providence (2001)
48. Li, J., Liu, C.-C.M., Liu, K., Zhou, J.: A mathematical theory of the topological vertex. Geom. Topol. **13**(1), 527–621 (2009)
49. Lerche, W., Mayr, P.: On N=1 mirror symmetry for open type II strings (2001). arXiv:hep-th/0111113
50. Lian, B.H., Liu, K., Yau, S.-T.: Mirror principle I. Asian J. Math. **1**(4), 729–763 (1997)
51. Lian, B.H., Liu, K., Yau, S.-T.: Mirror principle III. Asian J. Math. **3**(4), 771–800 (1999)
52. Liu, C.-C.M.: Moduli of J-holomorphic curves with Lagrangian boundary conditions and open Gromov-Witten invariants for an S^1-equivariant pair (2002). arXiv:math/0211388
53. Liu, C.-C.M., Liu, K., Zhou, J.: A proof of a conjecture of Mariño-Vafa on Hodge integrals. J. Differ. Geom. **65**(2), 289–340 (2003)
54. Liu, C.-C.M., Liu, K., Zhou, J.: A formula of two-partition Hodge integrals. J. Am. Math. Soc. **20**(1), 149–184 (2007)
55. Mariño, M.: Open string amplitudes and large order behavior in topological string theory. J. High Energy Phys. **2008**(3), 060, 34 (2008)

56. Maulik, D., Oblomkov, A., Okounkov, A., Pandaripande, R.: Gromov-Witten/Donaldson-Thomas correspondence for toric 3-folds. Invent. Math. **186**(2), 435–479 (2011)
57. Mayr, P.: $\mathcal{N} = 1$ mirror symmetry and open/closed string duality. Adv. Theor. Math. Phys. **5**(2), 213–242 (2001)
58. Mulase, M., Zhang, N.: Polynomial recursion formula for linear Hodge integrals. Commun. Number Theory Phys. **4**(2), 267–293 (2010)
59. Norbury, P., Scott, N.: Gromov-Witten invariants of $\mathbb{P}^1$ and Eynard-Orantin invariants. Geom. Topol. **18**(4), 1865–1910 (2014)
60. Ross, D.: On GW/DT and Ruan's conjecture in all genus for Calabi-Yau 3-orbifolds (2014). arXiv:1409.7015v2
61. Ross, D., Zong, Z.: The Gerby Gopakumar-Mariño-Vafa formula. Geom. Topol. **17**(5), 2935–2976 (2013)
62. Ross, D., Zong, Z.: Cyclic hodge integrals and loop schur functions. Adv. Math. **285**, 1448–1486 (2015)
63. Teleman, C.: The structure of 2D semi-simple field theories. Invent. Math. **188**(3), 525–588 (2012)
64. Tseng, H.-H.: Orbifold quantum Riemann-Roch, Lefschetz and Serre. Geom. Topol. **14**(1), 1–81 (2010)
65. Zhou, J.: Topological recursions of Eynard-Orantin type for intersection numbers on moduli spaces of curves. Lett. Math. Phys. **103**(11), 1191–1206 (2013)
66. Zhou, J.: Arithmetic properties of moduli spaces and topological string partition functions of some Calabi-Yau threefolds. Thesis (Ph.D.), Harvard University (2014), 155 pp. ISBN: 978-1321-02505-7
67. Zhou, J.: Local mirror symmetry for one-legged topological vertex (2009). arXiv:0910.4320
68. Zhou, J.: Local mirror symmetry for the topological vertex (2009). arXiv:0911.2343
69. Zong, Z.: Generalized Mariño-Vafa formula and local Gromov-Witten theory of orbi-curves. J. Differ. Geom. **100**(1), 161–190 (2015)
70. Zong, Z.: Equviariant Gromov-Witten theory of GKM orbifolds (2016). arXiv:1604.07270

The Total Ancestor Potential in Singularity Theory

Todor Milanov

Contents

Abstract This is an extended version of a long lecture given on the workshop "Pedagogical workshop on B-model" held at the University of Michigan, Ann Arbor on 3–7 March 2014. The main goal is to prove that the total ancestor potential in singularity theory depends analytically on the deformation parameters.

2000 Mathematics Subject Classification 14D05, 14N35, 17B69

1 Introduction

Motivated by quantum cohomology and Gromov–Witten theory Dubrovin invented the notion of a Frobenius manifold [4]. Furthermore, he noticed that the Frobenius manifolds satisfying certain semi-simplicity condition play a key role in the theory of integrable hierarchies. This lead to the remarkable discovery that every semi-simple Frobenius manifold gives rise to an integrable hierarchy [5]. Partially motivated by Dubrovin's work, Givental discovered a certain higher-genus reconstruction formalism in Gromov–Witten (GW) theory which lead him to introduce

T. Milanov (✉)
Kavli IPMU (WPI), The University of Tokyo, Kashiwa, Chiba, Japan
e-mail: todor.milanov@ipmu.jp

E. Clader, Y. Ruan (eds.), *B-Model Gromov-Witten Theory*, Trends in Mathematics,
https://doi.org/10.1007/978-3-319-94220-9_6

the notion of a total ancestor potential in the abstract settings of an arbitrary semi-simple Frobenius manifold S (see [8]). The potential is defined for each semi-simple point $s \in S$ and it has the form

$$\widetilde{\mathcal{A}}_s(\hbar; \mathbf{t}) = \exp\Big(\sum_{g=0}^{\infty} \hbar^{g-1} \widetilde{\mathcal{F}}_s^{(g)}(\mathbf{t})\Big)$$

where $\widetilde{\mathcal{F}}_s^{(g)}(\mathbf{t})$ is a formal power series in some formal vector variables $\mathbf{t} = (t_0, t_1, t_2, \dots)$. Let us denote by $B \subset S$ the subset of non-semisimple points. It is known that B is an analytic hypersurface in S and that the coefficients of the formal power series $\widetilde{\mathcal{F}}_s^{(g)}(\mathbf{t})$ depend analytically in s for all $s \in S \setminus B$.

Givental conjectured that if S is the quantum cohomology of some compact Kahler manifold, then under the semi-simplicity assumption, the total ancestor potential of the Frobenius structure is a generating function for the so called ancestor GW invariants (see Sect. 2.4 for more details). Givental's conjecture was proved by Teleman [21] in the more general settings of semi-simple Cohomological Field Theories (CohFT).

On the other hand, most of the CohFT that we would like to compute satisfy the semi-simplicity condition only after we deform them, so in order to use Givental's higher genus reconstruction it is important to determine whether the total ancestor potential $\mathcal{A}_s(\hbar; \mathbf{t})$ of a given semi-simple Frobenius structure extends analytically through the non-semisimple locus. For example, if S is the orbit space of the Weyl group of a non-simply laced simple Lie algebra (i.e., types B, C, F, or G), then there is a natural Frobenius structure on S (see [4, 19]), but the total ancestor potential *does not* extend analytically. It is a very interesting question to determine whether the total ancestor potential of the Frobenius structures in that case has a geometric origin, i.e., it is related in some way to some CohFT of Fan-Jarvis-Ruan-Witten (FJRW) [7]. In fact, some progress in this direction was recently made by Liu-Ruan-Zhang [15].

One of the most important examples of a semi-simple Frobenius structure, that plays a crucial role in mirror symmetry, is Saito's flat structure [18]. Motivated by the classical theory of period integrals, K. Saito introduced the notion of a *primitive form*. Let S be the base of the universal unfolding of the germ of a holomorphic function $f \in \mathcal{O}_{\mathbb{C}^{n+1},0}$ with an isolated critical point at $0 \in \mathbb{C}^{n+1}$. A primitive form is the germ of a holomorphic volume form on $\mathbb{C}^{n+1}$, possibly depending on the deformation parameters $s \in S$, with some very special properties. Spelling out the precise definition is quite difficult, but the main idea is that a primitive form and its covariant derivatives with respect to the Gauss–Manin connection, provide a frame for the vanishing cohomology bundle in which the Gauss–Manin connection turns into a Dubrovin's connection. In particular, the base S inherits a natural Frobenius structure, which is always semi-simple, because the critical values provide canonical coordinates (see [11, 20]). The goal in these notes is to prove the following theorem.

Theorem 1 *Let S be the base of the universal unfolding of some $f \in \mathcal{O}_{\mathbb{C}^{n+1},0}$ equipped with a Frobenius structure corresponding to a primitive form; then the coefficients of the total ancestor potential $\mathcal{A}_s(\hbar; \mathbf{t})$ in front of the monomials in $\mathbf{t}$ and $\hbar$ extend analytically across B to analytic functions on the entire Frobenius manifold S.*

Theorem 1 motivates the following question. Given a singularity $f \in \mathcal{O}_{\mathbb{C}^{n+1},0}$ and a primitive form, can we identify the total ancestor potential of the singularity with the generating function of invariants of some CohFT. For example, if the germ f can be represented by an invertible weighted-homogeneous polynomial and the primitive form is chosen appropriately, then there is a conjecture that the appropriate CohFT is a FJRW-CohFT [7].[1]

The proof of Theorem 1 follows the argument from [17]. We will try to keep the exposition as self-contained as possible. In particular, up to some linear algebra exercises, we give an introduction to Givental's higher-genus reconstruction, define and prove the properties of the so called propagators from [2], and finally give a proof of the local Eynard–Orantin recursion [6, 16]. The only requirements for reading this text is the knowledge of a Frobenius structure (see [4]). However, it might be useful also to refer from time to time to Givental's work [10], where the period integrals were introduced and some of their most fundamental properties were established.

2 Givental's Total Ancestor Potential

Let S be a complex semi-simple Frobenius manifold and $B \subset S$ be the analytic hypersurface consisting of non-semisimple points. Motivated by Gromov–Witten theory, Givental has defined the total ancestor potential $\mathcal{A}_s(\hbar; \mathbf{q})$ of the Frobenius manifold S for every semi-simple point $s \in S \setminus B$. The goal in this section is to recall Givental's construction.

2.1 Givental's Symplectic Loop Space Formalism

Let H be a complex vector space equipped with a non-degenerate bi-linear pairing $(\ ,\ )$ and with a distinguished vector $\mathbf{1} \in H$. By definition, Givental's symplectic loop space $\mathcal{H} = H((z^{-1}))$ is the space of formal Laurent series in z^{-1} with coefficients in H, equipped with the following symplectic structure:

$$\Omega(f(z), g(z)) = \mathrm{Res}_{z=0}(f(-z), g(z))dz,$$

the residue is interpreted formally as the coefficient in front of z^{-1}.

[1]One of the pleasant outcomes of the workshop was that this conjecture was confirmed by generalizing the approach of [14].

The vector space $\mathcal{H}$ viewed as an abelian Lie algebra has a natural central extension $\mathcal{H} \oplus \mathbb{C}$ in which the symplectic form coincides with the cocycle defining the extension,

$$[v_1, v_2] := \Omega(v_1, v_2), \quad v_1, v_2 \in \mathcal{H}.$$

Since $\mathcal{H}\oplus\mathbb{C}$ is a Heisenberg Lie algebra it has a standard Fock space representations. In our case the construction is as follows. Let us fix bases $\{\phi_a\}_{a=1}^N$ and $\{\phi^a\}_{a=1}^N$ of H dual with respect to $(\ ,\ )$, then

$$\Omega(\phi^a(-z)^{-n-1}, \phi_b z^m) = \delta_{a,b}\delta_{n,m}.$$

Let us fix a sequence of formal vector variables $\mathbf{q} = (q_0, q_1, q_2, \dots)$, where $q_k = \sum_{a=1}^N q_{k,a}\phi_a$. We will be interested in the Fock space of formal power series

$$\mathbb{C}_\hbar[\![q_0, q_1 + \mathbf{1}, q_2, \dots]\!],$$

where $\mathbb{C}_\hbar$ is the field of formal Laurent series in $\hbar^{\frac{1}{2}}$. The shift by $\mathbf{1}$ is known as the *dilaton shift*. The linear operator of the Fock space representing $v \in \mathcal{H} \oplus \mathbb{C}$ will be denote by $\widehat{v}$ or $(v)^\wedge$. The representation of the Heisenberg algebra on the Fock space is uniquely defined by

$$(\phi_a z^m)^\wedge := \hbar^{\frac{1}{2}} \partial_{q_{m,a}}, \quad (\phi^a(-z)^{-m-1})^\wedge := -\hbar^{-\frac{1}{2}} q_{m,a},$$

where $1 \le a \le N$ and $m \ge 0$.

2.1.1 Quantization of Quadratic Hamiltonians

Note that the map

$$\mathcal{H} \to \mathcal{H}^*, \quad v \mapsto \Omega(\ , v)$$

induces an isomorphism of Lie algebras

$$\mathcal{H} \oplus \mathbb{C} \cong \mathcal{H}^* \oplus \mathbb{C},$$

where the RHS is the vector space of constant and linear functions on $\mathcal{H}$ and the Lie bracket is the Poisson bracket corresponding to the symplectic form Ω. On the other hand, a linear operator A on $\mathcal{H}$ is an infinitesimal symplectic transformation if and only if the map $v \mapsto Av$ is a Hamiltonian vector field. Moreover, the Hamiltonian is given by the quadratic function $h_A(v) = \frac{1}{2}\Omega(Av, v)$. Put

$$p_{m,a} = \Omega(\ , \phi_a z^m), \quad q_{m,a} = -\Omega(\ , \phi^a(-z)^{-m-1}),$$

then h_A is a quadratic expression in $p_{m,a}$ and $q_{m,a}$. We define the quantization $\widehat{A} := \widehat{h}_A$ by

$$(p_{m,a}p_{n,b})^\wedge = \hbar \partial_{q_{m,a}}\partial_{q_{n,b}},$$
$$(p_{m,a}q_{n,b})^\wedge = (q_{n,b}p_{m,a})^\wedge = q_{n,b}\partial_{q_{m,a}},$$
$$(q_{m,a}q_{n,b})^\wedge = \hbar^{-1}q_{m,a}q_{n,b}.$$

We leave it as an exercise to verify the following properties

$$[\widehat{A}, \widehat{v}\,] = (Av)^\wedge, \quad \{h_A, h_B\} = h_{[A,B]},$$

for all $v \in \mathcal{H}$ and all infinitesimal symplectic transformations A and B, where $\{\ ,\ \}$ is the Poisson bracket.

2.1.2 Quantization of Symplectic Transformations

Let us assume that the operator series

$$R(z) = 1 + R_1 z + R_2 z^2 + \cdots, \quad R_k \in \mathrm{End}(H)$$

is a symplectic transformation. It will be convenient to identify the sequence $\mathbf{q} = (q_0, q_1, \dots)$ with the series $q_0 + q_1 z + q_2 z^2 + \cdots$, then the natural action of $R(z)$ on $H[z]$ induces an action on the formal sequence: $\mathbf{q}(z) \mapsto R(z)\mathbf{q}(z)$, or in components

$$q_n \mapsto R_0 q_n + R_1 q_{n-1} + \cdots + R_n q_0.$$

Let us also define $V_{k\ell} \in \mathrm{End}(H)$ by the identity

$$\sum_{k,\ell=0}^{\infty} V_{k\ell} z^k w^\ell = \frac{R^T(z)R(w) - 1}{z + w}, \tag{1}$$

where ^{T} is transposition with respect to the bi-linear form $(\ ,\ )$. We can define formally $A(z) = \log R(z)$, so that $R(z) = e^{A(z)}$. By definition the quantization $\widehat{R} := e^{\widehat{A}}$. The action of $\widehat{R}$ on the Fock space is not well defined in general. We have the following Lemma.

Lemma 2 *If $\mathcal{F}(\hbar; \mathbf{q})$ is a formal power series in the Fock space and $\widehat{R}\mathcal{F}$ is well defined, then*

$$\widehat{R}\mathcal{F}(\hbar; \mathbf{q}) = \left(e^{\hbar V(\partial,\partial)/2}\mathcal{F}\right)(\hbar; R^{-1}\mathbf{q}),$$

where $V(\partial,\partial)$ is the following 2nd order differential operator

$$V(\partial,\partial)=\sum_{k,\ell=0}^{\infty}\sum_{a,b=1}^{N}(\phi^a, V_{k\ell}\phi^b)\partial_{q_{k,a}}\partial_{q_{\ell,b}}.$$

2.1.3 Tame Asymptotical Functions

We will be interested in the so called *tame* functions. To define them let us introduce first another sequence of formal vector variables $\mathbf{t}=(t_0,t_1,t_2,\ldots)$, so that $t_k=q_k+\delta_{k,1}\mathbf{1}$. Formal power series in the Fock space of the type

$$\mathcal{A}(\hbar;\mathbf{q})=\exp\Big(\sum_{g=0}^{\infty}\hbar^{g-1}\mathcal{F}^{(g)}(\mathbf{t})\Big),\quad \mathcal{F}^{(g)}\in\mathbb{C}[[\mathbf{t}]],\quad \mathcal{F}^{(g)}(0)=0$$

are called formal asymptotical functions. We say that a formal asymptotical function is tame if its Taylor's coefficients satisfy the $3g-3+r$-jet condition

$$\left.\frac{\partial^r\mathcal{F}^{(g)}}{\mathrm{d}t_{k_1,a_1}\cdots\mathrm{d}t_{k_r,a_r}}\right|_{\mathbf{t}=0}=0\quad\text{if}\quad k_1+\cdots+k_r>3g-3+r.$$

Let us recall the following result from [10].

Lemma 3 *If $\mathcal{A}$ is a tame asymptotical function, then $\widehat{R}\mathcal{A}$ is a well defined tame asymptotical function.*

2.1.4 Symplectic Loop Space for a Frobenius Manifold

We fix a flat coordinate system $\{t_a\}_{a=1}^N$ on S, s.t., the Euler vector field takes the form

$$E=\sum_{a=1}^{N}d_a\,t_a\partial_a+\sum_{b:d_b=1}r_b\partial_b,$$

where $d_1,d_2,\ldots,d_N$ and r_b $(b:d_b=1)$ are some constants and $\partial_a:=\partial/\partial t_a$. For simplicity, let us assume that S is simply connected, so that the flat vector fields give a trivialization of the tangent and the cotangent bundle. More precisely, let us denote by H the tangent space at some reference point, then we have

$$T^*S\cong TS\cong S\times H,$$

where the first isomorphism is via the non-degenerate bi-linear form of the Frobenius structure and the second one is via parallel transport with respect to the corresponding Levi–Civita connection. The vector space H can be viewed also as the vector space of flat vector fields on S. Note that $\phi_a = \partial_a$ and $\phi^a := dt^a$ form bases of H dual with respect to the Frobenius pairing. Recalling Givental's symplectic loop space formalism applied for H with $(\ ,\ )$ being the Frobenius pairing, and **1** the unit vector field we get a symplectic loop space and a Fock space equipped with a representation of the Heisenberg algebra and a *projective* representation of the Poisson algebra of quadratic Hamiltonians on H.

2.2 Canonical Coordinates

Let us fix a semi-simple point $s \in S \setminus B$. By definition, there exists a coordinate system $\{u_i\}_{i=1}^N$ defined locally near s in which both the Frobenius multiplication and the flat metric are diagonal, i.e.,

$$\partial_{u_i} \bullet \partial_{u_j} = \delta_{i,j} \partial_{u_j}, \quad (\partial_{u_i}, \partial_{u_j}) = \frac{\delta_{i,j}}{\Delta_j},$$

where $\{\Delta_j\}_{j=1}^N$ are some functions analytic with no zeros in a neighborhood of s. Coordinates $\{u_i\}$ with the above properties are called *canonical*. They are unique up to permutation and a constant shift.

Let us denote by U the diagonal matrix of size $N \times N$ whose diagonal entries are $U_{i,i} = u_i$. We need also the $N \times N$ matrix Ψ corresponding to the linear map

$$\Psi : \mathbb{C}^N \to T_s B \cong H, \quad e_i \mapsto \sqrt{\Delta_i} \partial_{u_i}.$$

The matrix of Ψ is constructed by using the standard basis $\{e_i\}_{i=1}^N$ of $\mathbb{C}^N$ and the flat basis $\{\phi_a\}_{a=1}^N$ of H, so that the entries of Ψ are

$$\Psi_{a,i} = \sqrt{\Delta_i} \frac{\partial t_a}{\partial u_i}, \quad 1 \le a, i \le N.$$

Let us summarize some of the basic properties of the matrix Ψ. The proofs follow immediately from the definitions, so they will be left as an exercise.

Proposition 4 *The matrix Ψ has the following properties:*

(1) *If $g = (g_{a,b})$, $g_{a,b} = (\phi_a, \phi_b)$ is the matrix of the flat pairing, then*

$$\Psi\, \Psi^T = g^{-1},$$

where ^{T} is the usual transposition of matrices.

(2) *Let* $A = \sum_{a=1}^{N} A_a dt_a$ *be the connection 1-form on* S *where* A_a *is the linear operator of Frobenius multiplication by* ∂_a, *then*

$$\Psi^{-1} A \Psi = dU.$$

(3) *The Euler vector field has the form* $E = \sum_{i=1}^{N} u_i \partial_{u_i}$. *In particular,*

$$\Psi^{-1}(E\bullet)\Psi = U,$$

where $E\bullet$ *is the linear operator of multiplication by the Euler vector field* E.

Recall the Dubrovin's connection ∇ on the trivial bundle $S \times \mathbb{C}^* \times H \to S \times \mathbb{C}^*$. In flat coordinates

$$\nabla = d - Az^{-1} + \Big(-\theta z^{-1} + (E\bullet)z^{-2}\Big)dz,$$

where θ is the so called *Hodge grading operator* defined by

$$\theta : H \to H, \quad \theta(\phi_a) = \Big(\frac{D}{2} + d_a - 1\Big)\phi_a, \quad 1 \le a \le N.$$

Proposition 5 *Dubrovin's connection has an irregular singularity at* $z = 0$ *and it has a unique formal asymptotical solution of the form*

$$\Psi(1 + R_1 z + R_2 z^2 + \cdots)e^{U/z}. \tag{2}$$

Proof Using Proposition 4 we get

$$\Psi^{-1}\nabla\Psi = d + \Psi^{-1}d\Psi - dUz^{-1} + (Vz^{-1} + Uz^{-2})dz,$$

where $V := -\Psi^{-1}\theta\Psi$. The asymptotical series (2) is a solution to the Dubrovin's connection if and only if $\{R_k\}_{k=0}^{\infty}$ (we set $R_0 = 1$) satisfies the following system of differential equations

$$dR_k + (\Psi^{-1}d\Psi)R_k = [dU, R_{k+1}], \quad \forall k \ge 0 \tag{3}$$

and

$$kR_k + [U, R_{k+1}] = -VR_k, \quad \forall k \ge 0. \tag{4}$$

We have to prove that the above system has a unique solution. Arguing by induction on ℓ we will prove that there is a unique sequence $R_1, \ldots, R_\ell$ satisfying (3) and (4) for all $k \le \ell - 1$, the diagonal part of (4) for $k = \ell$, and $E(R_k) = -kR_k$ for all $k \le \ell$.

Let us first prove the statement for $\ell = 1$. Using (3) with $k = 0$ and comparing the (i, j)-th entries of the matrices with $i \neq j$ we get

$$(\Psi^{-1} d\Psi)_{i,j} = (du_i - du_j)(R_1)_{i,j}.$$

The flatness of ∇ implies that $[dU, \Psi^{-1}d\Psi] = 0$. In particular, $(du_i - du_j) \wedge (\Psi^{-1}d\Psi)_{i,j} = 0$, which by the de Rham lemma implies that $(\Psi^{-1}d\Psi)_{i,j} = \alpha_{i,j}(du_i - du_j)$ for some function $\alpha_{i,j}$ analytic in a neighborhood of s. Hence $(R_1)_{i,j} = \alpha_{i,j}$. Comparing the diagonal entries in (4) for $k = 1$ we get

$$(R_1)_{i,i} = -\sum_{p \neq i} V_{i,p}(R_1)_{p,i},$$

so R_1 is uniquely determined. Let us check that the R_1 satisfies (4) with $k = 0$. We need only to compare the off-diagonal entries. Fix $i \neq j$, then by definition we have

$$(\Psi^{-1}\partial_{u_p}\Psi)_{i,j} = 0, \quad p \neq i, j,$$

and

$$(R_1)_{i,j} = (\Psi^{-1}\partial_{u_i}\Psi)_{i,j} = -(\Psi^{-1}\partial_{u_j}\Psi)_{i,j},$$

hence

$$[U, R_1]_{i,j} = (u_i - u_j)(R_1)_{i,j} = (\Psi^{-1}E(\Psi))_{i,j},$$

where $E = \sum_{i=1}^N u_i \partial_{u_i}$ is the Euler vector field. Since by definition $\mathrm{Lie}_E(\, , \,) = (2 - D)(\, , \,)$ we get that $E(\Delta_i) = D\,\Delta_i$ and

$$E(\Psi_{a,i}) = \Big(\frac{D}{2} + \deg(t_a) - 1\Big)\Psi_{a,i} = \theta_{a,a}\Psi_{a,i}.$$

In other words $\Psi^{-1}E(\Psi) = \Psi^{-1}\theta\Psi = -V$. Finally, note that $E(U) = U$ and $E(V) = 0$, so the identity $[U, R_1] = -V$ implies that $E(R_1) = -R_1$.

Assume that we have constructed $R_1, \ldots, R_\ell$. We would like to construct $R_{\ell+1}$ so that the inductive assumption holds. Note that since ∇ is flat we have

$$\left(d + \Psi^{-1}d\Psi\right)^2 = \Psi^{-1}d^2\Psi = 0, \quad [dU, d + \Psi^{-1}d\Psi] = 0,$$

so

$$[dU, dR_\ell + \Psi^{-1}d\Psi R_\ell] = (d + \Psi^{-1}d\Psi)[dU, R_\ell] = (d + \Psi^{-1}d\Psi)^2 R_{\ell-1} = 0.$$

Now the same argument that we used to construct R_1 can be used to construct $R_{\ell+1}$. The details are straightforward and will be left as an exercise. □

2.3 The Total Ancestor Potential

Let us begin first with the case when $S = \mathbb{C}$ is equipped with the natural Frobenius structure corresponding to the standard multiplication of complex numbers and the pairing is $(1, 1) = 1$. The total ancestor potential in this case is defined through the intersection theory on the Delign–Mumford moduli spaces $\overline{\mathcal{M}}_{g,n}$ of stable curves. Let us denote by ψ_i the 1st Chern class of the orbifold line bundle on $\overline{\mathcal{M}}_{g,n}$ corresponding to the cotangent lines at the i-th marked points. Put

$$\langle \psi_1^{k_1}, \dots, \psi_n^{k_n} \rangle_{g,n} := \int_{\overline{\mathcal{M}}_{g,n}} \psi_1^{k_1} \cdots \psi_n^{k_n}. \tag{5}$$

The Witten-Kontsevich tau-function is a formal series in $\mathbf{t} = (t_0, t_1, \dots)$ defined by

$$\widetilde{\mathcal{A}}_{\mathrm{pt}}(\hbar; \mathbf{t}) = \exp\Big(\sum_{g=0}^{\infty} \hbar^{g-1} \widetilde{\mathcal{F}}^{(g)}(\mathbf{t}) \Big),$$

where the *genus-g potential*

$$\widetilde{\mathcal{F}}^{(g)}(\mathbf{t}) = \sum_{n=0}^{\infty} \frac{1}{n!} \langle \mathbf{t}(\psi_1), \dots, \mathbf{t}(\psi_n) \rangle_{g,n}$$

is defined as follows. We identify $\mathbf{t}$ with the formal series $\mathbf{t}(z) := t_0 + t_1 z + \cdots$ and the n-point genus-g correlator is expanded multilinearly in $t_0, t_1, \dots,$ so that the correlators are reduced to expressions of the type (5). The total ancestor potential $\mathcal{A}_{\mathrm{pt}}$ is obtained from $\widetilde{\mathcal{A}}_{\mathrm{pt}}$ via the dilaton shift: $\mathbf{t}(z) = \mathbf{q}(z) + z$, or in components $t_k = q_k + \delta_{k,1}, \quad k = 0, 1, 2, \dots,$ i.e.,

$$\mathcal{A}_{\mathrm{pt}}(\hbar; \mathbf{q}) = \widetilde{\mathcal{A}}_{\mathrm{pt}}(\hbar; \mathbf{q}(z) + z) \quad \in \quad \mathbb{C}_\hbar[\![q_0, q_1 + 1, q_2, \dots]\!].$$

Note that in this case $B = \emptyset$ and that by definition $\mathcal{A}_{\mathrm{pt}}$ is independent of the choice of a semi-simple point.

If S is an arbitrary simply connected semi-simple Frobenius manifold, then we fix a reference tangent space H with a basis $\{\phi_a\}_{a=1}^N$ that gives rise to a flat coordinate system $t = \{t_a\}_{a=1}^N$. In a neighborhood of a fixed semi-simple point $s \in S \setminus B$ we pick canonical coordinates $\{u_i\}_{i=1}^N$ and fix a branch of $\sqrt{\Delta_i}$, so that the matrix Ψ is uniquely defined. The total ancestor potential is defined by

$$\mathcal{A}_s(\hbar;\mathbf{q}) := \widehat{\Psi}\,\widehat{R}\,e^{(U/z)^\wedge}\prod_{i=1}^{N}\mathcal{D}_{\rm pt}(\hbar\Delta_i;{}^i\mathbf{q}(z)\sqrt{\Delta_i}) \in \mathbb{C}_\hbar[\![q_0,q_1+\mathbf{1},q_2\dots]\!], \tag{6}$$

where ${}^i\mathbf{q}(z) = \sum_{k=0}^{\infty} {}^i\mathbf{q}_k z^k$. The expression preceding $\widehat{\Psi}$ is a formal series in the variables ${}^i\mathbf{q}_k$ $(1\le i\le N,\ k\ge 0)$. The quantization $\widehat{\Psi}$ is interpreted as the change of variables

$$\sum_{i=1}^{N} {}^i q_\ell \Psi(e_i) = \sum_{a=1}^{N} q_{\ell,a}\phi_a,$$

i.e., $\widehat{\Psi}$ transforms a formal series in ${}^i q_\ell$ into a formal series in $q_{\ell,a}$ via the substitution

$${}^i q_\ell = \sum_{a=1}^{N} (\Psi^{-1})_{i,a} q_{\ell,a}.$$

Proposition 6 *The coefficients in the formal series expansion of $\mathcal{A}_s(\hbar;\mathbf{q})$ as a series in $q_0, q_1+1, q_2,\dots$ are Laurent series in $\hbar$, whose coefficients extend analytically to the open subset $S\setminus B$ of semi-simple points.*

Proof In order to prove that the coefficients extend analytically along any path in $S\setminus B$, it is enough to prove that the canonical coordinates u_i have this property. Let us denote by $L\subset T^*S$ the characteristic variety of the Frobenius multiplication. Namely, L is defined as the zero locus of the sheaf of ideals $\mathcal{I}$ on T^*S generated by the kernel of the map

$$\mathrm{Sym}(\mathcal{T}_S)\to\mathcal{T}_S,\quad v_1\dots v_k\mapsto (v_1\bullet\cdots\bullet v_k). \tag{7}$$

Here we are using that there is a natural map $\pi^*\,\mathrm{Sym}(\mathcal{T}_S)\to\mathcal{O}_{T^*S}$, where $\pi: T^*S\to S$ is the projection, so that the kernel of the map (7) can be mapped to $\mathcal{O}_{T^*S}$ and it makes sense to define the ideal $\mathcal{I}$ generated by the image.

If s is a semi-simple point, then we can choose canonical coordinates $(u_1,\dots,u_N)$ around s and fiberwise coordinates $x_1,\dots,x_N$ on T^*S, so that all 1-forms in a neighborhood of s are given by $\sum_{i=1}^N x_i du_i$. In the local coordinates $(u_1,\dots,u_N,x_1,\dots,x_N)$ the characteristic variety L is given by the equations

$$x_ix_j-\delta_{i,j}x_j=0,\quad 1\le i,j\le N.$$

It follows that over a neighborhood of s the subvariety L is a N-sheet covering and the N sections of T^*S that define L are precisely the 1-forms du_i $(1\le i\le N)$.

It is not hard to see from here that the projection π induces a branched covering $L \to S$ of degree N and moreover the set B of non-semi-simple points coincides with the branching locus, i.e., with the support of the sheaf of relative differentials $\Omega^1_{L/S}$. Since L induces a regular covering on $S \setminus B$ the differential forms du_i extend along any path in $S \setminus B$, which proves that u_i also extends.

The analytic continuation along a closed loop in $S \setminus B$ acts as a permutation on the sequence $(u_1, \ldots, u_N)$, while on the sequence $(\sqrt{\Delta_1}, \ldots, \sqrt{\Delta_N})$ the action is given by the same permutation, but with possible sign changes of $\sqrt{\Delta_i}$. It remains only to check that formula (6) is independent of the choices of signs in $\sqrt{\Delta_i}$ and invariant under the permutations of the canonical coordinates. This follows easily from the definitions. □

2.4 The Ancestor Correlators

In order to motivate our definition of correlators, let us first recall the definition in the geometric settings, following [9]. For a given projective manifold V, let us denote by $\overline{\mathcal{M}}_{g,n}(V,d)$ the moduli space of degree-d stable maps from a genus-g nodal Riemann surface, equipped with n marked points, to V. The ancestor correlator functions are defined by the following intersection numbers:

$$\langle \phi_{a_1}\overline{\psi}_1^{k_1}, \ldots, \phi_{a_n}\overline{\psi}_n^{k_n}\rangle_{g,n}(t) := \sum_{m=0}^{\infty}\sum_{d}\frac{Q^d}{m!}\int_{[\overline{\mathcal{M}}_{g,n+m}(V,d)]^{\mathrm{virt}}} \mathrm{ev}^*(\phi_{a_1}\otimes\cdots\otimes\phi_{a_n}\otimes t^{\otimes m})\prod_{a=1}^{n}\overline{\psi}_a^{k_a},$$

where the notation is as follows. The classes $\{\phi_{a_s}\}_{s=1}^n$ and t are cohomology classes on V, the 2nd sum is over all effective curve classes $d \in H_2(V;\mathbb{Z})$ and Q^d is an element of the Novikov ring. Furthermore, evaluating the stable map at the marked points gives rise to the evaluation map

$$\mathrm{ev}: \overline{\mathcal{M}}_{g,n+m}(V,d) \to V^{n+m},$$

while the operation forgetting the last m marked points, the stable map, and stabilizing (i.e. contracting the unstable components) gives a map ft : $\overline{\mathcal{M}}_{g,n+m}(V,d) \to \overline{\mathcal{M}}_{g,n}$. The cohomology classes $\overline{\psi}_s := \mathrm{ft}^*(\psi_s)$ $(1 \le s \le n)$. Finally, $[\overline{\mathcal{M}}_{g,n+m}(V,d)]^{\mathrm{virt}}$ is the virtual fundamental cycle. Let us point out that if $\overline{\mathcal{M}}_{g,n}$ is empty, i.e., $2g-2+n \le 0$, then the ancestor correlator is by definition 0. The total ancestor potential of V has the form

$$\widetilde{\mathcal{A}}_t(\hbar;\mathbf{t}) = \exp\Big(\sum_{g=0}^{\infty}\hbar^{g-1}\widetilde{\mathcal{F}}_t^{(g)}(\mathbf{t})\Big), \tag{8}$$

where $t \in H := H^*(V;\mathbb{C})$, $\mathbf{t} = \{t_{k,a}\}$ is a set of formal variables and

$$\widetilde{\mathcal{F}}_t^{(g)}(\mathbf{t}) = \sum_{n=0}^{\infty} \frac{1}{n!} \langle \mathbf{t}(\overline{\psi}_1), \ldots, \mathbf{t}(\overline{\psi}_n)\rangle(t)_{g,n}$$

is the so called genus-g ancestor potential, where $\mathbf{t}(z) = \sum_{k,a} t_{k,a}\phi_a z^k$ and the definition of the correlator is extended mult-linearly.

Let us return to the settings of an abstract semi-simple Frobenius manifold. It can be proved that the ancestor potential (6) still has the form (8). Motivated by Gromov–Witten theory we would like to define the analogues of the ancestor correlator functions, so that the ancestor potential can be written in the same way. Put

$$\langle \phi_{a_1}\psi^{k_1}, \ldots, \phi_{a_n}\psi^{k_n}\rangle_{g,n}(s;\mathbf{t}) := \partial_{t_{k_1,a_1}} \cdots \partial_{t_{k_n,a_n}} \widetilde{\mathcal{F}}_s^{(g)}(\hbar;\mathbf{t})$$

and

$$\langle \phi_{a_1}\psi^{k_1}, \ldots, \phi_{a_n}\psi^{k_n}\rangle_{g,n}(s) := \langle \phi_{a_1}\psi^{k_1}, \ldots, \phi_{a_n}\psi^{k_n}\rangle_{g,n}(s;0), \tag{9}$$

then by the Taylor's formula we have

$$\widetilde{\mathcal{A}}_s(\hbar;\mathbf{t}) = \exp\Big(\sum_{g,n=0}^{\infty} \frac{\hbar^{g-1}}{n!} \langle \mathbf{t}(\psi), \ldots, \mathbf{t}(\psi)\rangle_{g,n}(s) \Big),$$

where by extending multi-linearly the definition (9) we allow the insertions of the correlator to be any formal power series from $H[\![\psi]\!]$. We will refer to (9) as the *ancestor correlators* of the Frobenius structure. According to Proposition 6 they are analytic functions in $s \in S \setminus B$.

3 The Local Eynard–Orantin Recursion

Let us assume that S is a semi-simple Frobenius manifold. The goal in this section is to derive a recursion for the ancestor correlators.

3.1 Virasoro Constraints for the Point

Recall that the Witten–Kontsevich tau-function is a vacuum vector for a certain representation of the Virasoro algebra. The representation can be constructed as follows. Put

$$I_{A_1}^{(k)}(u,\lambda) = (-1)^k \frac{(2k-1)!!}{2^{k-1/2}} (\lambda - u)^{-k-1/2}, \quad k \geq 0$$

$$I_{A_1}^{(-k-1)}(u,\lambda) = 2\frac{2^{k+1/2}}{(2k+1)!!} (\lambda - u)^{k+1/2}, \quad k \geq 0.$$

These functions are known to be the periods of the A_1-singularity. They satisfy the following crucial property

$$\partial_\lambda I_{A_1}^{(n)}(u,\lambda) = I_{A_1}^{(n+1)}(u,\lambda). \tag{10}$$

We form the generating series

$$\mathbf{f}_{A_1}(u,\lambda;z) = \sum_{n\in\mathbb{Z}} I_{A_1}^{(n)}(u,\lambda)\,(-z)^n$$

and define

$$L_{A_1}(u,\lambda) := \frac{1}{4} : (\partial_\lambda \mathbf{f}_{A_1}(u,\lambda;z)^\wedge (\partial_\lambda \mathbf{f}_{A_1}(u,\lambda;z)^\wedge : + \frac{1}{16}(\lambda-u)^{-2} =: \sum_{m\in\mathbb{Z}} L_{A_1,m}(\lambda-u)^{-m-2},$$

where : : is the normal ordering which means that all differentiation operations precede all multiplication ones. The operators $L_{A_1,m}$ form a representation of the Virasoro algebra (with central charge 1) on the Fock space of the Frobenius manifold $S = \mathbb{C}$. The first few of them have the form

$$L_{A_1,-1} = \frac{q_0^2}{2\hbar} + \sum_{k=0}^{\infty} q_{k+1}\partial_{q_k},$$

$$L_{A_1,0} = \frac{1}{16} + \frac{1}{2}\sum_{k=0}^{\infty}(2k+1)q_k\partial_{q_k},$$

$$L_{A_1,1} = \frac{\hbar}{8}\frac{\partial^2}{\partial q_0{}^2} + \frac{1}{4}\sum_{k=0}^{\infty}(2k+3)(2k+1)q_k\frac{\partial}{\partial q_{k+1}},$$

$$L_{A_1,2} = \frac{3\hbar}{8}\frac{\partial^2}{\partial q_0 \partial q_1} + \frac{1}{8}\sum_{k=0}^{\infty}(2k+5)(2k+3)(2k+1)q_k\frac{\partial}{\partial q_{k+2}}.$$

It was conjectured by Witten [22] and proved by Kontsevich [13] that $\widetilde{\mathcal{D}}_{\mathrm{pt}}$ is a tau-function of the KdV hierarchy. In addition, $\widetilde{\mathcal{D}}_{\mathrm{pt}}$ satisfies the string equation. According to Kac and Schwarz [12] there exists a unique tau-function of KdV satisfying string equation, which can be characterized also as the vacuum vectors for the Virasoro algebra. In our notation the Virasoro constraints take the form

$$L_{A_1,m}\,\mathcal{A}_{\mathrm{pt}}(\hbar;\mathbf{q}) = 0, \quad m \geq -1.$$

3.2 Virasoro Constraints for the Total Ancestor Potential

Fix a neighborhood of a generic semi-simple point, so that the canonical coordinates $(u_1,\ldots,u_N)$ are pairwise distinct, i.e., $u_i \neq u_j$ for $i \neq j$. Let us fix a sufficiently small disk D_i near each u_i, s.t, $D_i \cap D_j = \emptyset$ for $i \neq j$. Put

$$\mathbf{f}_i(s,\lambda;z) := \Psi\, R(z)\, e^{U/z}\mathbf{f}_{A_1}(0,\lambda;z)e_i = \Psi\, R(z)\,\mathbf{f}_{A_1}(u_i,\lambda;z)e_i,$$

where for the 2nd equality we used the translation property (10). Expanding in the powers of z we get

$$\mathbf{f}_i(s,\lambda;z) = \sum_{n\in\mathbb{Z}} I_i^{(n)}(s,\lambda)(-z)^n,$$

where each $I_i^{(n)}(s,\lambda)$ makes sense as a formal Laurent series in $\lambda - u_i$. However, using that $\Psi R e^{U/z}$ is a solution for the Dubrovin's connection, it is easy to prove that $I_i^{(n)}(s,\lambda)$ is a solution to the following system of ODEs

$$\begin{aligned}
\partial_a I^{(n)}(s,\lambda) &= -\phi_a \bullet I^{(n)}(s,\lambda) \\
\partial_\lambda I^{(n)}(s,\lambda) &= I^{(n+1)}(s,\lambda) \\
(\lambda - E\bullet)\partial_\lambda I^{(n)}(s,\lambda) &= \Big(\theta - n - \frac{1}{2}\Big) I^{(n)}(s,\lambda).
\end{aligned} \tag{11}$$

Equation (11) has regular singularities at $\lambda = u_i$ $(1 \leq i \leq N)$, which implies that the Laurent series representing $I_i^{(n)}(s,\lambda)$ is convergent for all $\lambda \in D_i$ and moreover we can analytically extend in λ along any path in $\mathbb{C}\setminus\{u_1,\ldots,u_N\}$.

After a direct computation using Lemma 2 we get the following Lemma.

Lemma 7 *The following identities hold:*

$$(\mathbf{f}_i(s,\lambda;z))^\wedge\,\widehat{\Psi}\widehat{R} = \widehat{\Psi}\widehat{R}(\mathbf{f}_{A_1}(u_i,\lambda;z)e_i)^\wedge, \quad 1 \leq i \leq N.$$

The symplectic vector space $\mathcal{H} = \mathcal{H}_+ \oplus \mathcal{H}_-$, where $\mathcal{H}_+ = H[z]$ and $\mathcal{H}_- = H[\![z^{-1}]\!]z^{-1}$ are Lagrangian subspaces. We denote by $\mathbf{f} \mapsto \mathbf{f}^+$ and $\mathbf{f} \mapsto \mathbf{f}^-$ the corresponding projections.

Lemma 8 *The symplectic pairing*

$$\Omega(\mathbf{f}^+_{A_1}(u_i,\lambda;z),\mathbf{f}^-_{A_1}(u_i,\mu;z)) = 2\log(\lambda-\mu) - 4\log((\lambda-u_i)^{1/2} + (\mu-u_i)^{1/2}),$$

where the RHS is expanded into a Laurent series at $\mu = u_i$, while keeping λ as a parameter.

The proof is an easy computation using the explicit formulas for $\mathbf{f}_{A_1}$. The proof of next Lemma is also a direct computation.

Lemma 9 *The symplectic pairing $\Omega(\mathbf{f}^+_i(s,\lambda;z),\mathbf{f}^-_i(s,\mu;z))$ coincides with*

$$\Omega(\mathbf{f}^+_{A_1}(u_i,\lambda;z),\mathbf{f}^-_{A_1}(u_i,\mu;z)) + \sum_{\ell',\ell''=0}^{\infty} (-1)^{\ell'+\ell''}(V_{\ell',\ell''}e_i,e_i)\, I^{(-\ell''-1)}_{A_1}(u_i,\lambda) I^{(-\ell'-1)}_{A_1}(u_i,\mu),$$

where $V_{\ell',\ell''} \in \mathrm{End}(\mathbb{C}^N)$ are defined in terms of R by (1).

Let us define the *propagator*

$$P_{i,i}(s,\lambda,\mu) := \partial_\lambda\partial_\mu \Omega(\mathbf{f}^+_i(s,\lambda;z),\mathbf{f}^-_i(s,\mu;z)). \tag{12}$$

where the RHS is interpreted as a Laurent series in $(\mu-u_i)$ whose coefficients are Laurent series in $(\lambda-u_i)$. In fact, using Lemmas 8 and 9 we get that the propagator has the form of a singular term $2(\lambda-\mu)^{-2}$ plus a Laurent series in $(\lambda-u_i)$ and $(\mu-u_i)$. Furthermore, we define

$$P^{(0)}_{i,i}(s,\lambda) := \frac{1}{2!}\,\partial^2_\mu\Big((\lambda-\mu)^2 P_{i,i}(s,\lambda,\mu)\Big)\Big|_{\mu=\lambda}.$$

It is convenient to define

$$\phi_j(s,\lambda;z) := \partial_\lambda \mathbf{f}_j(s,\lambda;z), \quad \widehat{\phi}_j(s,\lambda) := (\phi_j(s,\lambda;z))^\wedge.$$

Put

$$L_i(s,\lambda) := :\widehat{\phi}_i(s,\lambda)^2: + P^{(0)}_{i,i}(s,\lambda).$$

Proposition 10 *The following formula holds*

$$L_i(s,\lambda)\,\widehat{\Psi}\widehat{R} = 4\,\widehat{\Psi}\widehat{R}\,L_{A_1}(u_i,\lambda).$$

Proof Put

$$P_{A_1,A_1}(u_i,\lambda,\mu) := \partial_\lambda\partial_\mu \Omega(\mathbf{f}_{A_1}(u_i,\lambda;z)_+,\mathbf{f}_{A_1}(u_i,\mu;z)_-),$$

and define

$$P^{(0)}_{A_1,A_1}(u_i,\lambda) := \frac{1}{2!}\,\partial_\mu^2\Big((\lambda-\mu)^2 P_{A_1,A_1}(u_i,\lambda,\mu)\Big)\Big|_{\mu=\lambda} = \frac{1}{4}(\lambda-u_i)^{-2}.$$

Note that

$$4L_{A_1}(u_i,\lambda) =: \widehat{\phi}_{A_1}(u_i,\lambda)^2 : + P^{(0)}_{A_1,A_1}(s,\lambda).$$

After this observation the proof is straightforward. Namely, according to Lemma 7 we have

$$\widehat{\phi}_i(s,\lambda)\widehat{\phi}_i(s,\mu)\,\widehat{\Psi}\widehat{R} = \widehat{\Psi}\widehat{R}\,\widehat{\phi}_{A_1}(u_i,\lambda)\widehat{\phi}_{A_1}(u_i,\mu). \tag{13}$$

On the other hand

$$\widehat{\phi}_i(s,\lambda)\widehat{\phi}_i(s,\mu) = :\widehat{\phi}_i(s,\lambda)\widehat{\phi}_i(s,\mu): + P_{i,i}(s,\lambda,\mu)$$

and

$$\widehat{\phi}_{A_1}(u_i,\lambda)\widehat{\phi}_{A_1}(u_i,\mu) = :\widehat{\phi}_{A_1}(u_i,\lambda)\widehat{\phi}_{A_1}(u_i,\mu): + P_{A_1,A_1}(u_i,\lambda,\mu).$$

Also

$$P_{i,i}(s,\lambda,\mu) = \frac{2}{(\lambda-\mu)^2} + P^{(0)}_{i,i}(s,\lambda) + O(\lambda-\mu)$$

and

$$P_{A_1,A_1}(u_i,\lambda,\mu) = \frac{2}{(\lambda-\mu)^2} + P^{(0)}_{A_1,A_1}(u_i,\lambda) + O(\lambda-\mu).$$

Hence after subtracting the singular term $2(\lambda-\mu)^{-2}$ from both sides in (13) we can set $\mu=\lambda$. We get precisely the identity that we wanted to prove. □

Corollary 11 *Let*

$$L_i(s,\lambda) = \sum_{m\in\mathbb{Z}} L_{i,m}(\lambda-u_i)^{-m-2}$$

be the Laurent series expansion at $\lambda = u_i$*; then*

$$L_{i,m}\,\mathcal{A}_s(\hbar;\mathbf{q}) = 0, \quad 1\le i\le N, \quad m\ge -1.$$

3.3 The Local Eynard–Orantin Recursion

By definition, the ancestor potential does not have non-zero correlators in the unstable range $(g, n) = (0, 0), (0, 1), (0, 2)$ and $(1, 0)$. It is convenient however, to extend the definition in the unstable range as well in the following two cases:

$$\left\langle \phi_j^+(s,\lambda;\psi_1), \mathbf{t} \right\rangle_{0,2} := \Omega(\mathbf{t}(z), \phi_j^-(s,\lambda;z)), \tag{14}$$

$$\left\langle \phi_j^+(s,\lambda;\psi_1), \phi_j^+(s,\lambda;\psi_1) \right\rangle_{0,2} := P_{j,j}^{(0)}(s,\lambda). \tag{15}$$

Theorem 12 *The ancestor correlators satisfy the following recursion*

$$\left\langle \phi_a \psi_1^m, \mathbf{t}, \dots, \mathbf{t} \right\rangle_{g,n+1}(s) =$$

$$\frac{1}{4}\sum_{j=1}^{N} \mathrm{Res}_{\lambda=u_j} \frac{\Omega(\phi_a z^m, \mathbf{f}_j(s,\lambda;z)_-)}{(I_j^{(-1)}(s,\lambda), \mathbf{1})} \times \left(\left\langle \phi_j^+(s,\lambda;\psi_1), \phi_j^+(s,\lambda;\psi_2), \mathbf{t}, \dots, \mathbf{t} \right\rangle_{g-1,n+2}(s) + \right.$$

$$\left. \sum_{\substack{g'+g''=g \\ n'+n''=n}} \binom{n}{n'} \left\langle \phi_j^+(s,\lambda;\psi_1), \mathbf{t}, \dots, \mathbf{t} \right\rangle_{g',n'+1}(s) \left\langle \phi_j^+(s,\lambda;\psi_1), \mathbf{t}, \dots, \mathbf{t} \right\rangle_{g'',n''+1}(s) \right),$$

for all stable pairs $(g, n+1)$, *i.e.*, $2g - 2 + n \geq 0$, *where all unstable correlators on the RHS are set to* 0, *except for the ones of the type (14) and (15).*

Proof We will prove that the recursion is equivalent to the Virasoro constraints stated in Corollary 11. To begin with let us write the generating series $L_j(s,\lambda)$ explicitly as the sum of the following three terms

$$\sum_{k',k''=0}^{\infty} \sum_{a,b=1}^{N} (-1)^{k'+k''} (I_j^{(k'+1)}(s,\lambda), \phi^a)\, (I_j^{(k''+1)}(s,\lambda), \phi^b)\, \hbar \partial_{q_{k',a}} \partial_{q_{k'',b}}, \tag{16}$$

$$\sum_{k',k''=0}^{\infty} \sum_{a,b=1}^{N} 2(-1)^{k''+1} (I_j^{(-k')}(s,\lambda), \phi_a)\, (I_j^{(k''+1)}(s,\lambda), \phi^b)\, q_{k',a} \partial_{q_{k'',b}}, \tag{17}$$

$$P_{i,i}^{(0)}(s,\lambda) + \sum_{k',k''=0}^{\infty} \sum_{a,b=1}^{N} (I_j^{(-k')}(s,\lambda), \phi_a)\, (I_j^{(-k'')}(s,\lambda), \phi_b)\, \hbar^{-1} q_{k',a}\, q_{k'',b}. \tag{18}$$

Note that the double sum in (18) is analytic at $\lambda = u_j$, so the sum does not contribute to the Virasoro constraints, which means that it can be ignored. Let us undo the dilaton shift, i.e., switch to the variables $t_{k,a} = q_{k,a} - \delta_{k,1}\delta_{a,1}$, where for simplicity we assume that $\phi_1 = \mathbf{1}$. Note that the only term affected by the change is (see (17) when $k' = a = 1$)

$$\sum_{k=0}^{\infty} 2(-1)^{k+1} (I_j^{(-1)}(s,\lambda), \mathbf{1})\, (I_j^{(k+1)}(s,\lambda), \phi^b)\, (t_{1,1}+1)\partial_{t_{k,b}}.$$

Now we need the following identity:

$$\sum_{j=1}^{N} \mathrm{Res}_{\lambda=u_j}\, (I_{\beta_j}^{(k')}(s,\lambda), \phi_a)\, (I_{\beta_j}^{(k'')}(s,\lambda), \phi^b)\, d\lambda = 2(-1)^{k'}\delta_{a,b}\delta_{k'+k'',0},$$

for all $k', k'' \in \mathbb{Z}$ and $a, b = 1, 2, \ldots, N$. The proof follows from the definitions, so it is left as an exercise. Fix $m \ge 0$ and $a \in \{1, 2, \ldots, N\}$, then

$$\frac{1}{4}\sum_{j=1}^{N} \mathrm{Res}_{\lambda=u_j} \frac{(I_j^{(-m-1)}(s,\lambda), \phi_a)}{(I_j^{(-1)}(s,\lambda), \mathbf{1})} L_j(s,\lambda) d\lambda =$$

$$\frac{\partial}{\partial t_{m,a}} + \frac{1}{4}\sum_{j=1}^{N} \mathrm{Res}_{\lambda=u_j} \frac{(I_j^{(-m-1)}(s,\lambda), \phi_a)}{(I_j^{(-1)}(s,\lambda), \mathbf{1})} \Big(L_j(s,\lambda)\big|_{\mathbf{q}=\mathbf{t}} \Big) d\lambda. \tag{19}$$

The Virasoro constraints for the ancestor potential can be stated equivalently as $L_j(s,\lambda)\mathcal{A}_s(\hbar;\mathbf{q})$ is analytic at $\lambda = u_j$ for all $j = 1, 2, \ldots, N$. Hence the operator (19) annihilates $\widetilde{\mathcal{A}}_s(\hbar;\mathbf{t})$. Comparing the coefficients in front of the monomial expressions in $\mathbf{t}$ and $\hbar$ of fixed degree n and genus $g-1$ we get the recursion that we wanted to prove. □

Remark 13 The recursion in Theorem 12 is the same as the local Eynard–Orantin recursion introduced in [6].

4 Analyticity of the Total Ancestor Potential in Singularity Theory

Let us assume now that S is the base of the universal unfolding F of some function $f \in \mathcal{O}_{\mathbb{C}^{n+1},0}$ with an isolated critical point at 0. We may assume that S is a small ball in $\mathbb{C}^N$ with center at 0, where N is the multiplicity of the critical point. Furthermore, we may arrange that the domain of F is an appropriate small contractible Stein domain $X \subset S \times \mathbb{C}^{n+1}$, s.t., $(0,0) \in X$ and $F(0,x) = f(x)$ (see [1] for some background on singularity theory). Let us fix a primitive holomorphic volume form $\omega \in \Omega^{n+1}_{X/S}(X)$, so that S becomes a Frobenius manifold (see [11, 20]). Moreover, it can be proved that the critical values provide a canonical coordinate system and the non semi-simple points B are precisely those $s \in S$ for which at least one of the critical points of $F(s, x)$ is not of type A_1.

4.1 Period Integrals

The map

$$\varphi : X \to S \times \mathbb{C}, \quad (s, x) \mapsto (s, F(s, x)),$$

gives rise to a smooth fibration on $(S \times D)'$, called the *Milnor fibration*. Here $D \subset \mathbb{C}$ is a sufficiently small disk with center 0 and $'$ denotes removing the points (s, λ) for which the fiber $X_{s,\lambda} := \varphi^{-1}(s, \lambda)$ is singular. Let us fix a reference point $(s_0, \eta_0) \in S \times D$ and denote by $\mathfrak{h} := H_n(X_{s_0,\eta_0}\mathbb{C})$ and by $\Delta \subset \mathfrak{h}$ the set of vanishing cycles. The definition of a primitive form implies that the functions $I_j^{(k)}(s, \lambda)$ introduced in Sect. 3.2 can be identified with period integrals of the following type:

$$I_\alpha^{(k)}(s, \lambda) := -d\Big((2\pi)^{-\ell}\partial_\lambda^{k+\ell} \int_{\alpha_{s,\lambda}} d^{-1}\omega\Big) \quad \in \quad T_s^*S \cong H,$$

where $\ell := n/2$ (by stabilizing the singularity if necessary we may assume that n is even), $\alpha \in \mathfrak{h}$ is a cycle, and $d^{-1}\omega$ denotes an arbitrary n-form $\widetilde{\omega}$, holomorphic in a neighborhood of $X_{s,\lambda}$, s.t., $\omega = d\widetilde{\omega}$. The period is a multi-valued function on $(S \times D)'$ and its value depends on the choice of a path from the reference point to (s, λ). In particular, we denoted by $\alpha_{s,\lambda}$ the parallel transport of α along the path. If $s \in S \setminus B$ is semi-simple and λ is in a neighborhood D_j of the critical value u_j, then let us choose $\alpha \in \Delta$ to be a vanishing cycle and fix the path in such a way that $\alpha_{s,\lambda}$ becomes the cycle vanishing over $\lambda = u_j$, then the period integral coincides with $I_j^{(k)}(s, \lambda)$ (see [10]).

For each fixed $s \in S$, the period vectors $I_\alpha^{(n)}(s, \lambda)$ satisfy Fuchsian differential equation in λ with singularities only at the critical values of $F(s, x)$ and $\lambda = \infty$. Hence using analytic continuation we may assume that the period integrals are define on $(S \times \mathbb{C})'$. Equivalently, the cohomology groups $H^n(X_{s,\lambda}; \mathbb{C})$, $(s, \lambda) \in (S \times D)'$ form a vector bundle equipped with a flat Gauss–Manin connection and the primitive form determines an extension of this bundle to a vector bundle on $S \times \mathbb{P}^1$, s.t., the Gauss–Manin connection has a logarithmic singularity at $\lambda = \infty$.

Finally, let us discuss the so called *primitive direction*. The flat identity of the Frobenius structure is a vector field δ_1, called *primitive*, s.t., $\delta_1 F = 1$. We denote by $s \mapsto s + \lambda \mathbf{1}$ the time-λ flow of δ_1. Note that if $(s, \lambda) \in S \times D$ is such that $s - \lambda\mathbf{1} \in S$, then $X_{s,\lambda} = X_{s-\lambda\mathbf{1},0}$, so the periods have the following translation symmetry

$$I_\alpha^{(n)}(s, \lambda) = I_\alpha^{(n)}(s - \lambda\mathbf{1}, 0).$$

Therefore we can extend the Frobenius structure in the primitive direction as well, i.e., we may assume that S is invariant under the translations $s \mapsto s + \lambda \mathbf{1}$ for all $\lambda \in \mathbb{C}$.

4.2 Propagators and the Monodromy Representation

Recall the propagators (12). In this section we prove that they can be extended analytically along any path in $(S \times \mathbb{C})'$ and moreover the analytic extension is compatible with the monodromy action. To begin with let us introduce the following terminology. Given cycles $\alpha, \beta \in \mathfrak{h}$ we define a *propagator* on $(S \times \mathbb{C})'$ from α to β to be a Laurent series

$$P_{\alpha,\beta}(s, \lambda, \mu) = \frac{(\alpha|\beta)}{(\lambda - \mu)^2} + \sum_{k=0}^{\infty} P^{(k)}_{\alpha,\beta}(s, \mu)(\lambda - \mu)^k,$$

where $(\alpha|\beta)$ up to the sign $(-1)^{\ell}$ is the intersection pairing of the cycles α and β, satisfying the following properties.

(1) For every $(s, \mu) \in (S \times \mathbb{C})'$, the radius of convergence of the series is non-zero.
(2) The functions $P^{(k)}_{\alpha,\beta}(s, \mu)$ extend analytically along any path in $(S \times \mathbb{C})'$ and the analytic continuation is compatible with the monodromy representation.
(3) If (s, μ) is such that s is semi-simple, μ is sufficiently close to $u_i(s)$, and β is the cycle vanishing over $\mu = u_i(s)$, then

$$P_{\beta,\beta}(s, \lambda, \mu) = P_{i,i}(s, \lambda, \mu).$$

Property (2) means that if C is a closed loop in $(S \times \mathbb{C})'$ based at (s, μ) and w is the monodromy transformation of $\mathfrak{h}$ corresponding to the parallel transport of cycles along C, then the analytic continuation of $P^{(k)}_{\alpha,\beta}(s, \mu)$ along C is $P^{(k)}_{w(\alpha),w(\beta)}(s, \mu)$.

The key to constructing propagators is the so called *phase 1-form* (see [2, 10])

$$\mathcal{W}_{\alpha,\beta}(s, \xi) = I^{(0)}_{\alpha}(s, \xi) \bullet I^{(0)}_{\beta}(s, 0) \quad \in \quad T^*_s S,$$

where the period vectors are interpreted as elements in $T^*_s S$ and the multiplication in $T^*_s S$ is induced by the Frobenius multiplication via the natural identification $T^*_s S \cong T_s S$. The dependence on the parameter ξ is in the sense of a germ at $\xi = 0$, i.e., we will be interested in the Taylor's series expansion about $\xi = 0$. The phase form is a power series in ξ whose coefficients are multivalued 1-forms on $S' := (S \times \{0\})'$.

Lemma 14 *We have*

$$(\alpha|\beta) = -\iota_E \mathcal{W}_{\alpha,\beta}(s,0) = -(I^{(0)}_\alpha(s,0), E \bullet I^{(0)}_\beta(s,0)).$$

This is a well known fact due originally to K. Saito [18].

Lemma 15 *The phase form is weighted-homogeneous of weight* 0*, i.e.,*

$$(\xi\partial_\xi + L_E)\mathcal{W}_{\alpha,\beta}(s,\xi) = 0,$$

where L_E is the Lie derivative with respect to the vector field E.

Proof Note that

$$\mathcal{W}_{\alpha,\beta}(s,\xi) = (I^{(0)}_\alpha(s,\xi), dI^{(-1)}_\beta(s,0)).$$

It is easy to check that $\mathcal{W}_{\alpha,\beta}$ is a closed 1-form, so using the Cartan's magic formula $L_E = d\iota_E + \iota_E d$, where ι_E is the contraction by the vector field E, we get

$$L_E\mathcal{W}_{\alpha,\beta} = d(I^{(0)}_\alpha(s,\xi), (\theta+1/2)I^{(-1)}_\beta(s,0)) = -d((\theta-1/2)I^{(0)}_\alpha(s,\xi), I^{(-1)}_\beta(s,0)).$$

We used that θ is skew-symmetric with respect to the residue pairing and that

$$\iota_E dI^{(-1)}_\beta(s,0) = EI^{(-1)}_\beta(s,0)) = (\theta+1/2)I^{(-1)}_\beta(s,0),$$

where the last equality comes from the differential equation (11) with $n=-1$ and $\lambda = 0$. Furthermore, using the Leibnitz rule we get

$$-((\theta-1/2)dI^{(0)}_\alpha(s,\xi), I^{(-1)}_\beta(s,0)) - ((\theta-1/2)I^{(0)}_\alpha(s,\xi), dI^{(-1)}_\beta(s,0)).$$

The first residue pairing, using the skew-symmetry of θ and the differential equation $dI^{(0)}_\alpha = -AI^{(1)}_\alpha$ becomes

$$(AI^{(1)}_\alpha(s,\xi), (\theta+1/2)I^{(-1)}_\beta(s,0)) = -(AI^{(1)}_\alpha(s,\xi), E\bullet I^{(0)}_\beta(s,0)). \tag{20}$$

Similarly, the 2nd residue pairing becomes

$$((\xi\partial_\xi + E)I^{(0)}_\alpha(s,\xi), dI^{(-1)}_\beta(s,0)) = \xi\partial_\xi\mathcal{W}_{\alpha,\beta}(s,\xi) + (E\bullet I^{(1)}_\alpha(s,\xi), AI^{(0)}_\beta(s,0)). \tag{21}$$

On the other hand, recall that $A = \sum_i(\phi_i\bullet)dt_i$ and that the Frobenius multiplication is commutative. In particular, $[A, E\bullet] = 0$, so the terms (20) and (21) add up to $\xi\partial_\xi\mathcal{W}_{\alpha,\beta}(s,\xi)$. The lemma follows. □

Given cycles $\alpha, \beta \in \mathfrak{h}$, then we define

$$P_{\alpha,\beta}(s,\lambda,\mu) := \partial_\lambda \partial_\mu \int_{s_0}^{s-\mu\mathbf{1}} \mathcal{W}_{\alpha,\beta}(s',\lambda-\mu), \tag{22}$$

where the integration is along a path C in $(S\times\mathbb{C})'$, s.t., s_0 is a generic point on the discriminant and the cycle $\beta_s \in H_n(X_{s,0};\mathbb{C})$ vanishes along C.

Proposition 16 *The integral in definition* (22) *is convergent, independent of the choice of path along which β vanishes, and the Laurent series expansion at $\lambda=\mu$ of $P_{\alpha,\beta}(s,\lambda,\mu)$ defines a propagator from α to β on $(S\times\mathbb{C})'$.*

Proof The integral can be computed explicitly in terms of the period integrals, because according to Lemma 15 we have

$$\partial_\lambda \mathcal{W}_{\alpha,\beta}(s',\lambda-\mu) = -d\Big(\frac{1}{\lambda-\mu}\,\iota_E \mathcal{W}_{\alpha,\beta}(s',\lambda-\mu)\Big),$$

which by definition is

$$d\Big(\frac{1}{\lambda-\mu}\,(I^{(0)}_\alpha(s',\lambda-\mu),(\theta+1/2)I^{(-1)}_\beta(s',0))\Big).$$

Using that $I^{(-1)}_\beta(s',0)$ vanishes as $s'\to s_0$, we get

$$P_{\alpha,\beta}(s,\lambda,\mu) = \partial_\mu\Big(\frac{1}{\lambda-\mu}\,(I^{(0)}_\alpha(s,\lambda),(\theta+1/2)I^{(-1)}_\beta(s,\mu))\Big).$$

The above series has a Laurent series expansion at $\lambda=\mu$ with a pole of order 2 and no residue. The leading order term is

$$\begin{aligned}&\frac{1}{(\lambda-\mu)^2}\,(I^{(0)}_\alpha(s,\mu),(\theta+1/2)I^{(-1)}_\beta(s,\mu))\\ &= \frac{1}{(\lambda-\mu)^2}\,(I^{(0)}_\alpha(s,\mu),(\mu-E\bullet)I^{(0)}_\beta(s,\mu)) = \frac{(\alpha|\beta)}{(\lambda-\mu)^2},\end{aligned}$$

where the last equality follows from Saito's formula (see Lemma 14).

It remains only to prove that if s is a semi-simple point, λ and μ are sufficiently close to a critical value $u_i(s)$, and $\alpha=\beta$ is vanishing cycle vanishing over $\lambda=u_i(s)$, then $P_{\alpha,\beta}(s,\lambda,\mu)=P_{i,i}(s,\lambda,\mu)$. Since in definition (22) we can choose the generic point s_0 and the integration path as we wish, let us pick $s_0 = s-u_i(s)\mathbf{1}$ and integrate along the straight segment $[s_0, s-\mu\mathbf{1}]$. Using integration by parts together with

$$(I_\alpha^{(k)}(s',\lambda-\mu), dI_\beta^{(-k-1)}(s',0)) =$$
$$d(I_\alpha^{(k)}(s',\lambda-\mu), I_\beta^{(-k-1)}(s',0)) - (I_\alpha^{(k+1)}(s',\lambda-\mu), dI_\beta^{(-k-2)}(s',0))$$

it is easy to see that the integral in (22) coincides with the Laurent series expansion at $\mu = u_i(s)$ of the symplectic pairing

$$\Omega(\mathbf{f}_\alpha^+(s,\lambda;z), \mathbf{f}_\beta^-(s,\mu;z)) = \sum_{k=0}^{\infty}(-1)^{k+1}(I_\alpha^{(k)}(s,\lambda), I_\beta^{(-k-1)}(s,\mu)).$$

This completes the proof. □

Note that in the course of the proof we derived the following explicit formulas for the the coefficients $P_{\alpha,\beta}^{(k)}(s,\mu)$ of the Laurent series expansion in $(\lambda-\mu)$ of the propagator:

$$\frac{1}{(k+1)!}(I_\alpha^{(k+1)}(s,\mu), (\theta+1/2)I_\beta^{(0)}(s,\mu)) + \frac{1}{(k+2)!}(I_\alpha^{(k+2)}(s,\mu), (\theta+1/2)I_\beta^{(-1)}(s,\mu)).$$

In particular,

$$P_{\alpha,\beta}^{(0)}(s,\mu) = \frac{1}{2}\Big((\mu - E\bullet_s)I_\alpha^{(1)}(s,\mu), I_\beta^{(1)}(s,\mu)\Big) = \frac{1}{2}\Big((\theta-1/2)I_\alpha^{(0)}(s,\mu), I_\beta^{(1)}(s,\mu)\Big). \tag{23}$$

Note that the propagator $P_{\alpha,\beta}^{(0)}(s,\mu)$ is symmetric with respect to α and β.

4.3 Twisted Representation of the Heisenberg VOA

Let us denote by $\mathcal{F} = \mathrm{Sym}(\mathfrak{h}[\zeta^{-1}]\zeta^{-1})$. Given $a \in \mathfrak{h}$ it is convenient to put $a_{(-n-1)} := a\zeta^{-n-1}$, then every element in $\mathcal{F}$ is a linear combination of elements of the type

$$a = \alpha^1_{(-k_1-1)}\cdots\alpha^r_{(-k_r-1)}, \quad \alpha^i \in \mathfrak{h}, \quad k_i \geq 0.$$

Following [2] we define differential operators acting on the Fock space as follows. First we define

$$X_{s,\lambda}(\alpha) := \widehat{\phi}_\alpha(s,\lambda), \quad \alpha \in \mathfrak{h}, \tag{24}$$

where we identify $\alpha \in \mathfrak{h}$ with $\alpha_{(-1)} \in \mathcal{F}$ and put $\widehat{\phi_\alpha}(s,\lambda) = (\partial_\lambda \mathbf{f}_\alpha(s,\lambda;z))^\wedge$, then we set

$$X_{s,\lambda}(a) = \sum_J \Big(\prod_{(i,j)\in J} \partial_\lambda^{(k_j)} P^{(k_i)}_{\alpha^i,\alpha^j}(s,\lambda)\Big) :\Big(\prod_{l\in J'} \partial_\lambda^{(k_l)} X_{s,\lambda}(\alpha^l)\Big):, \tag{25}$$

where $\partial_\lambda^{(k)} := \frac{\partial_\lambda^k}{k!}$ and the sum is over all collections J of disjoint ordered pairs $(i_1, j_1), \dots, (i_s, j_s) \subset \{1,\dots,r\}$ such that $i_1 < \cdots < i_s$ and $i_l < j_l$ for all l, and $J' = \{1,\dots,r\} \setminus \{i_1,\dots,i_s,j_1,\dots,j_s\}$. Although we are not going to use the theory of vertex algebras here, let us point out that formula (25) is obtained by the axioms of vertex operator algebra representations. Namely, the vector space $\mathcal{F}$ has a standard structure of a Heisenberg Vertex Operator Algebra (VOA) and the fields (24) are known to be local to each other. It was proved in [2], that the definition (24) extends uniquely to a σ-twisted representation of $\mathcal{F}$, where σ is the classical monodromy corresponding to a big loop that goes around the discriminant.

For $(s,\lambda) \in (S\times\mathbb{C})'$ and $c_1,\dots,c_r \in \mathfrak{h}$ we define

$$\Omega^{(g)}_{c_1\cdots c_r}(s,\lambda;\mathbf{t}) \quad \in \quad \mathbb{C}[\![t_0,t_1,\dots,]\!]$$

by the following equation

$$X_{s,\lambda}(c_1\cdots c_r)\mathcal{A}_s(\hbar;\mathbf{q}) =: \sum_{g=0}^{\infty} \hbar^{g-\frac{r}{2}} \Omega^{(g)}_{c_1\cdots c_r}(s,\lambda;\mathbf{q})\mathcal{A}_s(\hbar;\mathbf{q}), \tag{26}$$

where in order to define $\Omega^{(g)}_{c_1\cdots c_r}(s,\lambda;\mathbf{t})$ we replace $\mathbf{q}$ by $\mathbf{t}$ *without* using the dilaton shift. If we denote by $W \subset \mathrm{GL}(\mathfrak{h})$ the monodromy group, then W acts naturally on $\mathcal{F}$ via $w(a_{(-n-1)}) := (w(a))_{(-n-1)}$. Since both the generating fields (24) and the propagators are compatible with the monodromy representation we get that the analytic continuation of $X_{s,\lambda}(a)$ along a closed loop C in $(S\times\mathbb{C})'$ is $X_{s,\lambda}(w(a))$, where $w \in W$ is the monodromy transformation corresponding to the loop C. In particular, the analytic continuation in (s,λ) along C of $\Omega^{(g)}_{c_1\cdots c_r}(s,\lambda;\mathbf{t})$ is $\Omega^{(g)}_{w(c_1)\cdots w(c_r)}(s,\lambda;\mathbf{t})$.

4.4 *Extension Through a Generic Non-semisimple Point*

Let $b_0 \in B$ be a generic point, so that $F(b_0,x)$ has $N-2$ critical points of type A_1 and 1 critical point of type A_2. The critical values corresponding to the A_1-critical points will be denoted by $u_i(b_0)$ $(1 \le i \le N-2)$ and we will assume that they are pairwise distinct. All points $b \in B$ that do not satisfy the above property are points in some codimension 2 analytic subvariety of S. Therefore, according to Hartogue's extension theorem, in order to prove that the ancestor potential extends analytically, it is enough to prove that it extends analytically at $s = b_0$.

4.4.1 Fixing a Neighborhood of b_0

We fix pairwise disjoint sufficiently small disks D_i, $1 \le i \le N-1$, in $\mathbb{C}$, s.t., the center of D_i is the critical value $u_i(b_0)$, where $u_{N-1}(b_0)$ is the critical value of the A_2-critical point. Put

$$S_0 = \{s \in S \mid u_i(s) \in D_i (1 \le i \le N-2),\ (u_{N-1}(s), u_N(s)) \in (D_{N-1} \times D_{N-1})/\mathbb{Z}_2\},$$

where $\mathbb{Z}_2 = \mathbb{Z}/2\mathbb{Z}$ acts on $D_{N-1} \times D_{N-1}$ by permuting the coordinates. Note that $S_0 \subset S$ is an open neighborhood of b_0 homeomorphic to

$$D_1 \times \cdots \times D_{N-2} \times (D_{N-1} \times D_{N-1})/\mathbb{Z}_2.$$

Furthermore, if we restrict the Milnor fibration to $(S_0 \times D_0)'$, where $D_0 = \sqcup_{i=1}^{N-1} D_i$, then the vanishing cycles Δ_0 of the new fibration form a disjoint union

$$\Delta_0 = \Delta_1 \sqcup \cdots \sqcup \Delta_{N-1},$$

where Δ_i $(1 \le i \le N-1)$ is the set of cycles vanishing respectively over $\lambda = u_i(b_0)$ $(1 \le i \le N-1)$ along some path in $(S_0 \times D_0)'$. Note that Δ_{N-1} is a A_2-root system, while the remaining Δ_i's are A_1-root systems.

4.4.2 Extending the Recursion

Let us rewrite the Eynard–Orantin recursion in terms of the operators (25). By definition

$$X_{s,\lambda}(\beta_i^2) =: \widehat{\phi}_{\beta_i}(s, \lambda)^2 : + P^{(0)}_{\beta_i,\beta_i}(s, \lambda).$$

Let us denote by $\widehat{\phi}^{\pm}_{\alpha}(s, \lambda)$ the quantization of $\phi^{\pm}_{\alpha}(s, \lambda; z)$, then the above operator becomes

$$\widehat{\phi}^{+}_{\beta_i}(s, \lambda)^2 + 2\widehat{\phi}^{-}_{\beta_i}(s, \lambda)\widehat{\phi}^{+}_{\beta_i}(s, \lambda) + \widehat{\phi}^{-}_{\beta_i}(s, \lambda)^2 + P^{(0)}_{\beta_i,\beta_i}(s, \lambda). \tag{27}$$

Recalling the definition (26), we get that $\Omega^{(g)}_{\beta_i,\beta_i}(s, \lambda; \mathbf{t})$ can be written as a sum of 4 type of terms corresponding to the 4 summands in (27). Let us compare $\Omega^{(g)}_{\beta_i,\beta_i}(s, \lambda, \mathbf{t})$ with the sum of correlators that appear in the big brackets on the RHS of the local Eynard–Orantin recursion in Theorem 12. The contributions of the 1st summand in (27) coincide with the sum of all stable correlators, the 2nd summand in (27) corresponds to the sum of all products of an unstable correlator of type (14) and a stable correlator, the third summand depends analytically on $\lambda - u_i$ so it does not contribute to the residue, and finally the 4th summand corresponds to the contribution of the unstable correlator (15). Hence, the local Eynard-Orantin recursion stated in Theorem 12 can be written conveniently in the following way

$$\langle \phi_a \psi^m \rangle_{g,1}(s;\mathbf{t}) = -\frac{1}{4}\sum_{i=1}^{N} \mathrm{Res}_{\lambda=u_i} \frac{\Omega(\phi_a z^m, \mathbf{f}^-_{\beta_i}(s,\lambda;z))}{y_{\beta_i}(s,\lambda)} \Omega^{(g)}_{\beta_i,\beta_i}(s,\lambda;\mathbf{t})\, d\lambda, \tag{28}$$

where β_i is a cycle vanishing over $\lambda = u_i$ and

$$y_\beta(s,\lambda) := (I^{(-1)}_\beta(s,\lambda), 1).$$

Let $\{\alpha, \beta\}$ be a basis of simple roots for Δ_{N-1}. Put

$$\chi_1 = \frac{2}{3}\alpha + \frac{1}{3}\beta, \quad \chi_2 = -\frac{1}{3}\alpha + \frac{1}{3}\beta, \quad \chi_3 = -\frac{1}{3}\alpha - \frac{2}{3}\beta.$$

We refer to these as 1-point cycles. Note that the root system Δ_{N-1} consists of all differences $\chi_i - \chi_j$ for $i \neq j$. Motivated by the construction of Bouchard–Eynard [3] we introduce the following integral

$$-\frac{1}{2\pi\sqrt{-1}} \oint \sum_{c_1,\ldots,c_r} \frac{1}{(r-1)!} \frac{\Omega(\phi_a z^m, \mathbf{f}^-_{c_1}(s,\lambda;z))}{\prod_{k=2}^{r} y_{c_k-c_1}(s,\lambda)} \Omega^{(g)}_{c_1\cdots c_r}(s,\lambda;\mathbf{t})\, d\lambda, \tag{29}$$

where the integral is along a closed loop in D_{N-1} that goes once counterclockwise around the critical values $u_{N-1}(s)$ and $u_N(s)$ and the sum is over all $r = 2, 3$ and all $c_1, \ldots, c_r \in \{\chi_1, \chi_2, \chi_3\}$ such that $c_i \neq c_j$ for $i \neq j$. Note that the integrand is monodromy invariant (see Sect. 4.3), hence a single valued analytic 1-form in $D_{N-1} \setminus \{u_{N-1}(s), u_N(s)\}$, so the integral makes sense.

Theorem 17 *The integral* (29) *coincides with the sum of the last two summands in* (28) *corresponding to the residues at* $\lambda = u_{N-1}, u_N$.

4.4.3 Proof of Theorem 17

The proof relies on a certain identity that we would like to present first. Let us denote by $\mathfrak{h}_{\Delta_i}$ $(1 \le i \le N)$ the vector subspace of $\mathfrak{h}$ spanned by the root system Δ_i (we assume that $\Delta_N = \Delta_{N-1}$). Let u_i and u_j $(1 \le i, j \le N)$ be two of the critical values, $\beta := \beta_j$ be the cycle vanishing over u_j, and $a \in \mathfrak{h}_{\Delta_i}$. Let us fix some Laurent series

$$f(\lambda,\mu) \in (\lambda - u_i)^{1/2}\mathbb{C}((\lambda - u_i, \mu - u_j)) + \mathbb{C}((\lambda - u_i, \mu - u_j))$$

where $\mathbb{C}((\lambda - u_i, \mu - u_j))$ denotes the space of formal Laurent series. We will have to evaluate residues of the following form:

$$\mathrm{Res}_{\lambda=u_i}\,\mathrm{Res}_{\mu=u_j}\sum_{\text{all branches}}\frac{\Omega(\phi_a^+(s,\lambda;z),\mathbf{f}_\beta^-(s,\mu;z))}{y_\beta(s,\mu)}\,f(\lambda,\mu)\,d\mu, \tag{30}$$

where the sum is over all branches (2 of them) of the multivalued function that follows.

Lemma 18 *If $f(\lambda,\mu)$ does not have a pole at $\lambda = u_i$; then the residue* (30) *is non-zero only if $i = j$ and in the latter case it equals to*

$$(a|\beta)\,\mathrm{Res}_{\lambda=u_i}\sum_{\text{all branches}}\frac{f(\lambda,\lambda)}{y_\beta(s,\lambda)}\,d\lambda.$$

Proof Put $a = a' + (a|\beta_i)\beta_i/2$; then a' is invariant with respect to the monodromy around $\lambda = u_i$. From this we get that $\phi_{a'}^+(s,\lambda;z)$ is analytic at $\lambda = u_i$, so it does not contribute to the residue. In other words, it is enough to prove the lemma only for $a = \beta_i$. Let us assume that $a = \beta_i$. Recall that by definition $\beta = \beta_j$, then we get

$$\Omega(\phi_a^+(s,\lambda;z),\mathbf{f}_\beta^-(s,\mu;z)) = \Omega(\mathbf{f}_{\beta_j}^+(s,\mu;z),\phi_{\beta_i}^-(s,\lambda;z)) + \Omega(\phi_{\beta_i}(s,\lambda;z),\mathbf{f}_{\beta_j}(s,\mu;z)).$$

The first symplectic pairing on the RHS does not contribute to the residue, because $\phi_a^-(s,\lambda;z)$ has a pole of order at most $\frac{1}{2}$ so after taking the sum over all branches, the poles of fractional degrees cancel out and hence the 1-form at hands is analytic at $\lambda = u_i$. For the second symplectic pairing, recalling that $\mathbf{f}_{\beta_k}(s,\lambda;z) = \Psi R\mathbf{f}_{A_1}(u_k,\lambda;z)$ for $k = i, j$, we get

$$\Omega(\phi_{A_1}(u_i,\lambda;z)e_i,\mathbf{f}_{A_1}(u_j,\mu;z)e_j) = 2\delta_{i,j}\frac{(\mu-u_j)^{\frac{1}{2}}}{(\lambda-u_i)^{\frac{1}{2}}}\,\delta(\lambda-u_i,\mu-u_j),$$

where

$$\delta(x,y) = \sum_{n\in\mathbb{Z}} x^n y^{-n-1}$$

is the formal δ-function. It is an easy exercise to check that for every $f(y) \in \mathbb{C}((y))$ we have

$$\mathrm{Res}_{y=0}\,\delta(x,y)\,f(y) = f(x).$$

The lemma follows. □

The integral (29) can be written as a sum of two residues: $\mathrm{Res}_{\lambda=u_{N-1}}$ and $\mathrm{Res}_{\lambda=u_N}$. We claim that each of these residues can be reduced to the corresponding residue in the sum (28). Let us present the argument for $\lambda = u_{N-1}$. The other case is completely analogous.

Let $\alpha = \beta_{N-1}$ be the cycle vanishing over u_{N-1}. The summands in (29) for which $r = 2$ and $c_1, c_2 \in \{\chi_1, \chi_2\}$ give precisely

$$\mathrm{Res}_{\lambda=u_{N-1}} \frac{\Omega(\phi_a z^m, \mathbf{f}^-_{\chi_1-\chi_2}(s,\lambda;z))}{y_{\chi_1-\chi_2}(s,\lambda)} \Omega^{(g)}_{\chi_1,\chi_2}(s,\lambda;\mathbf{t})\, d\lambda.$$

On the other hand, using that $\alpha = \chi_1 - \chi_2$ we get

$$\Omega^{(g)}_{\chi_1,\chi_2}(s,\lambda;\mathbf{t}) = -\frac{1}{4}\,\Omega^{(g)}_{\alpha,\alpha}(s,\lambda;\mathbf{t}) + \frac{1}{4}\,\Omega^{(g)}_{\chi_1+\chi_2,\chi_1+\chi_2}(s,\lambda;\mathbf{t})$$

Since $(\chi_1 + \chi_2|\alpha) = 0$, the form $\Omega^{(g)}_{\chi_1+\chi_2,\chi_1+\chi_2}(s,\lambda;\mathbf{t})$ is analytic at $\lambda = u_{N-1}$, so it does not contribute to the residue. Therefore we obtain precisely the $(N-1)$-st residue in (28). It remain only to see that the remaining summands with $r = 2$ cancel out with the summand with $r = 3$.

There are two types of quadratic summands: $c_1, c_2 \in \{\chi_1, \chi_3\}$ and $c_1, c_2 \in \{\chi_2, \chi_3\}$. They add up respectively to

$$\frac{\Omega(\phi_a z^m, \mathbf{f}^-_{\chi_1-\chi_3}(s,\lambda;z))}{y_{\chi_1-\chi_3}(s,\lambda)} \Omega^{(g)}_{\chi_1,\chi_3}(s,\lambda;\mathbf{t})\, d\lambda \tag{31}$$

and

$$\frac{\Omega(\phi_a z^m, \mathbf{f}^-_{\chi_2-\chi_3}(s,\lambda;z))}{y_{\chi_2-\chi_3}(s,\lambda)} \Omega^{(g)}_{\chi_2,\chi_3}(s,\lambda;\mathbf{t})\, d\lambda. \tag{32}$$

By definition

$$\sum_{g=0}^{\infty} \hbar^{g-3/2}\, \Omega^{(g)}_{\chi_i,\chi_3}(s,\lambda;\mathbf{t})\mathcal{A}_s = \hbar^{-1/2}\Big(:\widehat{\phi}_{\chi_i}(s,\lambda)\widehat{\phi}_{\chi_3}(s,\lambda): + P^{(0)}_{\chi_i,\chi_3}(s,\lambda)\Big)\, \mathcal{A}_s. \tag{33}$$

We claim that the propagators $P^{(0)}_{\chi_i,\chi_3}(s,\lambda)$ in (33) do not contribute to the residue at $\lambda = u_{N-1}$. Indeed, their contribution is given by the residue at $\lambda = u_{N-1}$ of the following function

$$\frac{\Omega(\phi_a z^m, \mathbf{f}^-_{\chi_1-\chi_3}(s,\lambda;z))}{y_{\chi_1-\chi_3}(s,\lambda)} P^{(0)}_{\chi_1,\chi_3}(s,\lambda) + \frac{\Omega(\phi_a z^m, \mathbf{f}^-_{\chi_2-\chi_3}(s,\lambda;z))}{y_{\chi_2-\chi_3}(s,\lambda)} P^{(0)}_{\chi_2,\chi_3}(s,\lambda).$$

Note that the above expression is invariant with respect to the local monodromy around $\lambda = u_{N-1}$ and that the coefficients in front of $P^{(0)}_{\chi_i,\chi_3}(s,\lambda)$ do not have a pole at $\lambda = u_{N-1}$. Recalling formula (23) we get that $P^{(0)}_{\chi_i,\chi_3}(s,\lambda)$ has a pole of order at mots 1/2, which implies that the entire expression is analytic at $\lambda = u_{N-1}$.

The normally ordered product on the RHS of (33) is by definition

$$\widehat{\phi}_{\chi_3}(s,\lambda)\,\widehat{\phi}^+_{\chi_i}(s,\lambda) + \widehat{\phi}^-_{\chi_i}(s,\lambda)\widehat{\phi}_{\chi_3}(s,\lambda). \tag{34}$$

Since $(\chi_3|\alpha) = 0$ the field $\widehat{\phi}_{\chi_3}(t,\lambda)$ is analytic at $\lambda = u_{N-1}$. In addition $\widehat{\phi}^-_{\chi_i}(t,\lambda)$ has a pole of order at most $\frac{1}{2}$ at $\lambda = u_{N-1}$. It follows that the second summand in (34) does not contribute to the residue and therefore it can be ignored. For the RHS of (33) we get

$$\sum_{g=0}^{\infty} \hbar^{g-1}\,\widehat{\phi}_{\chi_3}(s,\lambda)\,\langle \phi^+_{\chi_i}(s,\lambda;\psi)\rangle_{g,1}(t;\mathbf{t})\,\mathcal{A}_s,$$

which after recalling the local recursion (28) becomes

$$-\frac{1}{4}\sum_{j=1}^{N} \mathrm{Res}_{\mu=u_j} \frac{\Omega(\phi^+_{\chi_i}(s,\lambda;z),\mathbf{f}^-_{\beta_j}(s,\mu;z))}{y_{\beta_j}(s,\mu)}\,\widehat{\phi}_{\chi_3}(s,\lambda)\,X^{u_j}_{s,\mu}(\beta_j^2)\,d\mu\,\mathcal{A}_s,$$

where $X^{u_j}_{s,\mu}(a)$ is the Laurent series expansion of $X_{s,\mu}(a)$ in $(\mu-u_j)$. Therefore we need to compute the residues $\mathrm{Res}_{\lambda=u_{N-1}}\,\mathrm{Res}_{\mu=u_j}$ of the following expressions

$$-\frac{1}{4}\sum_{i=1,2} \frac{\Omega(\phi_a z^m,\mathbf{f}^-_{\chi_i-\chi_3}(s,\lambda;z))}{y_{\chi_i-\chi_3}(s,\lambda)}\,\frac{\Omega(\phi^+_{\chi_i}(s,\lambda;z),\mathbf{f}^-_{\beta_j}(s,\mu;z))}{y_{\beta_j}(s,\mu)}\,\widehat{\phi}_{\chi_3}(s,\lambda)\,X^{u_j}_{s,\mu}(\beta_j^2)\,d\mu\,\mathcal{A}_s.$$

The operator $\widehat{\phi}_{\chi_3}(s,\lambda)\,X^{u_j}_{s,\mu}(\beta_j^2)$ can be written as

$$:\widehat{\phi}_{\beta_j}(s,\mu)^2\,\widehat{\phi}_{\chi_3}(s,\lambda): + 2[\widehat{\phi}^+_{\chi_3}(s,\lambda),\widehat{\phi}^-_{\beta_j}(s,\mu)]\,\widehat{\phi}_{\beta_j}(s,\mu) + P^{(0)}_{\beta_j,\beta_j}(s,\mu)\widehat{\phi}_{\chi_3}(s,\lambda). \tag{35}$$

Since $(\chi_3|\alpha) = 0$ the operator $\widehat{\phi}^+_{\chi_3}(s,\lambda)$ is regular at $\lambda = u_{N-1}$. It follows that the commutator

$$[\widehat{\phi}^+_{\chi_3}(s,\lambda),\widehat{\phi}^-_{\beta_j}(s,\mu)] \in \mathbb{C}((\lambda - u_{N-1},\mu - u_j))$$

and therefore we may recall Lemma 18. The above residue is non-zero only if $j = N-1$. In the latter case we get

$$-\frac{1}{4}\,\mathrm{Res}_{\lambda=u_{N-1}} \sum_{i=1,2} (\chi_i|\alpha)\, \frac{\Omega(\phi_a z^m, \mathbf{f}^-_{\chi_i-\chi_3}(s,\lambda;z))}{y_{\chi_i-\chi_3}(s,\lambda)\, y_\alpha(s,\lambda)}\, \widehat{\phi}_{\chi_3}(s,\lambda)\, X^{u_{N-1}}_{s,\lambda}(\alpha^2)\, d\lambda\, \mathcal{A}_s. \tag{36}$$

Note that

$$[\widehat{\phi}^+_{\chi_3}(s,\lambda), \widehat{\phi}^-_{\beta_j}(s,\mu)] = \iota_{\lambda-u_{N-1}}\, \iota_{\mu-u_{N-1}}\, P_{\chi_3,\beta_j}(s,\lambda,\mu),$$

where $\iota_{\lambda-u_{N-1}}$ is the Laurent series expansion at $\lambda = u_{N-1}$. Hence

$$\widehat{\phi}_{\chi_3}(s,\lambda)\, X^{u_{N-1}}_{s,\lambda}(\alpha^2) = \iota_{\lambda-u_{N-1}} X_{s,\lambda}(\chi_3\alpha^2).$$

By definition

$$-\frac{1}{4}\alpha^2 = \chi_1\,\chi_2 - \frac{1}{4}\chi_3^2$$

and since χ_3 is invariant with respect to the local monodromy around $\lambda = u_{N-1}$, the field $X_{s,\lambda}(\chi_3^3)$ does not contribute to the residue. We get the following formula for the residue (36):

$$\mathrm{Res}_{\lambda=u_{N-1}} \sum_{i=1,2} (\chi_i|\alpha)\, \frac{\Omega(\phi_a z^m, \mathbf{f}^-_{\chi_i-\chi_3}(s,\lambda;z))}{y_{\chi_i-\chi_3}(s,\lambda)\, y_\alpha(s,\lambda)}\, X^{u_{N-1}}_{s,\lambda}(\chi_1\chi_2\chi_3)\, d\lambda\, \mathcal{A}_s.$$

Using that $\alpha = \chi_1 - \chi_2$, $(\chi_1|\alpha) = 1$, and $(\chi_2|\alpha) = -1$ we get

$$\mathrm{Res}_{\lambda=u_{N-1}} \Bigg(\frac{\Omega(\phi_a z^m, \mathbf{f}^-_{\chi_1}(s,\lambda;z))}{y_{\chi_2-\chi_1}(s,\lambda)\, y_{\chi_3-\chi_1}(s,\lambda)} + \frac{\Omega(\phi_a z^m, \mathbf{f}^-_{\chi_2}(s,\lambda;z))}{y_{\chi_1-\chi_2}(s,\lambda)\, y_{\chi_3-\chi_2}(s,\lambda)}$$
$$+ \frac{\Omega(\phi_a z^m, \mathbf{f}^-_{\chi_3}(s,\lambda;z))}{y_{\chi_1-\chi_3}(s,\lambda)\, y_{\chi_2-\chi_3}(s,\lambda)} \Bigg) \times \sum_{g=0}^{\infty} \hbar^{g-3/2}\Omega^{(g)}_{\chi_1\chi_2\chi_3}(s,\lambda;\mathbf{t})\, d\lambda\, \mathcal{A}_s.$$

This sum cancels out the contribution to the residue at $\lambda = u_{N-1}$ of the cubic terms (i.e. the terms with $r = 3$) of the integral (29). □

Note that in the integral (29) we may choose the integration contour to be the boundary of the disk D_{N-1}. Since the integrand in (29) has singularities only at the critical values $u_{N-1}(s)$ and $u_N(s)$, which are inside the disk D_{N-1} for all $s \in S_0$, we get that the integral (29) depends analytically on $s \in S_0$. Using Theorem 17 we can set up a recursion that produces functions analytic in a neighborhood of any generic point $b_0 \in B$. For example, let us write the recursion for $\langle \phi_a \rangle_{1,1}(s; 0)$. Since

$$\Omega^{(1)}_{c_1,c_2}(s, \lambda; 0) = P^{(0)}_{c_i,c_j}(s, \lambda), \quad \Omega^{(1)}_{c_1,c_2,c_3}(s, \lambda; 0) = 0,$$

we have

$$\begin{aligned}\langle \phi_a \rangle_{1,1}(s; 0) = \frac{1}{4}\sum_{i=1}^{N-2} \frac{1}{2\pi\sqrt{-1}} \oint_{C_i} \frac{(I^{(-1)}_{\beta_i}(s, \lambda), \phi_a)}{(I^{(-1)}_{\beta_i}(s, \lambda), \mathbf{1})} P^{(0)}_{\beta_i,\beta_i}(s, \lambda) d\lambda \\ - \frac{1}{2\pi\sqrt{-1}} \oint_{C_{N-1}} \sum_{1\leq i<j\leq 3} \frac{(I^{(-1)}_{\chi_i-\chi_j}(s, \lambda), \phi_a)}{(I^{(-1)}_{\chi_i-\chi_j}(s, \lambda), \mathbf{1})} P^{(0)}_{\chi_i,\chi_j}(s, \lambda) d\lambda,\end{aligned}$$

where C_i is the boundary of the disk D_i, $1 \leq i \leq N - 1$. By definition if $(s, \lambda) \in S_0 \times C_i$, then (s, λ) is not a point on the discriminant. Note that $I^{(-1)}_{\varphi}(s, \lambda) \neq 0$ if φ is a vanishing cycle and (s, λ) is not on the discriminant, because according to Lemma 14

$$2 = (\varphi|\varphi) = (I^{(0)}_{\varphi}(s, \lambda), (\theta + 1/2) I^{(-1)}_{\varphi}(s, \lambda)).$$

All integrals depend analytically on $s \in S_0$, so the correlator $\langle \phi_a \rangle_{1,1}(s; 0)$ is analytic in the entire neighborhood S_0 of $b_0 \in B$. Using induction on the lexicographical order of the pairs (g, n), where g is the genus and n is the number of insertions, we can prove by induction that all ancestor correlators are analytic in the neighborhood S_0. Theorem 1 follows from the Hartogues extension theorem.

Acknowledgements I am thankful to Y. Ruan for organizing the workshop and creating an inspiring environment. This work is supported by JSPS Grant-In-Aid 26800003 and by the World Premier International Research Center Initiative (WPI Initiative), MEXT, Japan.

References

1. Arnol'd, V.I., Gusein-Zade, S.M., Varchenko, A.N.: Singularities of Differentiable Maps. Vol. II. Monodromy and Asymptotics of Integrals. Monographs in Mathematics, vol. 83, Birkhäuser, Boston (1988)
2. Bakalov, B., Milanov, T.: W-constraints for the total descendant potential of a simple singularity. Comps. Math. **149**, 840–888 (2013)
3. Bouchard, V., Eynard, B.: Think globally, compute locally. J. High Energy Phys. **2013**(2), 143 (2013)
4. Dubrovin, B.: Geometry of 2D topological field theories. In: Integrable Systems and Quantum Groups (Montecatini Terme, 1993). Lecture Notes in Mathematics, vol. 1620, pp. 120–348. Springer, Berlin (1996)
5. Dubrovin, B., Zhang, Y.: Normal forms of hierarchies of integrable PDEs, Frobenius manifolds and Gromov-Witten invariants, 295 pp. A new 2005 version of arXiv:math/0108160v1

6. Dunnin–Barkowski, P., Orantin, N., Shadrin, S., Spitz, L.: Identification of the Givental formula with the spectral curve topological recursion procedure. Commun. Math. Phys. **328**(2), 669–700 (2014)
7. Fan, H., Jarvis, T., Ruan, Y.: The Witten equation, mirror symmetry, and quantum singularity theory. Ann. Math. **178**(2), 1–106 (2013)
8. Givental, A.: Semisimple Frobenius structures at higher genus. Int. Math. Res. Not. **2001**(23), 1265–1286 (2001)
9. Givental, A.: Gromov–Witten invariants and quantization of quadratic Hamiltonians. Mosc. Math. J. **1**, 551–568 (2001)
10. Givental, A.: A_{n-1} singularities and nKdV Hierarchies. Mosc. Math. J. **3**(2), 475–505 (2003)
11. Hertling, C.: Frobenius Manifolds and Moduli Spaces for Singularities. Cambridge Tracts in Mathematics, vol. 151, x+270 pp. Cambridge University Press, Cambridge (2002)
12. Kac, V., Schwarz, A.: Geometric interpretation of the partition function of 2D gravity. Phys. Lett. B **257**, 329–334 (1991)
13. Kontsevich, M.: Intersection theory on the moduli space of curves and the matrix Airy function. Commun. Math. Phys. **147**, 1–23 (1992)
14. Li, C., Li, S., Saito, K., Shen, Y.: Mirror symmetry for exceptional unimodular singularities. J. Eur. Math. Soc. **19**(4), 1189–1229 (2017)
15. Liu, S.-Q., Ruan, Y., Zhang, Y.: BCFG Drinfeld–Sokolov hierarchies and FJRW-theory. Invent. Math. **201**(2), 711–772 (2015)
16. Milanov, T.: The Eynard–Orantin recursion for the total ancestor potential. Duke Math. J. **163**(9), 1795–1824 (2014)
17. Milanov, T.: Analyticity of the total ancestor potential in singularity theory. Adv. Math. **255**(2), 217–241 (2014)
18. Saito, K.: On periods of primitive integrals I. RIMS (1982, preprint)
19. Saito, K.: On a linear structure of the quotient variety by a finite reflection group. Publ. RIMS Kyoto Univ. **29**, 535–579 (1993)
20. Saito, K., Takahashi, A.: From primitive forms to Frobenius manifolds. Proc. Sympos. Pure Math. **78**, 31–48 (2008)
21. Teleman, C.: The structure of 2D semi-simple field theories. Invent. Math. **188**(3), 525–588 (2012)
22. Witten, E.: Two-dimensional gravity and intersection theory on moduli space. In: Surveys in Differential Geometry, pp. 243–310, Lehigh University, Bethlehem (1991)

Lecture Notes on Bihamiltonian Structures and Their Central Invariants

Si-Qi Liu

Contents

Abstract In these lecture notes, we give an introduction to the classification theorem of semisimple bihamiltonian structures, with as much details as possible. The equivalence classes of this classification problem are characterized by the so-called central invariants. In the last section, two examples are given to illuminate the applications of central invariants in cohomological field theories.

1 Introduction

Let X be a toric Fano variety. It is well known that the quantum cohomology $QH^*(X)$ is a semisimple Frobenius manifold, and the generating function of all its Gromov-Witten invariants, which is usually called the total descendant potential of X, is given by Givental's quantization formula (see [25] for more details):

$$Z_X = \tau_I(u)\hat{S}_u^{-1}(z)\hat{\Psi}_u\hat{R}_u(z)e^{(U/z)\hat{}}\left(\prod_{i=1}^{n} Z_{pt}^{(i)}\right).$$

S.-Q. Liu (✉)
Department of Mathematical Sciences, Tsinghua University, Beijing, People's Republic of China
e-mail: liusq@tsingha.edu.cn

E. Clader, Y. Ruan (eds.), *B-Model Gromov-Witten Theory*, Trends in Mathematics,
https://doi.org/10.1007/978-3-319-94220-9_7

Givental also proved the Virasoro conjecture in this case, that is, Z_X satisfies the Virasoro constraints

$$L_m Z_X = 0, \quad m \geq -1,$$

where $\{L_m\}_{m\in\mathbb{Z}}$ is a set of linear differential operators satisfying the Virasoro commuting relations [17, 22].

The preprint version of Givental's [25] was released in Aug 2001. In the same month, Dubrovin and Zhang put another preprint [13] on arXiv, which showed that the total descendant potential of a semisimple Frobenius manifold is uniquely determined by its genus zero part and the Virasoro constraints. Let $F = \log Z_X$ be the free energy of X, then expand F with respect to the string coupling constant $\hbar$

$$F = \sum_{g\geq 0} \hbar^{g-1} F_g.$$

Dubrovin and Zhang derived a series of differential equations for F_g from the Virasoro constraints, whose generating function is called the loop equation for X, and showed that one can obtained F_g recursively from these equations. In particular, they gave an explicit formula of F_2 for an arbitrary semisimple Frobenius manifold, which is not easy to obtain from Givental's quantization formula.

According to Dubrovin-Zhang's uniqueness theorem, their approach is equivalent to Givental's quantization formula. Givental's formula has drawn much attention, while Dubrovin-Zhang's approach is less well known. One possible reason is that Dubrovin-Zhang's preprint [13] is too long: it contains more than 180 pages, whose first 150 pages are about an axiomatic framework for integrable systems that may govern a Gromov-Witten theory. Their loop equation appears in the last 30 pages, and the main results are also proved in this last part. It seems that to understand their main results one must read the first 150 pages, which is indeed a tough work for people not working on integrable systems. But in my personal opinion, the last 30 pages of Dubrovin-Zhang's preprint is almost independent of the first 150 pages, so one can read it directly.

In an informal workshop on Landau-Ginzburg B-model held in University of Michigan, March 10–14, 2014, I gave a short introduction to Dubrovin-Zhang's loop equation, especially on the case with $X =$ point. I planned to give more details for general cases in the present lecture notes. But Zhang told me that Dubrovin and he have been working on a similar introductory paper for months, and there is also a good introduction to this subject in Dubrovin's new paper [8], so I decide to talk about something else—something on the first 150 pages of Dubrovin and Zhang's preprint [13].

Saying one can skip the first 150 pages of [13] doesn't mean that this part is not important. Instead, this part is more general, so it includes not only the cases in which Givental's formula or Dubrovin-Zhang's loop equation work but also the cases make these two approaches fail. For example, Dubrovin-Zhang's axiom system consists of four axioms (see [13] for more details):

- BH = Bihamiltonian structure
- QT = Quasi-triviality
- TS = Tau structure
- VS = Linearizable Virasoro symmetries

If an integrable system satisfies all these axioms, the corresponding total descendant potential must be given by Givental's formula or Dubrovin-Zhang's loop equation. But, if it satisfies all but the last axiom, one can also define its total descendant potential, and this potential is not equivalent to Givental's one in general. Recently, Wu showed that the Drinfeld-Sokolov hierarchies of BCFG types are integrable systems of this kind [34]. Then Ruan, Zhang and I show that the generating functions of FJRW invariants of boundary singularities of BCFG type gives tau functions of these integrable systems [30]. In particular, we show that the BCFG Drinfeld-Sokolov hierarchies must be not equivalent to Dubrovin-Zhang's hierarchies, so the generating function of BCFG FJRW invariants must be not given by Givental's formula.

To show that two integrable systems are not equivalent is highly nontrivial. One need to find out the orbits of a class of integrable systems under the action of a certain transformation group. Such a classification problem is first precisely stated in Dubrovin-Zhang's [13] for the integrable systems satisfying the BH axiom. We introduced the concept of *central invariants*, which can be regarded as coordinates on the orbit space, and answered the uniqueness part of this classification problem [14, 27]. As a byproduct, we also show that the QT axiom is a corollary of the BH axiom, which is also conjectured and partially proved in [13]. The existence part of the above classification problem is also resolved recently. In [29], we founded a new framework for the computation of the cooresponding bihamiltonian cohomologies, and proved the existence theorem for the simplest case, that is the bihamiltonian structure of the Korteweg-de Vries hierarchy. We planned to consider the general cases in [16] by using a similar argument. This is not an easy generalization, because our computation method, even for the simplified one, is still very complicated. In a recent preprint [2] (c.f. [1]), Carlet, Posthuma, and Shadrin developed some new computing techniques based on our approach, several interesting spectral sequences, and some homotopy formulae, then proved the existence theorem for the general cases.

The central invariants of a bihamiltonian structure are a set of functions of one variable. For the integrable systems satisfying Dubrovin-Zhang's four axioms, all central invariants must be $1/24$. On the other hand, we computed the central invariants for the bihamiltonian structure for Drinfeld-Sokolov hierarchies [15]. For the BCFG cases, their central invariants are unequal constants, so they are not equivalent to Dubrovin-Zhang's integrable hierarchies.

In these lecture notes, I will give an introduction to our results with as much details as possible. In Sect. 2, I recall some basic facts of finite dimensional Poisson geometry. We introduce the Schouten-Nijenhuis bracket in an unusual way, which can be also used in the infinite dimensional case. Then we give the definition of Hamiltonian structures for partial differential equations in Sect. 3. In Sects. 4 and 5, we prove some results on the relation between classification problems of (bi)hamiltonian structures and their cohomologies. We also prove a Darboux theorem for certain Hamiltonian structures. Then we introduce the notion of central invariants of a semisimple bihamiltonian structure in Sect. 6. In the last subsection, we give an introduction to the Drinfeld-Sokolov bihamiltonian structure and their central invariants.

2 Finite Dimensional Poisson Geometry

2.1 Basic Definition

Let M be a smooth manifold of dimension n, and $\mathcal{A}_0 = C^\infty(M)$ be the algebra of smooth functions on M (we will explain why we use this notation in the next section). A Poisson bracket on M is, by definition, a bilinear map $\{\,,\} : \mathcal{A}_0 \times \mathcal{A}_0 \to \mathcal{A}_0$ satisfying the following conditions:

$$\text{Skew-symmetry: } \{f, g\} + \{g, f\} = 0, \tag{2.1}$$

$$\text{Jacobi identity: } \{\{f, g\}, h\} + \{\{g, h\}, f\} + \{\{h, f\}, g\} = 0, \tag{2.2}$$

$$\text{Leibniz's rule: } \{f \cdot g, h\} = f \cdot \{g, h\} + \{f, h\} \cdot g, \tag{2.3}$$

where $f, g, h \in \mathcal{A}_0$, and $\cdot$ is the multiplication of $\mathcal{A}_0$. The manifold M is called a Poisson manifold if it is equipped with a Poisson bracket.

The condition (2.1) and (2.2) show that $(\mathcal{A}_0, \{\,,\})$ forms a Lie algebra, and the condition (2.3) implies (by using Hadamard's Lemma) that the Poisson bracket is locally given by [1]

$$\{f, g\} = P^{\alpha\beta}(u) \frac{\partial f}{\partial u^\alpha} \frac{\partial g}{\partial u^\beta}, \tag{2.4}$$

where $(u^1, \ldots, u^n)$ is a set of local coordinates on M. The functions $P^{\alpha\beta}(u)$ are actually given by

$$\{u^\alpha, u^\beta\} = P^{\alpha\beta}(u),$$

and they are called the components of the Poisson bracket $\{\,,\}$ in the local coordinates system $(u^1, \ldots, u^n)$.

[1] In this paper, summation over repeated *Greek* indexes is always assumed, and we don't sum over *Latin* indexes.

The formula (2.4) shows that we can introduce a bivector, i.e. a skew-symmetric tensor of (2, 0) type,

$$P = P^{\alpha\beta}(u)\frac{\partial}{\partial u^\alpha} \wedge \frac{\partial}{\partial u^\beta}, \tag{2.5}$$

and then write the Poisson bracket as the following form

$$\{f, g\} = \langle P, \mathrm{d}f \wedge \mathrm{d}g\rangle,$$

where $\langle\, ,\rangle$ is the standard pairing between tensors of $(2, 0)$ and $(0, 2)$ types. The tensor P is called the Poisson tensor or Poisson structure of the Poisson manifold $(M, \{\, ,\})$.

The condition (2.2) of the Poisson bracket $\{\, ,\}$ is equivalent to the following condition on the components of P:

$$\frac{\partial P^{\alpha\beta}}{\partial u^\sigma} P^{\sigma\gamma} + \frac{\partial P^{\beta\gamma}}{\partial u^\sigma} P^{\sigma\alpha} + \frac{\partial P^{\gamma\alpha}}{\partial u^\sigma} P^{\sigma\beta} = 0. \tag{2.6}$$

This condition also has a coordinate-free form, which requires the notion of Schouten-Nijenhuis bracket.

The Schouten-Nijenhuis bracket is a bilinear operation defined on the space $\Lambda^* = \Gamma(\wedge^* T(M))$ of polyvectors. There are several equivalent ways to define this operation. We give two of them, which can be easily generalized to the infinite-dimensional case.

2.2 *Nijenhuis-Richardson Bracket*

Let $P \in \Lambda^p$ be a p-vector. We define its action on p smooth functions $f_1, \ldots, f_p \in \mathcal{A}_0$ as follow:

$$P(f_1, \ldots, f_p) = \langle P, \mathrm{d}f_1 \wedge \cdots \wedge \mathrm{d}f_p\rangle,$$

so P can be regarded as a linear map from $\wedge^p \mathcal{A}_0$ to $\mathcal{A}_0$.

Let $\mathcal{V}^* = \mathrm{Hom}(\wedge^* \mathcal{A}_0, \mathcal{A}_0)$, whose elements are called generalized polyvectors. In particular, we have $\mathcal{V}^0 = \Lambda^0 = \mathcal{A}_0$, and $\mathcal{V}^{<0} = 0$. We regard Λ^* as a subspace of $\mathcal{V}^*$, and it is easy to see that $P \in \mathcal{V}^p$ belongs to Λ^p if and only if

$$P(f \cdot g, f_2, \ldots, f_p) = f \cdot P(g, f_2, \ldots, f_p) + P(f, f_2, \ldots, f_p) \cdot g \tag{2.7}$$

for all $f, g, f_2, \ldots, f_p \in \mathcal{A}_0$.

Theorem 1 ([28])

(a) There exists a unique bilinear map $[\,,]: \mathcal{V}^p \times \mathcal{V}^q \to \mathcal{V}^{p+q-1}$ *satisfying the following conditions:*

$$[P, f](f_2, \ldots, f_p) = P(f, f_2, \ldots, f_p), \tag{2.8}$$

$$[P, Q] = (-1)^{pq}[Q, P], \tag{2.9}$$

$$[[P, Q], f] + (-1)^{qp}[[Q, f], P] + [[f, P], Q] = 0, \tag{2.10}$$

where $P \in \mathcal{V}^p$, $Q \in \mathcal{V}^q$, *and* $f, f_2, \ldots, f_p \in \mathcal{A}_0$. *It is called the Nijenhuis-Richardson bracket of the generalized polyvectors.*

(b) The Nijenhuis-Richardson bracket satisfies the following graded Jacobi identity:

$$(-1)^{pr}[[P, Q], R] + (-1)^{qp}[[Q, R], P] + (-1)^{rq}[[R, P], Q] = 0, \tag{2.11}$$

where $P \in \mathcal{V}^p$, $Q \in \mathcal{V}^q$, *and* $R \in \mathcal{V}^r$.

Proof (a) We prove uniqueness first. Let $P \in \mathcal{V}^p$, $Q \in \mathcal{V}^q$. When $p = q = 0$, $[P, Q]$ must vanish, since $\mathcal{V}^{-1} = 0$. When $(p, q) = (1, 0)$, then the property (2.8) implies that $[P, Q] = P(Q)$. The (0, 1) case is similar, due to the property (2.9). When $(p, q) = (1, 1)$, take an $f \in \mathcal{A}_0$, then we have

$$\begin{aligned}[P, Q](f) &= [[P, Q], f] = [[Q, f], P] - [[f, P], Q] \\ &= P(Q(f)) - Q(P(f)).\end{aligned}$$

In general, take $f, f_2, \ldots, f_{p+q-1}$, we have

$$\begin{aligned}[P, Q](f, f_2, \ldots, f_{p+q-1}) &= [[P, Q], f](f_2, \ldots, f_{p+q-1}) \\ &= -\left((-1)^{qp}[[Q, f], P] + [[f, P], Q]\right)(f_2, \ldots, f_{p+q-1}),\end{aligned}$$

so the bracket defined on $\mathcal{V}^p \times \mathcal{V}^q$ is determined by the brackets defined on $\mathcal{V}^{p-1} \times \mathcal{V}^q$ and $\mathcal{V}^p \times \mathcal{V}^{q-1}$. Since we have shown the uniqueness for the $0 \le p, q \le 1$ cases, it also holds true for general cases. The uniqueness is proved.

To prove the existence, we recall the product $\bar{\wedge} : \mathcal{V}^p \times \mathcal{V}^q \to \mathcal{V}^{p+q-1}$ defined in [31]:

$$P \bar{\wedge} Q(f_1, \ldots, f_{p+q-1}) = \sum_{I \in S_{p,q}} (-1)^{|I|} P(Q(f_{i_1}, \ldots, f_{i_q}), f_{i_{q+1}}, \ldots, f_{i_{p+q-1}}),$$

where $S_{p,q}$ is the following subset of the symmetry group S_{p+q-1}:

$$S_{p,q} = \left\{ I = (i_1, \ldots, i_{p+q-1}) \in S_{p+q-1} \,\middle|\, \begin{array}{c} i_1 < \cdots < i_q \\ i_{q+1} < \cdots < i_{p+q-1} \end{array} \right\},$$

and $|I|$ is the parity of the permutation I.

The bracket [,] can be defined as

$$[P, Q] = (-1)^{(p+1)q} P \bar{\wedge} Q + (-1)^p Q \bar{\wedge} P.$$

We need to show that this bracket satisfies the conditions (2.8)–(2.10). The condition (2.8) and (2.9) are easy to verify. In particular, if $P \in \mathcal{V}^p$, $f, f_2, \ldots, f_p \in \mathcal{A}_0$, we have

$$[P, f](f_2, \ldots, f_p) = P \bar{\wedge} f(f_2, \ldots, f_p) = P(f, f_2, \ldots, f_p).$$

We denote $i_f(P) = [P, f]$, then one can show that

$$i_f(P \bar{\wedge} Q) = P \bar{\wedge} i_f(Q) + (-1)^{q+1} i_f(P) \bar{\wedge} Q,$$

which implies the condition (2.10). The existence is proved.

(b) We prove the identity by induction on $p + q + r$. When $r = 0$, it is just the condition (2.10). When $r > 0$, we assume that the identity (2.11) holds true for any p', q', r' satisfying $p' + q' + r' < p + q + r$. Let $P \in \mathcal{V}^p$, $Q \in \mathcal{V}^q$, $R \in \mathcal{V}^r$, and take an $f \in \mathcal{A}_0$, one can show that

$$i_f([[P, Q], R]) = [[i_f(P), Q], R] + (-1)^p [[P, i_f(Q)], R] + (-1)^{p+q} [[P, Q], i_f(R)].$$

Then by using the induction assumption, we obtain

$$i_f\left((-1)^{pr}[[P, Q], R] + (-1)^{qp}[[Q, R], P] + (-1)^{rq}[[R, P], Q]\right) = 0,$$

which implies the identity (2.11). The theorem is proved. □

Remark 2 The above theorem only used the fact that $\mathcal{A}_0$ is a linear space. In next section, we will replace $\mathcal{A}_0$ by another linear space to define the corresponding bracket operation on that space.

Proposition 3 *The Nijenhuis-Richardson bracket can be restricted onto the subspace Λ^*, that is, if $P \in \Lambda^p$, $Q \in \Lambda^q$, then $[P, Q] \in \Lambda^{p+q-1}$. The restricted bracket* [,] *is called the Schouten-Nijenhuis bracket of polyvectors.*

Proof We prove the proposition by induction on $p + q$.

When $(p,q) = (0,0), (1,0), (0,1), (2,0), (0,2)$, the proposition is trivially true. When $(p,q) = (1,1)$, take $f, g \in \mathcal{A}_0$, we have

$$\begin{aligned}
&[P,Q](f\cdot g)\\
=&P(Q(f\cdot g)) - Q(P(f\cdot g))\\
=&P(f\cdot Q(g) + g\cdot Q(f)) - Q(f\cdot P(g) + g\cdot P(f))\\
=&(f\cdot P(Q(g)) + P(f)\cdot Q(g) + g\cdot P(Q(f)) + P(g)\cdot Q(f))\\
&-(f\cdot Q(P(g)) + Q(f)\cdot P(g) + g\cdot Q(P(f)) + Q(g)\cdot P(f))\\
=&f\cdot [P,Q](g) + g\cdot [P,Q](f),
\end{aligned}$$

so $[P,Q] \in \Lambda^1$. From now on we can assume $p+q \geq 3$.

Suppose the proposition holds true for any p', q' satisfying $p'+q' < p+q$, take $f, g, f_2, \dots, f_{p+q-1} \in \mathcal{A}_0$, we have

$$\begin{aligned}
&[P,Q](f\cdot g, f_2, \dots, f_{p+q-1})\\
=&([i_{f_2}(P), Q] + (-1)^p[P, i_{f_2}(Q)])(f\cdot g, f_3, \dots, f_{p+q-1}).
\end{aligned}$$

Note that $i_{f_2}(P) \in \Lambda^{p-1}$, $i_{f_2}(Q) \in \Lambda^{q-1}$, so we have

$$\begin{aligned}
&[i_{f_2}(P), Q](f\cdot g, f_3, \dots, f_{p+q-1})\\
=&f\cdot [i_{f_2}(P), Q](g, f_3, \dots, f_{p+q-1}) + g\cdot [i_{f_2}(P), Q](f, f_3, \dots, f_{p+q-1})\\
&[P, i_{f_2}(Q)](f\cdot g, f_3, \dots, f_{p+q-1})\\
=&f\cdot [P, i_{f_2}(Q)](g, f_3, \dots, f_{p+q-1}) + g\cdot [P, i_{f_2}(Q)](f, f_3, \dots, f_{p+q-1}),
\end{aligned}$$

so we have

$$\begin{aligned}
&[P,Q](f\cdot g, f_2, \dots, f_{p+q-1})\\
=&f\cdot [P,Q](g, f_2, \dots, f_{p+q-1})) + g\cdot [P,Q](f, f_2, \dots, f_{p+q-1})).
\end{aligned}$$

The proposition is proved. □

Lemma 4 *Let $P \in \Lambda^2$ be a bivector, the following conditions are equivalent*

i) P gives the Poisson tensor of a Poisson bracket $\{\,,\}$;
ii) $[P,P] = 0$;
iii) The map $d_P : \Lambda^* \to \Lambda^{*+1}$, $Q \mapsto [P,Q]$ satisfies $d_P^2 = 0$.

Proof For any $P, Q \in \mathcal{V}^2$ and $f, g, h \in \mathcal{A}_0$, we have

$$\begin{aligned} &[P, Q](f, g, h) \\ =&P(Q(f, g), h) + P(Q(g, h), f) + P(Q(h, f), g) \\ &+ Q(P(f, g), h) + Q(P(g, h), f) + Q(P(h, f), g). \end{aligned}$$

Define $\{f, g\} = P(f, g)$, then we have

$$\{\{f, g\}, h\} + \{\{g, h\}, f\} + \{\{h, f\}, g\} = \frac{1}{2}[P, P](f, g, h).$$

The equivalence of i) and ii) is proved.

For any $Q \in \Lambda^q$, we have

$$[[P, P], Q] + [[P, Q], P] + [[Q, P], P] = 0,$$

which implies that

$$[P, [P, Q]] = -\frac{1}{2}[[P, P], Q].$$

The equivalence of ii) and iii) is proved. □

2.3 Odd-Symplectic Bracket

The above axiomatic definition of Schouten-Nijenhuis bracket is not very convenient for computation, so we also need another one.

Let $\hat{M} = \Pi(T^*(M))$ be the cotangent bundle of M with fiber's parity reversed, that is, the fiber $T_p^*(M)$ at $\forall p \in M$ is regarded as a super vector space of dimension $(0|n)$. Suppose $(u^1, \dots, u^n)$ is a set of local coordinates on M, and $(\theta_1, \dots, \theta_n)$ be the coordinates on fibers with respect to the basis $\mathrm{d}u^1, \dots, \mathrm{d}u^n$. It is easy to see that, if we change the local coordinate system to another one, say $(\tilde{u}^1, \dots, \tilde{u}^n)$, the transformation $\theta \mapsto \tilde{\theta}$ is given by the following formula:

$$\tilde{\theta}_\alpha = \frac{\partial u^\beta}{\partial \tilde{u}^\alpha}\theta_\beta, \tag{2.12}$$

which is same with the transformation formula for $\frac{\partial}{\partial u^\alpha}$. Denote by $\hat{\mathcal{A}}_0 = C^\infty(\hat{M})$ the superalgebra of smooth functions on $\hat{M}$.

Lemma 5 *There is an isomorphism* $\jmath : \hat{\mathcal{A}}_0 \to \Lambda^*$.

Proof The superalgebra $\hat{\mathcal{A}}_0$ can be decomposed as

$$\hat{\mathcal{A}}_0 = \bigoplus_{p=0}^{n} \hat{\mathcal{A}}_0^p,$$

where $\hat{\mathcal{A}}_0^p$ is the subspace consisting of functions which have the following form in a local coordinate system:

$$P = P^{\alpha_1 \cdots \alpha_p} \theta_{\alpha_1} \cdots \theta_{\alpha_p},$$

where $P^{\alpha_1 \cdots \alpha_p}$'s are components of a skew-symmetric tensor of $(p, 0)$ type. In particular, $\hat{\mathcal{A}}_0^0 = \mathcal{A}_0$.

We regard Λ^* as the subspace of $\mathcal{V}$ whose elements obey the Leibniz's rule (2.7), and then define the isomorphism $\jmath$ as follow:

$$\jmath : \hat{\mathcal{A}}_0^p \to \Lambda^p, \quad P \mapsto \jmath(P),$$

where the action of $\jmath(P)$ on $f_1, \ldots, f_p \in \mathcal{A}_0$ is given by

$$\jmath(P)(f_1, \ldots, f_p) = \frac{\partial^p P}{\partial \theta_{\alpha_p} \ldots \partial \theta_{\alpha_1}} \frac{\partial f_1}{\partial u^{\alpha_1}} \cdots \frac{\partial f_p}{\partial u^{\alpha_p}}.$$

Then it is not hard to show that $\jmath$ is an isomorphism. □

The cotangent bundle $T^*(M)$ has a canonical symplectic structure, so $\hat{M}$ has a canonical odd-symplectic structure. The corresponding odd-Poisson bracket can be written as

$$[P, Q]_{\hat{\mathcal{A}}_0} = \frac{\partial P}{\partial \theta_\alpha} \frac{\partial Q}{\partial u^\alpha} + (-1)^p \frac{\partial P}{\partial u^\alpha} \frac{\partial Q}{\partial \theta_\alpha}, \tag{2.13}$$

where $P \in \hat{\mathcal{A}}_0^p$, $Q \in \hat{\mathcal{A}}_0^q$. Note that this bracket has other variants (see [24] for example). Here we choose the one that is equivalent to the Schouten-Nijenhuis bracket introduced in the last section.

Proposition 6 *We have the following identity:*

$$\jmath([P, Q]_{\hat{\mathcal{A}}_0}) = [\jmath(P), \jmath(Q)]. \tag{2.14}$$

Proof We only need to show that $[\,,\,]_{\hat{\mathcal{A}}_0}$ also satisfies the conditions (2.8)–(2.10). This is not a hard task, so we left it to readers. The proposition is proved. □

From now on, we can identify $\hat{\mathcal{A}}_0$ and Λ^*, then write $[\,,\,]_{\hat{\mathcal{A}}_0}$ as $[\,,\,]$. A bivector (2.5) can be written as the following form:

$$P = \frac{1}{2} P^{\alpha\beta} \theta_\alpha \theta_\beta.$$

It is a Poisson structure if and only if $[P, P] = 0$. Here the bracket $[\,,\,]$ can be computed by using (2.13).

If $X = X^\alpha \frac{\partial}{\partial u^\alpha}$ is a vector field on M, we can identity it with $X = X^\alpha \theta_\alpha$. Let $H \in \mathcal{A}_0$, the Hamiltonian vector field X_H of H is defined as $X_H = [P, H]$, then we have

$$[X_F, X_G] = X_{\{F,G\}}.$$

In local coordinates, we have

$$X_H = X_H^\beta \theta_\beta, \quad \text{where } X_H^\beta = P^{\alpha\beta} \frac{\partial H}{\partial u^\alpha},$$

so the corresponding ODE can be written as

$$u_{t_H}^\beta = X_H^\beta = \{H, u^\beta\}.$$

3 Infinite Dimensional Poisson Geometry

3.1 Jet Bundles and Differential Polynomials

In this section, we will define the notion of Hamiltonian structure for an evolutionary partial differential equation of the following form:

$$u_t^\alpha = X^\alpha(u, u', u'', \ldots), \quad \alpha = 1, \ldots, n, \tag{3.1}$$

where $u^\alpha(x, t)$ are n smooth functions of real variables x and t, and X^α are certain functions of $u = (u^1, \ldots, u^n)$, $u' = (u_x^1, \ldots, u_x^n)$, $\ldots$, and so on.

A significant difference between the above equation and usual evolutionary PDE is that it can contain higher derivatives of u^α of any orders, because integrable systems arising from Gromov-Witten theories often take this form. For example, if $X = \mathbb{P}^1$, it is well known that the corresponding integrable system is the Toda lattice hierarchy [12, 23, 32, 35], whose first nontrivial member can be written as

$$u_t^1 = \frac{1}{\varepsilon}\left(e^{u^2(x+\varepsilon)} - e^{u^2(x)}\right) = e^{u^2}u_x^2 + \sum_{\ell\geq 1}\varepsilon^\ell X_\ell^1(u, u', \dots, u^{(\ell+1)}), \tag{3.2}$$

$$u_t^2 = \frac{1}{\varepsilon}\left(u^1(x) - u^1(x-\varepsilon)\right) = u_x^1 + \sum_{\ell\geq 1}\varepsilon^\ell X_\ell^2(u, u', \dots, u^{(\ell+1)}). \tag{3.3}$$

Here $\varepsilon = \sqrt{\hbar}$, and X_ℓ^α, which are the Taylor coefficients of the left hand side, are certain polynomials of $u_x^\alpha, \dots, u_{(\ell+1)x}^\alpha$ whose coefficients are smooth functions of u^α. If we introduce the following gradation

$$\deg f(u) = 0, \quad \deg u_{\ell x}^\alpha = \ell,$$

then $\deg X_\ell^\alpha = \ell + 1$. To describe functions X^α with these properties, we need to introduce the notion of infinite jet spaces and the algebra of differential polynomials on them.

Let $\hat{N}$ be a super manifold of dimension $(n|m)$. For any integer $k \geq 0$, we define the k-th jet bundle $J^k(\hat{N})$ of $\hat{N}$ as follow: the base manifold of the bundle is $\hat{N}$; the fiber manifold is $\left(\mathbb{R}^{n|m}\right)^k$; the bundle map is denoted by $\pi_{k,0} : J^k(\hat{N}) \to \hat{N}$. Suppose $(z^1, \dots, z^{n+m})$ is a set of coordinates over an open set U of $\hat{N}$, the corresponding coordinates on the fiber are denoted by

$$\{z^{\alpha,s} \mid \alpha = 1, \dots, n+m,\ s = 1, \dots, k\}.$$

In particular, we also take $z^{\alpha,0} = z^\alpha$, then the coordinates for the corresponding open set $\pi_{k,0}^{-1}(U)$ of $J^k(\hat{N})$ can be written as

$$\{z^{\alpha,s} \mid \alpha = 1, \dots, n+m,\ s = 0, \dots, k\}.$$

If we turn to another open set $\tilde{U}$ with coordinates $(\tilde{z}^1, \dots, \tilde{z}^{n+m})$, then the transition functions of the bundle $J^k(\hat{N})$ are given by

$$\begin{aligned}
\tilde{z}^{\alpha,1} &= z^{\beta,1}\frac{\partial \tilde{z}^\alpha}{\partial z^\beta},\\
\tilde{z}^{\alpha,2} &= z^{\beta,2}\frac{\partial \tilde{z}^\alpha}{\partial z^\beta} + z^{\beta_1,1}z^{\beta_2,1}\frac{\partial^2 \tilde{z}^\alpha}{\partial z^{\beta_2}\partial z^{\beta_1}},\\
\tilde{z}^{\alpha,3} &= z^{\beta,3}\frac{\partial \tilde{z}^\alpha}{\partial z^\beta} + 3z^{\beta_1,2}z^{\beta_2,1}\frac{\partial^2 \tilde{z}^\alpha}{\partial z^{\beta_2}\partial z^{\beta_1}}\\
&\quad + z^{\beta_1,1}z^{\beta_2,1}z^{\beta_3,1}\frac{\partial^3 \tilde{z}^\alpha}{\partial z^{\beta_3}\partial z^{\beta_2}\partial z^{\beta_1}}, \quad \dots
\end{aligned}$$

The rule for these transition functions is very simple: if $z^{\alpha,s}$ gives the s-th derivative of a curve $\gamma : (-\epsilon, \epsilon) \to \hat{N}$ in the local chart U, then $\tilde{z}^{\alpha,s}$ should be the same derivatives in the local chart $\tilde{U}$. In general, we have

$$\tilde{z}^{\alpha,s+1} = \sum_{t=0}^{s} z^{\beta,t+1} \frac{\partial \tilde{z}^{\alpha,s}}{\partial z^{\beta,t}}. \tag{3.4}$$

Note that these transition functions are not linear in $z^{\alpha,s}$, so jet bundles are not vector bundle, thought their fibers are vector spaces.

Definition 7

(a) A function $f \in C^\infty(J^k(\hat{N}))$ is called a differential polynomial if it is a polynomial of jet variables.

More precisely, let U be an open set of $\hat{N}$ with coordinates $(z^1, \ldots, z^{n+m})$, and $\pi_{k,0}^{-1}(U)$ be the corresponding open set of $J^k(\hat{N})$ with coordinates

$$\{z^{\alpha,s} \mid \alpha = 1, \ldots, n+m, \ s = 0, \ldots, k\},$$

then we have

$$f|_{\pi_{k,0}^{-1}(U)} \in C^\infty(U)[z^{\alpha,s} \mid \alpha = 1, \ldots, n+m, \ s = 1, \ldots, k].$$

This definition is independent of the choice of the open set U because of the definition of transition functions (3.4).

All differential polynomials form a subalgebra of $C^\infty(J^k(\hat{N}))$. We denote this subalgebra by $\mathring{\hat{\mathcal{A}}}^{(k)}(\hat{N})$.

(b) We define

$$\deg f(z) = 0 \text{ if } f(z) \in C^\infty(\hat{N}), \quad \deg z^{\alpha,s} = s \text{ if } s \geq 1,$$

and extend it to the whole $\mathring{\hat{\mathcal{A}}}^{(k)}(\hat{N})$, then $\mathring{\hat{\mathcal{A}}}^{(k)}(\hat{N})$ becomes a graded ring.

For any $f \in \mathring{\hat{\mathcal{A}}}^{(k)}(\hat{N})$, we can uniquely decompose it as follow

$$f = f_{d_{\min}} + f_{d_{\min}+1} + \cdots + f_{d_{\max}},$$

where $f_{d_{\min}}, f_{d_{\max}} \neq 0$ and $\deg f_d = d$. The number $d_{\min}$ is called the valuation of f, which is denoted by $\nu(f)$. (The number $d_{\max}$ can be called the degree of f, but we never use this notion.)

(c) We define a distance function over $\mathring{\hat{\mathcal{A}}}^{(k)}(\hat{N})$:

$$dist(f, g) = e^{-\nu(f-g)}, \quad \forall f, g \in \mathring{\hat{\mathcal{A}}}^{(k)}(\hat{N}).$$

Then denote by $\hat{\mathcal{A}}^{(k)}(\hat{N})$ the completion of $\mathring{\hat{\mathcal{A}}}^{(k)}(\hat{N})$ with respect to *dist*.

More precisely, let $f \in \hat{\mathcal{A}}^{(k)}(\hat{N})$, and U be an open set of $\hat{N}$, then we have

$$f|_{\pi_{k,0}^{-1}(U)} \in C^\infty(U)[[z^{\alpha,s} \mid \alpha = 1, \dots, n+m, \ s = 1, \dots, k]].$$

Here the formal power series ring $C^\infty(U)[[z^{\alpha,s}]]$ is completed by using the distance function *dist*.

We are only interested in $\hat{\mathcal{A}}^{(k)}(\hat{N})$, and will never use the notation ${}^\circ\hat{\mathcal{A}}^{(k)}(\hat{N})$ and the distance function *dist*. So, to abuse of language, we will call elements of $\hat{\mathcal{A}}^{(k)}(\hat{N})$ differential polynomials from now on, though they are actually formal power series in general. To indicate the degrees of every homogeneous components, we may write $f \in \hat{\mathcal{A}}^{(k)}(\hat{N})$ as

$$f = f_0 + f_1 + f_2 + \cdots = f_0 + \varepsilon f_1 + \varepsilon^2 f_2 + \cdots, \quad \text{where } \deg f_d = d.$$

Then the topology on $\hat{\mathcal{A}}^{(k)}(\hat{N})$ is just the ε-adic topology.

For $k \geq l \geq 0$, there is a projection map $\pi_{k,l} : J^k(\hat{N}) \to J^l(\hat{N})$, which just forgets the coordinates $z^{\alpha,s}$ with $s > l$. Jet bundles and the projection maps among them form an inverse system

$$\left(\{J^k(\hat{N})\}_{k\geq 0}, \ \{\pi_{k,l}\}_{k\geq l\geq 0}\right).$$

We denote the inverse limit of this inverse system by $J^\infty(\hat{N})$, and name it the infinite jet space of $\hat{N}$.

The projection $\pi_{k,l}$ ($k \geq l$) induces a pullback map $\pi_{k,l}^* : \hat{\mathcal{A}}^{(l)}(\hat{N}) \to \hat{\mathcal{A}}^{(k)}(\hat{N})$. The differential polynomial algebras and the pullback maps among them form a direct system

$$\left(\{\hat{\mathcal{A}}^{(k)}(\hat{N})\}_{k\geq 0}, \ \{\pi_{k,l}^*\}_{k\geq l\geq 0}\right).$$

We denote the direct limit of this direct system by $\hat{\mathcal{A}}(\hat{N})$, and name it the differential polynomial ring of $\hat{N}$.

Note that the maps $\pi_{k,l}^*$ are all injective, so every $\hat{\mathcal{A}}^{(k)}(\hat{N})$ can be regarded as a subalgebra of $\hat{\mathcal{A}}(\hat{N})$. These subalgebras define a filtration on $\hat{\mathcal{A}}(\hat{N})$:

$$\hat{\mathcal{A}}^{(0)}(\hat{N}) \subset \hat{\mathcal{A}}^{(1)}(\hat{N}) \subset \hat{\mathcal{A}}^{(2)}(\hat{N}) \subset \cdots \subset \hat{\mathcal{A}}(\hat{N}).$$

The maps $\pi_{k,l}^*$ preserve the gradation on $\hat{\mathcal{A}}^{(k)}(\hat{N})$, so $\hat{\mathcal{A}}(\hat{N})$ also has a gradation

$$\hat{\mathcal{A}}(\hat{N}) = \bigoplus_{d\geq 0} \hat{\mathcal{A}}_d(\hat{N}), \quad \hat{\mathcal{A}}_d(\hat{N}) = \{f \in \hat{\mathcal{A}}^{(d)}(\hat{N}) | \deg f = d\},$$

which is called the standard gradation of $\hat{\mathcal{A}}(\hat{N})$. In particular, $\hat{\mathcal{A}}_0(\hat{N}) = C^\infty(\hat{N})$.

Let M be a smooth manifold of dimension n, and $\hat{M} = \Pi\,(T^*(M))$ be the odd-symplectic cotangent bundle introduced in the last section. We can define $J^\infty(M)$ and $J^\infty(\hat{M})$ as above, whose differential polynomial algebras are denoted by $\mathcal{A} = \hat{\mathcal{A}}(M)$ and $\hat{\mathcal{A}} = \hat{\mathcal{A}}(\hat{M})$ respectively. Their local coordinates are written as

$$\{u^{\alpha,s} \mid \alpha = 1, \ldots, n,\ s \geq 0\}$$

and

$$\{u^{\alpha,s},\ \theta_\alpha^s \mid \alpha = 1, \ldots, n,\ s \geq 0\}$$

respectively. The algebra $\mathcal{A}$ can be identified with the subalgebra of $\hat{\mathcal{A}}$ whose elements don't depend on any θ_α^s. The superalgebra $\hat{\mathcal{A}}$ has another gradation

$$\hat{\mathcal{A}} = \bigoplus_{p\geq 0} \hat{\mathcal{A}}^p, \quad \hat{\mathcal{A}}^p = \{f = \sum_{s_1,\ldots,s_p\geq 0} f^{\alpha_1,\ldots,\alpha_p}_{s_1,\ldots,s_p}\theta^{s_1}_{\alpha_1}\cdots\theta^{s_p}_{\alpha_p} \mid f^{\alpha_1,\ldots,\alpha_p}_{s_1,\ldots,s_p} \in \mathcal{A}\},$$

which is called the super gradation of $\hat{\mathcal{A}}$. We also use the notation $\hat{\mathcal{A}}^p_d = \hat{\mathcal{A}}^p \cap \hat{\mathcal{A}}_d$. In particular, we have $\hat{\mathcal{A}}^0 = \mathcal{A}$, $\hat{\mathcal{A}}^0_0 = \mathcal{A}_0 = C^\infty(M)$. This explains the notations we used in the last section.

3.2 Evolutionary Partial Differential Equations

We can define evolutionary PDEs of the form (3.1) now. Let us prove two lemmas first.

Lemma 8 *The following operator*

$$\partial_{\hat{N}} = \sum_{s\geq 0} z^{\alpha,s+1}\frac{\partial}{\partial z^{\alpha,s}}$$

defines a global vector field on $J^\infty(\hat{N})$, and it also defines a derivation of $\hat{\mathcal{A}}(\hat{N})$.

Proof According to the definition (3.4) of transition functions of the bundle $J^\infty(\hat{N})$, we have

$$\partial_{\hat{N}} = \sum_{s\geq 0} z^{\alpha,s+1}\frac{\partial}{\partial z^{\alpha,s}} = \sum_{s\geq 0} \tilde{z}^{\alpha,s+1}\frac{\partial}{\partial \tilde{z}^{\alpha,s}}.$$

The lemma is proved. □

When $\hat{N} = \hat{M}$ (or M), $\partial_{\hat{N}}$ has the following expression

$$\partial_{\hat{M}} = \sum_{s\geq 0}\left(u^{\alpha,s+1}\frac{\partial}{\partial u^{\alpha,s}} + \theta_\alpha^{s+1}\frac{\partial}{\partial \theta_\alpha^s}\right) \quad \left(\text{or } \partial_M = \sum_{s\geq 0}\left(u^{\alpha,s+1}\frac{\partial}{\partial u^{\alpha,s}}\right)\right),$$

Note that $\mathcal{A} = \hat{\mathcal{A}}^0$, and $\partial_M = \partial_{\hat{M}}|_{\hat{\mathcal{A}}^0}$, so we denote them by $\partial = \partial_{\hat{M}} = \partial_M$ to abuse of notation.

Lemma 9 *Let $X : \hat{\mathcal{A}}(\hat{N}) \to \hat{\mathcal{A}}(\hat{N})$ be a continuous derivation such that $[X, \partial] = 0$, then we have*

$$X = \sum_{s\geq 0}\partial^s(X^\alpha)\frac{\partial}{\partial z^{\alpha,s}}, \tag{3.5}$$

where $X^\alpha \in \hat{\mathcal{A}}(\hat{N})$.

Proof Without loss of generality, we can assume that X is homogeneous with respect to the super gradation of $\hat{\mathcal{A}}(\hat{N})$, that is $X(\hat{\mathcal{A}}^p(\hat{N})) \subset \hat{\mathcal{A}}^{p+|X|}(\hat{N})$, where $|X| \in \mathbb{Z}$ is called the super degree of X. Then a derivation is a linear map $X : \hat{\mathcal{A}}(\hat{N}) \to \hat{\mathcal{A}}(\hat{N})$ such that

$$X(f \cdot g) = X(f)\cdot g + (-1)^{|X||f|} f \cdot X(g),$$

where $f \in \hat{\mathcal{A}}^{|f|}(\hat{N})$, $g \in \hat{\mathcal{A}}(\hat{N})$.

If $f \in \hat{\mathcal{A}}^{(n)}(\hat{N})$ for some $n \in \mathbb{N}$, then it is easy to see that

$$X(f) = \sum_{s=0}^{n} X(z^{\alpha,s})\frac{\partial f}{\partial z^{\alpha,s}} = \sum_{s\geq 0}\partial^s(X^\alpha)\frac{\partial f}{\partial z^{\alpha,s}},$$

where $X^\alpha = X(z^\alpha) \in \hat{\mathcal{A}}(\hat{N})$.

If f doesn't belong to any $\hat{\mathcal{A}}^{(n)}(\hat{N})$,

$$f = \sum_{d\geq 0} f_d, \quad \text{where } f_d \in \hat{\mathcal{A}}_d \subset \hat{\mathcal{A}}^{(d)}(\hat{N}),$$

then we have

$$\begin{aligned} X(f) &= X\left(\lim_{n\to\infty}\sum_{d=0}^{n} f_d\right) \\ &= \lim_{n\to\infty} X\left(\sum_{d=0}^{n} f_d\right) \quad (\Leftarrow X \text{ is continuous}) \end{aligned}$$

$$= \lim_{n\to\infty}\left(\sum_{s\geq 0}\partial^s(X^\alpha)\frac{\partial}{\partial z^{\alpha,s}}\right)\left(\sum_{d=0}^{n} f_d\right) \quad (\Leftarrow \sum_{d=0}^{n} f_d \in \hat{\mathcal{A}}^{(n)})$$

$$= \sum_{s\geq 0}\partial^s(X^\alpha)\frac{\partial f}{\partial z^{\alpha,s}}.$$

The last equality holds true because $\partial^p(X^\alpha)\frac{\partial}{\partial z^{\alpha,s}} : \hat{\mathcal{A}}(\hat{N}) \to \hat{\mathcal{A}}(\hat{N})$ is continuous for all p, and the summation $\sum_{p\geq 0}\partial^p(X^\alpha)\frac{\partial}{\partial z^{\alpha,s}}$ is uniformly convergent, so the summation itself is also continuous. □

If we have an evolutionary PDE (3.1) with $X^\alpha \in \mathcal{A}$, then for any $f \in \mathcal{A}$, we have

$$f_t = \sum_{s\geq 0}\left(u^{\alpha,s}\right)_t\frac{\partial f}{\partial u^{\alpha,s}} = \sum_{s\geq 0}\partial^s\left(X^\alpha\right)\frac{\partial f}{\partial u^{\alpha,s}},$$

which is just $X(f)$ with X given by (3.5), so we have the following definition.

Definition 10

(a) We denote by $\mathrm{Der}(\hat{N})$ the Lie algebra of continuous derivations over $\hat{\mathcal{A}}(\hat{N})$, and define

$$\hat{\mathcal{E}}(\hat{N}) = \mathrm{Der}(\hat{N})^\partial = \{X \in \mathrm{Der}(\hat{N}) \mid [X, \partial] = 0\},$$

whose elements are called evolutionary vector field on $J^\infty(\hat{N})$.

(b) According to Lemma 9, an element $X \in \hat{\mathcal{E}}(\hat{N})$ always takes the following form:

$$X = \sum_{s\geq 0}\partial^s(X^\alpha)\frac{\partial}{\partial z^{\alpha,s}}.$$

We denote it by $X = (X^\alpha)$ for short. The differential polynomials X^α's are called the components of X.

(c) We denote $\mathcal{E} = \hat{\mathcal{E}}(M)$ and $\hat{\mathcal{E}} = \hat{\mathcal{E}}(\hat{M})$.

It is easy to see that $\mathcal{E}$ is a Lie algebra, and $\hat{\mathcal{E}}$ is a graded Lie algebra.

3.3 Conserved Quantities

To develop a Hamiltonian formalism for Eq. (3.1), we still need the notion of *conserved quantity*. Roughly speaking, a conserved quantity for (3.1) is a functional

$$I[u] = \int_{\mathbb{R}} f(u, u', u'', \ldots, u^{(N)})\mathrm{d}x$$

such that if $u(x,t)$ is a solution for (3.1), then

$$\frac{dI}{dt} = \int_{\mathbb{R}} f_t \, dx = \int_{\mathbb{R}} X(f) \, dx = 0.$$

This definition is not very convenient, because we need some conditions on u and f to ensure that the integrations are convergent. A better choice is to replace $\mathbb{R}$ by $S^1 = \mathbb{R}/\mathbb{Z}$, and assume that $u(x) = (u^1(x), \dots, u^n(x))$ is actually the coordinates of a smooth map $\phi : S^1 \to M$.

Let $\mathcal{L}(M) = C^\infty(S^1, M)$ be the loop space of M. For any $\phi \in \mathcal{L}(M)$, we can lift it to a map $\phi^k : S^1 \to J^k(M)$ for all $k = 1, 2, \dots, \infty$. Then for any $f \in C^\infty(J^\infty(M))$, we can define a smooth function $(\phi^\infty)^*(f) : S^1 \to \mathbb{R}$, $x \mapsto f(\phi^\infty(x))$, and then define the following functional:

$$I_f[\phi] = \int_{S_1} (\phi^\infty)^*(f)(x) dx.$$

Lemma 11 *Let* $^\circ\mathcal{F}$ *be the linear space of functionals of the form* I_f, *and* $I : C^\infty(J^\infty(M)) \to {}^\circ\mathcal{F}$ *be the map* $f \mapsto I_f$. *Then* $\mathrm{Ker}(I) = \mathrm{Im}(\partial)$, *hence we have an isomorphism* $^\circ\mathcal{F} \cong C^\infty(J^\infty(M))/\partial(C^\infty(J^\infty(M)))$.

Proof By definition,

$$C^\infty(J^\infty(M)) = \varinjlim_k C^\infty(J^k(M)),$$

if $f \in C^\infty(J^\infty(M))$, then there exists $k \in \mathbb{N}$ such that $f \in C^\infty(J^k(M))$.

If $f \in \mathrm{Im}(\partial)$, there exists $g \in C^\infty(J^k(M))$ such that $f = \partial g$, so we have

$$I_f[\phi] = \int_{S^1} (\partial g)(\phi^\infty(x)) dx = g(\phi^\infty(x))\big|_0^1 = 0,$$

that is, $\mathrm{Im}(\partial) \subset \mathrm{Ker}(I)$.

Conversely, if $I_f[\phi] = 0$ for any $\phi \in \mathcal{L}(M)$, we need to construct $g \in C^\infty(J^k(M))$ such that $f = \partial g$.

Suppose M is connected (otherwise, we can do the following for each of M's connected component), fix a point $P_0 \in J^k(M)$, and take $Q_0 = \pi_{k,0}(P_0) \in M$. For any $P \in J^k(M)$, let $Q = \pi_{k,0}(P) \in M$. There exists a path $\gamma : [0, 1/2] \to M$ such that

$$\gamma(0) = Q_0, \quad \gamma(1/2) = Q, \quad \gamma^k(0) = P_0, \quad \gamma^k(1/2) = P,$$

where $\gamma^k : [0, 1/2] \to J^k(M)$ is the lifted map. Then define

$$g : J^k(M) \to \mathbb{R}, \quad P \mapsto g(P) = \int_0^{1/2} f(\gamma^k(x))\mathrm{d}x.$$

This definition is independent of the choice of γ (because $f \in \mathrm{Ker}(I)$), and it is easy to see that $\partial g = f$. The lemma is proved. □

The above lemma shows that, even the loop space is not necessary: we can define $\mathcal{F}$ as the cokernel of ∂. Inspired by this fact, we give the following definition.

Definition 12

(a) We define $\hat{\mathcal{F}}(\hat{N}) = \hat{\mathcal{A}}(\hat{N})/\partial\hat{\mathcal{A}}(\hat{N})$, whose elements are called local functionals on $\hat{N}$. A local functional $f + \partial\hat{\mathcal{A}}(\hat{N})$ is usually denoted by $\int f\mathrm{d}x$, and the representative f is called a density of this functional.
(b) The space $\hat{\mathcal{F}}(\hat{N})$ has an natural $\hat{\mathcal{E}}(\hat{N})$-module structure,

$$\left(X, F = \int f\mathrm{d}x\right) \mapsto X(F) = \int X(f)\mathrm{d}x.$$

A local functional $F \in \hat{\mathcal{F}}(\hat{N})$ is called a conserved quantity of $X \in \hat{\mathcal{E}}(\hat{N})$, if $X(F) = 0$.
(c) We denote $\mathcal{F} = \hat{\mathcal{F}}(M)$ and $\hat{\mathcal{F}} = \hat{\mathcal{F}}(\hat{M})$. Note that ∂ preserves the two gradations on $\hat{\mathcal{A}}$, so there are induced standard gradation and super gradation on $\hat{\mathcal{F}}$. We denote them by

$$\hat{\mathcal{F}} = \bigoplus_{d\geq 0}\hat{\mathcal{F}}_d = \bigoplus_{p\geq 0}\hat{\mathcal{F}}^p,$$

and $\hat{\mathcal{F}}_d^p = \hat{\mathcal{F}}_d \cap \hat{\mathcal{F}}^p$. In particular, $\mathcal{F} = \hat{\mathcal{F}}^0$.

Lemma 13 *Let $X = (X^\alpha) \in \hat{\mathcal{E}}(\hat{N})$, $F = \int f\mathrm{d}x \in \hat{\mathcal{F}}(\hat{N})$, then we have*

$$X(F) = \int \left(X^\alpha \frac{\delta F}{\delta z^\alpha}\right)\mathrm{d}x,$$

where

$$\frac{\delta F}{\delta z^\alpha} = \sum_{s\geq 0}(-\partial)^s \frac{\partial f}{\partial z^{\alpha,s}}$$

is the variational derivative of F with respect to z^α.

Proof In the space $\hat{\mathcal{F}}(\hat{N})$, we still have integration by parts, so

$$X(F) = \int \left(\sum_{s\geq 0} \partial^s (X^\alpha) \frac{\partial f}{\partial z^{\alpha,s}} \right) \mathrm{d}x$$

$$= \int \left(X^\alpha \sum_{s\geq 0} (-\partial)^s \frac{\partial f}{\partial z^{\alpha,s}} \right) \mathrm{d}x.$$

The lemma is proved. □

3.4 Hamiltonian Structures

We are ready to define Hamiltonian structures for the evolutionary PDE (3.1). Similar to the finite-dimensional case, a Hamiltonian structure on M is a Lie bracket over the space of local functionals (i.e. $\mathcal{F}$) whose action is given by certain differential operations in a local chart.

Definition 14

(a) Let $\mathcal{V}^* = \mathrm{Hom}(\wedge^* \mathcal{F}, \mathcal{F})$, whose elements are called generalized variational polyvector. According to Theorem 1, there is a unique bracket operation $[\,,\,] : \mathcal{V}^p \times \mathcal{V}^q \to \mathcal{V}^{p+q-1}$ satisfying the condition (2.8)–(2.10) with $\mathcal{A}_0$ replaced by $\mathcal{F}$ and the condition (2.11). We still call it the Nijenhuis-Richardson bracket.

(b) A generalized variational p-vector $P \in \mathcal{V}^p$ is called a variational p-vector, if its action on $F_1, \ldots, F_p \in \mathcal{F}$ is given by

$$P(F_1, \ldots, F_p) = \int \left(\sum_{s_1,\ldots,s_p \geq 0} P^{\alpha_1,\ldots,\alpha_p}_{s_1,\ldots,s_p} \partial^{s_1} \left(\frac{\delta F_1}{\delta u^{\alpha_1}} \right) \cdots \partial^{s_p} \left(\frac{\delta F_p}{\delta u^{\alpha_p}} \right) \right) \mathrm{d}x, \tag{3.6}$$

where $P^{\alpha_1,\ldots,\alpha_p}_{s_1,\ldots,s_p} \in \mathcal{A}$. The space of variational p-vectors is denoted by Λ^p. We denote by $\Lambda^* = \bigoplus_{p\geq 0} \Lambda^p$, which is a subspace of $\mathcal{V}^*$.

(c) A variational bivector $P \in \Lambda^2$ is called a Hamiltonian structure, if $[P, P] = 0$.

We have an infinite-dimensional analogue of Proposition 3.

Proposition 15 *If $P \in \Lambda^p$, $Q \in \Lambda^q$, then $[P, Q] \in \Lambda^{p+q-1}$.*

The definition (3.6) of variational polyvectors is very complicated. It is not easy to determine whether a generalized variational polyvector $P \in \mathcal{V}^*$ belongs to Λ^*, so we cannot prove the above proposition directly. In what follows, we will give another description of Λ^*, then prove the proposition by using the odd-symplectic bracket on $\hat{\mathcal{F}}$.

Lemma 16 *Define a map* $\jmath : \hat{\mathcal{F}}^p \to \Lambda^p$,

$$P = \int \tilde{P} \mathrm{d}x \mapsto \jmath(P)(F_1, \ldots, F_p)$$
$$= \int \left(\sum_{s_1,\ldots,s_p \geq 0} \frac{\partial^p \tilde{P}}{\partial \theta_{\alpha_p}^{s_p} \cdots \partial \theta_{\alpha_1}^{s_1}} \partial^{s_1} \left(\frac{\delta F_1}{\delta u^{\alpha_1}} \right) \cdots \partial^{s_p} \left(\frac{\delta F_p}{\delta u^{\alpha_p}} \right) \right) \mathrm{d}x.$$

Then $\jmath(P)$ *is independent of the choice of the density* $\tilde{P}$, *and* $\jmath$ *is surjective with* $\mathrm{Ker}(\jmath) = \mathbb{R}\omega \subset \hat{\mathcal{F}}^1$, *where* $\omega = \int (u^{\alpha,1}\theta_\alpha)\mathrm{d}x$. *So we have the isomorphisms* $\Lambda^p \cong \hat{\mathcal{F}}^p$ $(p \neq 1)$, *and* $\Lambda^1 \cong \hat{\mathcal{F}}^1/\mathbb{R}\omega$.

We have to omit the proof of this lemma because of its length. In [28], we proved a generalization of this lemma in §2.3. One can easily reduce that proof to the present case.

Define the action of $P \in \hat{\mathcal{F}}^p$ on $F_1, \ldots, F_p \in \mathcal{F}$ by

$$P(F_1, \ldots, F_p) = \jmath(P)(F_1, \ldots, F_p).$$

Then we have the following lemma.

Lemma 17 *For* $P \in \hat{\mathcal{F}}^p$, $Q \in \hat{\mathcal{F}}^q$, *define*

$$[P, Q] = \int \left(\frac{\delta P}{\delta \theta_\alpha} \frac{\delta Q}{\delta u^\alpha} + (-1)^p \frac{\delta P}{\delta u^\alpha} \frac{\delta Q}{\delta \theta_\alpha} \right) \mathrm{d}x,$$

then the operation $[\,,\,]$ *satisfies the condition* (2.8)–(2.10) *with* $\mathcal{A}_0$ *replaced by* $\mathcal{F}$ *and the condition* (2.11), *hence* $(\hat{\mathcal{F}}, [\,,\,])$ *forms a graded Lie algebra. In particular, its center is given by* $\mathbb{R}\omega$.

Proof Suppose $P = \int \tilde{P}\mathrm{d}x \in \hat{\mathcal{F}}^p$, $F \in \mathcal{F}$, then

$$[P, F] = \int \left(\frac{\delta P}{\delta \theta_\alpha} \frac{\delta F}{\delta u^\alpha} \right) \mathrm{d}x = \int \left(\sum_{s \geq 0} \frac{\partial \tilde{P}}{\partial \theta_\alpha^s} \partial^s \left(\frac{\delta F}{\delta u^\alpha} \right) \right) \mathrm{d}x,$$

so we have

$$[P, F](F_2, \ldots, F_p)$$
$$= \int \left(\sum_{s_2,\ldots,s_p \geq 0} \frac{\partial^p}{\partial \theta_{\alpha_p}^{s_p} \cdots \partial \theta_{\alpha_2}^{s_2}} \left(\sum_{s_1 \geq 0} \frac{\partial \tilde{P}}{\partial \theta_{\alpha_1}^{s_1}} \partial^{s_1} \left(\frac{\delta F}{\delta u^{\alpha_1}} \right) \right) \right.$$
$$\left. \partial^{s_2} \left(\frac{\delta F_2}{\delta u^{\alpha_2}} \right) \cdots \partial^{s_p} \left(\frac{\delta F_p}{\delta u^{\alpha_p}} \right) \right) \mathrm{d}x$$

$$
\begin{aligned}
&= \int \left(\sum_{s_1,\dots,s_p\geq 0} \frac{\partial^p \tilde{P}}{\partial \theta_{\alpha_p}^{s_p} \cdots \partial \theta_{\alpha_1}^{s_1}} \partial^{s_1} \left(\frac{\delta F_1}{\delta u^{\alpha_1}} \right) \cdots \partial^{s_p} \left(\frac{\delta F_p}{\delta u^{\alpha_p}} \right) \right) \mathrm{d}x \\
&= P(F, F_2, \dots, F_p).
\end{aligned}
$$

The identity (2.8) is proved.

Suppose $P \in \hat{\mathcal{F}}^p$, $Q \in \hat{\mathcal{F}}^q$, then we have

$$
\begin{aligned}
[P, Q] &= \int \left(\frac{\delta P}{\delta \theta_\alpha} \frac{\delta Q}{\delta u^\alpha} + (-1)^p \frac{\delta P}{\delta u^\alpha} \frac{\delta Q}{\delta \theta_\alpha} \right) \mathrm{d}x \\
&= \int \left((-1)^{(p-1)q} \frac{\delta Q}{\delta u^\alpha} \frac{\delta P}{\delta \theta_\alpha} + (-1)^{p+p(q-1)} \frac{\delta Q}{\delta \theta_\alpha} \frac{\delta P}{\delta u^\alpha} \right) \mathrm{d}x \\
&= (-1)^{pq} \int \left(\frac{\delta Q}{\delta \theta_\alpha} \frac{\delta P}{\delta u^\alpha} + (-1)^q \frac{\delta Q}{\delta u^\alpha} \frac{\delta P}{\delta \theta_\alpha} \right) \mathrm{d}x \\
&= (-1)^{pq} [Q, P].
\end{aligned}
$$

The identity (2.9) is proved.

The identity (2.10) is a special case of (2.11), so we only need to prove the latter one. For any $P \in \hat{\mathcal{F}}^p$, we define an operator $D_P : \hat{\mathcal{A}} \to \hat{\mathcal{A}}$

$$
D_P = \sum_{s\geq 0} \left(\partial^s \left(\frac{\delta P}{\delta \theta_\alpha} \right) \frac{\partial}{\partial u^{\alpha,s}} + (-1)^p \partial^s \left(\frac{\delta P}{\delta u^\alpha} \right) \frac{\partial}{\partial \theta_\alpha^s} \right), \tag{3.7}
$$

then it is easy to see that $D_P(\hat{\mathcal{A}}^q) \subset \hat{\mathcal{A}}^{p+q-1}$, $[D_P, \partial] = 0$, and $[P, Q] = \int D_P(Q) \mathrm{d}x$ for any $Q \in \hat{\mathcal{F}}^q$. The identity (2.11) is equivalent to the following identity:

$$
(-1)^{p-1} D_{[P,Q]} = D_P \circ D_Q - (-1)^{(p-1)(q-1)} D_Q \circ D_P, \tag{3.8}
$$

which is a corollary of the following identity:

$$
\frac{\delta}{\delta u^\alpha} [P, Q] = D_P \left(\frac{\delta Q}{\delta u^\alpha} \right) + (-1)^{pq} D_Q \left(\frac{\delta P}{\delta u^\alpha} \right), \tag{3.9}
$$

$$
(-1)^{p-1} \frac{\delta}{\delta \theta_\alpha} [P, Q] = D_P \left(\frac{\delta Q}{\delta \theta_\alpha} \right) - (-1)^{(p-1)(q-1)} D_Q \left(\frac{\delta P}{\delta \theta_\alpha} \right). \tag{3.10}
$$

To prove the identity (3.9), (3.10), we introduce the following operators

$$\delta_{\alpha,s} = \sum_{t\ge 0}(-1)^t \binom{t+s}{s}\partial^t \frac{\partial}{\partial u^{\alpha,s}},$$

$$\delta_s^{\alpha} = \sum_{t\ge 0}(-1)^t \binom{t+s}{s}\partial^t \frac{\partial}{\partial \theta_\alpha^s},$$

which are called the higher Euler operators. In particular,

$$\delta_{\alpha,0} = \frac{\delta}{\delta u^\alpha}, \quad \delta_0^\alpha = \frac{\delta}{\delta\theta_\alpha},$$

and they satisfy the following identities:

$$\delta_{\alpha,0}(f\cdot g) = \sum_{t\ge 0}(-1)^t\left(\delta_{\alpha,t}(f)\partial^t(g) + \partial^t(f)\delta_{\alpha,t}(g)\right),$$

$$\delta_{\alpha,t}\delta_{\beta,0} = (-1)^t \frac{\partial}{\partial u^{\beta,t}}\delta_{\alpha,0}, \quad \delta_{\alpha,t}\delta_0^\beta = (-1)^t\frac{\partial}{\partial\theta_\beta^t}\delta_{\alpha,0}.$$

Then we have

$$\begin{aligned}
&\delta_{\alpha,0}[P,Q]\\
=&\delta_{\alpha,0}\left(\delta_0^\beta(P)\delta_{\beta,0}(Q) + (-1)^p\delta_{\beta,0}(P)\delta_0^\beta(Q)\right)\\
=&\sum_{t\ge 0}(-1)^t\left(\delta_{\alpha,t}(\delta_0^\beta(P))\partial^t(\delta_{\beta,0}(Q)) + \partial^t(\delta_0^\beta(P))\delta_{\alpha,t}(\delta_{\beta,0}(Q))\right)\\
&+(-1)^p\sum_{t\ge 0}(-1)^t\left(\delta_{\alpha,t}(\delta_{\beta,0}(P))\partial^t(\delta_0^\beta(Q)) + \partial^t(\delta_{\beta,0}(P))\delta_{\alpha,t}(\delta_0^\beta(Q))\right)\\
=&\sum_{t\ge 0}\left(\frac{\partial(\delta_{\alpha,0}(P))}{\partial\theta_\beta^t}\partial^t(\delta_{\beta,0}(Q)) + \partial^t(\delta_0^\beta(P))\frac{\partial(\delta_{\alpha,0}(Q))}{\partial u^{\beta,t}}\right.\\
&\left.+(-1)^p\left(\frac{\partial(\delta_{\alpha,0}(P))}{\partial u^{\beta,t}}\partial^t(\delta_0^\beta(Q)) + \partial^t(\delta_{\beta,0}(P))\frac{\partial(\delta_{\alpha,0}(Q))}{\partial\theta_\beta^t}\right)\right)\\
=&D_P(\delta_{\alpha,0}(Q)) + (-1)^{pq}D_Q(\delta_{\alpha,0}(P)).
\end{aligned}$$

The identity (3.9) is proved. The identity (3.10) can be proved similarly.

Suppose $Q = \int \tilde{Q} \mathrm{d}x \in \hat{\mathcal{F}}^q$, we have

$$
\begin{aligned}
[\omega, Q] &= \int \left(\frac{\delta \omega}{\delta \theta_\alpha} \frac{\delta Q}{\delta u^\alpha} - \frac{\delta \omega}{\delta u^\alpha} \frac{\delta Q}{\delta \theta_\alpha} \right) \mathrm{d}x \\
&= \int \left(u^{\alpha,1} \frac{\delta Q}{\delta u^\alpha} + \theta_\alpha^1 \frac{\delta Q}{\delta \theta_\alpha} \right) \mathrm{d}x \\
&= \int \sum_{s \geq 0} \left(u^{\alpha,s+1} \frac{\partial \tilde{Q}}{\partial u^{\alpha,s}} + \theta_\alpha^{s+1} \frac{\partial \tilde{Q}}{\partial \theta_\alpha^s} \right) \mathrm{d}x \\
&= \int (\partial(Q)) \, \mathrm{d}x = 0,
\end{aligned}
$$

so ω is in the center of the graded Lie algebra $(\hat{\mathcal{F}}, [\,,\,])$.

Suppose $P \in \hat{\mathcal{F}}^p$ is in the center of $(\hat{\mathcal{F}}, [\,,\,])$, then for any $F \in \mathcal{F} = \hat{\mathcal{F}}^0$ we have $[P, F] = 0$. Consider the action of $[P, F]$ on $F_2, \ldots, F_p \in \mathcal{F}$,

$$[P, F](F_2, \ldots, F_p) = \jmath(P)(F, F_2, \ldots, F_p) = 0,$$

so $P \in \mathrm{Ker}(\jmath) = \mathbb{R}\omega$. The lemma is proved. □

For more properties of the higher Euler operators and their generalizations, please refer to [24, 28] and the references therein.

Proof of Proposition 15 Suppose $P \in \Lambda^p$, $Q \in \Lambda^q$, take $P' \in \hat{\mathcal{F}}^p$, $Q' \in \hat{\mathcal{F}}^q$ such that

$$P = \jmath(P'), \quad Q = \jmath(Q'),$$

then define $[P, Q]' = \jmath([P', Q'])$. According to the above two lemmas, this definition is independent of the choice of P' and Q'. Lemma 17 shows that the operation $[\,,\,]'$ must coincides with the Nijenhuis-Richardson bracket $[\,,\,]$, so we have $[P, Q] \in \Lambda^{p+q-1}$. The proposition is proved. □

Lemma 16 shows that $\hat{\mathcal{F}}^p$ and Λ^p can be identified, except the $p = 1$ case. When $p = 1$, Lemma 13 shows that $\Lambda^1 \cong \mathcal{E}/\mathbb{R}\omega$, so we can identify $\mathcal{E}$ and $\hat{\mathcal{F}}^1$ as follow

$$X = (X^\alpha) \in \mathcal{E} \quad \leftrightarrow \quad X = \int (X^\alpha \theta_\alpha) \mathrm{d}x \in \hat{\mathcal{F}}^1.$$

It is easy to see that the action of $X \in \mathcal{E} = \hat{\mathcal{F}}^1$ on $F \in \mathcal{F} = \hat{\mathcal{F}}^0$ is exactly given by $[X, F]$. From now on, we will always working with $\hat{\mathcal{F}}$, and forget about $\mathcal{F}$, $\mathcal{E}$, $\mathcal{V}$, and Λ^*.

Definition 18 An element $X \in \hat{\mathcal{F}}^1$ is called an evolutionary PDE. An element $F \in \hat{\mathcal{F}}^0$ is called a conserved quantity of X if $[X, F] = 0$. An element $P \in \hat{\mathcal{F}}^2$

is called a Hamiltonian structure if $[P, P] = 0$. An evolutionary PDE X is called Hamiltonian if there is a Hamiltonian structure P and a conserved quantity F such that $X = [P, F]$.

4 Hamiltonian Structures

4.1 Presentations and Examples

Let $P = \int \tilde{P} \mathrm{d}x \in \hat{\mathcal{F}}^2$ be a variational bivector, then $\tilde{P}$ satisfies the following homogeneous condition

$$\tilde{P} = \frac{1}{2} \sum_{s \geq 0} \theta_\alpha^s \frac{\partial \tilde{P}}{\partial \theta_\alpha^s},$$

so we have

$$P = \frac{1}{2} \int \left(\sum_{s \geq 0} \theta_\alpha^s \frac{\partial \tilde{P}}{\partial \theta_\alpha^s} \right) \mathrm{d}x = \frac{1}{2} \int \left(\theta_\alpha \frac{\delta P}{\delta \theta_\alpha} \right) \mathrm{d}x.$$

Suppose

$$\frac{\delta P}{\delta \theta_\alpha} = \sum_{s \geq 0} P_s^{\alpha\beta} \theta_\beta^s = \left(\sum_{s \geq 0} P_s^{\alpha\beta} \partial^s \right) \theta_\beta, \tag{4.1}$$

then

$$P = \frac{1}{2} \int \left(\theta_\alpha \left(\sum_{s \geq 0} P_s^{\alpha\beta} \partial^s \right) \theta_\beta \right) \mathrm{d}x, \tag{4.2}$$

so a variational bivector corresponds to a matrix differential operator

$$\mathcal{P} = \left(\mathcal{P}^{\alpha\beta} \right) = \left(\sum_{s \geq 0} P_s^{\alpha\beta} \partial^s \right). \tag{4.3}$$

By computing the variational derivative of both side of (4.2) with respect to θ_α, one can show that

$$\mathcal{P} + \mathcal{P}^\dagger = 0, \tag{4.4}$$

where

$$\mathcal{P}^{\dagger} = \left((\mathcal{P}^{\dagger})^{\alpha\beta}\right) = \left(\sum_{s\geq 0}(-\partial)^s P_s^{\beta\alpha}\right).$$

It is easy to see that the variational bivectors are one-to-one corresponding to the matrix differential operators (4.3) satisfying the condition (4.4), so we have the following definition.

Definition 19 Let $P \in \hat{\mathcal{F}}^2$ be a Hamiltonian structure, the matrix differential operator $\mathcal{P}$ defined by (4.1) is called the Hamiltonian operator of P.

In literatures, a Hamiltonian structure is often given by its Hamiltonian operator.

Let $P \in \hat{\mathcal{F}}^2$ be a Hamiltonian structure, then the bracket operation

$$\{\,,\}_P : \mathcal{F} \times \mathcal{F} \to \mathcal{F}, (F, G) \mapsto \{F, G\}_P = P(F, G)$$

is a Lie bracket, whose action can be computed explicitly:

$$\{F, G\}_P = P(F, G) = [[P, F], G] = \int \left(\frac{\delta f}{\delta u^\alpha}\mathcal{P}^{\alpha\beta}\left(\frac{\delta G}{\delta u^\beta}\right)\right) \mathrm{d}x.$$

If we enlarge the space $\mathcal{F}$ to contain functionals of the following form

$$u^\alpha(y) = \int_{S^1} u^\alpha(x)\delta(x-y)\mathrm{d}x,$$

then we have

$$\begin{aligned}\{u^\alpha(y), u^\beta(z)\}_P &= \int_{S^1} \delta(x-y)\mathcal{P}^{\alpha\beta}(u(x))\delta(x-z)\mathrm{d}x\\ &= \sum_{s\geq 0} P_s^{\alpha\beta}(u(y))\delta^{(s)}(y-z).\end{aligned}$$

This is another common way to present a Hamiltonian structure. We can call it the coordinate presentation.

Example 20 Suppose $M = \mathbb{R}$, so $n = 1$. We can omit the α index.

Let $P = \frac{1}{2}\int g(u)\theta\theta^1 \mathrm{d}x \in \hat{\mathcal{F}}^2$, then we have

$$\begin{aligned}\frac{\delta P}{\delta u} &= \frac{1}{2}g'(u)\theta\theta^1,\\ \frac{\delta P}{\delta \theta} &= \frac{1}{2}\left(g(u)\theta^1 + \partial\left(g(u)\theta\right)\right) = g(u)\theta^1 + \frac{1}{2}g'(u)u^1\theta,\end{aligned}$$

so we have

$$[P, P] = 2\int \frac{\delta P}{\delta\theta}\frac{\delta P}{\delta u}\mathrm{d}x = 0,$$

so P is a Hamiltonian structure. The Hamiltonian operator reads

$$\mathcal{P} = g(u)\partial + \frac{1}{2}g'(u)u^1,$$

and the coordinate presentation reads

$$\{u(y), u(z)\} = g(u(y))\delta'(y-z) + \frac{1}{2}g'(u(y))u'_y\delta(y-z).$$

Consider a deformation of P:

$$\tilde{P} = P + c\int \theta\theta^3\mathrm{d}x,$$

then

$$[P, P] = 2\int \frac{\delta P}{\delta\theta}\frac{\delta P}{\delta u}\mathrm{d}x = 2c\int g'(u)\theta\theta^1\theta^3\mathrm{d}x.$$

It is easy to see that $[P, P] = 0$ if and only if $g''(u) = 0$. So we obtain a family of Hamiltonian operators with three parameters a, b, c:

$$\mathcal{P}_{a,b,c} = (a\,u + b)\partial + \frac{a}{2}u^1 + c\,\partial^3.$$

In particular, the operators

$$\mathcal{P}_1 = \mathcal{P}_{0,1,0} = \partial,$$

$$\mathcal{P}_2 = \mathcal{P}_{1,0,\hbar/8} = u\partial + \frac{1}{2}u^1 + \frac{\hbar}{8}\partial^3$$

give the two Hamiltonian structures of the Korteweg-de Vries equation:

$$u_t = u\,u_x + \frac{\hbar}{12}u_{xxx}.$$

And the operators

$$\mathcal{P}_1 = \mathcal{P}_{0,1,-1} = \partial - \partial^3,$$

$$\mathcal{P}_2 = \mathcal{P}_{1,0,0} = u\partial + \frac{1}{2}u^1$$

give the two Hamiltonian structures of the Camassa-Holm equation:

$$u_t - u_{xxt} = 3\,u\,u_x - 2\,u_x\,u_{xx} - u\,u_{xxx}.$$

Example 21 Let $M = \mathbb{R}^2$, we denote

$$u^1 = u, \quad u^2 = v, \quad \theta_1 = \theta, \quad \theta_2 = \phi.$$

Define a series of shift operators

$$\mathcal{S}^k = e^{k\varepsilon\partial}, \quad k \in \mathbb{Z},$$

and denote by $a^\pm = \mathcal{S}^{\pm 1}(a)$, $a^{[k]} = \mathcal{S}^k(a)$ for $a \in \hat{\mathcal{A}}$, $k \in \mathbb{Z}$. The Toda equation (3.2) and (3.3) can be written as

$$u_t = e^{v^+} - e^v, \quad v_t = u - u^-.$$

Here we take $\varepsilon = 1$ for convenience.

The second Hamiltonian structure of the Toda equation can be written as

$$P_2 = \int \left(e^{v^+}\theta\theta^+ + u\theta(\phi^+ - \phi) + \phi\phi^+\right) \mathrm{d}x.$$

Its variational derivatives read

$$\frac{\delta P_2}{\delta u} = \theta(\phi^+ - \phi),$$
$$\frac{\delta P_2}{\delta v} = e^v\theta^-\theta,$$
$$\frac{\delta P_2}{\delta \theta} = e^{v^+}\theta^+ - e^v\theta^- + u(\phi^+ - \phi),$$
$$\frac{\delta P_2}{\delta \phi} = u\theta - u^-\theta^- + \phi^+ - \phi^-.$$

Here we used the identity:

$$\frac{\delta F}{\delta z} = \sum_{k\in\mathbb{Z}} \mathcal{S}^{-k}\frac{\partial f}{\partial z^{[k]}},$$

where $F = \int f \mathrm{d}z \in \hat{\mathcal{F}}$, $z = u, v, \theta, \phi$.

Then, by using the following fact

$$\int a\,\mathrm{d}x = \int a^{[k]}\,\mathrm{d}x, \quad \text{for all } a \in \hat{\mathcal{A}},\ k \in \mathbb{Z},$$

we obtain

$$\begin{aligned}
&\frac{1}{2}[P_2, P_2] \\
=&\int \left(\left(e^{v^+}\theta^+ - e^{v}\theta^- + u(\phi^+ - \phi)\right)\theta(\phi^+ - \phi)\right. \\
&\left.\left(u\theta - u^-\theta^- + \phi^+ - \phi^-\right)e^{v}\theta^-\theta\right) \mathrm{d}x \\
=&\int \left(e^{v^+}\theta^+\theta\phi^+ - e^{v^+}\theta^+\theta\phi + e^{v}\theta^-\theta\phi - e^{v}\theta^-\theta\phi^-\right) \mathrm{d}x \\
=&\int \left(e^{v}\theta\theta^-\phi - e^{v}\theta\theta^-\phi^- + e^{v}\theta^-\theta\phi - e^{v}\theta^-\theta\phi^-\right) \mathrm{d}x \\
=&0,
\end{aligned}$$

so P_2 is indeed a Hamiltonian structure.

The first Hamiltonian structure of the Toda equation can be written as

$$P_1 = \int \left(\theta(\phi^+ - \phi)\right) \mathrm{d}x.$$

One can show its hamiltonianily by using a similar method.

The two Hamiltonian operators read

$$\mathcal{P}_1 = \begin{pmatrix} 0 & \mathcal{S} - 1 \\ 1 - \mathcal{S}^{-1} & 0 \end{pmatrix},$$

$$\mathcal{P}_2 = \begin{pmatrix} \mathcal{S}e^{v} - e^{v}\mathcal{S}^{-1} & u(\mathcal{S} - 1) \\ (1 - \mathcal{S}^{-1})u & \mathcal{S} - \mathcal{S}^{-1} \end{pmatrix}.$$

The coordinate presentations can be also written down by acting the above operators on δ-functions.

4.2 Miura Transformations

Consider the follow equations:

$$\text{KdV}: \quad u_t - 6uu_x + u_{xxx} = 0,$$

$$\text{mKdV}: \quad v_t - 6v^2v_x + v_{xxx} = 0.$$

Miura found that if v is a solution to the mKdV equation, then $u = v^2 + v_x$ gives a solution of the KdV equation.

In general, for an evolutionary PDE

$$u_t^\alpha = X^\alpha, \quad \text{where } X^\alpha \in \mathcal{A},$$

we can transform it to another equation by using transformations of the following form:

$$u^\alpha \mapsto \tilde{u}^\alpha = F^\alpha(u) + Y^\alpha$$

where $F^\alpha(u)$ is a local diffeomorphism, and $Y^\alpha \in \mathcal{A}_{>0}$. We also call them Miura transformations.

Miura transformations are important for Gromov-Witten theory. For example, when considering the target space $\mathbb{P}^1$, the corresponding integrable system is the extended Toda hierarchy, whose equations (like (3.2) and (3.3)) contain $\varepsilon = \sqrt{\hbar}$. On the other hand, the free energy and two-point functions of this model should be formal Laurent series of $\hbar$, so we need to perform certain Miura transformations to eliminate the terms containing odd powers of ε.

It is easy to see that any Miura transformation can be written as the composition of a local diffeomorphism and a Miura transformation of the following form:

$$u^\alpha \mapsto \tilde{u}^\alpha = u^\alpha + Y^\alpha.$$

Local diffeomorphisms are just coordinates transformation on the manifold M, which is easy to deal with. For Miura transformations of the above form, we have the following lemmas.

Lemma 22 *For any $Y^\alpha \in \mathcal{A}_{>0}$ $(\alpha = 1, \ldots, n)$, there exists a variational vector $Z \in \hat{\mathcal{F}}^1_{>0}$ such that*

$$\tilde{u}^\alpha = u^\alpha + Y^\alpha = e^{D_Z}(u^\alpha),$$

where D_Z is the derivation defined by (3.7)*. Z is called the generator of this Miura transformation.*

Proof Let $\nu = \min\{\nu(Y^\alpha) \mid \alpha = 1, \ldots, n\} > 0$. Write Y^α as sum of its homogeneous components

$$Y^\alpha = Y_\nu^\alpha + Y_{\nu+1}^\alpha + \cdots.$$

Take $Z_{(1)} = \int \left(Y_\nu^\alpha \theta_\alpha\right) dx \in \hat{\mathcal{F}}^1_\nu$, then we have

$$\begin{aligned}&e^{-D_{Z_{(1)}}}(u^\alpha + Y^\alpha)\\ =&u^\alpha + Y_\nu^\alpha + Y_{\nu+1}^\alpha + \cdots\end{aligned}$$

$$- \left(Y_\nu^\alpha + D_{Z_{(1)}}(Y_{\nu+1}^\alpha) + \cdots\right) + \cdots$$
$$= u^\alpha + \tilde{Y}^\alpha,$$

where $\nu(\tilde{Y}^\alpha) \geq \nu + 1$.

For $\tilde{Y}^\alpha$, we can take a $Z_{(2)} \in \hat{\mathcal{F}}^1_{\nu+1}$, such that

$$e^{-D_{Z_{(2)}}}(u^\alpha + \tilde{Y}^\alpha) = u^\alpha + \tilde{\tilde{Y}}^\alpha,$$

where $\nu(\tilde{\tilde{Y}}^\alpha) \geq \nu + 2$.

So we obtain a series of variational vector $Z_{(1)}, Z_{(2)}, \cdots \in \hat{\mathcal{F}}^1$ such that

i) $\nu(Z_{(1)}) < \nu(Z_{(2)}) < \cdots$,
ii) $u^\alpha + Y^\alpha = e^{D_{Z_{(1)}}} e^{D_{Z_{(2)}}} \cdots (u^\alpha)$.

Then, by using the Baker-Campbell-Hausdorff formula and the commuting relation (3.8), one can show that there exist $Z \in \hat{\mathcal{F}}^1_{>0}$ such that $u^\alpha + Y^\alpha = e^{D_Z}(u^\alpha)$. The lemma is proved. □

Lemma 23 *Let $u^\alpha \mapsto \tilde{u}^\alpha = u^\alpha + Y^\alpha$ be a Miura transformation with generator $Z \in \hat{\mathcal{F}}^1$, then this Miura transformation transforms $P \in \hat{\mathcal{F}}^p$ to $e^{-\mathrm{ad}_Z}(P)$. We name Miura transformations of this form gauge transformations.*

This lemma depends on a transformation formula of variational derivatives, whose proof cannot be given here, so we omit it. One can find a full proof in §2.5 of [28].

In Poisson geometry, Darboux theorem plays an important role, which classifies the equivalence classes of Poisson structures modulo local coordinates transformations. We have a similar problem for the infinite dimensional case. Let $\mathcal{H}$ be the set of Hamiltonian structures

$$\mathcal{H} = \{P \in \hat{\mathcal{F}}^2 \mid [P, P] = 0\},$$

and $\mathcal{G}$ be the group of gauge transformations

$$\mathcal{G} = \{e^{\mathrm{ad}_Z} \mid Z \in \hat{\mathcal{F}}^1_{>0}\},$$

then $\mathcal{G}$ acts on $\mathcal{H}$, and the corresponding Darboux theorem is a certain description of the quotient space $\mathcal{H}/\mathcal{G}$.

A classification problem is often converted to a deformation problem. For a Hamiltonian structure $P \in \mathcal{H}$, let $\nu = \nu(P)$, and write P as

$$P = P_0 + Q, \quad \text{where } P_0 \in \hat{\mathcal{F}}^2_\nu,\ \nu(Q) > \nu,$$

then P_0 must be a Hamiltonian structure. We call it the leading term of P. Then the equation $[P, P] = 0$ can be written as

$$d_{P_0}(Q) + \frac{1}{2}[Q, Q] = 0, \tag{4.5}$$

where $d_{P_0} = \mathrm{ad}_{P_0}$. Equation (4.5) is called the Maurer-Cartan equation for P_0, and a solution to it is called a Maurer-Cartan element for P_0.

Let $\mathcal{MC}(P)$ be the set of Maurer-Cartan elements for a homogeneous Hamiltonian structure $P \in \hat{\mathcal{F}}^2_\nu$:

$$\mathcal{MC}(P) = \{Q \in \hat{\mathcal{F}}^2_{>\nu} \mid d_P(Q) + \frac{1}{2}[Q, Q] = 0\},$$

then $\mathcal{G}$ also acts on $\mathcal{MC}(P)$:

$$(e^{\mathrm{ad}_Z}, Q) \mapsto \tilde{Q} = e^{\mathrm{ad}_Z}(P + Q) - P.$$

The deformation problem is just to ask the structure of the quotient space $\mathcal{MC}(P)/\mathcal{G}$.

The following definition and lemma are very standard in deformation theory, so we omit their proof.

Definition 24 Let $P \in \hat{\mathcal{F}}^2_\nu$ be a homogeneous Hamiltonian structure.

(a) $Q \in \hat{\mathcal{F}}^2_{>\nu}$ is called a infinitesimal deformation of P if $d_P(Q) = 0$.
(b) Two infinitesimal deformation Q_1, Q_2 are called equivalent if there exists $Z \in \hat{\mathcal{F}}^1_{>0}$ such that $Q_1 - Q_2 = d_P(Z)$.
(c) An infinitesimal deformation Q is called trivial if it is equivalent to 0.
(d) The triple $(\hat{\mathcal{F}}, [\,,\,], d_P)$ forms a differential graded Lie algebra (DGLA). Its cohomology is defined as

$$H(\hat{\mathcal{F}}, d_P) = \mathrm{Ker}(d_P)/\mathrm{Im}(d_P).$$

Note that P is homogeneous, so we have the following decomposition

$$H(\hat{\mathcal{F}}, d_P) = \bigoplus_{p\geq 0}\bigoplus_{d\geq 0} H^p_d(\hat{\mathcal{F}}, P),$$

where

$$H^p_d(\hat{\mathcal{F}}, d_P) = \frac{\mathrm{Ker}(d_P : \hat{\mathcal{F}}^p_d \to \hat{\mathcal{F}}^{p+1}_{d+\nu})}{\mathrm{Im}(d_P : \hat{\mathcal{F}}^{p-1}_{d-\nu} \to \hat{\mathcal{F}}^p_d)}.$$

Lemma 25 *Let $P \in \hat{\mathcal{F}}^2_\nu$ be a homogeneous Hamiltonian structure.*

(a) *The space of equivalence classes of infinitesimal deformations of P is given by $H^2_{>\nu}(\hat{\mathcal{F}}, d_P)$. In particular, every deformation of P is trivial if and only if $H^2_{>\nu}(\hat{\mathcal{F}}, d_P)$ vanishes.*
(b) *Let ν' be the lowest degree of classes in $H^2_{>\nu}(\hat{\mathcal{F}}, d_P)$. If $H^3_{\geq 2\nu'}(\hat{\mathcal{F}}, d_P)$ vanishes, then every infinitesimal deformation can be extended to a genuine deformation, and the space of equivalence classes of deformations of P is just $H^2_{>\nu}(\hat{\mathcal{F}}, d_P)$.*

Example 26 Let $P = \frac{1}{2}\int \left(P^{\alpha\beta}(u)\theta_\alpha\theta_\beta\right) \mathrm{d}x \in \hat{\mathcal{F}}^2_0$ be a Hamiltonian structure. Then it is easy to see that $(P^{\alpha\beta}(u)\frac{\partial}{\partial u^\alpha} \wedge \frac{\partial}{\partial u^\alpha})$ gives a Poisson structure on the manifold M. We assume that $\det(P^{\alpha\beta}) \neq 0$, then, according to the Darboux theorem in finite dimensional symplectic geometry, there exists a local coordinate system $(u^1, \dots, u^n)$ such that $(P^{\alpha\beta})$ is a constant matrix. Let $\theta^\alpha = P^{\alpha\beta}\theta_\beta$, then

$$\frac{\delta P}{\delta u^\alpha} = 0, \quad \frac{\delta P}{\delta \theta_\alpha} = \theta^\alpha.$$

The operator D_P (see (3.7)) reads

$$D_P = \sum_{s\geq 0} \partial^s\theta^\alpha \frac{\partial}{\partial u^{\alpha,s}}.$$

If we write $\partial^s\theta^\alpha$ as $\mathrm{d}u^{\alpha,s}$, then D_P is just the de Rham differential of $J^\infty(M)$. In particular, $D_P^2 = 0$, so we have a complex $(\hat{\mathcal{A}}, D_P)$.

By definition, the following sequence of complex morphisms is exact

$$0 \to (\hat{\mathcal{A}}/\mathbb{R}, D_P) \xrightarrow{\partial} (\hat{\mathcal{A}}, D_P) \xrightarrow{\int} (\hat{\mathcal{F}}, d_P) \to 0,$$

so we have a long exact sequence of cohomologies

$$\begin{aligned} \cdots &\to H^p_{d-1}(\hat{\mathcal{A}}/\mathbb{R}, D_P) \to H^p_d(\hat{\mathcal{A}}, D_P) \to H^p_d(\hat{\mathcal{F}}, d_P) \\ &\to H^{p+1}_{d-1}(\hat{\mathcal{A}}/\mathbb{R}, D_P) \to H^{p+1}_d(\hat{\mathcal{A}}, D_P) \to H^{p+1}_d(\hat{\mathcal{F}}, d_P) \to \cdots. \end{aligned}$$

Define a map

$$F : [0, 1] \times J^\infty(\hat{M}), \quad (t, (u^{\alpha,s}, \theta^s_\alpha)) \mapsto (t^s\, u^{\alpha,s}, t^s\, \theta^s_\alpha),$$

which induces a homotopy equivalence from the complex $(\hat{\mathcal{A}}, D_P)$ to the de Rham complex $(\Omega^*(M), \mathrm{d}_{dR})$ of M, so we have

$$H^p_d(\hat{\mathcal{A}}, D_P) \cong \begin{cases} H^p_{dR}(M), & d = 0; \\ 0, & d > 0. \end{cases}$$

Similarly,

$$H_d^p(\hat{\mathcal{A}}/\mathbb{R}, D_P) \cong \begin{cases} H_{dR}^0(M)/\mathbb{R}, & p = 0,\ d = 0; \\ H_{dR}^p(M), & p > 0,\ d = 0; \\ 0, & d > 0. \end{cases}$$

So we have

$$H_d^p(\hat{\mathcal{F}}, d_P) \cong \begin{cases} H_{dR}^p(M), & d = 0; \\ H_{dR}^{p+1}(M), & d = 1; \\ 0, & d \geq 2. \end{cases}$$

In particular, if $H_{dR}^3(M) \cong 0$, every deformation of P is trivial.

If $H_{dR}^3(M) \not\cong 0$, there are non-trivial infinitesimal deformations, which can be always extended to a genuine deformation since $H_{\geq 2}^3(\hat{\mathcal{F}}, d_P) \cong 0$. For example, if G is a simple compact Lie group, and $M = T^*G$, then M has the canonical symplectic structure, and $H_{dR}^3(M) \cong H_{dR}^3(G) \not\cong 0$, so there is a non-trivial infinitesimal deformation with degree one. The Drinfeld-Sokolov Hamiltonian structure can be regarded as a reduction of this deformation.

4.3 Hydrodynamic Hamiltonian Structures

In this subsection, we consider homogeneous Hamiltonian structures with degree one.

Lemma 27 ([9]) *Let $P \in \hat{\mathcal{F}}_1^2$ be a variational bivector, the corresponding matrix differential operator reads*

$$\mathcal{P}^{\alpha\beta} = g^{\alpha\beta}(u)\partial + \Gamma_\gamma^{\alpha\beta}(u)u^{\gamma,1}.$$

Suppose $\det(g^{\alpha\beta}) \neq 0$, *then P is a Hamiltonian structure if and only if the following two conditions hold true:*

i) $g = (g_{\alpha\beta}) = (g^{\alpha\beta})^{-1}$ *is a flat (not necessary positive definite) metric on M.*
ii) $\Gamma_{\alpha\beta}^\gamma = -g_{\alpha\sigma}\Gamma_\beta^{\sigma\gamma}$ *give the Christoffel symbols of the Levi-Civita connection of* g.

Proof The bivector $P \in \hat{\mathcal{F}}_1^2$ reads

$$P = \frac{1}{2}\int \left(g^{\alpha\beta}\theta_\alpha\theta_\beta^1 + \Gamma_\gamma^{\alpha\beta}u^{\gamma,1}\theta_\alpha\theta_\beta\right) \mathrm{d}x.$$

The skew-symmetry condition $\mathcal{P} + \mathcal{P}^\dagger = 0$ gives

$$g^{\alpha\beta} = g^{\beta\alpha}, \tag{4.6}$$

$$\Gamma^{\alpha\beta}_\gamma + \Gamma^{\beta\alpha}_\gamma = \frac{\partial g^{\alpha\beta}}{\partial u^\gamma}. \tag{4.7}$$

The variational derivatives of P read

$$\begin{aligned}
\frac{\delta P}{\delta u^\sigma} &= \Gamma^{\beta\alpha}_\sigma \theta_\alpha \theta^1_\beta + \frac{1}{2}\left(\frac{\partial \Gamma^{\alpha\beta}_\gamma}{\partial u^\sigma} - \frac{\partial \Gamma^{\alpha\beta}_\sigma}{\partial u^\gamma}\right) u^{\gamma,1}\theta_\alpha\theta_\beta,\\
\frac{\delta P}{\delta \theta_\sigma} &= g^{\sigma\beta}\theta^1_\beta + \Gamma^{\sigma\beta}_\gamma u^{\gamma,1}\theta_\beta.
\end{aligned}$$

Let $W = \frac{1}{2}[P, P]$, then we have

$$W = \int \left(A^{\alpha\beta\gamma}\theta_\alpha\theta^1_\beta\theta^1_\gamma + B^{\alpha\beta\gamma}_\sigma u^{\sigma,1}\theta_\alpha\theta_\beta\theta^1_\gamma + C^{\alpha\beta\gamma}_{\sigma_1\sigma_2} u^{\sigma_1,1}u^{\sigma_2,1}\theta_\alpha\theta_\beta\theta_\gamma\right) dx$$

where

$$\begin{aligned}
A^{\alpha\beta\gamma} &= g^{\gamma\sigma}\Gamma^{\alpha\beta}_\sigma,\\
B^{\alpha\beta\gamma}_\sigma &= \frac{1}{2} g^{\gamma\delta}\left(\frac{\partial \Gamma^{\alpha\beta}_\sigma}{\partial u^\delta} - \frac{\partial \Gamma^{\alpha\beta}_\delta}{\partial u^\sigma}\right) + \Gamma^{\delta\alpha}_\sigma \Gamma^{\gamma\beta}_\delta,\\
C^{\alpha\beta\gamma}_{\sigma_1\sigma_2} &= \frac{1}{2}\Gamma^{\gamma\delta}_{\sigma_2}\left(\frac{\partial \Gamma^{\alpha\beta}_{\sigma_1}}{\partial u^\delta} - \frac{\partial \Gamma^{\alpha\beta}_\delta}{\partial u^{\sigma_1}}\right).
\end{aligned}$$

If $W = 0$ then $\frac{\delta W}{\delta \theta_\alpha} = 0$ for all $\alpha = 1, \ldots, n$, so we have

$$\begin{aligned}
0 &= \frac{\partial}{\partial \theta^2_\beta}\frac{\delta W}{\delta \theta_\alpha} = \frac{\partial}{\partial \theta^2_\beta}\left(\frac{\partial \tilde{W}}{\partial \theta_\alpha} - \partial \frac{\partial \tilde{W}}{\partial \theta^1_\alpha}\right) = \frac{\partial^2 \tilde{W}}{\partial \theta^1_\alpha \partial \theta^1_\beta}\\
0 &= \frac{\partial}{\partial u^{\beta,2}}\frac{\delta W}{\delta \theta_\alpha} = \frac{\partial}{\partial u^{\beta,2}}\left(\frac{\partial \tilde{W}}{\partial \theta_\alpha} - \partial \frac{\partial \tilde{W}}{\partial \theta^1_\alpha}\right) = -\frac{\partial^2 \tilde{W}}{\partial \theta^1_\alpha \partial u^{\beta,1}},
\end{aligned}$$

where $\tilde{W}$ is the density given above. The above two identities imply that

$$A^{\alpha\beta\gamma} = A^{\alpha\gamma\beta}, \tag{4.8}$$

$$B^{\alpha\beta\gamma}_\sigma = B^{\beta\alpha\gamma}_\sigma. \tag{4.9}$$

Equation (4.6) shows that g can be regarded as a metric. Equation (4.7) shows that the metric g is invariant with respect to the connection defined by $\Gamma^{\gamma}_{\alpha\beta}$. Equation (4.8) shows that this connection is torsion-free, so it must be the Levi-Civita connection of g. The last Eq. (4.9) is equivalent to the flatness of this connection.

Conversely, if g and Γ satisfy the condition i) and ii), we can choose a system of flat coordinates such that g is a constant metric and Γ vanish, then it is easy to show that P is a Hamiltonian structure. The lemma is proved. □

Definition 28 A Hamiltonian structure $P \in \hat{\mathcal{F}}^2_1$ is called of hydrodynamic type if it satisfies the conditions in Lemma 27.

According to Lemma 27, we can always choose a coordinate system such that

$$P = \frac{1}{2}\int \left(\eta^{\alpha\beta}\theta_\alpha\theta^1_\beta\right) \mathrm{d}x, \tag{4.10}$$

where $(\eta^{\alpha\beta})$ is a constant symmetric non-degenerate matrix.

From now on, we assume that M is connected and contractible, then consider the deformation problem of (4.10). The computation is similar to the degree zero case. The variational derivatives read

$$\frac{\delta P}{\delta u^\alpha} = 0, \quad \frac{\delta P}{\delta \theta_\alpha} = \eta^{\alpha\beta}\theta^1_\beta.$$

We denote $\theta^{\alpha,s} = \eta^{\alpha\beta}\theta^s_\beta$, then the operator D_P reads

$$D_P = \sum_{s\geq 0}\theta^{\alpha,s+1}\frac{\partial}{\partial u^{\alpha,s}}.$$

The algebra $\hat{\mathcal{A}}$ can be decomposed as $\hat{\mathcal{A}} = \hat{\mathcal{A}}' \otimes \hat{\mathcal{A}}''$, where

$$\begin{aligned}\hat{\mathcal{A}}' &= \mathcal{A} \otimes \wedge^*\left(\mathrm{Span}_{\mathbb{R}}\left\{\theta^{\alpha,s} \mid \alpha = 1,\dots,n;\ s \geq 1\right\}\right),\\ \hat{\mathcal{A}}'' &= \wedge^*\left(\mathrm{Span}_{\mathbb{R}}\left\{\theta^{1,0},\dots,\theta^{n,0}\right\}\right).\end{aligned}$$

Note that $D_P(\hat{\mathcal{A}}'') = 0$, so we have $(\hat{\mathcal{A}}, D_P) = (\hat{\mathcal{A}}', D_P) \otimes \hat{\mathcal{A}}''$, and

$$H^*(\hat{\mathcal{A}}, D_P) = H^*(\hat{\mathcal{A}}', D_P) \otimes \hat{\mathcal{A}}''.$$

On the other hand, if we replace $\theta^{\alpha,s+1}$ by $\mathrm{d}u^{\alpha,s}$, then $(\hat{\mathcal{A}}', D_P)$ is again the de Rham complex of $J^\infty(M)$, so we have

$$H^p_d(\hat{\mathcal{A}}', D_P) = \begin{cases}\mathbb{R}, & (p,d) = (0,0);\\ 0, & (p,d) \neq (0,0),\end{cases}$$

which imply

$$H_d^p(\hat{\mathcal{A}}, D_P) = \begin{cases} \wedge^p(\mathbb{R}^n), & d = 0; \\ 0, & d > 0. \end{cases}$$

Then it is easy to see

$$H_d^p(\hat{\mathcal{A}}/\mathbb{R}, D_P) = \begin{cases} \wedge^p(\mathbb{R}^n), & p > 0, d = 0; \\ 0, & \text{otherwise}. \end{cases}$$

Finally, by using the long exact sequence

$$\begin{aligned} \cdots \to H_{d-1}^p(\hat{\mathcal{A}}/\mathbb{R}, D_P) \to H_d^p(\hat{\mathcal{A}}, D_P) \to H_d^p(\hat{\mathcal{F}}, d_P) \\ \to H_d^{p+1}(\hat{\mathcal{A}}/\mathbb{R}, D_P) \to H_{d+1}^{p+1}(\hat{\mathcal{A}}, D_P) \to H_{d+1}^{p+1}(\hat{\mathcal{F}}, d_P) \to \cdots, \end{aligned}$$

we obtain

$$H_d^p(\hat{\mathcal{F}}, d_P) = \begin{cases} \wedge^p(\mathbb{R}^n) \oplus \wedge^{p+1}(\mathbb{R}^n), & d = 0; \\ 0, & d > 0. \end{cases}$$

In particular, $H^2_{>0}(\hat{\mathcal{F}}, d_P) \cong 0$, so there is no non-trivial deformation of P. This gives the Darboux theorem for Hamiltonian structures of hydrodynamic type.

Theorem 29 *Let $P \in \hat{\mathcal{F}}_1^2$ be a Hamiltonian structure of hydrodynamic type, then for any deformation $\tilde{P} = P + Q$, there exists a gauge transformation e^{ad_Z} such that $e^{\mathrm{ad}_Z}(\tilde{P}) = P$.*

It is interesting to ask whether there are Darboux theorems for Hamiltonian structures with degrees ≥ 2. For example, a degree two Hamiltonian operator has the following general form

$$\mathcal{P}^{\alpha\beta} = g^{\alpha\beta}\partial^2 + \Gamma_\gamma^{\alpha\beta} u^{\gamma,1}\partial + \left(P_\gamma^{\alpha\beta} u^{\gamma,2} + Q_{\xi\zeta}^{\alpha\beta} u^{\xi,1} u^{\zeta,1}\right).$$

We can assume that $g = \left(g^{\alpha\beta}\right)$ is non-degenerate, then g^{-1} is a symplectic structure on M. One can show that $\Gamma_\gamma^{\alpha\beta}$ is given by a symplectic connection of g^{-1}, and it should satisfy a certain flatness condition. But we know nothing about P and Q.

In [6], De Sole and Kac computed certain cohomology groups similar to $H^*(\hat{\mathcal{F}}, d_P)$ for $\mathcal{P}^{\alpha\beta} = g^{\alpha\beta}\partial^N$ with $g^{\alpha\beta}$ being constant and $\det(g^{\alpha\beta}) \neq 0$. Their definition is slightly different from ours, but the result is quite comparable (see [5] for details).

5 Bihamiltonian Structures

5.1 Definition and Semisimplicity

A bihamiltonian structure (P_1, P_2) is a pair of Hamiltonian structures such that $[P_1, P_2] = 0$.

Lemma 30 *Let $P \in \hat{\mathcal{F}}^2$ be a Hamiltonian structure, if there is an $X \in \hat{\mathcal{F}}^1$ such that $[X, [X, P]] = 0$, then $(P, [P, X])$ is a bihamiltonian structure. Bihamiltonian structures obtained by this way are called exact bihamiltonian structures.*

Proof Let $Q = [P, X]$, then $[P, P] = 0$, $[P, Q] = 0$, and

$$[Q, Q] = [[P, X], Q] = -[[X, Q], P] - [[Q, P], X] = 0.$$

The lemma is proved. □

Example 31 The KdV equation has two Hamiltonian structures

$$P_1 = \int \theta\theta^1 \mathrm{d}x, \quad P_2 = \int \left(u\theta\theta^1 + \frac{\hbar}{8}\theta\theta^3\right) \mathrm{d}x.$$

Let $X = \int \theta \mathrm{d}x$, then $P_1 = [P_2, X]$, and $[P_1, X] = 0$, so (P_1, P_2) is indeed a bihamiltonian structure, and it is exact.

Example 32 The Toda equation has two Hamiltonian structures

$$P_1 = \int \theta(\phi^+ - \phi)\mathrm{d}x,$$

$$P_2 = \int \left(e^{v^+}\theta\theta^+ + u\theta(\phi^+ - \phi) + \phi\phi^+\right) \mathrm{d}x.$$

Let $X = \int \theta \mathrm{d}x$, then $P_1 = [P_2, X]$, and $[P_1, X] = 0$, so (P_1, P_2) is also a bihamiltonian structure, and it is exact.

Example 33 The Camassa-Holm equation has two Hamiltonian structures:

$$P_1 = \int \theta(\theta^1 - \theta^3)\mathrm{d}x, \quad P_2 = \int \left(u\theta\theta^1\right) \mathrm{d}x.$$

We have shown in the last section that any linear combination of P_1 and P_2 is a Hamiltonian structure, which implies that $[P_1, P_2] = 0$, so (P_1, P_2) is a bihamiltonian structure. Note that this bihamiltonian structure is not exact.

Let (P_1, P_2) be a bihamiltonian structure, if both P_1 and P_2 are of hydrodynamic type, then (P_1, P_2) is also called of hydrodynamic type. According to Lemma 27, there exists a pair of flat metric g_1 and g_2, such that

$$P_a = \frac{1}{2}\int \left(g_a^{\alpha\beta}(u)\theta_\alpha\theta_\beta^1 + \Gamma_{\gamma,a}^{\alpha\beta}u^{\gamma,1}\theta_\alpha\theta_\beta\right) \mathrm{d}x,$$

where $a = 1, 2$, and $g_a^{\alpha\beta}$ and $\Gamma_{\gamma,a}^{\alpha\beta}$ are given by the contravariant metric and the connection coefficients of g_a. In general, one cannot find a coordinate system such that both g_1 and g_2 are constant.

Definition 34 Let (P_1, P_2) be a bihamiltonian structure of hydrodynamic type, whose contravariant metric are $g_1^{\alpha\beta}(u)$ and $g_2^{\alpha\beta}(u)$. If the roots

$$\lambda^1(u), \dots, \lambda^n(u)$$

of the characteristic equation

$$\det\left(g_2^{\alpha\beta}(u) - \lambda g_1^{\alpha\beta}(u)\right) = 0$$

are not constant and distinct, the bihamiltonian structure (P_1, P_2) is called semisimple. The roots $\lambda^1, \dots, \lambda^n$ are called the canonical coordinates of (P_1, P_2).

Theorem 35 ([20]) *Let (P_1, P_2) be a semisimple bihamiltonian structure. Its canonical coordinates can serve as local coordinates near any point on M. Furthermore, the two metric has the following form in the canonical coordinates*

$$g_1^{ij} = \delta^{ij} f^i(\lambda), \quad g_2^{ij} = \delta^{ij} \lambda^i f^i(\lambda).$$

Note that we don't sum over repeated Latin indexes i, j.

In canonical coordinates, the two Hamiltonian structures have the following forms:

$$P_1 = \frac{1}{2}\int \left(\sum_{i=1}^n f^i(\lambda)\theta_i\theta_i^1 + \sum_{i,j=1}^n A^{ij}\theta_i\theta_j\right) \mathrm{d}x,$$

$$P_2 = \frac{1}{2}\int \left(\sum_{i=1}^n g^i(\lambda)\theta_i\theta_i^1 + \sum_{i,j=1}^n B^{ij}\theta_i\theta_j\right) \mathrm{d}x,$$

where $g^i(\lambda) = \lambda^i f^i(\lambda)$, and

$$A^{ij} = \frac{1}{2}\left(\frac{f^i}{f_j}\frac{\partial f^j}{\partial \lambda^i}\lambda^{j,1} - \frac{f^j}{f_i}\frac{\partial f^i}{\partial \lambda^j}\lambda^{i,1}\right),$$

$$B^{ij} = \frac{1}{2}\left(\frac{g^i}{f_j}\frac{\partial f^j}{\partial \lambda^i}\lambda^{j,1} - \frac{g^j}{f_i}\frac{\partial f^i}{\partial \lambda^j}\lambda^{i,1}\right).$$

Note that $f^i \neq 0$, $g^i \neq 0$, and A^{ij}, B^{ij} are skew-symmetric.

Example 36 ([7]) Let M be a Frobenius manifold, then we have a pair of compatible metric

$$g_1^{\alpha\beta} = \eta^{\alpha\beta}, \quad g_2^{\alpha\beta} = E^\gamma c_\gamma^{\alpha\beta},$$

which define a bihamiltonian structure (P_1, P_2) of hydrodynamic type. If M is semisimple, then (P_1, P_2) is also semisimple, and the canonical coordinates of (P_1, P_2) coincide with the ones of M. Bihamiltonian structures of Frobenius manifolds are always exact, because $P_1 = [P_2, e]$, where e is the unit vector field.

5.2 *Bihamiltonian Cohomology*

In this subsection, we consider the deformation problem of a semisimple bihamiltonian structure.

Let (P_1, P_2) be a semisimple bihamiltonian structure, denote by $d_a = \mathrm{ad}_{P_a}$ $(a = 1, 2)$, then they satisfy

$$d_1^2 = 0, \quad d_1 d_2 + d_2 d_1 = 0, \quad d_2^2 = 0,$$

so we have a double complexes $(\hat{\mathcal{F}}^2, d_1, d_2)$.

A deformation of (P_1, P_2) is a bihamiltonian structure of the following form

$$(\tilde{P}_1, \tilde{P}_2) = (P_1 + Q_1, P_2 + Q_2),$$

where $Q_a \in \hat{\mathcal{F}}^2_{>1}$ $(a = 1, 2)$. According to the results given in Sect. 4.3, there is a gauge transformation e^{ad_Z} such that $e^{\mathrm{ad}_Z}(\tilde{P}_1) = P_1$, so we can take $Q_1 = 0$, and rename Q_2 to Q. Then $(\tilde{P}_1, \tilde{P}_2) = (P_1, P_2 + Q)$ is a bihamiltonian structure if and only if

$$d_1(Q) = 0, \quad d_2(Q) + \frac{1}{2}[Q, Q] = 0.$$

A bivector Q satisfying the above conditions is called a Maurer-Cartan element for (P_1, P_2), and we denote the set of Maurer-Cartan elements by $\mathcal{MC}(P_1, P_2)$:

$$\mathcal{MC}(P_1, P_2) = \{Q \in \hat{\mathcal{F}}^2_{>1} \mid d_1(Q) = 0,\ d_2(Q) + \frac{1}{2}[Q, Q] = 0\}.$$

Two deformations are equivalent if there exists a gauge transformation that convert one to another. Note that our P_1 is fixed, so the gauge transformations should preserve P_1, we denote such gauge transformations as $\mathcal{G}(P_1)$:

$$\mathcal{G}(P_1) = \{e^{\mathrm{ad}_Z} \mid Z \in \hat{\mathcal{F}}^1_{>0},\ d_1(Z) = 0\}.$$

The deformation problem for the bihamiltonian structure (P_1, P_2) is just to ask the structure of the quotient space $\mathcal{MC}(P_1, P_2)/\mathcal{G}(P_1)$.

Definition 37

(a) $Q \in \hat{\mathcal{F}}^2_{>1}$ is called a infinitesimal deformation of (P_1, P_2) if $d_1(Q) = 0$, $d_2(Q) = 0$.
(b) Two infinitesimal deformations Q_1, Q_2 are called equivalent if there exists $Z \in \hat{\mathcal{F}}^1_{>0}$ such that $d_1(Z) = 0$, $d_2(Z) = Q_1 - Q_2$.
(c) An infinitesimal deformation Q is called trivial if it is equivalent to 0.
(d) The bihamiltonian cohomologies of (P_1, P_2) are defined as

$$BH^p_d(\hat{\mathcal{F}}, d_1, d_2) = \frac{\hat{\mathcal{F}}^p_d \cap \mathrm{Ker}(d_1) \cap \mathrm{Ker}(d_2)}{\hat{\mathcal{F}}^p_d \cap \mathrm{Im}(d_1 d_2)}.$$

The following lemma is quite standard, so we omit its proof.

Lemma 38 *Let (P_1, P_2) be a semisimple bihamiltonian structure.*

(a) The cohomology group $BH^2_{>1}(\hat{\mathcal{F}}, d_1, d_2)$ gives the space of equivalence classes of infinitesimal deformations of (P_1, P_2).
(b) Let v' be the lowest degree of classes in $BH^2_{>1}(\hat{\mathcal{F}}, d_1, d_2)$. If

$$BH^3_{\geq 2v'}(\hat{\mathcal{F}}, d_1, d_2) \cong 0,$$

then every infinitesimal deformation of (P_1, P_2) can be extended to a genuine deformation, and $BH^2_{>1}(\hat{\mathcal{F}}, d_1, d_2)$ actually gives the space of equivalence classes of deformations.

In [27] and [14], we proved the following theorem.

Theorem 39 *Let (P_1, P_2) be a semisimple bihamiltonian structure, then*

$$BH^2_{d\geq 2}(\hat{\mathcal{F}}, d_1, d_2) \cong \begin{cases} \bigoplus_{i=1}^{n} C^\infty(\mathbb{R}), & d = 3; \\ 0, & d = 2, 4, 5, \ldots. \end{cases}$$

The classes in $BH^2_3(\hat{\mathcal{F}})$ are actually parameterized by n functions of canonical coordinates $\{c_1(\lambda_1), \ldots, c_n(\lambda_n)\}$. We will discuss their definition and properties in the next section.

In [29], we prove the following theorem.

Theorem 40 *Let* $\left(P_1 = \int (\theta\theta')\,\mathrm{d}x,\ P_2 = \int (u\theta\theta')\,\mathrm{d}x\right)$ *be the leading term of the bihamiltonian structure of KdV equation, then*

$$BH^3_{d\geq 4}(\hat{\mathcal{F}}, d_1, d_2) \cong 0.$$

The proofs of these two theorems are very long, so they cannot be given here. Combining the above theorems and lemma, we obtain the following corollaries.

Corollary 41 *Let* (P_1, P_2) *be a semisimple bihamiltonian structure. For any deformation* $(\tilde{P}_1, \tilde{P}_2)$ *of* (P_1, P_2)*, one can define n functions*

$$c_1(\lambda_1), \ldots, c_n(\lambda_n),$$

which are called the central invariants of $(\tilde{P}_1, \tilde{P}_2)$*, such that*

(a) Two deformations are equivalent if and only if their central invariants coincide.
(b) Write the deformation $(\tilde{P}_1, \tilde{P}_2)$ *as the sum of homogeneous components*

$$\tilde{P}_a = P_a + \sum_{k\geq 1} \varepsilon^k P_a^{[k]}, \quad a = 1, 2,$$

where $P_a^{[k]} \in \hat{\mathcal{F}}^2_{k+1}$*, then there exists a gauge transformation* e^{ad_Z} *such that* $(e^{\mathrm{ad}_Z}(\tilde{P}_1), e^{\mathrm{ad}_Z}(\tilde{P}_2))$ *doesn't contain odd powers of* ε*.*
(c) If (P_1, P_2) *is the leading term of the bihamiltonian structure of the KdV hierarchy, then for any smooth function* $c(u)$ *there exists a deformation whose central invariant is given by* $c(u)$*.*

Part (a) is called the uniqueness theorem of the deformation problem. Part (b) is important for Gromov-Witten theory, because it ensure that the corresponding integrable hierarchy can be always written as a formal power series of $\hbar$. Part (c) is called the existence theorem of the deformation problem. We conjecture that it is true for arbitrary semisimple bihamiltonian structure.

Recently [2] (c.f. [1]), Carlet, Posthuma, and Shadrin proved the following theorem, which showed that our conjecture is true.

Theorem 42 ([2]) *Let* (P_1, P_2) *be a semisimple bihamiltonian structure of hydrodynamic type, then* $BH^p_d(\hat{\mathcal{F}}, d_1, d_2)$ *vanishes for most* (p, d)*. In particular,* $BH^3_{d\geq 5}(\hat{\mathcal{F}}, d_1, d_2) \cong 0$*, which implies that the existence of a full dispersive deformation of* (P_1, P_2) *starting from any its infinitesimal deformation.*

The proof of this theorem is sophisticated, so we cannot give it here. Please refer to [2] for details.

5.3 Bihamiltonian Vector Fields

Let (P_1, P_2) be a bihamiltonian structure, $X \in \hat{\mathcal{F}}^1$ is called a bihamiltonian vector field, if there exists $I, J \in \hat{\mathcal{F}}^0$ such that $X = d_1(I) = d_2(J)$. Suppose (P_1, P_2) is semisimple, and $(\tilde{P}_1, \tilde{P}_2)$ is a deformation of (P_1, P_2). In this subsection, we will consider their bihamiltonian vector fields.

Lemma 43 *The space of bihamiltonian vector fields of (P_1, P_2) is given by $BH^1_{\geq 1}(\hat{\mathcal{F}}, d_1, d_2)$.*

Proof Let X be a bihamiltonian vector field of (P_1, P_2). Note that $\deg(P_a) = 1$ $(a = 1, 2)$, so $\nu(X) \geq 1$.

The bihamiltonian cohomology $BH^1_{\geq 1}(\hat{\mathcal{F}}, d_1, d_2)$ is defined as

$$BH^1_{\geq 1}(\hat{\mathcal{F}}, d_1, d_2) = \{X \in \hat{\mathcal{F}}^1_{\geq 1} \mid d_1(X) = 0,\ d_2(X) = 0\},$$

so every bihamiltonian vector field belongs to $BH^1_{\geq 1}(\hat{\mathcal{F}}, d_1, d_2)$.

On the other hand, if $X \in BH^1_{\geq 1}(\hat{\mathcal{F}}, d_1, d_2)$, then there must exist $I, J \in \hat{\mathcal{F}}^0$ such that $X = d_1(I) = d_2(J)$, because $H^1_{\geq 1}(\hat{\mathcal{F}}, d_a) \cong 0$ $(a = 1, 2)$. □

Lemma 44 *We have $BH^1_{\geq 2}(\hat{\mathcal{F}}, d_1, d_2) \cong 0$.*

Proof Suppose $X = d_1(I) = d_2(J) \in \hat{\mathcal{F}}^1_d$ $(d \geq 2)$, where

$$I = \int p\mathrm{d}x, \quad J = \int q\mathrm{d}x,$$

and $p, q \in \mathcal{A}^{(N)}$, $1 \leq N \leq d$. We are to show that one can always choose another pair of density $p', q' \in \mathcal{A}^{(N-1)}$ such that $I = \int p'\mathrm{d}x$, $J = \int q'\mathrm{d}x$. Then the theorem can be proved by induction on N.

Let $Z = d_1(I) - d_2(J) = \int (Z^\alpha \theta_\alpha)\,\mathrm{d}x$. It is easy to see that $Z^\alpha \in \mathcal{A}^{(2N+1)}$. We introduce a notation $a_{(i,s)} = \frac{\partial a}{\partial \lambda^{i,s}}$ for $a \in \mathcal{A}$. Then one can obtain that

$$Z^i_{(j,2N+1)} = (-1)^{N+1} f^i \left(p_{(i,N)(j,N)} - \lambda^i q_{(i,N)(j,N)}\right) = 0,$$

so $(\lambda^i - \lambda^j) q_{(i,N)(j,N)} = 0$. Since $\lambda^i \neq \lambda^j$ $(i \neq j)$, we have $q_{(i,N)(j,N)} = 0$ $(i \neq j)$. Denote by $r^i = q_{(i,N)(i,N)}$, then $q_{(i,N)(j,N)} = \delta^{ij} r^i$, $p_{(i,N)(j,N)} = \delta^{ij}\lambda^i r^i$.

Next, compute $Z^i_{(j,2N)}$:

$$0 = Z^i_{(j,2N)} = (-1)^{N+1} \left(\left(N + \frac{1}{2}\right) f^i r^i \lambda^{i,1} \delta^{ij} + (\lambda^i A^{ij} - B^{ij}) + \right.$$
$$\left. f^i \left(p_{(i,N)(j,N-1)} - p_{(j,N)(i,N-1)}\right) - g^i \left(q_{(i,N)(j,N-1)} - q_{(j,N)(i,N-1)}\right) \right).$$

Take $i = j$, we obtain $r^i = 0$, so p and q are linear in $\lambda^{\alpha,N}$. For $i \neq j$, we obtain

$$p_{(i,N)(j,N-1)} = p_{(j,N)(i,N-1)}, \quad q_{(i,N)(j,N-1)} = q_{(j,N)(i,N-1)},$$

which imply that one can choose $\tilde{p}, \tilde{q} \in \mathcal{A}^{(N-1)}$ such that $p' = p - \partial(\tilde{p})$ and $q' = q - \partial(\tilde{q})$ belong to $\mathcal{A}^{(N-1)}$. The lemma is proved. □

The above proof can be regarded as a demo version of the proofs for Theorem 39 and 40. In the latter cases, we also use an induction on N for $\mathcal{A}^{(N)}$. This computation method can be translated to the language of spectral sequence. In Carlet, Posthuma, and Shadrin's new preprints [1, 2], they introduce more spectral sequences, which help them to compute almost all the bihamiltonian cohomologies $BH_d^p(\hat{\mathcal{F}}, d_1, d_2)$.

The above lemma shows that bihamiltonian vector fields of (P_1, P_2) must have degree one.

Corollary 45 *Let*

$$X = \int \left(X^\alpha \theta_\alpha\right) \mathrm{d}x$$

be a bihamiltonian vector fields of (P_1, P_2)*, then* X *must be diagonal hydrodynamic, i.e.* $X^i = V^i(\lambda)\lambda^{i,1}$.

Proof Suppose $X = d_1(I) = d_2(J)$, where $I = \int p\mathrm{d}x$, $J = \int q\mathrm{d}x$, $p, q \in \mathcal{A}_0$, then X^i takes the following form:

$$X^i = \sum_{j=1}^{n} V_j^i(\lambda)\lambda^{j,1},$$

where the coefficients read

$$V_j^i = -f^i \mathcal{D}_{ij}(p) = -g^i \mathcal{D}_{ij}(q),$$

and $\mathcal{D}_{ij}$ is the following linear differential operator

$$\mathcal{D}_{ij} = \frac{\partial^2}{\partial\lambda^i \partial\lambda^j} + \frac{1}{2}\frac{\partial \log f^i}{\partial\lambda^j}\frac{\partial}{\partial\lambda^i} + \frac{1}{2}\frac{\partial \log f^j}{\partial\lambda^i}\frac{\partial}{\partial\lambda^j}.$$

Note that $\mathcal{D}_{ij}$ is symmetric, so we have $(\lambda^i - \lambda^j)\mathcal{D}_{ij}(q) = 0$, which implies $V_j^i = 0$ if $i \neq j$. The corollary is proved. □

Corollary 46 *If* X_1, X_2 *are bihamiltonian vector fields of* (P_1, P_2)*, then* $[X_1, X_2] = 0$.

Proof Let $Y = [X_1, X_2]$, then $d_1(Y) = 0$, $d_2(Y) = 0$, so $Y = 0$, since $Y \in BH_2^1(\hat{\mathcal{F}}, d_1, d_2) \cong 0$. □

Now let us consider the bihamiltonian vector fields of $(\tilde{P}_1, \tilde{P}_2)$. Let $X \in \hat{\mathcal{F}}^1$ be such a vector field, then $\nu(X) \geq 1$. We expand it with respect the standard gradation

$$X = X_1 + X_2 + \ldots, \quad X_d \in \hat{\mathcal{F}}^1_d,$$

then it is easy to see that X_1 must be a bihamiltonian vector field of (P_1, P_2). We call X_1 the leading term of X.

Theorem 47

(a) If X_1, X_2 are bihamiltonian vector fields of $(\tilde{P}_1, \tilde{P}_2)$, then $[X_1, X_2] = 0$. If they have the same leading term, then $X_1 = X_2$.
(b) For any bihamiltonian vector field X_1 of (P_1, P_2), there exists a bihamiltonian vector field X of $(\tilde{P}_1, \tilde{P}_2)$ such that X's leading term is just X_1.

Proof For Part (a), we only need to show that if the leading term of a bihamiltonian vector field X of $(\tilde{P}_1, \tilde{P}_2)$ vanishes, then $X = 0$. Expand X as

$$X = X_1 + X_2 + X_3 + \cdots, \quad X_1 = 0, \; X_d \in \hat{\mathcal{F}}^1_d.$$

We also expand $(\tilde{P}_1, \tilde{P}_2)$ in the same way:

$$\tilde{P}_1 = P_1 + \sum_{k\geq 1} P_1^{[k]}, \quad \tilde{P}_2 = P_2 + \sum_{k\geq 1} P_2^{[k]}.$$

Then the condition $[\tilde{P}_a, X] = 0$ $(a = 1, 2)$ implies that

$$d_a(X_d) + \sum_{k=1}^{d-2} [P_a^{[k]}, X_{d-k}] = 0, \quad a = 1, 2.$$

When $d = 2$, we obtain $d_1(X_2) = 0$, $d_2(X_2) = 0$, so we have $X_2 = 0$. Then, by induction on d, one can show that $X_d = 0$ for $d = 2, 3, \ldots$, so $X = 0$.

To prove Part (b), we also expand X, P_1, and P_2 as above. We need to show that if X_1 satisfies $d_1(X_1) = 0$, $d_2(X_1) = 0$, then there exist $X_2, X_3, \ldots$ such that

$$d_a(X_d) + \sum_{k=1}^{d-1} [P_a^{[k]}, X_{d-k}] = 0, \quad a = 1, 2. \tag{5.1}$$

Without loss of generality, we can assume that $P_a^{[1]} = 0$ (see Part (b) of Corollary 41), then we can take $X_2 = 0$ directly.

The existence of X_d ($d \geq 3$) can be proved by induction on d. Suppose we have obtained $X_2, \ldots, X_{d-1}$, and we are to find X_d. Denote by

$$W_a = -\sum_{k=1}^{d-1}[P_a^{[k]}, X_{d-k}], \quad a = 1, 2,$$

then X_d satisfy $d_1(X_d) = W_1$ and $d_2(X_d) = W_2$.

We assert that $d_1(W_1) = 0$. By using the Jacobi identity, we have

$$d_1(W_1) = \sum_{k=1}^{d-1}\left([d_1(P_1^{[k]}), X_{d-k}] + [P_1^{[k]}, d_1(X_{d-k})]\right).$$

Note that $\tilde{P}_1$ is a Hamiltonian structure, so we have

$$d_1(P_1^{[k]}) + \frac{1}{2}\sum_{j=1}^{k-1}[P_1^{[j]}, P_1^{[k-j]}] = 0.$$

From the above identity and (5.1) with X_d replaced by X_{d-k}, one can show that $d_1(W_1) = 0$. Similarly, we have $d_2(W_2) = 0$.

Since $H^2_{d+1}(\hat{\mathcal{F}}, d_1) \cong 0$, there exists $Y \in \hat{\mathcal{F}}^1_d$ such that $W_1 = d_1(Y)$, then the general solution to $d_1(X_d) = W_1$ can be written as $X_d = Y + d_1(Z)$ for arbitrary $Z \in \hat{\mathcal{F}}^0_{d-1}$. Then the equation $d_2(X_d) = W_2$ becomes $d_1 d_2(Z) = Q$, where $Q = d_2(Y) - W_2$.

It is easy to see that $d_2(Q) = 0$. One can also show that $d_1(Q) = 0$ by using the condition $[\tilde{P}_1, \tilde{P}_2] = 0$, so $Q \in \hat{\mathcal{F}}^2_{d+1} \cap \mathrm{Ker}(d_1) \cap \mathrm{Ker}(d_2)$. Note that $d + 1 \geq 4$, so $BH^2_d(\hat{\mathcal{F}}, d_1, d_2) \cong 0$, so there must exist $Z \in \hat{\mathcal{F}}^0_{d-1}$ such that $Q = d_1 d_2(Z)$. The existence of X_d is proved. □

6 Central Invariants

6.1 *Definition and Properties*

In this subsection, we explain how to compute the central invariants of a deformed semisimple bihamiltonian structure.

Let (P_1, P_2) be a semisimple bihamiltonian structure, $(\tilde{P}_1, \tilde{P}_2)$ be a deformation of (P_1, P_2), and $\mathcal{P}_a, \tilde{\mathcal{P}}_a$ ($a = 1, 2$) be the corresponding matrix differential operators in canonical coordinates. Expand $\tilde{\mathcal{P}}_a$ ($a = 1, 2$) with respect to the standard gradation

$$\tilde{\mathcal{P}}_a^{\alpha\beta} = \mathcal{P}_a^{\alpha\beta} + \sum_{s\geq 1}\left(\sum_{t=0}^{s+1} P_{s,t,a}^{\alpha\beta}\partial^t\right),$$

where $a = 1, 2$, $P^{\alpha\beta}_{s,t,a} \in \mathcal{A}_{s+1-t}$. It is easy to see that $P^{\alpha\beta}_{s,s+1,a}$ is a tensor on M. The *central invariants* of $(\tilde{P}_1, \tilde{P}_2)$ are defined as

$$c_i(\lambda) = \frac{1}{3\left(f^i\right)^2}\left(P^{ii}_{2,3,2} - \lambda^i P^{ii}_{2,3,1} + \sum_{k\neq i}\frac{\left(P^{ki}_{1,2,2} - \lambda^i P^{ki}_{1,2,1}\right)^2}{f^k(\lambda^k - \lambda^i)}\right), \tag{6.1}$$

where $i = 1, \ldots, n$, λ_i's are the canonical coordinates, and f^i's are the diagonal entries of the first metric (see Definition 34 and Theorem 35). Note that the semisimplicity of (P_1, P_2) plays a crucial role in the definition of central invariants: (1) λ_i's are not constants, so we can use them as coordinates; (2) they are distinct, so the denominator in the above formula never vanish.

Theorem 48

(a) The central invariants are invariant under gauge transformations.
(b) The i-th central invariant $c_i(\lambda)$ only depends on λ^i.
(c) The cohomology class corresponding to the infinitesimal deformation of $(\tilde{P}_1, \tilde{P}_2)$ has a representative

$$Q = d_2 d_1\left(\int\left(\sum_{i=1}^{n} c_i(\lambda^i)\lambda^{i,1}\log\lambda^{i,1}\right)\mathrm{d}x\right) \in \hat{\mathcal{F}}^2_3.$$

The proof of this theorem is simple but tedious [14], so we omit it.

In Part (c), we give Q in the form $d_2 d_1(J)$. This expression looks confusing, since elements of the form $d_2 d_1(J) = -d_1 d_2(J)$ should be exact in the cohomology group $BH^2_3(\hat{\mathcal{F}}, d_1, d_2)$. But Q is indeed not trivial, because the density of the local functional J given above is *not* a differential polynomial, so $J \notin \hat{\mathcal{F}}^0$. This expression shows that if we enlarge the group of gauge transformation, then there is no nontrivial infinitesimal deformations. This result is called the quasi-triviality theorem [14].

Theorem 49 *Denote by $\mu = \prod_{i=1}^n \lambda^{i,1}$, $\tilde{\mathcal{A}} = \hat{\mathcal{A}}[\mu^{-1}]$, $\tilde{\mathcal{F}} = \tilde{\mathcal{A}}/\partial\tilde{\mathcal{A}}$.*

(a) For any deformation $(\tilde{P}_1, \tilde{P}_2)$ of a semisimple bihamiltonian structure (P_1, P_2), there exists $Z \in \tilde{\mathcal{F}}^1_{>0}$, such that $(e^{\mathrm{ad}_Z}(\tilde{P}_1), e^{\mathrm{ad}_Z}(\tilde{P}_2)) = (P_1, P_2)$.
(b) Let $X \in \hat{\mathcal{F}}^1$ be a bihamiltonian vector field of $(\tilde{P}_1, \tilde{P}_2)$ with leading term $X_1 \in \hat{\mathcal{F}}^1_1$, then $e^{\mathrm{ad}_Z}(X) = X_1$.

This theorem implies that Dubrovin-Zhang's QT Axiom is a corollary of the BH Axiom, so the QT Axiom can be removed from their construction.

6.2 Example: Frobenius Manifolds

Let (P_1, P_2) be the bihamiltonian structure associated to a semisimple Frobenius manifold (see Example 36). In [10], Dubrovin and Zhang constructed a genus one deformation of (P_1, P_2) satisfying their VS Axiom [11]. Note that a genus one deformation is exactly an infinitesimal deformation of degree 3. So it is natural to ask: what are its central invariants?

By checking the expressions given in [10], the tensors used in (6.1) read

$$f^i = \frac{1}{\psi_{i1}^2}, \quad P^{ki}_{1,2,1} = 0, \quad P^{ki}_{1,2,2} = 0,$$

$$P^{ii}_{2,3,1} = \frac{1}{12\psi_{i1}^4} \sum_{j\neq i} \gamma_{ij} \left(\frac{\psi_{i1}}{\psi_{j1}} + \frac{\psi_{j1}}{\psi_{i1}}\right),$$

$$P^{ii}_{2,3,2} = \frac{1}{72\psi_{i1}^4} + \frac{\lambda^i}{12\psi_{i1}^4} \sum_{j\neq i} \gamma_{ij} \left(\frac{\psi_{i1}}{\psi_{j1}} + \frac{\psi_{j1}}{\psi_{i1}}\right),$$

then we immediately obtain the central invariants

$$c_1 = \cdots = c_n = \frac{1}{24}.$$

In [36], Zhang showed that if a deformation $(\tilde{P}_1, \tilde{P}_2)$ admits a tau function, then its central invariants must be constant. In this case, the genus one free energy has the form

$$F_1 = \sum_{i=1}^{n} c_i \log(\lambda^{i,1}) + G(\lambda).$$

When $c_i = 1/24$ $(i = 1, \ldots, n)$, we obtain the well-known formula for genus one free energy of a semisimple cohomological field theory

$$F_1 = \frac{1}{24} \log\left(\prod_{i=1}^{n} \lambda^{i,1}\right) + G(\lambda).$$

We conjecture that the converse propositions of the above results are also true.

Conjecture 50 Let $(\tilde{P}_1, \tilde{P}_2)$ be a deformation of (P_1, P_2) with central invariants $c_1, \ldots, c_n$.

(a) If c_i $(i = 1, \ldots, n)$ are all constant, then the corresponding integrable hierarchy admits a tau structure.

(b) if $c_i = 1/24$ $(i = 1, \dots, n)$, then the corresponding integrable hierarchy has linearizable Virasoro symmetries.

If these conjectures hold true, then Dubrovin-Zhang's TS Axiom and VS Axiom can be replaced by the above conditions on central invariants.

6.3 Example: Drinfeld-Sokolov Hierarchy

Let $\mathfrak{g}$ be a simple Lie algebra of dimension m and rank n, and $\mathrm{u}^1, \dots, \mathrm{u}^m$ be a set of basis. Suppose

$$[\mathrm{u}^\alpha, \mathrm{u}^\beta] = C^{\alpha\beta}_\gamma \mathrm{u}^\gamma.$$

Let $M = \mathfrak{g}^*$, and $\mathrm{v}_1, \dots, \mathrm{v}_m$ be dual basis of $\mathrm{u}^1, \dots, \mathrm{u}^m$, then any element $q \in M$ can be written as

$$q = u^\alpha \mathrm{v}_\alpha, \quad u^\alpha \in \mathbb{R}.$$

The bracket

$$\{u^\alpha, u^\beta\} = C^{\alpha\beta}_\gamma u^\gamma$$

defines a Poisson structure on M, which is called the Lie-Poisson structure.

The Lie-Poisson structure defines a Hamiltonian structure $P_0 \in \hat{\mathcal{F}}^2_0$:

$$P_0 = \int \left(C^{\alpha\beta}_\gamma u^\gamma \theta_\alpha \theta_\beta \right) \mathrm{d}x.$$

Its action on $F, G \in \hat{\mathcal{F}}^0$ is given by

$$\{F, G\}_{P_0} = \int \left(C^{\alpha\beta}_\gamma u^\gamma \frac{\delta F}{\delta u^\alpha} \frac{\delta G}{\delta u^\beta} \right) \mathrm{d}x. \tag{6.2}$$

Note that $C^{\alpha\beta}_\gamma u^\gamma = \langle q, [\mathrm{u}^\alpha, \mathrm{u}^\beta] \rangle$, where $\langle \, , \rangle$ is the pairing between $\mathfrak{g}^*$ and $\mathfrak{g}$. If we introduce a notation

$$\mathrm{grad}(F) = \frac{\delta F}{\delta u^\alpha} \mathrm{u}^\alpha \in \mathcal{A} \otimes \mathfrak{g},$$

then the Poisson bracket (6.2) becomes

$$\{F, G\}_{P_0} = \int \langle q, [\mathrm{grad}(F), \mathrm{grad}(G)] \rangle \mathrm{d}x.$$

Let $\langle\,,\rangle_{\mathfrak{g}}$ be a invariant non-degenerate symmetric bilinear form on $\mathfrak{g}$, we can identify $\mathfrak{g}^*$ and g such that

$$\langle q,\cdot\rangle = \langle q,\cdot\rangle_{\mathfrak{g}}.$$

We always assume this identification, then the Poisson bracket (6.2) can be also written as

$$\{F,G\}_{P_0} = \int \langle \mathrm{grad}(F), [\mathrm{grad}(G), q]\rangle_{\mathfrak{g}} \mathrm{d}x.$$

Define $\eta^{\alpha\beta} = \langle \mathrm{u}^\alpha, \mathrm{u}^\beta\rangle_{\mathfrak{g}}$. The Hamiltonian structure P_0 admit a degree one deformation

$$P = \int \left(C^{\alpha\beta}_{\gamma} u^\gamma \theta_\alpha\theta_\beta - \eta^{\alpha\beta}\theta_\alpha\theta^1_\beta\right) \mathrm{d}x.$$

The action of P on $F, G \in \hat{\mathcal{F}}^0$ reads

$$\{F,G\}_{P} = \int \langle \mathrm{grad}(F), [\mathrm{grad}(G), \partial + q]\rangle_{\mathfrak{g}} \mathrm{d}x.$$

Here we assume that $[\partial, a] = -[a,\partial] = \partial(a)$ for $a \in \mathcal{A}\otimes\mathfrak{g}$.

Let $X_0 = \int \left(u_0^\alpha\theta_\alpha\right)\mathrm{d}x$, where $u_0^1, \ldots, u_0^m \in \mathbb{R}$ are some fixed constants. Then it is easy to see that $[X_0, [X_0, P]] = 0$, so $([X_0, P], P)$ forms an exact bihamiltonian structure. We rename $P_1 = [X_0, P]$, $P_2 = P$. The bihamiltonian structure (P_1, P_2) is called the Zakharov-Shabat bihamiltonian structure.

The second component of the Zakharov-Shabat bihamiltonian structure can be regarded as a reduction of the deformed Hamiltonian structure mentioned in Example 26. The Drinfeld-Sokolov bihamiltonian structure is a further reduction of the Zakharov-Shabat one. A detailed description of the Drinfeld-Sokolov bihamiltonian structure would make the present lecture notes too long, so we only give the final result.

Theorem 51 ([15]) *The Drinfeld-Sokolov bihamiltonian structure (Q_1, Q_2) is an exact bihamiltonian structure on a submanifold $V \subset M$ with* $\dim V = n$.

(a) The leading term of (Q_1, Q_2) coincides with the bihamiltonian structure associated to the Frobenius structure on the orbit space of the Weyl group of $\mathfrak{g}$. In particular, it is semisimple.
(b) The central invariants of (Q_1, Q_2) are given by (up to a rearrangement)

$$c_i = \frac{\langle \alpha_i^\vee, \alpha_i^\vee\rangle_{\mathfrak{g}}}{48}, \quad i = 1, \ldots, n,$$

where $\{\alpha_1^\vee, \ldots, \alpha_n^\vee\}$ is a collection of simple coroots of $\mathfrak{g}$.

(c) If we choose $\langle\,,\,\rangle_{\mathfrak{g}}$ *to be the normalized one*

$$\langle\,,\,\rangle_{\mathfrak{g}} = \frac{1}{2h^{\vee}}\langle\,,\,\rangle_{\mathrm{K}},$$

where $h^{\vee}$ *is the dual Coxeter number of* $\mathfrak{g}$, *and* $\langle\,,\,\rangle_{\mathrm{K}}$ *is the Killing form, then the central invariants for* $\mathfrak{g}$ *of type* X_n *is given by the following table.*

$$\begin{array}{ll} X_n & c_1 \;\dots\; c_{n-1} \;\; c_n \\ A_n & \frac{1}{24} \;\cdots\; \frac{1}{24} \;\; \frac{1}{24} \\ B_n & \frac{1}{24} \;\cdots\; \frac{1}{24} \;\; \frac{1}{12} \\ C_n & \frac{1}{12} \;\cdots\; \frac{1}{12} \;\; \frac{1}{24} \\ D_n & \frac{1}{24} \;\cdots\; \frac{1}{24} \;\; \frac{1}{24} \\ E_n,\ n=6,7,8 & \frac{1}{24} \;\cdots\; \frac{1}{24} \;\; \frac{1}{24} \\ F_n,\ n=4 & \frac{1}{24} \;\; \frac{1}{24} \;\; \frac{1}{12} \;\; \frac{1}{12} \\ G_n,\ n=2 & \frac{1}{8} \qquad\quad \frac{1}{24} \end{array} \tag{6.3}$$

When $\mathfrak{g}$ is of ADE type, the central invariants are all equal to $1/24$, so the Drinfeld-Sokolov bihamiltonian structure is equivalent to Dubrovin-Zhang's deformation [10, 13], and the total descendant potential coincides with the one given by Givental's formula. Recently, Fan, Jarvis and Ruan rigorously define the Landau-Ginzburg A-model for a quasi-homogeneous singularity, which is called the FJRW theory. They also proved that the total descendant potential of FJRW theory for an ADE singularity is given by Givental's formula, so it is a tau function of the corresponding Drinfeld-Sokolov hierarchy. This result is called the ADE Witten conjecture. Please see [18, 19, 21, 26, 30, 33, 34] for more details.

When $\mathfrak{g}$ is of $BCFG$ type, the central invariants are constant, but not all equal to $1/24$. Define $R = 24\sum_{i=1}^{n} c_i$, then we have

$\mathfrak{g}$	B_n	C_n	F_4	G_2
R	$n+1$	$2n-1$	6	4

It is well-known that a simple Lie algebra of B_n type can be embedded into a simple Lie algebra of D_{n+1} type as the fixed locus of an order two automorphism. Similarly, C_n can be embedded into A_{2n-1}, F_4 can be embedded into E_6, and G_2 can be embedded into D_4. So the number R gives exactly the rank of the *ambient*

Lie algebra. This observation suggests us how to prove the generalized Witten conjecture of $BCFG$ type [30].

Remark 52 The above two examples both have constant central invariants. There also exist bihamiltonian structures possessing non-constant central invariats. For example, the bihamiltonian structure of the Camassa-Holm hierarchy (see Example 20) has $c(\lambda) = \frac{\lambda}{3}$. Its two-component generalization (see [27], [3]) has

$$c_1(\lambda_1) = \frac{\lambda_1^2}{24}, \quad c_2(\lambda_2) = \frac{\lambda_2^2}{24}.$$

We also considered its multi-components generalization in [4], and more complicated central invariants arose there.

The Camassa-Holm equation and its generalization are very popular recently in the area like PDE analysis or hydrodynamic, because they often have interesting weak solutions and wave-breaking phenomena. They are also the main source of our work [28], whose results play an important role in the present paper. However, there seems no direct connection between such integrable hierarchies and Gromov-Witten theories.

References

1. Carlet, G., Posthuma, H., Shadrin, S.: Bihamiltonian cohomology of KdV brackets. Commun. Math. Phys. **341**(3), 805–819 (2016)
2. Carlet, G., Posthuma, H., Shadrin, S.: Deformations of semisimple Poisson pencils of hydrodynamic type are unobstructed. J. Diff. Geom. **108**(1), 63–89 (2018)
3. Chen, M., Liu, S.-Q., Zhang, Y.: A two-component generalization of the Camassa-Holm equation and its solutions. Lett. Math. Phys. **75**, 1–15 (2006)
4. Chen, M., Liu, S.-Q., Zhang, Y.: Hamiltonian structures and their reciprocal transformations for the r-KdV-CH hierarchy. J. Geom. Phys. **59**, 1227–1243 (2009) (preprint). arXiv: 0809.0397
5. De Sole, A., Kac, V.G.: Essential variational Poisson cohomology. Commun. Math. Phys. **313**(3), 837–864 (2012)
6. De Sole, A., Kac, V.G.: The variational Poisson cohomology. Jpn. J. Math. **8**(1), 1–145 (2013)
7. Dubrovin, B.: Flat pencils of metrics and Frobenius manifolds. In: Integrable Systems and Algebraic Geometry (Kobe/Kyoto, 1997), pp. 47–72. World Science Publishing, River Edge (1998)
8. Dubrovin, B.: Gromov-Witten invariants and integrable hierarchies of topological type. In: Topology, Geometry, Integrable Systems, and Mathematical Physics. Amer. Math. Soc. Transl. Ser. 2, vol. 234. Adv. Math. Sci., vol. 67, pp. 141–171. Amer. Math. Soc., Providence, RI (2014)
9. Dubrovin, B.A., Novikov, S.P.: Hamiltonian formalism of one-dimensional systems of the hydrodynamic type and the Bogolyubov-Whitham averaging method. (Russian) Dokl. Akad. Nauk SSSR **270**(4), 781–785 (1983)
10. Dubrovin, B., Zhang, Y.: Bi-Hamiltonian hierarchies in 2D topological field theory at one-loop approximation. Commun. Math. Phys. **198**(2), 311–361 (1998)
11. Dubrovin, B., Zhang, Y.: Frobenius manifolds and Virasoro constraints. Selecta Math. (N.S.) **5**(4), 423–466 (1999)
12. Dubrovin, B., Zhang, Y.: Virasoro symmetries of the extended Toda hierarchy. Commun. Math. Phys. **250**(1), 161–193 (2004)

13. Dubrovin, B., Zhang, Y.: Normal forms of hierarchies of integrable PDEs, Frobenius manifolds and Gromov-Witten invariants (2001, preprint). arXiv: math/0108160
14. Dubrovin, B., Liu, S.-Q., Zhang, Y.: On Hamiltonian perturbations of hyperbolic systems of conservation laws. I. Quasi-triviality of bi-Hamiltonian perturbations. Commun. Pure Appl. Math. **59**(4), 559–615 (2006)
15. Dubrovin, B., Liu, S.-Q., Zhang, Y.: Frobenius manifolds and central invariants for the Drinfeld-Sokolov bi-Hamiltonian structures. Adv. Math. **219**(3), 780–837 (2008)
16. Dubrovin, B., Liu, S.-Q., Zhang, Y.: Bihamiltonian Cohomologies and Integrable Hierarchies II: The General Case. https://doi.org/10.1007/s00220-018-3176-y
17. Eguchi, T., Hori, K., Xiong, C.-S.: Quantum cohomology and Virasoro algebra. Phys. Lett. B **402**(1–2), 71–80 (1997)
18. Fan, H., Jarvis, T., Ruan, Y.: The Witten equation, mirror symmetry, and quantum singularity theory. Ann. Math. (2) **178**(1), 1–106 (2013)
19. Fan, H., Jarvis, T., Ruan, Y.: The Witten equation and its virtual fundamental cycle (2011, preprint). arXiv: 0712.4025
20. Ferapontov, E.V.: Compatible Poisson brackets of hydrodynamic type. Kowalevski workshop on mathematical methods of regular dynamics (Leeds, 2000). J. Phys. A **34**(11), 2377–2388 (2001)
21. Frenkel, E., Givental, A., Milanov, T.: Soliton equations, vertex operators, and simple singularities. Funct. Anal. Other Math. **3**(1), 47–63 (2010)
22. Getzler, E.: The Virasoro Conjecture for Gromov-Witten invariants. In: Algebraic Geometry: Hirzebruch 70 (Warsaw, 1998). Contemporary Mathematics, vol. 241, pp. 147–176. American Mathematical Society, Providence (1999)
23. Getzler, E.: The Toda conjecture. Symplectic geometry and mirror symmetry (Seoul, 2000), pp. 51–79. World Scientific Publishing, River Edge (2001)
24. Getzler, E.: A Darboux theorem for Hamiltonian operators in the formal calculus of variations. Duke Math. J. **111**(3), 535–560 (2002)
25. Givental, A.B.: Gromov-Witten invariants and quantization of quadratic Hamiltonians. Dedicated to the memory of I. G. Petrovskii on the occasion of his 100th anniversary. Mosc. Math. J. **1**(4), 551–568, 645 (2001)
26. Givental, A.B., Milanov, T.E.: Simple singularities and integrable hierarchies. In: The Breadth of Symplectic and Poisson Geometry. Progress in Mathematics, vol. 232, pp. 173–201. Birkhäuser, Boston (2005)
27. Liu, S.-Q., Zhang, Y.: Deformations of semisimple bihamiltonian structures of hydrodynamic type. J. Geom. Phys. **54**(4), 427–453 (2005)
28. Liu, S.-Q., Zhang, Y.: Jacobi structures of evolutionary partial differential equations. Adv. Math. **227**(1), 73–130 (2011)
29. Liu, S.-Q., Zhang, Y.: Bihamiltonian cohomologies and integrable hierarchies I: a special case. Commun. Math. Phys. **324**(3), 897–935 (2013)
30. Liu, S.-Q., Ruan, Y., Zhang, Y.: BCFG Drinfeld-Sokolov hierarchies and FJRW-theory. Invent. Math. **201**(2), 711–772 (2015)
31. Nijenhuis, A., Richardson, R.W. Jr.: Deformations of Lie algebra structures. J. Math. Mech. **17** 89–105 (1967)
32. Okounkov, A., Pandharipande, R.: The equivariant Gromov-Witten theory of $\mathbb{P}^1$. Ann. Math. (2) **163**(2), 561–605 (2006)
33. Wu, C.-Z.: A remark on Kac-Wakimoto hierarchies of D-type. J. Phys. A **43**(3), 035201, 8 pp. (2010)
34. Wu, C.-Z.: Tau functions and Virasoro symmetries for Drinfeld-Sokolov hierarchies. Adv. Math. **306**, 603–652 (2017)
35. Zhang, Y.: On the $\mathbb{CP}^1$ topological sigma model and the Toda lattice hierarchy. J. Geom. Phys. **40**(3–4), 215–232 (2002)
36. Zhang, Y.: Deformations of the bihamiltonian structures on the loop space of Frobenius manifolds. Recent advances in integrable systems (Kowloon, 2000). J. Nonlinear Math. Phys. **9**(suppl. 1), 243–257 (2002)

The manufacturer's authorised representative in the EU is Springer Nature Customer Service Centre GmbH, Europaplatz 3, 69115 Heidelberg, Germany. If you have any concerns regarding our products, please contact ProductSafety@springernature.com

Printed and bound by CPI Group (UK) Ltd, Croydon, CR0 4YY
30/09/2026
02225228-0020